How Modern Science Came Into The World

HOW MODERN SCIENCE CAME INTO THE WORLD

Four Civilizations, One 17th-Century Breakthrough

H. Floris Cohen

Amsterdam University Press

Cover illustration: Yves Tanguy, There Motion Has Not Yet Ceased (Là ne finit pas encore le mouvement), 1945, oil on canvas, 71,1 x 55,5 cm. Solomon R. Guggenheim Museum, New York, Bequest, Richard S. Zeisler, 2007, 2007.47 / © c/o Pictoright Amsterdam 2010
Cover design: Studio Jan de Boer, bno, Amsterdam
Lay-out: ProGrafici, Goes

ISBN   978 90 8964 239 4
e-ISBN   978 90 4851 273 7
NUR   685

In memory of Rob Wentholt, mentor and friend

TABLE OF CONTENTS

PREFACE XIII

PROLOGUE: SOLVING THE PROBLEM OF THE SCIENTIFIC REVOLUTION XV
The historiography of the Scientific Revolution: Past and present state XVII
The toolkit XX
Major questions here resolved XXIX
Users' guide XXXIII
    Notes on literature used XXXIX

PART I NATURE-KNOWLEDGE IN TRADITIONAL SOCIETY 1

I GREEK FOUNDATIONS, CHINESE CONTRASTS 3
A tale of two cities 4
Athens and Alexandria compared 15
Athens and Alexandria: Rare efforts at unification 23
Greek knowledge of nature: Upswing and downturn 27
Chinese knowledge of nature 33
Chinese and Greek nature-knowledge compared 44
Theory – a latent developmental potential and conditions for its realization 47
    Notes on literature used 50

II GREEK NATURE-KNOWLEDGE TRANSPLANTED: THE ISLAMIC WORLD 53
Upswing 54
Downturn 64
On the threshold of a Scientific Revolution? 70
    Notes on literature used 74

III GREEK NATURE-KNOWLEDGE TRANSPLANTED IN PART: MEDIEVAL EUROPE 77
Upswing 77
Downturn 89

VIII   *Nature-knowledge in Islamic civilization and in medieval Europe: A comparison*   90
      Notes on literature used   97

IV   GREEK NATURE-KNOWLEDGE TRANSPLANTED, AND MORE: RENAISSANCE EUROPE   99
Athens replayed in full   99
Alexandria: A replay with a difference   100
Europe's coercive empiricism   113
At the dawn of the Scientific Revolution   141
      Notes on literature used   152

PART II THREE REVOLUTIONARY TRANSFORMATIONS   157

V   THE FIRST TRANSFORMATION: REALIST-MATHEMATICAL SCIENCE   159
Johannes Kepler   161
Galileo Galilei   178
The knowledge structure of Alexandria-plus   195
Causes of the first transformation   201
      Notes on literature used   216

VI   THE SECOND TRANSFORMATION: A KINETIC-CORPUSCULARIAN PHILOSOPHY
OF NATURE   221
Revival continued   222
Beeckman and the transformation of ancient atomism   224
The Cartesian variety of kinetic corpuscularianism   226
Causes of the second transformation   238
      Notes on literature used   243

VII   THE THIRD TRANSFORMATION: TO FIND FACTS THROUGH EXPERIMENT   245
Bacon's vision   245
Bacon's proposed practice: The natural history of sound   247
Gilbert: Lodestone and amber treated the Baconian way   249
Harvey: Bodily processes revised   252
Van Helmont: Paracelsianism reformed   255
Theory: Experimentation, theorizing, and background worldview   258
Causes of the third transformation   260
      Notes on literature used   268

VIII   CONCURRENCE EXPLAINED   271   IX

       An explanatory overview   271
       Causal gaps identified   272
       An underlying sense of values shared across the culture   273
       An upswing luckily not interrupted   276
       The achievement still at risk   278
            Notes on literature used   279

IX   PROSPECTS AROUND 1640   281
       Chronology and continuities   281
       Dynamics of the revolution, in brief   284
            Notes on literature used   287

PART III DYNAMICS OF THE REVOLUTION   289

X   ACHIEVEMENTS AND LIMITATIONS OF REALIST-MATHEMATICAL SCIENCE   291
       The classic Alexandrian subjects absorbed   293
       Scholastic concepts mathematized   307
       Craft techniques mathematized   309
       Mathematical instruments   327
       Analogies of motion   335
       From Euclidean ratios to the calculus   346
       Power and pull of realist-mathematical science   360
            Notes on literature used   368

XI   ACHIEVEMENTS AND LIMITATIONS OF KINETIC CORPUSCULARIANISM   373
       Musical sound as moving corpuscles: An explanatory sample   375
       Rapid adoption   382
       Modifications of the doctrine   388
       Whirlpools for a revolving Earth   394
       Another sort of power; another sort of pull   397
            Notes on literature used   401

XII   LEGITIMACY IN THE BALANCE   403
       Strangeness: Three successive clashes   403
       Strangeness: Against common sense   415
       Sacrilege: Three successive clashes   417
       Strangeness and sacrilege: A looming crisis of legitimacy   426
            Notes on literature used   441

X   XIII   ACHIEVEMENTS AND LIMITATIONS OF FACT-FINDING EXPERIMENTALISM   445
            Facts collected and categorized   446
            Instrument-driven fact-finding   448
            Subject-driven fact-finding   463
            Craft techniques improved through experimental science   472
            Problems with facts and how to ascertain them   483
            Pooling of efforts   494
            Power and pull of fact-finding experimentalism   499
                Notes on literature used   506

    XIV   NATURE-KNOWLEDGE DECOMPARTMENTALIZED   509
            Whence the breakdown of barriers?   511
            Quantities and corpuscles   515
            Revolutionary fusion in the making   517
                Notes on literature used   519

    XV   THE FOURTH TRANSFORMATION: CORPUSCULAR MOTION GEOMETRIZED   521
            Motion, four principal ways   522
            Anomalous refraction revisited   539
                Notes on literature used   546

    XVI   THE FIFTH TRANSFORMATION: THE BACONIAN BREW   549
            Kinetic corpuscularianism crosses the Channel   549
            Spirit and active principles in kinetic corpuscularianism   552
            The Baconian Brew   554
                Notes on literature used   564

    XVII LEGITIMACY OF A NEW KIND   565
            The edges off controversy, and a shift of center Europe-wide   566
            Strangeness mitigated   568
            Sacrilege insulated   572
            Utility sanctioned in the Baconian Ideology   578
            Sustenance for nature-knowledge in monotheist surroundings –
            a comparative summing up   590
                Notes on literature used   595

XVIII   NATURE-KNOWLEDGE BY 1684: THE ACHIEVEMENT SO FAR   599   XI

    Predecessors on their way out   600

    17th-century props   606

    Advances on many fronts: the big picture   611

    The force knot   621

        Notes on literature used   634

XIX   THE SIXTH TRANSFORMATION: THE NEWTONIAN SYNTHESIS   637

    The Newton knot   637

    Toward the Principia   640

    Principia   658

    Toward the Opticks   678

    Opticks   696

    Two fragments and one whole   702

        Notes on literature used   716

EPILOGUE: A DUAL LEGACY   719

    Expanding modern science   722

    Science and values   731

        Notes on literature used   740

ENDNOTES   743

NAME INDEX   767

SUBJECT INDEX   779

PREFACE

Once upon a time 'the Scientific Revolution of the 17th century' was an innovative and inspiring concept. It yielded what is still the master narrative of the rise of modern science. The narrative has meanwhile turned into a straitjacket – so often events and contexts just fail to fit in. In the classroom we make the best of the situation; in our researches most of us prefer just to drop the concept altogether, regarding it as beset by truly unmanageable complexity. And yet, neither the early, theory-centered historiography nor present-day contextual and practice-oriented approaches compel us to drop the concept altogether. Instead, in the present book I provide a narrative restructured from the ground up, by means of a comprehensive approach, sustained comparisons, and a tenacious search for underlying patterns.

Key to my analysis is a vision of the Scientific Revolution as made up of six distinct yet tightly interconnected revolutionary transformations, each of some twenty-five to thirty years' duration. This vision equally enables me to explain how modern science could come about in Europe rather than in Greece, China, or the Islamic world.

In the prologue that follows I set forth all this at greater length, with proper attention given to the present state of the historiography and to the theoretical views that have served me as guidelines for my effort at fresh conceptualization.

I began writing this book in 1994. Since then I have been fortunate to receive numerous benefits.

Regarding institutions, I owe a great deal to the Dibner Institute, where (under the highly appreciated co-directorship of Jed Buchwald and Evelyn Simha) I was a fellow from February through June 1995; to the history-of-science group at the University of Twente with its frequent, no-holds-barred but also friendly discussions of work-in-progress; to the Department of Humanities of Utrecht University for a generous grant and (at the same university) to the Descartes Centre for the History and Philosophy of the Sciences and the Humanities, which under the visionary leadership of Wijnand Mijnhardt has since 2006 provided me with a most congenial working environment.

As for individuals who helpfully commented on portions of the book, I beg forgiveness from those I may fail to list against my best intentions. Critical readers of the entire book in its various stages include John Heilbron, John Henry, Peter Pesic, Bert Theunissen, and (last but definitely not least) Bob Westman; also, with her familiar hawk's eye, Pamela Bruton. I

XIV   further received useful comments on ideas, sections, and chapters from Klaas van Berkel, Domenico Bertoloni Meli, Mario Biagioli, Michel Blay, Rens Bod, Henk Bos, Geoffrey Cantor, Peter Dear, Fokko Jan Dijksterhuis, Mark Elvin, Peter Engelfriet, Moti Feingold, Rivka Feldhay, Dan Garber, Stephen Gaukroger, Penelope Gouk, Jehane Kuhn, Dick van Lente, David Lindberg, Frans van Lunteren, Nancy Nersessian, Lodewijk Palm, Larry Principe, Jamil Ragep, John Schuster, William Shea, Nathan Sivin, Noel Swerdlow, Edith Sylla, Steve Turner, Tomas Vanheste, John Walbridge, the late Sam Westfall, Catherine Wilson, Joella Yoder, and also, with friendly incisiveness, from Rob Wentholt, who died as this book went to the press. I dedicate it to his memory.

Dear friends and colleagues, I am truly grateful for how all of you have helped me in your various ways. Thanks to you, this has become a better book.

And thanks to Marita's loving support, the author has become a better man.

PROLOGUE

SOLVING THE PROBLEM OF THE SCIENTIFIC REVOLUTION

> The affairs of the Empire of letters are in a situation in which they never were and
> never will be again; we are passing now from an old world into the new world,
> and we are working seriously on the first foundation of the sciences.[1]
> Dom Robert Desgabets OSB, 18 September 1676

Around 1600 the pursuit of nature-knowledge was radically transformed. This happened in Europe over the course of a few decades, and our modern science is what grew out of the event. The transformation and its immediate aftermath have for quite some time been known as the Scientific Revolution of the 17th century. In a book published in 1994, *The Scientific Revolution: A Historiographical Inquiry*, I subjected to critical scrutiny some sixty views on the event selected from the vast literature for their boldly creative, interpretive sweep. I now present my own view. It has taken shape in critical dialogue with those sixty and several more-recent interpretations and also with many more narrowly focused studies. In good measure, it also rests upon firsthand familiarity with the subject. In the final chapter of my historiographical book I presented a preliminary sketch of my own budding view. That sketch has served me well as a stepping-stone, but my thinking has taken many a new turn in the meantime. I hereby discard that final chapter, with thanks for the encouragement it once gave me.

In tracing over time a range of events which culminated in 17th-century Europe, I seek answers to two basic questions. The first is: How did modern science come into the world, and (as part of that question) why did this happen in Europe rather than in China or in Islamic civilization? The other question is: Why did this 17th-century breakthrough in the pursuit of knowledge about nature instigate the as-yet-unbroken chain of scientific growth that we are wont to take for granted in our own time, four centuries later? Why did it not peter out, as every previous period of florescence suggests it very well might have? In short, the questions this book claims to resolve are *whence the onset, and whence the original staying power, of modern science*?

On the first question, the principal point I shall be concerned to make is that modern science came into the world by way of a threefold transformation. Transformed in revolutionary fashion were three mutually very different and also very much separate modes of acquiring knowledge about nature. The *mathematical* portion of the Greek corpus of

XVI   nature-knowledge, after several centuries of reception and enrichment in Islamic civilization and then in Renaissance Europe, was unpredictably turned by Galileo and by Kepler
into the beginnings of an ongoing process of mathematization of nature experimentally
sustained. Another portion of the Greek corpus, the *speculative,* contained four distinct,
rival systems of natural philosophy, with Aristotle's paramount. It was replaced, at the instigation of Descartes and other corpuscularian thinkers, by a natural philosophy of atomist
provenance that was decisively reinforced by a novel conception of motion broadly similar
to Galileo's. Thirdly, a specifically European-colored mode of investigation intent upon *accurate description and practical application* that had arisen by the mid-15th century began
to consolidate around 1600, under the aegis of Francis Bacon's calls for a general reform of
nature-knowledge, into a fact-finding, practice-oriented mode of experimental science.

Thus, the onset of the Scientific Revolution yielded three distinct modes of nature-
knowledge of a kind that the world had never seen. If we wish to understand how modern
science could *arrive* in the world, we must ask how, around 1600, these three almost simultaneous transformations could come about. Most answers to be given here are specific to
each distinct case of revolutionary transformation. Insofar as answers pertain to the question of 'why in Europe and not elsewhere?' they hinge on a comparison between Greek and
Chinese nature-knowledge and on a historical theory of upswing, downturn, and chances
for refreshment yielded by feats of cultural transplantation – all of which I unfold in a foundational first chapter. If, next, we wish to understand how kernels of "recognizably modern
science"[2] managed to *stay* in the world once they had arrived there, we ought to note first
that their very survival was a close call – by midcentury the revolutionary movement was
undergoing a veritable crisis of legitimacy. But, rather than losing momentum for good, a
new political climate and the emergence, by the early 1660s, of an ideology for innovative
nature-knowledge allowed the movement to regain pace. Three distinct driving forces propelled it forward. One was a specific dynamics built into the 17th-century practice of mathematization of nature experimentally sustained; another was a similar yet characteristically
different dynamics built into the 17th-century practice of fact-finding experimentation. And
I shall make a case for a midcentury event that has so far not been conceptualized at all. This
is the unprecedented breaking down of barriers between the Galilean, the Cartesian, and
the Baconian modes of nature-knowledge, leading in the 1660s to mid-1680s to three more
revolutionary transformations marked by hugely productive mutual interaction. In sum, the
Scientific Revolution of the 17th century may fruitfully be regarded, not as one monolithic
event, but rather as being made up of six distinct revolutionary transformations, each of
some twenty to at most thirty years' duration.

So much by way of an outline of the *argument* to be unfolded in the present book in the
format of an ongoing, chronological *narrative.*

To any reader – professional historian of science or not, student or not – the book has a *story* to tell. Between ancient Greece and Newton's *Principia* and *Opticks* it covers the major episodes and the major figures (also numerous minor ones). I take pains to avoid jargon and to present successive issues as clearly and simply as I can. Details not directly relevant to the story line (e.g., biographical data that go beyond the brief characterization or the telling anecdote) are left out. Even so, the book builds its message up from a long concatenation of topics, and my detailed treatment of certain issues may tax the patience of the nonprofessional. For over and above its providing a story, this book is meant to be an *argument*. It is directed in the first place at convincing my co-professionals in the history of science that the conception here unfolded of how modern science came into the world is worth considering in earnest. In view of how history-of-science writing has developed over the past decades, this may not be an easy task – it runs up against an ingrained skepticism concerning the very questions I now claim to have resolved.

The historiography of the Scientific Revolution: Past and present state

The concept of the Scientific Revolution was coined in the 1930s, as one product of a major historiographical overhaul that took place between the mid-1920s and the early 1950s. It was meant to identify a period in European history that covers roughly the second half of the 16th and almost all of the 17th century (i.e., between Copernicus and Newton) as marking a uniquely radical, conceptual upheaval out of which modern science emerged essentially as we still know it. This view quickly began to be articulated in the budding profession of historians of science. It did so in ways that turned the previously customary listing of one heroic scientific achievement after another into the careful reconstruction of such conceptual knots as those individuals who brought the Scientific Revolution about actually faced and strove to disentangle.

Out of the concept-focused mode of history writing thus emerging came a range of pathbreaking narratives. These shared, at a minimum, a focus on how the once self-evident conception of our Earth as a stable body at the center of the cosmos gave way to the core of the modern worldview – the Earth and the other planets placed in a solar system, itself a tiny portion of an infinite universe. The major vehicle to bring about this fundamental reversal, along with several other major, closely related accomplishments (a new conception of motion; deliberate creation of void space), was held to be a quickly expanding process of 'mathematization of nature'. By this was meant the subjection of increasing ranges of empirical phenomena to mathematical treatment in ways suitable as a rule to experimental testing. Key figures in the process were held to be Copernicus, as the man first to compute down to the required detail planetary trajectories in a Sun-centered setting; Kepler, as the man first to turn Copernicus' setup into a previously unthinkable 'celestial physics' leading to his discovery of the planets' elliptical paths; Galileo, as the man first to mathematize with success a significant terrestrial phenomenon (falling and projected bodies) in an effort to counter major objections to Copernicus' setup; Descartes, as the man first to conceive of the universe and of the particle-governed mechanisms at work in it mathematically; Newton, as the man to cap the whole development by uniting terrestrial and celestial physics in his mathematically exact, empirically sustained conception of universal gravitation. Not that these men and their principal accomplishments were taken to represent all there really was to the Scientific Revolution. Still, for decades historians were inclined to treat other noteworthy attainments of a modern-scientific nature, such as Harvey's discovery of the circulation of the blood or Boyle's chemical-testing procedures, as by-products, somehow, of the major development.

Starting in the 1960s, a range of perspectives was introduced that led to a widening of this 'master narrative'. Our history-of-science forebears unreflectively identified the present-day definition and classification of scientific disciplines with their apparent 17th-century counterparts. This habit has been given up in favor of a still-increasing awareness that what we now call 'mechanics', for example, scarcely had a counterpart in the early 17th century, so different, and differently aligned, was the intellectual context in which problems of motion used to be considered from the ancient Greeks onward. Even 'science' as a general expression is on its way out. It carries too many associations far removed from 17th-century realities (e.g., the professional identity of 'scientist' is a term of the 19th century, not earlier).

Further, research subjects and/or people previously left wholly or partly in the margins have come to be included in the narrative. Examples are subjects that (at the time) were non-mathematical and chiefly descriptive, like magnetism and illness; subjects that are scarcely practiced anymore, like musical science, and/or are held under grave suspicion, like alchemy; but also previously neglected contributors not of the first or even quite the second rank (e.g., hosts of ably experimenting Jesuits).

Most important of all, the goal of putting the history of scientific ideas in institutional and other sociocultural contexts has become a fixture of most articles and books on the subject. History writing in the vein of 'this major thinker brought this particular conceptual breakthrough about, then that thinker that one' has not come to an end, but a sense has emerged that a proper understanding of scientific accomplishment requires an awareness of how it was situated in time and place. In this way, for example, we have learned of the considerable extent to which practitioners depended upon Europe's patronage market. Also, an influential argument has been made for a constitutive link between the contested viability of instrument-aided experimentation per se as articulated in Boyle's and Hobbes' early-1660s dispute over the void and the politics of the Stuart Restoration. As a result, local particularity has in recent decades been gaining the upper hand over the universal validity claimed with ever diminishing vigor for the most seminal outcomes of the Scientific Revolution. One genuine accomplishment of this context-oriented approach is a greater concern

for the day-to-day practice of experimental research and for the trustworthiness of results
so attained. Another has been a heightened sense that there is room for contingency in the
story – not everything that happened was bound to happen, or was bound to happen the
way it did happen. Historians of science have further become aware that there were more
significant reasons for contemporary perceptions of modern-science-in-the-making as innately strange and disturbing than sheer backwardness and/or superstition.

With regard to the concept of the Scientific Revolution, the net effect of this plurality
of mostly productive, novel viewpoints has been resignation. What numerous historians of
science have in the meantime given up is not, to be sure, the ongoing production of novel interpretations of episodes in 17th-century science *but the very idea that, deeply underneath the
surface of individual events, something identifiable holds so complex a series of events together.*
And it is certainly true that a once-enlightening yet too one-sided formula like 'Scientific
Revolution = mathematization of nature' can no longer be accepted. But is this conclusion
tantamount to giving up the quest for underlying coherence altogether?

In everyday practice, it surely is. True, publishers keep inviting authors to produce texts
for the classroom. The dozen or so up-to-date surveys to result from such requests are of as
great a use to the students taught therefrom as they are vital for the health of the profession.
But inevitably they obey a format that precludes a concentrated effort to seek an underlying
coherence.

The reigning atmosphere of skeptical resignation does not, to be sure, stem solely from
the manifold perspectives brought to bear upon the Scientific Revolution over the past decades or solely from a despair-inducing sense of the ever more apparent complexity of the
event. Resignation stems in perhaps equal measure from the apparent elusiveness of all those
big 'why?' questions once raised about the origins of modern science. The causal adventure,
enthusiastically embarked upon in the 1930s, has ended in failure and disillusion. Starting in
the 1980s it has gradually petered out. But should it have?

The difficulty with much causal debate at the time rested neither in its vivacity nor in
the vital nature of the questions asked but rather in the peculiar habit historians of science
acquired to investigate them. Efforts at explanation almost invariably took shape as a thesis,
usually named after the historian to put it forward, about *the* one and only, all-encompassing cause of *the* Scientific Revolution. The Zilsel thesis explained the Scientific Revolution
through an alleged, early-17th-century closing of the perennial gap between scholars and
craftsmen. The Merton thesis was made by adherents and opponents alike to explain the Scientific Revolution through the contemporary adoption of Puritan values in a capitalist setting. The Yates thesis explained the Scientific Revolution as the next step both extending and
opposing a magical worldview of Hermeticist origin. The Duhem thesis explained the Scientific Revolution as a 14th-century revolt against Aristotle. The Eisenstein thesis explained the
Scientific Revolution through Europe's move from script to print. And such piling up of ex-

xx   planations went on and on, without much exchange taking place between adherents over the respective merits of these theses, each of which was shown by acute critics to fail to clinch the all-encompassing case made for it. It is small wonder that explanatory habits so unsubtle have in the end induced a sense of resignation. True, Thomas S. Kuhn once proposed to cut such causal theses down to size by restricting their scope to one identifiable portion of the Scientific Revolution rather than to the whole of innovative 17th-century science. And in my historiographical book I broadened his constructive proposal into a plea for "judicious combination and cross-fertilization leading to consciously applied transformation" of available conceptions of the Scientific Revolution.[3] But the drift of history writing has gone in another direction – that of giving up the causal quest altogether.

The message that such a posture of resignation manifestly conveys to outside scholarship is that the advent of modern science – a decisive event in world history, really the most outstanding among prime motors of our modern world – was in effect due to chance. But, as the great pioneer of cross-culturally comparative history of science Joseph Needham remarked in another context, "to attribute the origin of modern science entirely to chance is to declare the bankruptcy of history as a form of enlightenment of the human mind".[4] He made the point half a century ago. As globalization gains pace by the day, scholars from other venues want to understand even more than they already have why modern science arose in Europe rather than in any of the other great civilizations of the past. Scholars from other venues have a manifest and rightful interest in being presented with accounts of how modern science arose and fared in Europe that seek and find coherent pattern and order in the event other than by arranging an array of episodes to meet the needs of the classroom.

Nor are historians of science compelled by any necessity to keep their fellow scholars waiting. A great deal of material, as well as numerous partial interpretations of penetrating profundity, is ready to hand to seek solutions to the big questions involved. No more is required for bringing these materials and these interpretations to bear upon them than a determination to shake off the reigning sense of resignation and to rethink from scratch ways and means to go about such a quest. No return to the ways of the forefathers is desired or even possible – who cares any longer for a monolithic answer? Instead, we must inspect our toolkit all over again and refresh it from the bottom up. Here is how I have done it.

The toolkit

First and foremost, this book is meant to be *analytical*. I have sought to make storytelling and analysis go well together; but wherever I had to choose, analysis has prevailed. What, then, does the writing of what may be called 'analytical history' actually entail?

Not only in the natural but also in most human sciences as also in philosophy, clear-cut conceptualization is a cardinal requirement. A failure to define one's concepts sharply and to delineate one's theories and hypotheses carefully causes the investigation to go astray. But historians tend to avoid clear-cut conceptualization. We seem to be happiest keeping our concepts a bit fuzzy. Many academics – philosophers and sociologists at the forefront – uncharitably ascribe what they regard as our mindless fact-grinding to a congenital lack of capability for abstract thought ('history is sociology with the brains left out'). There are some good reasons for the apparently lax habit, though. The prime commodity we are uniquely dealing in is change over time. But change over time cannot be captured well by means of fixed concepts imposed upon a subject in flux. If the concepts selected to handle it are just modern ones projected back upon the past, the investigation tends to end in presentism (anachronistic accounts written as if history were destined to move toward some preordained end finally attained in modern times). Also, a gap between the facts reported and the concepts invoked to capture them is bound to remain. If, on the other hand, the concepts selected are confined to those employed by one's protagonists themselves (so-called 'actors' categories'), the historian robs himself of the enhanced understanding that flows from his privilege to look beyond the temporal horizon to which his subjects were inevitably confined. Indeed, maintaining some conceptual fuzziness provides the middle way needed here.

But conceptual fuzziness is not a virtue in itself. One may easily slip into giving one's key terms a multiplicity of meanings, which is as unreflective as it is confusing. Insofar as conceptual fuzziness exceeds the exigencies of the historian's inevitable predicament, then, it ought to be combated. That is, we should seek to make our concepts as clear-cut as our sense of the past – its manifold complexity, the sheer richness and endless variety of human life over time – allows us.

To the extent that historians are sensitive to philosophers' and sociologists' condescension toward our ingrained methods of dealing with concepts, they often seek to mimic those other disciplines, borrow a conceptual apparatus there, adopt it wholesale, and inform accordingly the historical tale to be told. One major example is how social-constructivist conceptions that in the 1980s emerged from work in the philosophy and the sociology of present-day scientific knowledge were adopted in certain accounts of 17th-century episodes in nature-knowledge. Thus imported, the conceptual apparatus served as a useful searchlight that nonetheless turned history-of-science writing into a battle scene for ideological contests of questionable relevance to the core business of the historian – to make sense of past events in their otherness from, *and* in their likeness to, our present-day concerns.

Sensitive likewise to a preference for restricting fuzziness of concept formation and theory building to the minimum that our predicament as historians dooms us to, I have sought in the present book for a solution other than parachuting a ready-made conceptual apparatus smack into the premodern world. Avoiding presentist categories as much as an a priori limitation to actors' categories while also refraining from squeezing the facts of history into

XXII concepts imported wholesale from elsewhere, I have found myself exploring a process perhaps best called 'conceptualization as we move along'.

In the writing of history, pattern must be discerned, not imposed. Wishing neither to press my causal accounts into preset conceptual schemata nor, on the rebound, to refrain from conceptualization altogether, *I have coined my concepts and conceived the historical theories that bind them together as I went along, in ongoing dialogue with the empirical material I found myself handling at every stage.* 'Dialogue' is the key word here. This ongoing process of concept formation and theory building has not been inductive only – as if facts could ever speak for themselves. The process has been deductive as well, insofar as I have selected the facts to go into the making of these concepts and theories with such broad conceptions in mind as work on my historiographical book had alerted me to. Also, in the course of my historiographical inquiries I have found certain fertile conceptualizations of numerous authors (e.g., Joseph Ben-David, Thomas S. Kuhn, Richard S. Westfall) that with a little adaptation proved well suited for pressing my analyses further.

As far as I can tell, the procedure here followed is an uncommon one, and it seems advisable to mark it. Wherever in the book that follows I have solidified my factual material into concepts and theories by means of this partly inductive, partly deductive process, I have put the reader on the alert by inserting the term 'Theory' in the titles of the relevant sections. Chapter I, where I discuss the problem of 'decline' and develop one of the key concepts of part I, 'cultural transplantations', abounds with conceptualizations of this kind.

Part of the effort at reconceptualization requires our aligning anew the world history of science. It is time to replace the still customary, Eurocentric periodization of the history of science. In a globalizing world it is growing more obsolete by the day. It also stands squarely in the way of resolving the question of why modern science emerged in Europe and not elsewhere.

Here is the easy answer to that hoary question. The adventure of science started in the West, namely, in ancient Greece, and (but for a holding action by the Arabs) the adventure has always *remained* of the West, namely, of Europe, where what the Greeks began was destined to come to full fruition. This broad view of things has been reiterated often, and there is no easy way to refute it squarely. Still, that something is amiss with the West-centered view so readily taken for granted by preceding generations is suggested by the one apparent anomaly it contains – suddenly nature-knowledge turns up in Islamic civilization and then just as suddenly vanishes from the story without leaving a trace. Thus, the West-centered view implies that the entire development from the Greeks up to and including the rise of modern science was of the West, *yet somehow not quite*. This chink in the armor is accompanied by another: the broad picture rests upon an underlying presupposition about Greece itself, which is that ancient Greece is where 'the West' took shape. And indeed, values and viewpoints that go back to ancient Greece have gone into the making of European civilization right from its inception in Carolingian times. Yet ever since the 19th century, not only XXIII has some sort of *identification* with Greece been made in European, especially in German and British, historical thought, but this identification has been carried one step further: the Greeks came to look very much like 19th-century Germans or Britons. *Almost*, the Greeks were already us, and the sole task left to Europeans was to take the final step and become *truly* us. Historians of ancient Greece have since broken with this still deeply ingrained picture and have sought to draw a different one that stresses, not so much the extent to which ancient Greece was already like Western Europe, but rather its radical *otherness*.

However one-sided and overdrawn that new picture of Greece was, it did provide a healthy corrective to a conception of 'the West' that still lingers in historians' everyday thinking, the thinking of historians of science definitely included. The obvious way to overcome it is to reconceive the issue in a world-historical setting. Two book-length efforts have so far been made to provide such a setting.

In *The Geography of Science* (1991) Harold Dorn outlines a world-historical panorama that in effect underwrites the standard view. Science, he argues, used to be cultivated in one of two possible ways. The state-run, narrowly utilitarian type of science is characteristic of civilizations that depended on techniques of water management and on the organized mass labor force needed to keep it going. The contrasting type of science is largely autonomous, curiosity driven, and unconnected to any state or statelike structure. It marks those rare civilizations that are set in temperate climes with regular and moderate rainfall leading to an even spread of sufficient water – the principal incentive for the power monopoly of a central state is lacking here. Before modern times, Dorn argues, the bureaucracy-dominated type of science was the rule. In this alignment Greece and Western Europe appear as the sole cases of the latter type, against Egypt, Mesopotamia, China, Islamic civilization, India, the empire of the Khmer, and a range of pre-Columbian civilizations.

In *The Rise of Early Modern Science: Islam, China, and the West* (1993), Toby E. Huff tries out another alignment of the world history of science. He treats the science of China, of Islamic civilization, and of medieval Europe as three by and large independent units of comparative analysis to show how the indispensable conditions for modern science that first emerged in medieval Europe grew to full maturity in early-modern Europe.

Both alignments miss the profound similarity of nature-knowledge in Islamic civilization and Renaissance Europe – in both it stemmed from what survived of the Greek corpus, subsequently to be enriched in quite similar ways. So my own alignment as unfolded in the present book runs as follows. I start with the Greeks (whose approach I contrast with that of the Chinese, albeit from a more content-oriented point of view than Dorn's treatment). In books dedicated to the Scientific Revolution Greece provides an unusual point of departure. E.J. Dijksterhuis' *Mechanization of the World Picture* (1950) is the only exception; most others confine themselves to 17th-century Europe or take their point of departure in the European

XXIV Middle Ages or in 1543 (in view of Copernicus' and Vesalius' revolution-inducing books appearing in that year). In the alignment that governs my account, Greece brings forth a definitely nonmodern corpus of nature-knowledge that (unlike its Chinese counterpart) was enabled by certain military events to be revived and enriched three times in succession: in Islamic civilization, in medieval Europe, *and all over again* (a crucial insight that I owe to Noel M. Swerdlow and to P.L. Rose) *in Renaissance Europe*. In this alignment, medieval Europe no longer serves as the preparatory stage, a role still customarily ascribed to it. Medieval Europe appears instead as the exceptional case of the three, in that this time the revival of the Greek corpus was a highly curtailed one and, as such, not only unrepresentative but also doomed to remain locked in its own framework. I shall further demonstrate that although the two other recipients, Islamic civilization and, later, Renaissance Europe, adopted, appropriated, and then creatively enriched the Greek corpus almost independently, the two have enough in common to make for fruitful historical comparison.

In thus replacing a unidirectional alignment with a concatenation of episodes of structural comparability, I throw open the gates to a full-scale comparative approach. As "the motor of historical thinking",⁵ comparison is indispensable for coming to grips with the big questions I seek to resolve. The historical comparisons consistently undertaken here come in three kinds.

I compare how nature-knowledge was pursued in a variety of civilizational settings. For example, I make a range of comparisons between certain well-circumscribed aspects of nature-knowledge in Islamic civilization and in Renaissance Europe.

I further compare the outcomes of certain pursuits of nature-knowledge considered over time within a single civilization (Europe in particular). For instance, I compare Newton to Huygens and Hooke on a range of concerns shared between them, with a view to explaining what enabled Newton in the end to transcend conceptual boundaries still maintained by his farthest-seeing elder rivals.

But I also compare roughly simultaneous events with each other. One example is a sustained comparison between three notorious 17th-century cases of perceived sacrilege and consequent persecution – Galileo versus the Inquisition and the pope; Descartes versus the university and the city council of Utrecht; French Cartesians versus their king and their archbishop. The comparison has enabled me to disclose a specific pattern of contested authority amounting to a looming crisis of legitimacy for nature-knowledge all over the Continent.

Meanwhile the question is, can the outcomes of such comparisons still satisfy the rules of empirical scholarship, or are they too speculative for that?

As historians – guardians of the unique and the unrepeatable – we tend to view comparative approaches to the past with deep and abiding suspicion. We do so in particular when the comparison is between Western and non-Western civilizations. We do so even more when the comparison is between what did happen (e.g., the Scientific Revolution in 17th-century Europe) and what did not happen but allegedly might have (a broadly similar Scientific Revolution elsewhere, at another time).

All kinds of objections have been raised against comparative research, particularly of the first, cross-cultural type. Some are methodological; for example, how 'broadly similar' may an event be allowed to be for it still to count as 'similar' at all? Or, what can one possibly learn from events that never took place? How can one argue that some events might have taken place even though – probably with good reason – in historical reality they didn't? In other words, is there room in history for the category of the possible-yet-unreal? Other objections are cultural; for example, does not the very act of raising the question of why only the West originated modern science already bespeak an undue sense of Western superiority? Still other objections are about the nature of science; for example, can one properly speak of non-Western scientific traditions? Or (with the thrust reversed), what is so special about Western science in the first place? Other criticisms are just practical; for example, how many languages and how many scientific traditions must one master before being entitled to compare non-Western and Western science with each other? That is, must one not be well-nigh superhuman to address such questions in a productive manner? And to be even more practical: is it not the case that the very failure of precisely such an almost superhuman figure – to wit, Joseph Needham – to come up in the end with a convincing answer to his self-styled question of why Europe created modern science whereas China did not, show the futility of such questions?

In my historiographical book I seized on Needham's and others' pioneering work in cross-cultural comparative history of science to ponder such skeptical questions at length. In the present book I prefer to demonstrate rather than argue all over again that (despite all the traps that Needham-the-bold-pioneer fell into) there are ways to make cross-cultural comparisons produce viable insights – *insights, moreover, that can be gained by no other means.* Not that I fancy my own results to be definitive in any way. No one individual's can be. Not even to a historian of Needham's astonishing breadth is it given to attain a sufficiently deep knowledge of so many culturally and linguistically distinct traditions. This is especially true since specialists in traditions such as the Chinese and the Islamic have abstained so far from providing nonspecialists with handy, chronologically arranged, reliably sources-based survey histories and also tend to disagree among themselves rather forcefully, making it hard for the sympathetic outsider to decide between specific interpretations. So what we need is not just one but a plurality of historians to do the comparing – the more, the better. For so much is certain that (as Toby Huff phrased it) "viewed from a comparative and civilizational point of view, the rise of modern science appears quite different than it does when seen exclusively as an intra-European movement."[6]

In regard to my second and third types of historical comparison I further add that the comparative method forms one effective antidote to the historian's venial sin of presentism. It

XXVI makes it possible to determine, again and again, what was new and different, in a manner that does not look back from the present but rather builds things up from a deeper past which then serves as the proper touchstone. Finally, and most importantly, such determination of what, at any given time and place, was truly new and different is indispensable if one wishes not just to *analyze* things but also to *explain* them.

One brief textbook excepted (John Henry's *The Scientific Revolution and the Origins of Modern Science* of 1997), there is no book-length treatment of the Scientific Revolution in which description and interpretation are intertwined with explanations of what is being described and interpreted. In the few books where attempts at explanation are undertaken at all, this is done by inserting a causal chapter that, inevitably, falls into the trap of seeking to explain the Scientific Revolution whole, as one monolithic unit. Equally inevitably, no cause or causes then appear to match so sizable and complex an event, thus contributing to the sense of resignation I have deplored above.

In the present book I interweave analysis and explanation in the following manner. As just set forth, the analysis proceeds by way of comparison, and this is what yields my *explananda* – those components of the full story that require explanation. Next, I seek with care and precision to match each distinct *explanandum* with its specific explanation or explanations. The operative causal mechanism invoked to link the *explanandum* to the explanation must be shown in each case to be causal indeed – that is, not just linked arbitrarily to the *explanandum* but able to clarify how what did happen could happen. One example of how to refrain from monolithic theses without missing what light these may still shed upon specific events is that I do not invoke the replacement of script by print as a catch-all cause of the Scientific Revolution, in the way pioneered in 1979 by Elisabeth L. Eisenstein. Rather, I ask at each stage of the Scientific Revolution and its Renaissance prehistory whether events were such as to require the printing press, or whether these might just as well have taken place in a script culture. Here sustained comparison with Islamic civilization provides ready-made empirical material to explore. I similarly treat the accomplishment of the pioneers of the Scientific Revolution in a vein of sustained comparison with the extent to which their immediate predecessors had prepared the way. This, and this alone, enables the historian to determine with sufficient precision what actually was transformed and how radical the transformation was, as an indispensable preliminary to the causal inquiry.

The multicausal plurality that goes with my partly inductive, one-at-a-time procedure runs the risk of leading to undue fragmentation. To alleviate it I have at several places (pp. 271, 284, 599) taken care to draw my findings together. The outcome confirms my hunch that some underlying coherence to the Scientific Revolution can indeed be realistically detected.

Operative causal mechanisms may come in several varieties – not just *one* mode of explanation is being applied here. Notably, the causal mechanism central to my argument in parts I and II rests in a conception of latent developmental possibilities unfolding under the impact of certain events that I have dubbed 'cultural transplantations' – the transfer, that is, of an entire body of knowledge to a civilization not previously touched by it (p. 45). Behind the scene of my frequent invocation of this particular mode of explaining historical events and episodes lurks a definite conception of history. Three passages (one quoted from a great literary author, the others from two great historians of the 20th century) express or at least intimate the conception of history that seems to me most appropriate for doing justice to the flow of human events:

> [Milan Kundera (1992)]: Man is bound to advance through the fog. But when he looks back to judge those who lived before him, he fails to detect any fog on their path. From his present, which was their far-away future, the path looks to him entirely clear, visible over its full extension. He can see the path, he can see the people who advance, but the fog is no longer there.[7]
> [François Furet (1995):] No understanding of our age is possible if we do not liberate ourselves from the illusion of necessity: The [20th] century is not explicable (insofar as it is at all) if we fail to acknowledge all that was unforeseeable about it.[8]
> [Hugh Trevor-Roper (1998):] History is not merely what happened: it is what happened in the context of what might have happened. Therefore it must incorporate, as a necessary element, the might-have-beens.[9]

In the view of history that I adopt here, the past, as also the present, displays a blend of situational logic and contingency. There are linkages rigorously concatenated over time and unfolding with apparent inexorability; but what looks inexorable may at unforeseeable moments be overthrown by something itself unforeseeable. Events are neither *wholly* predetermined nor *just* a matter of chance. As for the Scientific Revolution, we must keep in mind that it was neither miraculous (i.e., beyond explanation) nor foreordained (i.e., bound to happen regardless). I shall even argue (p. 71) that one particular Scientific Revolution (or at least the onset thereof) that never occurred may revealingly be incorporated in my account as, indeed, a might-have-been.

To the extent that the actual Scientific Revolution and the fog-covered path toward it can be subjected to explanatory analysis, then, we must define with care what our unit of analysis is going to be. For this is the one piece of conceptual apparatus that cannot be developed as we go along, in dialogue with our factual material. It must be established at the outset; that is, right here.

Ongoing historization of the past of science has made nearly unfeasible a practice that we were once accustomed to apply without even pausing: organizing our historical accounts in accordance with present-day disciplines. But Aristotle and Galileo were not two experts in mechanics, with the main difference between them that the former had it mostly wrong and xxviii the latter had it almost right. Even though moving objects were a significant shared concern, the assumption that they worked in the same discipline is bound to mislead rather than enlighten – Aristotle was a philosopher and, as such, sought to grasp the totality of the world, whereas Galileo, in the mathematician's customary way, deliberately operated in a piecemeal fashion. Prior to the 19th century, disciplines in their present-day alignment can hardly ever be employed as viable units of analysis. Neither would my ends in this book be served by selecting research subjects or individuals or successive time segments for my unit of analysis. To do so would obscure rather than help uncover what underlying coherence the Scientific Revolution possessed. The same is true of the six 'styles' that A.C. Crombie identified. He presented these in his monumental *Styles of Scientific Thinking in the European Tradition* (1994) as timeless entities amenable neither to transformation nor to mutual interaction, thus missing the very dynamics that have propelled scientific advance from early in the 17th century onward.

Instead, the unit of analysis I have in the end found myself working with is *modes of nature-knowledge*. By this I mean consistent ranges of distinct approaches to natural phenomena, which may differ in several dimensions. Their scope may have been comprehensive, with a view to deriving the whole wide world from first principles, or deliberately partial. The way in which knowledge was attained may have been predominantly empiricist or chiefly intellectualist. If any practices went with a given mode of nature-knowledge, these may have been observational, experimental, instrumental, etc. Knowledge may have been sought for its own sake or with a view to achieving certain practical improvements. Exchange may or may not have taken place between practitioners of distinct modes of nature-knowledge that were pursued at the same time and place.

Of particular concern in distinguishing a variety of modes of nature-knowledge is what I label their 'knowledge structure'. By this I mean something that the principal difference between Aristotle and Galileo illustrates. For instance, we shall find in due course that in the 17th century much conceptual confusion emerged from the different handling of seemingly similar or even overlapping conceptions (notably those of motion and force) in different modes of nature-knowledge (those of Aristotle and Galileo, but also those of Galileo and Descartes). Was knowledge organized wholesale or rather piecemeal? How was knowledge conceived to be oriented in time – did practitioners see themselves as working toward an open future, or as reconstructing past perfection, or as personally constructing the truly definitive schema of all possible knowledge? At what level of abstraction did they seek to capture nature's phenomena? How were empirical facts handled – in their own right or made to serve some a priori schema and, if the latter, by way of illustrative confirmation or for a posteriori checking?

I treat these modes of nature-knowledge as dynamic entities. What turns them into viable instruments of historical analysis is the additional category of *transformation*. Modes of

nature-knowledge need not remain fixed over time. At least potentially they were subject to   xxix
being transformed, in ways that varied from enrichment within a given framework to such
revolutionary transformations as came in time to mark the Scientific Revolution.

The entities here distinguished as 'modes of nature-knowledge' do not form actors' categories; rather, I present them as a historian's constructs. Still, they are not just arbitrary
categories which happen to suit my ends. Distinguishing between diverse modes of nature-
knowledge captures patterns of thought and behavior that did manifest themselves in the
reality of the past. As a rule, distinct modes of nature-knowledge were practiced in almost
watertight separation from one another until, in the mid-17th century, an unprecedented
breakdown of barriers occurred. Before that time, even if one individual practiced two
modes of nature-knowledge at a time (as for instance Ibn Sina, João de Castro, and Thomas
Harriot did), his thought and practice in the one had scarcely an impact upon the other.

Consequently, in this book I routinely assign *this* practitioner to *this* specific mode of nature-knowledge (e.g., al-Biruni to mathematical science in the Alexandrian tradition), *that*
one to another (e.g., Clavius to natural philosophy in the Athenian tradition). This is not
a pointless game of squeezing events and people into preset categories. Only the unfolding
story can determine whether the use of 'modes of nature-knowledge' as my principal unit of
analysis serves more than taxonomic purposes and enables us to catch both the conditions
under which the Greek corpus could be revolutionized and the specific dynamics that then
propelled the Scientific Revolution forward.

Major questions here resolved

> Truth will sooner emerge from error than from confusion.
>
> Francis Bacon, *Novum organum* (1620)[10]

This, indeed, is the prize I aim for in this book: to resolve those big questions that for so long
used to animate the profession of historians of science and that remain as valid as ever. I do
not claim *fully* to have resolved the two biggest questions – how modern science came into
the world and how it managed to stay there. As the challenge is taken up and the big issues
are engaged afresh, new viewpoints will reveal novel aspects. But limits are also set to historical explanation per se. There is wisdom in an idea that, in 1948, one of the great historians of
science, Alexandre Koyré, expressed thus:

> all explanations, however plausible they may be, ultimately turn around in a circle. Which,
> after all, is not a scandal for the human mind. It is fairly normal in history – even in the history
> of the human mind – that there are inexplicable events, irreducible facts, absolute beginnings.
> . . . It is impossible, in history, to empty the fact, and to explain everything.[11]

xxx   It is for this reason especially that I indicate at appropriate places where, in my own view, I have reached the limit of what can be explained and where elements of contingency take over or where certain facts stubbornly refuse to be 'reduced' any further.

Among major issues resolved in the present book I count the following.

- The question of *decline* has frequently come up with regard to nature-knowledge in ancient Greece but also in Islamic civilization. Proper definition followed by comparative handling of the issues involved reveals the presence in premodern times of a general pattern of upswing and downturn of nature-knowledge (pp. 28, 64, 90).

- Needham was the first to investigate on the grand scale why the Scientific Revolution occurred in Europe and not in China. In my historiographical book I figured out how he asked the question, reviewed his diverse answers, and examined critically his practice of cross-cultural comparison. I now place myself on the shoulders of a man who, for all his idiosyncrasies, was an intellectual giant, in hopes of seeing farther. My comparison between structural components of Chinese nature-knowledge and its Greek counterpart yields two answers of a novel kind to 'Needham's Grand Question' (p. 44).

- For decades historians of science were in the habit of debating whether the Scientific Revolution was marked by continuity with preceding events (thus calling into question the very phrase 'Revolution') or rather by discontinuity. Caricatures have often taken the place of more historically responsible distinctions between events smoothly running their course and events taking a sudden, unexpected turn, thus making a more or less drastic (though never, *pace* Koyré, an absolute) break with the past. Here I have seized upon a suggestion made in 1992 by David C. Lindberg – to consider the possibility of a differential rate of discontinuity for different segments of 16th- and 17th-century nature-knowledge. Lindberg focused on optical science as displaying more continuity over time than astronomy did. My own conception of the revolutionary transformation of modes of nature-knowledge carries within itself the rejection of the idea of an absolute beginning – it is always something that already exists that finds itself transformed, whether in a more or a less revolutionary fashion. And this is indeed what I have come to conclude. Of the three revolutionary transformations that together mark the onset of the Scientific Revolution, the third (transition from 'coercive empiricism' to 'fact-finding experimental science') turns out to display far more continuity with its immediate predecessor than the other two (p. 260).

The issue of continuity came up first when, in the 1910s, Duhem argued that modern science really emerged in the 14th century, at the hands of certain Parisian scholastics whose forgotten manuscripts he rescued from undeserved oblivion. On this view, Galileo and Descartes et al. just put finishing touches on work essentially accomplished already. No historian of science accepts Duhem's conception wholesale any more; but the underlying conceptualization of 'medieval science' remains a lingering presence. I here break fully with those lingering remains. This makes it also possible to show the specific manner in which the Aristotelian innovations that Duhem's favorite scholastics certainly did bring about, after a quarter-millennium-long deadlock became instantly productive upon infusion with the revolutionary transformation that Galileo accomplished (p. 307).

There is another element of discontinuity that has not been conceptualized earlier. Modern mathematical scientists are accustomed to resolving unexplained phenomena by drawing large-scale analogies with better-understood phenomena. Here I show that, once Galileo turned 'motion' into a subject of mathematical science for the first time, the method of modeling thus-far-uninvestigated types of motion after types already known formed one element of radical discontinuity in the unfolding Scientific Revolution (p. 336).

- Above, I remarked that for the most seminal outcomes of the Scientific Revolution, local particularity has been gaining the upper hand over universal validity. In cultural history generally, it often seems as if research spent on the local and time-bound circumstances under which, by definition, each and every cultural product comes into the world pre-cludes acknowledgment of the universal value some of them may over time attain. No such opposition seems warranted, however. Attention ably given to what is situation bound in even the greatest and most enduring human accomplishments is of course a boon. Movies like *Amadeus* are particularly apt vehicles for reenacting the historical cir-cumstances under which timeless masterpieces came into being and were originally per-ceived. The power that such movies possess to enchant us comes from how they illustrate the very point here at issue. Opera buffs all over the world still enjoy the *Nozze di Figaro* two centuries after it was composed – in *Amadeus* we watch it born amid a meanwhile senseless, courtly power struggle over the respective merits of a German versus an Italian libretto. Here the tension that responsible historians face between careful reconstruction of the local and proper acknowledgment of the universal is made visible in a quite cogent manner.

In the history of science the tension makes itself felt with particular urgency. This is so in view of the claim to truth so often made for science. By the 1980s the claim ceased in many circles to be taken as self-evident. Two consequences followed. In its extreme ver-sion, nature itself has been defined as nothing but the outcome of negotiations between scientists that reflect, not any state of things objectively there, but only the momentary power balance between these scientists in their ongoing struggle for recognition and funding. Few historians of science have fully gone along with this reductionist message from the 'sociology of scientific knowledge'. Still, in their everyday practice many have come to regard the issue as a minefield better shunned. The rationale for this preference not to engage an issue of crucial importance runs roughly thus: much past historiogra-phy of science was unpalatably triumphalist, in that much of it used to be portrayed as a seemingly inexorable march toward our present-day truths; in order, therefore, to get rid of triumphalism we just bracket the claim to truth. The logic of this syllogism leaves something to be desired – there are other ways to accomplish the aim of nonpresentist history writing than by throwing out the baby with the bathwater.

I have dealt with the tension between the universal and the local in two ways.

In parts I and II, I make the following distinction. Certain premodern modes of nature-knowledge turned out soon after their emergence to be transferable to very different localities, thus acquiring distinct traits of universality. This applies notably to the transfer of mathematical science and natural philosophy from Greece to Islamic civilization and beyond. In contrast, certain other components of premodern nature-knowledge were marked far more by the culture, such as notably a 15th- and 16th-century movement of 'coercive empiricism' that was profoundly colored by peculiarities of European civilization. So paths from local to universal may be uneven, and only the evidence of history can help us make proper distinctions in this regard. As a consequence, the Scientific Revolution will be viewed in what follows, not as a one-time-only triumph of the universal, but rather as a historically situated blend of rapidly universalizing and forever local components.

From another point of view as well, the Scientific Revolution is not equated here with universal verities. For every mode of nature-knowledge under scrutiny I have investigated empirically how practitioners actually sought in everyday practice to make their knowledge claims clinching. Notably, the pioneers of the Scientific Revolution were well aware that the claim to indubitable certainty that was customarily made for the ruling philosophy of nature, Aristotle's, could no longer be maintained. So they went about finding out how far their everyday practice allowed them to stretch their own vivid sense of establishing something definitely valid about the constitution of nature. How to steer clear of arbitrariness, or the "fansying" that Newton still found to be rampant among his contemporaries?[12] For 17th-century fact-finding experimental science, historians like, notably, Shapin and Schaffer have made a beginning by listing ways and means actually taken by practitioners to make claims to knowledge as clinching as the case seemed to allow. Here I have been able to build on the literature and expand its main findings (ch. 13). But for mathematical science and for natural philosophy the question has not even been asked. For those two modes of revolutionary nature-knowledge I have made a comparative investigation into the means used by practitioners to attain maximum validity for their knowledge claims (chs. 10 and 11).

· Another recurrent topic that is closely connected to the Scientific Revolution and that is still largely governed by preconceptions is the one known as 'the scholar and the craftsman'. Here, too, by empirically investigating the subject I have sought to cut through a maze of positions taken largely on a priori grounds. In chapters 10 and 13 I examine for

each case of alleged improvement of craft practice by revolutionary science whether or xxxiii
not the promise was actually realized during the 17th century. For example, did organ
builders benefit from the mathematical schemes for temperament peddled at the time
by Huygens and others? If not, why not? Rather to my surprise, the conclusion, reached
at the end of an item-by-item survey of the literature, turns out to be that with very few
exceptions such promises did not even begin to be fulfilled until far into the 18th century.
This in its turn confirmed (or rather radicalized) my suspicion that ideology-fostered
expectations rather than actual *feats* of science-based craftsmanship helped give the new
modes of nature-knowledge a prominence and social anchoring without any precedent
in world history. It also served as the basis for an across-the-board investigation of what
contemporary impediments stood in the way of making reality match the promise of the
times.

· Finally, all those proliferating, monolithic causal theses that made further efforts to explain the Scientific Revolution look like a sheer waste of energy find a place in my account. Their authors were not stupid, and most of their theses, when transformed and/or
cut down to proper size, turn out to have a role to fulfill. Not only have I woven them into
the account, but for some (such as the Zilsel thesis and the Merton thesis) I point out in
the 'Notes on Literature Used' section appended to every chapter what in their original
makeup I find worth preserving.

Users' guide

> ... history requires every bit as much attention to detail as does science –
> and the history of science perhaps twice as much.
> Carl B. Boyer (1959)[13]

This book may well be used as a survey, but it is not set up that way. In particular, no historical facts are reported here on the mere ground that they happened. Just about every
fact related forms one part of the chain of my argument, be it as a small link or by way of
illustration of the point being made. Often a fact mentioned in passing at one place finds an
'echo' someplace else. For instance, in chapter 4 I mention that around 1540 the Portuguese
admiral João de Castro had his own reading of the compass checked by a caulker ("a practical man of great experience", as Castro went so far as to add). This is echoed in chapter 13,
where I cite Robert Boyle's lordly distrust in the ability of divers to produce trustworthy
reports. The 'echo' serves to remind us that the social distance between elite and plebeian
circles was not fixed but varied with time and place.

xxxiv The amount of evidence required to sustain an argument is not determined beforehand. Although exhaustiveness has not been my aim, with certain issues it proved imperative to treat all pertinent cases (as with my determination of how much improvement of 17th-century craft practice was attained by science of a new kind). But even at those many places where exhausting the issues would have been pointless or even self-defeating, an author may confine the evidence to be selected to the bare minimum needed to make his points, or he may catch it in a more tightly woven net. In this book I have opted for the latter. The selection of evidence has been governed throughout by a desire to enable the reader to enter into dialogue with my argument, so as to provide her or him with material suitable for possible disagreement. I do not want readers to feel constrained by the argument. Past facts can always be interpreted in more than one way, and the material that is presented should reflect that capacity. Together with the widely acknowledged complexity of the subject, this goes far to explain the length of the book.

Even so I could have made it longer still. I have not sought to include everything worthy of note that happened in the pursuit of nature-knowledge in 17th-century Europe. Much remained the same throughout the period – my concern has been with the emergence of revolutionary novelty against the background of what remained the same. Much novelty made for dispute in ever-widening audiences, for example, debates over witchcraft and over the portentous meaning of comets, in ways that seem to foreshadow the coming Enlightenment. I have neglected to discuss specifically such debates, focusing instead on the general midcentury crisis of legitimacy that early disputes helped engender and later ones helped resolve. Further, I have simplified much, at times perhaps beyond what specialists may be prepared to endure. For instance, at the end of the section 'Toward the *Principia*' I write that in the spring of 1685 Newton, the groundwork for his dynamics established and three major obstacles toward universal gravitation overcome, was now ready to write the *Principia* (p. 658). That way I tacitly bracket as needlessly cumbersome for the reader Newton's slaving at 'Lectiones de motu' which, through ongoing expansion and revision, gradually took shape as the *Principia*.

In the 'master narrative' of the Scientific Revolution, privileged treatment was given to mathematical science. More directly empirical and/or experimental subjects used to receive comparatively short shrift and were treated in any case as subordinate to the main story line. By the same token, conceptual development used to prevail in historians' treatment over the everyday practice of experimentation. On the rebound, experimental practice has over the past decades far outweighed purely conceptual development in mathematical science in most current surveys of the Scientific Revolution. I have here made an effort not only to give each their proper due but also to compare their respective ways of advancing – their dynamics and the means by which practitioners sought to make their conclusions stick. With fact-finding experimental science as also with speculative natural philosophy

this makes for straightforward reading happily interspersed with some hilarious episodes.   xxxv
With mathematical-experimental science this is not always so. Mathematical work must be
presented the mathematical way. I have done my best to clarify, to illustrate, and responsibly
to simplify. Even so, some elementary geometry on the reader's part is required for following
the argument in full.

For more than half a century the historiography of science has been plagued by an alleged antithesis between 'internal' and 'external' approaches. The distinction is not pointless.
Over the past decades ways and means have been found to treat conceptual development
in a sociocultural context. Still, present-day contextualism goes only part of the way toward replacing the somewhat artificial opposition with a more balanced treatment, in that
it often remains confined to local circumstances. In the present book relevant portions of
world history provide the principal sociocultural context, allowing me to show under what
conditions nature-knowledge could flourish and take the course it did. As a consequence,
each chapter mixes aspects of the development of ideas with their circumstances and effects.
In some chapters the latter predominate (notably chs. 12 and 17 on the mid-17th-century
crisis of legitimacy and how it was overcome), and in others the former, but even where
conceptual development is preponderant, questions of historical conditioning and historical impact are addressed at the spots proper to them.

Big-picture presentations like this one require the points successively made to be illustrated by example rather than by the full empirical material available. I have regularly taken
my examples from a domain that was once my own specialty, the history of what we now
call musical acoustics. Its frequent usage for illustrative purposes is not meant as an implicit claim that the grand developments I discuss were somehow driven by what happened
in musical acoustics. Rather, I employ them because of their freshness and also because
they confirm what few historians of science are prepared to recognize – that issues in musical acoustics form as much a legitimate and (at the time) fully recognized part of nature-
knowledge as planetary trajectories or magnets did and do. Something similar is true of two
other topics that are often neglected and that happen to reflect my personal background as
a Dutchman and a onetime museum curator – as the account proceeds, scientific instruments and Dutch practitioners appear more frequently than usual. More generally speaking,
since the Second World War the geographic center for the history of science has moved to
the Anglo-Saxon world, and there has been a tendency to focus disproportionately on what
happened in England at the expense of the Continent. England certainly has a special place
in the story, which I go to great lengths to point out. Even so I have aimed for a balanced
treatment of European nature-knowledge wherever it flourished.

In present-day historiography care is increasingly taken to avoid our modern terminology if it fails to match 17th-century realities, such as 'science' if meant as a general expression.
There is less agreement over what to use instead. Throughout the present book I have re-

xxxvi placed 'science' as a generic term with the deliberately bland expression 'nature-knowledge' (which can be found at times in 17th-century parlance). That way I enable myself to use more pointed expressions for its various components over time (notably for 'natural philosophy', here used in a sense that I take care to define, and also to distinguish from broader 'worldviews', on p. 9; 258). I reserve 'science', used in a sense that goes beyond individual disciplines, for two cases only. Right from the start I speak of 'mathematical science' without more ado. And as the Scientific Revolution advances, occasions begin to present themselves where the likeness to present-day science is so great as to make it sheer pedantry still to speak of 'nature-knowledge' or any other artificial equivalent.

Unless it becomes too awkward to do so, I have likewise avoided the names of present-day disciplines. I have scholars ponder light rays and vision rather than contribute to optics. This restriction applies a fortiori to 'mechanics'. I never employ the term before Newton in the *Principia* creates the discipline known to him and us as 'mechanics' (now for the first time including the 'dynamics' that in earlier times was missing from the 'statics' that went under the name of 'mechanics'). I also avoid the adjective 'mechanical', following a recommendation of E.J. Dijksterhuis, who once listed six distinct meanings in current use among historians of science. I have refrained in particular from adopting the expression 'mechanical philosophy', even though it was coined by a 17th-century 'actor', Boyle. Throughout the historiography of the Scientific Revolution the expression has given rise to undue identification (or at least association) of this speculative, wholly qualitative philosophy of nature with the mathematical science of (modern) mechanics. To the extent that the two ought to be associated indeed, this must follow from careful scrutiny of the historical data rather than be imposed by sheer terminological overlap. I have coined instead the expression 'kinetic corpuscularianism'. It may sound awkward, yet it points directly at the one feature that fundamentally distinguished this philosophy of nature from both ancient and contemporary atomism – its concern with how corpuscles move.

Two more terms eschewed are 'physics' and 'technology'. The latter term invites undue association with the science-based type of technology that we at present are accustomed to – not until the 18th century did isolated portions thereof begin to emerge from the Scientific Revolution. The phrase 'the arts and crafts' provides a ready-made replacement. The case of 'physics' is more complicated. Until the 17th century the term was used as fully equivalent to 'natural philosophy' in the sense here employed. One event among those that mark the Scientific Revolution as truly revolutionary is the change of meaning that 'physics' underwent in the course of the 17th century – more and more did it take on something close to its present-day meaning. An early example is what Kepler meant when entitling his first masterpiece 'New Astronomy, or Celestial Physics'. What he specifically meant by the expression is a newly dynamic treatment of the hoary problem of planetary trajectories; that is, he considered the force or forces in play in the solar system. In the expressions that I employ when rendering Kepler's work I have sought to convey the transition from 'natural philosophy' to something closer to our modern idea of what physics is about.

I abstain throughout from the expression 'early modern'. It carries with it an undue sense that the coming-into-being of our modern world was already preordained by the 16th century or thereabouts. In my view Europe was not *destined* to give rise to our modern world prior to the actual event, which I join a few global historians in situating in the early 19th century, when the Industrial Revolution began to run at full steam (p. 729).

Finally, I had to come to grips with 'rational' and its derivative 'rationality'. Tacitly at the background of much current denigration of the very idea of the Scientific Revolution is the suspicion that proponents of the idea mean to convey thereby the unique rationality of science. In his influential books on the Scientific Revolution, A. Rupert Hall portrayed the event as the triumph, long in the making, of rationality over magic and what he took to be other forms of obscurantism and superstition; others, in not quite so explicit a vein, have followed suit. In the Epilogue I dissociate myself from much in their interpretation. There I seek to define the *specifically pointed kind of rationality* that modern science embodies, in ongoing and indeed inevitable rivalry with other varieties of rationality that legitimately mark other human pursuits (p. 734). So as not to prejudge the issue by means of the sheer choice of terminology, I have generally abstained from using either 'rational' or 'rationality' before reaching the Epilogue. I have opted instead for 'intellectualist' when dealing with the classic contrast between 'rationalist' and 'empiricist' approaches, while using circumlocutions like 'reason-based' in the few cases where I had to distinguish some given effort from enterprises drawing on other resources (notably, religious revelation).

Big-picture thinking of the kind here undertaken requires a mind-set a little different from how one approached one's material in an earlier scholarly life. Also, the material itself differs in scope as well as in kind. Only in cases of prior personal acquaintance or of an urgent need to resolve a stubborn knot of major importance to the argument can one afford to address the sources themselves. Decades-long participation in the profession ought to give a sense of where to find some of the most reliable and most up-to-date literature on any given topic. In the 'Notes on Literature Used' appended to every chapter I have listed those books and articles subject by subject, with some further comments inserted. When dealing with literature treated previously in my 1994 *The Scientific Revolution. A Historiographical Inquiry*, I just refer to *SRHI,* followed by section number. Further, I have often consulted but not, as a rule, listed separately entries in the sixteen volumes of *Dictionary of Scientific Biography,* which Charles C. Gillispie edited in 1970–1980 (abbreviated *DSB*) and which is currently being electronically updated.

Passages that I quote are referenced in consecutively numbered endnotes. More than once in the historiography of science have misinterpretations (on occasion quite authoritative ones) emerged from faulty translation or even from entirely spurious sources. I there-xxxviii fore find it imperative to give the reader a chance to check the accuracy of the translations here rendered. On www.hfcohen.com I have placed them side by side with the originals.

NOTES ON LITERATURE USED

Conceptualization in history of science is a tricky business. For articulation of my intuitions on how to give concepts maximal sharpness in a historically responsible manner I have learned much from how it is done in Joseph Ben-David's *The Scientist's Role in Society. A Comparative Study*. Englewood Cliffs (NJ): Prentice Hall, 1971, and how it is both advocated and done in Nancy J. Nersessian, 'Faraday's Field Concept'. In: D. Gooding & F.A.J.L. James (eds.), *Faraday Rediscovered: Essays on the Life and Work of Michael Faraday*. London: Macmillan, 1985; pp. 175-187.

Surveys of the Scientific Revolution to come out before 1994 have been treated at length in part I of my *SRHI,* and briefly in an entry 'Scientific Revolution' that I contributed to W. Applebaum (ed.), *Encyclopedia of the Scientific Revolution from Copernicus to Newton*. New York/London: Garland, 2000; p. 589-593. Surveys to come out since are Steven Shapin, *The Scientific Revolution*. Chicago: University of Chicago Press, 1996; John Henry, *The Scientific Revolution and the Origins of Modern Science*. Hampshire/London: Macmillan, 1997 (3rd ed.: Houndsmill: Palgrave Macmillan, 2008); Rienk Vermij, *De wetenschappelijke revolutie*. Amsterdam: Nieuwezijds, 1999; James R. Jacob, *The Scientific Revolution. Aspirations and Achievements, 1500–1700*. Amherst: Prometheus, 1999; Michel Blay, *La naissance de la science classique au XVIIe siècle*. Paris: Nathan, 1999; Lisa Jardine, *Ingenious Pursuits: Building the Scientific Revolution*. New York: Nan Talese, 1999; Paolo Rossi, *The Birth of Modern Science*. Oxford: Blackwell, 2000; Peter Dear, *Revolutionizing the Sciences. European Knowledge and Its Ambitions, 1500–1700*. Basingstoke: Palgrave, 2001.

Also relevant here are two books covering larger periods: Stephen Gaukroger, *The Emergence of a Scientific Culture. Science and the Shaping of Modernity 1210-1685*. Oxford: Clarendon Press, 2006, and Peter Dear, *The Intelligibility of Nature. How Science Makes Sense of the World*. Chicago: University of Chicago Press, 2006. For the books by Dorn and Huff discussed in the main text the full data are as follows: Harold Dorn, *The Geography of Science*. Baltimore: Johns Hopkins UP, 1991 (the alignment there probed has been extended in James E. McClellan III & Harold Dorn, *Science and Technology in World History. An Introduction*. Baltimore: Johns Hopkins UP, 1999), and Toby E. Huff, *The Rise of Early Modern Science. Islam, China, and the West*. Cambridge UP, 1993.

David C. Lindberg's helpful discussion of the issue of continuity and discontinuity is in the final chapter (pp. 355-368) of his *The Beginnings of Western Science*. Chicago: University of Chicago Press, 1992 (2nd ed.: 2008).

Absent quest for underlying coherence. Katharine Park and Lorraine Daston, in their editorial 'Introduction: The Age of the New' to idem (eds.), *Early Modern Science*. Cambridge UP, 2006 (vol. 3 of the series 'The Cambridge History of Science'), pp. 1-17, characteristically explain their decision to leave 'The Scientific Revolution' out of the title by stating on p. 13 that "it is no longer clear that there was any coherent enterprise in the early modern period that can be *identified* with modern science" (my italics). Rather, they regard the concept of the Scientific Revolution as "a *myth* about the *inevitable* rise to global domination of the West" (p. 15; my italics). They do not explain why a renewed search for coherence (not really undertaken before, as all surveys since the mid-1980s have been meant primarily for the classroom) would be doomed at the outset, nor why the concept necessarily implies historical inevitability.

Terminology. E.J. Dijksterhuis' distinction between six different meanings of 'mechanical' occurs at the end of his 'Die Mechanisierung des Weltbildes'. In: *Veröffentlichungen der Gesellschaft für internationale Wissenschaftsgeschichte, Sitz Bremen*, 1952, pp. 5-31. Jack A. Goldstone's article 'The Problem of the 'Early Modern' World'. *Journal of the Economic and Social History of the Orient* 41, 3, 1998, has persuaded me never again to use the expression 'early modern'. Rob Wentholt's work (notably his typescript 'Selfish, Unselfish, And Much More Besides; A Treatise On The Nature Of Human Nature') has alerted me to how important it is in scholarship generally to refrain from setting up arguments that on investigation prove to derive from sheer terminological confusion.

PART I

NATURE-KNOWLEDGE IN TRADITIONAL SOCIETY

I

GREEK FOUNDATIONS, CHINESE CONTRASTS

The 'Old World', not our familiar modern world, is what I shall address. The various modes of nature-knowledge that constitute our subject emerged and were transformed long before our modern world came into being. The modern world did not begin to take shape until the 19th century, induced in good part by a new kind of technology based in its turn upon nature-knowledge of a radically new kind – recognizably modern science.

In the Old World things were done by tradition. Practices of the previous generation were passed on largely unchanged to the next. But for the effects of large-scale war or epidemic illness, life was much as one's parents' had been and one's children's was to be. Novelty, rarely absent entirely, was the exception rather than the routine phenomenon it has become for us. We shall focus on the incidental appearance and (with luck) consolidation of inherently transitory novelty as more germane to our quest than traditional verities and practices and ways of being. We ought to keep in mind all the same that tradition was always the backdrop of the novel developments that form our proper subject.

It was in the Old World that human beings in a variety of civilizations began to perceive and examine regularities in certain natural phenomena. Babylonian astronomy, the Mayan calendar, and the minute observations underlying Polynesian seafaring are cases in point. However perceptive, intricate, and accurate such pursuits might be in execution and outcome, they were narrowly specialized. A Babylonian stargazer by, say, 700 BCE would record solar eclipses and planetary oppositions, but he did not examine any other natural phenomena in so sustained and orderly a fashion. No overall conception lurked behind such achievements other than that events in the world around us were ascribed to specific, unmediated, ongoing actions of task-oriented gods or godlike figures.

Two civilizations to start and then maintain a tradition of their own in the pursuit of nature-knowledge failed to obey the rule of one-domain specialization. In China and in some Greek city-states practitioners widened their coverage. Not only did their researches range from celestial events to properties of musical sound, of living bodies, of magnetic stones, and of many more phenomena, but also these investigations took place against the background of a comprehensive view of how the world is broadly constituted. Pattern and order in the world were sought for and given expression in ways that, while not as a rule excluding the divine, did entail attribution of the generality of natural phenomena to agents or forces operating without immediate divine intervention. 'Naturalist' has become the standard label for such conceptions.

4   In China, the outlines of a naturalist worldview that encompassed a budding understanding of a wide variety of natural phenomena emerged during the Warring States period, which preceded the unification of 221 BCE. This worldview came about on its own, without stimuli or borrowings from the outside. In contrast, the Greeks owed quite a debt, of which they were well aware, to the civilizations of Egypt and Babylonia. At trade centers where Greeks mingled with foreigners, the early pre-Socratics raised an already impressive achievement to a new plane of understanding, in a profound process of transformation misleadingly known of old as 'the birth of science and philosophy'. In the one case, then, the sustained pursuit of nature-knowledge was undertaken from scratch, whereas in the other we have a large-scale expansion and elaboration of a range of specialized ideas and findings adopted and ingeniously adapted from elsewhere.

Not only were the foundations different, but so were the outcomes. The Chinese endeavor yielded in the end one unitary, comprehensive worldview. It animated, and gave coherence to, a vast range of specialized conceptions and discoveries. The Chinese attained these by creatively examining heaps of empirical data (p. 33; 45). The Greek endeavor was grounded in an intellectualist approach rather than in the laborious collection of sense data. Not unitary at all, it was made up of two noncommunicating components. One component, centered in Athens, contained not one but four distinct worldviews, each in the specific format of a natural philosophy. The other, centered in Alexandria, consisted of bits and pieces of highly abstract mathematical science. A millennium and a half later, in late Renaissance Europe, the Scientific Revolution was to emerge in large part from their radical transformation. 'Athens' found itself transformed into 'Athens-plus'; 'Alexandria' into 'Alexandria-plus'. And about halfway through the Scientific Revolution, by the 1660s, Alexandria-plus and Athens-plus began to cross-fertilize – the first time in history for two distinct modes of nature-knowledge to do so. No explanation of these revolutionary transformations is at all possible without a prior understanding of (1) the outlines of Athenian natural philosophy, (2) the outlines of Alexandrian mathematical science, and (3) the fundamental thesis here maintained that both intellectually and institutionally a vast chasm yawned between the two, leaving what little interaction took place between them sparse and inconsequential. We take up these three topics in turn.

A tale of two cities

Athens.   The founding of schools of natural philosophy began with Plato early in the 4th century BCE. By the end of the founding period in the 3rd century BCE, four philosophies of nature had been established in the city. Those of Plato and his pupil Aristotle arose when Athens was still an independent city-state; atomism and the Stoa arose and became established after Alexander's conquests upset the entire order of the eastern Mediterranean. The four philosophies shared the following features. Natural philosophy was no more than part of the founders' efforts to explain the world in its totality – in each, political and ethical concerns were paramount. Each felt that human beings can attain secure knowledge of how the world is broadly constituted. Each provided an account of how such knowledge is actually attained. Each took its starting point in problems and doctrines first raised by the pre-Socratics. Each of the founders established a school of his own, where basic teachings were handed down over the centuries and discussed with pupils under the guidance of, first, the founder and, later, a range of hand-picked successors. All four had their conclusions and their ways of reaching them challenged at a fundamental level by the antiphilosophy of the skeptics, who founded a school of their own by the end of the creative period of Athenian philosophy.

Plato's philosophy took its starting point in the question of how best to live. His answer was: 'in justice, to be acquired through a visionary insight into the world of Ideas or Perfect Forms; in particular, into the Idea of the Good'. The conception of a realm of Ideas resting behind the phenomena that appear to our senses represents Plato's solution to a fundamental problem raised by Parmenides. This pre-Socratic thinker had maintained the permanence of things in a world only seemingly in flux. To come to terms with the challenge involved in his radical assertion, Plato posited the existence of a realm – to be grasped only by the illuminated mind – of eternal, unchanging Ideal Forms. Each corresponds to, and is itself the source of, its own imperfect manifestation in the world of the senses. The conception drew heavily on another pre-Socratic attainment – the elevation of mathematics from the concreteness of Egyptian and Babylonian computation rules to the abstract plane of proof given for every statement, even for properties that seem self-evidently true. Plato argued roughly thus. Angles of a triangle add up to 180° – this is a proven property. But measurement of a triangle drawn in the sand will yield some value close to, yet more or less subtly different from, 180°. The property is rigorously valid only for a mentally imagined triangle. Just so are all trees here on earth no more than imperfect embodiments of the one, perfect, and eternal Ideal Form of the Tree. Similarly, all examples of good conduct are weaker or stronger reflections of the Perfect Goodness to which all humans should aspire.

This is not a mathematical philosophy in the sense of mathematical entities themselves entering the picture. They do so only in Plato's sole excursion into more detailed natural philosophy, his *Timaios* dialogue with its speculations about the constitution of the material world out of tiny triangles of various shapes. Even this is quite unlike what mathematical science has come to mean since, with not so much the shapes of things but rather their mutual relations finding mathematical expression. With Plato, instead, mathematics as transformed by the early pre-Socratics provided his epistemic model – learning mathematics serves as a preparation for grasping the doctrine of Ideal Forms. In the first all-around philosophy to appear in the Greek tradition, then, mathematics held an elevated status. It formed one cornerstone of a comprehensive philosophy which had as its primary function to teach human beings how to live for themselves and how to live together. Unlike with the predominantly naturalist pre-Socratics, the understanding of natural phenomena is here pushed to the margin.

This and more came out quite differently in the even more comprehensive and systematic philosophy that Aristotle conceived in critical dialogue with his master's teachings. Aristotle's reigning passion was knowledge for its own sake. He sought it first of all in a sustained effort to understand life-forms, notably the parts of animals and the functions thereof. Inspired by such researches, he reconstructed Plato's leading ideas so that he could give more ample room to the world of sense. The essence of the phenomena that we find in the world does not reside elsewhere, in an autonomous realm behind appearances, but rather in the phenomena themselves. There is no Ideal Tree, but there are only trees, sharing an essential nature which we must identify and bear in mind while examining the ongoing changes to which every single tree is subjected. These changes are purposeful; they are directed toward definite ends. It lies in the nature of the tree to realize – to bring out into the open as it grows from seed to mature tree – those essential properties which have rested in it from the start as potentialities. In defining Becoming as the realization of what is potentially there, Aristotle solved in his own way the problem first raised by Parmenides of how to account for change.

In Aristotle's scheme of things the place assigned to mathematics is far less elevated than in his teacher's. There are four categories of change: change of being, or generation and decay; change in properties, or qualitative change; change in degree, or quantitative change; and change of place, or motion. With Aristotle mathematics is neither less nor more than the branch of learning that may assist us in grasping quantitative changes. Its status is on a par with intellectual tools that are meant to help us understand other, more consequential kinds of change. There is a domain of quality and there is a domain of quantity, and the two barely touch. As Aristotle insisted, the realm of the qualitative is irreducible to number.

This approach to natural phenomena seems at first sight to leave much room for observation and for a mind open to unexpected phenomena nature may offer us. In Aristotle's specialty as a scholar, comparative analysis of the parts of animals and their behavior, this is indeed how he proceeded. Yet the fundamental principles he derived from comparative analysis (notably, change conceived as the actualization of what is potentially there) lead one to assign to sense data their fitting place inside an overall scheme of things. Phenomena, taken at the outset as what they appear to be, are quickly assigned their proper place in a comprehensive system which has an explanation for everything that lends itself to explanation. For instance, in watching a branch fall to the ground we notice that the heavier the branch, the faster it descends, and also that it does not attain full speed until some time has elapsed. On further reflection we recognize in free fall one manifestation of a tendency shared by all heavy bodies in the terrestrial sphere: to get as close as possible to the center of the universe, which is their 'natural place' – the ultimate purpose of their existence. Fall thus belongs to the class of natural motion; natural motion is one subclass of motion as such (as distinct from violent motion, which is all those movements not directed toward natural place); motion is one subclass of change; change, or becoming, finally, is the realization of what a thing potentially is. In this manner even such a humdrum event as the fall of a branch to the ground is shown to be linked, in a multilevel, logically coherent system, to one core problem in Greek philosophy – that of Being and Becoming.

Aristotle had imbibed the teachings of Plato that his own philosophy sought to reconstruct at the school that Plato founded, the Academy. On Plato's death Aristotle, a foreigner to Athens, was passed by for the succession, so in due time he founded his own school, the Lukeion ('Lyceum'; 335 BCE). In the late 4th and early 3rd centuries BCE, when Athens lost its independence first to Alexander and his successors and then to the Romans, three more schools of philosophy came to be founded there. A selection of pre-Socratic teachings not employed by Plato and Aristotle reappeared in the atomist school (founded by Epikouros) and in the Stoa (named after the building in Athens where the founder, the Cypriot Zenôn, set up his school). The existence of no fewer than four rival conceptions of the world, each assured of the truth of its knowledge claims, gave rise to sustained doubt about the very possibility of knowledge. Such doubt took shape in a fifth system of philosophy or, more properly speaking, an antiphilosophy: skepticism (from Greek *skepto*, 'to judge'). Here is an outline of the main teachings of the atomist and stoic schools.

Atomists focused on the discrete. They held the world to consist solely of void space and of atoms moving through it in straight lines. Every now and then, however, atoms are diverted to an oblique 'swerve'. In natural philosophy, the swerve that Epikouros attributed to their motions served to produce whatever coherence a world of moving particles can possess; in ethics, it created room to attribute free will to human beings. One lengthy exposition of the doctrine has been preserved: in the 1st century BCE a great Roman poet, Lucretius, was attracted to it and expounded on it in metered detail. His *De rerum natura* (On the Nature of Things) lists numerous illustrative facts taken from everyday life. One example is fine dust particles dancing in a sunbeam that enters a darkened room; another, the invisible yet, over time, unmistakable wearing away to which, for example, rocks and metal rings are subject. How may one account for this other than that the material disappears particle by particle, due to the ongoing action of the sand or our inadvertent rubbing? Such everyday phenomena were used to convey their capacity to be brought into line with a speculative doctrine already settled in advance. Epikouros left his disciples free to choose the details of the specific explanation best suited to enhance the individual tranquility of mind (*ataraxia*) that formed the final objective of atomist philosophy.

In sharp contrast to the discreteness of a world populated by atoms, Zenôn and his disciples in the early Stoa worked out a no less speculative conception of the natural world in which the *continuity* of things provided the main theme. Chrysippos took the lead in ascribing the generality of natural phenomena to tensions in the *pneuma*, an aether-like blending of air and fire that he held to pervade the cosmos. He distinguished ordinary translations from those motions that are carriers of disturbances. For example, the propagation of sound is imagined to take place by comparing the phenomenon to the ripples that occur when a stone is thrown into a quiet pond (where the ripples carry forward, not the water itself, but only the original disturbance). Consequently, whatever happens in the world reverberates all through it, ready to be picked up someplace else, the way a spider senses whatever touches its web be it ever so subtly. Things and events are connected and may affect one another in ways that depend upon the nature of the correspondences between them. Those governed by mutual sympathy lead to attraction; those governed by antipathy bring about repulsion.

In each of these natural philosophies – Platonic, Aristotelian, atomist, stoic – arguments pro or con a certain explanation are drawn from two resources exclusively. They stem either from abstract reasoning or from everyday examples that were invoked to lend empirical plausibility to any given system of natural philosophy. In practice, they stemmed from a mixture of the two. Take the objection that one Aristotelian, Alexander of Aphrodysias, raised against so fundamental a stoic point as the tension exerted by the *pneuma*. Alexander was bound to reject the *pneuma* in view of Aristotle's own aether, which remains confined to the immutable celestial realm and is not therefore thought of as a carrier of dynamic action. Alexander's objection appealed to the alleged tenuousness of the *pneuma* – how can such subtle stuff possess a capacity to make far more solid things cohere? His stoic opponent was then bound to retort that *pneuma* owes that capacity to its characteristic blending of the properties of air, which is elastic, and of fire, which is active.

What we have by the mid-3rd century BCE, then, is four distinct systems of philosophy. Each has disciples attached to a school in the literal sense (Academy, Lyceum, garden of Epikouros, Stoa). Each looks plausible on the face of it. Each is locked in ongoing, at times strident, rivalry. Yet none is in a position to clinch *its* particular claim of possessing with indubitable certainty the one and only *truly* true conception of the world. Indeed, their ongoing rivalry gave rise to fundamental doubts about what humans can know at all. In a fifth school of philosophical thought founded likewise in Athens, a range of skeptical arguments was developed, the sum total of which was meant to show that all our presumed knowledge is a sham. Our senses as well as our intellect may deceive us in a large variety of ways, and in reality we know nothing. According to the majority of skeptics, following Pyrrhôn, we cannot know even that, so the only way open to us is a suspension of judgment about whatever goes beyond the immediately apparent. And with this message of utter epistemic despair the creative upswing of Greek philosophical thought came to an end.

Theory: the concepts of 'philosophy' and 'worldview' distinguished. Each founder of a school appealed to core ideas of some pre-Socratic predecessor for his own philosophy of nature. Both Plato and Aristotle sought to resolve the core problem raised by Parmenides of how, in an intellectually satisfactory manner, to understand the very possibility
of change. In seeking likewise to balance the apparent presence of change *and* permanence
in the natural world, Epikouros exploited an idea first raised by Demokritos (his conception
of immutable atoms, as yet without swerve), just as Chrysippos did with Herakleitos (his
idea of a dynamic world-in-flux held together by fire). The big difference in all these cases
is that what with the pre-Socratics had taken shape as a broad worldview now found itself
systematized into a full-fledged philosophy. All four Athenian systems of philosophy are
marked by a knowledge structure of first-principle thinking. Whether the empirical content
of one or another philosophy of nature be smaller or larger, the knowledge structure pervading it was one of comprehensiveness, of an aimed-for totality, of prestructured patterns
emerging from very general, basic principles, with an anchoring in the phenomenal world
being sought in assorted bits and pieces of apparently well fitting empirical evidence. *This
peculiar mode of attaining and organizing knowledge is the one feature that all four Athenian
natural philosophies, however different otherwise, held in common until far into the 17th century.* Throughout the present book I take this to be the defining characteristic. And wherever
I use the expression 'natural philosophy', this and nothing else is what I mean by it.

Taken in this strict sense, then, philosophy appears as the more pointed variety of something broader but also less clear-cut, a distinctive conception of how the world hangs together, or *worldview* for short. The pre-Socratics produced a large variety of worldviews.
What emerged at roughly the same time in Chinese civilization shall likewise prove to be a
range of worldviews, gradually solidifying after unification of the Chinese polity under one
emperor into one broad, consensual worldview.

Whether a worldview be tightened into the strict format of a natural philosophy or not,
there are always two sides to it. It helps to organize phenomena and is able to accommodate
certain phenomena not yet perceived or included in the grand scheme of things. But it also
closes off other ways to organize experience, not recognizing certain phenomena outside its
scope or not admitting that phenomena within its scope could be viewed in quite different
ways. It depends in good part on the quality of a worldview and on how it is handled by
successors whether the 'widening' or the 'closing-off' aspect prevails. Or rather, one way to
judge the quality of a worldview is the extent to which it displays a creative tension between
these two aspects.

Most often such tension remains hidden under the surface of the worldview. The best
chance for it to come into the open and reveal its creative potentialities resides in the confrontation with a rival worldview. But chances for productive confrontation, never large
in the premodern world, were bound to vanish as soon as first-principle thinking, with its

definite claim to definitive truth, came to freeze the relative fluidity of a worldview into the rigidity of a natural philosophy. As the objection Alexander of Aphrodysias raised against stoic *pneuma* exemplifies, clashes between philosophies of nature were bound to remain sterile. But this leaves open what might happen in the different situation of a rival mode of attaining knowledge of natural phenomena arising, not *inside* such first-principle thinking, but rather from the *outside*. Indeed, in the late 4th century BCE such a rival emerged. In Alexandria certain phenomena of nature were tackled, not as subordinate portions of an overarching conception of the world, but as relatively autonomous, far more disjointed pieces of knowledge.

Alexandria. Alexander's conquests turned Greece at one unpredictable stroke from a sideshow on the stage of world history into the cultural center of the civilized Mediterranean world. Once again a large-scale admixture with foreigners provided a powerful incentive toward vigorous cultural exchange and consequent flourishing. Besides Athens, the center of philosophy in the sense just discussed, the new-founded capital city of Hellenist Egypt, Alexandria, quickly became the intellectual center for nature-knowledge of an in many ways quite different kind. Its focal point was the Mouseion (House of the Muses) founded there on the orders of Alexander's successor in Egypt, his erstwhile general Ptolemy. In this environment a range of natural phenomena was subjected to mathematical treatment over the course of some two centuries (c. 320–120 BCE).

In surveying the range I omit almost all technical detail. To the extent that such details were to prove instrumental in the revolutionary transformation of the Greek corpus that took place in Europe around 1600, I explicate them in later chapters. For now it is the big picture of mathematical science in Greek antiquity that we must grasp.

In geometry, Euclid turned into one coherent system a range of theorems and proofs accumulated over several generations by certain pre-Socratics and by mathematicians in Plato's circle. From five axioms (unprovable, yet apparently self-evident propositions) Euclid derived by means of rigorous proof all known theorems in plane and solid geometry and in the theory of proportions. The most important mathematical work done besides this did not take place in Alexandria itself but on Rhodes and in Syracuse, with Alexandria still serving as clearinghouse or at least as a point of intellectual convergence. This concerns Apollonios' treatment of conic sections and Archimedes' virtuoso usage of rigorous methods for determining areas and volumes of various kinds of irregular, mostly curved figures. The procedure employed by Archimedes is known as the method of exhaustion. It was meant to circumvent in a rigorously demonstrative manner certain dilemmas of the infinite and the infinitesimal (by the early 17th century European mathematicians began to handle these in looser fashion by means of the limiting methods of what then became the calculus; p. 346).

Rigorously deductive geometry was used to analyze certain empirical phenomena: musical sound, equilibrium states, light rays, and planetary trajectories. The prime example that Alexandrian geometers followed in treating these empirical phenomena in mathematical fashion stemmed from one particular pre-Socratic school, the Pythagoreans. The key discovery has been attributed to Pythagoras himself. There are a few cases in which musical notes, when sounded together or in quick succession, appear to blend well. These are the consonant intervals, and they are produced by vibrating strings whose lengths are to each other in ratios made up of the first few integers. The unison appears to be given by 1:1, the octave by 1:2, the fifth by 2:3, and the fourth by 3:4. All other, dissonant intervals are represented by ratios with less neat numbers (e.g., the major third, which is a dissonance in the Pythagorean scheme of things, by 64:81). Hence, the natural phenomenon of consonance is limited to the four first integers. To Pythagoreans 1 + 2 + 3 + 4 (the *tetraktys*) was a sacred number, which served to explain why there are only so many consonant intervals.

The discovery has the startling implication that a certain regularity in nature (we hear certain musical sounds as combining well) appears to correspond not roughly but exactly to something on a quite different plane – an abstract mathematical regularity. In retrospect, the road toward our mathematical physics may seem to lie open.

But only in retrospect. The frame and context of pre-Socratic thought led Pythagoras and his followers to ignore almost fully the real-world aspect of their setup – those divisions of the vibrating string. Rather, they went on to treat the ratios for the consonances, not as prime markers of to-and-fro motion, but as starting points for intricate exercises with abstract numbers. And so it was with those Alexandrians, like Euclid, who worked out these Pythagorean findings in systematic fashion and in greater detail.

A similarly abstract approach was taken by Archimedes when he set out to examine equilibrium states of solids and fluids by means of mathematical reasoning. His proof for an already well-known empirical rule concerning the force applied to a balance beam and the distance covered that way (known since as the law of the lever) is rendered in Figure 1.1.[14]

Archimedes did not use a numerical example. He gave the proof first for any ratio of integral numbers, and then for ratios with any numbers at all. The validity of the proof has been questioned. The distribution of the weights presupposes that the common center of gravity of two weights suspended from a beam remains undisturbed by the distribution – a presupposition which in its turn presupposes a theory of centers of gravity not even mentioned by Archimedes. Also, is not the second axiom, properly considered, equivalent to the very proposition that Archimedes set out to prove?

Archimedes also found, reputedly in his bath, that the apparent loss of weight of his or any other body immersed in a fluid equals the weight of the fluid displaced thereby. As with the proof of the law of the lever, which omits all reference to friction or to resistance of the air, Archimedes' presentation took shape as a process of mathematical abstraction – not justified by him in any way but tacitly handled as if it were self-evident. He considered

The proof (in a simplified rendition here simplified further by means of a numerical example) runs as follows. Two axioms state that: 1) a lever loaded symmetrically is in balance; 2) a body suspended from a lever may be replaced by one of the same weight if the distance from the point of suspension to the fulcrum remains the same.

Let be given a beam with fulcrum O and weights W_1 and W_2 suspended at A_1 and A_2.

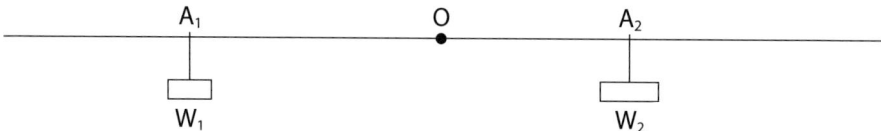

Proof is required that the lever remains in balance if the weights are inversely proportional to the respective distances of their points of suspension to the fulcrum, i.e., if W_1:W_2 = OA_2:OA_1. The weights are represented by integral numbers, say, 3 and 4 ounces, respectively. Hence, the distances OA_1 and OA_2 (measured in, say, inches) are as 4 and 3. Now find points B and C on the lever such that both BA_1 and CA_1 equal OA_2 (hence, BC = 6, and CA_2 = OA_1 = 4); likewise, find point D such that DA_2 equals OA_1 (hence, DA_2 = CA_2 = 4 and OB = OD = 7).

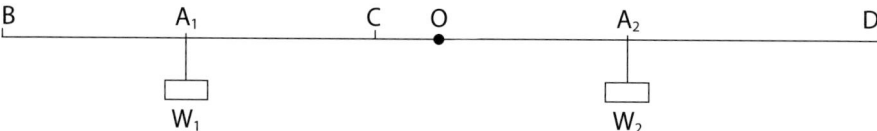

As a result, BC comprises 6 inches, and CD 8. Now divide the beam between B and D into 14 segments of 1 inch each.

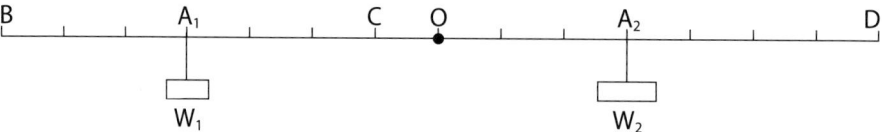

Replace weight W_1 by 6 weights of ½ oz. each, and distribute these such that each is suspended at midpoint of a segment; weight W_2 is likewise replaced by 8 weights of ½ oz. each.

These replacements are allowed by the second postulate and result together in a symmetrical loading of the beam; hence, given the first postulate, the lever is in balance (which was to be proven).

Figure 1.1: Archimedes' proof of the law of the lever

everyday phenomena like immersed and floating bodies or suspended weights, not in their 13
full concreteness, but rather as divested of their materiality, thus rendering them susceptible
to mathematical treatment.

Archimedes exploited his theoretical findings by devising a range of ingenious machines
of mostly legendary yet on occasion actual construction. Later, several Alexandrian practitioners, including Ktesibios and Herôn, demonstrated that the five 'simple machines' then
in use (lever, pulley, wedge, windlass, and screw) all obey the law of the lever.

In similarly mathematical fashion the paths of light rays in reflection were subjected to
geometric treatment by a variety of authors (Euclid among them), most of whom assumed
that vision occurs by way of the eye emitting, not receiving, rays of light.

Most importantly, the course of the heavenly bodies was taken up in mathematical fashion. This was not a simple matter – if one follows their trajectories over a range of successive nights, numerous irregularities turn up, most drastically so, of course, with the planets.
How to get a handle on those irregularities? Here Plato's influence upon the mathematizing
trend in Greek nature-knowledge of the 3rd and 2nd centuries was most in evidence. In accordance with the doctrine of Ideal Forms, he allegedly required the irregular movements
of all celestial bodies to be analyzed in terms of circular motions uniformly traversed. The
requirement derived from the ideal regularity and perfection of the circle, but later mathematical astronomers saw as an added boon that of all curves the circle is defined by a
uniquely limited number of points. With those three points established a priori or determined through observation, the whole figure is given and if, in addition, a uniform angular
velocity is posited, times passed can with relative ease be computed from spaces traversed or
the other way round. Hipparchos invented some ingenious devices to reduce the complications – the 'inequalities' – of planetary motion to the relative simplicity of combinations of
uniformly traversed circles. For example, he accounted for the inequality of the duration of
the seasons by means of a model in which the Sun revolves around the Earth such that the
Earth is placed at a slight distance from the center of the circle. That is, heavenly order is
restored by taking the circle to be *eccentric* (Figure 1.2).[15]

The eccentric was one of three devices that made it possible for the mathematical astronomer to do his job properly. Hipparchos also invented the epicycle, and three centuries
later Ptolemy came up with the equant point (both shown on p. 107, where I discuss the dissatisfaction of Ptolemy's greatest intellectual heir, Copernicus, with one of them). The three
devices make their joint appearance in Ptolemy's *Almagest* (the book's Arabic title slightly
Latinized, stemming from *Megistè syntaxis,* or Grand Synthesis). This book, written c. 150
CE, provides a detailed, technical synthesis of Greek mathematical astronomy, with one
model for each heavenly body in apparent motion – Moon, Sun, Mercury, Venus, Mars, Jupiter, Saturn. The book's degree of originality cannot be assessed with precision, since almost
no original work by Hipparchos or other preceding mathematical astronomers survives.

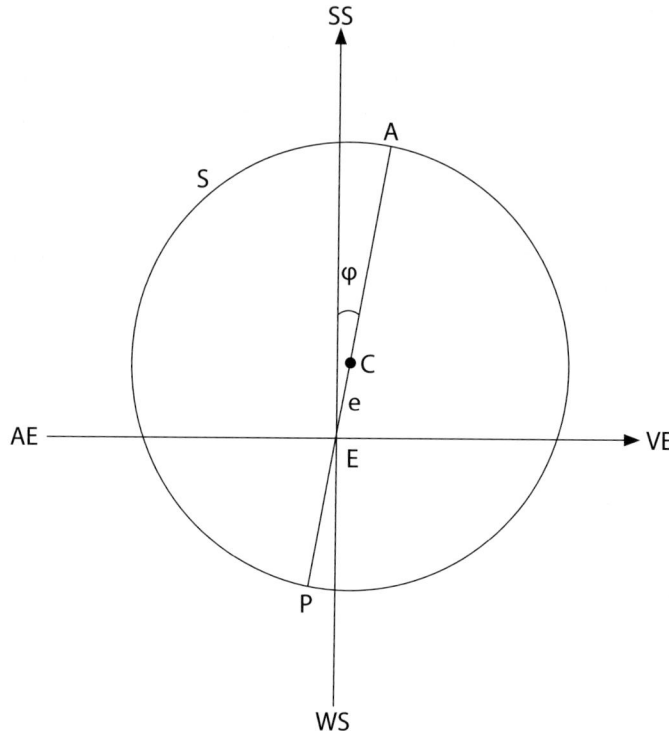

Since assuming the Sun to traverse the ecliptic with nonuniform angular velocity would violate the requirement of uniformly traversed, circular orbits, Hipparchos took the Earth not to be situated quite at the center C of the Sun's circular orbit, but rather at E.

The Sun was known to take 94½ days from vernal equinox VE to summer solstice SS, and 92½ days to get from there to autumnal equinox AE, thus using 187 days for spring + summer and leaving 178 days for fall + winter. For the hypothesis of an eccentric circle to fit these quantitative data, Hipparchos had to determine the distance CE, expressed as a fraction e (for eccentricity) of the diameter AP (constructed by extending CE to point A, for apogee, where the Sun is farthest from the Earth, and P, for perigee, where it is closest), and the angle φ determining the position of AP (called the line of apsides) in relation to the seasonal points.

Hipparchos arrived in the end at a value of 1/24 for e and of 24°30' for φ.

Figure 1.2: Hipparchos' account of the inequality of the seasons

Not only Hipparchos and Ptolemy but every single Greek astronomer but one operated within the framework of a geocentric universe. The exception is another mathematical scientist, Aristarchos. He did not, however, work out in any detail his idea of a Sun-centered

universe, with the Earth a planet like the five others. Certainly in the absence of mathematical detail, the idea had little if anything to commend it – plain common sense and everyday
observation clearly show us otherwise (the arguments pro and con for placing the Sun at the
center will be discussed in connection with Copernicus: p. 111). On this near-undisputed,
geocentric basis, and with use made in addition of Babylonian observations and computation methods, Greek mathematical astronomy had by the end of its creative period attained
a great deal. With knowledge acquired of the spherical shape of the Earth, the what and
when of eclipses, the positions of numerous stars, and so on, Greek astronomers were able
to predict the positions of the Sun, the Moon, and the other wandering stars with an accuracy amounting to c. 10 minutes of arc (not to be surpassed until, in the late 16th century,
Tycho Brahe narrowed it down to 2 minutes, the very limit in pretelescopic observation;
p. 129).

Athens and Alexandria compared

It is time to take stock and compare. There was, on the one hand, a plurality of comprehensive philosophies. Their first principles went beyond issues of political action and individual
conduct to encompass the constitution of the natural world. The principles found empirical
support in a selection of well-fitting observations, fairly small with Platonism and the Stoa,
larger with Aristotelianism and atomism. All this was centered in Athens. There were, on the
other hand, a few pockets of mathematized science, in tandem with a well-organized body
of deductive geometry, all of which had Alexandria as intellectual center. In sustained comparison between these two distinct modes of Greek nature-knowledge – systems of natural
philosophy and pieces of mathematical science, respectively – I shall first consider what
place each occupied in society. I shall then list the numerous ways in which the structure of
Athenian knowledge differed from its Alexandrian counterpart. Finally, I shall examine the
how and why of the wide gap between them.

An anchoring in society.   Throughout this book I confine the expression 'natural
philosophy' to first-principle speculation designed to render a causal account of the totality of natural phenomena. Not that such an entity has ever existed *as such*. Wherever
we encounter natural philosophy, we encounter it as part of an overarching philosophy
directed as well or more so at overtly human concerns. This has always been philosophy's
mainstay in society. The philosopher was seen as a person skilled in public argument and
in spreading wisdom about the world and how to conduct oneself in it. A philosopher as a
rule earned his living by instructing others in these matters, whether they be aspiring poli-
ticians (as in Greek city-states) or a broader audience in search of directions for achieving

16 individual happiness by understanding the world one found oneself living in (as in the Hellenist world or, later, the Roman Empire). Philosophy could also be the principal concern of gentlemen of leisure or the part-time occupation of men gainfully employed in other ways, as in public oratory. Further, philosophy could on occasion benefit from overlap with other occupational areas. In the Greek, Hellenist, and Roman worlds a range of learned doctrines of healing that were eventually collected in the Hippocratic corpus served to some extent as a cognate occupational domain. When, in the 2nd century CE, the Roman physician Galen codified in writing the medical knowledge of his time, he based certain teachings upon current ideas about natural phenomena. For example, he grounded the idea of the four bodily humors that he encountered in Hippokrates' teachings in Aristotle's doctrine of the four elements.

Philosophy, then, managed to remain a presence in the ancient world until its very end. In contrast, mathematical science proved dependent for its support in society upon one particular institution, royal patronage. Frustratingly little is known for certain about the support a succession of kings of Hellenist Egypt (all named Ptolemy, after the founder of the dynasty) gave to the cultivation of mathematical science in their newly built capital, Alexandria. We do not know how or to what extent practitioners depended on its central building, the Mouseion and its library housing thousands of scrolls, or the intensity of their exchanges or what motivated their royal protectors (although the furtherance of astrological forecasting would seem a safe bet). Nor do we know for how long patronage was extended by the dynasty beyond the first three kings. It can nonetheless be maintained that the specialized pursuit of mathematical science in antiquity stood or fell with royal support – those practitioners we know of were with few exceptions connected either to the Alexandrian court or to a secondary one (Pergamon, Rhodes, Syracuse) in close touch with it.

The structure of knowledge in philosophy and in mathematical science. The ways in which knowledge was attained in philosophy and in mathematical science shared one major characteristic: their overriding *intellectualism*. Both Athens and Alexandria (I shall use the city names as shorthand for the two distinct modes of nature knowledge) provided a priori constructions, with concepts predominant and empirical phenomena assigned a subordinate place in the overarching conceptual edifice. Nevertheless, the nature of the concepts and the nature of the subordination differed greatly between them. Here is how they differed.

In Athens the central operation was explanation through the positing of first principles; in Alexandria, description in mathematical terms. First principles of various kinds were put forward by a range of Athenian thinkers; what these first principles had in common was, indeed, their being *posited*, with a blend of inner self-evidence and external, empirical illustration serving to underwrite their validity. Validity was held in each case to be warranted by the very nature of the principles – but for the level of details, knowledge was not just probable but established once and for all. Alexandrian thought had no use for any such first principles. Practitioners took the basics for granted. Their sole aim was to establish mathematical regularities without explanatory pretensions or underlying ontology. Still, they also laid claim to indubitably certain knowledge, albeit attained quite another way, by means of mathematical proof sought and given for each successive theorem.

In Athenian thought, empirical phenomena are used as examples, selected primarily for their capacity to illustrate the validity of the first principles posited. In Alexandrian thought, selected phenomena are employed as individual points of departure for mathematical analysis. Each school of natural philosophy was ideally capable of explaining each and every phenomenon in terms of its first principles, which after all apply to the whole world. In practice, empirical evidence served primarily to enhance the intuitive plausibility of the first principles. In Alexandrian thought, just as a vibrating string gave occasion to observe the numerical regularity of the consonances, other objects of sense, like balance beams or mirrors or lenses or planetary positions, could give rise to mathematical analysis if they proved susceptible to such treatment in the first place.

Athenian thought was comprehensive; Alexandrian, piecemeal. The aim of Athenian thinkers was to grasp the whole, to explain the world or at the very least to understand that which gives the world the inner coherence they assumed it possessed. In Alexandrian thought, investigators went about their researches one at a time, without positing or even seeking any necessary coherence between them. Here the sole common thread was the mode of investigation they applied: the judicious application of mathematical relations (congruence, equality, proportion, and so on) and of known theorems.

Athenian thought solidified in four rival and alternately dominating schools, plus one school skeptical of the very possibility of ascertained knowledge. The leading names attached to each school are naturally different. In debates between adherents, any educated person could take part. In Alexandrian thought, to contribute to research required highly specialized skills possessed by relatively few, whose expertise most often extended to several subject areas. For instance, Euclid wrote not only his celebrated *Stoicheia* (Elements) but also mathematical treatises on light rays and on the consonances.

In Athenian thought, finally, the aim, and the claim, were to gain a solid grasp of reality; in Alexandrian thought, those real phenomena from which an investigation took its point of departure quickly vanished behind a process of ever increasing abstraction. The kind of reality that Athenian thought was after was everyday reality considered from a special point of view (this is true even of Platonism, so concerned to transcend everyday reality). As distinct from this, in Alexandrian thought the empirical starting point got ever more abstract as its mathematical handling proceeded. *This will turn out to be a point crucial to the onset of the Scientific Revolution* (chs. 4 and 5), and therefore I shall now show for each individual branch of mathematical science how, and how far, the process of abstraction ran.

18 *Degrees of abstraction in Alexandrian mathematical science.* The process of abstraction was carried the farthest with balance problems and with consonant intervals. Archimedes' proof of the law of the lever does not apply to those real balance beams, with real weights suspended from them, that he used as a point of departure. Rather, the analysis is all about right lines to which numbers denoting weights have been assigned. In the end, there is no exact quantitative correspondence to reality but rather a cloud of approximation. For instance, due to friction and air resistance the proportion in the case of a real balance beam is no more than roughly inverse. It is quite the same with the entities imagined to float or sink in a fluid (p. 11). And with the consonances, once their integer ratios were established, no one bothered to examine the nature of the vibrations that the string produces at various lengths.

With light rays and with planetary positions the leftover link with reality was a little greater. Since light rays instantiate straight lines so well, pertinent investigations retained ipso facto a larger measure of reality content. Still, Euclid's and Ptolemy's geometric ray tracing involved abstraction from some crucial components of reality. For instance, in tracing reflected or refracted rays it was not asked what happens at the boundary plane between the air through which a light ray moves and the mirror that reflects it or the water that refracts it.

In mathematical astronomy, empirical observations played a larger part than in any other area of Alexandrian-type research. Here observed positions of heavenly bodies served as constraints upon the accuracy of the predictions that were generated by the geometric models devised to derive them. Still, models like the eccentric explicated in Figure 1.2 (p. 14) neither could nor did lay claim to existence as such in the reality of the heavens. In contemporary parlance, they were meant 'to save the phenomena'.

To sum up, then, natural philosophy was about reality, grasped (with few exceptions) qualitatively; mathematical science about fundamentally abstract entities treated with exactitude.

Near-absent interaction. If now considered conjointly, the one common marker (a shared intellectualism) and the numerous contrasts just listed invite us to ask how these two modes of nature-knowledge stood in relation to one another. I seek to answer the question, not in terms of formal pronouncements or of a priori conceptions, but empirically, according to what actual practice proves to have been like. When the issue is addressed in this sense, the bottom line turns out to be that *Athens and Alexandria remained far apart.* Not only is this true in the obvious sense of geographic separation, but intellectual exchange between the two proves also to have been minimal.

If taken in the broadest sense, the assertion is less than fully true. Archimedes' and others' efforts at handling empirical phenomena like floating bodies or light rays or planetary trajectories the mathematical way are likely to have benefited from the elevated epistemic status assigned to mathematics by Plato. Also, practitioners of such mathematical sciences operated with certain assumptions at the back of their minds which were adopted, deliberately or not, from one or another philosophy of nature. For example, an Aristotelian conception of heaviness seems to govern some of Archimedes' tacit assumptions in his demonstration of the law of the lever. Yet on the level of *specific* phenomena and of *specific* theorizing this is hardly so – what happened in mathematical science remained strikingly insulated from what was going on in natural philosophy. This assertion resembles Pierre Duhem's idea of an absolute, programmatically sustained split in Greek astronomy between a cosmological and a mathematical variety. This idea, by now a century old, has largely been discredited in the meantime. In the way Duhem put it forward it also carried a great deal of ideological baggage. No Duhem-like preconceptions disfigure the three characteristic examples that I will now invoke to illustrate what was really going on. They show that, on the operational level we are addressing here, there was indeed some connection between natural philosophy and mathematical science, but not much. There was at any rate a good deal less of it than was attained, for the first time, in Europe in the mid-17th century, when barriers between diverse modes of pursuit of nature-knowledge began to break down (p. 509).

The first example concerns *mathematics and problems of the infinite*. The very existence of such problems had in the 5th century BCE been shown by Zeno of Elea. His well-known paradoxes (Achilles and the tortoise, the flying yet resting arrow, etc.) were meant to demonstrate that his teacher Parmenides had been right – all apparent change is really a sham. Here is how Zeno's contemporary, Demokritos, the early atomist, showed the pertinence of such paradoxes of the infinite in everyday run-of-the-mill mathematics. Suspecting that the volume of a cone is exactly one-third of the volume of a cylinder with equal base and height, he perceived the paradox involved:

> If a cone were cut by a plane parallel to the base, what must we think of the surface of the sections? Are they equal or unequal? For, if they are unequal they will make the cone irregular, as having many indentations, like steps, and unevennesses; but if they are equal, the sections will be equal, and the cone will appear to have the property of the cylinder and to be made up of equal, not unequal circles, which is quite absurd.[16]

In principle, there are two ways out of such an inherently paradoxical effort to grasp the continuous by means of steps that, albeit divided more and more finely, still remain finite. One is the 'method of exhaustion'. Its inventor, Eudoxos, used it to prove with impeccable rigor Demokritos' theorem, among others, and Archimedes was to achieve even more with it. The method strikes us as needlessly cumbersome, but it is quite in keeping with the decision, firmly made by Greek geometers ever after Zeno, to keep the infinite at bay. Our

20 modern, looser method comes down to a decision to tame the infinite by approximating it ever more closely, that is, by making it converge to a limit. We do not think of a static distance between the two sections, however minute; we rather think of it as a dynamic continuum, as belonging to an infinite series of sections approximating the given section as the distance between each member of the series and the fixed section approximates zero. And this is the other possible way out of the paradox. The procedure is as questionable from a logical point of view as, ever since the rise of the calculus, it has proven heuristically fruitful. Yet, and this is the point here, the idea of convergence to a limit *did* make a fleeting appearance in Greek thought. Archimedes' younger contemporary Chrysippos, an early stoic, sought to resolve Demokritos' paradox by arguing that "the surfaces will neither be equal nor unequal; the bodies, however, will be unequal, since their surfaces are neither equal nor unequal".[17] That is, we watch Chrysippos here in the act of deriving from the general conception of dynamic continuity that was central to his stoic philosophy a broad idea of convergence to a limit. The very way in which he phrased his argument betrays both his struggling with the idea and the inability of his wholly qualitative conceptual apparatus to give it more adequate shape. So we have here two distinct lines of thought – the Athenian one of convergence to the limit qualitatively groped after and the Alexandrian one of tackling mathematical problems alternatively resolvable along that very path. But neither then nor later in the history of Greek thought were the two combined. If they had been, some other, no less searching conception would have emerged that might in the end have led to the calculus; even if not, it would at the very least have given occasion for further thought in its general direction. To observe (not by way of posthumous recrimination but as a piece of historical analysis) that this did not happen illustrates the point that natural philosophy and mathematical science remained far apart, not only geographically but also conceptually. There were retrospectively identifiable chances for cross-fertilization, but they were not taken up.

Another telling example arises from the mathematical treatment of *consonant intervals*. As noted, the Pythagorean doctrine in which these are linked up with the quantitative regularity of numerical ratios had at its very inception been abstracted away from its empirical origin, the vibrating string. The account was expanded and formalized in Alexandrian hands, particularly so in the Euclidean *Katatome kanonos* (Division of the Scale), yet never again in antiquity was it to be linked back to anything empirical. Nor did this happen when a seemingly ready opportunity presented itself in two rival accounts of sound that emerged in the atomist and the stoic traditions. Neither the idea of the emission of sound corpuscles in the former nor the conception of a wavelike propagation of sound in the latter was brought to bear by anyone upon the mathematical regularity observable in the production of consonant intervals. Once again, the very same step was to be taken with conspicuous smoothness, quite in passing really, in late-Renaissance Europe (p. 144).

A final example concerns *planetary theory*. The job, as the mathematical astronomer conceived it, was to devise geometric models aimed at predictive accuracy so as to 'save the phenomena'. This was not entirely distinct from the natural philosopher's job of presenting a unified outline of the overall constitution of the heavens, based upon first principles fit to serve as an explanation by means of a selection of crudely observed celestial phenomena. An example of the latter approach is Plato's cosmology as expounded in *Timaios*. Another is Aristotle's view of the world as neatly divided into a sublunar (i.e., terrestrial), imperfect domain of four elements undergoing incessant change and a supralunar realm of unchanging, ethereal planets (including Sun and Moon) moving uniformly in perfect circles below the outermost sphere of the fixed stars (Figure 1.3).

Here, too, there were shared features of a general kind. Mathematical astronomers allowed themselves to be bound by a philosophical requirement ascribed to Plato (to work with nothing but circles uniformly traversed served their ends well in a more technical sense, too). There was further the utterly self-evident, commonsense idea that the Earth stands at rest at the center of the universe. Even so natural philosophers and mathematical astronomers pursued distinct branches of learning. The former aimed at an explanatory, comprehensive picture without bothering much about quantitative accuracy. The latter were concerned with the resolution of all known irregularities in the motions of the Sun, the Moon, and the planets without regard for how the mathematical models employed could be made to correspond with any possible reality out there.

All this leads to a conclusion regarding the general status of the mathematical sciences in Greek thought. Whereas in our modern era the big problem is to preserve quality in a world of quantity, in antiquity the issue (definable, obviously, in retrospect only) was quite the reverse. Not only in Greece, but everywhere humans were living in what has been called the "world of the more-or-less".[18] In that world, so hard for us to recapture nowadays, the problem for the mathematical sciences was rather how to find a place for quantity in a world of quality (in a world, that is, conceived in the qualitative way). For consider once again how extremely tenuous, even in these few portions of mathematical science, the connection with reality actually was. The Pythagorean theory of consonance, derived at the outset from experiences in actual music making, had been cut loose from these at an early stage. Nor was it clear to what celestial realities the calculated models of the mathematical astronomer might be thought to correspond. Hence, the only solid points of connection between the empirical world and its mathematical treatment were the mirror, the five simple machines known to obey the law of the lever, and regularly shaped bodies floating in water. With so little quantity introduced into so relentlessly qualitative a world, it is no surprise that no breakthrough toward mentally conceiving a world of quantity occurred at this point. Utterly marginal, the mathematical branches of Greek nature-knowledge formed little pockets of exactitude inside a world of the more-or-less without there being any significant occasion to think

Schema huius præmissæ diuisionis Sphærarum.

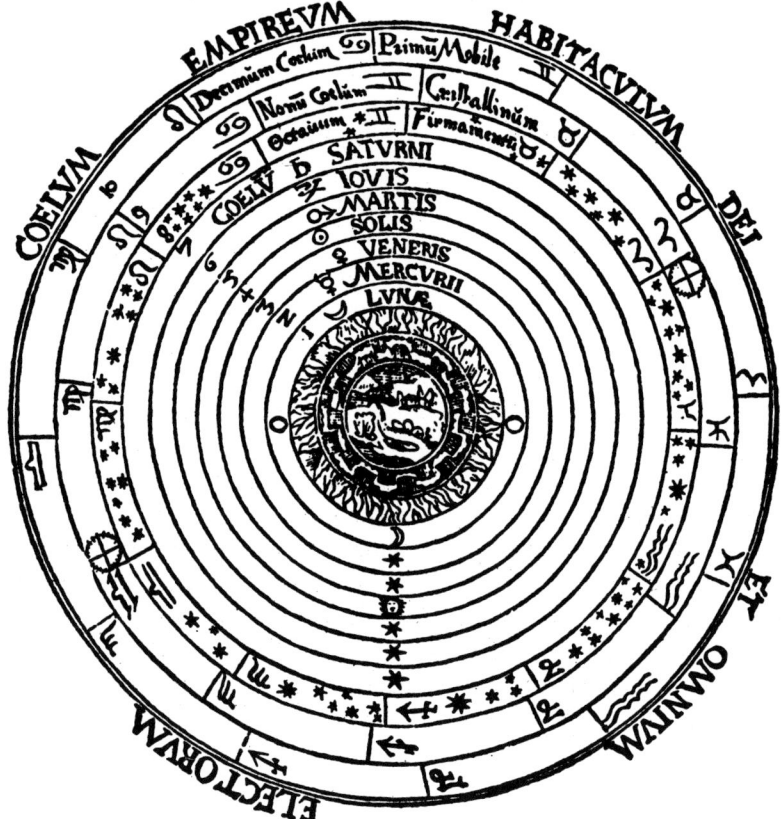

This diagram from Petrus Apianus' Cosmographicus liber of 1529 is often used to show Aristotle's conception of the cosmos. Notice, from the inside out, the Earth itself, then a layer of water (the oceans and seas), a layer of air (the atmosphere), and a layer of fire, thus rounding off the sublunar world with these four elements in a state of ongoing flux. The supralunar, ethereal, perfectly unchanging realm successively encompasses the spheres of the Moon, Mercury, Venus, the Sun, Mars, Jupiter, Saturn, and the fixed stars.

Two more spheres and the Empyrean Heaven are later, medieval accretions arising from both theological and technical problems with the original schema. The order of the planets is the one Ptolemy opted for (p. 25); planets beyond Saturn and the satellites of several of the planets cannot be seen with the naked eye and were therefore unknown (p. 275).

Figure 1.3: The Aristotelian universe

that they might be turned into kernels of a new, entirely unheard-of "universe of precision". 23
Guided by hindsight we can say that the Alexandrian corpus was inherently capable of such
an outcome. We cannot say that such an outcome was bound to occur *either then or at any
later time.*

Athens and Alexandria: Rare efforts at unification

How wide was the terrain still to be covered for such an outcome to be realized? History provides a measure to estimate its width. Two distinct efforts were undertaken in antiquity to
fuse components of Athenian *and* Alexandrian nature-knowledge into a larger whole. One
such effort was the construction of a coherently astrological conception of the world. The
other was Ptolemy's effort to enhance the reality content of each branch of mathematical
science that he addressed.

An astrological worldview.   Astrological forecasting had never been far from mathematical astronomers' concern. The heavenly bodies were by no means considered as just
geometric figures or arithmetical units to be reckoned with. Heavenly bodies were regarded
by both Babylonian and (under their direct influence) Greek stargazers as omens, as por-
tenders of things to come and of fates settled in advance. In Alexandria in the 2nd century
BCE, the ongoing search for accuracy in predicting positions of heavenly bodies culminated
in the construction of a peculiar, predominantly astrological worldview. Assorted pieces of
both Alexandrian mathematical science and Athenian natural philosophy found a place in it.
This first effort at unification took shape as a picture of the world in which the central tenet
of astrology – 'as above, thus below' – held the various parts together:

> Astrology grew out of a union of aspects of advanced Babylonian celestial divination with Ar-
> istotelian physics [i.e., natural philosophy] and Hellenistic astronomy; this union … occurred
> in Egypt in the second century BC. The product was the supreme attempt made in antiquity
> to create in a rigorous form a causal model of the *kosmos*, one in which the eternally repeating
> rotations of the celestial bodies, together with their varying but periodically recurring inter-
> relationships, produce all changes in the sublunar world of the four elements that, whether
> primary, secondary, or tertiary effects, constitute the generation and decay of material bodies
> and the modifications of the parts or functions of the rational and irrational souls of men,
> animals, and plants. In other words, ancient Greek astrology in its strictest interpretation was
> the most comprehensive scientific theory of antiquity, providing through the application of
> the mathematical models appropriate to it predictions of all changes that take place in a world
> of cause and effect.[19]

24   The purported author of the texts in which this worldview was recorded was 'Hermes Trismegistos' (the thrice great Hermes). It is hard to say how far the impact of this worldview extended through the Roman Empire and how long it lasted. In early-2nd-century CE Alexandria it was still alive with certainty – an astrological treatise that was composed then and there fed right into it.

Ptolemy as an early bridge builder.   Ptolemy, the author of that treatise, confined himself in his *Almagest* to producing a range of planetary models which, in being two-dimensional, laid no claim to representing any supposedly true state of the heavens. From there on, however, he sought to build bridges to celestial reality. The point bears stronger phrasing: Ptolemy provides the one and only case of a mathematical scientist expertly and more than fleetingly concerned to construct the very kind of specific linkages between Alexandrian and Athenian approaches that I have demonstrated were by and large absent from Greek nature-knowledge. Here is how he did it, in work on heavenly bodies, on light rays, on consonant intervals, and on the determination of place on Earth.

With *heavenly bodies* Ptolemy sought in three different ways to enlarge upon the geometric two-dimensional models presented in the *Almagest*. At the head of that book he placed six 'hypotheses' (points of departure), for example, 'that the Earth makes no motion involving change of place'. He drew support for these hypotheses from empirical phenomena if he could and from natural philosophy (Aristotelian or stoic) if he had nothing else to draw on. The procedure stemmed from his stated view of natural philosophy as just "guesswork".[20] It marks a characteristically 'opportunist' stance toward philosophy that Copernicus, Galileo, and many a later mathematical scientist adopted – a posture of 'use whatever philosophical assertion suits you to close incidental gaps in your mathematical argument'.[21] Further, in his astrological treatise, the *Tetrabiblos*, Ptolemy explained how the full certainty attainable mathematically for the realm from the Moon up ought to be supplemented with a more fallible science of how celestial motion affects events in the world below. This book also contributed its part to the astrological worldview just discussed. Finally, in his *Planetary Hypotheses* he sought to resolve, or at least to lessen, innate tensions between the mathematical and the natural-philosophical way of doing astronomy. That is, he sought to bridge the gap between accurate prediction and broad cosmology, the latter of the Aristotelian variety, to which Ptolemy here, too, ascribed with his customary opportunism in matters philosophical. One such source of tension appeared intractable, however. In qualitative cosmology, the Earth is naturally at the center of the universe. So it is in quantitative astronomy – but not quite. The inequality of the seasons was mathematically successfully, yet philosophically inexplicably, 'saved' by assuming the Earth to be slightly eccentric to the path that the Sun traverses around it (p. 13). Ptolemy cared little for the discrepancy, and at any rate, he managed for the rest to devise a quantitative cosmology that was to remain persuasive to most experts for fifteen centuries to come. He took one Aristotelian doctrine – that no empty spaces separate the ethereal spheres of the planets from one another – and joined it to several known geometric devices for measuring cosmic distances. In this manner he determined the measure of the universe from its center, the Earth, to its outer bounds, the sphere of the fixed stars. He computed the radius of the universe to be 20,000 Earth radii (c. 50 million modern miles); he calculated the Sun's diameter to be 5½ times as large as that of the Earth. By means of some further, in part arbitrary, assumptions a suitable order for the planets emerged as well:

> the Ptolemaic System of cosmic dimensions was an ingenious mélange of philosophical tenets, geometric demonstrations with spuriously accurate parameters, planetary theories, and naked-eye estimates. It was a speculative by-product of the first complete system of mathematical astronomy. Although in retrospect the distances were in error by an order of magnitude, the values found by the nesting sphere principle were confirmed by the distance of the Sun found in the Almagest and also by Aristarchus's ratio of solar to lunar distance. The naked-eye estimates of the apparent diameters of fixed stars and planets seemed reasonable enough. The resulting scheme of sizes and distances was, therefore, as plausible as it was ingenious, and it came to occupy an important place in cosmological thought among Moslem astronomers and then the Christian schoolmen.[22]

So much for Ptolemy's threefold bridge building in astronomy. In similar fashion, in his treatise on *light and vision* Ptolemy built on previous, geometric analyses of the paths of light rays. He did so in an ingenious yet retrospectively defective attempt to find a rule for how, in refraction, the angles of incidence and of refraction are related. He did so, too, in his analysis of how vision takes place. Presumably using available insights into the parts and functioning of the eye, he aimed to reconcile the geometrically conceived 'extramission' view current in the mathematical tradition and Aristotle's qualitative conception of light rays reaching the eye from the luminous object rather than being emitted by the eye. A reconciliation along such lines was not only attempted but actually attained nine centuries later by Ibn al-Haytham (p. 59).

Ptolemy's treatise on the *consonant intervals* displays a similar effort at bridging the divide. Pythagorean musical arithmetic claimed that the consonances (the very building blocks of music) were marked by exact ratios. Thus, it maintained the primacy in music of harmonic relations. Against this, a pupil of Aristotle's, Aristoxenos, argued for the primacy of the flow of the melody as an irreducible entity in all music making and, hence, in musical analysis also. Euclid and other thinkers in the Alexandrian mold merely extended Pythagorean musical analysis without bothering about Aristoxenos' attack, of which they might have been unaware. But Ptolemy sought to strike a workable compromise between these two, ultimately irreconcilable points of view. He sought to show that the very interval relations which undeniably display the numerical regularity found empirically by Pythagoras do govern the rules for melodic progression that sheer hearing reveals to us and that Aristoxenos had sought to codify without regard for any such underlying regularity. Musical judgments rendered by 'sense' and by 'reason' (to use contemporary parlance) are not really opposed in Ptolemy's view; the one follows quite naturally the exigencies imposed by the other. Later scholars were to become ever more doubtful about the extent to which the testimony of the ear in imbibing the dynamic flow of music can fruitfully be subjected to analysis in terms of essentially static, mathematically determined interval relations (p. 144). Even so, the historical significance of Ptolemy's effort rests in his conviction that a balance could be struck here in the first place.

Finally, Ptolemy's concern, in his *Geographia*, with the *determination of place on Earth* could not but take a somewhat different guise. Here he sought, not to enhance the reality content of Alexandrian topics but, conversely, to impose state-of-the-art geometry upon a topic drawn straight from the real world. He took from one recent predecessor's work available data about the location of places all over the Mediterranean, listed these by what seemed ascertainable about their latitude and longitude, and then crossed over to Alexandrian terrain by working out two alternative, highly sophisticated methods for map projection meant to render the known world with minimal distortion.

All in all, then, Ptolemy's work displays a manifest concern to put highly abstract, mathematical science into closer touch with reality. If we judge the outcome by modern standards, numerous shortcomings become quickly apparent. Not only did he conveniently paper over a range of defects that were also defects in terms of what he himself knew and strove for (e.g., a far too large value for the apparent diameter of the Moon; no mathematical grounds given for certain melodic progressions). Not only were the syntheses thus arising to prove untenable in the long run. More significantly, the very plane on which he undertook the reconciliation appears misconceived from the viewpoint of how, around 1600, mathematical science was to be infused with reality by two pioneers of the Scientific Revolution, Kepler (who deliberately placed himself in Ptolemy's tradition) and Galileo.

But are these the proper standards by which to judge Ptolemy's consistent effort at bridge building? The very effort shows that one late thinker in the Alexandrian mold had become aware of a need to break out of the apparent limits set to mathematical science. Nor was he to remain alone in perceiving so much – in the 11th century Ibn al-Haytham was to take Ptolemy's lead on the subject of light rays, followed in the 16th by Zarlino on musical consonance. The proper standard by which to judge the effort of these rare bridge builders is rather this: their attempts show us how profoundly nonobvious the method of infusing mathematical science with reality that arose in early-17th-century Europe truly was.

Greek knowledge of nature: Upswing and downturn

The argument so far has concerned Athenian natural philosophy and Alexandrian mathematical science as two distinct, largely separate entities. I have surveyed their contents and then laid bare their principal features and the contrasts between them. I have argued that, despite some overlaps (notably, a shared intellectualism and a commonly held conviction of a centrally fixed Earth), overall they stood far apart. The astrological synthesis alone excepted, little interaction took place between the two distinct modes of nature-knowledge. This assertion stands confirmed by the circumstance that retrospectively obvious opportunities for interaction remained unexploited. And Ptolemy's lonely bridge-building effort has served to demonstrate *a contrario* the presence and the vast width of the chasm.

From here on in the present chapter I will no longer consider Athens and Alexandria separately but will discuss Greek nature-knowledge as a whole. My concern is threefold. First, I show how the chronological order of events reveals a pattern of upswing and downturn that helps resolve what is known in the literature as 'the problem of decline'. Next, I compare Greek nature-knowledge with its Chinese counterpart to bring out the contrasts between them and also resolve what is known in the literature as 'Needham's Grand Question': 'Why did modern science not originate in China?' Finally, I develop conceptual tools to explain how Greek nature-knowledge, its upswing and downturn completed, could in later times and elsewhere be resurrected and enriched and even transformed in revolutionary ways.

A brief chronology.   Among first-rate practitioners of nature-knowledge in Greek antiquity, Ptolemy (fl. c. 150 CE) was definitely a latecomer. The age of the pre-Socratics with whom it all started is customarily taken to cover the 6th and 5th centuries BCE. These men who transformed mathematics and came up with a veritable proliferation of worldviews numbered only about half a dozen practitioners in every generation, or even fewer. The next stage of Greek nature-knowledge was marked by the founding of the first schools of philosophy, Plato's Academy around 387 BCE and Aristotle's Lyceum in 335 BCE. Soon after the latter's death in 322 and around the time of Alexander's conquests, the number of practitioners exploded. This is when the atomist, the stoic, and the skeptical schools were founded in Athens; this is when, in and around Alexandria, mathematical science came into its own. All this took place between late-4th and mid-2nd century BCE. And then 'decline' set in.

Decline is an elusive topic of historical discourse, hard to attain sufficient clarity about yet indispensable for our core problem. Conceptualization of what I mean when speaking of the decline of nature-knowledge must do justice to all historical cases under scrutiny. In seeking to resolve the question of how modern science arose and then survived in Europe, and not elsewhere, we meet not just one but a number of cases of 'decline' of nature-knowledge in some Greece-inspired variation (Greece itself, Islamic civilization, medieval

28 Europe) or at least the onset of decline (Europe about halfway the Scientific Revolution). The first step is to distinguish between what may be asserted of them all and the specific, very civilization-dependent guises decline took in each case individually.

Theory – 'decline' of nature-knowledge.   Even without undue widening, the problem is large enough. It is the decline of *nature-knowledge* I am addressing here, not the decline of the state, of economic prosperity, or even of civilization as such. To start with a deliberately general definition, I am concerned with the recurrent historical phenomenon of a failure to maintain the intensity and overall quality of the pursuit of nature-knowledge over the full course of a civilization. Hellenist civilization, Islamic civilization, and European civilization lived on while the pursuit of nature-knowledge undertaken in each suffered one or more serious setbacks, and the question is, Why?

Or is that really the proper question? What, on reflection, induces us to expect some built-in driving force powering without cease the pursuit of such knowledge? Is decline not, on second thought, rather the natural thing to expect?

We at present are familiar with the broad continuity of the scientific enterprise. Few endeavors seem less likely nowadays to break down than ongoing scientific discovery. We do not worry about the contingency because we take its modern props for granted, notably the common perception that science-based technology enhances our prosperity in major ways. But science-based, prosperity-enhancing technology with its institutional support system is a modern phenomenon. In the Old World very little even came close to this state of affairs. So we must not project the presence of our modern props back to times prior to the Industrial Revolution.

Upon thus reversing the question of *why* nature-knowledge was subject to decline in any of the civilizations where it flourished for a while – the short answer being 'What else would you expect?' – we return to the *how* question. To answer it we need a less bland, analytically more productive definition of what we mean when speaking of the decline of nature-knowledge in times prior to the late-18th-century rise and subsequent institutional anchoring of science-based technology. The definition must be provisional until all pertinent cases have been examined (p. 148). Common to all the cases of decline was a loss of momentum, marked by a leveling off in the sense of the nonrecurrence of a unique, uncommonly dense constellation of luminaries at work on a high-quality level of performance, not to be surpassed in the sequel unless incidentally, by one or another lone practitioner. Or, to put it simpler, *in each case an upswing takes place that within two to three centuries culminates in a relatively short-lived 'Golden Age', and then a steep downturn occurs that is nevertheless punctuated by some rare, individual achievements at a level of quality far above what has in the meantime become standard.*

Naturally the general pattern, even if it be provisionally granted to cover the principal facts in all our cases, still leaves much room for civilization-specific differences.

In the first place, stimuli for the upswing differed. The Greek case is marked by a mix of original undertakings (the rise of 'naturalist' worldviews with the pre-Socratics) and radical transformation of practices adopted from elsewhere (shift from rule-governed to demonstrative mathematics). By contrast, in Islamic civilization as in medieval and then again in Renaissance Europe, what stood at the origin of the upswing was a large-scale translation effort (the body of texts to be translated in each case being, of course, the one created by the Greeks). Further, the substantive content and overall level of achievement attained in the upswing were to some extent stamped with the mark of the particular civilization where it was attained. In the Greek case an adversarial style of argument and the lively, cosmopolitan culture that sprang from Alexander's conquests were particularly strong markers of such a kind, whereas the cultural markers that were in due time to stamp the creative pursuit of nature-knowledge in Islamic civilization and in Europe derived from their respective varieties of monotheist faith.

So much for the upswing. In the absence of our modern props, its momentum was bound to subside sooner or later. The question of *what* made it occur *when* it did may likewise yield a variety of answers. So may the respective aftermaths. These, too, may differ widely, in duration and in numerous other respects that I shall examine now for the Greek case after its Golden Age came to an end.

Specifics of the downturn in the Greek case. A marked acceleration and intensification in the pursuit of nature-knowledge took place between c. 340 and c. 120 BCE. The period, studded with the names of Apollonios and Archimedes and Aristarchos and Chrysippos and Epikouros and Euclid and Hipparchos and Pyrrhôn (to name only the greatest innovators), must count as a Golden Age by any standard. With few exceptions, the achievement of each constituted an endpoint. In natural philosophy a reshuffling of conceptions definitively established by the end of this period, rather than innovative addition to them, became the rule for the remainder of antiquity. In mathematical science most individual achievements (those of Archimedes and Apollonios in particular) remained endpoints, too. Two exceptional practitioners still to operate on a level with their Golden Age predecessors in Alexandria and even to surpass them in certain respects were Ptolemy in the 2nd century CE and, in the 3rd, Diophantos, who used advanced arithmetic to solve problems in what later became algebra.

Why the Golden Age came to an end when it did, roughly by the mid-2nd century BCE, is a question hard to answer with assurance in view of the scantiness and unreliability of the evidence. Still, two major causal factors that we shall meet in our later cases can confidently be ruled out for Greece. At the time, about halfway between Alexander and the armed establishment of Roman supremacy, no invasion or military conquest was so destructive as to disrupt the entire culture. And of sacrilege, in the sense of a widely shared perception of current

30 nature-knowledge trespassing religious boundaries, there was very little question. Nor was any further handling of the various branches of mathematical science bound to be fruitless. If Ptolemy's achievement is considered apart from the observational data that went into it, it might just as well have followed upon Hipparchos' work at once rather than after an interval of almost three centuries. What Ptolemy achieved could very well have been the immediate prolongation of the Golden Age rather than an incidental, one-man effort undertaken amid its steady decline. If, therefore, there is a viable answer to the question of 'why *then*, rather than earlier or later?' it must be sought in what the very Alexandria-centeredness of the entire enterprise of mathematical science in antiquity suggests – royal support first given, then withheld. But we do not know whether the dynasty of the Ptolemies extended its patronage beyond the period of the Golden Age, nor whether perhaps such patronage was provided by the governor of Roman Egypt during the time of Ptolemy, the astronomer. So this is where speculation must end. For natural philosophy it is easier to believe that contemporary practitioners considered its developmental possibilities to be exhausted. The emergence of the skeptical school of thought points in that direction. What sort of innovation could still have taken place in natural philosophy after four rival schools had come into being and one anti-school had just subjected their shared epistemological foundations to a critique so searching yet so unproductive in its net outcome?

Speculation aside, there remains the plain historical fact of a steep downturn all around. What marked the prolonged aftermath of the Golden Age of Greek nature-knowledge was not so much the end of all creativity as rather its flow into other channels. Excepting such incidental, late-Alexandrian luminaries as Ptolemy and Diophantos, the aftermath was characterized by *dogmatic assertion*, *codification*, the writing of *commentaries*, *simplification*, and *syncretism*. All these second-order activities took place against the background of an ever more *spiritual*, religion-imbued conception of things. I shall address them in turn.

The adversarial style of Greek thinking about the natural world had furthered the ongoing creation of a plurality of powerful, imaginative, and well-construed worldviews and natural philosophies. Beyond the Golden Age, the style began to display the liabilities that might come with its assets. The communication mode of public disputation lost its creative potential and turned repetitive, polemical-for-its-own-sake, and stale, giving way in the end to sheer dogmatic assertion. It is always hazardous to infer from a deliberate caricature the reality behind it. Still, a sketch like 'Philosophers for Sale', written in the 2nd century CE by the wittiest of skeptics, Loukianos of Samosata, indicates what dispute between philosophers had meanwhile degenerated into. Their striving for eloquence displaced the search for new truths that in earlier times had been the purpose of such disputation.

More productive than ongoing dispute, and of decisive importance for the very survival of the corpus of Greek nature-knowledge in an age of script and perishable papyrus, were the countless acts of preservation undertaken on behalf of what the luminaries of a bygone age had put down in writing. Out of centuries of copying, cutting, and pasting came the hodgepodge we know as the Greek corpus. In some rare cases the greater part of someone's life's work was preserved, as with Plato's dialogues and Ptolemy's treatises; in other cases, we have only a more narrow selection, as with Euclid, whose *Elements* survived wholesale whereas his writings on light rays and on consonant intervals survived only in didactic summaries. Sometimes a faithful rendition by epigones made it through the ages, as with Lucretius for Epikouros' teachings or with the skeptical battery of arguments set forth in tiresome detail by one late adherent, Sextus Empiricus. Cases of survival on the large scale are exceptional, though. What we know of Hipparchos we know through Ptolemy, and the natural philosophy of the stoics had to be reconstructed from tantalizing fragments. In mathematical science codification took more creative guises than in natural philosophy, as for instance with a range of advanced geometric problems collected by Pappos in 4th-century CE Alexandria.

Commentaries offered another flourishing venue for preservation. Meant in the first place for textual exegesis in a didactic setting, they provided the format best suited to satisfy any remaining striving for originality. High points in this regard are a Christian-inspired, profound commentary on principles underlying geometry by Proklos (head of Plato's Academy in the 5th century) and Johannes Philoponos' highly critical commentaries on several works by Aristotle (6th century). It is in the nature of such comments or marginal glosses to be directed to one partial issue after another; they can be of little help in building up an independent argument. Even so they could acquire a life of their own, as happened with Philoponos' objections to Aristotle's account of projectile motion (p. 57; 84).

Simplification is a highly commendable attribute of teaching at its best. In the first sustained act of translation to befall Greek nature-knowledge it took special guises. Translation is never a culture-neutral act. It cannot even be so when the translator's sole aim is a maximally faithful rendition of a text in another language. That, however, was not the primary aim of those Romans who, starting with Cicero in the 1st century BCE, sought to render selected pieces of Greek scholarship into Latin. Seizing on Greek compilations, themselves products already of considerable simplification, they rewrote these materials, thereby divesting them of their original, intricate structure. In philosophy the 'natural' portions often fell by the wayside. Either they were left out of expositions of some philosophical system altogether, or they found themselves assembled in collections of facts and factoids compiled for their curiosity value. Such enumerations of phenomena divested of the natural-philosophical frame and structure they stemmed from reached a high point in the elder Pliny's *Naturalis historia* (1st century CE). This vast compilation of factual material taken from hundreds of authors covers, in a succession more apparent than real, the heavens, the Earth, humans, animals, and vegetable and mineral products. Something similar happened with advanced mathematical argument in Alexandrian style. In its Latin rendition little more remained as a

32 rule than the qualitative outcome, illustrated by some crude, numerical data. Or so at least it
was with mathematical astronomy (the branch of Alexandrian science that evoked the most
vivid interest, chiefly because of its uses in astrology). In certain other branches, notably
in the mathematical analysis of consonant intervals, highly competent compilations were
prepared by 'the last Roman', the 6th-century senator Boethius.

Not only could the teachings of the various schools of philosophy be recorded one by
one (as with Lucretius for atomism) or expounded one by one (as with Philoponos for Aristotelianism) or jointly ransacked for their empirical content alone (as with Pliny), but they
were also subjected to processes of fusion at the level of their first principles:

> pieces [are] borrowed from all systems under the primary condition of their sharing a certain
> spiritual state... . So as to endow this compromise with the solidity that it can no longer draw
> from rigorous argument (which would be fatal to it), it is covered with a vague religiosity that
> must cement and dogmatize the whole. ... Prolific to the point of overflowing, the artisans of
> this blend, Poseidonios [1st century BCE] and Philo [1st century CE] give their responses to
> everything, satisfy everyone, handle every subject, soothe every controversy, draw from each
> system what suits them, and reconcile all dogmas with each other.[23]

The philosophical systems to merge into such syncretic concoctions were those of Plato,
Aristotle, and the Stoa – those where (unlike in atomism) soul and spirit were conceived of
as exalted far above mere common matter. In embodying a predominantly spiritual outlook
on life, syncretic Greek philosophy both gave expression to and fostered in its turn an other-
worldly climate of thought on the rise throughout the period.

This shift in climate is expressed likewise in how in the late Roman Republic and then
in imperial times the Stoa first, then Platonism in a creative adaptation, got the upper hand
over their more nature-oriented rivals, Aristotelianism and atomism (the latter hampered
in addition by its reputed atheism). It is characteristic of the ascendancy of the Stoa in the
Roman world between the 1st century BCE and the 2nd century CE that its natural-philosophical portion (the dynamic tension of *pneuma*) fell increasingly by the wayside. The
subsequent predominance of a highly spiritualized variety of Platonism laid down in the
3rd century by Plotinos went even further in the same direction. Rarely has the deprecation
of matter and the elevation of the soul been carried to farther extremes than in his *En-
neads*. The rise to prominence of Christianity also contributed its part. If we consider that
numerous Christians took their faith to be directly antagonistic to the very enterprise of
philosophy, the picture of a wholesale shift away from whatever might still smack of natural
philosophy seems complete.

Christendom, however, came in many varieties. Adherents to those conceptions of
Christ's true nature that eventually lost out in the grand theological battles of the 4th–6th

centuries were to make themselves the very bearers of the Greek corpus in both natural 33
philosophy and mathematical science. Chased from the lands of the resident emperor in
Constantinople, Nestorian and Monophysite clergymen and laymen at the eastern outskirts
of the empire (Syria) or even beyond (Persia) set out to examine the corpus resulting from
seven to eight centuries of assiduous codification and commentary. They translated large
portions into Syriac or Persian. Naturally, they undertook their translation effort for purposes that fitted in with the local culture of Persian Nisibis and Jundishapur. But in comparison with the extensive simplification that, in Latin translation, had befallen the Greek corpus
in the West, the 'Eastern' translation effort remained far more faithful to the structure, the
intricacies, and the advanced quality of the body of work to come out of the Golden Age.

This completes the account of nature-knowledge in Greek antiquity. It was not the only
large-scale body of nature-knowledge that originated around the 6th century BCE and held
its own for a long time. The Chinese produced one, too.

Chinese knowledge of nature

As with the Greeks, the Chinese pursuit of nature-knowledge started with an 'overture',
in the course of which themes destined to become predominant at a later stage were first
sounded out. As with the pre-Socratics in several Greek city-states, in the Warring States of
ancient China naturalist ideas about the constitution of the world began to proliferate in the
6th and 5th centuries BCE. Over the course of some three centuries a whole range of text
lineages came into being, each of which referred back to ancient founders like Kung Fu Tzu
(Confucius) or Mo Ti or Lao Tzu.

As with the Greeks, the overture was followed by a period of sifting the early ideas and
systematizing those that survived the sifting. In Greece the process led to the establishment
of four schools of natural philosophy and one independent current of mathematical science.
In the Chinese case sifting and systematizing led to a radically different outcome. Instrumental in this regard was the unification of China, in 221 BCE, under the First Emperor.
Upon the speedy downfall of his brutal regime the first long-lived imperial dynasty (the
Han, 206 BCE–221 CE) oversaw the production of a large-scale synthesis out of a range of
leading ideas first explored during the earlier period. The Han synthesis both expressed and
consolidated a budding consensus among the literati. Under later dynasties it was greatly
elaborated and extended. It was not, however, to be questioned at the level of its fundamentals until the encounter with European science got under way in earnest (p. 726).

I shall first outline the basic constituents of the Han synthesis and the events that brought
it about. Next, I supply some examples of how the abstract conceptions of the synthesis were
made operative in the everyday pursuit of nature-knowledge, with attention focused upon

34 two high points: the achievement of Shen Kua in the 11th century and a large-scale reform of the 'season-granting system' in the 13th. Prior to making a sustained comparison between the Chinese and the Greek pursuits of nature-knowledge I also call attention to the Mohist approach, which derives its special interest from a dual circumstance: it resembles Greek thinking more closely than any of its Chinese rivals do, and it never made it into the Han synthesis.

Basic constituents of the Han synthesis. The core issue in Chinese thought right from the start was how to attain and/or maintain a stable social order. The fundamental prerequisite for preventing society from falling apart was that the social order must be in harmony with human nature, which in its turn reflects the harmonic order of the cosmos. Society, the individual, the world around us display an infinite multiplicity of phenomena, customarily called 'the myriad things'. What underlying coherence these possess resides in their mutual dependence:

> The regularity of natural processes is conceived of, not as a government of law, but of mutual adaptations to community life. ... [The way of the Chinese was] to systematise the universe of things and events into a structural pattern which conditioned all the mutual influences of its different parts. ...
>
> For the ancient Chinese, things were connected rather than caused. ... The universe is a vast organism, with now one component, now another, taking the lead at any one time, with all the parts co-operating in a mutual service which is perfect freedom.
>
> In such a system as this, causality is not like a chain of events. ... The concept of causality where the idea of succession was subordinated to that of interdependence dominated Chinese thinking.[24]

The world is an infinitely subtle fabric. Each thread, however tiny, is interwoven with every other. The nature of the fabric can be grasped by means of correlations, so as to get at the various ways in which the mutual coherence of all things makes itself felt in some given phenomenon. Four concepts exemplify and express this correlative way of thinking. Together, they form the core of what would eventually be turned into *the* Chinese view of the world. These concepts are *tao* (the Way), *chi* (matter/energy), *wu-hsing* (the five phases), and *yin-yang*.

In the sense here used, 'Tao' does not refer specifically to the Taoist religion. Its beliefs and practices developed over the centuries in close connection with one of those venerable text lineages from which the Chinese worldview emerged – the one for which Lao Tzu and Chuang Tzu are counted as the founders. 'The Way' in a more general sense is the end toward which each quest in each of the diverse text lineages aspired. Confucius (6th/5th century)

began to endow the customary term for 'way', or 'path', with a more abstract meaning: as   35
the Way that accords with the individual and with his society, in harmony also with the
fabric of nature. The ancient sages followed this Way of their own accord, but the Way has
meanwhile been lost, and it is about time to find it again. Confucius thought that to observe
proper ritual would lead back to the Way. The natural order as such did not concern him
much, and the contribution that his immediate followers made to nature-knowledge was
relatively insignificant. Lao Tzu and Chuang Tzu, in a manner quite different from Confucius, distinguished between a Way which can be spoken of, the ever transitory Way of natural
processes, and the unchanging Way which cannot be spoken of because (as in all varieties
of mysticism) words by their very nature fall short of expressing ultimate meanings. Many
more 'Ways' were put forward as rivals during the Warring States period. Not until the first
century of the Han dynasty, with the Chinese polity solidly unified, was the diversity of Ways
lumped into one conception which henceforth coincided with the cosmic order in which the
new, centralized empire was grounded.

How and to what extent can we know the cosmic order? The intellect alone is not sufficient to grasp it; intuition, contemplation, and imagination are required likewise:

> Study was one of several kinds of self-cultivation. It provided understanding and useful
> knowledge of the world (which was one aspect of the Way). The deeper aspect of reality (the
> nameless Way) is so subtle that one can penetrate to it only through non-cognitive means.
>
> The *Book of the King of Huai-nan* puts it cogently: 'What the feet tread does not take up
> much space; but one depends on what one does not tread in order to walk at all. What the
> intellect knows is limited; one must depend on what it does not know in order to achieve il-
> lumination.'[25]

Just as with the Way, each of the other core concepts, too, went through a development of its
own before, in the Han synthesis, they found themselves jointly transformed into the mental
equipment required to come to grips with the world fabric.

Ch'i can hardly be translated into some equivalent modern concept; experts render it
as 'matter/energy'. Originally *ch'i* stood for a wide diversity of phenomena that can be observed but not touched, like air, breath, smoke, fog, or clouds. From there the meaning of
ch'i expanded into physical vitality and then into the forces of climate and cosmos which
co-determine health.

'Yin' and 'yang', as the familiar image shows, form polar opposites which are also comple-
mentary. Even contraries cannot make do without each other; their interconnectedness is
built into them. In principle, every configuration and every process that take place over time
are subject to an interpretation in terms of yin-yang, be it action and reaction or male and
female or growth and decay.

36 In those cases when a division into two makes itself apparent, yin-yang offers a ready-made frame for interpretation. When facing more complex divisions, another concept stood ready to be invoked, *wu-hsing,* or the 'five phases'. Albeit rendered in terms of matter alone – Water, Fire, Wood, Metal, and Earth – *wu-hsing* really denotes both the concrete substances and the more abstract processes that regulate the course of things. For example, Earth stands for vegetative processes, Metal for those fixed forms that can turn into another state by melting or by evaporation. The explanatory power of the five phases resides in the distinction made between various cycles. The production cycle Wood ⇨ Fire ⇨ Earth ⇨ Metal ⇨ Water indicates the order in which one process gives rise to another. The conquest cycle Wood ⇨ Metal ⇨ Fire ⇨ Water ⇨ Earth marks the order in which one process prevails over another. Whether such processes are conceived concretely or in the abstract, they do not run linearly but, indeed, in cycles. After the wooden spade has conquered the earth, the metal hatchet has carved the piece of wood, the fire has melted the metal object, and water has extinguished the fire, the conquering water is in its turn dammed and canalized by earth. The cyclical course of things is built into the cosmic order:

> [Chinese thinkers] made sense of the momentary event by fitting it into the cyclical rhythms of natural process, for the life-cycle of an individual organism – birth, growth, maturity, decay, and death – had essentially the same configuration as those more general cycles which went on eternally and in regular order, one fitting inside the other: the cycle of day and night which regulated the changes of light and darkness, the cycle of the year which regulated heat and cold and the farmer's growing seasons, and the greater astronomical cycles. ... In order to make the Tao of a particular thing intelligible, its life-cycle needed to be located with respect to the greater periods. The different parts of a cycle could be analysed in terms of [the core concepts just listed], for instance the Yin and Yang, which were the passive and active phases through which any natural cycle must pass.[26]

So much for the core concepts of Chinese thought. In the 2nd and 1st centuries BCE they were jointly taken up in an all-embracing synthesis. What occasioned the creation of the synthesis was a political crisis of major proportions.

The making of the Han synthesis.   Unlike the Greek city-states, China prior to unification consisted of a decreasing number of medium-large to large states, entangled in ongoing warfare against each other with a view to attaining definitive hegemony. All this strife furthered rather than halted the emergence of a healthy plurality of views, making the age of the Warring States, when concepts like those just sketched took shape in a range of text lineages, so creative a period for Chinese thought. The pioneers and adherents of each text lineage competed with each other to sell their own Way on the market of ideas. That 'market' was formed by the rulers of the various states at war with each other. Thinkers circulated from one court to the next in an unceasing effort to persuade one ruler after another that their particular Way was best suited as a guideline for political action.

The unification of Chinese civilization by the state of Ch'in was decisive for the shape the Chinese worldview was to take. The First Emperor established his rule in rigorous fashion. He founded his claim to strict obedience on nothing but his military victory. The only expert knowledge that he needed at court concerned the techniques of maintaining raw power. He had no time for any more refined ideas, and only a few of the various text lineages that were created during the Warring States period managed to survive the vast destruction (or at any rate the loss) of manuscripts that marked the years of the Ch'in dynasty. As a result, most thinkers of the Warring States period left little to posterity other than their names and, with luck, some scattered passages. Among the victims was a range of texts in the tradition of the pacifist Mo Ti (about which more below).

Rule over a large population cannot be maintained over the long run by just naked force or the threat thereof. No one could be more sure of this truism than the man who successfully rebelled against the heirs of the First Emperor and, in 206 BCE, founded a new dynasty, the Han. This dynasty recognized that if it wished to maintain its rule for longer than the fifteen years of its savage predecessor, it had to proclaim and disseminate a central conception of the world that would cause its subjects to accept the authority of the dynasty as genuine and legitimate and therefore obey it of their own accord, not just because of the threat of sheer force. Here the remnants of the text lineages of old proved useful. Under the Han, a consensus on the principal features of Chinese thinking about the constitution of the world was produced that, at least in gross outline, lasted until the onset of modernization.

Central to the consensus was the idea of harmony. There is a celestial harmony, which for as long as things go well is reflected in the harmonious course of events in the realm at large. And whether or not things go well depends on the emperor, whose supreme task rests in maintaining and confirming harmony. He rules by virtue of the 'Mandate of Heaven' – dynasty change takes place when the emperor forsakes the Mandate. All kinds of natural phenomena, like earthquakes, floods, and eclipses, may be harbingers and tokens of the Mandate of Heaven being forsaken by the ruling dynasty.

Natural phenomena, however, are more than portents; it is also possible to investigate in some detail how they hang together. Their prime coherence rests in the harmony between them, and the way to find out more about that harmony is to consider the interplay of the basic concepts outlined above.

The Han synthesis has been given the enlightening label 'organic materialism'. It is a worldview where material processes that also have a spiritual component are combined into one organic fabric:

38   The Chinese cosmos is a constant flux of transformation, always regenerating itself as its constituents spontaneously change. *Ch'i* is matter, transformative matter, always matter of a particular kind, matter that incorporates vitality.

  Yin-yang and the five phases had, by the end of the first century B.C., a consistent, dynamic
character as part of the *ch'i* complex. Anything composed of or energized by *ch'i* is yin or yang
not absolutely but with reference to some aspect of a pair to which it belonged and in relation
to the other members. ... Yin-yang provided a flexible language well suited to discussing the
balance of opposites. This was a balance not of quantity but of the dynamic quality of each
in interacting domains – for instance, something could be yang in its activity and yin in its
receptivity. When the focus was not on a binary opposition, however, but on more complex
sequences of growth and decay or conquest and subjugation within a larger process, the various sequences of the five phases came readily into play.[27]

What did all this mean more concretely? How did a Chinese scholar go about investigating
a given range of natural phenomena? The full panoply of core concepts was not always invoked in a specific case. The scattered literature on the history of nature-knowledge in China
(at least to the Western scholar, no chronological account is yet available) suggests that three
main characteristics may be distinguished: (1) an inclination to undertake researches with
some practical end in mind, more often than not such as derived from court priorities; (2) a
tendency to categorize things according to preformed schemata like the ones just listed; (3) a
highly variable measure of interplay between, on the one hand, nature-knowledge gained
the empirical way and, on the other, the predominant worldview that had taken shape and
gained general assent over the preceding centuries.

Examples of research practice and its high points.   Broadly speaking, each area of
possible study, such as the phenomena of sound or the processes that appeared to display or
speed up the transmutation of metals,

    was determined by an original demarcation of a field of observation and experience, defined
    by imposition of the common natural philosophy ['worldview'], and developed partly by
    working deductively through the various permutations of particular facts.[28]

One example of how all this played out in workaday research is the discovery, in the 1st
century CE, that there is a close connection between the tides and the phases of the Moon.
The spectacular alternation between ebb and flood in the Yangtse River delta and the waxing
and waning of the Moon provided the observational data. What, so early in the history of
thought, could bring an observer to correlate such seemingly unrelated empirical phenomena? The answer lies in Chinese scholars' sensitization to coherence at so deep a level by a

background worldview according to which things form an organic fabric and events run in cycles.

Or take sympathetic resonance. On sounding a string or blowing a pipe, another string or pipe at some distance away may seemingly spontaneously begin to sound the same note. The phenomenon was used for the fine-tuning of bells. But practical usage was not all that was aimed for; the phenomenon was also to be explained. The concept ready to be invoked was *ch'i*, which was taken to blow through the pipe like some sort of cosmic wind or to help spread the sound produced by the string. Time and again the conception of mutual harmony in the fabric of nature assisted the senses in noticing such subtle phenomena. Indeed, with musical instruments all kinds of delicately different timbres were distinguished ('timbre' is everything in a musical note of given pitch and loudness that may still sound different, as with organ pipes in different stops). Here categorization took place according to the material of which the instruments were made but also to the eight directions of the wind (north, northwest, etc.). The underlying connection between the various timbres was once again conceived to be ruled by *ch'i*.

Or take the pursuit of alchemy. It was engaged in on a large scale and over many centuries, most often in a Taoist setting. It involved many practices, such as distillation, sublimation, and precipitation, that were in due time carried to high degrees of refinement. It also involved various beliefs, notably in the growth of metals in the earth and in the possibility of prolonging life or even achieving immortality. Unlike with other cosmic cycles captured by means of the trusted concepts of *ch'i*, yin-yang, and the five phases, in alchemical maturation or the achievement of longevity the cyclical course of events is in the end overcome:

> Although the perfection of the elixir is the result of a repeated cyclical process, at each step of the treatment the intermediary product is not the same, but rather progressively exalted. Thus superimposed upon the cycle is a progressive upward tendency, which does not revert itself.[29]

With beliefs like these went an ease in moving between, on the one hand, numbers obtained from carefully weighing the various reactants and, on the other, numbers derived from purported correspondences between the reactions examined and their proper places in the cosmos. For instance, nothing but painstakingly precise measurement can have taught Chen Shao-Wei early in the 8th century that distillation of 16 ounces of high-quality cinnabar (mercury sulfide) yields between 13 and 14 ounces of mercury (the modern figure is 13.8). He also gave exact figures for the net yields of mercury from ores that contain cinnabar in varying degrees of purity. However, modern analysis shows that he derived those exact figures, not or not primarily from observation, but rather from an a priori ordering of assumed degrees of maturation of the metal in the earth.

By way of a more-extended illustration of how research and worldview intertwined, here

40   is a passage from an 11th-century Taoist tract that gives the flavor of these investigations and how they were reported. To elucidate its meaning, I quote next the straightforward paraphrase that the foremost student of these matters, Nathan Sivin, has appended to it (all parenthetical insertions are also his):

> Jupiter is the Wood planet, the vital animus of the sun and the essential *chi* of Water. This animus is scarlet, because (it corresponds to) Fire. Fire gives birth to Wood. In response to the *chi* of Mars (the Fire planet), cinnabar is born. Cinnabar holds within it the Yin *chi* of Wood, and thus contains quicksilver. Quicksilver is called the Caerulean Dragon; and the Caerulean Dragon belongs to Wood (w–W–F).
>
> ... Mars is Fire, and the seminal essence of Fire. It receives the *chi* of the Wood planet (Jupiter) and also transmits the animus of the sun. Its flowing seminal essence enters Earth (or the earth) and gives birth to cinnabar. The animus (of cinnabar) belongs to Fire and so it is born out of Wood. Since within it there is Yin, it gives birth to mercury. Fire gives birth to Earth. Earth contains the Balanced Yang, and gives birth to realgar, the sapidity of which is sweet. (W-F-E).
>
> We can ... reduce ... the text to a straightforward assertion: 'There exists the genetically related binary system mercury/cinnabar, of which mercury, corresponding to Wood, is the young (i.e., immature) Yang phase and cinnabar, corresponding to Fire, is the mature Yang phase.'[30]

Note that the invocation of the names of the planets does not signify any idea of an influence exerted by them upon the growth of metals in the earth or upon anything else – they are associated with the five phases by way of symbols only.

Speaking now more generally, the planets and their irregular trajectories did not in the Chinese view have the same prominent place that the Greeks assigned to them. This came to the fore in particular during two major efforts initiated by the imperial court to reform the calendar. Every year the emperor would issue a calendar, a kind of almanac which listed significant future events:

> Successfully predicting a phenomenon fitted it into the dynamic cosmic and social order that the emperor maintained on behalf of his people. Unpredictable phenomena and failed predictions were either good or bad omens. Each challenged the established order, and each had a meaning. Bad omens warned that the emperor's mediating virtue, which maintained concord between the cosmic and political orders, was deficient. Good portents generally signalled, and approved, a step in a new direction. Successfully predicting celestial events neutralized their ominousness, preserving the charisma of the ruling dynasty.
>
> The annual calendar (or almanac) issued by the throne was thus important in the solemn

ceremonies that asserted the emperor's authority, as the ancient motto put it, to 'grant the seasons'.

Granting the seasons involved ongoing technical refinement or at least the semblance there-of: "Astronomical reinforcement of the Mandate of Heaven called forth endless attempts to improve constants."[31] Solar and lunar phenomena far more than planetary ones held impli-cations for the calendar. It was not until c. 1075, with the appearance at court of a learned man by the name of Shen Kua, that an interest arose in including planetary trajectories in a large-scale reform of the season-granting system. Shen Kua (1031–1095) rose to high office on the coattails of a major reformer. He was appointed by a young emperor out to enhance his power and authority by means of a far-reaching tax reform and a bold military effort to chase invading nomads back beyond the Great Wall. Shen Kua took an active part in getting the appropriate measures executed, but he also made detailed plans for a major overhaul of the season-granting system in current operation.

Shen Kua's reform project included an unprecedented effort to come to grips with one major irregularity in the trajectories of the planets (retrograde motion; p. 13; 107) and to predict their positions. He also improved the design of observation instruments, at the ser-vice of a wholesale program for restructuring the observational basis of the calendar. For the planets, but also for the Moon (the many data available had accumulated large errors over the centuries), he proposed to read "exact coordinates ... three times a night for five years".[32] He further proposed to confront the astronomical phenomenon at the heart of imperial authority – solar eclipses. These were hard to capture with sufficient precision by means of the algebraic (not, as with the Greeks, geometric) tools that predominated in Chi-nese astronomy. More than anything else did eclipses run a serious risk of failed prediction. However, an inability to make his inert staff cooperate in his bold and well-prepared reform projects, along with the sudden fall from grace of his immediate patron, caused Shen Kua's efforts to come to naught. He spent the remainder of his days at Dream Brook, an exquisite garden estate that had appeared to him in a dream years before he sought for and purchased it.

Almost two centuries later, an even vaster and also more successful reform of the season-granting system was undertaken at the behest of Kubilai Khan, the first Mongol emperor, who established the Yuan dynasty. In a move meant to reconcile with his rule both his newly acquired subjects in northern China and those in the south whom he planned to subjugate next, he assembled a tightly knit group of Chinese literati at court and ordered them to pre-pare and install a new season-granting system by and large from scratch. In 1276 the reform group, which counted more than a hundred officials altogether, was given authority to by-pass the existing directorates of the imperial Astronomical Bureau and set up a directorate of its own. The group devised new, ingenious interpolation techniques, new measuring in-

42 struments of unprecedented size (e.g., two gnomons each ten feet high), and new techniques for making measurements. For instance,

> teams of observers ... carried out a great latitude survey with portable equipment at 27 locations scattered from Siberia southward for 3600 miles. They recorded for each place time differences of eclipse observations, variations in the ratio of day length to night length, and changes in the latitude of the sun, the moon, and possibly the planets.[33]

To test the accuracy of the new measurements, the new directorate made a vast inventory of previous observations as recorded under earlier dynasties over many centuries and kept in the imperial archives ever since. Kubilai's astronomers

> included in their tests a series of early eclipse observations that specified not only the day but the time of the eclipse, often to the nearest quarter of an hour. This was not the norm. Those records were, by and large, quite accurate. The Yuan group used them systematically and profusely, not only because of its high standard, but because its funds permitted it to do that.[34]

Indeed, generous funding made possible certain things but inhibited others. On the one hand, there was Kubilai's manifest interest in enhancing, in a way the Chinese population could recognize as authentic, the accuracy of celestial prediction for the determination of the seasons, for divination, and for similar purposes. On the other hand, his interest was bound up with positions and eclipses of the Sun and the Moon, not of the planets. Just as with Shen Kua's failed reform two centuries earlier, no efforts were undertaken to direct the new techniques to determining planetary trajectories with enhanced precision.

Let us now return to Shen Kua in the late 11th century. He spent his enforced retirement writing up the experiences and insights of a lifetime in a text entitled *Brush Talks at Dream Brook*. His sole remaining companion, his brush, served him to jot down notes of a few lines to a full page on a variety of subjects, among them poetry, omens, comments on the classics, or the careful observation and interpretation of certain rock formations. These 'Brush Talks', although they did not have much of an impact upon the further course of events, have in our day caused Shen Kua to be regarded as China's greatest premodern thinker about natural phenomena. He perceptively noted, for instance, that the peculiar shape of those rock formations stemmed from erosion patterns created long ago by streams that had in the meantime dried up – an effect that he went on to recognize in hills, "miniatures of the Yentang mountains, but in earth rather than in stone".[35] Another example is his observation of a peculiar spring with water that when heated turned into bitter alum. If he kept heating it, it yielded copper, and if he heated the alum for a long time in an iron pan, the pan itself turned into copper. The phenomenon (a well-known displacement reaction in modern chemistry)

reminded him of the conquest cycle of the five phases – Metal conquers Water. Clearly, the 43
very existence of that conceptual apparatus made the investigator receptive to observations
of such a subtle kind.

Something similar is true of Shen Kua's researches involving magnets. Building on the
doctrine of geomancy (*feng shui*, the determination of places propitious for building), he
made magnetic needles float on water and found that there is always a slight displacement
vis-à-vis the north pole. To establish and measure magnetic declination he made trials with
a suspended needle but also determined with great accuracy how far the polestar is removed
from true north.

So much by way of an outline of the Chinese conception of nature and of paths taken
to investigate its myriad details. Also during the Sung dynasty (960–1279), by and large simultaneous with Shen Kua, the Han synthesis underwent its boldest extension ever. It was
due to five successive thinkers collectively known by the label 'neo-Confucian', even though
their commentaries on the classics were also tinged by Taoist and even Buddhist ideas. They
were concerned to refine a conceptual understanding of how precisely the myriad things
cohered in the ceaseless alternation of cyclical processes. Their speculations culminated in a
newly forged concept, *li*, or 'organic pattern-principle' in the closest English equivalent. On
occasion neo-Confucian thinkers extended the principle to more mundane matters. For instance, Cheng I rebuked run-of-the-mill physicians for not recognizing that, when a certain
yellow substance (a tannine) is added to another given substance that is white (an iron salt),
and the two mix to form black stuff (a pigment used for ink), the latter should be regarded
as an essentially new substance. He wrote:

> if we have c and continue to look for a and b in it, if we have black and persist in looking for
> yellow and white in it, then we are failing to understand the nature of things. (This is why)
> the ancient (sages) investigated to the utmost the organic pattern-principles of things; they
> studied tastes, smelt odours, differentiated between colours, and acquired knowledge of what
> substances will mix or combine together.[36]

The Mohist exception.   Throughout China's premodern history, the myriad natural
phenomena were treated in the broad manner just elucidated. Surely variations of many
kinds were tried out over time, and in many ways nature-knowledge under the Sung looks
different from that of the Han period. Even so, the basic setup of correlative pattern seeking
in a broadly empiricist vein remained by and large the same: explanations were sought in
terms of the Tao, *ch'i*, yin-yang, and *wu-hsing* throughout premodern times. But there is one
genuine exception to this rule – an approach to phenomena undertaken on a fundamentally
different plane. This concerns a collection of writings stemming from the text lineage that
emerged with Mo Ti. The surviving fragments are known as the *Canon and Expositions* and

44 date from the Warring States period (c. 300 BCE). The approach taken in these texts is not so much marked by the correlative mode of thought that goes with the conception of the cosmos as an organically cohering fabric. Tellingly, the texts invoke neither yin-yang nor the five phases nor even *ch'i*. Ideas and concepts are developed in more rigorous fashion, not nearly so associative but more by way of logical inference, directed toward if/then relations rather than the and/and of all-encompassing, mutual interdependence:

> The advantage of the Mohists over the Yin and Yang and Five Elements schools (which laid the foundation of traditional Chinese science) is that their logic gives them a clear conception of what they find acceptable as an adequate explanation. "One uses explanations to bring out reasons" ... and *ku* 'reasons' are of two kinds, the necessary conditions without which something "necessarily will not be so", and the necessary and sufficient conditions having which it "necessarily will be so". ... The Mohist's fundamental objection to the Five Elements type of explanatory principle is that it lacks the necessity he finds in causal explanation. But ... he looks for causal relations only in specific phenomena. ... His problems arise from the manipulation of mirrors, balances, ladders or masonry. Why is the image in a concave mirror upside down? And why only if the object reflected is outside the centre of curvature? The problems, although suggested by practical situations, are purely theoretical; one point not mentioned is the practical function of the concavity, to make a burning-mirror.[37]

The *Canon and Expositions*, then, focuses on questions of motion and force, of light and shadow. Beside the distinction made between cases when the image in a concave mirror appears reduced and right side up and when it appears enlarged and inverted, specific answers are sought to such questions as "'Why does a cross-bar not bend under a weight?' or 'Why does the longer arm of a beam go down when equal weights are placed on both sides?'"[38] After the emergence of the Han synthesis and its organic materialism, topics like these were seldom investigated – *what little has been preserved of the Mohist text lineage was never again called back into life.*

Chinese and Greek nature-knowledge compared

Why in the end was modern science grafted upon the body of knowledge to come out of the Greek, and not the Chinese, quest for nature-knowledge? Might not the outcome, but for some quirk of history, just as well have been otherwise? Or may a comparison between the two help us detect certain causal factors at work that precluded the alternative outcome?

Two preliminaries are indispensable. Any attempt to approach either quest without the benefit of hindsight must impress upon us the sheer boldness and grandeur of the only two

efforts ever undertaken to face the world of nature afresh and make coherent sense of it on   45
the truly large scale. To speak of 'failure' in either case is to gauge their respective accomplishments against a measure foreign to them. Nor does it make sense to attribute the (in
the end) world-shaking difference in outcome to some alleged, native irrationality on the
part of the Chinese. From the point of view of how we at present know (yes, dear reader, we
know) the natural world to be constituted, *both* had it mostly wrong. But likewise, as soon
as that point of view is dropped as indeed it should be, *both* can be seen to have succeeded
well in the sole objective sensibly to be labeled 'rational' in this regard for times prior to the
rise of our modern science. Both aimed to make coherent sense of the world in naturalist
ways capable of capturing a large variety of readily observable phenomena. Of course, the
Greeks and the Chinese undertook their respective quests in different spirits – one befitting
the 'Masters of Truth'; the other, the 'Possessors of the Way'.[39] Of course, the spectacle of the
Greeks and the Chinese breaking the world up in different, really incommensurable ways is
just what a contemporary observer from Mars ought to have expected. The true historical
problem lies elsewhere, in the questions posed in the previous paragraph. In seeking to answer them, we reap from the preceding accounts some illuminating major contrasts.

Empiricism versus intellectualism.   Early in the 17th century, Francis Bacon used his
fabulous gift for the catchy metaphor to elucidate what has since become a classic distinction
between modes of knowledge acquisition. Empiricists are like *ants*, who build their hills by
bringing in one load after another; *spiders*, who weave their cobwebs out of their own bodies, are intellectualists; whereas *bees*, in turning their patiently collected nectar into honey,
take the middle way Bacon himself advocated. In this sense, then, the Greeks (whether Athenian or Alexandrian) were on the whole very much spider-like (which is precisely what
Bacon held against them); the Chinese were much more like ants. The contrast is far from
absolute. Greek thinkers, Parmenides alone excepted, did attach at least some independent
value to the sheer givenness of phenomena; the Chinese were hardly dull collectors of facts
but categorized and ordered these in light of their broad worldview. Still, compared to the
very intricately bee-like structure of modern science (with its actually rather un-Baconian
interplay between mathematical models and experimental feedback mechanisms), neither
the Chinese nor the Greek achievement was definitely placed on the high road toward modern science. Why, then, did the one get there in the end, but not the other?

Part of the answer must be delayed pending a more detailed comparison between Chinese nature-knowledge and a similarly empiricist current that arose and flourished in Renaissance Europe (p. 137). But another part of the answer can be given right away.

Theory – the historical force of cultural transplantations.   Civilizations may clash;
they may also cross-fertilize. Processes of transformation and of what I shall from here on

46   label 'cultural transplantation' offer the single most potent boost to novelty and creativity history knows of. An influx of foreign people, foreign ideas, foreign practices may under propitious circumstances greatly enhance chances for novel things to happen to ideas or habits worn out or petrified or grown stale in their original setting. Already in its creative upswing Greek nature-knowledge went through the process twice, at its very inception in Ionian centers of interethnic commerce and then as a result of Alexander's conquests, which was one major stimulus toward the Golden Age. *Due primarily to the vicissitudes of military-political history, such stimulating and transformative experiences were to keep happening to the Greek corpus of nature-knowledge but never to its Chinese counterpart.* Unlike the Roman Empire, China's imperial dynasties kept their civilization unscathed by their barbarian enemies, be these held at arm's length beyond the borders or rapidly sinicized once inside. At least until the arrival of the Jesuits in the late 16th century, and in most respects far beyond that time, the Chinese pursuit of nature-knowledge (for all its branching out to Korea and Japan) was to remain encapsulated inside itself. Consequently, whatever developmental possibilities might be hidden in the Chinese corpus of nature-knowledge missed the best chance history has to offer to come into the open.

This is true in the first place of the organic materialism that, with its increasingly sharp articulation culminating under the Sung, formed the core of the Chinese worldview from Han times onward. It is highly doubtful whether this enticing worldview and the 'correlative' way of thinking that it expressed contained within themselves the potential to be transformed into something sufficiently close to modern science to usher in an alternative Scientific Revolution. Fleeting speculations once made along such lines fail to carry conviction. Decisive, however, is that organic materialism, in remaining locked up inside itself, never got a chance to prove its mettle in a refreshing act of cultural transplantation.

The China-Greece contrast does not apply solely to the bodies of nature-knowledge central in each (one broad worldview in the former; a range of rival philosophies in the latter). It had consequences, too, for what remained on the margin of both. Mathematical science in the Alexandrian mode, which from Euclid to Ptolemy was pursued in the same, highly abstract mold, was to gain new developmental opportunities upon being taken up in fresh environments (first Islamic civilization, then Europe). There is a telling contrast here with the potentially no less fruitful approach embodied in the Mohist *Canons and Expositions*. These texts, not so mathematical but with a closer connection to natural reality, remained just as marginal in China as the Alexandrian mathematical texts did in the Hellenist world. However, the one setback that the Mo Tzu text lineage received – its large-scale destruction under the Ch'in – sufficed to make it wither away for good, whereas the no less marginal Alexandrian approach, despite quite comparable, if not more severe, setbacks, was over time to get a second, a third, and even a fourth chance *to realize its latent developmental potential in fresh cultural environments.* Primarily responsible here was the world-historically unique

ability of the Chinese polity to keep, if not always its political then at least its cultural integrity intact. As one unintended consequence, then, nature-knowledge in China remained
closed to the kind of reinvigoration that the procedures and conceptions of Archimedes and
Ptolemy, in particular, were repeatedly to receive, with world-shaking consequences in the
end.

The comparison, then, comes down to this. The organic/correlative view of the world
that, as the Mo Tzu text lineage faded away, came to permeate the Chinese corpus of nature-
knowledge may or (much more likely) may not have had the potential that its Greek counterpart possessed in retrospect to be transformed in such a way as to lead to the emergence
of recognizably modern science. In terms of actual accomplishment neither corpus was inherently superior. The decisive difference is that the one, but not the other, was to meet with
opportunities for the latent developmental potential contained within it to become manifest
over time.

Theory – a latent developmental potential and conditions for its realization

In close connection with the point just made about the transformative force of cultural
transplantations, my central explanatory category rests in this idea of latent developmental
potential. I owe it to the historian David Landes. He has used it for comparing the European
mechanical clock invented in the early 14th century with a big, ingenious water clock that
Su Sung (a contemporary of Shen Kua) constructed c. 1094. It kept time far more accurately than its mechanical counterpart, yet it ran into what Landes has called "a magnificent
dead-end".[40] It was a typical end product – due to the specifics of its setup, Su Sung already
squeezed out of his clock all accomplishment it was inherently capable of. The mechanical
clock, in contrast, was endowed with an as yet latent developmental potential. Not only did
it prove to lend itself well to miniaturization and to easy repair – eventually it proved subject
to transformation into a precision pendulum clock (p. 329).

Landes' concept of latent developmental potential opens wide venues for historical explanation. Applied to the present case, it has led me to distinguish a threefold 'if'. A body of
expert knowledge, having taken shape in the vigor of its original upswing,
(a) may or may not still contain major developmental possibilities – if it does not, it is bound
   to run into 'a magnificent dead-end', whereas if it does,
(b) it may or may not be subjected to processes of cultural transplantation – if it is not, it is
   likely to remain self-contained and true to itself in all its fundamentals, whereas if it is,
(c) it may or may not be radically transformed.
In the first place, the point at which the original upswing comes to be reversed is crucial.

48 Whether or not at that particular point a given body of expert knowledge still contains fertile developmental possibilities is best regarded as a gift of history, ascertainable only in retrospect. Further, the occurrence or nonoccurrence of cultural transplantations is a function of geography, of opportunities for interethnic trade, and (above all) of large-scale military conquest. Finally, my entire investigation will culminate in the question of whether or not transplantation induced transformation, be it of a more or a less revolutionary kind, with my specific conceptions of upswing and downturn central to the search for an answer.

What does all this now mean for the Greek body of nature-knowledge in the more or less garbled state it had reached when Greco-Roman civilization had definitely come to an end?

Certainly neither Senator Boethius in 6th-century Rome nor Bishop Sebokht near 7th-century Nisibis could have suspected that the Greek learning they helped preserve was richly gifted indeed with a significant developmental potential. Nor could either man have foreseen that the early caliphs' conquests would upset so thoroughly the ordinary course of events east and south of the Mediterranean, followed by the civil war that resulted in the overthrow of the Umayyad court by a new dynasty, the Abbasids. Nor, farther down the line, could they have foreseen the profuse interethnic mingling that marked 12th-century Toledo and Sicily, or the fall of Byzantium in 1453. The raw fact of history that three successive cultural transplantations were actually induced by all this could not possibly have been predicted by them either. If duly informed of these future events, however, the senator and the bishop would have been in the best possible position to infer from past experience what conditions would be favorable to turning transplantation to foreign soil into a success story. They might have acknowledged four such conditions. Would further acts of translation take place, along the lines of their own efforts? In what ways would translated texts be communicated? Would Ptolemy's effort to infuse Alexandria with a measure of reality be pursued further? Would there be chances for the pursuit of nature-knowledge to become anchored in society?

Translation of the Greek corpus of nature-knowledge could be done with purposes in mind that tended toward selectivity and extensive simplification, as in the Latin West. Alternatively, translation policies might involve, as in the Syriac East, adoption of texts wholesale, preservation of frame and structure, and something approaching literal rendition. For any future, large-scale translation effort it mattered whether all texts were considered for translation and whether those who undertook the task of translation considered faithfulness toward the original important. Would translators also be willing and able to master the source and the target languages? Would they in addition acquire the skills needed to deal with technical terms like *pneuma* or mathematical concepts like 'proportion' and the ways of proof that went with them? It would depend on the fulfillment of requirements like these whether or not the Greek corpus of nature-knowledge, in taking root in the soil of a budding civilization, would be fit to serve as a carrier of universal values transcending the Greco-Roman civilization from which it sprang.

Once this condition was fulfilled, whether the body of knowledge thus transplanted would
remain a dead letter or be granted new life in a creative spirit was bound to depend on
prevailing modes of *transmission*. Public dispute in an adversarial spirit had proven a two-
edged sword in the Greek case. At first it stimulated creative rivalry and well-constructed
argument, but in the end it lapsed into stale polemics and idle eloquence. Now that the
body of Greek nature-knowledge was no longer being created but formed one completed
whole, accessible to be handled in ways less or more creative, *educational practice* was likely
to become decisive – would it be attuned to learning by rote, or would it proceed in a critical
spirit directed at improvement?

Once this condition was fulfilled as well, whether *the enhancement of the reality-content
of the Alexandrian portion* of the Greek corpus of nature-knowledge would be attempted
might depend on whether Ptolemy's pertinent efforts were recognized, be it in their own
right or as an example worth following.

Underlying the two last-mentioned conditions, in particular, was the presence or absence
of viable channels into which *support* for the pursuit of Greece-born nature-knowledge
from outside the relatively narrow circle of its practitioners could flow. For the flourishing
of mathematical science the Greek experience powerfully suggests that court patronage was
indispensable. Here a slight extension might suggest in addition that a plurality of major
courts would be better than just the one in Alexandria (plus a few short-lived, derivative
courts) that Greek mathematical scientists had to make do with. For philosophy, its claim
to both worldly and otherworldly wisdom gave it opportunities for gaining a wider audience. Also, a variety of related occupational areas might be available (as, in Greece, oratory and medical practice). However, whereas philosophy was practiced for as long as the
Greco-Roman world existed, an increasingly spiritual climate of thought caused its natural-
philosophical portion to be attended to less and less. *Clearly, the more this-worldly the general atmosphere of the culture in which the Greek corpus of nature-knowledge would find itself
transplanted, the better the chances for its being pursued at all.*

Nothing in the above implies that, even if all such conditions were fulfilled, the Greek
corpus was *bound* in the end to usher in recognizably modern science. Only, in discussing
as we are about to do what successively happened to it in the course of no fewer than three
major feats of cultural transplantation, we should keep in mind the conditions just listed,
while remaining on the alert for still others to emerge from the inquiry as it proceeds.

NOTES ON LITERATURE USED

Traditional society. An early conceptualization of traditional as distinct from modern societies is due to Max Weber (in his posthumous *Wirtschaft und Gesellschaft*. Tübingen: Mohr/Siebeck, 1921).

'Non-Western' knowledge of nature. Helaine Selin (ed.), *Encyclopaedia of the History of Science, Technology, and Medicine in Non-Western Cultures*. Dordrecht: Kluwer, 1997, contains numerous enlightening entries.

- **Egypt, Babylonia**. B.L. van der Waerden, *Science Awakening*. New York: Wiley, 1963; chs. 1–3. An up-to-date view of later Babylonian-Greek exchanges in mathematical astronomy is in an essay review by Francesca Rochberg, 'The Historical Significance of Astronomy in Roman Egypt'. *Isis* 92, 4, December 2001, pp. 745-748.

- **China**. No straightforward history of pre-modern Chinese nature-knowledge is available (at least to those who, like myself, do not read Chinese). I have learned most from data supplied in work by Joseph Needham (*SRHI*, 6.5., 6.6.), and from various writings by Nathan Sivin: a book cowritten with Geoffrey E.R. Lloyd, *The Way and the Word. Science and Medicine in Early China and Greece*. New Haven: Yale UP, 2002; a chapter 'The Theoretical Background of Elixir Alchemy' that Sivin contributed to Joseph Needham *et al.*, *Science and Civilisation in China*. Vol. 5, Part 4; Cambridge UP, 1980; pp. 210-298; a lemma 'Shen Kua' in *DSB* 12, pp. 369-393 (a slightly revised version is on Sivin's personal website); ch. 1 'Astronomical Reform and Occupation Politics' of his *Granting the Seasons: The Chinese Astronomical Reform of 1280, with a Study of Its Many Dimensions and an Annotated Translation of Its Records*. Berlin: Springer, 2008, and a chapter cowritten with A.C. Graham: 'A Systematic Approach to the Mohist Optical Propositions'. In: N. Sivin & S. Nakayama (eds.), *Chinese Science: Explorations of an Ancient Tradition*. Cambridge (Mass.): MIT Press, 1973; pp. 105–152 (this chapter sums up some major points later set forth in A. C. Graham, *Later Mohist Logic, Ethics and Science*. Hongkong: Chinese University Press, 1978).

- **India**. For reasons set forth in *SRHI*, 6.1., I have decided against undertaking an effort to include nature-knowledge in India on a par with China and Greece – it seems to fall somewhere in between the two categories of one-domain specialization and comprehensive coverage.

Greek nature-knowledge. Conceptualization prepared in *SRHI*, 4.2. (discussion of works by Dijksterhuis, Farrington, Clagett, Sambursky, Lloyd, and Ben-David). For a handy overview and a factual check, chs. 1–7 of David C. Lindberg, *The Beginnings of Western Science*. Chicago: University of Chicago Press, 1992 (second, revised edition: 2008) have been very useful.

- **Pre-Socratics**. Jonathan Barnes, *Early Greek Philosophy*. Harmondsworth: Penguin, 1987. TRANSFORMATION OF MATHEMATICS: ch. 4 of van der Waerden's *Science Awakening*. PYTHAGOREAN MUSICAL THEORY: B.L. van der Waerden, *Die Pythagoreer. Religiöse Bruderschaft und Schule der Wissenschaft*. Zürich: Artemis, 1979.

- **Athens and Alexandria distinguished**. More than a century ago Pierre Duhem distinguished between 'realist' and 'instrumentalist' conceptions of science. He grounded the distinction historically in an interpretation of Greek astronomy that has been widely disputed. My own argument, which covers wider terrain in a mostly different manner, is not to be identified with Duhem's distinction. It carried a heavy ideological load, and I join the majority of his critics in finding it simplistic and mostly misconceived.

- **Athens**. Five philosophical schools: Jean-François Revel, *Histoire de la philosophie occidentale, I: De l'Antiquité à la Renaissance*. Paris: Stock, 1969. Plato and Aristotle: E.J. Dijksterhuis, *The Mechanization of the World Picture*. Oxford UP, 1961; sections I, II D-E. For the triangular constituents of matter in Plato's *Timaios*: Judith V. Field, *Kepler's Geometrical Cosmology*. Chicago: University of Chicago Press, 1988, ch. 1. atomism: A. Stückelberger (ed., intr., transl.), *Antike Atomphysik*. München: Heimeran, 1979. Stoa: Samuel Sambursky, *Physics of the Stoics*. London: Routledge & Kegan Paul, 1959.

- **Alexandria**. E.J. Dijksterhuis' book cited above, sections I, III A-C. Various entries in the *Companion Encyclopedia of the History and Philosophy of the Mathematical Sciences* edited by I. Grattan-Guinness (2 vols.; London: Routledge, 1994). Lindberg's book cited above. Reviel Netz, *Ludic Proof. Greek Mathematics and the Alexandrian Aesthetic*. Cambridge UP, 2009.

- **Athens and Alexandria compared**. In *The Physical World of Late Antiquity*. London: Routledge, 1963; p. ix-x, Samuel Sambursky broached the idea of absent integration between Greek natural philosophy and mathematical science.

- **World of more-or-less vs. universe of precision**. Alexandre Koyré, 'Du monde de l'"à-peu-près" à l'univers de la précision'; in: idem, *Études d'histoire de la pensée philosophique*. Paris: Armand Colin, 1961 (pp. 341-361 of the 1971 reprint by Gallimard); discussed at length in *SRHI*, starting on p. 86.

- **Late unifications**. Astrology: Tamsyn Barton, *Ancient Astrology*. London: Routledge, 1994. Ptolemy: G.J. Toomer (transl.), *Ptolemy's Almagest*. London: Duckworth, 1984. Albert Van Helden, *Measuring the Universe*. Chicago: University of Chicago Press, 1985. Liba Chaia Taub, *Ptolemy's Universe. The Natural Philosophical and Ethical Foundations of Ptolemy's Astronomy*. Chicago & LaSalle: Open Court, 1993. Andrew Barker, 'Plato and Aristoxenus on the Nature of μελος'. In: C. Burnett, M. Fend, P. Gouk (eds.), *The Second Sense. Studies in Hearing and Musical Judgement From Antiquity to the Seventeenth Century*. London: Warburg Institute, 1991; pp. 137-169.

- **Decline**. general: *SRHI*, 4.2.3.; 6.2.5. Greece specifically: Sambursky's *Physical World of Late Antiquity*. Geoffrey E.R. Lloyd, *Greek Science After Aristotle*. New York: Norton, 1973. Joseph Ben-David, *The Scientist's Role in Society. A Comparative View*. Englewood Cliffs (NJ): Prentice Hall, 1971, ch. 3 'The Sociology of Greek Science'. Scott L. Montgomery, *Science in Translation. Movements of Knowledge through Cultures and Time*. Chicago: University of Chicago Press, 2000.

Greek nature-knowledge and the Scientific Revolution. *SRHI*, 4.2.3.

- **China and Greece compared**. The argument reflects what I have learned from the views of Needham and his foremost critics (discussed in *SRHI*, 6.5.4., 6.6.; I have summed up and extended conclusions drawn there in 'Joseph Needham's Grand Question, and How to Make It Productive

52   for Our Understanding of the Scientific Revolution'. In: A. Arrault and C. Jami (eds.), *Science and Technology in East Asia*, vol. 9: *The Legacy of Joseph Needham*. Turnhout: Brepols, 2001; pp. 21-31). I have further benefited from points made since by Lloyd & Sivin in their joint book cited above. Francis Bacon's metaphor of the ants, spiders, and bees is in his *Novum organum*, aphorism 95 (*Works* 1, p. 201; *Works* 4, p. 93).

- **Cultural transplantation**. The idea that encounters with foreign habits and ideas are forceful motors of creativity and large-scale historical change is central to William H. McNeill's *The Rise of the West*. Chicago: University of Chicago Press, 1963 (reprinted 'with a retrospective essay' in 1991). The idea that such encounters may work with particular cogency in cases when the native civilization has perished and, with the umbilical chord severed, another picks up its legacy and enriches it in ways out of reach to the native civilization itself has been suggested by a Dutch author, Karel van het Reve in his *Een dag uit het leven van de reuzenkoeskoes*. Amsterdam: Van Oorschot, 1979; pp. 225-226. NB Ancient China did open itself to cultural exchanges in one case, with India. The importation of Buddhism has been widely studied; other exchanges have recently begun to receive serious attention. Amartya Sen's account in 'Passage to China'. *New York Review of Books* 51, 19; 2004, pp. 61-65, seems to imply that the 'nature-knowledge' component in these exchanges was small.
- **Latent developmental potential**. David S. Landes, *Revolution in Time. Clocks and the Making of the Modern World*. Cambridge (Mass.): Harvard UP, 1983; ch. 1 (*SRHI*, pp. 436-437).

II

GREEK NATURE-KNOWLEDGE TRANSPLANTED:
THE ISLAMIC WORLD

Three questions govern the account that follows of nature-knowledge in Islamic civilization. What happened to Athenian natural philosophy and Alexandrian mathematical science, and to the vast gap between them? To what, if any, extent did Islamic civilization conform to the pattern of upswing and downturn that characterized the Greek case? And to what extent was the body of nature-knowledge that was transplanted also transformed thereby?

Underlying the latter question is the claim that, for transformation to occur at all, transplantation is a fundamental precondition. That claim is paradoxically confirmed by the case of Byzantium, which I shall address first.

Portions of the Greek corpus of nature-knowledge were subjected, as we have seen, to sustained processes of translation into Latin, Syriac, and Persian. Greek manuscripts holding whatever survived of the corpus were preserved in many former centers of Hellenism but chiefly in and near Constantinople. The state of Byzantium into which by the early 7th century the Eastern Roman Empire was refashioned held the documentary record until the Ottomans conquered it in 1453. Thus, Byzantine civilization had almost a millennium to absorb and enhance its inheritance. What did it make of its opportunity?

The answer reinforces the idea of cultural transplantations serving as major incentives toward innovation. As a rule, the material transmitted has to be acquired first, by sustained efforts at translation of difficult texts into target languages lacking vocabulary for it. That was not the case with Byzantium, which inherited Greek nature-knowledge as children inherit from their parents – without any effort of their own. So much the more telling, then, that Byzantium sat on its manuscript treasure and, apart from bouts of intensive copying and some reshuffling, did nothing with it. Other, adjacent civilizations were to make productive use of manuscripts resting in Byzantium's palaces and monasteries. Caliphs sent emissaries to take copies for translation in Baghdad. Seven centuries later, the fall of Byzantium set free a stream of documents that in crossing the Adriatic enabled the incipient Italian Renaissance to expand far beyond arts and belles lettres (p. 99). The sense of excitement created by these transplantations was palpable. Byzantium, by contrast, faced no such challenge. If modern science was ever to emerge from the Greek corpus of nature-knowledge, learned Byzantines were not even in a position to make it happen.

Upswing

After wresting the caliphate from the Umayyads, who had made Damascus their capital, the Abbasid caliphs had a new capital, Baghdad, built in the 760s (AH 140s) on the model of Alexandria. The new dynasty made a determined effort over several centuries to acquire manuscripts of Greek learning from all over the realm and beyond. They had them translated (at first mostly from Syriac or Persian, later directly from the Greek) into the language of the Quran. Emerging from a civil war, and badly in need of legitimation as the newly ruling dynasty, these late descendants of Mohammed's uncle Abbas hit upon the project of large-scale book translation as one way to accommodate their many Persian subjects, the learned among whom they needed to run their administration. Caliph al-Mansur (714–775; AH 95–158), who initiated the movement, was keen to "incorporate the translation culture of the Sasanids [the last Persian dynasty before the Arab conquest] as part of his overall imperial ideology."[41] According to Sasanid legend the Greek corpus had been taken by Alexander from the Persian king he vanquished. Now it would be restored to its rightful place. Another major motive to embark on a sustained effort of translation was to acquire arguments for disputing religious issues with their numerous Zoroastrian, Manichean, Christian, and Jewish subjects. Philosophy (*falsafa*) could help them correct their lack of sophistication, or so the early Abbasid caliphs thought. Thus, out of the whole of the Greek literary corpus, which included poetry and drama and historiography, the Abbasids chose philosophical writings for translation and examination. Still other motives came to the fore later, as Islamic culture devoted extensive effort to mathematical science, in large measure for its practical applications in astrological forecasting but also for its mental challenge. Al-Mamun (al-Mansur's great-grandson) gave the movement a great push as part of his anti-Byzantine policy. It allowed him to pose as the true lover of an ancient Greek wisdom forsaken by its direct descendant.

We examine now how the Greek corpus fared in its new environment. Given the present state of historiography, I had to compose the account out of numerous disjointed pieces, some much disputed and/or in flux.

From translation to enrichment.   Openness toward the wider world marked the translation movement right from the start. Al-Mansur shaped his new capital on the cosmopolitan example of ancient Alexandria, and for several centuries his successors helped to keep that spirit alive. The realm not only encompassed large portions of the earlier Hellenist world and Persia but also Central Asia, bordered by both India and China. A commercial free zone extending from Cordoba to the Chinese border made wide and instructive traveling comparatively easy. Texts, no longer written on papyrus or parchment but on cheap paper, spread widely and were preserved all over the realm in libraries opened to the public for the first time in history. These developments in the possibilities of travel, preservation, and communication go some way toward explaining how the number zero and the decimal place system could be adopted from India, how the Taoist idea of achieving longevity through the *elixir* could reach the Islamic world, and how in mathematical astronomy not only Greek texts but also Persian and Indian ones were translated.

The technical competence of translators grew with experience. So did the inherent interest they took in what had come about as a state enterprise. At first, perceived utility was the principal ground for selection of texts. Chosen for translation on this basis were Aristotle's *Topics* on how to conduct an orderly argument and texts on astrology, invoked by al-Mansur for the building of Baghdad and a part of Abbasid ideology ever after. Later, more abstract subjects were chosen as well. As technicalities were mastered and a suitable vocabulary took shape, the best translators began to conceive of their craft as a creative enterprise in its own right. This is how the translation of texts and the enrichment of their contents became two sides of one and the same coin:

> [al-Kindi, in his Euclid-inspired study of light and vision:] We wish to complete the mathematics and set forth therein what the ancients have transmitted to us, and increase that which they began and in which there are for us opportunities of attaining all the goods of the soul.[42]
> [Thabit ibn Qurra, in the foreword to his own edition of an earlier translation of the *Almagest*:] The work was translated from the Greek into the Arabic language by Ishaq ibn Hunayn ibn Ishaq [etc.] for Abu s-Saqr Ismail ibn Bulbul and was corrected by Thabit ibn Qurra from Harran. Everything that appears in this book, wherever and in whatever place or margin it may occur, whether it constitute commentary, summary, expansion of the text, explanation, simplification, explication for the sake of clearer understanding, correction, allusion, improvement, and revision, derives from the hand of Thabit ibn Qurra al-Harrani.[43]

When around 1000 (AH 400) the translation movement came to an end, it was because what had meanwhile been translated was deemed adequate and because the pool of suitable texts appeared depleted. Indeed, the two greatest mathematical scientists in the half-century to follow, al-Biruni and Ibn al-Haytham (Latinized, Alhazen), raised ongoing enrichment to another plane no longer mixed with translation activity of any kind. And at that high point the creative pursuit of nature-knowledge in Islamic civilization lost its momentum and fell into decay, albeit not for good.

Why momentum was lost when it did, *what* might have ensued if it had not been, and *how* the downturn was later halted and (up to a point) reversed are questions to be addressed below (p. 64). I shall first survey developments in natural philosophy and list how the various branches of mathematical science were enriched.

56 *Aristotelian predominance.*   In the late Roman Republic and early Roman Empire the Stoa gained supremacy over its Athenian rivals; in the late empire, Platonism in Plotinos' highly spiritualized variety took over. Now, in Islamic civilization, it was Aristotelianism's turn, which had gained a head start with the early translation of the *Topics*. Its lasting predominance was neither exclusive nor untainted. In the eastern parts of the Islamic world, al-Kindi, al-Farabi, and Ibn Sina (Avicenna) systematized, interpreted, and on occasion criticized Aristotelian doctrine from a Platonic point of view. Their readings reserved a special place for Plato's doctrine of Ideal Forms and for Plotinos' vision (sometimes ascribed to Aristotle) of the gradual unfolding of the soul through a range of successive emanations. Later, a purer variety of Aristotelianism was adopted in al-Andalus – the westernmost part of the Islamic world, with Cordoba as its center. There Ibn Rushd (Averroes) strove to divest Aristotle's teachings of their Platonist accretions.

In these developments, whether they concerned supremacy, accretion, or expurgation, the philosophy of nature was not in and of itself of decisive import. Rather, the philosophic enterprise in Islamic civilization was largely determined by issues presented by the faith. Just as in the Roman Empire, but now driven by a different problematic, the portion of *falsafa* dedicated to nature-knowledge was of lesser significance than political philosophy, ethics, logic, or epistemology. Issues of great concern were the nature of the soul; the proper role of the ruler as (at least in theory) both spiritual and worldly leader; the created or eternal nature of the Quran; how revealed religion stood with respect to outcomes of speculative investigation; the uses of logic in law and in *kalam* (a system of reasoned defense of the faith); and so on. It was in such contexts that the other, non-Aristotelian systems of Athenian provenance came into play. Platonism proved attractive for its vision of creation and of the soul. The Stoa held on offer a rival account of logic. Atomism was taken up in the frame, not of *falsafa*, but of *kalam*, for its account of the composition of matter, with a wholly speculative debate going on over the number of dimensionless atoms minimally required to produce a body in length, width, and depth. Skepticism, finally, came into play for its critical point that the very contradictoriness of philosophical argument between schools belied the claim to indubitable certainty made on behalf of each. It was this powerful objection from the skeptical armory that al-Ghazali employed in his influential treatise *The Incoherence of the Philosophers*. His stated aim was to help clear the way for the mystical path to certainty he favored himself. Indeed, Ibn Rushd's no less famous rebuttal, *The Incoherence of 'The Incoherence'*, was meant to vindicate all over again the indubitable certainty of the teachings of one philosophical school in particular, Aristotle's.

The ongoing systematization, exposition, clarification, and occasional criticism of Aristotle's frequently obscure teachings thus took shape by way of a succession of treatises written most often in a format pioneered in antiquity by commentators faithful to the original doctrine – Alexander Aphrodisias, Simplikios, or (of more independent mind) Johannes

Philoponos. Although directed toward exposition and explanation in the first place, commentaries by al-Kindi, al-Farabi, Ibn Sina, and others also came up with objections directed at specific doctrinal points. Their various criticisms might lead to adjustment, not so much of the philosophical whole, but of individual tenets found to be out of keeping with the faith, with the overall tenor of the system, or occasionally with empirical facts thought to contradict them. One example of the latter is Aristotle's account of projectile motion, with its worrisome implication of the air sustaining, rather than impeding, the flight of a projectile (p. 84).

Mathematical science enriched. Greek mathematics was overwhelmingly geometric. Diophantos' range of solutions to special problems in arithmetic was as far as Greek mathematics went in the direction of algebra. Mathematicians in the Islamic world made a new start, which was aided by the decimal place system (including usage of a sign for zero) developed in previous centuries in India. But for incidental advances like the examination of irrational coefficients, the new algebraic methods that were worked out in Islam did not allow the solution of previously insoluble problems. Their principal merit rested rather in the development of general techniques for how to manipulate equations and polynomials of, mostly, the first and second degree, as in the very first book on algebra, written by al-Khwarizmi in the early 9th (3rd) century. He distinguished between six basic forms (e.g., $ax = b$, or $ax^2 + bx = c$, all expressed in words) and gave methods for reducing any given linear or quadratic equation to one of these standard types. The forms and (in arithmetic) the notational language in which mathematical problems were thus being cast opened venues that had remained closed to the Greek approach. Even so, the developmental possibilities contained in these new forms (in the end, the full *equivalence* of algebra and geometry; p. 351) were as yet being tapped on an incidental basis only. The best example is Umar al-Khayyami's geometric solution of certain cubic equations, for which he used the intersection of two conic sections determined by the coefficients of the particular equation to be solved.

In geometry itself, as knowledge of works by Euclid, Apollonios, and Archimedes became available, several issues were developed further. Certain books of the *Elements* gave rise to profound inquiries into foundational issues, notably, Euclid's theory of proportion and his postulate about parallel lines. Also, the *Conics* was used for the construction of figures not constructible by compass and ruler, like the regular heptagon, or for the trisection of angles. Finally, the Indian trigonometric concept of the *sine* led to the systematic replacement of Ptolemy's equivalent, yet more cumbersome, chord.

To mention trigonometry (including its spherical branch, which was further elaborated in Islam) is to point at planetary astronomy, the branch of mathematical science for the benefit of which the subject was developed in the first place. As in Greece, astronomy was the mathematical science practiced on larger scale than any other. Ptolemy's *Almagest* became and remained the central text. It was no longer held to be inviolable in its many details, but with one exception its fundamentals were never questioned. Innovative astronomical thought and practice, then, took shape by way of a more or less radically executed confrontation with details of Ptolemy's legacy:

> Ptolemy had saved selected phenomena spread over some centuries in Antiquity. When extrapolated to the ninth century his tables were no longer in step with contemporary phenomena. ... Trying to reconcile the data in the *Almagest* with the evidence of fresh observations created long-term problems of an unexpected complexity at the very root of astronomical science. The Arabic astronomers could not confirm the *Almagest*'s values of basic parameters, naturally thought of as constants of nature. These included the duration of the (tropical) solar year, the obliquity of the ecliptic (its inclination to the celestial equator) and the rate of precession (the slow displacement of the vernal equinox among the fixed stars). To save phenomena that spanned a period of more than a millennium proved to be anything but trivial. Either one had to trust the old data, and accordingly establish a theory for the secular variation of the basic parameters, or one had to be satisfied with producing, on the basis of contemporary data, tables of an admittedly ephemeral validity.[44]

The challenge involved in opting for the more daring, first horn of the dilemma was taken up in the eastern part of the realm, where (unlike later in al-Andalus) no Aristotelian near monopoly could stifle Alexandrian pursuits. Most of this work was highly sophisticated, and we will meet some of it when discussing how Copernicus dealt with the *Almagest* (p. 105). All we need here is to get the drift of what advances were made. Among early solutions, Thabit ibn Qurrah found that, not the Sun's annual trip from one vernal equinox to the next, but rather the 'sidereal year', which measures it against the background of the fixed stars, has a constant value. Al-Battani, in the 9th (3rd) to early 10th century, derived from some thirty years' worth of observation new values for the eccentricity of Venus and the Sun and discovered that the obliquity of the ecliptic is not a constant value and that the place where the distance of the Sun from the Earth is greatest (its 'apogee') varies over time. Also, while adopting as generally unproblematic Ptolemy's grand schema of the distances and sizes and resulting order of the planets, he aimed at better attunement by correcting several of their distances.

Mathematical musical theory did not develop far beyond the synthesis wrought in Ptolemy's *Harmonics*. The main difference between the principal theorists was that al-Kindi went deeply into Pythagorean speculations on the harmony of the spheres, whereas al-Farabi and Ibn Sina confined themselves to the theory of consonance and its corollary, the division of the octave. Remarkably, actual music making in Arab culture, which was founded upon tone scales set up quite differently from the Greek, scarcely made an impact upon musical theory in the Greek vein or vice versa. The exception is the tuning of the *ud,* or four-stringed lute, which was done by means of ratios obtained from the Pythagorean monochord.

In the mathematical treatment of equilibrium states first probed by Archimedes, whose works reached Islamic civilization only in rather garbled forms, some extension of his results was achieved. Examples are efforts, undertaken by Thabit ibn Qurrah and others, to extend the law of the lever to balance beams with a weight of their own and the establishment, by the most versatile of mathematical scientists in Islam, al-Biruni, of a measure akin to the modern concept of specific gravity for numerous substances.

The most drastic advances in mathematical science to be attained in Islamic civilization concerned problems of light and vision. Ptolemy's quantitative results on reflection were extended for a range of variously curved mirrors; his search for a quantitative law of refraction was crowned with success in the hands of Ibn Sahl, whose discovery remained unknown until our own day and had to be done all over again by Harriot, Snel, and Descartes more than six centuries later (p. 202). Above all, Ptolemy's effort to bring together Aristotle's broad conception of human vision and Euclid's mathematical tracing of light rays was extended and, in the process, altered greatly by Ibn al-Haytham in the first half of the 11th (5th) century. Ibn al-Haytham built on work by Ibn Sahl; he made a clear-cut distinction for the first time between how we see and how light propagates; he used his physician's knowledge of Galen's teachings on the parts and function of the eye, and he followed Ptolemy's bridge building to work all this and more into a grand synthesis of light and vision. Part of his effort consisted in devising ranges of experiments, of a kind probed before but now undertaken on larger scale. Thus, in the course of a systematic effort to demonstrate

> that light radiates from every point on the surface of an object, whether self-luminous or illuminated, along every straight line that can be imagined to extend from it in all directions. … [He placed] an oil lamp, with a particularly wide wick to provide an intense and constant source of light … in front of the opening of a tube where it pierced the centre of a copper sheet to enable the light to be projected via the tube. A screen was placed facing the other end of the tube. When the light was rotated around the aperture (of the tube), the spot projected onto the screen remained unaltered. With the narrowing of the opening, the luminous spot, fainter and smaller, still continued to appear. In this way he demonstrated that light is emitted equally from all parts of the wick and dispersed radially.[45]

Ibn al-Haytham's synthesis of light and vision, together with the work of his contemporary al-Biruni, formed the high point of mathematical science in Islamic civilization. While subjected in later times to incidental improvement and/or extension, its competence, scope, and broad vision remained unsurpassed until c. 1600 Kepler came along. Therefore, it is

60 the more remarkable that in the Islamic world the *Optics* practically disappeared from view soon after its appearance in the eleventh century until, in the beginning of the fourteenth century, the Persian scholar Kamal al-Din composed his great critical commentary on it.[46]

Such sudden disappearance, followed centuries later by resurrection in the guise of critical commentary, actually exemplifies a pattern of grander proportions – a loss of momentum, followed by an aftermath both quite similar to and markedly different from the Greek experience.

Athens and Alexandria.　Before addressing the much-disputed topic of upswing and downturn, we need to explore more deeply what happened to the Athens and Alexandria modes of nature-knowledge as a result of their first cultural transplantation. I shall examine (1) the broad frame conceived for each, (2) their place in the classification of learning in Islamic civilization, and (3) what, if any, interaction was made to take place between them.

For nature-knowledge in its philosophical as also in its mathematical guise *the framework established by the Greeks was left intact.* Critical questioning and creative enrichment, however insightful at their best, kept being directed at individual issues and lines of argument, not at the knowledge structure itself, which remained speculative in philosophy and, in mathematical science, divested of reality in the varied senses I have taken care to distinguish (p. 18).

For natural philosophy, ongoing preservation of its traditional knowledge structure meant that first-principle thinking continued to reign supreme. Practitioners agreed that, whatever its other merits, *falsafa* was needed for proper understanding of the faith. Debate between them turned on the question of whether or not Aristotle's first principles could stand on their own or were in need of support by Plato's. Also, just as with the Greeks, to the extent that empirical facts were invoked these still served to bolster outcomes of speculative argument. However much a spirit of constructive criticism was able to accomplish, there remained a basic continuity with the attitude toward the work of the pioneers that was taken in the long aftermath of the first Golden Age. No transformation took place, but rather a shifting or blending of allegiances and emphases, with Aristotelianism in particular coming out the richer.

Similarly, the account of how mathematical science was enriched shows that it remained confined to the highly abstract procedures of the ancient pioneers. This is true, too, of al-Biruni, with his penchant for attaining maximal precision through "observational methods that yielded direct results, as against techniques requiring extensive reduction by computation".[47] Not even he went beyond such a helpful yet limited extension of Alexandrian standard practice. Nor did the number of mathematized subjects increase beyond the given range of planetary trajectories, musical consonances, light rays, the lever, and floating bodies, with some attention to mapping techniques. Even al-Biruni, who mastered more of these subjects than anyone else, and whose curiosity led him to examine and record with sympathy the state of learning in Hindu India, did not move beyond those five topics. To conclude that the frame in which mathematical science was pursued in Islam remained largely unchanged does certainly not mean that the sum total was a mere rehash of the Greek performance. Surely Alexandria found itself enriched in these new surroundings, in an often critical spirit conducive to taking certain novel approaches (notably in arithmetic), as also to the attainment of numerous new findings, some of outstanding merit. The achievement was far more than the mere holding action so often ascribed to 'the Arabs'. Even so, the lively, creative pursuit of mathematical science did not transcend the scope attained or the knowledge structure established many centuries before in and around Alexandria.

The analytical distinction here maintained between Athens and Alexandria finds no match in the kind of classifications that were actually made in Islamic civilization by practitioners and adversaries alike. The prime division was between what in the modern literature most often goes by the labels 'foreign' (*awail*) and 'Arabic' learning. The latter comprised study of the Quran (whether approached in an intellectualist, *kalam*-oriented or in a spiritual/mystical manner), of *hadith* (sayings handed down from the times of the Prophet), and of *sharia* (law). The former comprised Greek learning, which thus found itself categorized as one single unit, with Aristotle, Ptolemy, and Galen serving as its major sages. Subdivisions into distinct branches of *awail* learning were made by a variety of scholars in a variety of ways, and rarely if ever do philosophy and mathematical science appear as distinct entities in such formal taxonomies. A significant contrast did nonetheless make itself felt at the level of their pursuit in actual practice, which was marked by a familiar *absence of interaction.*

In Greece, mathematical scientists were versatile men, who contributed as a rule to more than just one Alexandrian subject. So it was in Islamic civilization – in a branch-by-branch survey certain names, like Thabit ibn Qurrah and al-Biruni, recur time and again. In Greece, mathematical scientists and philosophers were different men. In Islamic civilization this was not always so. Some men whose claim to fame rests properly with their philosophical writings, notably al-Kindi, al-Farabi, and Ibn Sina, also produced work (more expository than original, to be sure) in one or more of the mathematical sciences. Remarkably, they performed these activities with, as it were, different parts of the brain. *In no way did methods practiced or conclusions reached in the one affect those in the other.*

Another sign of the persistence of the Greek split between Athens and Alexandria in Islamic civilization was the so-called Andalusian revolt against Ptolemy. In the western part of the Islamic world those active in the restoration of Aristotelianism sought to replace the *Almagest* by some alternative planetary arrangement in which all planetary motions, in their full irregularity, were directly referred to an impeccably central Earth. They had two grounds for their wholesale rejection of what mathematical planetary science had wrought. One was to vindicate Aristotle in view of the eccentric – the mathematical tool (p. 14) by means of which Ptolemy had violated the absolute centrality of the Earth required by Aristotle's doctrine of natural place. But the principal complaint was made on behalf of natural philosophy as such. In Ibn Rushd's succinct wording of the issue:

> The astronomical science of our days [with which Ibn Rushd was well acquainted] surely offers nothing from which one can derive an existing reality. The model that has been developed in the times in which we live accords with the computations, not with existence.[48]

Although this is an extreme statement of the Athens/Alexandria split in planetary astronomy, it does match the split as originally constructed in antiquity. Nor were, in Islamic civilization, natural philosophers alone in thus conceiving it – at the other side of the divide, too, mathematical scientists took it very much to heart. The 'hypotheses' Ptolemy put at the head of the *Almagest* provide a telling example. If, but only if, he felt that he had run out of mathematical argument or sufficiently persuasive empirical evidence or both, he would find refuge in stopgap arguments taken from one or another philosophy of nature (p. 24). By contrast, al-Biruni preceded a number of later mathematical astronomers in his flat refusal to have truck with the introduction into mathematical arguments of any natural-philosophical doctrine whatsoever, even if this were done only in Ptolemy's highly selective, 'opportunist' manner.

Nevertheless, one mathematical scientist in Islamic civilization did respond to Ptolemy's across-the-board efforts to infuse mathematical science with some realism drawn from suitable components of natural philosophy (p. 24). Ibn al-Haytham's synthesis of light and vision unmistakably emulates Ptolemy's bridge building. Quite as ingenious and quite as retrospectively mistaken as Ptolemy's own efforts in this regard, Ibn al-Haytham's synthesis reminds us that there was no obvious or easy way to make abstract mathematical science embrace the real world in a truly valid manner.

Cultural markers. The adoption-with-enrichment of the Greek corpus did not exhaust the pursuit of nature-knowledge in Islamic civilization – to some extent that pursuit touched base with the culture as well. Unlike with musical theory, which retained Pythagorean categories and neither affected nor was affected by native scales, some other enterprises were profoundly colored by their cultural environment. This concerns notably the search for transmutation of metals, the proper direction and times of prayer, and certain institutional arrangements with a religious background.

Among prominent contributors to alchemical practice were 'Jabir ibn-Hayyan' (a collection of authors known under that name) and al-Razi. Alchemy as adopted from a few original practitioners in Alexandria was enriched in the 8th–9th (2nd–3rd) centuries with an account of the composition of solid metal out of sulphur and mercury and with a range of

distinct techniques (distillation, calcination, solution, sublimation, filtration, etc.) meant to
produce the elusive *elixir*. The goal of this esoteric exercise was to make possible the transformation of base metals into the higher ones, silver and gold. The *elixir* was believed to be
also the ultimate remedy for the fatally sick or (influenced by Taoist conceptions) a means to
prolong life. Alchemy resonated with the culture most specifically in the idea that the laborious purification of metals was an exercise in the purification of the soul.

The Quran enjoins believers to face Mecca when at prayer (a direction known as the *qibla*). To lay out the building of mosques at places far away in strict obedience to this requirement constituted an advanced problem in spherical trigonometry. It took mathematicians
in Islam about two centuries to solve (Figure 2.1).

Early Quranic tradition further prescribes five prayers a day, the timing of which has to
do with dawn and dusk and shadow length. Much astronomical expertise is required to make
exact determinations, and Muslim mathematical scientists devoted as much attention to this
problem as to the *qibla*. The Quran has strict rules for the division of legacies, and a good deal

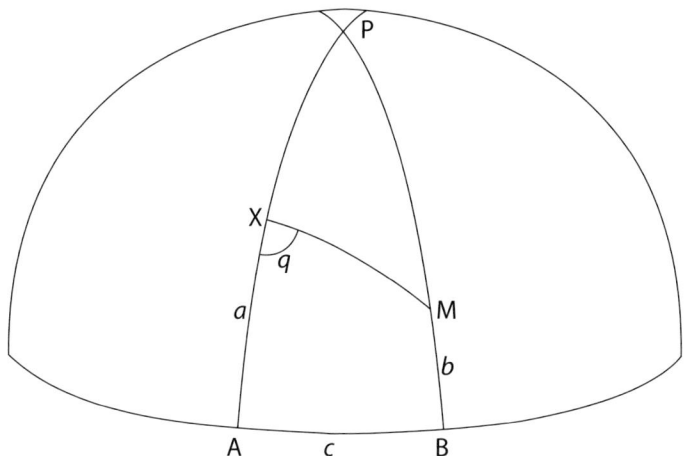

On the Earth's Northern Hemisphere here shown, M is Mecca, X is any given place in the Islamic
world, AB is the equator, and P is the North Pole. If one knows the latitudes of Mecca (MB = b) and
of X (XA = a), and also the longitude difference between Mecca and X (AB = c), then the problem
is to calculate the qibla, measured by the angle AXM (=q), from these data. Equivalents to the
modern formula, $q = \cot^{-1}\frac{\sin a \cos c - \tan b}{\sin c}$ were found from the early 9th (3rd) century onward
by using and extending Greek mathematical knowledge and techniques. This made it possible to
develop tables for a variety of places like Baghdad or Toledo or Cairo.

Figure 2.1: The qibla

64 of effort in arithmetic was spent on working these out in practical detail, as, for example, in the very same book that al-Khwarizmi devoted to the algebraic systematization of equations.

Finally, great importance was attached to keeping the community of believers in good health. Accessible hospitals were duly established all over the realm; public concern for the practice of healing extended to its theory, and that is why, unlike in Greece, numerous renowned philosophers and/or mathematical scientists had a medical background. Ibn Sina and Ibn al-Haytham provide the best-known examples of this combination, with the latter using it to particular advantage in working the structure of the eye into his grand synthesis of light and vision. This is the context in which Galen could be revered as the third Greek sage along with Aristotle and Ptolemy. This is also why in Pythagoras- or Aristoxenos-inspired treatises on musical theory, otherwise conceding so little to the peculiarities of Arab music making, the argument was often accompanied by hints and rules on how to use music as a cure for all kinds of mental ailments.

Downturn

My account of the upswing now faces a fiercely contested question. The 'decline' of nature-knowledge in Islamic civilization has been subject to an ongoing, emotionally charged debate filled at times with grand generalizations of questionable relevance. There are notable exceptions, acknowledged in the contribution to the debate that follows. I undertake the effort without passion and from the sidelines (p. 263). My aim is to throw light on the issue by means of the comparative approach taken throughout the present book. I consider the 'decline' of nature-knowledge here, not for the case of Islamic civilization alone, but as part of a larger historical problem which equally involves nature-knowledge in Greece, in medieval Europe, and in Europe halfway the Scientific Revolution. For the case of Greece I have enunciated a preliminary pattern (p. 28). I shall now show that (one major, time- and place-dependent variation notwithstanding) the 'decline' of nature-knowledge in Islamic civilization displays much the same pattern.

The Golden Age comes to an end.   As with Greece, we start from an empirical consideration about which there is broad agreement. Even if we leave out of the picture those men whose chief occupation was translation, between c. 815 (AH 200; reign of al-Mamun) and c. 1050 (430) a dense constellation of luminaries manifested themselves. The names of al-Battani, al-Biruni, al-Farabi, Ibn al-Haytham, Ibn Sina, al-Khwarizmi, al-Kindi, al-Razi, Thabit ibn Qurra stand out in every account of the period. *No such across-the-board outburst of concentrated creativity was to repeat itself in Islamic lands.* The three greatest, Ibn Sina, al-Biruni, and Ibn al-Haytham, all died in the mid-11th century (mid-5th century). And then,

just as in Greece some 1200 years earlier, momentum was lost and a steep downturn set in. As
with Greece, I consider the question of why momentum was lost at all to be a nonquestion
in view of this being the standard phenomenon in the premodern world (p. 28). Rather, the
question to ask is why it occurred *when* it did.

65

Note first that, although most of these luminaries and their numerous, lesser colleagues
operated in Abbasid Baghdad or in places under its immediate influence, the three greatest
did not. These three were eager, on completion of the translation effort, to reap what earlier
practitioners had sown, and they gave a powerful boost to already ongoing enrichment in
regions far removed from where it had all started: Ibn Sina and his longtime rival al-Biruni
in or to the east of Persia; Ibn al-Haytham in Egypt. The dynasties they served (the Fatimids
in Egypt; six short-lived dynasties in the far east) were not the Abbasids (who at this time
were mere puppets controlled by their military commanders) but independent rulers, badly
in need of legitimation and ready therefore to serve as patrons. *One generation later, the zest
had gone out of the enterprise.* Although their work held the promise of further advances to
be made by the next generation in the same or, just possibly, in hitherto untried directions,
that promise remained unfulfilled. For instance, no one took serious notice of Ibn al-Haytham's work on light and vision for centuries to come.

Why did this fading away occur? Although the wave of large-scale invasions from far less
civilized quarters that started at about this time and that continued unabated for some three
hundred years did not put an end to the pursuit of nature-knowledge in Islam once and for
all, it did take the fire out of its ongoing enrichment *at a retrospectively critical moment.*

In invading the world of Islam at a variety of times across a variety of borders, Berbers,
Mongols, Banu-Hillal, and also (to a lesser extent) Seljuk Turks and crusading Europeans
went to such lengths of destruction as no finely structured civilization can endure for any
length of time:

> Thus the Islam of 1300 was very different from that of 1000. The free, tolerant, inquiring and
> 'open' society of Omayyad, Abbasid and Fatimid days had given place, under the impact of
> devastating barbarian invasions and economic decline, to a narrow, rigid and 'closed' society.

It is easy to see how, when prolonged, massively destructive fighting occurs, the civilization
tends to turn inward. In 1241 Europe owed its escape from a similar fate to the fortuitous circumstance that the invading Mongols were called back home to attend the election of a new
khan. In large parts of the Islamic world, with everything seeming to fall apart, a remnant
of security was sought by clinging to a literalist faith. Spiritual values gained the upper hand
over the outward-looking curiosity of the Golden Age. What all this meant for the pursuit of
nature-knowledge is that, when and where such circumstances prevailed, *longstanding objections to 'foreign' learning could now gain a vigor and effectiveness quite new to them.* The point

66   is not so much that the distinction between 'foreign' and 'Arabic' learning was maintained over the full duration of premodern Islamic civilization (in some quarters, up to the present day). The point is rather that peaceful coexistence between these two varieties of learning gave way to a faith-saturated atmosphere in which 'foreign' learning came to be perceived as superfluous or perhaps even sacrilegious.

Objections of this kind had been raised almost from the start of the translation movement. In Baghdad in the late eighth (mid-2nd) century, Ibn Qutaiba leveled a fierce critique at 'foreign' learning, not just on behalf of the faith but also in view of its apparent uselessness for everyday life. He complained that even lowly placed government clerks felt an urge to master foreign learning – so deeply had the Greek corpus and ongoing translation thereof taken root in the civilized circles of early Abbasid Baghdad. Not until the open-minded mentality attacked in vain by Ibn Qutaiba and his ilk dissipated in those later waves of destructive invasion did such attacks begin to fall on fertile soil.

What happened to al-Ghazali's treatise *The Incoherence of the Philosophers* serves well to illustrate the profound change. The author's intentions centered on the issue of indubitable certainty and where to attain it (his ultimate answer being, not in philosophy, but in mystical revelation). In a move far exceeding those well-circumscribed intentions, his treatise now came to be perceived as a symbol for what it was not: a wholesale attack on foreign learning as incompatible with the faith.

Or take the *qibla* and the daily times of prayer and mathematicians' laborious efforts to provide the complex problems involved in these issues with exact solutions. How little such sophisticated scholarship was appreciated by the clergy appears from the telling fact that those very triumphs of religiously inspired mathematical science were generally ignored. Undisturbed, the *ulamas* kept declaring the sun set when they saw it and laying out their mosques in the general direction of where they felt Mecca must roughly be located. As Islamic civilization turned inward, such obscurantism in the presence of refined 'foreign' knowledge prevailed with ease.

Or take the establishment all over the Islamic world of schools of higher learning (*madrasas*). These privately endowed institutions began to spread at just about the period we are now addressing, to serve as strongholds for the revival of a literal-minded, exclusive spiritualism. The higher learning there on offer was confined as a rule to Quran, *hadith*, and law – the three principal branches of 'Arabic' learning.

In cultural environments subject to change of this sort, the lingering background question could become far more pressing than before: what is 'foreign' learning good for in the face of the certainty of revealed belief? Significantly, no self-confident answer to the question was offered by the practitioners of these suspect branches of learning. Unlike in Europe by the mid-17th century, the outward-looking curiosity of the Abbasid epoch, while evidently compatible with the faith, *was not, however, religiously laden in any positive sense.* Hence,

those turning it upside down were unlikely to meet with much faith-based resistance. I reserve further discussion of pertinent differences between the Islamic and Latin-Christian
varieties of monotheism for several later chapters (starting on p. 93), and confine myself
here to the outcome alone. Whether the literal-minded traditionalists who now had much
of the say over these matters regarded nature-knowledge of Greek provenance as just superfluous or as downright damaging, they did manage to rob freely flowing investigation of
nature's phenomena of much of the attractiveness required to pursue it, or to patronize it.

This is not however where the story ends – the bleak picture thus emerging does not
cover the entire aftermath of the Golden Age. At various times in different regions open-
minded, outward-looking environments emerged in which nature-knowledge of Greek
provenance could prosper once again, *albeit in a fashion decisively different from before the
fire had been taken out of it.*

Regional reversal and a protracted, high-level aftermath.   As in Greece, the aftermath was of far longer duration than the upswing itself had taken (see p. 28 for how we
have defined these concepts). As in Greece, in the aftermath a few individuals came to display outstanding achievement, not in anything like the dense clustering that had marked the
Golden Age but rather like a rare hilltop rising high over a vast plain. What is true of Ptolemy
and Diophantos in mathematical science and of Proklos and Philoponos in natural philosophy is just as true of Ibn Rushd in natural philosophy and of Nasir ed-Din al-Tusi and Ibn
as-Shatir in mathematical science. Their achievements, on a level at least with that of their
Golden Age predecessors, were separated from those of their predecessors by centuries and
accomplished in a context marked by lower-grade achievement all around. The big *difference*
from the Greco-Roman aftermath rests solely in the attainment, in between such rare peaks,
of a level of *average* achievement considerably higher in Islamic civilization. Or so it was in
certain regions at certain times under certain circumstances, which came about as follows.

The partial reversal of the downturn was due to two major feats. Upon settling down, destruction and conquest accomplished, the rulers of three invading peoples – Berbers, Mongols, Ottoman Turks – set up courts whose largesse included at times a renewed patronage
of nature-knowledge. The revival was reinforced by the establishment, or the revitalization,
of certain institutional strongholds without precedent in the Greco-Roman world – hence
the overall higher average level of whatever nature-knowledge was still being pursued in the
respective aftermaths. This is the broad picture, now to be filled in court by court.

In al-Andalus, two successive Berber dynasties regarded patronage of *falsafa* as a means
to enhance their hold on the throne and also bring them certain more narrowly political
benefits. This was the environment in which Ibn Rushd and several others operated. It was
in line with these rulers' specific objectives that, as *falsafa* prospered, mathematical science
was more and more ignored – a circumstance fraught with consequences for the distinctive
pursuit of nature-knowledge that soon took hold in the Latin West (p. 77).

68   In Persian lands in the early 13th (7th) century, the Mongol conqueror of late-Abbasid Baghdad, Hulagu Khan, founded the first large-scale observatory in history at his capital city Maraghah. As with the reform of the Chinese 'season-granting system' instigated at about the same time by his elder brother Kubilai (p. 41), one major purpose was to provide the new ruler with observationally improved horoscopes. Almost two centuries later and still farther east, another great conqueror's grandson, Ulugh Bey, set up at Samarqand another well-equipped observatory. "With few exceptions the royal observatories of Islam lived no longer than their royal founders".[50] But this did not prevent mathematical astronomers on the staff of (or in close connection with) the Maraghah observatory from taking things up where al-Battani and Thabit ibn Qurrah and al-Biruni and others had left off. In ways even bolder than their predecessors they returned to inspecting, adjusting, and correcting Ptolemy's observational data and/or technical devices. For example, al-Urdi subjected Ptolemy's grand schema of planetary order to a more severe critique than al-Battani's (p. 58). Out of his effort came a somewhat altered order of the planets, generally greater distances, and increased diameters (actual rather than apparent), all the while leaving untouched the conceptual pillars upon which the schema rested. At about the same time al-Urdi's boss at the observatory, Nasir ed-Din al-Tusi, went so far as to reject one geometric device invented by Ptolemy and employed throughout the *Almagest*. He replaced it with a more sophisticated device known since its rediscovery in the 1950s as the 'Tusi couple'. One century later Ibn as-Shatir (timekeeper at the grand mosque of Damascus) saw that the couple could be exploited on a grander scale for absolving Ptolemy of his own violation of strictly uniform angular velocity. Whereas both the Tusi couple itself and Ibn as-Shatir's models remained without consequence in their own civilization, centuries later Copernicus employed the same techniques in his own planetary modeling (p. 106). These devices and models figure prominently in historians' debates over the extent to which the Scientific Revolution came about in Europe due to nature-knowledge imported from elsewhere (an issue I shall address on p. 207).

    Finally, by the mid-15th century the Ottoman Turks had come close to reuniting what had once been the realm of the early Abbasids. Successive sultans helped sustain a tradition of patronizing nature-knowledge that remained essentially unbroken until almost the end of the 19th century. Here, too, certain institutional strongholds came to reinforce their patronage, notably Istanbul's short-lived observatory and numerous *madrasa*s. In the more secular spirit now reigning once again, these *madrasa*s adopted more *awail* learning in their curricula than was the case at their original foundation. This was an environment in which 'foreign' learning could once again be pursued with self-assurance and sophistication, and yet, *the excitement and vigor with which more than half a millennium earlier it had all started were gone for good*. Ongoing examination of dust-covered manuscripts may of course disclose unexpected finds, but so far the picture seems to be of a certain ossification, in the sense of a tradition turned inward on itself, of comments heaped upon comments in a fashion similar to the aftermath in the Greco-Roman world. There was no apparent desire even to inspect the manuscripts filled with Greek learning that in Byzantium had been preserved and copied over almost a millennium. The impact that these texts made upon their release occurred hundreds of miles away, in Italy, *not* on the spot in Istanbul. Instead, scholars in the Ottoman Empire maintained continuity with another glorious past – the one they acknowledged as their own. There was a readiness to examine the products of the Golden Age of nature-knowledge in Islam all over again and a capacity to come up with fresh thoughts on this or that point of running scholarly debate. But such fresh thoughts, even if highly promising in retrospect, remained encapsulated inside an unchanging framework and were not subjected to questioning, let alone to replacement.

One example is a protracted argument among mathematical scientists, with roots in al-Biruni's work, on whether or not there are observable phenomena which on or near the Earth would appear differently if the Earth underwent daily rotation. Al-Qushji, who thought not, also went along with al-Biruni's refusal to follow Ptolemy and introduce considerations from natural philosophy to settle the case. Thus, al-Qushji freed his mind from any objection to the idea of a moving Earth. Tellingly, however, he did not take the retrospectively obvious next step of finding out how far such a supposition might get him in terms of his own concern to improve known models in planetary astronomy. Also in the course of that debate, al-Birjandi suggested in passing the possibility that circular motion on Earth might persist forever. No less tellingly, he did not pursue this thought any further in the direction taken about a century later by Galileo (p. 184). Historians are tempted to regard such hunches and passing thoughts as 'anticipations', which indeed they are. Analytically it is more fruitful to note that *ossification within a given framework is what prevented the developmental possibilities contained in such hunches from becoming manifest.*

Later events confirm this diagnosis. Not that any Ottoman 'failure' to instigate a Scientific Revolution is at stake here – with respect to the nonoccurrence of such a feat, it does not even make sense to speak in terms of 'failure'. The rigidity of the Ottoman approach to nature-knowledge is demonstrated rather from how its practitioners, who were in a position to benefit from commercial and diplomatic contacts with neighboring Europe, dealt with the new science once, by the late 17th century, they were exposed to it. They took it in a wholly utilitarian vein, helping themselves to measurements rightly deemed more accurate than their own while ignoring the radically new modes in which nature-knowledge was being pursued in Europe (p. 725).

Schema 1 visualizes the pattern of upswing and downturn charted across ancient and Islamic civilizations. It displays numerous telling similarities, as well as three italicized, no less telling disparities.

Schema 1: Upswing and downturn compared in Greece and Islamic civilization

	Greece	*Islamic civilization*
UPSWING	creation + (in mathematics) transformation	*translation ⇨ enrichment*
• Golden Age	Plato to Hipparchos	al-Kindi to Ibn Sina, al-Biruni, and Ibn al-Haytham
DOWNTURN	c. 150 BCE	c. 1050 CE (AH 430)
1) Why?	'normal'	'normal'
2) Why then?	skeptical crisis? cessation of patronage?	*invasions ⇨ inward turn*
3) Aftermath	steep, protracted; generally low level	steep, protracted; followed by *partial, higher-level reversal* sustained by court proliferation and by institutional strongholds (observatory, later *madrasas*)
• marked by	codification, commentary, syncretism ...	commentary mostly
• still outstanding	Ptolemy, Diophantos; Proklos, Philoponos	Nasir ed-Din al-Tusi, Ibn as-Shatir; Ibn Rushd

On the threshold of a Scientific Revolution?

Does it follow from this account that the pursuit of nature-knowledge in Islamic civiliza-tion might have culminated in an event sufficiently like the Scientific Revolution to produce broadly the same outcome?

Theory – 'what-if history' and a dispassionate view of the past. To conceive of possible, alternative routes in history need not be a frivolous pastime. Although 'what-if history' may degenerate into a silly game, if handled with some care and good sense it can serve as a useful tool to discipline our causal thinking about those events that actually did happen. Among events in the stream of the past we ought to distinguish three kinds: those that happened (e.g., Newton's derivation of universal gravitation from Galileo's and Kepler's laws of motion); those that neither did nor possibly could have happened (e.g., derivation of universal gravitation from Philoponos' innovative conceptions of free fall and projectile motion); and those that, indeed, did not happen but which realistically might have hap-pened (e.g., derivation by Huygens of universal gravitation from Galileo's and Kepler's laws of motion). The only irrefutable objection to this threefold distinction is a strictly determinist point of view – possibly true, yet utterly sterile for purposes of historical analysis. After all, in everyday life we make comparisons all the time between what we did and what we might (most often: wish we would) have done instead. In 'big' history it is no different. In the latter case the historian is well advised to keep comparisons between 'possible' events as close as possible to those that did happen and to choose those similarities that display the attainable maximum of mutual coherence. For example, Huygens and Newton were similar thinkers and doers in numerous pertinent respects (p. 531), whereas Philoponos and Newton were not. Never forget meanwhile that room for uncertainty always remains, and that the whole purpose of the exercise rests in throwing into sharper relief the nature and causes of those events that actually did happen.

Two more preliminaries are in order. First, our exercise in 'what-if history' can only be provisional. Not until we have considered the first stage of the Scientific Revolution can we really begin to give due weight to the other side of the comparison. Second, recall the comparison between Greek and Chinese nature-knowledge, undertaken from the perspective of their own, not of some later, accomplishments. The answer to the question of how the Scientific Revolution could in the end be grafted upon the former, not the latter (p. 44), turned out to rest in the developmental potential hidden in the Greek corpus and the chances for transformation handed it by the vagaries of world history. Throughout my handling of the issue I have abstained from making claims of superiority based on what can in retrospect be seen to have come out of the one corpus of nature-knowledge but not the other. Just so, in presently drawing a conclusion from my survey of how the Greek corpus fared due to the first transplantation it underwent, I take a similarly dispassionate view of the matter – neither praise nor blame ought to be the objective, but only reaping the full benefit of sustained comparison.

Caught in the framework; high promise cut off.   It may seem, then, as if on two particular occasions, widely separated in time, the Islamic world stood poised to transform radically at least the Alexandrian component of the Greek corpus. At the *second* of these occasions, Ottoman practitioners came up with assertions remarkably close in retrospect to some prominent novelties that mark the onset of the European Scientific Revolution – notably al-Qushji's removal of objections to the idea of an Earth in daily rotation and al-Birjandi's idea of persisting circular motion. If so much was possible, why should the next step, so obvious in retrospect, have been out of reach? The answer, already hinted at, rests in the inclination to emulate the Golden Age that, naturally enough, reigned supreme in Ottoman nature-knowledge. As in the Greek aftermath, commentaries were heaped upon commentaries, incidentally producing finds of great promise yet in the grand scheme of things leading to unredeemed ossification. In short, the pursuit of nature-knowledge in Islamic civilization got caught in its own framework.

72 In contrast, on the *earlier* occasion a revolutionary transformation might have been possible indeed. If in our historical imagination we take the Golden Age to have lasted for one or two generations longer, with no destructive invasions intervening, 'Alexandria' might conceivably have been transformed into what I shall in chapter 5 dub 'Alexandria-plus' (i.e., mathematical science turned realist). For might not that transformation have been achieved in a variety of ways? With such an outcome residing as an unrealized possibility in the Greek corpus of nature-knowledge, there is no need to assume that the European way of bringing it into the open was the only one possible. More than a century ago E.C. Sachau, the translator of al-Biruni's book on India, observed that

> the fourth [tenth] century is the turning point in the history of the spirit of Islam. ... But for Al Ashari and Al Ghazali the Arabs might have been a nation of Galileos, Keplers, and Newtons.[51]

As it stands, the assertion is untenable. Not even authors as influential as al-Ashari (the founder of a long-lived tradition in Islamic theology in the late 9th [3rd]–early 10th [4th] century) and al-Ghazali (the author of *The Incoherence of the Philosophers*) could on the force of their books alone effect so major a turnabout. Still, there is historical substance to Sachau's hunch. The names of these two men have come to symbolize the large-scale inward turning that followed the political fragmentation of the caliphate under the impact of the barbarian invasions (which al-Ghazali experienced firsthand). *The forward zest of fresh, ongoing enrichment born of the excitement of the translation movement was cut off from the outside at a moment of high promise.* Whether, absent those invasions, the promise would have been fulfilled – that, indeed, is bound to remain anybody's guess. The crucial point is that there is nothing a priori inconceivable in some Galileo- and/or Kepler-like figure coming forward in the generation following that of men of such stature, achievement, and wide vision and scope as al-Biruni and Ibn al-Haytham.

One definite and three preliminary conclusions. Pending a range of comparisons to be made between Islamic possibilities and European actualities, I confine myself for now to baldly stating four conclusions:

(1) As just argued, there is no decisive and *inherent* reason why the achievement of the Islamic Golden Age of nature-knowledge was not crowned by one or more Galileo/Kepler-like figures.

(2) Even if that had happened indeed, we must still abandon Sachau's "Newtons". There is no conceivable way in which the new science thus arising in Islam could have survived the crisis of legitimacy that would inevitably have ensued.

(3) Conclusion (2) pertains to the circumstance that the pursuit of nature-knowledge in Islamic civilization reflected values underlying the culture (in a sense defined on p. 134) to a far lesser extent than in Renaissance Europe.

(4) Conclusion (3) also has to do with the circumstance that the religion of Islam belonged to the mainstream of monotheism, whereas Latin Christendom uniquely developed a thoroughgoing, this-worldly orientation. That orientation was to prove decisive in that it provided a way out of the profound crisis of legitimacy that revolutionary science underwent at a critical moment halfway the 17th century (chs. 12 and 17; p. 590 especially).

NOTES ON LITERATURE USED

Byzantium. Rare overviews are in Edward Grant, *The Foundations of Modern Science in the Middle Ages. Their Religious, Institutional, and Intellectual Contexts.* Cambridge UP, 1996; pp. 186-191 (section 'The Other Christianity: Science and Natural Philosophy in the Byzantine Empire'), and in A. Jones, entry 'Later Greek and Byzantine Mathematics'. In: Ivor Grattan-Guinness (ed.), *Companion Encyclopedia of the History and Philosophy of the Mathematical Sciences* (2 vols.; London: Routledge, 1994); vol. 1, pp. 64-69.

Translation. The vicissitudes and cultural contexts of all but one of the translation efforts successively directed at the Greek corpus of nature-knowledge (Greek to Latin, Greek to Syriac + Persian, Greek + Syriac + Persian to Arabic, Arabic to Latin) are surveyed jointly in Part I of Scott L. Montgomery, *Science in Translation. Movements of Knowledge Through Cultures and Time.* Chicago: University of Chicago Press, 2000. Regarding the sociocultural context of the translation movement in Baghdad I have further benefited greatly from Dimitri Gutas, *Greek Thought, Arabic Culture. The Graeco-Arabic Translation Movement in Baghdad and Early Abbasid Society (2nd–4th/8th–10th centuries).* London/ New York: Routledge, 1998. His views have the advantage of giving a coherent picture that also looks plausible from the point of view of the politics involved at the time (the obvious need for a newly established dynasty to gain legitimacy, and its need to appeal to learned Persian subjects). George Saliba, in chs. 1–3 of his *Islamic Science and the Making of the European Renaissance.* Cambridge MA, MIT Press, 2007, has made a case for a significant number of translations having already been prepared under the Umayyads.

Nature-knowledge in Islam. In *SRHI*, 6.2.1.-6. I examined and compared the following authors' views on 'decline' and several other issues: von Grunebaum, Sayili, Saunders, and Sabra. Unfortunately, no well-informed, even moderately up-to-date survey of the history of Islamic nature-knowledge exists that is longer than A.I. Sabra's entry 'Science, Islamic' in *Dictionary of the Middle Ages* (ed. J.R. Strayer), vol. 11, New York (Scribner's), 1988, pp. 81-89, and ch. 8 ('Science in Islam') in the second, much revised 2008 edition of David C. Lindberg, *The Beginnings of Western Science.* Chicago: University of Chicago Press, 1992. So I had to piece my account together from a wide variety of resources. That effort has been complicated by the following difficulties: practitioners tend to address the views of their fellow specialists at variance with their own in a remarkably polemical manner; they tend to approach out-siders' views with great, albeit to some extent understandable, suspicion, and they seem to persist in their belief that the time has not yet come for a synthetic (and, of course, as such necessarily imperfect) conception of where more than a century's worth of assiduous scholarship has gotten us. Here is a list of the resources I have employed:

- numerous entries in three fairly recent encyclopedias: *Companion Encyclopedia of the History and Philosophy of the Mathematical Sciences* cited above; *Encyclopedia of the History of Arabic Science* edited by Roshdi Rashed (3 vols.; London: Routledge, 1996), and *Encyclopaedia of the History of Science, Technology, and Medicine in Non-Western Cultures* edited by Helaine Selin (Dordrecht: Kluwer, 1997);

• some more-specialized literature – PHILOSOPHY: Magid Fakhry, *A History of Islamic Philoso-* 75
phy. New York: Columbia University Press, 1970, and a similarly entitled book in Dutch: Michiel
Leezenberg, *Islamitische filosofie. Een geschiedenis*. Amsterdam: Bulaaq, 2001; FAITH-GOVERNED
NATURE-KNOWLEDGE: David A. King, 'Science in the Service of Religion: The Case of Islam': ch. 1
of idem, *Astronomy in the Service of Islam*. Aldershot: Variorum, 1993; MEASURE OF THE UNIVERSE:
Albert Van Helden, *Measuring the Universe*. Chicago: University of Chicago Press, 1985, pp. 29-33;
MUSICAL THEORY: Fadlou Shehadi, *Philosophies of Music in Medieval Islam*. Leiden: Brill, 1995;
Amnon Shiloah's Introduction to his catalogue *The Theory of Music in Arabic Writings*. München:
Henle, 1979; and an entry 'Arabische Musik' by Alexis Chottin on pp. 131-152 of a selection of ency-
clopaedia entries *Außereuropäische Musik in Einzeldarstellungen*. München: dtv/Bärenreiter, 1980.
• work that explores what, after the Golden Age, happened to nature-knowledge in the eastern parts
of the Islamic world; in particular, F. Jamil Ragep, 'Tusi and Copernicus: The Earth's Motion in
Context'. *Science in Context* 14, March/June 2001; pp. 145-163, and several articles by Ekmeleddin
Ihsanoglu colllected in his *Science, Technology and Learning in the Ottoman Empire*. Aldershot:
Ashgate Variorum, 2004. NB The names of Ragep and Ihsanoglu must be added to those of von
Grunebaum, Sayili, Saunders, and Sabra for their likewise dispassionate approach to basic issues
in the history of nature-knowledge in Islamic civilization, as also Gutas and Montgomery in the
books on the translation movement cited above, and John Walbridge in an enlightening account
of the wide ramifications and present state of the debate, a review essay entitled 'Islam and Science'.
Islamic Studies 37, 3, 1998; pp. 395-410.

What-if history. While most historians are still skeptical of 'what-if history', more and more are applying it even without lengthy apologies. The argument I draw from everyday life occurs in a little-known
essay by Max Weber, 'Objektive Möglichkeit und adäquate Verursachung in der historischen Kausal-
betrachtung'. In: idem, *Gesammelte Aufsätze zur Wissenschaftslehre*. Tübingen: Mohr/Siebeck, 1922; pp.
266-290; key passage on pp. 279-280. More persuasive still than Weber's inevitably verbal argument is
a delightful British movie *Sliding Doors* (1994; directed by Peter Howitt, with Gwyneth Paltrow as the
moving heroine in two carefully synchronized roles). It makes the point in a realistic, everyday-life
manner that only visual means can accomplish. In the final scene it also dares address the vital, inherently insoluble question of whether alternative paths lead to ever more widely diverging outcomes or
rather converge in the end.

III

GREEK NATURE-KNOWLEDGE TRANSPLANTED IN PART: MEDIEVAL EUROPE

On the move from his birthplace in Cremona, a wandering scholar by the name of Gerard settled down in recently reconquered Toledo in or about 1140 and translated from Arabic into Latin, first, the *Almagest* and then, over four decades, seventy more works of Greek or 'Islamic' provenance. Until then, nature-knowledge in medieval Europe had been confined to texts stemming from the effort at simplified translation of selected portions of the Greek corpus that had been undertaken from Cicero to Boethius (p. 31). These texts survived in the few centers of learning that managed to hold their own in Western Europe once this partly Romanized, thinly populated, raiders-battered region was cut off by the early Islamic conquests from its center, the Mediterranean, and found itself willy-nilly compelled to set off on a path of its own.

Such nature-knowledge as had been preserved was organized in the *quadrivium*, which comprised *geometria*, *arithmetica*, *musica*, and *astronomia*. The men to keep alive quadrivial knowledge were clerics, the near-exclusive bearers of learning up to the 12th century. Not counting a few individuals with quite exceptional knowledge, the first stirrings of creative thought about nature on a more than incidental scale took place in the 12th century, when city life began to flourish and the center of education shifted from monasteries to schools attached to cathedrals and other urban institutions. With the help of Plato's cosmology as outlined in the *Timaios*, efforts were made to explain what happened during the six days of Creation by invoking natural agents. But the medieval pursuit of nature-knowledge did not acquire its distinctive character until Europe gained possession of a more substantial portion of the Greek corpus, due to the second round of transplantation – the one instigated on the grand scale by Gerard of Cremona.

Upswing

The pursuit of nature-knowledge in medieval Europe acquired its distinctive character from its Andalusian origins. The specific patronage needs of two successive dynasties of Berber origin (p. 67) were decisive. These led to the expurgation and correspondingly enhanced supremacy of the Aristotelian corpus and to the abandonment of the active cultivation of mathematical science. Aristotle's teachings were de-Platonized, and the mathematical sciences, albeit well preserved and well known in al-Andalus, were no longer pursued there

78   in a creative manner. Two consequences flowed from such curtailment of the full spectrum of extant nature-knowledge of Greek provenance. As the western part of the Islamic world shrank under the impact of the 'Reconquista' (Christian kings slowly but surely driving Islam out of Spain), the contemporaneous, step-by-step reunification of its eastern parts under the Ottomans sealed the split. Henceforth, what pursuit of nature-knowledge remained in the east tapped resources larger and more varied than in the west. It is telling that, of the three great men of the aftermath of the Islamic Golden Age, the philosopher, Ibn Rushd, was an Andalusian, whereas one of the two mathematical scientists, Nasir ed-Din al-Tusi, lived and worked in Persia and the other, Ibn as-Shatir, in nearby Damascus. The other consequence was decisive for the events to fill the present chapter – *it was in its western, hence, curtailed guise that the Greek corpus as enriched in the Golden Age of nature-knowledge in Islamic civilization came to be known in medieval Europe.*

Translation.   Once again, then, it was military conquest followed by enhanced, inter-ethnic commerce that opened up opportunities for cultural transplantation. Other than in Islam, the massive translation movement that came to center in Toledo soon after the city's reconquest was not state-sponsored (on Sicily, a secondary translation center, it was) but rather a matter of private enterprise, "as a profession without an institution".[52] Gerard of Cremona was but the most prolific of a dozen wandering scholars who between c. 1140 and 1180 settled in Toledo. With indispensable help from Islamic scholars and/or Jewish intermediaries at first, these men set out to turn manuscripts written in Arabic into at times slavishly literal but often more creatively handled Latin. Some works of theological interest (notably, the Quran) found themselves translated as well, yet nature-knowledge was the focus of translators' attention all along, as had been the case in Baghdad. On Sicily, slightly later, the same thing happened on a smaller scale, albeit with somewhat wider scope in that a number of manuscripts in Greek, procured from Byzantium on an incidental basis, were translated, too.

Not nearly the full corpus of learning in Islamic culture was thus transmitted. Texts in Arabic of Chinese or Indian provenance did not reach al-Andalus and thus remained unknown in Europe for centuries to come. By the same token, much original work done in the eastern parts of the Islamic realm never reached the west. For instance, not only al-Biruni's report on the state of nature-knowledge in India but his entire corpus of work has only over the past hundred years or so gradually become known to the world. Still, a considerable portion of texts in mathematical science and in Aristotelian philosophy, as also much knowledge about their various Islamic enrichments, had by the early 13th century been dropped into a Latin West thirsting for knowledge of a kind far more sophisticated and profound than available so far.

Aristotelian near monopoly.   The event decisive for how the nascent civilization of
the West would handle all this excitingly novel knowledge thus occurred right at the start.
Whereas in Islamic civilization Aristotelian doctrine had gained the upper hand over rival
philosophies of nature, in medieval Europe it secured almost uncontested supremacy, not
only within natural philosophy, where no rivals remained, but over mathematical science as
well. 'Near monopoly' does not imply that every medieval thinker subscribed to every single
tenet of Aristotle's – on the contrary, many a daring innovation was to be tried out before
the upswing was over. The point is rather that Aristotelianism, once accepted, served as the
frame inside which even the most striking departures from Aristotle's original teachings
were fitted and, hence, acquired their proper meaning.

Why was the specifically Andalusian constellation so well suited to medieval Europe,
to the point that it was adopted wholesale and even radicalized with respect to its built-in
one-sidedness? Three interconnected causes are in play here: (1) the *inherent attractions* of
Aristotelian doctrine, (2) chances exploited for *reconciliation* between Aristotelian teachings and Christian tenets seemingly at odds with them, and (3) the simultaneous rise of an
educational stronghold of rare homogeneity and staying power – the *university*. We address
these causes in turn.

More than any other Greek philosophies of nature, Aristotelian doctrine could account
in a thoroughly reasoned, logically satisfying, broadly coherent manner for whole ranges of
everyday natural phenomena (p. 6). This characteristic attracted a Dominican monk whose
wide scholarly interests included plant description and who is known to posterity as Albertus Magnus (Albert the Great). During the early second half of the 13th century he and his
pupil Thomas Aquinas wrote commentaries and comprehensive summaries of the full range
of Latinized Aristotelian texts. They interspersed these with theological interpretation and
justification. Indeed, even under such favorable conditions, the establishment of a near monopoly hinged on whether or not the Aristotelian conception of the world would be judged
by the clergy at large to be compatible with Christian tenets and points of view.

It was not an easy task. Numerous weighty obstacles presented themselves. For example,
how could the eternity of the world upheld by Aristotle be made compatible with a Christian
conception of time bounded between Creation and the Last Judgment? Or how to uphold belief in the life of the soul once it leaves the body, given that this prime Christian dogma seemed
to be ruled out by Aristotle's conception of the soul as a form inseparable from matter?

The issue at the root of the problem came in the end to be seen as one of explaining
how, if at all, the sovereignty of the biblical God, with His absolute freedom to deal with His
creation as He pleases, could be reconciled to the large range of causal necessities that Greek
philosophy generally, and Aristotle in particular, held to be implanted in nature. Thomas
Aquinas was the man to perceive how it could be done. His points of departure were not
new. In the 4th century St. Augustine had proclaimed what came to be known as the 'hand-

80   maiden' solution to the problem of how religion was related to philosophy: philosophy could render important services in the defense of as well as the proper understanding of religious truths. Thomas also took up an idea of Philo's, an early-1st-century CE syncretist philosopher (p. 32): there must be an underlying harmony between the truths of theology and philosophy, so that apparent conflicts can always be resolved. The crucial distinction that Thomas made was between God's absolute power and His ordained power. While absolutely free to act with respect to the created world as He pleases at any given moment, He has actually used His freedom for making a free decision to let the various regularities it has pleased Him to implant in His creation generally run their own course.

This is how Thomas mitigated the causal necessities built into the very foundations of Aristotelian doctrine to the point of making them compatible with divine freedom. Next, he adopted a range of distinct Aristotelian conceptions so as to show how these fitted in with Christian doctrine after all. Sometimes he employed adaptation, for example, by declaring Aristotle's form joined to the body to be imperishable. More often he used straightforward adoption, as when he identified Aristotle's 'prime mover' (the agent from which all motion in the world emerges) with the biblical God. One core feature of Aristotle's teachings that made such reconciliation both possible and attractive was his ongoing invocation of *purpose* in the universe. Over the bridge of God's ordained power His ends with the world could be identified with the final state that nature is unceasingly striving toward in the Aristotelian view of things.

Thomas' effort at reconciliation effectively persuaded the regular and secular clergy of Europe (still the men most concerned with intellectual matters at the time) to embrace Aristotelian doctrine. But the initial success does not fully explain the tenacious hold the doctrine was to exert for many centuries to come, particularly in view of the renewed clerical opposition to Thomas' teachings within half a century of their original enunciation. In 1277, at the instigation of the pope himself, the bishop of Paris condemned a range of tenets ascribed to Aristotle as incompatible with the Christian faith – in most cases, because he felt God's omnipotence to be undermined thereby. That is, he rejected the idea of God's ordained power as an inadmissible compromise. What made the condemnation both significant and influential is that, on the European continent, Paris was the center of Aristotelian teaching. Nevertheless, even in the short run, the decree did not succeed in banning the condemned tenets or the conception underlying them; as we shall presently see, its long-term effect was rather to make incipient innovation of these meanwhile settled teachings take one route rather than another.

What remains to be explained is why Aristotelianism, once established, proved enduring. This is where the simultaneous rise of universities as permanent bodies of autonomous learning comes in. Aristotelianism became rapidly entrenched in the leading universities of Paris and Oxford (founded by the 13th century) and in many others. Previously, schools of advanced learning had been marked by the specific doctrine they taught – for example, Platonism at the Academy and specific interpretations of the law at specific *madrasas*. But in medieval universities curricula looked very much alike. From Coimbra to Cracow and from Vienna to St. Andrews, Aristotelian doctrine and the *quadrivium* were taught as foundation courses in arts faculties by means of a uniform didactic system of questions and comments. 'Scholastic' is the adjective by which these educational practices came to be known.

Principally on these three grounds, then – inherent explanatory merit, perceived compatibility with the predominant cultural value, and institutional entrenchment – Aristotelian doctrine came to envelop almost every view held on the constitution and operations of nature. This state of intellectual affairs was without precedent. It was quite unlike the coexistence and mutual rivalry that had marked the pursuit of nature-knowledge in Hellenist and (albeit to a smaller extent) in Islamic civilization. The new state of affairs had important consequences, *both* for what happened in medieval Europe to existing domains of knowledge *and* for the directions innovation might take.

Here is what happened in the broadest of terms. To some limited extent the various Athenian rivals in natural philosophy had already made their appearance in medieval thought – several components of Platonism but also scattered fragments of stoic and even, in Carolingian times, of atomist thought. All these now vanished from the scene almost entirely, leaving Aristotle the undisputed "master of those who know" (as Dante in his *Commedia* was to call him).[53] As for the Alexandrian mathematical sciences, some branches were refashioned in a scholastic mold so as to fit the reigning philosophy, and others, although passed down the centuries in manuscript lineages of considerable continuity, nonetheless remained in the margin. We shall now examine these developments in more detail, starting with the threefold fate of the Alexandrian heritage: incorporation into the Aristotelian frame, or 'scholastization'; fringe existence; and incidental extension.

Alexandria scholasticized, submerged, and somewhat enriched.   Incorporation proper is what befell all but two of the Alexandrian subjects. Adoption of Boethius' account of *musica* involved a threefold division between *musica mundana* (harmony of the macrocosm), *musica humana* (harmony of the microcosm; i.e., of the human being as a miniature reflection of the macrocosm), and *musica instrumentalis* (music as sung and played according to the rules of harmony). The examination of the harmonic (consonant) intervals remained confined to calculations carried out with their ratios, so that in practice the teaching of *musica* came down to a variety of *arithmetica* in equally simplified forms. In *geometria* and *astronomia*, the translated texts of Euclid's *Elements* and Ptolemy's *Almagest* were simplified and scholasticized. They were clad in logical forms that were as proper to Aristotelian categories as they were far removed from their original structure of theorems and proofs. The mathematical and quantitative *finesse* of Ptolemaic planetary theory (including

82   his schema of cosmic distances) was subordinated in practice to a qualitative cosmology as expounded in the popular textbook *De sphaera* (On the Sphere) written by 'Sacrobosco' (John of Holywood). Discussion of cosmological details took place in the scholastic format of *quaestiones*. The mathematical tradition in equilibrium problems of solid bodies was similarly scholasticized, with treatment in accordance with Aristotelian categories replacing mathematical proofs in Archimedes' vein.

Only one Alexandrian domain was not affected by scholastization, at the price, however, of a life on the margins. Archimedes' work on floating bodies remained stuck in a text lineage all by itself. The text was ably translated in the late 13th century by Willem van Moerbeke from a Byzantine manuscript. Although it was repeatedly copied, its impact on medieval thought remained minimal. Archimedes' treatment of solid bodies immersed in fluids is all about the relative weights of bodies variously composed, yet it did not give rise to questioning the Aristotelian doctrine of the polar opposition between 'heavy' and 'light', such as was to happen indeed several centuries later in a decisively altered intellectual context (p. 404).

Regarding problems of light and vision, finally, the synthesis wrought by Ibn al-Haytham (Latinized, Alhazen; p. 57) was taken as authoritative from the outset. Under the name of *perspectiva* it was studied both inside and outside the *quadrivium*. Whether Alhazen's manner of grafting an Aristotelian account of vision upon the Euclid/Ptolemy tradition in the tracing of light rays was scholasticized wholesale (as by Roger Bacon, albeit without the far-reaching simplification customary in the other branches) or whether it remained insulated from the Aristotelian mainstream, this is what Europe received and, for all Bacon's and others' exertions, substantially adhered to. One novel finding was added to *perspectiva:* Dietrich von Freiberg's early-14th-century explanation of the rainbow. He showed in a qualitative manner how the rainbow is formed by the combined reflection and refraction of sunlight in water droplets in a cloud. He was not the only one to make this discovery. At the same time, yet independent of Dietrich, the first scholar in Islam to comment on Ibn al-Haytham's synthesis of light and vision and extend it, Qutb al-Din al-Shirazi, seized on his Ptolemy-inspired rules for refraction to explain the same phenomenon that Dietrich accounted for in similar fashion.

What further innovation took place inside the *quadrivium* was likewise small scale and incidental. *Geometria* and *astronomia* were so thoroughly scholasticized as to preclude any innovation in the sense of enriching, in the way of Ibn al-Haytham or Nasir ed-Din al-Tusi, what Euclid and Ptolemy had wrought. In *musica*, finally, an as yet inconsequential adaptation was due to the rise of a uniquely medieval brand of polyphonic music. Here singers employed the major third as a pure, consonant interval. Unlike in Islamic civilization, musical practice quickly began to affect musical theory, as quadrivial theorists came to realize that this major third could not possibly be the stridently dissonant interval 64:81 (two whole notes, 8/9 x 8/9) in the Pythagorean scheme of things. Rather, it is represented by 4:5 – a theoretical misfit in a Pythagorean scale as yet left intact otherwise.

Aristotelian teachings enriched.   Teachers at the arts faculties of Paris and Oxford took the lead in critically inspecting and creatively enriching what Albert and Thomas had set forth in their all-encompassing summaries. The accomplishment of these 14th-century scholastics represents without question the Golden Age of medieval nature-knowledge.

Novel viewpoints arose as a rule in the frame of commentaries made upon some Aristotelian text. Often these took the form of *quaestiones*, the written counterpart to the oral disputations by means of which knowledge was regularly passed on. Opinions were confronted by counteropinions and reasoned down to a final conclusion over convoluted stretches of logical argument. Jean Buridan's 'Questions upon the Eight Books of Aristotle's *Physics*' (second quarter of the 14th century) provides an instructive example. At one point in a lengthy argument Buridan wanted to know "whether in the motions of heavy and light bodies toward their natural places all succession arises from the resistance of the medium".[54] Sound Aristotelian doctrine holds that all bodies in the sublunar domain are endowed with a tendency to move toward their natural places. For heavy bodies in descent this is the center of the universe, which coincides therefore with the center of the Earth. The question is now whether or not their apparent failure to fall in just an instant of time is due to more than the resistance that is naturally exerted by the medium interposed between the body and its final destination. The point of asking the question is that the presence of other sources of resistance beyond the resistance exerted by the medium would invalidate Aristotle's proof of the impossibility of the void. That proof, in its turn, depends crucially on the consideration that a body falling through an infinitely rare medium (i.e., through void space) in not meeting with any resistance would move instantaneously not successively, which Aristotle deemed absurd. Not wishing to uphold the void other than through God's absolute power to create one, Buridan sought and found reasons for rejecting all conceivable sources of added resistance that previous commentators had suggested. For instance, may not a heavy body possess an internal resistance arising from the fact that only its midpoint can reach the center of the universe, not its sides, which therefore resist the motion toward it? To this Buridan objected that, of two contiguous parts of water, the one part does not seek a lower position than the other:

> Even if a sailor descends to the bottom of the sea so that he has hundred vessels of water upon his shoulders, he does not sense the weight of that water, as that water which is above him does not incline to be farther below.

But this example is not good enough, as water in its own natural place (at sea bottom) is weightless anyway. So a better case is this:

even if the water were not in its natural place, but very high in a vessel like on top of the tower of the Notre Dame, still one part of it would not incline to be below, so that if someone would be there in a bath with his leg on the bottom so that above that leg there would be a large quantity of water which in the air he would not be able to carry, he would still not feel the weight of that water. [55]

And this argument settles the issue – parts of heavy bodies do not indeed resist natural motion on account of their inability ever to arrive at the precise midpoint of the universe.

For a proper understanding of the role allotted to experience in natural philosophy this is a revealing passage. Most often in natural philosophy experience is invoked directly, with a view to rendering a certain doctrinal point more plausible (e.g., the argument about the elasticity of stoic *pneuma* reported on p. 8). Here, in contrast, it looks as if Buridan is appealing, as an early Galileo, to experimentally designed experience in an effort to settle which of two predicted alternatives actually holds in the real world. But that is not so – the 'experience' Buridan appeals to is wholly predetermined by the enveloping, theoretical structure. Not only is there no question of Buridan himself carrying a bath up the stairs of Notre Dame to check whether his seemingly empirical assertion holds: Does one *really* not feel the weight of the bathwater resting there upon one's leg? (Note in addition that Buridan, so well abreast of the Aristotelian literature, apparently has no inkling of what Archimedes had shown to go on in one's bathwater). The point is rather that such a check would be irrelevant to Buridan's apparent aim here. This is none other than didactic enlightenment using the most drastically extreme cases he can think of in order to display to his students the overall consistency of Aristotelian doctrine even in cases, like the present one, where Aristotle had overlooked a logically possible alternative.

It is only when a case presents itself where Buridan has become persuaded that consistency has not been provided by Aristotle himself but must be reinstated by his latter-day commentators that he moves toward innovation of a kind. One well-known example is Buridan's idea of *impetus*. Here he built on Avicenna's (Ibn Sina's) quarrel with Aristotle's account of projectile motion – a quarrel which went back in its turn to points first raised by Johannes Philoponos. Buridan argued that the agent which continues to move a projectile – say, a stone – after it has been thrown away is not the ambient air invoked by Aristotle but rather an internal motive force, which he called *impetus*. For this he invoked everyday experience of various sorts, for example, a boat with a sail does not stop moving once the sail is suddenly torn off and, thus, the ambient air is removed. He went on to apply his idea of internal motive force to two other cases. He argued that to explain the motion of the planets around the Earth there is no need to posit celestial 'intelligences' (supposed agents often invoked for the job and identified by some with angels). Rather, at Creation God endowed every planet with a certain amount of impetus. Further, the common observation that (up to a point) as objects fall their speed increases can easily be accounted for by supposing them to acquire a certain amount of impetus at every successive moment of their fall. This suggestion had the merit of addressing a highly problematic spot in Aristotle's account of local motion. Vertical descent itself seemed adequately explained by ascribing to terrestrial objects an inclination toward their natural place, but Aristotelian categories could not explain why the descent is not uniform but slower at the beginning than later on. So by means of his *impetus* account Buridan helped close a gap in an otherwise nicely coherent system, and later scholastics were eager to adopt it. But that is all there is to it – *impetus* was neither meant to nor did bring about a rupture with the Aristotelian framework.

Another innovative approach concerns method. Lengthy debates were devoted to re-finements in Aristotelian methodology, which concerned proper derivation of its first principles. Not the first principles themselves were questioned; it was rather argued by some that mathematics ought to be given a place in their derivation, whereas others pleaded for a larger role to be assigned to empirical observation.

The decree by means of which the bishop of Paris had in 1277 sought to ban numerous Aristotelian doctrines in view of their apparent incompatibility with God's omnipotence had a dual impact on the innovative trends of the 14th century. Since one may not consider God bound by what humans might find reasonable for Him to do with the created world, the decree inadvertently promoted examination of possible phenomena that Aristotle ruled out on a priori grounds. But what intellectual free room the decree gave with one hand it took away with the other, as the freedom thus gained was often of the type called at the time 'speculation according to the imagination'. Here are two influential examples, both taken from the work of Nicole Oresme, who taught at the University of Paris one generation after Buridan.

In his commentaries upon Aristotle's book on the heavens, Oresme considered the possibility that not so much the outer sphere of the heavens revolves in twenty-four hours around the Earth but that, instead, the Earth rotates around its own axis. He sought to demolish in orderly fashion both 'experiences' and 'reasons' first listed against that supposition and then went on to advance positive reasons for it. He argued that "motion can be perceived only if we can see that one body assumes a different position relative to another body",[56] so that if first the heavens were rotating and then the Earth we would not notice the difference. Also, an arrow shot up vertically would not fall back due west with the Earth turning away underneath; rather, the arrow is "moved toward the east very rapidly with the air through which it passes".[57] The supposition also allows a more economical explanation of heavenly appearances. Also, Bible passages like "God hath established the world which shall not be moved" need not be taken literally, since they are accommodated to everyday speech. But then, in the final passage, Oresme suddenly makes a U-turn: "But everyone maintains, and I think myself, that the heavens do move and not the Earth".[58] And now he invokes all over again the very Bible passage "God hath established the world" that, just a few folios previously, he has told his audience not to take literally, and says that all those persuasive reasons just advanced "are not evidently conclusive".[59] The same, however, is true of the opposite position, of which he has now shown sufficiently that it cannot be demonstrated either. His entire argument pro or con the Earth's axial rotation, so Oresme concludes, has been meant to serve as just an "intellectual exercise and a diversion".[60]

The same applies to 14th-century work exploring how far certain topics in Aristotle's natural philosophy could be quantified. The matter was pursued first at Merton College (Oxford University). A range of scholars collectively known as the 'calculators' were concerned to quantify whatever in the Aristotelian philosophy of nature seemed to allow it. For example, Thomas Bradwardine proposed to alter Aristotle's rule for heavy bodies pulled or pushed – speed increases with force applied. In the customary rendition of Aristotle's ambiguous text it increases proportionally – by doubling the force, you double the speed, etc. The difficulty with straightforward proportionality is that a minute force like a thumb pressing cannot displace a heavy object like a stone pillar. So a lower limit must be set, which disrupts neat proportionality. Bradwardine solved the problem by positing proportionality of another kind, one that he could express only verbally and that we can understand only when writing it up as a logarithmic function. More significantly, there were no empirical grounds for his rule whatever, his sole aim being to restore theoretical consistency at a point where it appeared lacking.

Another issue taken up by the calculators was Aristotle's insistence upon the irreducibility of quality to quantity. Cannot the extent to which a body possesses a certain quality be subjected to quantitative treatment after all? Not only may one body be more red than another, with the intensity of redness being liable to numerical expression, but it is also possible for one and the same body to be colored by different intensities of red. Such considerations led the calculators to the idea of 'intensification and remission of qualities'. Oresme not only adopted it but went on to invent a novel tool for it, graphic representation. His core idea was to express a given intensity by means of a straight line of corresponding length. If every successive intensity of a certain quality was drawn perpendicular to a baseline that represents that quality, a surface figure arises which Oresme called the *configuratio* of that quality. For example, the rectangle forms the configuration that belongs to a uniform distribution of intensities. Similarly, an intensity which *varies* uniformly has either a rectangular trapezium or a rectangular triangle for its configuration (Figure 3.1).

More complicated rates of change yield configurations bounded by curved lines of many different shapes. Oresme broke these down in an exhaustive categorization, with chapters 'on uniform and difform quality', 'on simple difform difformity', 'on four kinds of simple difform difformity', 'on composite difformity and how it has sixty-two species', etc. He went on to apply all this to one of Aristotle's four classes of change – change of place, or motion.

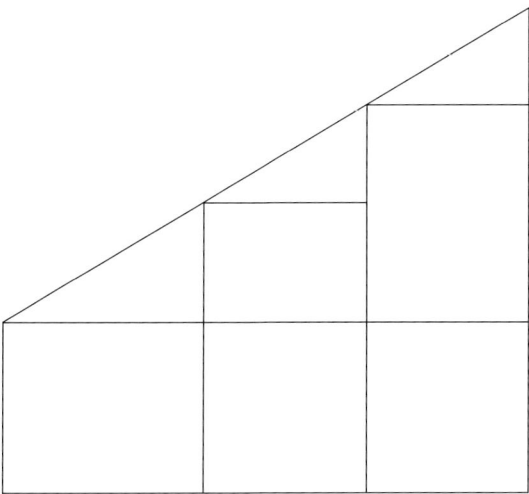

Figure 3.1:   Oresme's uniformly difform configuration

In this particular case, successive velocities of a moving body are marked against the flow of time to yield a 'total velocity' for configuration.

In our time most attention has been drawn to Oresme's treatment of uniform acceleration in moving bodies, because it seems to 'anticipate' one of Galileo's major feats (pp. 307, 358). However, Oresme made his explorations of uniform acceleration in the imagination only, and the crucial idea is missing that there might be such a motion in the real world. Furthermore, for the author himself the principal interest of the whole exercise resided in the status he assigned to his configurations. These he took to represent the thing or person affected with such and such a quality at such and such a varying rate of intensity. By means of the differences in configuration that he so minutely derived he sought to explain "the relative pungency of different substances, the attraction of iron by magnets, the natural hostility or friendship of people and animals, the resistance of certain people to common phenomena like lightning or snake bites, the remarkable properties of herbs and gems, and so on."[61] So much for the once-popular idea that with Oresme we are already halfway to the rise of modern science.

Cultural markers.   In medieval Europe as in Islamic civilization a few subjects were handled in ways that, rather than neatly fitting either Athens or Alexandria, were colored to a large extent by the culture. In medieval Europe these subjects were alchemy, the *computus* (calculation of Easter dates), falconry, and the magnet.

88   Inspiration for the transmutation of base metals into higher ones came from the translated works of the alchemical authorities in Islam, 'Jabir' and al-Razi (p. 62). Practitioners in medieval Europe adopted their account of the composition of metals out of 'sulphur' and 'mercury' (concepts encompassing far more than their respective modern equivalents). They also adopted, and refined, techniques like distillation and filtration, with gains in quantitative precision attained by means of the assayer's balance. The way in which 'Geber' and other alchemists handled the balance allows us in retrospect to recognize an idea of specific weight tacitly underlying much of their experimentation. It speaks to the almost airtight insulation obtaining between distinct modes of premodern nature-knowledge that at the same time but in other circles a better articulated yet wholly untried concept of specific gravity lived on just barely breathing in the text lineage of Archimedes' *On Floating Bodies*. It is even more telling that, in apparent ignorance of that text, teachers at the arts faculties kept handling Aristotle's account of heaviness in absolute, not specific terms (as with Buridan and his imagined bathtub on top of Notre Dame).

Medieval alchemists' efforts at purifying metals to prepare them for transmutation went in tandem with an ideal of purification of the soul as developed originally in Islam. Their dependence on the great Islamic authorities was too large as yet to do more than render in biblical language the luscious, naturally Islam-colored imagery in which 'Jabir' and al-Razi et al. had clad their accounts. Not until the Renaissance, with Paracelsus in particular (p. 132), was European alchemy to adopt fully the coloring of its cultural environment.

More thoroughly determined by the culture was the calculation of Easter. In 325 the Council of Nicaea made an effort to determine for every year prior to Judgment Day the biblically proper date to celebrate Easter. This involved reconciling the vast complications of a part-solar, part-lunar calendar. The rules eventually set were so complex in their effects as to require the working out and ongoing refinement of highly specialized calendrical reckoning. This became known at the time as the *computus*. Refinement continued over the entire medieval period and indeed far beyond (the greatest mathematician of the early 19th century, Gauß, still made a lasting contribution).

Likewise far removed from any university context, two distinct investigations were undertaken by the Sicily-based, German emperor Friedrich II von Hohenstaufen and by a Frenchman of whom almost nothing is known, Pierre de Maricourt. Their respective researches were neither mathematical nor dependent on philosophical first principles. Rather, they depended on the patient collection and critical sifting of empirical facts and the cautious drawing of conclusions from those facts. In about 1250 the emperor completed a lavishly illustrated manuscript treatise *De arte venandi cum avibus* (On the Art of Hunting with Birds). It displays "his intense curiosity about the particulars of nature, most unusual in an age that was forever seeking universals".[62] Not only did Friedrich extend in a critical spirit Aristotle's observations about the anatomy and the feeding, migration, and mating habits of birds of various kinds. He also displayed an inquisitive mind open to whatever phenomena turned out to be, irrespective of any a priori conceptions. For instance, he sent an expedition to the far north to check whether, as a persistent story still around in the 17th century had it, barnacle geese (never watched nesting) do or do not hatch from trees.

Two decades later, on 8 August 1269, Pierre de Maricourt completed a treatise on a subject that had not aroused much scholarly interest from the Greeks or in Islamic civilization: the capacity of stone of Magnesia to make iron move toward it. In his *Epistola de magnete* (Letter on the Magnet), he assembled and described accurately such topics as what magnetic poles are, how magnetized iron aligns itself if left free to do so, and in what sense magnets attract or repel each other. Pierre set his investigation in a context of practical application, directed at the improvement of the compass so as to assist sailors in their orientation when out of sight of the coast.

Downturn

Whether widely admired by contemporaries (Friedrich II's treatise) or not (Pierre's), such empiricist researches remained just as incidental and marginal to the whole of medieval nature-knowledge as did those branches of mathematical science that were not subjected to scholastization. Only Aristotelian natural philosophy flourished on a scale beyond the incidental and the marginal. Undertaken on a massive scale in the 13th century and creatively enriched in the 14th, the work of Jean Buridan, the calculators, and Nicole Oresme must count as the Golden Age of medieval nature-knowledge. Once again, the downturn was sudden and steep. Using the provisional criterion offered previously (after a dense constellation of luminaries, no next-generation follow-up; p. 28), we can conclude that with Oresme's death in 1382 the Golden Age came to an end.

Its end was more decisive still than in either Hellenist or Islamic civilization. Not only did the *average* level of performance remain devoid of just about all creativity, but also no incidental, high-quality achievers of the likes of Ptolemy or Ibn as-Shatir came along to take up where the men of the Golden Age had left off. Instead, the aftermath was marked by an unrelieved reshuffling of ever more hackneyed ideas and concepts in the persistent framework of 'Questions' and 'Commentaries'. It was also of comparatively short duration. In the mid-15th century, with the fall of Byzantium, Greek nature-knowledge underwent a *second* transplantation to European civilization. Numerous rival doctrines were restored, and Aristotelianism lost its near monopoly. The event compelled Europe's arts faculties to refresh their Aristotelian teachings – a feat they accomplished between roughly the mid-15th and the mid-16th century (p. 100). The medieval aftermath was over; in its newly revitalized state Aristotelian natural philosophy was ready to face the challenge of those rivals recovered from oblivion by the humanist movement of the Renaissance (see the next chapter).

90   *Caught in the framework.*   With regard to the medieval downturn only, the two distinct questions of 'why then?' and 'what was the aftermath like?' receive the same answer: By the late 14th century just about all opportunities for substantive renewal of Aristotelian natural philosophy had been exploited to the very limits. In what further directions might innovation still have been sought within just one conceptual framework, even one as flexible and spacious (comparatively speaking) as Aristotle's? No different in this regard from the Ottoman reversal of the 15th to 19th centuries, nature-knowledge of the medieval type found itself caught in its own framework, to the point of dooming an aftermath still swarming with busy scholastics to near-complete sterility.

This diagnosis is not contradicted by, but rather finds paradoxical confirmation in, the so-called enigma of Domingo de Soto. In a collection of questions and commentaries not otherwise marked by originality, this mid-16th-century scholastic added to his docile rendition of Oresme's doctrine of configurations one remarkable point. Quite in passing, and without elaborating any further, he mentioned that uniformly accelerated velocity does occur in nature, namely, in free fall. Half a century later the idea would serve Galileo for his point of departure (p. 192). How can it be that, put forward in a 1545 commentary, this idea was not picked up by anyone, not even by its own author? This is enigmatic only if we think of past ideas in isolation, without proper regard for the preconceptions from which they stem. In the commonsense frame of Aristotelian philosophy the idea makes little sense – our unmediated perception of objects falling shows that the descent quickly lapses into uniformity. For the idea to make proper sense the frame of some as yet nonexistent mode of nature-knowledge was required, where such observational refutation would not at once count against it. The entire development of medieval nature-knowledge, however, had gone in precisely the opposite direction – toward a near monopoly of one mode of nature-knowledge and one set of preconceptions. For as long as these remained unquestioned, no innovation that went substantially beyond Oresme's could occur.

Nature-knowledge in Islamic civilization and in medieval Europe: A comparison

We are now in a position to compare how the upswing/downturn cycle unfolded in Islamic civilization and in medieval Europe. Adding the latter to the earlier Schema 1 (p. 70) and once again italicizing the disparities yields Schema 2.

There are some major disparities between successive items in the two pertinent columns, but do these negate the overall validity here claimed for the upswing/downturn pattern of premodern nature-knowledge? Pending the next addition (Schema 3 on p. 149), it is clear already that all disparities on display in Schema 2 stem from one feature only: the uncommonly

Schema 2: Upswing and downturn compared in Greece, Islamic civilization, and medieval Europe   91

	Greece	Islamic civilization	Medieval Europe
UPSWING	creation + (in mathematics) transformation	translation ⇨ enrichment	translation ⇨ *Aristotelian* enrichment
• Golden Age	Plato to Hipparchos	al-Kindi to Ibn Sina + al-Biruni + Ibn al-Haytham	Albert the Great to Oresme
DOWNTURN	c. 150 BCE	c. 1050 CE (AH 430)	c. 1380
1) Why?	'normal'	'normal'	'normal'
2) Why then?	skeptical crisis? cessation of patronage?	invasions ⇨ inward turn	*possibilities for enrichment exhausted*
3) Aftermath	steep, protracted; generally low level	steep, protracted; followed by partial, higher-level reversal	steep, protracted; *unreversed*
• marked by	codification, commentary, syncretism ...	commentary	commentary
• still outstanding	Ptolemy, Diophantos; Proklos, Philoponos	Nasir ed-Din al-Tusi, Ibn as-Shatir; Ibn Rushd	*none*

narrow basis on which the pursuit of nature-knowledge during its medieval heyday rested. The 'Andalusian' curtailment of the full spectrum of nature-knowledge, first adopted and then reinforced on conceptual, theological, and educational grounds alike, was truly decisive.

A more-detailed comparison must now be made in terms of how the conceptual substance of Alexandria and Athens developed over two successive transplantations. I shall focus on similarities and differences with respect to (1) the idea of a Scientific-Revolution-like event waiting just around the corner, (2) institutional support, (3) nature-knowledge and faith, (4) a naturalist drift.

A Scientific Revolution just around the corner?   Both Ottoman work in mathematical science and medieval innovations in Aristotelian philosophy yielded incidental results that have struck commentators as harbingers of conceptions that were in due time to mark the Scientific Revolution. Oresme's graphic rendition of uniform acceleration and his argument about the Earth's daily rotation are cases in point, as also the freedom from both empirical and conceptual objections against he latter gained by al-Qushji, and the persistence of circular motion suggested by al-Birjandi. In all these cases, however, the frame in which they were set precluded any sustained follow-up or further extension. Some (notably Oresme's) were meant as intellectual play far removed from reality. But truly decisive is

92 that they all remained trapped in their own framework – in the Ottoman case, emulation of Golden Age mathematical science; in the medieval case, the Aristotelian near monopoly. No Scientific Revolution was around the corner in any conceivable way. The part that these surely promising, medieval innovations were destined to play in historical reality was quite different. Rather than prefiguring the Scientific Revolution, they were in due time absorbed into it. It took their being taken up in the frame of a drastically altered mode of nature-knowledge to reveal what potential merit lay hidden inside (p. 307).

Institutional support.   In Islamic civilization, the one exception to the rule of pursuit of nature-knowledge standing or falling with courtly patronage came late and remained without much consequence: the *madrasa*, which helped preserve an established tradition but did little if anything to refresh it. As al-Biruni, client to no fewer than six *dynasties* over a lifetime, inferred with some desperation from his own experience,

> The number of branches of learning is great, and it may be still greater if the public mind is directed towards them at such times as they are in the ascendancy and in general favour with all, when people not only honour learning itself, but also its representatives. To do this is, in the first instance, the duty of those who rule over them, of kings and princes. For they alone could free the minds of scholars from the daily anxieties for the necessities of life, and stimulate their energies to earn more fame and favour, the yearning for which is the pith and marrow of human nature.
>
> The present times, however, are not of this kind. They are the very opposite, and therefore it is quite impossible that new learning should emerge or a new beginning should arise in our days. What we have of learning is nothing but the scanty remains of bygone better times.[63]

We may leave aside both al-Biruni's somewhat premature despair and the specific motives he ascribed to the patrons of 'foreign' learning and their clients. But he did express the general rule for Islamic civilization: with patronage, the pursuit of nature-knowledge may flourish; take the patronage away, and it steeply declines. In antiquity that rule had been true of mathematical science alone; in Islamic civilization it was at least for a while true of *falsafa* proper as well.

In medieval Europe almost the exact opposite situation obtained. Court patronage remained confined to Friedrich II's sponsorship of a few second-generation translators. The Toledo translators, in contrast, thrived on private initiative, and then newly founded, statute-endowed, autonomous corporations of students and scholars took over to provide space for the examination of Aristotelian texts. Universities became the institutional embodiment of both the medieval upswing and its downturn. In the case of the mathematical sciences, the ongoing scholastization produced by these events leaves us with two possibilities. Was it stimulated by the absence of courtly interest in its pursuit, so that there was no place else to go for the aspiring mathematical scientist but the university, there to see the substance of his endeavor refashioned in Aristotelian categories? Or did the early scholastization of mathematical science preclude the rise of courtly interest in the first place? Whether the one or the other or a self-sustaining dynamics arising from their intertwinement, it is certain that in medieval Europe occupational opportunities for mathematical science became for the first time in the history of Alexandria nonexistent from start to finish.

With philosophy, as always, things were different. For philosophers in antiquity, several occupations beyond its proper job of spreading worldly and otherworldly wisdom stood open as well (notably, the teaching of eloquence). In Islamic civilization, *falsafa*, even if cultivated in tandem with medicine, was most often bound up with princely patronage. So, when the tables turned after the wave of invasions, *falsafa* proved vulnerable in a way unknown in the Greco-Roman world. Suspected to be a sacrilegious enemy of revealed religion, it underwent a major setback and took its 'foreign' fellow, mathematical science, down with it. In medieval Europe the possible sacrilege implied in philosophy (notably of the Aristotelian variety) was recognized from the start, as it had been in Islam, but on more massive scale, so as to provoke the concoction of a remedy unlike anything produced in Islam.

Nature-knowledge and faith.   Thomas Aquinas drew his remedy (God's ordained power) from Philo's idea of guaranteed harmony between the truths of philosophy and religion and from St. Augustine's conception of philosophy as the 'handmaiden' of theology (p. 79). The latter conception, of philosophy as a help in defending and clarifying the faith, reflects Christian practice in Augustine's time. But it reflects later Islamic practice as well. Both in late antiquity and in early Islamic civilization the conversion of intellectually skilled infidels proved again and again to be accomplished, not by revelation alone, but by offering reasons for the religion's purported verities. The big difference is that in Islam, unlike in Christianity from Augustine to Thomas and beyond, *the practice was not turned into established doctrine.* An effort in that direction was certainly made. Under Caliph al-Mamun's active sponsorship, and using Greek rules for how to conduct reasoned argument, certain theologians called 'mutazilites' undertook a full-scale intellectualization of the faith. The effort foundered on the vicissitudes of patronage as, in the end, all secular learning did. The *mutazila*, called in by al-Mamun to seal the religious authority he claimed for the caliph as Mohammed's not only worldly but also otherworldly heir, fell with the Abbasids. Their fall, and with it the end of the caliphate, left a void at the center of religious authority that was to remain unfilled in Islam forever. Henceforth, a multiplicity of interpretations of the Quran, the *hadith*, and the law could flourish in the absence of a center that could enforce one sole interpretation as valid and declare all others mistaken or even heretical.

94   Latin Christianity was in possession of such a center – the papacy and, for particularly weighty doctrinal issues, the council of bishops. Consequently, the challenge became to get God's ordained power and all that flowed from it accepted in Rome. At first, the opposite happened, with the condemnation of 1277 serving as one late, papally stimulated effort at a showdown. But soon enough the uniquely close intertwinement between Christian dogma and Aristotelian philosophy that Thomas had proposed in his *Summae* came to be widely recognized, in Rome as elsewhere, as a boon bought at a price well worth paying – the apparent need for *revealed* truth to find sustenance in *reasoned proof for it*. Those willing to pay the price were far-sighted indeed – the alliance between Jesus Christ and Aristotle was to hold for centuries, to their apparent mutual benefit. Who at the time could possibly have foreseen that centuries down the line a Scientific Revolution would make the alliance come unstuck within decades (p. 432)?

In the meantime, almost the entire cultural, social, and political elites of Europe over several hundreds of years were to pass through the arts faculties of its universities. Even the dullest student could hardly leave them without a smattering of quadrivial and Aristotelian knowledge, served up not as something out of the ordinary but as part and parcel of basic higher education. And that knowledge was sanctioned and taught by the very bearers of the most prominent cultural values around – those of the Christian faith. In medieval Christendom from roughly the mid-13th century onward, the Greek corpus, virtually equated as it was with its Aristotelian portion, had by and large managed to evade suspicion of heresy. Or rather, suspicion of heresy remained confined to the clerical fringe. The 'handmaiden' and 'God's ordained power' solutions to the problem of the relationship between secular learning and revealed religion proved to be durable. Whether a turning inward of the civilization similar to what happened in the Islamic world would have sufficed to call the doctrinal harmony thus established into renewed doubt is a question that neither can nor need to be answered. For the time being, Christian Europe was moving in the opposite direction, toward an ever more outer-directed, nature-oriented conception of things.

Naturalist drift.   The history of philosophy in the Greek tradition displays ongoing shifts in the place allotted within the whole of philosophy to the study of *nature*. Marginalized ever more in the Roman Empire, natural philosophy lived in Islamic civilization in a somewhat uneasily shifting balance with more spiritual concerns. In medieval Europe, by contrast, when philosophy became institutionally entrenched, it was with its nature-directed portion clearly predominant. The proportion of *quaestiones* devoted to issues in natural philosophy exceeds by far the amount of attention paid to natural philosophy in Islamic civilization. Also, the readiness to question specific Aristotelian tenets on grounds, not only of lack of inner consistency or of apparent conflict with the faith, but of apparent conflict with empirical phenomena was much greater in Europe than in Islamic civilization.

Outside philosophy proper, too, there are pointers in the direction of an enhanced natural-ism. Compare the extent to which Pythagorean musical theory was affected by the realities of music making in the two cases: hardly so in Islamic civilization and quickly so in medieval Europe. Or take those two directly faith-derived and, as such, straightforwardly comparable pieces of sophisticated nature-knowledge: the *qibla* and the *computus*. Compare how little the Islamic clergy bothered to follow the exact rules for orienting prayers and mosques to-ward Mecca that mathematical scientists had painstakingly worked out with the following account:

> when it became clear that the Lord's return would be delayed indefinitely, the Church shoul-
> dered the burden of planning for centuries and even millennia. To its everlasting glory, it raised
> the little Julian excess of eleven minutes a year into a major problem. If uncorrected, the excess
> might imperil the souls of those who, by Sosigenes' [Julius Caesar's calendrical adviser] error,
> were led to celebrate Easter on the wrong day; and it would certainly cause discord between
> segments of the Church that tried to correct for the error in their own ways.[64]

This does not of course mean that the *computus* was undertaken on any other grounds than those derived straightforwardly from religion or religious politics. The point is rather that in the Latin West churchmen found it worth their while, at the inevitable cost of making themselves dependent on those in possession of the secular knowledge required, to go into detail on an issue they might just as well have left a bit fuzzy. In Islamic civilization, faced with the same choice, the decision made was exactly opposite.

In Friedrich II's and Pierre de Maricourt's treatises on falconry and on the magnet, other telling symptoms of this uncommonly heightened naturalism (the possible causes thereof will occupy us later, p. 263) came to the fore. What makes these treatises unique is their position halfway between the high-level intellectualism of nature-knowledge of Greek prov-enance even in its most simplified renditions and such practice-oriented rules as emerged in a variety of skilled undertakings in trade and in the arts. Guido of Arezzo developed a practi-cal system of solmization ('ut/re/mi/fa/sol/la') for fixing musical notes. Cathedral builders used practice-derived rules to prevent collapse. Leonardo ('Fibonacci') of Pisa showed in his *Liber abaci* how merchants could benefit from the use of calculation devices in Arabic numerals as originally developed by al-Khwarizmi and now commodified and extended by Leonardo. At a level of abstraction *higher* than such wholly practice-oriented guidelines yet *lower* than the highfalutin' verities of the scholastics, Friedrich and Pierre practiced an inter-mediary kind of nature-knowledge without any known precedent in either Greco-Roman or Islamic civilization. In both treatises, empirical description that was aimed at painstakingly accurate observation rather than at exemplification of preset first principles was blended with a thoroughly practical objective: to improve hunting and to enhance the reliability of the mariner's compass.

96   The two treatises stand isolated in their own time. But by the mid-15th century the medieval trickle turned into a Renaissance stream. A third mode of nature-knowledge, marked by this spirit of practice-oriented empiricism, emerged as its two elder rivals, Athens and Alexandria, underwent transplantation one final time.

Translation. Once again, Scott L. Montgomery, *Science in Translation. Movements of Knowledge Through Cultures and Time*. Chicago: University of Chicago Press, 2000, ch. 4 offers an enlightening account, which I have supplemented with Charles Burnett, 'The Coherence of the Arabic-Latin Translation Program in Toledo in the Twelfth Century'. *Science in Context* 14, 2001; pp. 249-288.

Conceptualization. In *SRHI*, sections 2.2.4.-2.4.1., 4.3., and 4.4.2., I have charted debates on the significance of the medieval episode inspired by authors like Pierre Duhem, Anneliese Maier, E.J. Dijksterhuis, R. Hooykaas, Edward Grant, and Paul L. Rose. Rose's suggestions for a radical break with the conceptualization of the medieval episode instigated in the 1910s by Duhem have stimulated me to carry it out and extend it.

Nature-knowledge in medieval Europe. Among surveys also used are Part 2 of E.J. Dijksterhuis, *The Mechanization of the World Picture*. Oxford UP, 1961 (original Dutch: 1950); chs. 9-14 of David C. Lindberg, *The Beginnings of Western Science*. Chicago: University of Chicago Press, 1992 (2nd edition: 2008), and Edward Grant, *The Foundations of Modern Science in the Middle Ages: Their Religious, Institutional, and Intellectual Contexts*. Cambridge: Cambridge University Press, 1996.

- **Mathematical science**. IN GENERAL: the surveys just mentioned, plus entries in Ivor Grattan-Guinness (ed.), *Companion Encyclopedia of the History and Philosophy of the Mathematical Sciences* (2 vols.); London: Routledge, 1994. MUSICA: Paolo Gozza's editorial introduction to *Number to Sound. The Musical Way to the Scientific Revolution*. Dordrecht: Kluwer, 2000. FLOATING BODIES: P.L. Rose, *The Italian Renaissance of Mathematics. Studies on Humanists and Mathematicians from Petrarch to Galileo*. Geneva: Droz, 1975, pp. 80-82.
- **Enrichment of Aristotelian natural philosophy**. With the exception of the weight of bath-water on top of Notre Dame, the examples inserted in the main text are standard in the literature.
- **Culturally marked nature-knowledge**. ALCHEMY: R. Hooykaas, ch. 3, 'The Philosopher's Stone' of his *Fact, Faith and Fiction in the Development of Science*. Dordrecht: Kluwer, 1999 (a *précis* of ch. 3, 'De Latijnsche alchemie' ['Latin alchemy'] of his 1933 Utrecht Ph.D. thesis, *Het begrip element in zijn historisch-wijsgeerige ontwikkeling* ['The historical-philosophical development of the concept of element']). William R. Newman and Lawrence M. Principe, *Alchemy Tried in the Fire. Starkey, Boyle, and the Fate of Helmontian Chymistry*. Chicago: University of Chicago Press, 2002; pp. 34-49. COMPUTUS: John L. Heilbron, *The Sun in the Church. Cathedrals as Solar Observatories*. Cambridge (MA): Harvard UP, 1999: ch. 2, 'The Science of Easter'. FRIEDRICH II: Michael McVaugh, entry in *DSB* 5, p. 146-148. PIERRE DE MARICOURT: Edward Grant, entry in *DSB* 10, p. 532-540.

GREEK NATURE-KNOWLEDGE TRANSPLANTED, AND MORE: RENAISSANCE EUROPE

When in 1453 Byzantium fell to Sultan Mehmet II, vast numbers of manuscripts in Greek philosophy and mathematical science were mobilized. Ottoman scholars, focused as they were on the Golden Age of Islamic nature-knowledge, did not regard the presence of these texts as an opportunity. But Italian scholars did. In Florence and Rome a movement of renewal in arts and literature was already under way. I follow convention and refer to this movement as the 'Renaissance', and to the scholars who contributed to it as 'humanists'. They perceived a major difference between their own outlook and what they contemptuously referred to as 'Middle Ages' caught in the dark middle between the glories of the ancient world and its ongoing re-creation. All this was now reinforced by the arrival from Byzantium of those excitingly ancient manuscript sources in natural philosophy and mathematical science. Soon other parts of civilized Europe were involved in the translation and examination of these sources.

The third transplantation of the Greek corpus, then, occurred on the same soil that had received the second. Inevitably, the medieval aftermath displays some temporal overlap with the new movement of translation and enrichment of the Greek corpus – in certain academic quarters the dissemination of scholastic texts went on unabated for more than a century. Still, it is vital to keep the two transplantations analytically distinct. Translations were now routinely made from Greek to Latin directly, thus giving occasion to expurgate such accretions and errors as had kept accumulating since the Baghdad translations. The medieval near monopoly enjoyed by Aristotelian doctrine came to an end, and the spectrum of ancient nature knowledge was restored even more fully than it had been centuries earlier in Baghdad. The sense of excitement at bringing to light all these forgotten riches was as palpable as in that earlier case, and so was the opportunity thus regained to explore what more could still be made of them. We shall now first survey what recovery and ensuing enrichment meant for Athens and for Alexandria.

Athens replayed in full

The one Greek system of natural philosophy to survive antiquity in something approaching its original shape was Platonism. Medieval Europe had to make do with the *Timaios*. The recent availability of the full range of Plato's dialogues challenged the budding humanist movement to develop its philological techniques. Along with texts reflecting Pythagorean, stoic, atomist, and many other points of view, Plato's dialogues were published and distributed with unprecedented swiftness by means of a novel communication tool, the printing press. Vivid debates over the respective merits of these texts ensued all over Europe. Aristotelian doctrine, securely ensconced at the arts faculties, was affected by the movement, too. The substance of the doctrine was not given up, but its forms were made more flexible than in medieval times, so as to enable it to meet the challenge posed by the resurgence of rival systems of thought. It was refreshed by an injection with newly expurgated versions of the basic texts. It was recast didactically. It was reinforced with new arguments, new empirical evidence, and refined methodological disquisitions. About halfway through the 16th century a revitalized Aristotelianism stood ready to contribute its share to a philosophical debate that, on the whole, took shape in ways similar to exchanges held in the ancient world.

The similarity extends to the revival of skepticism. Being confronted by so many competing perspectives on the world, each marked by a claim to incontrovertible truth, led in Renaissance Europe to the same doubts as had arisen in Greece about the capacity of human beings to know anything at all with certainty. Under the aegis of Loukianos of Samosata, and with a view to avoiding dogmatic polarization in the early days of the Reformation, humanists like Desiderius Erasmus and Thomas More initiated the first skeptical revival. Their irreverently nonpartisan wit was reinforced when the full battery of skeptical arguments became known in the final quarter of the 16th century. It found a wide audience due to Michel de Montaigne, who gave attractive literary form to the printed arguments of one minor expositor whose work happened to survive, Sextus Empiricus.

With didactically recast Aristotelianism making itself heard at the universities, with a Platonist climate reigning among scholars at Italian courts especially, with skeptical rejection of dogmatic commitment setting the trend at many other centers of learning, and with all kinds of positions taken from other systems of philosophy being defended by humanist scholars all over Europe, the stage was set for a replay of a performance already given in ancient Greece. And indeed, little original substance was added to the Athenian portion of the Greek corpus during the period here considered. Whereas over the medieval period philosophers had been at the forefront of whatever enrichment of nature-knowledge was achieved, during the Renaissance it was rather mathematical scientists who surpassed what the Golden Age of Greek nature-knowledge had wrought.

Alexandria: A replay with a difference

Eleven years after the fall of Byzantium, in 1464, a young scholar called Johannes Müller of

Königsberg (Regiomontanus) came to Padua to give a lecture series. His first address was in praise of the mathematical sciences. Taking his cue from Ptolemy's preface to the *Almagest* (p. 24), he argued that

> the theorems of Euclid have the same certainty today as a thousand years ago. The discoveries
> of Archimedes will instill no less admiration in men to come after a thousand centuries than
> the delight instilled by our own reading.[65]

One aim that Regiomontanus hoped to attain with his celebration of the perennial certainty of mathematical reasoning concerned the philosophers. Their own claim to indubitable certainty, he asserted, is untenable; nonetheless, they are the ones who stand uppermost in the hierarchy of knowledge. As Ptolemy showed, philosophical arguments can never be more than plausible at best, and yet, philosophy is far more solidly entrenched at the universities than the very repository of incontrovertibly true knowledge, mathematical science. It is wrong, therefore, that philosophers' incomes are not only more secure but also much higher.

Over the century and a half that followed, Regiomontanus' complaint was to be reiterated time and again – the same situation still vexed Galileo as a university teacher in mathematical science (p. 417). In the meantime mathematical humanists worked hard at undoing what, on this view, had gone wrong throughout the Middle Ages: scholastization of the great majority of Alexandrian sciences, plus marginalization of the authentic remainder. An ongoing effort to get rid of all those scholastic accretions contributed to the zest with which originally Alexandrian texts were recovered, translated, and enriched over the period.

To set up a publication program for precisely such concerted action was the orator's other apparent concern. The scope and ambition of Regiomontanus' program were without precedent. He called for recovery of all resources that might serve to restore true mathematical learning as attained in antiquity. Eligible for publication in his view were not only manuscripts from Byzantium but also texts that had in earlier times reached Europe indirectly. Even so, his printing program reflects his conception of the purity of the ancients in that he ignored texts written by the Schoolmen almost entirely. By the same token he treated mathematical science in Islam as just a foreign interlude, with the Arabs temporarily preserving for Europe its Greek heritage – a periodization of the history of science that has survived to the present day (p. xxii).

Regiomontanus began his scholarly career as an early 'mathematical humanist' under the protection of Cardinal Bessarion. Born near Byzantium, Bessarion had converted to the Roman variety of the Catholic faith. When the Ottomans captured the city he turned himself into the leading patron in Italy to foster the importation and translation of Greek manuscripts. Johannes Müller proved his most brilliant catch. Eight years after his Padua oration, when Regiomontanus had availed himself of his own home printing press, he wrote

102 up his publication program in detail. He did not live to carry it out in person for more than the bare beginnings. It was nonetheless executed by like-minded scholars during the century or so that followed his premature death in 1476. By the end of the 16th century, Europe had put itself in possession of a very large portion of the Alexandrian corpus, but now divested of its scholastic overgrowth, restored to its text-critical purity, augmented with such improvements as had during earlier episodes accrued to it, and further enriched by many a mathematical humanist so hard at work to establish these texts in the first place.

What happened in Alexandrian science over the period of recovery varied from one branch to another.

In the subject area of light and vision hardly anything changed. Ibn al-Haytham's synthesis, as preserved and extended in medieval university circles, did not elicit original, published contributions until 1604, when Johannes Kepler took up the subject at the very points where Ibn al-Haytham and Roger Bacon and Witelo had left it, so as radically to transform it (p. 297).

In producing his synthesis, Ibn al-Haytham had followed the example of Ptolemy's bridge building between Alexandrian science and Athenian natural philosophy. What he had done for light and vision was now done likewise for the consonances. Humanists recovered ancient texts in musical theory like Euclid's and Ptolemy's, and the latter's effort at synthesis caught on with two musical scholars of wide learning and practical experience, Gioseffo Zarlino and Francisco Salinas. With the medieval rise of polyphonic music, which was built around the use of pure major thirds, the theoretical problem had emerged of how to account for those definitely non-Pythagorean thirds (4/5 rather than 64/81; see p. 82). Seizing upon a musical scale devised by Ptolemy, by the mid-16th century Zarlino and Salinas put together in one harmonious construction an account of music that encompassed the newly consonant thirds and sixths. They did so in ways satisfying not only the Pythagorean/Euclidean style of musical analysis but also the polyphonic practice of their time, all the while shoring up their accounts by means of a selection of arguments taken at suitable points from Plato or Aristotle. The integrity of cosmic harmony was restored, seemingly for good.

Apollonios and Archimedes enriched.   For the recovery of Greek texts that dealt with geometry or with planetary trajectories or balance problems treated the geometric way, substantive enrichment ran another course. Unlike in Baghdad seven centuries earlier, most scholars who did the translating turned themselves into experts capable of moving beyond the sheer recovery of proofs and theorems to the tentative reconstruction of some which seemed irretrievably lost, and from there toward even more tentatively undertaken improvement.

Take the surviving record in geometry. How incomplete and unreliable these texts really were became ever more apparent as translators' bilingual and technical proficiency grew and the publication of texts came closer to the point of exhaustion. A growing awareness that certain theorems, proofs, commentaries, or even entire collections thereof were wrong or incomprehensible or even missing altogether posed a challenge that the best practitioners were eager to take up. Aimed-for reconstitution could go hand in hand with effective innovation. For instance, the last four books of Apollonios' *Conics* were regarded as lost (they did not turn up until 1661), but Pappos' 4th-century comments on all eight books gave some inkling of their contents. No fewer than three times was it attempted with the help of those comments to reconstitute the lost books, at first around 1550 by a typical 'mathematical humanist', Francesco Maurolyco, and later by two first-rate mathematical innovators, François Viète and Pierre de Fermat.

Nothing is more revealing of how differently the Greek corpus was received in the Middle Ages and in the Renaissance than the fate that befell Archimedes' work on the lever and on equilibrium conditions in fluids. With the latter, Renaissance scholars even worked from the same text version that had already in the 13th century been prepared from a Byzantine manuscript by Willem van Moerbeke. What had not received any follow-up during the uncontested reign of the Schoolmen was now enthusiastically taken up and extended. This happened notably in late-16th-century Italy among a group of scholars centered at Urbino under the patronage of Guidobaldo dal Monte. Here, too, they sought to go beyond the sheer recovery and publication of ancient texts. Guidobaldo and others in his circle strove to extend Archimedean concepts and modes of attack beyond equilibrium problems (where due to mutual balancing no motion occurs) to bodies actually moving. But their efforts came to naught. The most brilliant attempt made in this regard was a set of manuscript treatises written by a young Florentine who in 1589 benefited from Guidobaldo's patronage to gain the chair in mathematics at the very same university in Pisa that he had only four years earlier left without a degree. In his 1591 manuscript *De motu* (On Motion) Galileo Galilei analyzed falling bodies in terms of hydrostatic concepts adopted from the man he unreservedly called there "the superhuman Archimedes" (also "the, as it were, most divine Archimedes").[66] Yet however often Galileo attacked the problem, he kept getting stuck, and when he left Pisa for Padua he also abandoned the unfinished manuscript. Antonio Favaro first brought it to light in 1890 as part 1 of Galileo's complete works, to enhance our understanding of how in his Padua years, between 1592 and 1610, Galileo took the revolutionary steps that were still out of reach to him in Pisa in 1591 (p. 193).

Simon Stevin's simultaneous, independently undertaken effort at extending the Archimedean analysis of equilibrium problems shows how, in the short term, less ambition met with more success. For instance, he devised an ingenious proof for the basic theorem on the equilibrium of a solid on an inclined plane (Figure 4.1.).

Another example of how Archimedes' work was both emulated and extended is Stevin's discovery and mathematical proof of what is still known as the hydrostatic paradox: for ves-

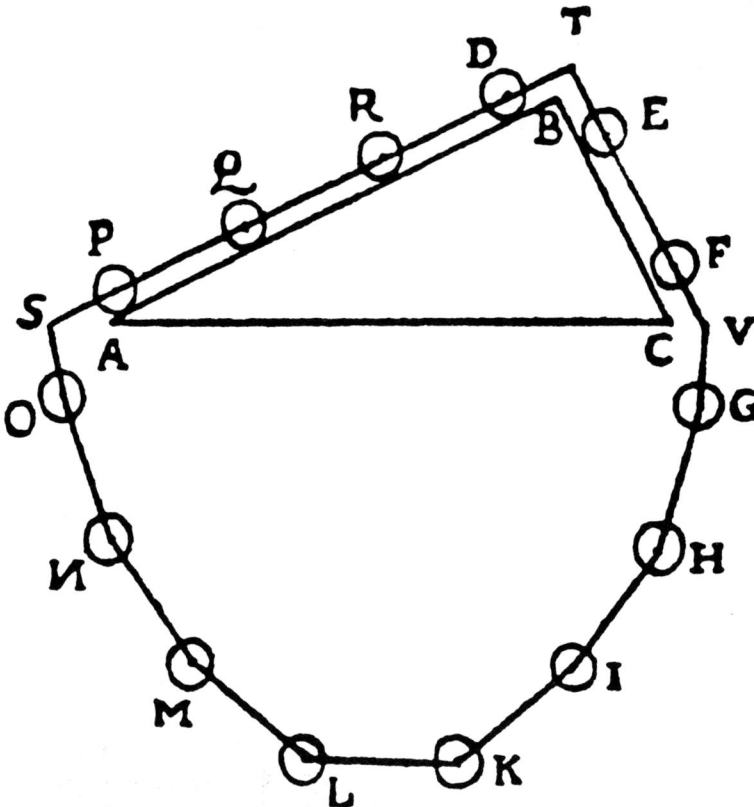

The property in need of proof is that two weights, connected to one another, will balance if they are proportional to the lengths of the slopes on which they rest. Over the sides AB and BC of a vertical triangle ABC a wreath is slung with equal spheres at equal distances, such that it can slide without impediment around the three fixed points S, T, and V situated at A, B, and C, respectively. Suppose segment AB of the wreath to be heavier than its counterpart BC, then the wreath starts moving but, as each sphere takes the place of its successor without the setup as such being changed thereby, the wreath will forever keep moving around – to Stevin, a self-evident absurdity. Hence, the wreath is at rest, nor will its rest be disturbed by taking away segment VS, which pulls equally at both sides (as ONML balances GHIK). Hence, segments AB and BC are at rest, too. For the case shown, where AB = 2BC, the four spheres over AB may be taken to be concentrated in one body with a weight 4/2 that of a similarly concentrated body at BC, which indeed is what had to be proven.

Figure 4.1: Stevin's 'wreath of spheres' proof

sels of given height and base area that contain a fluid, the pressure that the fluid exerts upon the base is independent of the shape of the vessel and, hence, of how much fluid is contained in it. The proof proceeds through a characteristically Archimedean effort at *reductio ad ab-surdum*, followed by an ingenious argument that is rendered in Figure 4.2.

The pattern underlying all these varied efforts may by now look familiar. In Renaissance Europe the Alexandrian corpus was recovered and in various ways enriched as it had been in Islamic civilization. In handling it thus, mathematical humanists remained within the boundaries maintained there, too: the original frame of highly abstract mathematical treatment was left intact. The question to be addressed now is whether the one remaining branch of the Alexandrian corpus, planetary trajectories, fits the same pattern or whether it went some way toward transcending it.

Ptolemy enriched; or, Copernicus restores the Almagest with a difference.

Planetary trajectories were the mathematical subject that Regiomontanus in 1464 exalted most and that he felt required the hardest labor to set the Greek record straight. His deepest interest in its expurgation was to gain possession of a secure basis for astrological prognostication. The man to take up his challenge half a century later, Nicolaus Copernicus from Torun/Thorn near Gdansk/Danzig, remained silent all his life about his motives in this particular regard. In his written work he presented himself as driven by a profound awareness that, over the centuries, the Ptolemaic account of the planetary trajectories had turned into what he termed a 'monster'. To emend that monster to fit the original, harmonious proportions that the heavens must display but that not even Ptolemy, in the *Almagest*, had quite been able to determine satisfactorily was the task Copernicus set himself in the first decade of the 16th century.

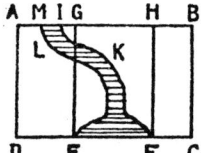 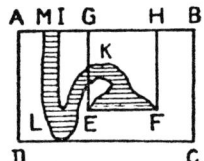 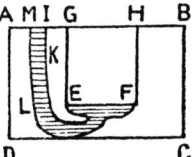

Imagine arbitrary portions of the fluid contained in ABCD to be solidified without altering its specific weight, thus leaving as fluid bodies only the shaded, irregular shapes IKFELM. In each case the pressure exerted upon EF remains equal to the weight of the fluid column EFGH.

Figure 4.2: Stevin's hydrostatic paradox

106   In view of the train of events that Copernicus' reform was to unleash about a century after he started work on it, it is tempting in hindsight to read into his considerable yet well-circumscribed achievement a good deal more than there was to it. Throughout my treatment of what (foreseen least of all by its original author) was in the end to turn into the 'Copernican revolution' we must distinguish between the many unintended consequences his reform turned out to entail and the historical frame in which his efforts make proper sense. That proper frame was the very replay of the Greek performance we are examining here – a replay with a difference, to be sure, yet one that was to remain nearly invisible for half a century to come.

Key to Copernicus' effort was his objection to the license Ptolemy had taken in violating uniform circular motion of the planetary spheres. The fundamental problem of planetary trajectories rested in two major, apparent irregularities. On observation over successive nights against the background of the stars, most planets in their respective orbits around the Earth display a varying brightness readily accounted for as due to their (over a full circuit) varying distance. This was called the first planetary 'inequality'. The second inequality, known as retrograde motion, was that each planet appears for a while to regress on its path. To account for these irregularities, Hipparchos invented two mathematical devices, the eccentric we met in Figure 1.2 (p. 14) and the epicycle on display in Figure 4.3. In the *Almagest* Ptolemy added a third, the equant point, which is also worked into Figure 4.3.[67]

The equant point was Ptolemy's way to break with Plato's allegedly strict requirement of uniform angular velocity without appearing to break with it – nominally, uniformity was maintained. Copernicus' objection lay precisely here. The issue of actual nonuniform motion of the planetary spheres had to be faced squarely and addressed by breaking it down into component spheres moving uniformly, Copernicus thought, rather than circumvented (as Ptolemy, on this view, had done) by positing an eccentric point inside the sphere from which the motion only looked, but actually was not, uniform. Similar objections had been voiced before by astronomers at the Maraghah observatory, with whose work Copernicus is likely to have been familiar (the ways of transmission, possibly oral only, are still unknown). Yet he went far beyond them, not so much in the specific solutions he sought to specific problems, many of which (like the 'Tusi couple') he adopted, but rather in the scope of his reform enterprise. What he decided to undertake was nothing less than writing a book on the same scale and in the same format as the *Almagest*, an improved *Almagest* really, which was to proceed from a novel, or rather from a very ancient, point of departure: the Earth is taken to orbit the Sun, not the Sun the Earth. When his book finally came out in 1543, Copernicus recalled in an early chapter that he had adopted the idea from the Pythagorean school, thus omitting here the name of his most obvious source, Aristarchos. The significant difference was that whoever else had previously put forward the idea of a Sun-centered universe had stopped short of elaboration in detail, so that in effect the

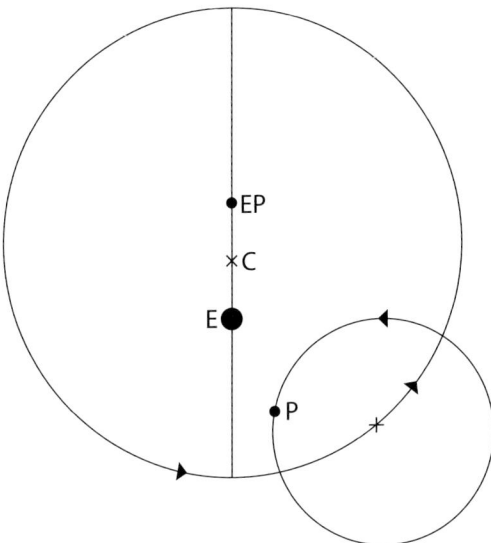

The device of the *epicycle* is meant to produce retrograde motion: planet P turns in circles around a midpoint that circles around center C. The Earth E does not coincide with C (*eccentric*). Considered, not from C but from the *equant point* EP, the angular velocity of the center of the epicycle remains constant.

Figure 4.3: Eccentric, epicycle, and equant point

proposal came down to an evident absurdity with little or no merit (astronomical or otherwise) to commend it. Instead, Copernicus set himself a dual task. He wished to turn what so far had been just an idea into the cornerstone of a full-fledged, reformed planetary astronomy. And he wished to show that this was not at all an absurdity in the *reality* of the heavens but rather an assertion subject to the very proof he hoped to be able to provide for it. However well he succeeded in the first task, he failed in the second, and he knew it.

What, then, were the terms on which Copernicus conceived of his primary task, to write an improved *Almagest*? He was aware (again, like astronomers in Islamic civilization, only more radically so) that since Ptolemy's day errors of observation, of computation, and of transmission had occurred and had had plenty of time to accumulate and thus to disfigure the integrity – such as it had – of the original whole. He was aware that the heliocentric hypothesis required a wholesale transposition of parameters employed by Ptolemy and his successors. This was comparatively easy to do in rough outline but required an enormous amount of fine-tuning when it came to detail, of which there were many thousands. He further became aware ever better as he went along that not all ancient observations were as

108   reliable as he had originally assumed, so he kept adding to the considerable stock of novel observations he had to make anyway during the decades that passed between his original conception and the final execution. Yet there was one thing he never ceased to take for granted as he went along – Ptolemy had provided the proper theoretical models with which to do the job. Whereas in retrospect Copernicus may appear to beckon a kind of modernity for which he really had no category of thought whatsoever, a more fitting picture in terms of his own stance in time is gained by joining several historians of the Copernican achievement in regarding him rather as Ptolemy's last and greatest heir. An intellectually unbroken tradition extending two thousand years back makes itself felt in the whole setup of Copernicus' life's work as well as in the way he addressed individual issues. Half a century later, Kepler perceptively observed that Copernicus' achievement had been to interpret Ptolemy rather than nature.

The achievement was put down in the end in a book entitled *De revolutionibus orbium coelestium* (On the Revolutions of the Heavenly Spheres). Books II through VI jointly carried the message that the job of devising a heliocentric planetary theory in accordance with all requirements accepted from ancient times down to Coperncus' own day could indeed be done. On closing the tome, readers with sufficient technical proficiency could rest assured that each and every pertinent heavenly motion had now been analyzed into its component parts of circles uniformly traversed, using the trusted means of epicycles and eccentrics, and employing freshly derived parameters in such a manner as to keep intact or, where need be, to restore to the best of his considerable ability the required match between positions computed and positions observed. And it is in this accomplishment (not without its occasional weaknesses, yet on the whole very ably executed) that contemporaries and at least one generation of successor astronomers acknowledged Copernicus' great merit to reside. Here is indeed where he almost wholly succeeded.

Where he failed was in carrying out to his own or to almost everybody else's satisfaction the task he set himself in the first, introductory book of *De revolutionibus*. In it, and in his preface addressed to the incumbent pope, he explained in sixteen pages why he had felt the need for a restoration of Ptolemaic astronomy; how he had arrived at the heliocentric hypothesis that provided him with the cornerstone of the reform; what reasons he had for thinking that the hypothesis represents reality in that the Earth really and truly orbits the Sun; and what the resulting picture of the universe looks like in broad outline. Yet the source of what soon proved to be endless controversy resides here, too. It was here that the author unequivocally demanded reality status for a claim that (if applied to the broad, didactically simplified outline of book I) defied all contemporary common sense, whereas (but now taken in terms of the details of planetary motion provided in books II through VI) the claim was clad in all the trappings of a by-and-large fictional model rather than of something anyone could possibly think to be straightforwardly there in the heavens. More than fifty circles were required by Copernicus to do the job of a fully elaborated, heliocentric astronomy; what possible heavenly machinery could be invoked to combine the component parts into the resultant, highly irregular motions? Although set up in three dimensions, the fully articulated model could not even be imagined to represent any conceivable, heavenly reality, nor did the established tradition of astronomy performed in the Alexandrian way yield stimuli in that direction. To bring the detailed model of books II through VI into line with the heliocentric hypothesis which made it tick, that hypothesis had to be declared no less fictitious than the model that in the end came out of it, and this is what the foreword to *De revolutionibus* strongly advised the reader to do. That foreword was not written by Copernicus, though. The vicissitudes of publication had put a Nuremberg vicar, Andreas Osiander, in a position to prefix it anonymously (and, hence, posing as from the author's own hands) to a book he had been assigned to supervise into print. Nothing but Copernicus' death within two months of the actual publication date precluded the effective correction that would otherwise in all likelihood have ensued. For Copernicus did not agree; even though aware that he had not made his realist claim truly clinching, he clung to it regardless, and he devoted a good part of the first book of *De revolutionibus* to giving reasons why.

Most striking about those reasons is their narrow focus compared to the size of the job. Copernicus has been called a narrow-minded astronomer in the technical sense, wearing as it were blinders which kept him from seeing what most readers were to see at once – that a whole world-picture was being disrupted here. As we shall see when dealing with Galileo (p. 186), there was no coherent way in which a thoroughgoing Aristotelianism could be reconciled with a universe in which planet Earth orbits the Sun and also rotates around its own axis. Yet in book I Copernicus sought to keep the Aristotelian views he had imbibed as a student in Bologna as little tainted with notions foreign to the doctrine as he could. Once again Ptolemy's 'hypotheses' at the head of the *Almagest*, and Ptolemy's 'opportunist' way of dealing with philosophy (p. 24), provided Copernicus with his model. Ptolemy had countered anticipated real-world objections by supplying certain empirical phenomena or, if no suitable ones were available, by an appeal to assorted bits and pieces of natural-philosophical argument. Since empirical evidence for a heliocentric setup was so far lacking altogether, Copernicus could not but take refuge in a similar effort at philosophy snatching. One example is his argument that heavy bodies once dropped do not tend toward the center of the universe, as Aristotle taught, but toward other bodies of the same stuff, as Plato held. The difficulty was that Ptolemy's geocentric setup had common sense on its side, so that not too much hinged on the quality of his natural-philosophical stopgap arguments. By contrast, Copernicus' corresponding medley was charged with an impossible task, and what arguments taken from natural philosophy he offered in favor of the reality of his heliocentric hypothesis were bound to leave the large majority of his readers unconvinced. What, then, were the *other* arguments, the mathematical-astronomical ones, by means of which

110   this mercifully blinkered, narrowly technical astronomer sought to make his case for the superiority of his own system over that of Ptolemy and successors? Or, expressed in terms adopted by Copernicus himself, in what sense had he restored the Ptolemaic 'monster' to its pristine harmony?

In the striking image Copernicus employed, the various parts of the body astronomical – the head, the feet, the arms, the trunk – failed to fit harmoniously together. Hence, for those ready to take a step backward as it were and look at the body whole, that body could not but give the appearance of a monster. There were two principal sources of this lack of harmony. One resided in the accumulation of errors and patchwork repair over time which only a wholesale approach as undertaken by Copernicus could hope to correct systematically. The other was at the heart of Ptolemy's, and everybody else's, geocentric point of departure. Copernicus went out of his way to explain as clearly and simply as he could the various respects in which his own heliocentric hypothesis yielded a more harmoniously satisfying picture of the universe than geocentrism could possibly produce. Even so, his considerations were inevitably somewhat technical. Here is an authoritative summing up, followed by further explication:

> The heliocentric theory – philosophical quibbles aside – gives the order and distances of the planets unambiguously and under the reasonable assumption that the equation of the anomaly shows the ratio of the radii of the planet's and earth's orbits. In so doing, it makes the planetary system into a single whole in which no parts can be arbitrarily rearranged. By contrast, in the geocentric theory the radii of the eccentrics and epicycles are known only relatively, one planet at a time, and only by additional assumptions, such as the contiguity of successive spheres, can the order and distances of the planets be determined. Certain features of planetary models that are unexplained in geocentric theory are understood as the necessary consequences of the transformation from a heliocentric to a geocentric arrangement: why the radii of the epicycles of the superior planets remain parallel to the direction from the earth to the mean sun; why the centers of the epicycles of the inferior planets lie in the direction of the mean sun; why planets closer to the earth's orbit, Mars and Venus, have longer synodic periods, longer periods of invisibility (at least for Mars), larger equations of the anomaly (i.e. relatively larger epicycles), and larger retrograde arcs (although this is actually more complicated) than planets farther from the earth's orbit, Jupiter and Saturn.[68]

That is, in 'unmasking' the five large epicycles that Ptolemy needed to account for one major irregularity in the observed motions of each of the planets (to wit, retrograde motion) as nothing but a projection arising from a moving Earth, Copernicus accomplished three major things. He simplified the system insofar as these five major epicycles could just be done away with. He could fix the true order of the planets without using Ptolemy's auxiliary assumptions (p. 25). And he could show why certain features that were arbitrary in Ptolemy's setup now reappeared as necessary consequences of his own rival construction.

This, then, is what Copernicus called turning a monster into a harmonious body, and here he had a point, his only point really. It is in this particular feature that the principal attraction resided which the heliocentric hypothesis *in the guise Copernicus gave it* could exert on his readership. By virtue of the very technicality of the point, however, none but knowledgeable astronomers were likely to stand ready to be persuaded. Even among these rare birds at the time, only those who were prepared to give less weight, on balance, to the plethora of *drawbacks* the heliocentric hypothesis entailed might seriously consider letting themselves be persuaded.

One drawback was of an equally technical nature. Copernicus had foreseen and sought to rebut it from the start. If the Earth truly orbits the Sun, an observer on Earth must see a star in spring under a different angle from a similar observation made in the fall, when the Earth is at the opposite side of its orbit; but such a difference ('parallax') had not been observed. Copernicus retorted that the sphere of the 'fixed' stars is so far away from us that, as seen from there, the Earth's orbit appears as a point; in other words, parallax would indeed manifest itself if it were not, alas, too small to be measured. This counterargument, which we know to be true, could hardly fail to appear weak and unfortunate at the time. Weak, in that it looks like a cheap evasion. Unfortunate, in that it enforced the assumption of a huge amount of void space between the sphere of Saturn and the fixed stars. Why assume such a waste of space if its only point was to save Copernicus from a lethal objection to a system inherently odd on many other counts as well? Again, in retrospect the road seems opened toward more modern ideas of infinite space, as indeed was to happen once the Copernican system came to look acceptable on other grounds. Meanwhile, the issue represented one more tough objection in a long list. The others on the list were less technical and more suitable therefore to such debate among nonexperts as the very absurdity of the proposal was almost bound to provoke. Prominent among these objections was a threesome we have already met in passing with Oresme (p. 85). One objection arose from common sense, another from Aristotle's conception of motion, a third from biblical exegesis.

The commonsense objection is obvious – does not Copernicus' hypothesis run counter to our failure to sense any motion of the Earth, as well as to our everyday observation that the Sun rises and sets and that the stars are seen moving all the time? Of course it does, and this is how one early adherent, Galileo, could name as the feature he admired most in Copernicus his willingness to go against common sense and adopt a counterintuitive point of view. But this is an admirable trait only if one feels at ease in a noncommonsense mode of pursuing nature-knowledge, which is precisely what Galileo was engaged in bringing about. In the 'world of the more-or-less' inhabited by just about everyone else at the time, such a violation of common sense constituted not so much a virtue as, rather, grounds for rejection out of hand.

Aristotle had placed commonsense ideas of motion in a coherently intellectualized frame, and that is how a whole range of objections that are nothing but logical consequences of those commonsense ideas could be raised in addition against the Copernican hypothesis. Ptolemy, who had indeed considered possible technical benefits to arise from a heliocentric point of departure, had nonetheless rejected it for this kind of reason. For instance, he provided an argument implying that on an Earth rotating around its own axis clouds would remain behind and birds would have to exert immense efforts solely to keep up (in the Ottoman Empire there had been a similar debate; p. 69). The argument can be varied endlessly, as indeed it was at the time. Here is one variety of particular cogency and historical significance: the argument from free fall. It may be represented in the format of a logically unobjectionable syllogism: If it were true that the Earth rotates, then when a person throws a stone vertically up, the stone would come down due west of the place from where it was sent upward, as the Earth would meanwhile have rotated away underneath; in reality, we observe no such deviation; hence, the supposition of a rotating Earth is mistaken. Copernicus had retorted that the stone follows the rotation of the Earth, but as a single statement, this made no sense within an Aristotelian framework he himself went on to accept without questioning. In retrospect we see a time bomb inadvertently set and ticking peacefully away here, yet its explosive potential could lead to actual detonation only if a whole range of conditions were fulfilled, one of these being a drastic improvement in the balance of pros and cons I am drawing up here. For the range of contemporary sets of objections is not completed yet – there was also the theological issue.

As Copernicus knew well, numerous biblical passages could be cited which either state or unmistakably imply the stability of the Earth. In a treatise to remain unknown until our own time, Copernicus' only disciple, Joachim Rheticus, explained under his master's supervision how the objection could be met. Like Oresme, Rheticus invoked as his principal argument a line of thought that St. Augustine had developed in the early fifth century. His objective had been to uphold the spherical shape of the Earth, much contested at the time by several other church fathers. Augustine had argued that the Bible should not be read as a textbook in astronomy; rather, its expressions are generally accommodated to conventional speech. Rheticus' appeal to the 'accommodation' mode of biblical exegesis was not an unusual way to defend the nonliteral interpretation of one or another biblical passage. Still, in a theological climate of 'interpretation according to the letter, unless ...', it left open the crucial question – not to be raised until seventy-two years after Copernicus died – of what compelling reasons there might be for reinterpreting the Bible in accordance with the heliocentric hypothesis. That is, how on balance to assess the merit of Copernicus' proposal?

In terms of contemporary presuppositions, nothing but some highly technical benefits to mathematical astronomy spoke in favor of the hypothesis, as against a compelling array of objections, some equally technical, but most of them readily accessible. Only one move appeared available by which to preserve the most alluring portion of Copernicus' achievement while insulating the hypothesis from all that was objectionable about it. This was the very move anonymously advocated in the foreword to *De revolutionibus:* to treat the Copernican hypothesis as fictional. Despite Copernicus' own conviction to the contrary, this was what happened indeed in the great majority of cases. Outsiders could ignore the heliocentric hypothesis or (like Martin Luther at the dinner table) shrug it off as a bad jest; experts could avail themselves of the finely elaborated models to which it had given rise. The full, detailed assembly of circles uniformly traversed that Copernicus presented in books II–VI of *De revolutionibus* held the promise of improved astronomical tables fit for improved predictions of the positions of the heavenly bodies, to be used in their turn for prognostication, calender reform, navigation, etc. And this is how Copernicus' work was made to fit seamlessly into the ongoing process of, for the third time in history, replaying the Greek performance in mathematical science and natural philosophy and enriching it somewhat in accordance with its own principles. A difference had now manifested itself in the replay, and that difference was gifted with a built-in potential to overturn a good deal of received knowledge about nature. Still, the difference could make itself felt if, and only if, the heliocentric hypothesis *as taken to represent reality* were, against many and powerful odds, to catch on. But why should it, ever?

A deep tension lay hidden inside Copernicus' book – a tension between the open fictionalism of Osiander's anonymous foreword, the implied fictionalism of books II–VI, and the explicit realism of book I. So far, the whole logic of the situation pointed in the direction of the fictionalist solution. That way, the difference in the replay of the Greek performance was bound to remain nearly invisible, even to its own author. Soundly put to rest in the very book that had raised it, the difference might well have rested there forever. Why in the end it did not is a question that shall occupy us at length in later chapters. We must first examine an excitingly new development in how nature-knowledge was pursued in Renaissance Europe.

Europe's coercive empiricism

The recovery and concomitant enrichment of the Greek corpus, accompanied by ongoing scholastic exposition of Aristotelian doctrines, do not nearly cover all 15th- and 16th-century European activity to acquire and preserve knowledge of nature, even though these were by far the most structured portions of that activity. What in both Islamic civilization and medieval Europe had taken shape in a minor way as incidental pieces of nature-knowledge marked by the culture from which they sprang (pp. 62; 88) multiplied in significance in Renaissance Europe. Here, too, but now in a major way, the appropriation of the Greek corpus was accompanied around the edges by a range of probings into more or less novel areas, *all*

114 *of which were colored by specific features of contemporary European civilization.* In principle, the developments analyzed so far in the present chapter might have taken place in any advanced civilization with sufficient maturity and readiness to receive the Greek corpus, as indeed quite similarly structured developments had taken place in Islamic civilization. This is not so with culturally marked pieces of nature-knowledge in Renaissance Europe. Their scope was so large and so varied and they display so much of a coherent, underlying knowledge structure that for analytical justice to be done to them they require categorization as a *third mode of nature-knowledge.* This third mode, then, emerged side by side with the ongoing recovery of Athens and Alexandria. To capture its accomplishments and characteristics I shall successively

· introduce this third mode through one early representative, Leonardo da Vinci;
· examine its varied pursuits in terms of their substantive content;
· investigate certain peculiarities of European civilization that are reflected in this third mode of nature-knowledge; and
· demonstrate that here, too, the customary, almost watertight compartmentalization between distinct modes of nature-knowledge remained as valid as ever.

Two components, exemplified by Leonardo.   In fragmentary passages spread over his private notebooks but also in sketches, in drawings, and even in his paintings, Leonardo pursued an extraordinary kind of nature-knowledge. It has often been interpreted in terms that are too modern. Here is how a perceptive art historian articulated what Leonardo sought to accomplish:

> [Leonardo's] science was to be founded on the shared principles behind the diverse phenomena of nature. But the principles themselves were never enough for Leonardo. Only when he had explained with the utmost rigour all the multitudinous varieties of observed effects could he rest content. Every tiny detail of anatomical structure must be explained in terms of its perfectly functioning design. Every vortex in turbulent water must be characterised in terms of the principles of impetus, percussion and revolving motion. Every light effect must be understood in terms of strength, distance, angle, reflection, colour, atmosphere, etc. Only when the investigator of nature had investigated these and all other natural phenomena, achieving a perfect match between causes and effects in every conceivable case, could he consider his quest accomplished. And only when the painter had achieved a complete understanding of natural form and function could a 'second world of nature' be constructed for any imagined situation in a way that matched Leonardo's ideals of perfection.[69]

Two drawings, selected almost at will, illustrate how Leonardo went about so perfectionist an effort to capture the essence of things by means of exhaustive descriptions of their countless details (Figures 4.4 and 4.5).[70]

Leonardo's drawing of a Star of Bethlehem served him as one plant study for his lost painting 'Leda and the Swan'. As Martin Kemp observes, it "breathe[s] the air of direct studies from nature, though it is doubtful if the leaves of the Star of Bethlehem adopt quite as emphatic a spiral arrangement as illustrated here. The procedure may be similar to that ... in which his method of observational analysis results in an emphasis upon the underlying organisational patterns."

Figure 4.4: Leonardo: Star of Bethlehem

The extraordinary lifelikeness of these drawings is by no means devoid of theoretical assumptions – of such concepts, like 'impetus' and 'reflection', that Leonardo used to guide him through the multitude of empirical phenomena. Even so, the relation that obtains here

Martin Kemp observes (and then quotes Leonardo) that this drawing "represents a synthesis of phenomena investigated separately in other drawings ... Leonardo's note carefully outlines the factors behind this synthesis:

> The motions of water which has fallen into a pool are of three kinds, and to these a fourth is added, which is that of the air being submerged in the water, and this is the first in action and it will be the first to be defined. The second will be that of the air which has been submerged, and the third is that of the reflected water as it returns the compressed air to the other air. Such water then emerges in the shape of large bubbles, acquiring weight in the air and on this account falls back through the surface of the water, penetrating as far as the bottom, and it percusses and consumes the bottom. The fourth is the eddying motion made on the surface of the pool by the water which returns directly to the location of its fall, as the lowest place interposed between the reflected water and the incident air. A fifth motion will be added, called the swelling motion, which is the motion made by the reflected water when it brings the air which is submerged within it to the surface of the water.

The obvious difficulty in tabulating the variables in a subject of this complexity is reflected in Leonardo's adding factors as he expounds his initial *tre spetie*. Although this is a synthetic drawing, deeply influenced by his theoretical framework of impetus and revolving motion, it does reflect many hours of patient scrutiny of water motions in natural, artificial and experimental situations, as recorded in [several] pocket-books."

Figure 4.5: Leonardo: Water falling in a pool

between what is seen and how the seen is accounted for differs in a major way from that in the Greek corpus. Leonardo's work does not display assorted pieces of real-life evidence adduced to shore up a set of all-encompassing principles established beforehand, as in Athenian natural philosophy. Nor does it display the mathematical handling of a restricted set of geometric or numerical figures thoroughly abstracted from a few handpicked natural phenomena, as in Alexandrian science. What Leonardo aimed at in the first place was, indeed, lifelikeness, factual accuracy, exhaustive description. Precisely this forms one major component of the third mode of nature-knowledge.

Another component comes in view once we realize that Leonardo's descriptive studies of nature were oriented, more often than not, toward practical ends. A well-known example is his examination of the flight of birds with a view to emulating it artificially (Figure 4.6).[71]

Leonardo's passion to get at the core of phenomena the empirical way did not stop short at phenomena given in nature. He also tried to gain a generalized understanding of man-made artifacts. In his time, as in the Old World generally, practical activities like mining or toolmaking were pursued by means of rules of thumb handed down from master to pupil, and if one sought to improve them at all, the preferred method was trial and error. Almost uniquely, Leonardo aimed at a thorough understanding before he went on to seek improvement. Around 1492, when he was appointed court engineer in Milan, he began systematically to analyze how machines work, with a view to optimizing their effective power. Never before had machines (e.g., water mills) been conceived as assemblies coherently made up of a quite limited number of constructive elements. Machines used rather to be handled as compounds of individual parts to which, if one wished to enhance their effect, one might add further parts at will. Leonardo perceived that in proceeding thus one usually loses more than one gains, due to excessive wear, often bad fit, and the introduction of new sources of friction. He went ahead to investigate under what systematically varied circumstances friction occurs and when and how it can be eliminated or at least minimized (Figure 4.7).[72] Thus, by experiment, he found that the amount of friction, while independent of the area of the surfaces in contact, increases in direct proportion to the load applied.

When, in the 18th century, books began to appear in which current engineering practice was subjected to theoretical investigation, no one could be aware that, in the privacy of his notebooks, Leonardo had preempted by centuries many an insight then emerging. Most often it makes little sense to call a person's accomplishment 'ahead of its time', and with Leonardo the expression has been abused particularly often. But his single-handed creation of the beginnings of a theory of engineering practice leaves us no choice.

In Leonardo, then, we have a case (however extraordinary in its individual execution) of one more feature characteristic of the third mode of nature-knowledge: its thorough-going *action-proneness* in the sense of an urge somehow to blend theory with practice. Not knowledge of natural phenomena for its own sake, or conceived as part and parcel of some

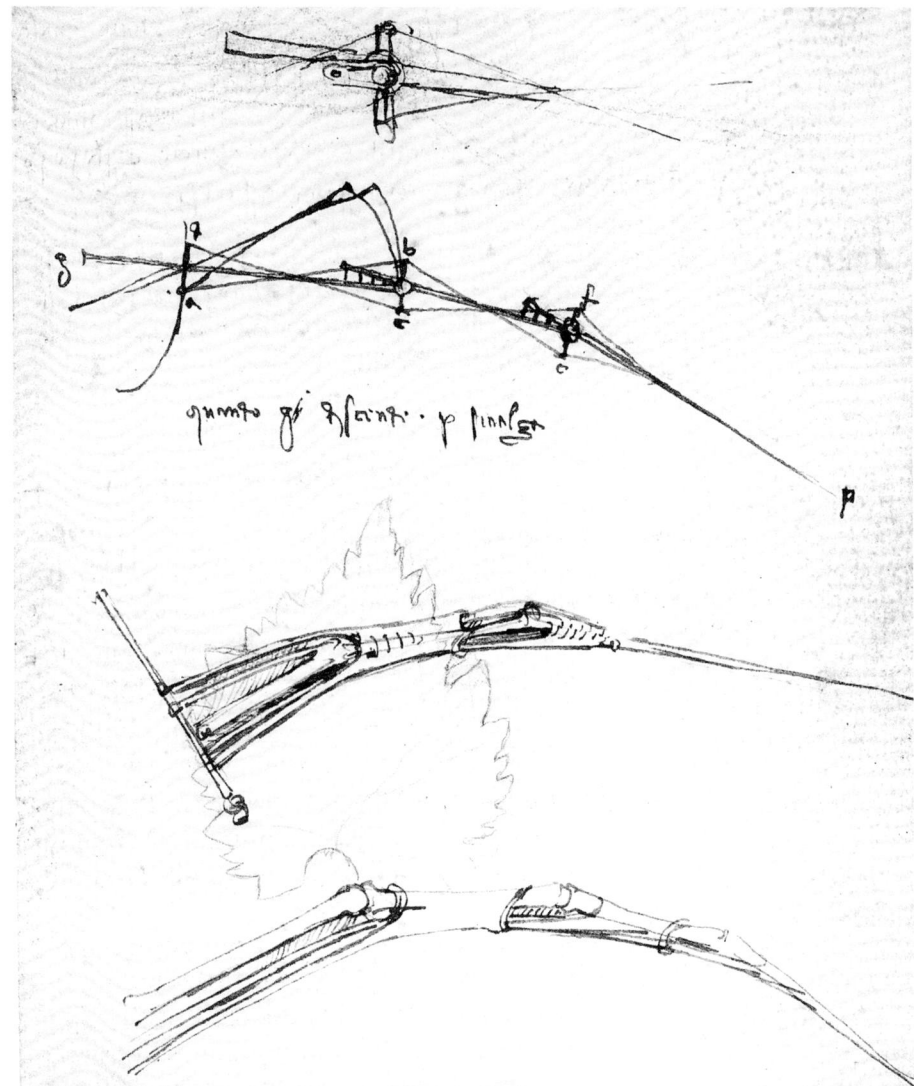

Below are two sketches of a tendon used to open and fold the wing; above are two sketches of how to imitate this by means of a system of pulleys. As usual with Leonardo, the text is in mirror writing; it reads "quando g discende p sinalza" (when g descends, p rises).

Figure 4.6: Leonardo: Bird flight examined

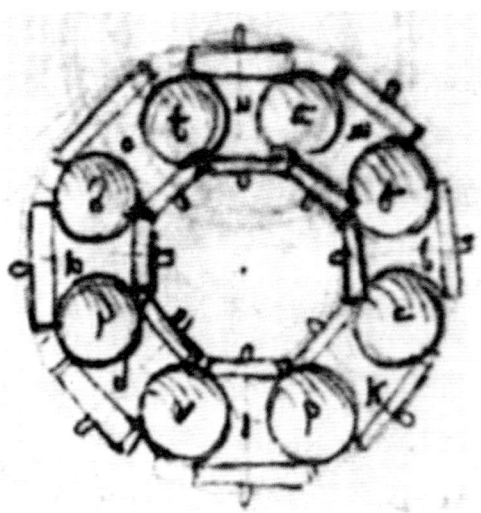

These anti-friction balls were "designed for a ball bearing (reinvented in 1920)."

Figure 4.7:   Leonardo: Antifriction balls

comprehensive conceptual structure, but knowledge directed toward practical ends is what characterizes this mode of nature-knowledge throughout – it marks its second component.

Accomplishments examined.   Leonardo's extraordinary gifts made the manner in which he went about his research program unique. But the program itself was not – it exemplifies what was at the heart of Europe's third mode of nature-knowledge. In the way Renaissance practitioners sought to carry it out, its two components (accurate description and practical improvement) were blended as a rule, albeit in varying proportions. In some researches the descriptive element was preponderant; in others the effort to improve current practices. We start our survey with those cases where accurate description was the weightier component.

Renaissance Europe experienced an outpouring of (often exquisitely illustrated) descriptions of unprecedented accuracy, whether of living plants and animals or of dead human bodies, whether of celestial phenomena or of lands and coastlines. Figures 4.8–11 display four samples, each juxtaposed to an antecedent counterpart.

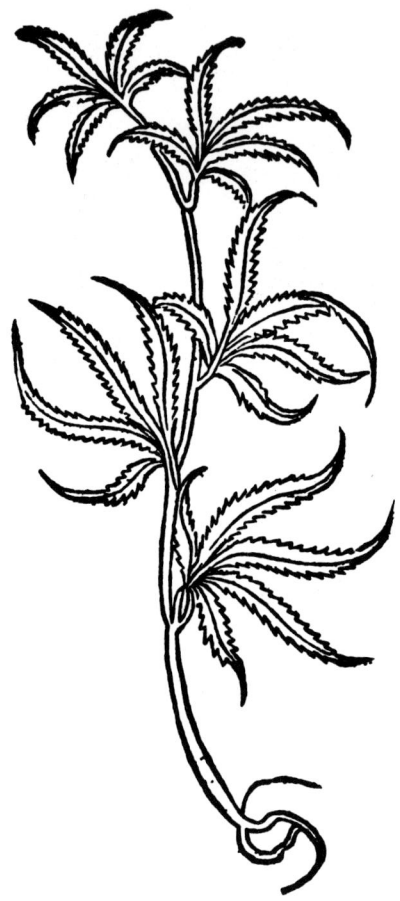

Depicted in Peter Schöffer's *Gart der Gesundheit* of 1485.

Figure 4.8a: Cannabis

CANNABIS
SATIVA

Zamer Hanff.

Depicted in Leonhard Fuchs' *De Historia Stirpium* of 1542.

Figure 4.8b: Cannabis

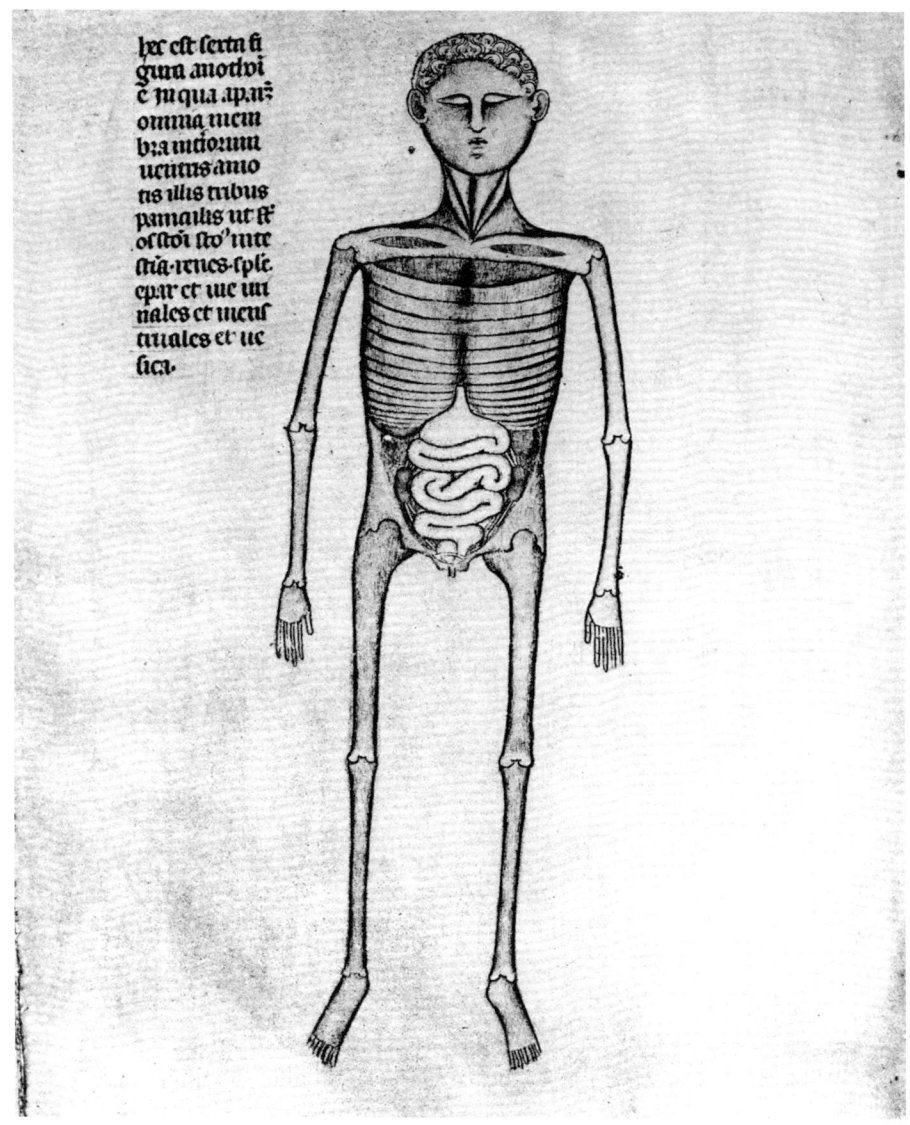

Skeleton and intestines as depicted in 1335 in an anatomical manuscript by Guido da Vigevano for the king of France.

Figure 4.9a: Man inside

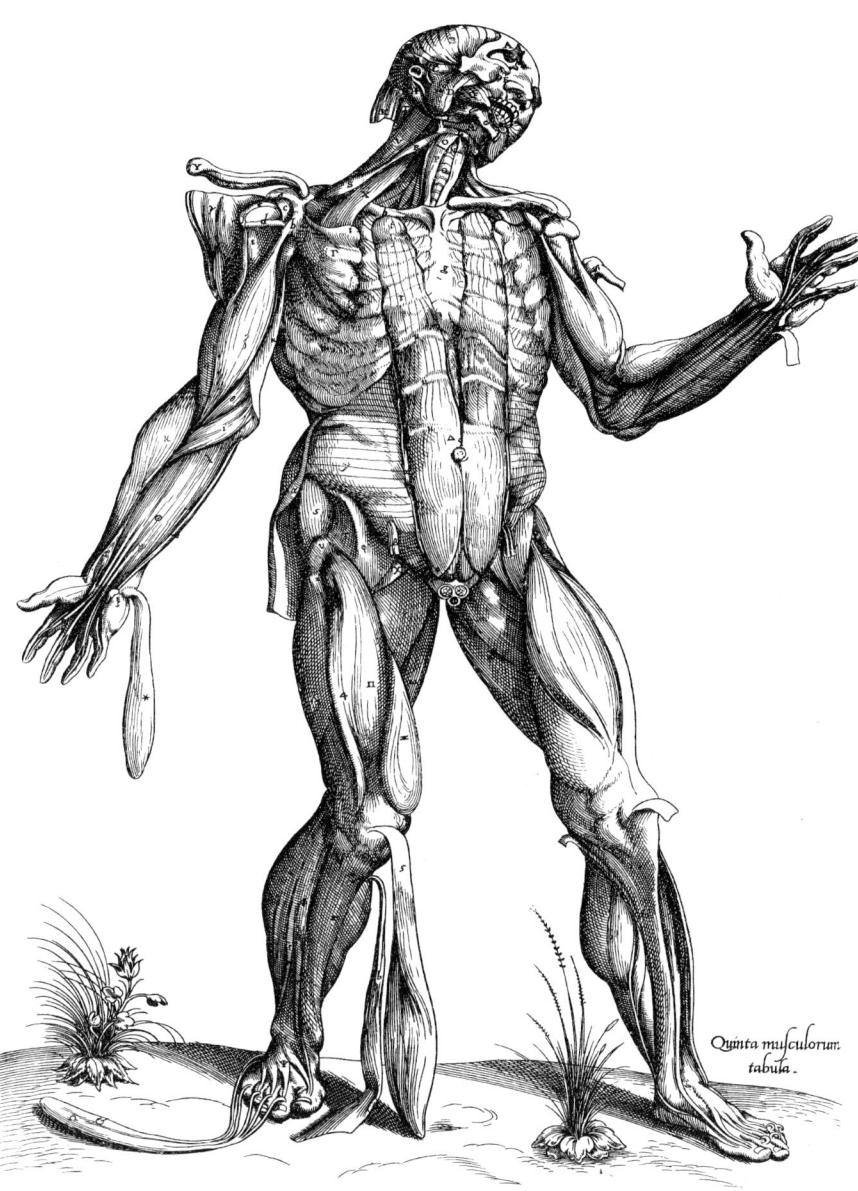

One of the 'muscle men' in Vesalius' 1543 *De Humani Corporis Fabrica*.

Figure 4.9b: Man inside

124

Cosmos from the Earth upward, in Conrad von Megenberg's 1475 *Buch der Natur.*

Figure 4.10a: The heavens

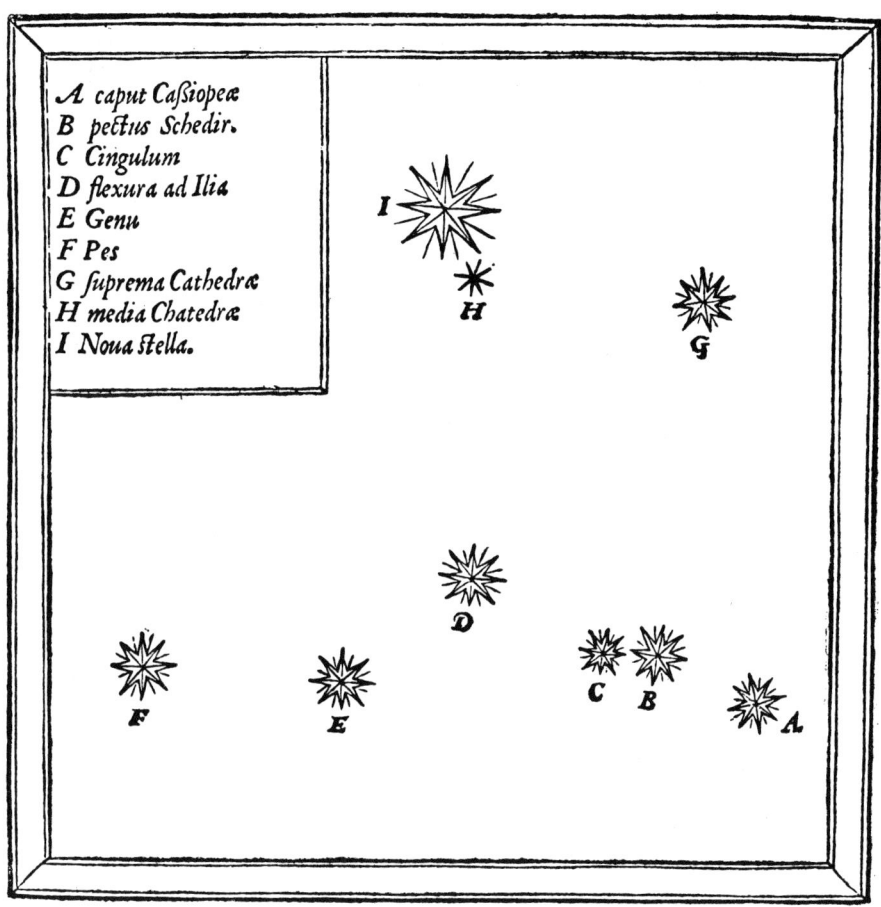

New star in the constellation Cassiopeia, as observed and investigated by Tycho Brahe in 1572.

Figure 4.10b: The heavens

Map 2 in Sebastian Münster's 1540 edition of Ptolemy's *Geographia* ('Map of the World as Described by Ptolemy').

Figure 4.11a: World maps

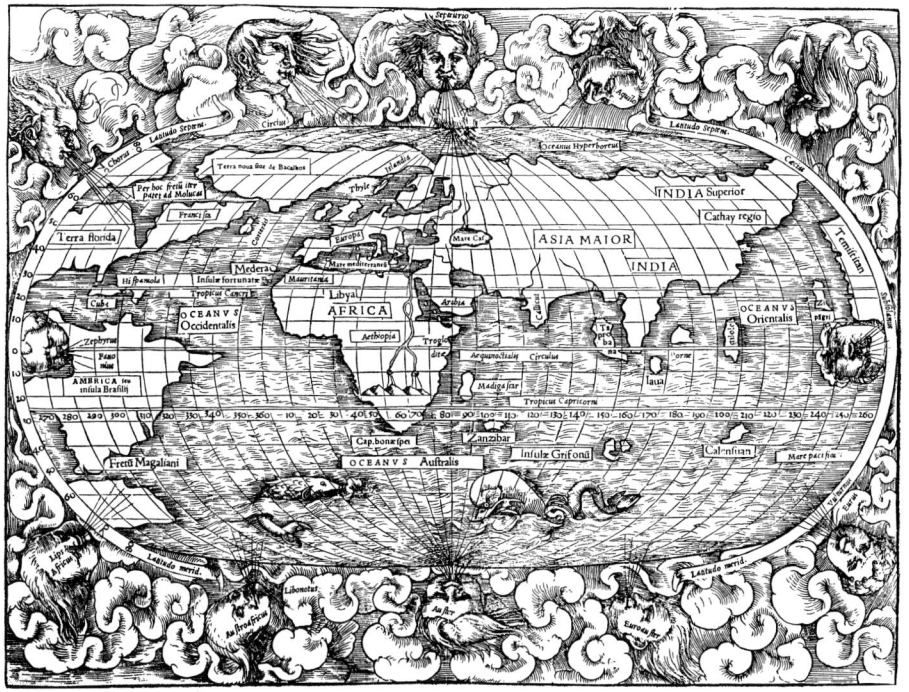

Map 1 in Sebastian Münster's 1540 edition of Ptolemy's *Geographia* ('Map of the Entire World').

Figure 4.11b: World maps

128　　If taken at face value, these juxtapositions would lead us astray. Neither the illustrations under (b) nor the verbal descriptions that accompany them are always so accurate and so devoid of hearsay, trust in the ancients, folklore, or sheer superstition as we like to associate with full-blown modernity. Also, the intellectual contexts out of which the antecedent samples arose were often quite dissimilar, thus rendering straightforward comparison unfair or even unenlightening. For example, Figure 4.8a is not so much an inept effort at lifelikeness but served rather as an aid to quick recognition of a medical drug. More often than not, medieval illustrations were meant to express the current doctrine of signatures, in which the symbolic meaning attached to things counts uppermost (e.g., a walnut is taken to stand for the human brain). This, however, is the very point to be made here: *Lifelikeness is precisely the new ideal those numerous practitioners of nature-knowledge colored the European way were after.*

This new ideal arose in close connection with the striving for lifelike depiction that started in Italy with Giotto and Duccio and that, over the course of the Renaissance, spread all over Europe. In 1533 already the movement had reached a point where Hans Holbein the Younger could inscribe his portrait of a certain Derich Born with a neat advertisement "Once you add the voice, you see Derich himself here, such that you wonder whether it was the painter or the Creator who made him."[73] As one consequence, previously strict boundaries between art and the pursuit of nature-knowledge began to dissolve. Leonardo's effective *identification* of the two is an extreme case – the Star of Bethlehem reproduced in Figure 4.4 (p. 115) is as much a botanical study as a preparation for a painting. In less extreme cases, artists collaborated with men at work to enhance the realist quality of what nature-knowledge they pursued. In the 1530s/1540s three German scholars by the names of Brunfels, Bock, and Fuchs produced herbals notable for their shared aim to provide accurate descriptions of the plants depicted therein. The pictures in Brunfels' and Bock's herbals were produced by first-rate draftsmen, Weiditz and Kandel, and their realist quality stemmed from the illustrations even more than from the accompanying texts.

Another case of close collaboration between artists and a man bent upon pursuing nature-knowledge in newly realist fashion resulted in the atlas of the human body published by Andries van Wesel (Andreas Vesalius) in 1543, *De humani corporis fabrica*. Even more than for his own written comments, this work is famous for the stunning lifelikeness on display in the authorized woodcuts prepared by students of Titian. Even though Vesalius' comments surpassed by far in accuracy any previous effort at describing the bones and muscles of the human body, his interpretation of what the woodcuts depicted was often close to Galen's teachings, which had been recently restored and expurgated.

An ideal of accurate description also reached the celestial realm. One generation after Copernicus, Tycho Brahe turned the latter's growing doubts about the trustworthiness of every single observation reported by the ancients into a huge, meticulously executed program in observational astronomy. Two or three events moved him toward undertaking it. When still a student he was struck by a mismatch between a predicted conjunction of Jupiter and Saturn and its actual occurrence a month earlier. In 1572 he was spurred on by the sighting of a new star of a kind literally overlooked for centuries in Europe (p. 274). In the way Tycho set up his program he may further have been stimulated by rumors about achievements at the earlier observatories in Islamic civilization. Whether or not he was, his program surpassed in thoroughness every earlier effort by far. Night after night, aided by fine measuring instruments of his own making and by assistants carefully groomed by him, he observed and recorded positions of stars and planets to a degree of accuracy that stretched naked-eye observation to the very limit. He brought the limit of accuracy down to c. 2 minutes of arc, as compared to the c. 10 maximally attained by Ptolemy and at best equaled since.

Tycho rejected Copernicus' heliocentrism, mostly for the vast amount of space wasted between Saturn and the fixed stars in view of absent parallax (p. 111). Instead, he devised a compromise system in which the Earth, left at the center, is orbited solely by the Moon and the Sun. This way he combined the virtues of a stationary Earth with advantages gained from having the planets orbit the Sun. It is characteristic of Tycho's stance, not in Alexandria but in the overwhelmingly empiricist approach to natural phenomena that marks the third mode, that he did not turn his compromise system into more than a rough sketch – unlike Copernicus he was above all an observer, not so much a mathematician. His thoroughgoing empiricism is also evident in his likewise decades-long sustained and likewise unprecedented program to record the weather on a daily basis in the vain hope of enabling him to predict it reliably.

In two further cases, the aims of accurate description and of attaining practical ends were about equally weighty: the making of maps and the collection of natural objects. Medieval Europe had given rise to two wholly unconnected map traditions, one edifyingly theological, one artisanal. Scholars pictured Jerusalem at the center of the world; mariners drew Mediterranean coastlines as they saw them. The former had concerns other than real-life descriptive accuracy; the latter lacked the knowledge properly to represent it. The two traditions began to fuse when Ptolemy's *Geographia* was translated into Latin from a Byzantine manuscript in 1406. Here scholars could learn the basic principles of map projection using coordinates. Yet no authentic maps survived to accompany the text, significant parts of western Europe inevitably failed to be of concern to Ptolemy, and by the end of the century ever more portions of the globe were being discovered of which Ptolemy had known nothing. The *Geographia* was printed in 1475 for the first time, and successive editions contained ever more novel and reliable details. Geography emerged as an art- and science-infused discipline in its own right. Borrowing from sailors' accounts, from astronomical tables, and from improved techniques in surveying and projective geometry, learned mapmakers like Mercator and Blaeu sought to attain a high degree of accuracy, thus meeting rising demands from the navigation business.

130 During the same period scholars began to collect rapidly increasing numbers of natural objects of ever increasing variety. Whether stuffed animals or surveying instruments or plants like tobacco or corn imported from the Americas, everything that could arouse curiosity might find a place in the houses of collectors and be shown on demand to high-ranking visitors. The ideal was to 'possess nature' by having on display in one's museum a specimen of every object in existence. Collecting went in tandem with preparation of museum catalogs that provided exhaustive descriptive accounts of all objects collected. These expanded into ever more unwieldy proportions as the collections they cataloged grew. Over the second half of the 16th century Ulisse Aldrovandi's house in Bologna occupied center stage, in size as well as in reputation, and his collection survived until far into the next century. Collecting need not remain an in-house activity. Carefully laid out botanic gardens, tended by a university's medical faculty for the preparation of herbal cures, flourished as well – here, too, description aimed at understanding and action aimed at practical objectives went harmoniously together.

There were, finally, numerous research subjects where accurate description was a lesser concern and practical utility provided the predominant objective. More often than not, the crafts were deeply involved. Craftsmanship and the pursuit of nature-knowledge were quite distinct activities as a rule, separated by a chasm at once social and intellectual. Exceptions were as yet few and far between, as with Ktesibios and Herôn applying the law of the lever to their pulleys, with Friedrich II's falconry and Pierre de Maricourt's work on magnetism and the compass, or with the practice and vision of alchemy. Artisans worked by experience gained from trial-and-error procedures with proven merit in past practice and orally passed on through the generations in the guise of 'how-to' rules rather than from rules extracted from an understanding of underlying principles. No theory of harmonics could guide the organ builder engaged in his workshop in intoning pipes for mixture stops (the delight of early organs). Similarly, no idea of the void or of the weight of air was around to show the way toward the construction of an engine that would be fit to marshal the force hidden in atmospheric pressure. There is retrospective irony in the situation. In iron and silver mines, technicians struggled in vain amid their groundwater-soaked ores to overcome the limitations to which the suction capacity of a water pump appeared liable in practice. At the same time, securely ensconced in their lecture halls, university professors like Jean Buridan held forth on the impossibility (but for possible divine intervention) of the very void these miners were inadvertently producing.

What happened in the course of the Renaissance was certainly not a full closure of the gap between two socially noncommunicating, intellectually opposite poles. The significant feat accomplished over the period was an ongoing exploration of the vast territory lying in between. For example, 16th-century humanists like Pierre de la Ramée (Petrus Ramus) or Luís Vives expressed their hopes for a general reform of the sciences to emerge from the quarters of the artisans. 'Get out of your armchairs, ye philosophers, and gain wisdom in the workshops' is what their insistent message came down to. Ramus found the kind of practical arithmetic he encountered in the marketplace far more significant for such a hoped-for reform than the books of Euclid revered by the more common run of mathematical humanists. A skilled potter and surveyor by the name of Bernard Palissy provides an example from the other direction. He collected fossils and theorized and lectured to Parisian audiences about their meaning as petrified animals and about other topics related to craft practice without any regard for Greek learning or apologies for his own lack of it. Instead, he argued pointedly that experience is at all times the mistress of the arts, and that visitors to his home collection of natural objects would there learn more in two hours than from forty years' worth of study in the books of the ancients.

In the same vein, abstract mathematics began to find some as yet well-circumscribed application. The striving for lifelike accuracy of representation that came to dominate Italian painting from the mid-14th century onward was much enhanced by the invention of linear perspective. Painters and geometers mingled to get the required geometry right and bring the budding practice to perfection. Nor was the practical use of available mathematical knowledge confined to problems in depiction. Architecture, fluvial engineering, and gunnery were affected, too. For instance, Niccolò Tartaglia rightly guessed without proving it that the elevation at which to attain maximal shooting range is 45°. The mathematics employed in such exercises, albeit far from trivial, also did not rise to the innovative or the profound. By the end of the 16th century, four domains of craft practice found themselves infused with mathematics – not only linear perspective but also navigation, mapmaking, and the construction of fortresses (p. 317).

This was the flourishing period of the artist-engineers. Men like Leon Battista Alberti, Leonardo da Vinci, and Francesco di Giorgio in Italy and Albrecht Dürer and Georg Bauer (Agricola) in Germany were engaged in the exploration of territory intermediate between nature-knowledge and the arts. One way to explore it was by seeking ways and means to apply rules of some degree of generality to practical undertakings. Or, the other way round, they might seek to extract certain underlying principles from their practical concerns in perspective or architecture or mining.

Unprecedented and partly successful as these activities were, their scope and impact were as yet highly circumscribed. An example of how achievement and limitation went together in this regard is Simon Stevin's effort at improving the construction of windmills the mathematical way. His aim was to increase the amount of water these could drain away per unit time. He drew on his theoretical understanding of the *Principles of Water Equilibrium* (the title of his Archimedes-style treatise of 1586). In the end, he took out patents on a range of improvements of two kinds – one qualitative and the other based on intricate calculations. The former proposal, a somewhat different shape for certain wheels that were used in

132 transmission, was widely adopted. Not so with the latter. It derived from a calculation – essentially sound even in retrospect – that suggested a substantial increase in the size of the blades of the paddle wheel while reducing their speed and number. That solution, however, turned out to be useless in practice. Note here that Stevin felt the match he had attained between mathematical theory and milling practice to be good enough to make economic exploitation of his ideas feasible. Indeed, throughout his career Stevin was concerned with making 'reflection' meet 'action'. So much the more telling that a man so hard at work at their interface still underestimated by far how difficult it is to make them match:

> He appears fully aware of the ideal character of the system [of Archimedean statics] thus constructed. Repeatedly he emphasizes the difference between what is valid 'when taken mathematically' and what is true 'when understood naturally', that is, in physical reality. The celebrated 'wreath of spheres' proof of the law of equilibrium on an inclined plane [p. 104] is conceived so fully to take place in an ideal domain that not one word is devoted to the apparent impossibility of physically realizing the situation described. Now, just like his great predecessor from Syracuse, Stevin was given not only to reflection but to action as well. Therefore, one might expect to find him paying attention to the question of the extent to which physical reality deviates from ideal theory. Curiously enough, this expectation is hardly fulfilled. To be sure, Stevin supplements his 'Art of Weighing' with a 'Practice of Weighing', where the theory explained in the preceding treatise is applied in practice. However, the machines discussed in the latter are kept in the domain of the ideal quite as much as is true of the windmill for which, in a separate work, he develops a quantitative theory – the first in history. True, Stevin repeatedly calls attention to the fact that, in dealing with real machines, things do not quite fit the theory taught in the 'Art of Weighing' so nicely. *The impression gained, however, is that Stevin seriously underestimates the extent of the deviation.*[74]

Today we are well aware of what emerged in the end – nothing less than the 18th-century onset of modern, science-based technology (p. 729). But we must not project this major fact of history back upon earlier times. Prior to the grand divide of the Scientific Revolution, the component of nature-knowledge present in the craft tradition even at its most sophisticated was bound to remain small and haphazard. From a contemporary point of view, unspoiled by hindsight, certain other practice-oriented manifestations of the 'third' mode of nature-knowledge loom much larger.

 Chemistry as practiced in Islamic civilization and in medieval Europe was in the course of the Renaissance placed in a markedly novel frame that went by the name of 'iatrochemistry' (medicine-oriented chemistry). In the first half of the 16th century Theophrastus Bombastus von Hohenheim, known to posterity as Paracelsus, turned well-proven components like Aristotle's conception of matter and the theory and practice of alchemy into something

distinctively different. He was hardly a systematic thinker. The vast influence that his posthumously published works were to exert for about a century after his death in 1541 is due to
their 'something-for-everyone' character, as well as to the near-universal scope Paracelsus
envisaged for chemistry. Paracelsians explained Creation in terms of God's "Halchymicall
Extraction, Seperation, Sublimation, and Coniunction".[75] They devised a doctrine of the
three elementary principles 'sulphur', 'mercury', and 'salt' and endowed it with religious (the
Trinity), theoretical, and practical significance. They held the Earth to be animated and regarded it as a big chemical workshop. Since the microcosm is a reflection of the macrocosm,
they regarded human beings likewise as spirit-infested bundles of chemical processes. They
went on from there to derive scores of mineral and metal recipes meant to restore the sick
to that precious chemical balance between the three elementary principles that expresses
itself in the state of health. The cultivation of iatrochemistry and alchemy thus ranged from
esoteric levels of lofty theoretical doctrine to thoroughly practical undertakings.

The theory as well as the practice of Paracelsianism made for close connections with one
more manifestation of Europe's third mode of nature-knowledge: a worldview pervaded
with magic.

If we understand magic to be concerned with influencing natural events in ways defy-
ing the ordinary course of things, the activity is surely of all times and places. In Renaissance Europe, under the aegis of the 'Hermes mistake', something more happened – magic
was turned into an elaborate worldview. By the late 15th century texts allegedly written by
Hermes Trismegistos (p. 24) were recovered in Florence's Platonizing circles. They were believed to antedate and, as it were, prefigure by many centuries Greek philosophy and also
Mosaic religion. Those who fell under their spell came to regard the universe as a network of
correspondences and occult forces acting upon one another at a distance by means of like-
nesses and sympathies (attractive forces) or of contrasts and antipathies (repulsive forces).
What we still call 'sympathetic resonance' – as one string is plucked, a similar one at some
remove may with apparent spontaneity begin to sound in unison with the plucked string –
offers as good an example as any of these mysterious natural actions and how they were accounted for in Hermes-inspired, magical thought. Magic conceived this way entailed more
than just a world*view* – it pointed the way toward tapping these 'occult', hidden forces so as
to put them to work for human ends. Here is a summing up of how, in a 1533 textbook of
magical theory and practice, Cornelius Agrippa von Nettesheim explained what made such
feats possible:

> The universe is divided into three worlds, says Agrippa, the elemental world, the celestial world,
> the intellectual world. Each world receives influences from the one above it, so that the virtue
> of the Creator descends through the angels in the intellectual world, to the stars in the celestial
> world, and thence to the terrestrial elements and all things composed of them. Magicians *think*

that they can make the same progress upwards, and draw the virtues of the upper world down by manipulating the lower ones. They try to discover the virtues of the elemental world by medicine and philosophy; the virtues of the celestial world by astrology and mathematics; and in regard to the intellectual world they study the holy ceremonies of religions.[76]

Nature's occult properties were not only examined and put to work on the lofty plane of an elaborate worldview. They also percolated down to the bookshop and the marketplace. Under the heading of 'secrets of nature', how-to rules were peddled all over Europe by conjurer-like 'professors of secrets', and books filled with natural prodigies or herbal lore or alchemical recipes reached wide audiences.

European peculiarities.   How do the varied undertakings that we have just surveyed compare with their counterparts in Islamic civilization several centuries earlier? Certainly in scope the difference is quite large. Whereas the Greek corpus was enriched in the two civilizations in roughly equal measure, what Europe in its pre-Scientific Revolution period added to that corpus out of resources of its own was a good deal richer. Nature-knowledge colored the European way covered far more subjects and counted many more contributors.

As before, neither praise nor blame is our objective; the question is rather why this should be so. We cannot even begin to seek answers until we have addressed the prior question of what enables us to speak here, too, of 'culturally marked' nature-knowledge the way we did with the Islamic precedent. In Islam, subjects like the *qibla*, the determination of times of prayer, and the division of legacies stemmed directly from its Holy Writ. In the specific orientation given to alchemy and in the prominence allotted to healing practices, values predominant in Islamic civilization came likewise to the fore, albeit not so directly. A similar distinction for Europe yields a different picture. The determination of Easter was the only subject of culturally marked nature-knowledge that derived from the faith in a straightforward manner. All other culturally marked subjects, from Pierre de Maricourt's magnets down to Tycho's celestial observations or Paracelsus' chemical cures, were inspired by the reigning faith in an indirect manner only.

We shall find later how this preponderance of pursuits expressive of values *ultimately* yet not *directly* grounded in the faith helps explain certain revolutionary turns that nature-knowledge was to take in 17th-century Europe (p. 263). But the prior question ought to be answered first: what in the wide spectrum of nature-knowledge here discussed allows us to label it 'marked by the culture'? What was so specifically European about the insistent pursuit of descriptive accuracy, the application of some current mathematics to artists' problems, the emergence of several other possible interfaces between nature-knowledge and the crafts, the universal claims raised for chemistry, and a magical worldview bent on the beneficial harnessing of occult forces?

The question ties in with an ongoing debate, instigated a century ago by Max Weber, over what peculiarities enabled European civilization to make the unprecedented leap to the modern world that we live in. Two classes of solutions had better be dismissed right away: appeals to a 'Faustian spirit' allegedly animating Europeans from the Renaissance onward, and efforts to treat certain medieval developments as predetermining Europe to feats of modernization still centuries away. A better pathway to viable solutions rests in cross-cultural comparison undertaken without the benefit of hindsight. Here is one particularly valiant attempt by Joseph Needham, who spent a lifetime comparing nature-knowledge in China and in Europe. He saw the issue in terms of "Europe['s] built-in quality of instability". He explained this quality

> in terms of the geography of what was in effect an archipelago, the perennial tradition of independent city-states based on maritime commerce and jostling military aristocrats ruling small areas of land, the exceptional poverty of Europe in the precious metals, the continual desire of Western peoples for commodities which they themselves could not produce (one thinks especially of silk, cotton, spices, tea, porcelain, and lacquer), and the inherently divisive tendencies of alphabetic script, which permitted the growth of numerous warring nations with centrifugal dialects or barbarian languages. By contrast China was a coherent agrarian land-mass, a unified empire since the third century B.C. with an administrative tradition unmatched elsewhere till modern times, endowed with vast riches both mineral, vegetable, and animal, and cemented into one by an infrangible system of ideographic script admirably adapted to her fundamentally monosyllabic language. Europe, a culture of rovers, was always uneasy within her boundaries, nervously sending out probes in all directions to see what could be got – Alexander to Bactria, the Vikings to Vineland, Portugal to the Indian Ocean. The greater population of China was self-sufficient, needing little or nothing from outside until the nineteenth century ... and generally content with only occasional exploration, essentially incurious about those far parts of the world which had not received the teachings of the Sage. Europeans suffered from a schizophrenia of the soul, oscillating forever unhappily between the heavenly host on the one side and the 'atoms and the void' on the other. ... It may well be that here, at this point of tension, lies some of the secret of the specific European creativeness when the time was ripe.[77]

In the picture here sketched, an archipelago-like geography, city autonomy, a perennial trade deficit, and alphabetic script are granted equal parts in endowing European society and culture with an extraordinary dynamism. As Needham pointedly observed, Europe's lack of self-sufficiency and its consequent restlessness and curiosity instigated wide-ranging explorations culminating in the Voyages of Discovery that the Portuguese pioneered. To be sure, Europeans were not the first to explore foreign lands; daring individuals in other civiliza-

136   tions had done so before. For instance, Ibn Battuta, a citizen of Tangier, traveled as far as Sudan and Sumatra and China, and upon his return to his native town he dictated an account of his experiences. Many more individuals in the Islamic world also traveled extensively, yet these trips were envisioned and undertaken by exceptional individuals. The European experience compares better with its Chinese counterpart – in both cases, exploration went beyond the individual and the incidental.

By the early 15th century the Portuguese set out to explore the Atlantic Ocean, moving ever farther south and then, on finding and rounding the Cape, due north along Africa's east coast. Some gray-haired longshoreman in Mombasa who in 1492 saw Vasco da Gama's fleet sail into the harbor might still recall how as a youth he had watched the vastly larger vessels of Admiral Chêng Ho anchor there. Both expeditionary forces were sent out on the initiative of the state. But there are also some telling contrasts. Chêng Ho ('the Three-Jewel Eunuch who went down into the West') headed a onetime effort, repealed within twenty years by the Ming imperial bureaucracy. Vasco da Gama's voyage to India forms one early link in a chain unbroken until, centuries down the line, the last white spot on the map of the Earth had been filled in. The attitude taken toward the peoples visited also differed. In keeping with their conception of a tributary world of barbarians beyond the borders of the Middle Kingdom, the Chinese blended avuncular benevolence with contemptuous indifference. The Portuguese, and in their wake the Spanish, the Dutch, and the British, soon turned to the attack, out of missionary zeal mixed with an urge for gold and spices. Their aggression, however, did not stand in the way of a genuine interest in the rules and habits and the natural surroundings of the peoples they sought to subject. Just as al-Biruni had reported on the state of nature-knowledge in India centuries earlier, just so did an apothecary by the name of Garcia de Orta publish *Colloquies on the Simples and Drugs of India* (1563).

Unlike with al-Biruni's individual effort, however, Garcia's treatise represents one link in an ever-lengthening chain. Portuguese sailors returned from their probings with reports much at variance with what, back home, leaders of the humanist movement were at the very same time finding asserted in Greek texts. Ptolemy had proved that all dry land is confined to the Northern Hemisphere – Portuguese navigators showed him to have been quite mistaken. And Aristotle, so given to explaining why what was not the case could not possibly be the case, had denied that human beings could live in the tropics. Learned scholars had parroted him over the centuries – "rude sailors", on passing the equator and coming on shore, found these regions to be teeming with humanity.[78] The message these adventurers brought home was that the world is full of phenomena as yet unknown, and that its diversity is worth being charted without the dead weight of preconceived ideas. In precisely this fashion did Garcia de Orta's encounter with a flora much at variance with that of his native land enable him to take an even more independent stance vis-à-vis ancient authority than that of the three German herbalists Brunfels, Bock, and Fuchs.

In other ways, too, these voyages impinged upon such accomplishments as mark Europe's third mode of nature-knowledge. Beyond the Canaries Portuguese sailors faced a tough struggle with unknown winds. To help them find their way along the west coast of Africa, Pedro Nunes in the early decades of the 16th century pioneered a geometric approach to the theory and practice of navigation and to mapping techniques that British and Dutch navigators were to adopt and extend later in the century. Further, among accomplishments in the third mode I have already mentioned how men like Aldrovandi assembled vast museum collections. Many specimens that came to fill these were colonial imports, such as tobacco, maize, and potatoes.

Feats like these served as an enduring source of inspiration and a powerful image to those engaged, not in the backward-oriented movement of restoring and on occasion enriching the Greek corpus, but rather in the forward-looking, novelty-hunting movement embodied in the third mode. For instance, when around 1600 Francis Bacon invoked the ongoing opening of the "regions of the material globe", he added pointedly that it would be a shame if, in contrast, "the intellectual globe should remain shut up within the narrow limits of old discoveries".[79]

In such varied ways did the Voyages of Discovery inform Europe's third mode of nature-knowledge. But the dynamism that singled out Europe among the world's civilizations came to the fore in other respects as well. A significant case in point is the creation of interfaces between nature-knowledge and the crafts. Here, too, a comparison between Europe and China is instructive. Earlier I contrasted the intellectualism that marked the Greek approach to nature with the empiricist and practice-oriented manner in which Chinese investigators went about their researches (p. 45). Greek nature-knowledge impinged only in the rarest of cases on craft practice, whether it be cultivated in its Athenian or in its Alexandrian mode and whether it be cultivated in Hellenist or Islamic civilization or in medieval or Renaissance Europe. Nature-knowledge of a kind intermediate between highly abstract theory and theoretically low-level craft practice could therefore emerge only outside the perimeter of the Greek tradition. We have seen how by the mid-15th century, around the edges of ongoing recovery and enrichment of the Greek corpus, the kind of research that Friedrich II and Pierre de Maricourt had probed first turned into a quickly widening stream – a mode of nature-knowledge in its own right. But an independent, parallel stream had already for centuries been flowing in China.

The parallelism goes quite far. Chang-heng arrived at seismological insights out of a felt need to guard against earthquakes, just as Paracelsus came to some chemical understanding out of a desire to cure the sick. Leonardo investigated material properties manifest in friction, just as, in the second half of the 11th century, Shen Kua examined directive properties of floating magnets used in geomancy (p. 43). Not only were such activities undertaken, more often than not, against a background of magic, but the principal likeness is that in both we

138 encounter, not grand theories in search of bits and pieces of empirical confirmation, but laborious exploration, undertaken with a view to attaining some practical end, of possible regularities in a multiplicity of diverse phenomena.

The parallelism does not go all the way, however. There is a subtle yet important difference in the *kind* of empiricism in each case. To borrow an expression coined by Goethe, Chinese nature-knowledge was marked throughout by the *zarte Empirie* ('tender empiricism') that he himself advocated for a truly scientific approach to nature's phenomena. In partial contrast, the empiricism that marked the European-colored mode of nature-knowledge was as a rule of another, ruder, and more interventionist kind, oriented toward control and domination as its ultimate objective. Whereas in mining and in land cultivation a Chinese could deal as ruthlessly with the integrity of the naturally given as any European could, the difference rests in whether or not nature *as such* was regarded as fair game for manipulation. Take Shen Kua as an example. He shared with all "literati thinking about Nature a millennium earlier" the accepted vision of nature as "an organismic system, its rhythms cyclic and governed by the inherent and concordant pattern uniting all phenomena". But as a court official he was also engaged in the large-scale mining of salt, which he curtly and tellingly defined as "a means to wealth, profit without end emerging from the sea".[80] Whatever the underlying unity that reconciles the apparent tension between these distinct attitudes, in Renaissance Europe no such tension existed. There, respect for the givenness of natural phenomena and the patient collection of data went together with a worldview that legitimated and fostered, rather than negated, the exploitation of nature.

The difference at stake here is between Leonardo's studies aimed at emulating bird flight and the Chinese infatuation with making huge kites fly with the winds. Joseph Needham has sought to capture the overall contrast in terms of the phrase *wu-wei*, which occurs so often in texts in the Lao Tzu and Chuang Tzu lineage. On this view, China's worldview of organic materialism was marked by an attitude of acceptance of the world. The ultimate aspiration of China's rule, the achievement of harmony in microcosm and macrocosm alike, could be attained best by abstaining from improper action. No mere passivity lay in the aim to live in harmony with nature; yet the Chinese posture was constrained by *wu-wei* – by a reluctance to 'go against the grain of nature'.[81] Behind the Chinese striving for harmony lay an acceptance of the world as it is given to us, in marked contrast to how European civilization aimed in action *and* in thought to intervene in the world and manipulate it for human ends. As diagnosed by Needham, Europe's singular dynamism is reflected in the peculiarly interventionist nature of its empiricist undertakings.

Unrelieved separation; a rare series of experiments. Conforming to a familiar pattern, the pursuit of nature-knowledge in this third mode was separated by a wide chasm from the ongoing recovery and reassessment of the Greek corpus – other people, other concerns. True, practitioners were wont likewise to take their point of departure in some ancient text. Vesalius started from Galen; Brunfels from Dioskorides; Nunes from Ptolemy; many a collecting naturalist from Pliny; Agrippa from legendary 'Hermes'; many an applied mathematician from the legendary Archimedes and the military feats ascribed to him of old. However, time and again events branched out from there in empiricist, forward-looking directions quite different from the efforts at reconstruction of ancient wisdom undertaken at the time by the philosophers and the mathematicians. Broadly speaking, the recovery and enrichment of Greek mathematical science were the concern of humanists enjoying court patronage, and the cultivation of natural philosophy was overwhelmingly an affair of university-seated scholastics. The empiricist search to enhance factual knowledge of the natural world and gain some measure of control over it, while enjoying court patronage more often than not, was carried out as a rule by scholars connected to the world of the arts and crafts. So the unmistakable lack of interaction between the three modes of nature-knowledge simultaneously present in Renaissance Europe (the two culture-transcending, Greek modes and the culture-bound, European mode) was furthered by the different social positions practitioners occupied. It was not, however, uniquely caused thereby. A suitable way to demonstrate that it was not is to consider the exceptional case of a man who took an active part in both the Athenian and the 'European' modes of nature-knowledge while characteristically managing to keep the two separated in his own head.

That man was a Portuguese scholar by the name of João de Castro. I use his case for another purpose as well – his achievement points at a significant potentiality inherent in Europe's control-oriented empiricism. As we saw earlier in the case of Leonardo, accurate, action-directed observation might on occasion consolidate into more narrowly targeted ranges of deliberate experimentation. It was not that the act of experiment was so novel per se – recall Pythagoras' consonance-producing monochord, Ibn al-Haytham's oil-lamp shining through a tube, Pierre de Maricourt's bisected yet still bipolar magnets, or all kinds of operations tried out over the centuries in alchemy. Truly new was experiments being carried out, not incidentally as before, but in well thought-out, progressive series. That is what, hidden from the world, Leonardo was occupied with: he was systematically setting up a coherent range of experiments to find out how materials behave under deliberately varied conditions of friction. In his own manner, Castro did similar things.

Around 1538, at the age of thirty-eight, he wrote *Tratado da esfera* – one of numerous cosmological treatises produced between the early 13th and the 17th century on the model of Sacrobosco's *De sphaera* (p. 82). Here Castro stuck to Aristotelian doctrine in its most orthodox variety, but for one peripheral exception. When treating Aristotelian themes like the allegedly small amount of dry land in proportion to the Earth's waters, he marked his disagreement by pointing at contrary evidence obtained by his countrymen in their maritime probings to the south and east. Castro was deeply engaged in these overseas adventures himself – he ended his career as viceroy of India from 1545 until his early death in 1548. Except for these few passages, however, nothing in the *Tratado* reflects the experience.

Little as Portugal's incipient empire building was able to direct Castro's mind in its Athenian mode, so much the more did it affect that same mind when engaged in writing up, between 1538 and 1541, the three consecutive *Roteiros* (travel reports) he kept on board ship from Lisbon to Goa, from Goa to Diu, and from Goa to Suez. In these notes a quite different spirit prevails – the coercively empiricist spirit of nature-knowledge colored the European way.

King João III charged Castro with making observations on the behavior of sea-water and magnets. The *Roteiros* display an unceasing, methodically exercised interest in both the unknown and the allegedly known. Castro gave a detailed eyewitness account of a basalt formation on the southwest coast of India. He interrogated Ethiopian chieftains to correct Pliny's tales about the sources of the Nile. He was on the lookout all the time for novel facts and described them in much accurate detail without any effort to squeeze them into philosophically prestructured categories. More often than not, Castro collected evidence with some assumed natural regularity in mind. A systematically executed series of compass readings convinced him of the untenability of the widespread conviction that there must be a 'true meridian' of variation zero in respect to the magnetic pole. More than half a century later a man of the stature of Simon Stevin would still erect his *Havenvinding* (The Haven-Finding Art) upon that hoped-for resolution of the problem of geographical longitude.

When facing new facts, Castro felt a persistent need to check their truth against the two sources of possible error he in effect identified: those arising from human fallibility and those that might flow from instrumental design and operation. He subjected his astrolabes and compasses to critical scrutiny. Anyone who happened to be around might be recruited to serve as a witness or to assist in the repetition of an observation. On one occasion

> Castro found a latitude of 16°, the contramestre 17 1/3°, a mariner 17 1/3° and the caulker – "a practical man of great experience" ["homem practico experimentado"] – 16 1/4°.[82]

Not until Castro had found that an observed yet incongruous fact withstood all serious criticism and had to be accepted as genuine did he seek for possible causes in unforeseen influences arising from some stuff or event nearby. That way "Castro [became] the first to discover that the capricious behaviour of the needle on board of a ship may be caused by iron objects."[83] More often than not, however, he was unable to identify such an element in the environment with sufficient plausibility. In such cases he consistently refused to take the customary way out, which was just to dream up some explanation. Instead, he left the solution to the admired Pedro Nunes, who had taught him the theory and practice of marine instruments, or he resorted to his pet phrase 'let Apollo solve it'.

Near coincident with Castro's death in 1548, the Portuguese, albeit the very people to set out on the discovery, exploration, and exploitation of foreign lands and their inhabitants, began to drop out of the further pursuit of the kind of nature-knowledge that went with these activities. The Spanish soon proved subject to this early case of 'imperial overstretch', too. Around 1600 other seafaring nations, notably the British and the Dutch, began to contribute to the advance of nature-knowledge of a descriptive, practice-oriented, and – in ways first explored by Leonardo and by Castro – occasionally experimental kind.

At the dawn of the Scientific Revolution

So far in this chapter we have examined how, each in its own manner, Athens, Alexandria, and Europe's newborn coercive-empiricist mode of nature-knowledge fared between c. 1450 and c. 1600. We have now at long last arrived at that fateful point in time, the turn of the 17th century. We shall ignore for a while the breakthrough that we know by hindsight to have been on the verge of taking place and, instead, remind ourselves of the state that nature-knowledge had reached in Europe at this particular point in time, so as to compare that state with the most comparable case by far – the state of nature-knowledge in Islamic civilization at the end of its most creative period. Or rather, we conjure up in our historical imagination a contemporary observer and charge her with making the comparison for us. We make a mental effort to cross back over the grand divide of the Scientific Revolution and place ourselves in or near the year 1600. We instruct our observer to report to us on where we now stand and what, *in view of past experience*, the future of nature-knowledge in Europe is most likely to have in store for us.

Theory – introducing a fictional observer.   What aim is served by charging an imaginary, detached, yet well-informed outside observer with reporting on the past and current state of nature-knowledge with a view to predicting its future chances? In employing this stratagem out of the historian's toolkit, I seek to highlight the amount of change brought about by the event we are trying to understand. Does the fictional procedure reveal a sizable contrast between what the imaginary observer would have every reason to expect at the time and what the future actually held in store? Or does it not, or hardly so? To estimate in this particular manner the size of the contrast may help us decide that perennial bone of contention among historians – the issue of continuity versus discontinuity. If our reporter is led to predict just about the exact opposite of what actually was going to happen, then clearly what actually happened betokens a radical break with past patterns.

Not that even a *radical* break can ever be an *absolute* break. There are always continuities; in history no slate is ever wiped clean. Not even the sudden evaporation of the human race

142 through full-scale thermonuclear war would involve the end of all life on earth, and for bacteria there would be no discontinuity worth mentioning at all. Even so some past pattern may be transformed more or less radically, and our fictional observer's report may serve as a measure for how drastic (perhaps even revolutionary) one or another transformation was.

The only viable pathway to prediction open to that observer involves the extrapolation of current trends in light of comparable precedents. This is the route customarily taken in reports meant to help modern policy makers find some firm ground and give their decisions a chance to reach beyond the erratic and the intuitive. Comparison, then, is at the heart of our reporter's exercise. The preceding cases of transplantation of the Greek corpus yield the sole data she can possibly go by. Her first task is to examine whether there are sufficient grounds for considering the point now reached as the Golden Age of Renaissance European nature-knowledge. And if yes, she is bound to consider whether that Golden Age is sufficiently like its predecessors to expect that the upswing will soon give way to the customary downturn.

Europe's Golden Age of nature-knowledge?   A glance back at Schema 2 (p. 91) reveals at once that what has happened in Renaissance Europe so far looks quite similar to previous episodes of flourishing nature-knowledge. Much more than the second, medieval, highly curtailed, and, hence, lopsided transplantation of the Greek corpus, however, it is the first one, to Islamic civilization, that displays a pattern of flourishing quite like what Renaissance Europe was now going through. The Golden Age of nature-knowledge in Islam was made up of three components: (1) translation and appropriation of substantial portions of the Greek corpus, (2) the partial enrichment of that corpus, in such a way as to preserve conformity with its broad frame and standard procedures, and (3) development of some pieces of nature-knowledge marked by the culture of Islam. *Between c. 1450 and c. 1600, the state of nature-knowledge in Europe looked overall quite similar*. True, in Renaissance Europe recovery and translation of manuscripts from the Byzantine treasure trove had restored the Greek corpus to even fuller measure, particularly in natural philosophy. True, too, pieces of nature-knowledge marked by European culture were far more numerous and also more coherent in expressing, not so much direct requirements of the faith as rather, indirectly, such outer-directed orientations as had come to perfuse it. Still, the broad pattern is quite alike – in both we find very similar processes of enrichment and, above all, a similar throng of highly gifted practitioners excited at the chance to master a novel world of learning. This ties in with the standard criterion for a Golden Age in nature-knowledge first defined on p. 28 – "a unique, uncommonly dense constellation of luminaries at work on a high-quality level of performance, not to be surpassed in the sequel unless incidentally, by one or another lone practitioner". Hence, our observer's report must sum up in its first part how far, by the end of the 16th century, these European luminaries had carried their reexamination of the Greek corpus.

In view of the high fences between the three distinct modes of nature-knowledge concerned
– Athens, Alexandria, and Europe's 'third mode' of coercive empiricism – it is impossible
for the report to focus on just one individual who encompasses all that had been attained
by the turn of the 17th century. The report must examine the achievement of several men,
each of whom represents the pinnacle of the specific mode of nature-knowledge in which
he was engaged. For Alexandria this is *Giovan Battista Benedetti*, customarily handled by
historians as Galileo's immediate 'forerunner'. For nature-knowledge colored the European
way *Vincenzo Galilei* (Galileo's father) is the man to select. For Athens it is harder to make a
choice. The principal new feat was recovery of all those philosophies (Platonism, stoicism,
atomism, skepticism) that had remained outside the realm of Aristotle-drenched, medieval
nature-knowledge. But what came out of the event was just restoration to their ancient integrity, without much by way of substantive enrichment taking place. Among Renaissance
Aristotelians, however, some managed to respond in a creative manner to the various challenges their doctrine found itself exposed to now that its near monopoly had come to an
end. Not only did scholastic disputations give way at many universities to more flexible,
humanism-inspired ways of teaching the same old doctrine, but a few far-sighted Aristotelians took a more daring path and sought to stretch the doctrine to its apparent limits in
an effort to come to terms with the two non-Athenian rivals newly emerging: Alexandrian
mathematical science on the one hand, and Europe's third mode of coercive empiricism on
the other. The man to seek to enrich Aristotle's doctrine with a dash of mathematical science
was *Christoph Clavius;* the man to seek to enrich it with a dash of empiricist magic was *Jean
Fernel*.

With the stage thus set for these four men whose accomplishments sum up the enrichment the Renaissance episode had brought to nature-knowledge over a century and a half,
we are ready to raise the curtain for the first, the 'upswing' part of our observer's report.

Giovan Battista Benedetti served as a court mathematician at Parma and then at Turin.
He taught mathematical science, drew horoscopes, gave counsel on public works both military and civil, and wrote treatises on sundials and on perspective. His innovative accomplishment does not reside in such court-inspired work of a practical bent but rather took its
departure in a search for alternative solutions to problems in Euclidean geometry. Soon he
turned to problems from outside mathematical science (most often from Aristotelian doctrine), which he then furnished with novel solutions, reached more often than not by applying Archimedean principles. His commitment was to Alexandria, not to Athens, and we do
not find him unduly disturbed by the circumstance that his exploitation of such principles
often caused him to contradict some Aristotelian tenet. One example of the kind of enrichment Benedetti brought to the Alexandrian corpus is an account of falling bodies by means
of Archimedes' concept of buoyancy. The outcome was at substantial variance with how
Aristotle accounted for the quickening of an object's descent as the density of the medium
decreases.

144 Benedetti took up each problem separately and solved it separately, without an effort made to seek for underlying coherence. What innovation any solution of his might contain remained confined therefore to the individual problem at hand. In that manner he produced an account of acceleration in free fall in terms of increments of impetus and an account of hydrostatic pressure. His partly novel account of consonance serves particularly well to show how far enrichment did and did not go in Benedetti's hands. Neither in antiquity nor later had the Pythagorean account of consonance been linked up with either the atomist or the stoic account of how sound propagates (p. 20). In summarily treating consonant intervals as arising from water-like airwaves, Benedetti began to give some real-world content to what had so far been handled as just a numerical regularity. Thus did Benedetti cross a boundary left untouched even by Zarlino and Salinas. The consequences were negligible nonetheless. Benedetti confined the conclusion that he drew from his lapidary account to just another play with numbers, rather than going for an all-out analysis of consonance in terms of vibrations of the air once moved by a string or pipe. Neither here nor elsewhere does his work display any awareness of the retrospectively promising lines of thought it was fraught with. It fell into immediate neglect, and only historians are in a position to perceive what genuine novelty its author so expertly hid from every reader's eyesight, his own included.

Vincenzo Galilei, a 16th-century lutanist and cloth merchant who married into the lower Tuscan nobility, exemplifies another pinnacle. In the 1580s he undertook a number of experiments to determine how pitch depends upon string tension and material. At issue was an ongoing debate over whether, in musical judgment, priority must go to 'reason' or to 'sense'. There was, on the one hand, the Pythagorean analysis of music in terms of the ratios of the consonant intervals that form the backbone of musical harmony. There was, on the other hand, Aristoxenos' principal point that our auditory experience of the flow of the melody cannot be reduced to any allegedly underlying principle. Ptolemy had sought a compromise solution (p. 25). From then on the center of gravity in musical analysis shifted increasingly from 'ratio' (focusing on mathematical regularity) to *sensus* (or, in contemporary parlance, the 'judgment of the ear'). The shift intensified by mid-16th century. What in effect comes down to the carving out of a distinctly aesthetic realm in musical theory was advocated with particular cogency by a range of 'musical humanists', members for the most part of a humanist Florentine circle well known among historians of early Baroque music as the Camerata. Vincenzo was drawn into the group, probably for his combination of practical expertise and extraordinary knowledge of musical theory.

Stimulated by this circle, Galilei became more and more aware of trouble with the grand musical-theoretical synthesis in which his teacher Zarlino had enveloped Ptolemy's compromise solution (p. 25). Zarlino had succeeded in incorporating the new, non-Pythagorean consonances in regular usage in European polyphony by deriving the full set of consonant intervals, no longer from Pythagoras' *tetraktys,* or 1–4 range of consonance-generating inte-

gers, but from the *senario,* or 1–6 range of consonance-generating integers. But that solution
appeared to entail a wholly unstable and therefore impracticable tonal scale. Vincenzo began
to realize that on this very point Aristoxenos could help out both the theorist and the musical practitioner. In vigorous, at times shrill polemic he set out to demolish Zarlino's way of
blending numerical ratios with the judgment of the ear. In the ensuing dispute he increasingly moved away from theoretical a priori pronouncements toward appeals to empirical
evidence, gained from the practice of singers and instrumentalists. Vincenzo did so under a
dual banner of humanist provenance: (1) Aristoxenos' recently printed text, which expressed
the author's views more freshly and extensively than Boethius' brief and rather partisan
summary had, and (2) a determination, of central concern to the antiquity-revering members of the Camerata, to re-create the sound of ancient Greek music (it is along this route
that the first operas emerged).

The pattern is familiar – here, too, a point of departure taken in an ancient text branches
out in newly empiricist directions. In Vincenzo's case this led in the end to an original, sustained, experimental effort to undercut the generally accepted basis of Zarlino's synthesis.
That basis was none other than the fundamental, Pythagorean observation, never contested
by anyone but Aristoxenos and his rare adherents, that our sensual experience of musical
intervals and chords is governed by their numerical ratios. In Zarlino's schema, the ratios
inside the *senario* represent the consonances (e.g., the major third 4:5); those outside, the
dissonances (e.g., the whole tone 8:9). What experimental discoveries, then, enabled Vincenzo to contradict all this hoary wisdom?

> Testing strings of various materials, Vincenzo found that to produce a true unison, two strings
> had to be made of the same material, of the same thickness, length, and quality, and stretched
> to the same tension. If any of these factors was absent, the unison would be only approximate.
> Moreover, he discovered that if a lute were strung with two strings, one of steel and one of
> gut, and these were stretched to the best possible unison, the tones produced by stopping the
> strings at various frets would no longer be in unison.[84]

And there was another way in which Vincenzo managed to sever the established link between 'reason' and 'sense' by showing up its less-than-universal validity. He discovered that
if one compares, not the *lengths* of strings, but rather their *tensions,* the traditional ratios
no longer hold. If one and the same string is successively stretched by different weights, in
order to get the octave one must suspend from the string, not a weight twice as heavy as the
first, but four times as heavy. Similarly, in order to get the fifth the weights must be in a 4:9
$= 2^2:3^2$ ratio. In terms of string tensions, then, the intervals are in a *squared* proportion to
the weights; hence, what ratios appear depends on what one decides to measure, and the
simplicity of both Pythagoras' *tetraktys* and Zarlino's *senario* dissolves (a few decades hence
to be reconstituted, in an ironic twist, by Vincenzo's own son; p. 301).

146   It is true that, in a vain hope to make his big find symmetrical, Vincenzo spoiled it by pronouncing an insufficiently checked, all-too-fanciful, *cubed* proportion for the volumes of
organ pipes. But never mind, the point to establish in bringing up Vincenzo's achievement
is not some flawless execution of whole ranges of novelty-directed experiments in the frame
of coercive empiricism but rather their naturally fallible, incipient emergence as its most
promising outgrowth.

We turn next to the two men who represented 16th-century Aristotelianism at its most
creative, Clavius and Fernel. *Christoph Clavius,* a widely respected author of widely read
textbooks was a Germany-born, Jesuit priest engaged between the 1560s and his death in
1612 in a successful campaign to give mathematics a more prominent place in the standard set of Jesuit courses. The education of Europe's elites with a view to strengthening
the Counter-Reformation was among the main objectives of the Jesuit Order. Therefore,
Clavius' campaign was of considerable significance for the future place of mathematics in
European thought. The Jesuit curriculum was grounded in Aristotle's teachings as interpreted by Thomas Aquinas. Aristotle's scheme of things contained at its margins a rather
obscure domain of scholarship known as 'mixed mathematics'. It stemmed from his distinction between four categories of change (p. 6). Quantitative change, which alone was held
to be suitable for mathematical treatment, was placed on a par with the other categories
('generation and corruption', qualitative change, and change of place). Natural philosophy
being concerned with the real world, quantitative change happened only to those relatively
few aspects of reality that lent themselves to computation; it was unthinkable for a quality
like, for instance, heat to be likewise quantified. Among the eldest components of 'mixed
mathematics' was a quite un-Archimedean treatment of the law of the lever found in an
early Aristotelian treatise on 'Mechanical Problems' and soon misattributed to Aristotle
himself (p. 336). In the heyday of scholasticism some further work in this vein served as
the most prominent case by far of 'mixed mathematics' (although the Alhazen tradition in
light and vision was sometimes reckoned to belong to it as well). To Clavius, by the 1560s, it
occurred that this did not nearly go far enough to meet the humanist challenge, which (as
we saw with Regiomontanus) involved an effort by mathematical scientists to regain intellectual and social territory lost during the reign of the Schoolmen. Clavius, an accomplished
mathematical humanist himself, sought to meet the challenge halfway by widening 'mixed
mathematics' and by making it more prominent. A case in point was his preparation of a
Latin edition of Euclid's *Elements.* As faithful Aristotelians, the Schoolmen had felt that syl-
logistic reasoning leads much more securely to indubitably true, empirical conclusions than
mathematical proof can. Therefore, they had refashioned Euclid's proofs to fit the mold of
the syllogism. Clavius now prepared an edition such that the two modes of proof came to
look equivalent. Another way to bring as much mathematical science as he dared under the
purview of natural philosophy rested in what has been called a "rigorous exploitation of

the quantitative characteristics of manifest experience".[85] These he encountered, not only in
balance problems and in the treatment of light and vision, but also in planetary theory, in
the astrolabe, in the calendar, in sundials, in perspective, and in musical consonance; all of
which subjects he expounded in voluminous textbooks. Thus was revived and expanded in
the grand manner an intermediary branch of learning meant henceforth to cover as much
of Alexandria as could be borrowed from it while preserving intact the Aristotelian variety
of Athens. That is, Clavius remained a philosopher, a person to whom the true state of things
in the world is revealed by way of secure possession of solidly true first principles, with the
mathematics added solely to fill in the quantitative details. All this is the very opposite of
what Benedetti was striving for, who was concerned with subjecting selected Athenian statements to rigorously Alexandrian critique. Two centuries earlier, Oresme and the Calculators
had stretched the category of quantitative change to the limit still within reach of Aristotelian doctrine in the absence of inspiration from a rival in mathematical science. With that
rival meanwhile making its way back at full speed from its previous, enforced retreat, Clavius
now set out for the limit of what Aristotelian natural philosophy could still absorb from that
rival without losing its defining character as natural philosophy in the full, Athenian sense.
In this way Clavius laid out a path, of proven attraction already during his lifetime, along
which to bring highly abstract Alexandria in somewhat closer touch with the real world.

Jean Fernel was not only a philosopher but also a medical man. He lived in the first half
of the 16th century and ended his career as physician to the king of France. In *On the Hidden Causes of Things* (1542, *De abditis rerum causis*) he extended his native Aristotelianism
in the direction, not of mathematics as with Clavius, but of natural magic. In Fernel's view
the doctrine of the four bodily humors, pioneered by Hippokrates and adopted by Galen in
view of its close correspondence to Aristotle's doctrine of the four elements (p. 16), was woe-
fully inadequate to account for diseases of three particular kinds: those arising from poison,
from contagion, and from pestilence. How in the world could their sudden, often lethal effects and (with the second and third) vast spread among the population be accounted for in
terms of one's bodily humors getting out of balance?

The idea to ascribe these three unexplained types of disease to occult causes was hardly
new at the time. But Fernel preferred to delve more deeply. He undertook a profound philosophical inquiry, with a view to giving the occult its deserved place in the predominant,
Aristotelian conception of the world. By means of a convoluted analysis of the twin notions
of 'matter' and 'form' at the very heart of Aristotle's philosophy of nature, Fernel sought to
pin down how form comes to inhere in matter. Along the way he invoked several remarks by
Aristotle on generation, all to the effect that for life to arise matter and form require a third
principle, variously called 'the Sun' or 'heaven ' or 'the First Mover'. Fernel took this to mean
that the process of in-formation of matter so as to constitute the 'substantial form' (or 'total
substance' in Galen's phrase) takes place from above, from the heavens. In Fernel's treatment

148   this was not just standard astrological fare, with its assumption of planetary angles or other tangible effects acting as the decisive vehicles of heavenly influence. Rather, it was based on a conception of hidden, heavenly forces busily at work through a network of hidden correspondences that operate through attraction and repulsion, sympathy and antipathy, to influence what happens on Earth (p. 133). Due to these heavenly influences, "some diseases of the total substance exist, which can corrupt the form of the body." Or, in paraphrase, "these diseases do not operate by disturbing the balance of the humours, they attack the body's form; that is to say they attack what makes the body what it is."

While granting that the occult qualities thus invoked are not liable to further explication in terms of the natural philosophy that carried his unswerving allegiance, Fernel nonetheless insisted upon their reality, which empirical research could and should help establish. In paraphrase, interspersed with literal quotation:

> Peony ... fastened to the neck of infants cures epilepsy. This must be due to its total substance because if it healed by drawing off phlegm, which is the supposed cause of epilepsy, it would be effective against other diseases where excess phlegm was the trouble. What's more, medicaments which do draw off phlegm are usually ineffective against epilepsy. Consequently, "This power [peony] possesses emanated not from the manifest qualities, but from a splitting up and repugnance of the total substance, by which it entirely antagonises not phlegm, not sluggishness of humour, but the essence of epilepsy, which is unseen and occult." Sympathies and antipathies are always regarded as highly specific in their operation, and Fernel's introduction of them into his discussion demonstrates his belief that, with regard to the occult diseases, humoral therapies are inadequate, and that what is required are specific remedies, capable of acting directly on specific diseases.[86]

What we have here, then, are all the trademarks of Europe's third mode – to investigate in the empirical way nature's manifest or hidden phenomena, with a view to benefiting from one's findings in practice. However, all this is coupled here to natural-philosophical analysis in the trusted, Aristotelian manner (or so Fernel thought). Fernel pioneered the approach; Fracastoro, Cardano, and the elder Scaliger were to follow in his wake later in the 16th century. The entire approach was in due time (the 1640s) to become part and parcel of a full-fledged, Jesuit worldview (p. 495; 517), complementing the other, mathematical enlarging upon sound Aristotelian doctrine that Christoph Clavius promoted.

Downturn ahead?   If we recall the vast translation effort that commenced after the fall of Byzantium and then survey the various enrichments its various products had by 1600 undergone at the hands of a vast collection of innovative practitioners – not only Benedetti and Vincenzo Galilei and Clavius and Fernel but also (again to list only the most prominent)

Bock, Brunfels, Castro, Copernicus, Fuchs, Garcia de Orta, Leonardo, Nunes, Paracelsus, Ra- 149
mus, Regiomontanus, Stevin, Tycho, Vesalius, Zarlino – the conclusion to be drawn from the
above report seems ineluctable. Yes, by 1600 Renaissance Europe was indeed experiencing
its Golden Age of nature-knowledge. The more difficult question to ask is, of course, What
next?

For clarity's sake, we have allowed our observer to sprinkle through her account a few
glimpses of things to come. But if we forget about these, so as to make our observer stare
into a (by 1600) wholly unknown future, what plausible course of future events is she to
predict in the closing section of her report? For her to make her guess as much an educated
one as she can, the only move open to her is to infer by past analogy and to prophesy by
extrapolation – that is, to anticipate more of the same along the lines reported on.

With the analogy of one more Golden Age of nature-knowledge thus sufficiently established, we complete it by means of one more extension of our familiar upswing/downturn
schema (n° 3).

Schema 3: Upswing and downturn compared: The whole picture

	Greece	*Islamic civilization*	*Medieval Europe*	*Renaissance Europe*
UPSWING	creation + (in mathematics) transformation	translation ⇨ enrichment	translation ⇨ Aristotelian enrichment	translation ⇨ enrichment
• Golden Age	Plato to Hipparchos	al-Kindi to Ibn Sina+al-Biruni+Ibn al-Haytham	Albert the Great to Oresme	Regiomontanus + Leonardo to Stevin + Vincenzo Galilei + Clavius
DOWNTURN	c. 150 BCE	c. 1050 (AH 430)	c. 1380	???
1) Why?	'normal'	'normal'	'normal'	'normal'
2) Why then?	skeptical crisis? cessation of patronage?	invasions ⇨ inward turn	possibilities for enrichment exhausted	possibilities for enrichment exhausted?
3) Aftermath	steep, protracted; generally low level	steep, protracted; followed by partial, higher-level reversal	steep, protracted; unreversed	---
• marked by	codification, commentary, syncretism ...	commentary	commentary	---
• still outstanding	Ptolemy, Diophantos; Proklos, Philoponos	Nasir ed-Din al-Tusi, Ibn as-Shatir; Ibn Rushd	none	---

150   Once again, the conclusion to be drawn from the analogy imposes itself. As in the two previous cases of transplantation of the Greek corpus of nature-knowledge, so now in Renaissance Europe, too, a fairly steep downturn is most likely to follow, this being after all the natural occurrence in traditional society. There is no inherent reason whatever for why the Renaissance-European upswing should in the end have escaped the destiny of every previous, large-scale endeavor to attain knowledge of nature, that destiny being to come to a standstill at some point. All previous experience ought to lead our observer to expect the Golden Age of European nature-knowledge still to continue for a while on its own, by now well-beaten track but sooner or later to come to an end as its Hellenist, its Islamic, and also its medieval predecessors had. Not so certain is, of course, precisely when the downswing was to set in; even less certain, *why* it was to happen *when* it was to happen. One possible answer to the latter question is suggested in Schema 3: the possibilities for further enrichment may have been exhausted, or almost. With respect to this suggestion, our observer's report is bound to be equivocal.

For as long as we make her consider Alexandria alone, the inference seems plausible enough. Some of the routes taken toward enrichment in Renaissance Europe differed from those taken in Islamic civilization, yet on the whole the outcome by 1600 was quite similar: a range of incidental, specific steps that went beyond where Euclid, Archimedes, and Ptolemy had left off. Even in Copernicus' case no break occurred in the familiar pattern of those five abstract mathematical sciences operating at several removes from reality. A contemporary feat – two highly gifted young men striving in the final decade of the 16th century to enrich the Alexandrian corpus in remarkably innovative ways – goes far to confirm this hunch of a dead end in the making. Indeed, suppose our observer's report to turn from Benedetti's mature accomplishment to a certain Florentine nobleman in the act of wrestling in vain with the problem of falling bodies treated the Archimedean way (1588–1591), or to how in 1595 a certain German/Austrian schoolmaster was fleshing out a Copernican universe with abstruse geometrical figures. If she included these young men in her report, she would have been bound to infer that, with young Galileo getting hopelessly stuck in an apparently impossible enterprise and young Kepler even more hopelessly flying off at a tangent on the wings of his excessive fancy, ongoing enrichment of Alexandria had come close to exhausting its developmental potential.

This much could be predicted with quite a measure of confidence, but a more hopeful tone of voice may be expected to fill the final pages of our observer's report. Ongoing stretching of Aristotelian doctrine in the direction of mathematical science, as pioneered by Clavius and taken up in the late 1590s already by some younger Jesuits, may well point toward an extended flourishing of 'mixed mathematics'. Just so, ongoing stretching of Aristotelian doctrine in the direction of natural magic, as pioneered by Fernel and then pursued by several others, has already been flourishing for some time without a natural end to such enrichment being anywhere near in sight. But what is bound to inspire our observer most of all is the third mode of coercive empiricism and its confident, forward-looking dynamism. If indeed by 1600 the Golden Age of European nature-knowledge is not over yet, ongoing discovery in this specifically European vein, interspersed every now and then with bouts of experimentation, is what may plausibly be expected to continue for some while.

This, then, is perforce how our observer's report ends, so perceptive in its drawing sensible conclusions from a large-scale analogy with the past of nature-knowledge, followed by cautious extrapolation from current trends. It is not her fault that these cautious, sensible conclusions turn out to be dead wrong. No contemporary observer, however perceptive we make her or him, could by 1600 have predicted what was to occur indeed – a veritable revolution or, rather, three revolutions or, rather, three revolutionary transformations. Nor could even the most perceptive of observers have foreseen the big paradox at the heart of these revolutionary transformations. For the largest break of them all occurred at the very spot least expected even to keep flourishing. It occurred inside a movement directed not forward but backward, not toward an unknown future confidently faced and actively prepared as it was in the third mode, but rather toward gradual extension along lines foredrawn by the Greeks – toward enrichment, that is, of the *Alexandrian* corpus. Indeed, the most radical transformation to mark the onset of the Scientific Revolution occurred at the hands of two men whose intellectual upbringing was profoundly rooted in the mathematical approach to phenomena undertaken in prominent fashion by their respective scientific forebears, Archimedes and Ptolemy. These men are none other than our two youngsters of the 1590s, Galileo Galilei and Johannes Kepler. Mathematical humanists themselves, they moved on from where we summarily left them in 1592 and 1595, respectively, and burst the limits of their Alexandrian heritage and turned it into something almost unrecognizable – into newly *realist*-mathematical science.

The third transplantation conceptualized. As Noel M. Swerdlow has insisted in his 'Science and Humanism in the Renaissance: Regiomontanus' Oration on the Dignity and Utility of the Mathematical Sciences'. In: P. Horwich (ed.), *World Changes. Thomas Kuhn and the Nature of Science*. Cambridge, Mass.: MIT Press, 1993, pp. 131-168, the restoration of the Greek corpus during the Renaissance is seldom treated as a distinct episode in the history of science. P.L. Rose, *The Italian Renaissance of Mathematics. Studies on Humanists and Mathematicians from Petrarch to Galileo*. Geneva: Droz, 1975, makes the same point more implicitly. Rose's book also gives the most detailed account yet available of the movement of translation from Greek into Latin, which Scott Montgomery (cited in previous Notes) failed to add to his successive accounts of Rome, Jundishapur, Baghdad, and Toledo. I have discussed Rose's book in *SRHI*, section 4.4.2., with much else in the present chapter going back to *SRHI*, section 4.4. As one consequence of this near-universal lack of conceptualization of 'Renaissance science' as a distinct episode in the history of the reception of the Greek corpus, I had to piece together the component parts of the present chapter from a large variety of resources, as follows:

Athens. *SRHI* 4.4.3.-5. for Renaissance Aristotelianism, Platonism, and skepticism.

Alexandria:

- **Regiomontanus' praise of the mathematical sciences:** account based upon Swerdlow's paper just cited.
- **Stevin**. E.J. Dijksterhuis, *Simon Stevin*. The Hague: Nijhoff, 1943 (this is still the standard biography in Dutch; a poorly translated, condensed version appeared posthumously in 1970 with the same publisher under the title *Simon Stevin. Science in the Netherlands around 1600*).
- **Copernicus.** My treatment was inspired most by two works in the voluminous literature that discuss his achievement from a deliberately nonmodernizing point of view: T.S. Kuhn, *The Copernican Revolution. Planetary Astronomy in the Development of Western Thought*. Cambridge, Mass: Harvard University Press, 1957 (discussed in *SRHI*, 4.4.1., and usefully assessed in view of later historiographical developments by R.S. Westman in his 'Two Cultures or One? A Second Look at Kuhn's *The Copernican Revolution*'. *Isis* 1994, 85, pp. 79-115) and N.M. Swerdlow and O. Neugebauer, *Mathematical Astronomy in Copernicus's De Revolutionibus*. New York: Springer, 1984; 2 vols.; pages 3-85 in particular (a neat summary of the principal technical problems that Copernicus faced is in N.M. Swerdlow, 'An Essay on Thomas Kuhn's First Scientific Revolution, *The Copernican Revolution*'. *Proceedings of the American Philosophical Society* 148, 1; March 2004; pp. 64-120). The implied fictionalism of books II–VI of *De Revolutionibus*, which is often ignored or glossed over in the nonspecialist literature, is stressed in section IV–1A of E.J. Dijksterhuis, *The Mechanization of the World Picture*. Oxford: Oxford University Press, 1961 (original Dutch: 1950). To what extent one may validly speak of 'fictionalism' in this regard is an issue illuminatingly treated by Nicholas Jardine, *The Birth of History and Philosophy of Science. Kepler's* A Defence of Tycho Against Ursus With Essays on Its Provenance and Significance. Cambridge UP, 1984.

Besides a daily rotation and an annual orbit Copernicus attributed to the Earth a third motion, which he thought he needed to keep its axis tilted parallel to itself and which he used to explain precession. With his new conception of motion Galileo showed it to be superfluous, and I have left it out of my account so as not to complicate treatment of Copernican heliocentrism still further.

Rheticus' theological defense of the Copernican hypothesis, long thought lost, was rediscovered by R. Hooykaas. He published it with an extensive commentary in *G.J. Rheticus' Treatise on Holy Scripture and the Motion of the Earth*. Amsterdam: North Holland, 1984.

Copernicus' 'monster' passage has been shown by Robert S. Westman (in his 'Proofs, Poetics, and Patronage: Copernicus's Preface to *De Revolutionibus*'. In: D.C. Lindberg & R.S. Westman (eds.), *Reappraisals of the Scientific Revolution*. Cambridge: Cambridge University Press, 1990; pp. 167-205) to be a rhetorical flourish common at the time. Still, a given expression need not lose its significance for being shopworn. Here is how Galileo used it in his *Dialogo*:

> there resulted a monstrous chimera composed of mutually disproportionate members, incompatible as a whole. Thus however well the astronomer might be satisfied merely as a calculator, there was no satisfaction and peace for the astronomer as a philosopher.

[NB: The translation is as given by Stillman Drake on p. 341 of his widely read, 1953 English translation of the *Dialogo*. In rendering *filosofo* as 'scientist' rather than 'philosopher' Drake obscured the very distinction Galileo was insisting upon here. This is just one example of a more general translation habit I discuss at greater length in endnote 101 and below on p. 219].

Conceptualization of Europe's 'third' mode of nature-knowledge. The two historians most intent to mark as distinct the mode of nature-knowledge here dubbed 'coercive empiricism' have been R. Hooykaas and Allen G. Debus (see below for more specific references). Over the past decade the episode has been receiving increasing attention, for instance, in Brian Ogilvie, *The Science of Describing. Natural History in Renaissance Europe*. Chicago: University of Chicago Press, 2006 (in his introduction he oddly claims priority for signalizing its thoroughgoing empiricism), or in Lissa Roberts, Simon Schaffer, Peter Dear (eds.), *The Mindful Hand: Inquiry and Invention From the Late Renaissance to Early Industrialisation*. Amsterdam: Edita, 2007.

· **Leonardo.** ARTIST/SCIENTIST: My views have been influenced most by those of the art historian Martin Kemp in an exhibition catalog he produced together with J. Roberts and P. Steadman, *Leonardo da Vinci*. Yale University Press, 1989, and also in a CD-ROM edition of Leonardo's 'Codex Leicester' in possession of the computer tycoon Bill Gates, coproduced under the 'Corbis' label. ENGINEER/SCIENTIST: Bertrand Gille, in his *Les ingénieurs de la Renaissance*. Paris: Hermann, 1964, provides a healthy corrective to all-too-inflated claims about Leonardo as an inventor of absolutely *sans pareil* ingenuity; Ladislao Reti in *DSB* 8, s.v. Leonardo da Vinci, pp. 206-214, details in what specific respects he nonetheless was one.

· **Herbals, bestiaries, anatomical atlases.** A.G. Debus, *Man and Nature in the Renaissance*. Cambridge UP, 1978.

154

- **Tycho.** Victor E. Thoren, *The Lord of Uraniborg. A Biography of Tycho Brahe.* Cambridge: Cambridge UP, 1990.
- **Mapmaking.** Brief synthetic surveys in ch. 20 'Ptolemy Revived and Revised' of Daniel J. Boorstin, *The Discoverers. A History of Man's Search To Know His World and Himself.* New York: Random House, 1983, and David N. Livingstone's 'Of Myths and Maps. Geography in the Age of Reconnaissance': ch. 2 of his *The Geographical Tradition.* Oxford: Blackwell, 1992.
- **Collections.** Paula Findlen, *Possessing Nature. Museums, Collecting, and Scientific Culture in Early Modern Italy.* Berkeley: University of California Press, 1994.
- **Nature-knowledge and the crafts prior to the Scientific Revolution.** *SRHI* 3.4.3., 5.2.1., 5.2.6. (referring to Olschki and Koyré).
- **Wisdom on the streets.** *SRHI* 5.2.6. (referring to Hooykaas and Rossi).
- **Practical mathematics.** *SRHI* 5.2.1. (referring to Olschki). On the complexities surrounding the origins of linear perspective I owe much to Jehane R. Kuhn's article 'Measured Appearances: Documentation and Design in Early Perspective Drawing', *Journal of the Warburg and Courtauld Institutes* 53, 1990, p. 114-132.
- **Paracelsian iatrochemistry.** Discussion based upon A.G. Debus, *The Chemical Philosophy. Paracelsian Science and Medicine in the Sixteenth and Seventeenth Centuries.* 2 vols. New York: Science History Publications (Watson), 1977, and idem, *Man and Nature in the Renaissance.* Cambridge: Cambridge UP, 1978. On the varied origins of Paracelsian iatrochemistry in Taoist, Alexandrian, Arabic, and medieval European thought and practice, there is a grand summary in J. Needham, *Science and Civilisation in China,* vol. 5, part 4, section 33 (i) 'Comparative Macrobiotics', with much on Chinese and Alexandrian alchemy on pp. 324-388; on Chinese and Arabic alchemy on p. 388-491, and on 'Macrobiotics in the Western World' on pp. 491-507.
- **Hermeticism and natural magic.** *SRHI*, 3.3.1.-5. and 4.4.4. (referring to Yates and her numerous critics and commentators); also a neat summing up of late Renaissance natural magic in John Henry, *The Scientific Revolution and the Origins of Modern Science.* Hampshire/London: Macmillan, 1997.
- **Secrets of nature.** William Eamon, *Science and the Secrets of Nature. Books of Secrets in Medieval and Early Modern Culture.* Princeton, NJ: Princeton University Press, 1994.
- **Renaissance art.** A direct, highly rewarding way to catch what was so special about the course uniquely taken in European art is to pay an extended visit to the Museum of Fine Arts in Boston, starting with the full range of its wonderful Asian collections and going on to compare these, first, with the European-medieval collection and then with what one finds in the hall of Renaissance painting and beyond. The former comparison brings out the profound similarities; the latter, the quickly intensifying differences.
- **Voyages of Discovery**. A summary of R. Hooykaas' numerous writings on the impact of the Portuguese Voyages is in *SRHI*, 5.8.2.; see also Onésimo T. Almeida, 'Portugal and the Dawn of Modern Science'. In: G.D. Winius (ed.), *Portugal, the Pathfinder. Journeys from the Medieval toward the Mod-*

ern World 1300 – ca. 1600. Madison, 1995. Needham's tendentious yet brilliant comparison between the Portuguese and the Chinese voyages is engagingly retold in Colin Ronan 's authorized *The Shorter Science and Civilisation in China. An Abridgement of Joseph Needham's Original Text*. Vol. 3 (A Section of Volume IV, Part 1 and a Section of Volume IV, Part 3 of the Major Series), Cambridge UP, 1986; pp. 128-152.

- **Goethe's 'tender empiricism'** is discussed at some length in P.F.H. Lauxtermann, *Schopenhauer's Broken World-View. Colours and Ethics Between Kant and Goethe*. Dordrecht: Kluwer Academic Publishers, 2000; especially on p. 36 and p. 63.
- **Castro**. Text based upon R. Hooykaas, *Science in Manueline Style. The Historical Context of D. João de Castro's Works*. Coimbra: Academia Internacional da Cultura Portuguesa, 1980 (separate edition of pp. 231-426 in vol. 4 of: A. Cortesão & L. de Albuquerque (eds.), *Obras Completas de D. João de Castro*, Coimbra, 1968-1980; 4 vols.). A brief account of this book was given by Onésimo T. Almeida in a lecture of March 18th, 1997, 'R. Hooykaas and his "Science in Manueline Style" – The Place of the Works of D. João de Castro in the History of Science'.

Fictitious report. The artifice here adopted of highlighting the unpredictability of a particular historical event by contrasting it with what an imagined, 'enlightened observer' would have had every reason to expect at the time has been effectively employed by Donald S.L. Cardwell on pp. 73-75 of his *Turning Points in Western Technology*. New York: Neale Watson, 1972.

- **Benedetti.** Treatment based chiefly upon Stillman Drake's lemma in *DSB* 1, pp. 604-609 and on ch. 7 of P.L. Rose's book cited above; further on my own contribution 'Benedetti's Views on Musical Science and Their Background in Contemporary Venetian Culture', in: *Cultura, scienze e technice bella Venezia del Cinquecento. Atti del Convegno Internazionale di Studio 'Giovan Battista Benedetti e il suo tempo'*. Venezia: Istituto Veneto di Scienze, Lettere ed Arti, 1987; pp. 301-310, and on other papers in that collection, notably Mario di Bono, 'L'astronomia Copernicana nell'opera di Giovan Battista Benedetti' (pp. 283-300).
- **Vincenzo Galilei.** In *Quantifying Music* (1984), pp. 34-45, I have explained the musical-theoretical problem underlying the dispute between Zarlino and Galilei. My own interpretation of the dispute is on pp. 78-85, with reference to D.P. Walker, 'Vincenzo Galilei and Zarlino', in idem, *Studies in Musical Science in the Late Renaissance*. London/Leiden: Warburg Institute / Brill, 1978, pp. 14-26, and to Claude V. Palisca, 'Scientific Empiricism in Musical Thought', in: H.H. Rhys (ed.), *Seventeenth Century Science and the Arts*. Princeton (NJ): Princeton UP, 1961, pp. 91-137. Further light has since been shed on the issue by Palisca in later work (summed up in his contribution 'Was Galileo's Father an Experimental Scientist?' to: V. Coelho (ed.), *Music and Science in the Age of Galileo*. Dordrecht: Kluwer, 1992, pp. 143-151), and by Michael Fend in 'The Changing Functions of *Senso* and *Ragione* in Italian Music Theory of the Late Sixteenth Century'; in C. Burnett, M. Fend, & P. Gouk (eds.), *The Second Sense. Studies in Hearing and Musical Judgement From Antiquity to the Seventeenth Century*. London: Warburg Institute, 1991, pp. 199-221. Fend's article caps a range of four consecutive papers jointly documenting the 'secular trend in musical judgment' I speak

156 of in the main text (cp. my review of the book in *Annals of Science* 51, 1994, pp. 427-429). A classic treatment of 'Musical Humanism' is in D.P. Walker's range of articles from 1941/2, reprinted in D.P. Walker (P. Gouk, ed.), *Music, Spirit and Language in the Renaissance*. London: Variorum, 1985. A no less classic treatment of the rise of the Baroque musical style out of the Camerata's mistaken efforts to re-create ancient Greek music is in chs. 1–2 of Manfred F. Bukofzer's *Music in the Baroque Era*. New York: Norton, 1947.

- **Clavius.** James M. Lattis, *Between Copernicus and Galileo. Christoph Clavius and the Collapse of Ptolemaic Cosmology*. Chicago: University of Chicago Press, 1994; Peter M. Engelfriet, ch. 2, 'Mathematics in Jesuit Context' of his *Euclid in China. The Genesis of the First Translation of Euclid's* Elements *in 1607 and its Reception up to 1723*. Leiden: Brill, 1998.

- **Fernel.** John M. Forrester & John Henry, 'Jean Fernel and the Importance of His *De Abditis Rerum Causis.*' Introduction to: Idem (ed., transl.), *Jean Fernel's On the Hidden Casuses of Things*. Leiden: Brill, 2005.

PART II

THREE REVOLUTIONARY TRANSFORMATIONS

V

THE FIRST TRANSFORMATION: REALIST-MATHEMATICAL SCIENCE

The unpredictable event to set the first of three revolutionary transformations in motion was that around 1600 the Copernican hypothesis began to catch on.

True, to a limited extent it had caught on during the previous half-century already. Between 1543 and 1600, numerous practitioners who did not take seriously the realist core Copernicus had intended for his hypothesis (p. 108), exploited it for computational purposes. Over the same period, a dozen men went so far as to adopt it as a real representation of the world. Some of the dozen were well-informed about astronomical issues inside a context of other, more pressing concerns (e.g., Benedetti or Stevin). Others were more-specialized, expert practitioners, like the German astronomer Michael Mästlin. Grounds and motives varied for adopting heliocentrism in its realist variety. Some adopted it wholesale; others in part only (e.g., daily rotation but not the annual orbit). But even if committed fully, none of these early Copernicans came even close to perceiving (to use a pointed expression coined by Kepler) 'how rich Copernicus was'.[87] Their acceptance of the Copernican hypothesis as a description of the real world or significant portions thereof remained in all but two cases inconsequential, being an 'opinion' to be held and defended but not (with the exception of the maverick outsider Giordano Bruno) something of striking import.

Two men changed this state of affairs (which might well have lasted indefinitely) in a decisive manner. They probed to the very depth the explosive potential of Copernican heliocentrism in its realist guise and proceeded to detonate the hidden time bomb, radically transforming ingrained habits of thought in the process. These men were, indeed, Galileo Galilei and Johannes Kepler. The former took it upon himself to make heliocentrism plausible in terms of terrestrial motion. The latter saw that a hidden simplicity was resting at the core of Copernicus' planetary theory, which if brought to light could perfect and make unassailable the harmonious structure of the solar system that Copernicus had perceptively claimed for it. Kepler (born 1571) set out on his journey in 1595, with its full momentum unleashed from 1600 on when he joined Tycho Brahe, through 1619 when he laid down to his own satisfaction the rules of harmony that govern the world, until 1621 when he finished publishing his textbook of a new astronomy; his journey essentially completed, he lived for nine more years. Galileo (born 1564) set out on his unwittingly parallel, yet differently aimed and differently pursued journey in about 1592, with high points reached in 1610 and 1632 and the culmination point – publication of the book that was to transform forever how we conceive of motion – not being arrived at until 1638, four years before he died. When these two men were done, they had delved up from the deep core components of a new astronomy and a new conception of motion on Earth, respectively. In thus bringing about an upheaval in two highly significant and visible branches of learning, they in effect turned Alexandria into 'Alexandria-plus'. That is, they brought about a revolutionary transformation of highly abstract mathematical science into still abstract but now also *realist* mathematical science. And in consequence of this already quite major feat, they laid the groundwork for one constitutive hallmark of our modern world – something that has been called 'the universe of precision'. The precise definition that the concept deserves must wait until we have watched it in actual operation over time. Suffice it to say here that by 'the universe of precision' I mean the ever-more-apparent capacity of abstract/realist-mathematical science to draw ever more domains of natural phenomena and of human experience into its own, ever-expanding realm (p. 724).

What Kepler and Galileo thus wrought constitutes the most radical turnabout in how to come to grips with nature and conceive of it since the early pre-Socratics. Unlike the pre-Socratics, however, they did not go about their investigations from scratch; instead, they were deeply rooted in the prehistory that I have reviewed in Part I. By 1600 Alexandrian mathematical science was experiencing a Golden Age for the third time, and now two individuals, upon fully absorbing its principal achievement, broke through its apparent limits. How do we measure their achievement, and how do we assess it?

Key to answering the first question is to determine with precision *what* was transformed and what was not. This is the burden of the two sections that follow. What conceptual innovations did Kepler and Galileo bring about; where exactly was the novelty located; how crooked or straightforward was the path they traveled to end up where they did? These two men gave a decisive twist to events – *what, then, did the twist consist of?* The full prehistory, as discussed in Part I, forms the baseline of the investigation – by c. 1600 so much had been accomplished already; in subsequent decades what was added to it? The prehistory presents us with the 'given', in ongoing *comparison* with which the extent of the revolutionary accomplishment can be measured. Only in setting forth what Galileo and Kepler did that al-Biruni or al-Birjandi or Benedetti or Copernicus did not can the measure of their achievement become clear.

But measurement alone is not enough; we also want to explain. Explanations in connection with the Scientific Revolution have been devised as a rule in too broadly comprehensive a manner (pp. xix; xxvi). Instead, we ought to make our explanations match as carefully as we can the historical phenomenon that stands to be explained, after first taking pains to define what that thing – each of three radical twists occurring in an already settled pattern – consisted in. The next step is to seek explanations. At first, these can be partial only, in that they must fit, not some undifferentiated whole, but rather each individual revolutionary

transformation distinctly. Not until we have exhausted the resulting range of partial expla-
nations will the time have come to find out what causal gaps may still appear to remain, so
as to end up with a residue of explanatory factors of more general scope (p. 205).

Johannes Kepler

Throughout his lifetime Kepler held the Greek corpus of nature-knowledge in high regard.
Traces of 'mathematical humanism' are certainly present in his work. He intended to publish
a Latin translation of Ptolemy's *Harmony*, and he did publish an attempted reconstruction
of three crucial but missing chapters of that book. Also, for all his vivid awareness of the
novelty of his accomplishment in planetary and harmonic theory, he regarded his own life's
work as a follow-up to the Greeks (Pythagoras, Plato, and most of all Ptolemy) and as attaining its significance in that setting.

Even so, Kepler's mathematical humanism was of a special kind. The broad conception
of the world to which it gave rise was pervaded by a peculiar set of Christian beliefs colored
yet not determined by his devout Lutheranism. God was the person central to Kepler's thinking throughout, in that he took it upon himself to decipher the code God had employed in
creating the world. From his student days onward Kepler embraced a Pythagorean conception of the harmony of the world along with a *Timaios*-like account of Creation. However,
he replaced Plato's demiurge, who *obeys* the geometric models of creation, with the biblical
God as an absolutely sovereign *handler* of such models. Here is how Kepler expressed the
basic layout of Creation:

> Geometry … , coeternal with God, and radiating in the divine Mind, provided God with the
> models according to which He could furnish the world such as to make it the best and the most
> beautiful and, in short, the most alike to its Creator.

> Geometry, from before the origin of things coeternal with the divine Mind, God himself …
> supplied God with the models to create the world, and together with the image of God it [i.e.,
> geometry] has passed into man.[88]

Since we are born in God's likeness we can rethink His thoughts and thus reconstruct what
those thoughts were at Creation; hence, we can find out how our world has been made and
therefore what it is like:

> Man is an image of his maker, and it may well be that on certain things that have to do with
> the adornment of the world his views are the same as God's. For the world partakes of quan-

tity, and there is nothing the human mind (a thing from above the world put into the world) understands better than those very quantities, for the grasping of which he has clearly been made.[89]

That is, God created the world according to geometric models, and He equipped us with an understanding of the geometry involved. This enables us to seek answers to two prime questions: *What* are those models, and in *what* empirical portions of nature has God hidden them?

The fundamentals of these answers were clear to Kepler right from the start. The geometric models God employed are the 'archetypes' of musical harmony, that is to say, the geometric figures and rules from which the consonant intervals ought to be derived. Once these harmonies have been reduced to their geometric origin, they can be found back in the world of phenomena at many levels and in many instances but most of all in the heavens. Geometry, musical theory, astrology, and astronomy are the four domains of investigation into which Kepler's mind was self-compelled to be drawn almost from the start. Together with some derivative problem areas (image formation in connection with astronomical observation, and chronology in connection with the problem of the duration of the world) these four branches of learning were to keep him occupied as a scholar throughout his life.

A Copernican with a difference.   The conceptions just outlined go far to explain with what irresistibility Kepler felt himself drawn to the Copernican hypothesis. If harmony, geometrically conceived, was the leading idea of God at Creation, then a planetary theory whose principal strength resided in its harmonious structure could not be far off the mark. Note that this rationale makes sense only for the Copernican hypothesis taken in the vein in which its originator had hoped to promote it – as a description of our real world. Only in its mostly decried, or ignored, realist sense could the Copernican hypothesis exert its momentous appeal on Kepler's mind.

There were additional considerations, to be sure. He was one of the few students to be exposed at the time to the Copernican doctrine in (more or less) realist fashion – the Arts Faculty at Tübingen University had Michael Mästlin as professor of astronomy. And the message fell on this budding theologian's willing ears in that, to Kepler's mind, the Sun held a significance far beyond its being just one heavenly body among others, albeit with the additional property of sending out light. In Kepler's view the Sun symbolized God-the-Father in an analogy with the Holy Trinity that extended to the outermost sphere of the fixed stars (God-the-Son) and to the space in between (the Holy Spirit). In a related view Kepler was soon to develop, the Sun served as a center of force (with the force radiating throughout this intermediary space) that held the planetary system together. Copernicus had waxed poetic over the central place that the Sun occupies in the simplified version of his hypothesis presented in book I of *De revolutionibus*:

In the middle of all sits Sun enthroned. For who in this most beautiful temple would place this lamp at another or better place than from where it can at the same time illuminate the whole?[90]

The patient reader of the detailed mathematical account of books II–VI would nonetheless find soon enough that, rather than the Sun itself, Copernicus placed the empty center of the Earth's eccentric orbit at the center of the universe, with the Sun doing nothing but lighting the whole from one side. Kepler's exalted view of the Sun made this subordinate, eccentric position from which to illuminate the cosmos quite inappropriate. So the first move in his long-drawn-out transformation of Copernicus' account of the heavens was to put the Sun back where he felt it properly belonged – at the very center of things celestial.

He did so in the first book he was to devote to this subject of perennial fascination to him – the structure of the universe. *Mysterium cosmographicum* (The Secret of the Universe) of 1596 is the book alluded to in the imaginary report on the state of European nature-knowledge around 1600 (p. 150). It accomplished many things at once. It presented the most cogent argument in favor of the Copernican hypothesis in its realist guise yet put forward. In so doing it almost tacitly set straight a number of points on which Copernicus, in this gifted beginner's view, had been mistaken, such as this matter of the center of the universe. It further raised three wholly new questions about the structure of the universe. It answered two of these in a tone of jubilant conviction at having at long last resolved the mystery of the world. And it went about the job of checking the figures in play against values derived from the observational tradition in a manner displaying some lingering doubt over how well they actually fitted.

Among young Kepler's corrections of Copernicus' own version of Copernicanism, the principal one besides placing the Sun back at the center was Kepler's recognition of the significance of the equant point (p. 107). Introduced by Ptolemy to get his calculations in order; rejected by Copernicus because it masked the effective introduction of nonuniform speed in a planetary world that should be thought of exclusively in terms of circles uniformly traversed, the equant point was reinstated into heliocentric astronomy by young Kepler because he realized that the heavenly bodies actually move nonuniformly – a fact of celestial life that must be faced squarely rather than circumvented or negated.

This second correction stemmed in no lesser measure than the first from a decisively different conception that young Kepler developed of things astronomical generally, which was marked by a *thoroughgoing realism*. To Kepler the mathematical representation of the motions of the heavenly bodies (i.e., the customary decomposition of actual orbits into supposed component circles) meant nothing if, below the surface of appearances, there was nothing real to shore the whole construction up. *Reality in this sense could be of three, by no means mutually exclusive, kinds: geometric, 'physical' (i.e., dynamic), or harmonic.*

Realism, mostly geometric.   In *Mysterium cosmographicum* Kepler revealed the geometric reality that he understood to be at work behind the complexities of the planetary orbits and the multiplicity of figures and parameters invoked to render them. It formed his own answer to two out of those three, wholly novel questions about the structure of the universe that had occurred to him: Why are there six planets, no more, no less, and why are their orbits vis-à-vis one another spaced as they are?

In the view of most contemporaries and of posterity alike, his answer was entirely fanciful. Even so, he was to stick to it for the rest of his life. It resided in the five regular polyhedra: the tetrahedron (with four equal faces), the cube (six), the octahedron (eight), the dodecahedron (twelve), and the icosahedron (twenty). Euclid had proved that no more than five such regular solids can exist, and it is here that Kepler found the explanation for the number of the planets: From the sphere of Saturn down to the sphere of Mercury one of the regular polyhedra is successively inscribed into that sphere (Figure 5.1).

Ideally, the whole space between Mercury and Saturn is thus filled up, but to make it so is no arbitrary matter. Kepler was well aware that each regular solid has its own, fixed ratio for the respective radii of its inscribed and circumscribed spheres – for the tetrahedron it is 1:3; for the cube and the octahedron, about 577:1000; for the dodecahedron and icosahedron, about 795:1000. Kepler's very first attempt to put each polyhedron at its fitting place, which he characteristically based on considerations of their respective dignities, proved to work out as well as it could. The fit, if considered as he thought it should be – from the proper center of the solar system, that is, from the Sun itself – happens to be remarkable indeed.

Still, some problems had to be resolved before the fit could be pronounced satisfactory by its proud inventor. There further proved to be some left-over problems, which were to occupy Kepler for much of the rest of his life, leading him to what we are compelled mostly against Kepler's own judgment to consider his lasting achievement. The main initial problem had to do with the notion of 'sphere' that a few lines above has been allowed to slip in. What were those 'spheres'?

There is no one answer to this question, as views among astronomers differed and also underwent a marked shift in Kepler's own time. The basic idea was that each planet was affixed to a material sphere, with its primary orbit marking a circle on the surface of that sphere. Opinions greatly differed on the substance that the spheres were made of, but it seems certain that most astronomers, Copernicus included, took the spheres to be solid and, hence, impenetrable. In what manner to imagine how the detailed machinery of eccentrics and epicycles and equants was connected to a planetary sphere was indeed a problem. Copernicus' prime reason for rejecting Ptolemy's equant point may even have rested in his rightly deeming it incompatible with the notion of solid spheres. Here resides the biggest change that Tycho's observation program brought about in how to conceive of the real constitution of the heavens. In 1577 his fine-tuned measurements demonstrated to any-

one willing to let himself be persuaded that the comet then in spectacular manifestation   165
was not so much a sublunary, atmospheric (as Aristotle, in dealing with comets, had been
self-compelled to uphold), as, rather, a supralunary, celestial phenomenon (Figure 5.2).[91]

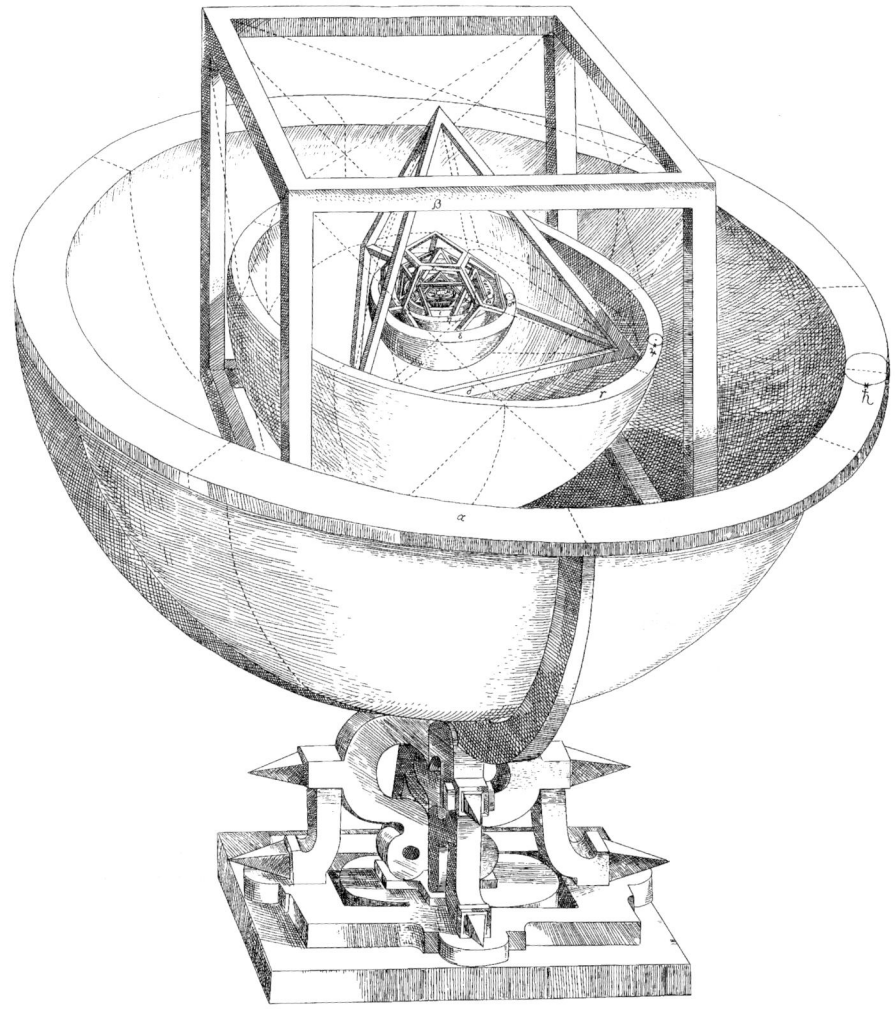

Between Saturn and Jupiter fits the cube; between Jupiter and Mars, the tetrahedron; between
Mars and the Earth, the dodecahedron; between the Earth and Venus, the icosahedron; between
Venus and Mercury, the octahedron.

Figure 5.1: Kepler's five regular polyhedra

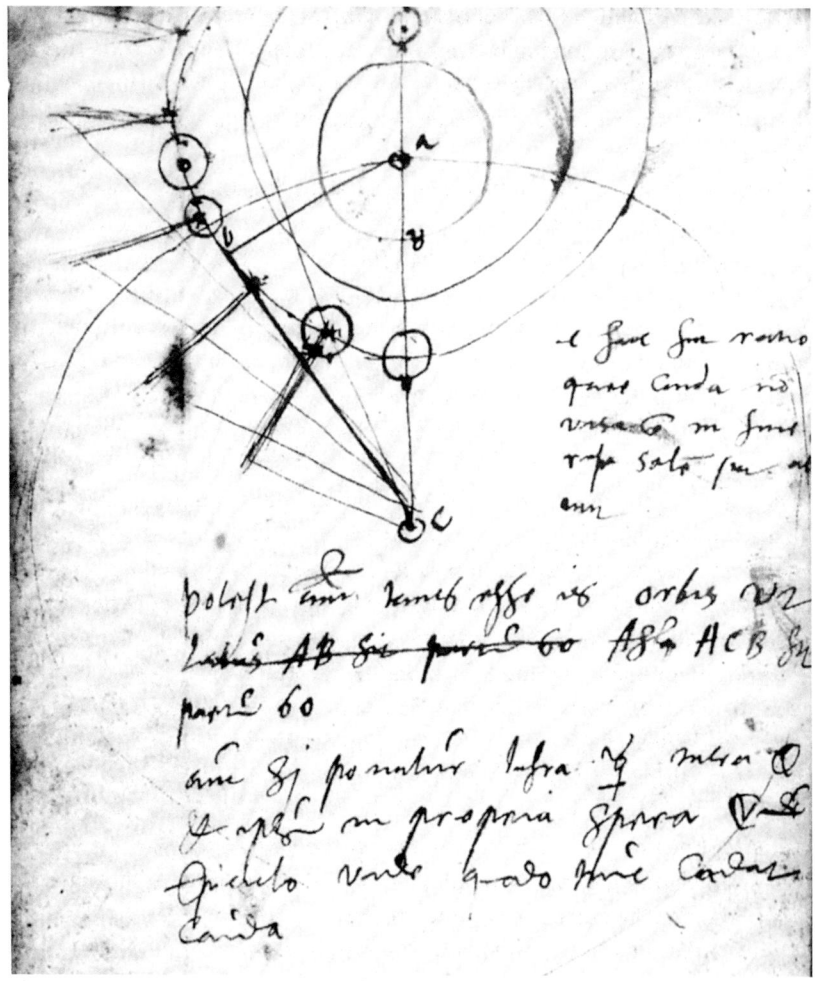

"Tycho's working hypothesis of the (retrograde) orbit of the comet of 1577 around the sun (a), and also around the orbits of Mercury and Venus."

Figure 5.2:   Tycho's comet of 1577

The planetary spheres could be penetrated, so whatever else they might be, they were not solid. For Tycho, this helped remove one barrier against promoting his compromise between pure geocentrism and pure heliocentrism (p. 129). Young Kepler, too, was among those to accept the penetrability and, therefore, the nonexistence of the solid spheres (thus allowing him to reinstate the equant point). Not only the spheres themselves but also the five regular polyhedra successively inscribed into them were, in Kepler's view, a mental construct, impressed by God when He set up the solar system, yet not in any sense tangibly there. Their reality lay rather in their expressing the model according to which God had created the world. But did they, really?

The problem was that discrepancies remained. With Mästlin's help, Kepler marked out needed portions of interplanetary space to take into account the epicyclic machinery of the planets by allowing each 'sphere' a certain thickness of its own. But discrepancies still remained. Kepler found that his calculations from the best publicly available tables of planetary eccentricities (based upon Copernicus' recomputation of planetary parameters) yielded a fit that was enticingly close yet still not perfect. What he needed, so he hinted at several places in his *Mysterium cosmographicum*, was more accurate data on planetary eccentricities to allow him neatly to round off his personal picture of the layout of the heavens.

On this note of slightly insecure optimism the book ran to its close. There are at least three ways to assess Kepler's *Mysterium*. For us, in hindsight (and equally so for Kepler, on looking back a quarter-century later), it already sounded all those themes that were to remain with him for a lifetime, leading him along crooked pathways to his great accomplishment: the three laws of planetary motion deservedly named after him. If he had died right after publication, though, the whole fantastic construction would equally deservedly have been shrugged off, probably for good. So it was by the one co-Copernican to whom Kepler had a gift copy sent, a certain Galilaeus Galilaeus at Padua. It took a thoroughly expert astronomer – as Galileo, in this sense, never was – to perceive underneath the surface of creative fantasy run rampant a brilliant theoretical mind at work. There was indeed an expert astronomer (the recipient of another free copy) who perceived so much. Tycho Brahe was in jealously guarded possession of the very kind of precise data on the planetary eccentricities that Kepler was beginning to realize might be of use to him. However, the one lived on an island in the Sont and the other in Graz, and there was as yet no way for the – almost penniless – latter to accept the invitation casually proffered by the former (always on the lookout for assistants) to look him up if he felt so inclined. Nor was the urge for accurate information on the eccentricities urgent enough yet. In both regards things were to change rather quickly.

Realism, mostly harmonic.   In the years following publication of the *Mysterium*, Kepler, in a couple of letters to Mästlin and to a noble protector, Herwart von Hohenburg, delved more deeply into his conception of the harmony of the world. He went along with Zarlino's authoritative view that there are eight consonant intervals; he took their ratios to form the true *logoi kosmopoietikoi*, or 'world-forming ratios', and he went on to ask three major questions about them:

168   (1) How can the whole set (unison 1:1, octave 1:2, fifth 2:3, fourth 3:4, major sixth 3:5, major third 4:5, minor third 5:6, minor sixth 5:8) be derived from one, intelligible, and *therefore* geometric rule or set of rules? Half a century earlier, Zarlino had answered the question arithmetically – the numbers that make up the ratios are all within the range of the first six integers (thus turning the minor sixth into an exception in need of some special pleading; p. 102). In hinging on just a play with numbers, Zarlino's criterion to distinguish consonance from dissonance could not satisfy Kepler. His own set of governing rules was to elude him for two more decades, to find a place eventually in book III of his *Harmonice mundi*.

(2) How have the *logoi* been worked by the Creator into the planetary aspects, that is, into the angles the planets form as seen from the Earth? This problem was to be resolved by Kepler in the fourth, astrological book of the same work.

(3) How have the *logoi* been worked by the Creator into the planetary orbits? Here the problem came up that had formed Kepler's third wholly new question in the *Mysterium:* Why is it that, if one considers the period that a planet requires for one orbit around the Sun, relative speed diminishes more and more as one moves from Mercury outward? (The point is that, if the period of Jupiter's orbit [twelve years] is taken as the standard for Saturn, which is almost twice as far removed from the Sun as Jupiter is, then Saturn's period would be about twenty-four years, whereas in reality it is near thirty.) That this is so had been suspected all along in geocentric quarters; Copernicus had confirmed it, and Kepler, in the *Mysterium*, had broken new ground by just asking, Why? Some hesitant considerations drawn from assorted pieces of natural philosophy had been all he was able to come up with at that time. Now the same question turned up in connection with the planetary harmonies he was searching for: What if the angular velocities of the planets are in consonant ratios to one another? Thus, by assigning velocity 3 to Saturn, 4 to Jupiter, 8 to Mars, etc., the consonant intervals could be seen indeed to be expressed in the heavens (e.g., Saturn and Jupiter would make a fourth together). If this were true, the conceptions of the five regular polyhedra and of the planetary harmonies combined would make it possible to derive the planetary eccentricities a priori. This was crucially important to Kepler because only an a priori derivation could confirm their being necessary consequences of constraints arising from God's geometric/harmonic layout of Creation. And all over again the question was: Do the values thus computed on the basis of theory actually match values derived from observation? For an answer, young Kepler had nowhere to turn but to Tycho, and he knew it.

Realism, mostly 'physical'.   The story of how, in 1597, the new king of Denmark caused a recalcitrant Tycho to leave the island Hven for (in the end) Prague and how the archduke of Austria made the unrepentant Protestant Kepler leave Graz on 1 January 1600 so as to seek and find refuge with the former must reluctantly be left untold here. So must the vicissitudes of their mercifully brief collaboration, if it can be called that. What concerns us, is what the move meant for Kepler's work. He had set out to obtain accurate values for the planetary eccentricities; within two years of his arrival in Prague, but now himself the Imperial Mathematician as Tycho's successor, he had inherited them all, sort of, and he went on to use them to advantage. At first he used them, not for patching up final details in his harmonic construction as he had intended, but rather to tackle the problem Tycho had assigned to him soon after his arrival – to find the true parameters of the orbit of the planet Mars. Because of its relatively large eccentricity the orbit of Mars was notoriously the toughest nut to crack in the framework of an astronomy bent on using nothing but circles. Within a year Kepler (no longer a beginner or anybody's assistant but himself a thoroughly proficient astronomer) had cracked it – or so he thought for a while.

Cracking the nut involved a good deal of clearing-away activity. Kepler let himself be guided by dynamic (i.e., force-related) considerations, all of which stemmed from his thoroughgoing realism. The dissolution of the solid spheres had led Kepler to embrace the equant point as a token of planetary pathways traversed truly nonuniformly. In his uniquely perceptive view it had also robbed astronomers of a plausible entity to *guide* these nonuniform and, on the face of it, highly irregular motions of planets along their orbits. What, if no solid sphere, could compel a planet to follow the course it did follow? Here Kepler's idea of a force emanating from the Sun came in – what else was there to keep a planet moving? Conversely, the introduction of a real (but, to us, after Newton, nonexistent) force steering the planets along their orbits served further to underscore a line Kepler was already taking in view of his own conception of the created world as geometric and harmonic and, therefore, beautifully simple. Any account of planetary motion, in sum, had to reduce apparent complexity to manifest simplicity (the very simplicity Kepler alone was aware lay hidden somewhere deeply inside Copernicus' unwitting riches). It had to do so in accordance with a plausible, force-directed, or (as he himself called it) 'physical' mechanism. It also had to make predicted and observed planetary positions match without a fault. None of these requirements was being met anywhere yet; in particular, no 'celestial physics' in the sense now envisaged by Kepler had ever been contemplated. Still, after about a year of strenuous thinking and computing Kepler had precisely such a model for Mars' orbit in his hands. Figure 5.3 displays its main features.

The task set by Tycho, then, had, to all appearances, been fulfilled.

But what followed from this point has ever since, and by Kepler himself in the first place, been regarded as the decisive event on his journey toward a 'new astronomy'. Up to now we have seen Kepler delicately poised between the flight of his theoretical fancy and the need for empirical confirmation. The search for accurate observational data on the planetary eccentricities had sent him to Tycho in the first place, but it was as yet undecided what he would

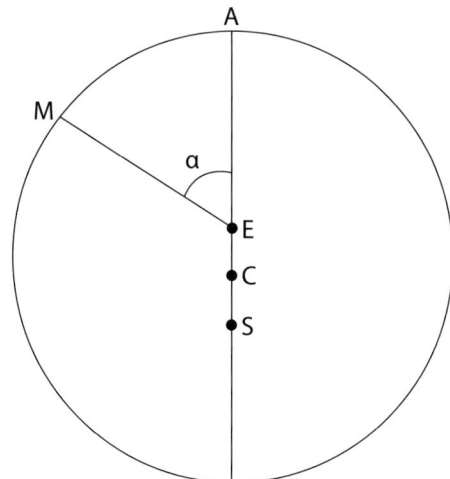

The first complete model Kepler designed for the orbit of Mars. M is Mars; C is the center of its orbit; S is the Sun; E is the equant point from which M's orbit appears uniform, so that angle α varies uniformly; A is aphelion (the point of Mars' orbit farthest removed from the Sun). CS=CE (this is called the bisection of the eccentricity). From Tycho's observations Kepler computed numerical values for the ratio CS:CA (the eccentricity) and for the position of A amid the stars.

Figure 5.3: Kepler's 'hypothesis vicaria'

do with them if they turned out to contradict his geometric and his harmonic constructs. Given Kepler's own brand of epistemology, according to which he was in the business of personally reading the message encoded by the Creator, chances were more than even that he would resolve any mismatch between theory and fact, particularly if it were a small one, in the direction of preserving the theory at the cost of the fact. Yet for the time being not quite so overpowering a theoretical commitment was at stake, for in his study of Mars he had invested a good year's worth of hard, pioneering work, not the leading idea of his life. Still, what would he do if a mismatch turned up?

We know the answer, for a mismatch did turn up. More than that, Kepler deliberately exposed his theory of Mars – the finest and most accurate ever yet put forward – to the most searching criticism he could possibly expose it to, which was a range of pertinent observations by Tycho on Mars that had not yet gone into the making of his novel theory. And indeed, the new observations fitted the theory beautifully. Or rather, the misfit of eight minutes of arc that he could not help noticing might well serve as a welcome confirmation in terms of the current standard of accuracy as upheld from Ptolemy to Copernicus, which was ten minutes of arc. However, Tycho had decreased the margin of error to two minutes of arc, and here is how Kepler, in those both moving and brilliantly rhetorical sentences that deservedly adorn many a historical treatise on the subject, announced how he planned to proceed:

> It behooves us, to whom God's benevolence has given that most diligent observer, Tycho Brahe, from whose observations the error of eight minutes in the calculation for Mars according to Ptolem[aic principles] is revealed, with gratitude to accept, and to exploit, this divine benefit. Let us, then, exert ourselves such that in the end (leaning upon the evidence by means of which our suppositions have been caught out as false) we hunt out the genuine form of the heavenly motions. And this is the way along which, in what follows, I myself shall precede others to the best of my ability. For if I had judged those eight minutes of longitude fit to be ignored, the correction already carried out (namely, by bisecting the eccentricity) of my hypothesis [on the orbit of Mars] would have sufficed. But since these eight minutes could not be ignored, they alone, therefore, have led the way toward the wholesale reformation of astronomy and have provided the matter for the larger part of this work.[92]

Theory – facts in science.   In our own day there is much talk about scientific facts being not so much objectively ascertainable as rather the subjective outcomes of *negotiations* between scientists in their ongoing struggle for recognition by their peers and the public. In the present case, though, the only person with whom Johannes Kepler can possibly be caught negotiating is Johannes Kepler, and it was up to him and no other to decide what the outcome of his self-imposed testing procedure implied. Nothing compelled him to take the objection arising from observation seriously, as the human mind can take refuge whenever it wishes in side-stepping strategies and, ironically, is sometimes well advised to do so. Given the makeup of Kepler's mind, the odds were rather against his taking seriously the facts as they presented themselves to him. Yet take them seriously he did, in a move that marks the radical transition in thought about nature we are examining here. And if we go on to ask what made him do so, rather than searching elsewhere for an answer, we need only listen carefully to his own testimony just quoted. For all his subjective belief that in his pursuit of the secret of the universe he was reading the mind of God, he rarely forgot that the Divine Mind had expressed Itself in the phenomena of nature. In being therefore no less God-given than the mental frame from which they sprang, phenomena accurately established deserved at least equal respect as such. It is true that phenomena as they present themselves can always be interpreted in different ways, with much of the business of science being devoted to that very task, and to this extent they are most certainly subject to our human subjectivity. Still, facts can in addition do one objective thing, which is to give us indispensable feedback by, at times, and if we are lucky and stand ready to accept the luck, contradicting precisely our

172   most cherished interpretations. More often than not the verdict of nature is spoken by other, mostly later participants to the ongoing debate of science. What is objective about the procedure is that in science, unlike any other human pursuit, broad, commonly accepted rules have been explored and worked out from the early 17th century onward on how to reach in the end (which at times is a long way off) a kind of agreement that, even though never wholly definitive, is nonetheless much more than just arbitrary and fashion-bound. Indeed, one major theme running through the present book is a painstaking examination of how in the 17th century numerous pioneers actually went about their search for ways and means to steer a course between the hoary rocks of arbitrary assertion, dogmatic certainty, and skeptical resignation. Remarkable about Kepler is that, with his particular mind-set almost predisposing him to go exactly the other way, he nonetheless recognized in the outcome of his own testing procedure the verdict of nature, and that he proceeded to obey it, creating in the end a truly new astronomy.

Realism, mostly physical (continued).   The steps successively made on this journey (which he re-created in rhetorical fashion in *Astronomia nova*, to placate Tycho's heirs and disarm possible critics) are for the most part quite technical. Only the broadest of outlines can be given here, by way of a ten-point listing. The reader may well prefer to skip even these ten points upon learning that the end results – Kepler's first two laws of planetary motion – came out of an ongoing interplay between partly old, partly new mathematical tools, on the one hand, and brand-new physical considerations, on the other. Note here that the meaning to be assigned to the term 'physical' thus began to shift from 'natural-philosophical' in the received sense in the general direction of what we nowadays understand a physicist to be concerned with.

(1)   Kepler's first move was to reinstate his rejected hypothesis on the orbit of Mars as a 'vicarious hypothesis', that is, as a temporary aid, since as yet he had nothing better to go by. From there on, he took a range of mostly exceedingly radical steps, in roughly the following order:

(2)   He subjected his own observation post (i.e., the moving Earth) to a searching scrutiny by imagining himself to be an observer on Mars. It then appeared that, contrary to Copernicus' idea, not even the Earth's orbit is traversed with uniform velocity – it, too, requires an equant point. More than that, at aphelion and perihelion (the points of the orbit where the Earth is farthest away from and closest to the Sun, respectively) velocity appeared to be inversely proportional to distance.

(3)   He sought a more general relation between the distance of a planet to the Sun and the planet's angular velocity.

(4)   He conceived in several stages the idea (and here his decision to match mathematical with 'physical' considerations came most prominently to the fore) that two distinct

forces, both emanating from the Sun, regulate the orbits of the planets. Like all his contemporaries with the sole exception of Galileo, Kepler never questioned the Aristotelian ground rule that force is required to keep a body in motion, so that the body ceases to move once the force stops acting. Therefore, he needed one force to keep the planet moving in its orbit and another to account for its eccentricity whatever the precise shape of the orbit. The primary force came from the Sun in its ongoing rotation around its own axis (posited by Kepler for theoretical reasons; confirmed in 1613 by means of observations made by Galileo and others) and was taken by Kepler to act the way light does. From this alone a circular motion would result, so in addition the Sun was taken by Kepler to be a giant magnet, attracting each planet over one half of its orbit and repelling it over the other half. For this 'physical' mechanism he borrowed insights from William Gilbert's book on the magnet published in 1600 (p. 249).

(5) To conceive of the Sun as a magnet went hand in hand with a grand, strictly illicit generalization that not just at aphelion and perihelion but at every position that the planet occupies its speed is inversely proportional to its distance from the Sun.

(6) On this bold generalization he based what for the time being he took to be no more than an approximation, which, however, has since come to be known as Kepler's area law. It holds that the time a planet needs to traverse any arc of its orbit is proportional to the area of the sector between the Sun and that arc (Figure 5.4).

If the shape of the curve is known, the rule makes it feasible to compute times passed from spaces traversed. That way the principal computational advantage previously enjoyed as a reward for using circles traversed uniformly (p. 13) was restored to some extent. But this very usage had in the meantime become ever more questionable to Kepler's mind.

(7) Armed with his area law, he tried for one final time to squeeze the data on Mars into a circular orbit. Failing once again, he went on to reject for good the Platonic requirement, twenty centuries old already and never even challenged before, to account for planetary orbits in terms of circles.

(8) Instead, he settled on an oval shape. However stubbornly he tried and kept trying, he could not determine its shape more closely. The problem was that whatever had looked precise before, including his area rule, had now suddenly become approximate, for which he lacked proper mathematical tools.

(9) Blending expert knowledge of the parameters involved, ongoing physical reasoning, total recall of Apollonios' *Conics*, and some good luck, he found in the end that his 'monstrous' oval must be symmetrical and could be represented in a satisfactory manner only by an ellipse.

(10) He rounded off his theory of the elliptic shape of the orbit of Mars (implicitly taken to be valid for each of the planets) by means of some additional, physical considerations.

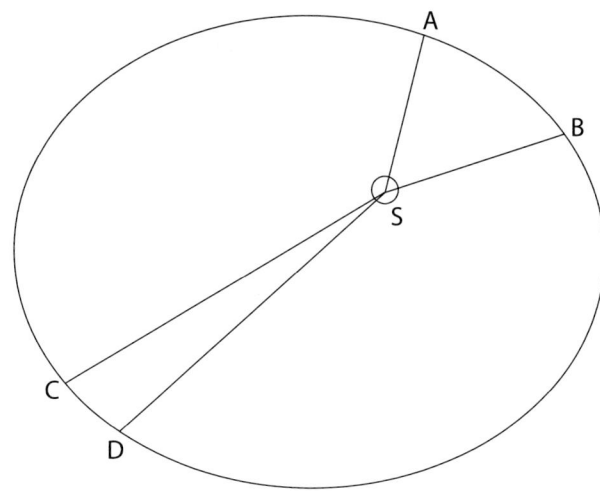

Areas ASB and CSD are equal – the nearer a planet is to the Sun S, the more it speeds up.

Figure 5.4: Kepler's area law

So short an account inevitably understates the immense labor the whole effort required. A large amount of theoretical ingenuity and heaps of dreary calculation went into it. Also, part of the mathematics involved was not readily available but had, as a child of necessity, to be invented along the way. Altogether Kepler's hunt for the true orbit of Mars took him some five, frequently interrupted years, from 1600 to 1605, with publication of the book that purported to tell the whole story, *Astronomia nova*, taking another four. The full title is even more revealing of what Kepler thought he had accomplished; it reads thus:

> *New Astronomy, based upon causes, or: Celestial Physics, treated by means of commentaries on the motions of the star Mars, from the observations of Tycho Brahe, Gent., by the order and at the expenses of Rudolph II, Emperor of the Romans etc., worked out at Prague in a tenacious study lasting many years by His Holy Imperial Majesty's Mathematician Joannes Kepler.*

Back to the harmonies.   Consider first the dimensions of Kepler's achievement as it appears to us. The deep tension present in Copernicus' *De revolutionibus* had not only been recognized but now found itself resolved in a beautifully simple picture of planetary motion. That picture blended predictive accuracy and mathematical exactitude with a plausible (albeit retrospectively misconceived) dynamic machinery. Instead of more than fifty circles loosely connected with one another yet devoid of any conceivable physical mechanism, Kepler had now summed it all up in the assertion that all six planets traverse an elliptical orbit of variously given eccentricity, with the Sun placed in one of the two focal points of the ellipse, and the radius vector between the planet and the Sun sweeping out equal areas in equal times.

And yet Kepler was not satisfied – but not for the one reason that might look plausible to us, which is that his celestial physics was wrong. Galileo's demonstration that the Aristotelian conception of motion requiring a force to sustain it was profoundly mistaken was still being worked out in the secrecy of his own home at Padua, not to be revealed to the world until two years after Kepler died. Hence, Kepler had no grounds whatever for thinking anything wrong with the first celestial physics ever conceived in the long history of astronomy. His misgivings were situated elsewhere: The deeper rationale for his two laws was lacking. The laws embodied empirical truths about our world in a neat, mathematical form, yet no compelling reason was in sight for why God had arranged the world that way. The elliptic shape of the planetary orbits did not fit, in any clear-cut sense, into the harmony of the world. It was, therefore, with a sense of relief that, right after completion of *Astronomia nova*, Kepler returned to those harmonies of his in which he delighted so much more: "May God," so he wrote in October 1605, "get me out of astronomy so that I can switch back and concern myself with my work on the harmony of the world."93

His elliptic detour mercifully behind him, Kepler recorded the final results of his harmonic labors in what he himself regarded as his true magnum opus, the *Harmonices mundi libri quinque* (Five Books on the Harmony of the World) of 1619. In this work all remaining riddles of *Mysterium cosmographicum* were finally resolved by means of his ingenious elaboration of the broad schema he had drawn up in 1599, on the eve of his enforced departure for Prague. Of the five books, the first two are geometric, the third is musical, the fourth astrological, the fifth (the crowning achievement of the whole effort) astronomical.

Book I is devoted to regular polygons. These are *not* the regular polyhedra, which are solid and of which there are five, but plane figures like the equilateral triangle or the square, of which there is an infinite variety. Kepler classified them in accordance with the relative degrees of commensurability they display between one side of a given regular polygon and the diameter of the circumscribed circle.

Book II, while a display of geometric ingenuity done mostly for the sheer fun of it, stands somewhat apart from the book's main argument in that it arranges the regular polygons according to criteria of congruence hardly employed further on.

Book III seizes upon the commensurability rules of book I to derive the eight consonant intervals. Kepler did so by means of seven successive divisions of circle arcs by successive regular polygons, with the outcomes governed by three specific rules devised for the purpose. The procedure yielded a neat reduction to geometric rule and order of the musical

176   harmonies, now shown to be the true *logoi kosmopoietikoi*. The reduction made it possible for Kepler to apply them in their turn to planetary theory. That is, he took it upon himself to find these 'world-forming ratios' back in those portions of the empirical world that mirror the *logoi* in the clearest and most significant fashion.

Book IV shows to what extent Kepler was prepared to go along with astrology, in the business of which the Imperial Mathematician was deeply involved. His basic idea was that the Earth has a soul, which perceives the rays coming from the planets and measures the angles these make with one another at any given moment. The angles occurring at nativity determine a person's character. That is, they do not so much determine a person's destiny as, rather, the general tenor of how he responds to events.

Book V, finally, returns to the third question Kepler had raised in *Mysterium cosmographicum*: the relation between the planets' angular velocities and their relative distances from the Sun. It does so in connection with the principal subject of the book, which is the harmony of the spheres. Well aware that Ptolemy preceded him in taking up this hoary, Pythagorean theme, Kepler regarded his own position as doubly favored in comparison. The true nature of the harmony of the cosmos could be disclosed only from a polyphonic viewpoint musically and from a heliocentric viewpoint astronomically, both out of Ptolemy's reach but fortunately at Kepler's disposal.

In the end Kepler found the celestial harmonies to reside in the minimal and maximal angular velocities of each planet, that is, in their speeds (measured in seconds of arc traversed daily) at aphelion and perihelion, respectively. The harmonies expressed themselves (rationally, not audibly, as Kepler insisted, because there is no air in the heavens) in four different respects, of which the most elementary and the most complicated one must suffice by way of examples. Saturn's speed at perihelion is 135, at aphelion 106, which by means of a small adjustment for which Kepler was not at a loss to adduce a rationale yielded 108:135, or 4:5, that is, the major third. This is the easy one, which can of course be computed not only for each planet but also for pairs of planets in a variety of ways. More complex harmonic configurations (chords, not just intervals) turn up when the whole range of speeds between aphelion and perihelion is taken into consideration. Is there going to be a moment in the history of the world when all six planets together sound ('rationally sound', that is) one big, consonant chord? The answer depends critically on the eccentricities of the six planets, with those of smallest eccentricity (Venus and Earth) setting the constraints. So here we are back at the problem of the planetary eccentricities, and Kepler is now close to the final step on his long journey – to explain why the planetary eccentricities are what they are.

In the penultimate chapter of book V of *Harmonice mundi* Kepler set forth in forty-eight rules what constraints following from the basic architecture that God had employed at Creation necessitated adjustments in the planetary harmonies as expressions of the planetary eccentricities. That is to say, given the broad celestial layout of the five regular polyhedra, and given the true nature of musical consonance, everything else in the design of the solar system followed by necessity and could therefore be explained a priori as well as empirically checked a posteriori. Not that the rigorous attitude toward deviant observation that Kepler had displayed in his hunt for the true orbit of Mars governed his work on the harmony of the spheres to quite the same extent, so much more was at stake for him here.

It is in this context that, with the proofs of part of book V already in, Kepler made one more discovery. He found the mathematical rule which expresses the tendency of planets to lag behind as they are placed at a larger distance from the Sun. The discovery came just in time to work the rule into his grand a priori listing of God's ways at Creation. We know it as Kepler's third law: the squares of the orbital periods of any two planets are as the cubes of their mean distances from the Sun.

The accomplishment made accessible.   How we have come to know the rule as Kepler's third law has less to do with the book in which he first published it than with the two obligations that, upon the book's triumphant completion, still remained to be fulfilled.

He set out first to explain the Kepler view of the universe in textbook format. This he did in seven books filled with questions and answers and entitled with deceptive modesty *Epitome astronomiae Copernicanae* (Summary of Copernican Astronomy; 1617–1621). To the quite limited extent that later astronomers wished to expose themselves to the intricacies of Kepler's universe, here is mostly where they found it.

By far the best known of Kepler's astronomical works, however, proved in the end to be the *Tabulae Rudolphinae* (Rudolphine Tables), which over years of drudgery he compiled from Tycho's raw observations and saw through the press in 1627. For, just as had happened to Copernicus, most astronomers in the next generation were to care far more for the opportunity for vastly improved prediction Kepler had given them by publishing the *Tables* than for the elaborate theoretical framework from which it derived its significance in the proud author's point of view. What to Kepler was a quite-unheard-of celestial physics–*cum*–harmonics–*cum*–realist geometry was reduced by most others in the next generation to the very kind of semifictional, purely mathematical astronomy Kepler had striven so hard to get rid of.

In one domain of nature-knowledge, then, abstract-mathematical science in the classic Alexandrian vein now began to be replaced by abstract/*realist*-mathematical science in a vein I label for short 'Alexandria-plus'. So vast was the step thus made, and so abstrusely wrapped in an idiosyncratic conception of God and Creation, that its immediate adoption would have been far more astonishing than the narrow, prediction-focused uses to which it was put by astronomers of the next generation. In a later chapter (p. 302) I shall examine those uses and how step-by-step they resulted in recognition of Kepler's lasting accomplishment. The prime task for now is to investigate the *other* domain of nature-knowledge where,

178   also in the early decades of the 17th century, abstract/realist-mathematical science broke through accustomed ways of proceeding. That domain was the motion of objects on Earth.

Galileo Galilei

Contrasts between Galileo and Kepler are endless and endlessly fascinating. Compare the paths the two men followed. If one takes what in effect was Kepler's research program as embodied in *Mysterium cosmographicum* and blends it with his cosmological/theological commitments, with the thoroughgoing realism these entailed, and with the singular perspicacity that guided him through his program, the path he followed is straightforward indeed. It looks oddly crooked only if considered from hindsight, as if it were meant by Kepler to prepare the way for Newton's *Principia*. On that account Kepler's research program not only fails badly but also begins to look unaccountably strange. By contrast, Galileo's path seems straightforward in retrospect, whereas if looked at without the benefit of hindsight its rapidly emerging ambiguities seem truly countless – an amazing feat for a man well-known for his ability to express himself with rarely surpassed clarity.

One ambiguity comes up as soon as we ask what turned Galileo into a Copernican. By 1597 at the latest he was fully committed to the doctrine – that much appears from the note in which he acknowledged receipt of the gift copy of Kepler's *Mysterium cosmographicum*. Much later he suggested that a lecture series by a certain German astronomer known to have died in 1588 made him take heliocentrism seriously for the first time. But we know nothing for certain about his reasons for adopting it, which is especially odd since no one was ever to put forward so many, and so compelling, arguments in its favor.

Many more things stand in the way of calling Galileo's path straightforward. I shall ignore in what follows one counterinformative obstacle: the range of Galileo legends still circulating among many outsiders (the leaning tower; 'eppur si muove'; the idea that Galileo was the first to advocate heliocentrism, etc.). To take these seriously is to complicate beyond hope of resolution the task Galileo's historians have taken upon themselves. But the major obstacles that remain must be taken into account, even though we are left with enduring uncertainties in the reconstruction of the principal outlines of his thought.

One uncertainty resides in the circumstance that, unlike Kepler, Galileo did not possess a broad conception of the world to cling to throughout his life and guide him – there are many, more or less fleeting commitments but nothing nearly so solid. Nor did Galileo set forth or follow one coherent research program in the nature of Kepler's.

Further, Kepler was in the habit of publishing his findings in the format of a record of how he had arrived at them. We know now that, as a historian of his own thinking, he was driven so much by rhetorical convention or (with *Astronomia nova*) by an urgent need to

placate Tycho's heirs that we cannot accept his own accounts without more ado. Still, his 179
well-preserved working notes allow for whatever correction proves necessary. Galileo, in
contrast, did nothing of the kind. Those private jottings that he did leave to posterity, although over past decades found to be informative indeed, nonetheless fail to allow the kind
of relatively straightforward reconstruction that Kepler's work and notes make possible once
presentist blinkers are removed.

Further, unlike with Kepler and his harmonic concerns, Galileo is the first thinker about
natural phenomena in all history whom modern scientists feel they can identify with. It
is not by chance that they feel that way. Much in Galileo's best work is indeed, in Stillman
Drake's apt phrase, "recognizably modern physics",[94] and this above all is what keeps alive the
vast interest it continues to evoke. But such a sentiment, if not marred by a determination
to take Galileo's own times as the standard against which to measure him, is treacherous indeed. Owing to the versatility of Galileo's mind, it may lead to perhaps the greatest impediment of all in the way of a straightforward reconstruction. Since, more than a hundred years
ago, the 'Galileo industry' began in earnest, he has been taken by serious historians for one
or more of the following: (1) an intuitive empiricist; (2) a worker in the medieval impetus
tradition; (3) the pioneer of the mathematization of nature; (4) a Platonist metaphysician;
(5) the man to continue what Archimedes began; (6) an Aristotelian methodologist; (7) the
pioneer of experimental verification and/or falsification; (8) the pioneer of fact-finding experiment; (9) a man steeped in craft practice; and even, incongruously, (10) a Taoist of sorts.
Please note that all these varied, partially overlapping attributions have been made with due
reference to portions of Galileo's own writing. How much in them is really in the eye of the
beholder? Inevitably, some of it is. One cannot write about Galileo (the pre-dominant figure,
in any case, in the story of how modern science came into the world) without informing the
account with one's own conceptions of what science ultimately is about, and it would be fu-
tile to pretend the present treatment to be exempt from such sources of possible distortion.

So much granted by way of preliminaries, let us begin at the beginning: a succinct listing
of Galileo's intellectual roots, which were many and varied indeed.

An Archimedean with a difference.   Up to the mid-1590s Galileo fitted neatly into
the recurrent pattern of mathematical humanism at the time – recovery of texts, leading to
efforts at reconstruction of lost pieces, leading in their turn to extension and enrichment
of the corpus. This is what kept him occupied as a member of a group of Archimedeans
centered on Guidobaldo dal Monte (p. 103). With Latin editions of pertinent Greek texts
already available to him thanks to his predecessors in the movement, Galileo's first contribution was an attempted reconstruction of Archimedes' reputed feat of determining the
alloy in the king of Syracuse's less-than-pure-gold crown by means of a special kind of balance. This helped him secure, through Guidobaldo's patronage, the mathematics chair at the

University of Pisa. Here he completed a treatise, *De motu* (On Motion), in which he went beyond pure Archimedeanism by grafting the law of the lever upon an account of free fall in terms of the gradual conquest of 'accidental lightness'. This concept fitted broadly into the Aristotelian conception of free fall. It was an addition made to it in the 2nd century BCE to explain the readily observable fact that, in starting to fall, objects at first move more slowly than later – the very phenomenon for which Jean Buridan in the 14th century invoked the related yet different idea of successive increments of impetus being added to the heaviness of the falling object (p. 85).

De motu is significant for three reasons. One is that in the end Galileo got stuck in this first attempt of his to treat a problem involving Aristotelian moving forces by means of a conceptual apparatus taken from the Archimedean treatment of equilibrium problems, and that he recognized as much, abandoning the manuscript as he abandoned Pisa in 1592. Further, he took much more from Archimedes than just a range of theorems on solid and fluid equilibrium. At places in the treatise recognizably real entities – in particular, objects falling along inclined planes – are not treated as such but rather as idealized abstractions. He assumed "a plane [being], as it were, incorporeal or at least very exactly polished and hard." Similarly, he took the "accidental resistance arising from roughness of the moving object or of the inclined plane, or from the shape of the object"[95] to be nothing but an *impediment* that, as such, ought not to be taken into account. This is the world of mathematical abstraction to which Archimedes had reduced his bodies in equilibrium. Galileo now sought to reduce falling bodies to it as well, in an attempt to treat these as somehow out of balance for as long as they fall with increasing speed, hence, as suitable for analysis by means of static concepts. As, in the end, Galileo himself was bound to acknowledge, the task cannot be accomplished. Nonetheless, and this is the third significant feature of *De motu*, the treatise as such testifies to his perspicacity in seeing that there were opportunities here ready to be exploited, even though he did not yet know quite how to avail himself thereof. He was aware, then, that there was an as-yet-unresolved problem of falling bodies, that it was important to resolve it in a coherent manner, and that the Archimedean approach in which he was steeped was necessary yet not in and by itself sufficient for accomplishing so much. And the question is what made him aware of these things in the first place. For answers, we must turn to other portions of the intellectual heritage that Galileo was in the end to put to unheard-of use.

There was, in the first place, Galileo's deep and detailed knowledge of Aristotle's teachings – strikingly so for the very man who was to engage in vehement controversy with staunch propounders of the doctrine and who was to caricature it with such remorseless wit, he knew exactly what he was talking about. Not only had he around 1584 composed an Aristotelian treatise *De universo* (On the Universe, no less) in hopes of securing a university position. He also got in touch with Clavius and other Jesuit scholars in Rome engaged in the effort – similar on the surface to his own – to enhance the status of mathematics within

Aristotelian doctrine. It further appears from course notes Galileo wrote out in the late 1580s and also from later statements that he was impressed with the logical rigor of the organization of Aristotelian argument and with the language in which Aristotelian methodology was clothed – a language he was to employ (in the enriched format he was to encounter somewhat later at Padua) to the end of his life. His respect for Aristotelianism was not to last – increasingly he came to see the doctrine as petrified beyond hope of reanimation. Yet his respect for the founder remained. This was so in view of the component of empirical open-mindedness in Aristotle's philosophy of nature that was in due time to serve Galileo as one secondary yet indispensable ingredient in the proper way to make sense of what was ultimately a mathematically imprinted universe. It was from this blend of Aristotelian motifs, then (however much subject to subsequent change), that Galileo's unusual, for an Archimedean, preoccupation with the problem of accelerated fall is likely to derive at least in part. Something similar, however, had been true of Benedetti, whose work Galileo appears not to have been aware of, and further components must be invoked to explain his interest, notably, inspiration drawn from Copernicanism.

Here we tread quite unstable ground, for it is not at all certain that Galileo, when writing 'De Motu', had already made up his mind about the virtues and vices of heliocentrism. In 'De Motu' there is talk of a rotating sphere at the center of the universe, which may be taken to indicate that Galileo subscribed to an advanced version of the Tychonic compromise (of which he may or may not have been aware at the time). Still, whether the Earth is taken to rotate around its own axis as a planet (as Copernicus had held) or rather at the stationary center of the universe (as Galileo here appeared to assume), a problem with rectilinear fall does present itself in either case (p. 112). Even though it is the cause of acceleration in free fall, not the problem of how vertical descent is at all possible on a rotating Earth, that occupied Galileo in 'De Motu', it is at least conceivable that Galileo's cosmological views served to enhance for himself the significance of this phenomenon, which by itself was of marginal interest in Aristotelian natural philosophy as a whole. In any case, and whatever it was that made him address it, the phenomenon of fall was never to lose its grip on Galileo for reasons that derive beyond doubt from his adherence (whenever it started) to the Sun-centered view of the universe. For it is certain that, after Pisa failed to extend his three-year appointment and Guidobaldo helped him secure the mathematics chair at Padua, Galileo kept working on the problem, yet began to put it on a novel, entirely unheard-of footing. This was to lead to the profound transformation of the very conception of motion that he was to achieve during his eighteen Padua years. But we cannot even begin to outline that revolutionary transformation until we examine two more roots of Galileo's mature thought.

While in his native Tuscany, Galileo had largely dealt in constructs at several removes from tangible reality – in the abstractions of Archimedes-style balance problems, in the formalities of Aristotelian methodology, in the first-principle, a priori manner of treating empirical phenomena that marked all natural philosophy. His move to Padua in the Veneto now exposed him to an engineering tradition very much alive in Venice.

To the limited extent that mathematical considerations went into craft practice at the time (p. 131), Galileo quickly became familiar with them. He owed this both to extracurricular teaching obligations he took upon himself to increase his meager professor's salary and to a workshop for mathematical instruments he set up for the same reason. A 'geometrical and military compass' (a computation aid of universal application that he designed in 1597–1599) exercised his manual skills. A course syllabus on the five simple machines ('Le Meccaniche'), which he composed around 1593 and reworked in 1601, shows him busily engaged in improving Archimedean applications of the law of the lever. His debt to two centuries of engineering literature in the Italian vernacular is evident at places in his later work. In at least one case, the limited capabilities of a suction pump owing to the as-yet-unwitting creation of a void space (p. 410), Galileo derived inspiration from visits he was in the habit of paying to the Venetian Arsenal. The turn toward the concreteness of manual practice that Galileo experienced at Padua is likely to have helped turn his mind as it did his hands toward the decisively novel approach to problems of motion on which he embarked there from about 1602 onward. The practical turn may have been enhanced by Galileo's early exposure to experiments undertaken in the late 1580s, when he was still a youth, by his father, Vincenzo, on musical consonance. In a creative extension of current practice with the monochord, Vincenzo had shown how consonance depends, not only on the various lengths into which a string can be divided, but also on such properties as its material and its degree of tension, thus (like Benedetti, but in a different way) going one step beyond the customary, purely arithmetical conception of consonance (p. 145).

What, in short, we have here is Galileo's absorption into, specifically, the Archimedean tradition and his thorough familiarity with two others (one Aristotelian, one in craft practice); all this was further enriched, mainly through some cross-fertilization, by a few novel elements. In other words, what we have here is some kind of improved Benedetti – nothing less, but not much more, either. In *what* respects, then, and *how* (and also, in a later section, *due to what*) did he manage to transcend these traditions held in common with many a contemporary of his? How did he, and he alone, manage to turn it all into something decisively different?

The mature doctrine of motion.   What scrap papers Galileo left behind do not provide a solid enough foundation for us to know with sufficient certainty how and in what order Galileo created his novel science of terrestrial motion. So in the following presentation of his achievement I shall leave even the pretense of chronological order behind. Instead, I shall review pertinent portions of his two masterpieces, the *Dialogo* of 1632 and the *Discorsi* of 1638, with occasional flashbacks to connect their contents with high and low points in his career as an observational astronomer and as a defender of Copernicanism, to which his doctrine of motion was indissolubly tied in many respects. To that end, let us take up the array of objections to the Copernican hypothesis understood as descriptive of reality that was listed on p. 111. We start with the objection from motion.

Whether the objection was phrased in terms of stones thrown upward or downward, of cannonballs shot due east and west or due north and south, of clouds left behind or of birds struggling to keep up, an elementary, commonsense conception of motion tacitly resided at the back of every objection of this type. Its core stands expressed in Aristotle's well-known assertion that 'everything that is in motion is moved by something'. That is to say, an object begins to move as some force makes it do so, and motion ceases as the force ceases to act upon the moving body. Two almost equally important corollaries of this ground rule in the context of Aristotelian doctrine are that the motion of an object is absolute, in the sense of taking place irrespective of what other moving objects do or do not do, and that a moving object can have one motion only.

These are commonsense notions. They appear on interrogation to be shared in essence by just about every person not educated in Newtonian mechanics. And this is precisely why they exerted an appeal far beyond those used to upholding their consequences in strictly Aristotelian fashion. One consequence is the impossibility of a stone falling back toward a rotating Earth to land at the same point from where it was thrown upward. It was, indeed, a matter of upholding *consequences*, for the core notions themselves were taken to be so self-evidently true as not even to be regarded as specific presuppositions. Galileo was the one who, during his Padua years, began to see that this is precisely what they are – presuppositions, with some limited validity in our rough-and-tumble world yet just false in a world from which all circumstances that disturb the actual properties of motion are removed, be it in practice or in thought. On the Second Day of the *Dialogo* Galileo pulled off the feat of making his own spokesman ('Salviati') bring the Aristotelian opponent ('Simplicio') very close to admitting a radically different conception of motion by gently guiding him alongside a range of precisely such phenomena as come nearest, in the world of everyday experience, to such 'disturbances-removed' situations. Here is how, in a well-known passage on the Second Day of his *Dialogo*, he introduced the idea of the relativity of motion:

> Motion, in so far as it is and acts as motion, to that extent exists relatively to things that lack it; and among things which all share equally in any motion, it does not act, and is as if it did not exist. Thus the goods with which a ship is laden leaving Venice, pass by Corfu, by Crete, by Cyprus and go to Aleppo. Venice, Corfu, Crete, etc. stand still and do not move with the ship; but as to the sacks, boxes, and bundles with which the boat is laden and with respect to the ship itself, the motion from Venice to Syria is as nothing, and in no way alters their relation among themselves. This is so because it is common to all of them and all share equally in it. If,

from the cargo in the ship, a sack were shifted from a chest one single inch, this alone would be more of a movement for it, in relation to the chest, than the two-thousand-mile journey made by all of them together.[96]

It testifies to both Galileo's marvelous sense of irony and his awareness of the slippery slope he is, with this one example, leading Simplicio toward that he makes the latter respond at once with: "This is good, sound doctrine, and entirely Peripatetic [i.e., Aristotelian]." For here we have precisely one of those real-life situations which in the clearest way, *once one's attention is drawn toward them*, display the defects of current ideas on motion. But what had drawn Galileo's attention to such real-life situations in the first place, and what had made him infer such radical conclusions? Had not previous thinkers, too, noticed contradictions between certain real-life phenomena and their Aristotelian explanations? Yes, they had, but one crucial difference is that men like Buridan had been content with local reparation of a marginal leakage, so to speak, whereas for Galileo an incongruity at the very center of things was at stake. Why central? The Copernican hypothesis, in its realist interpretation to be sure, had made it central. And the other difference with previous amendments made to this or that Aristotelian tenet on this or that phenomenon of motion was that there is something special about the kind of real-life situations Galileo invoked to sustain his own, far more radical revisions. For these real-life situations, such as the ship from Venice to Aleppo, were invariably made to act as bridges between two different levels of reality – the concrete world of ordinary experience and the world of geometric idealization. It is in this latter world (which Galileo knew how to display before his readers as almost real in the ordinary sense) that his new conception of motion made proper sense. Of that new conception, this idea of the relativity of motion was but one, albeit highly significant, corollary.

Central to the new conception is a notion that Salviati went on to outline for the benefit of both Simplicio and the third partner in the conversation, Sagredo, who in his utterances was made to blend the intelligent layman's comments with positions previously held by Galileo yet subsequently abandoned with greater or lesser regret. This central, wholly novel conception of motion is, in one word, its tendency to be *retained*. Once an object moves, it tends to keep moving without any outside influence being needed for its doing so. If in everyday life it invariably seems otherwise, this is only because everyday life puts so many impediments in the way of the true phenomenon having a chance to display itself. Galileo used many pages on that vitally important Second Day to introduce his astounding conception of motion tending to be retained rather than to return to rest. He made Salviati squeeze out of Simplicio grudging acknowledgment of three successive situations:

- A perfectly spherical ball made of bronze is put on an inclined plane "as smooth as a mirror and made of some hard material like steel":[97] upon release the ball will roll down with ever-increasing acceleration, yet more slowly the less inclined the plane.

- The same ball, when made to roll upward along the same incline, will in the end come to
a standstill, and the smaller the angle of the incline, the more time this process will take.
- Hence (and this is of course the crucial point), if the plane has no tilt at all but is parallel
to the horizon, a ball upon it would either lie still or, if even the smallest amount of motion were imparted to it, continue to move forever.

It is not known for sure precisely when, or how exactly, Galileo arrived at this most crucial point in his campaign to weed out commonsense conceptions of motion. He first mentioned it in public writing in 1614, in one of his 'Letters on Sunspots'. How entirely paradoxical such a notion was at the time may be gauged from the fact that even Kepler, who read those 'Letters', just overlooked the passage. Only by means of the kind of elaborate discussion Galileo was shrewd enough to provide in the *Dialogo*, zooming in on the idea from many different angles and invoking lively examples of the kind he was adept at thinking up, did his idea of horizontal motion being retained have a chance of moving from the paradox it was for all his contemporaries to the truism is has become since.

With one important reservation, however. To Galileo, 'horizontal' meant, literally, 'parallel to the horizon', that is to say, circular. Here as in numerous other instances Galileo's thought was bound by received notions to a larger degree than he stood ready to acknowledge, be it to others or to himself. At pains to combat Aristotle's opposition between natural and forced motion, he continued to the end of his days to operate with a similar dichotomy, even though defined differently. For Galileo terrestrial motion was 'natural' when always at equal distance from the center, and 'forced' when not. There is no way Galileo could at one stroke have enunciated the Newtonian principle of inertia, which is about rectilinear motion retained. No one can ever make a wholly clean break with his own past; it is up to successor generations to draw consequences that seem inevitable by hindsight only but that really require the benefit of being educated from the start into those mostly novel yet partly inherited views the original inventor had not, by definition, had a chance of being educated into. True, on the Fourth Day of the *Discorsi* Galileo came close to the persistence of rectilinear motion, yet he was at pains to point out that this was no more than approximately valid, permitted only in view of the small distances involved. Still, one sole message comes across loudly and clearly from the pages of the Second Day – Galileo has rightly been said to "instill in his readers an inertial feeling such as the ancients had never possessed, and which Kepler, too, still wholly lacked."[98]

From the introduction of this central idea Galileo moved on to lengthy discussions of its second corollary. This was his equally novel principle that, since motion is retained, an object can partake of several motions at once, just adding one more to the motion or motions it already had. If, and to the extent that, an observer is subject to the same motion or motions, he himself does not even notice it.

What we have here, in sum, is *a quite-unheard-of triplet of conceptions of motion to re-*

place their commonsensical counterparts in Aristotelian doctrine: The motion of an object is retained; it exists only relative to other objects in motion or at rest; and an object may be subject to a variety of motions at one and the same time. With this conceptual apparatus in place, Galileo was in possession of the means needed to dispose of the whole list of objections to heliocentrism arising from motion. No longer was he left to assert *that* a stone thrown upward does not cease to partake in the rotatory motion of the Earth and is therefore bound to fall back to the same place irrespective of whether the Earth rotates or stands still. He could now show *why* this is so. Other examples could be treated similarly – and Galileo was careful enough to go through each exercise in turn as if it were brand new – whether bullets, clouds, birds, balls falling alongside masts on sailing ships, and so on and so forth. The Second Day, then, is devoted in its entirety to eliminating one central set of objections to heliocentrism – yet *not* (as Galileo himself insisted) to adducing arguments counting positively for it.

The same goes for the First Day of the *Dialogo*. Galileo's objective on this Day was to show that the disruption of the Aristotelian cosmos implied in treating the Earth as a planet was not an unfortunate consequence of a move made for technically astronomical ends, and therefore a powerful argument against the acceptance of heliocentrism. Rather, it was the very move needed to accommodate an array of observations made in the sky over previous decades by Galileo himself (also to some extent by others, but Galileo was never willing to share his glory with anybody else). These discoveries had all been made with the telescope.

The disruption of the broad Aristotelian picture of the world.   Central to the Aristotelian conception of the universe was the strict separation between what is perfect, unchanging, naturally circular, and superlunary, and what is imperfect, subject to perennial change, naturally rectilinear, and sublunary (i.e., the terrestrial world; p. 21). To make the Earth one of the planets pulled the rug out from under such a rigorous dichotomy, as Copernicus' opponents realized far better than Copernicus himself. In book I of the *Dialogo* Galileo proceeded to make as accessible as he could what at the time of his original telescopic discoveries still remained subject to debate among the experts, namely, that the thrust of those discoveries was radically to undermine the distinction. In 1609 Galileo heard of the recent invention of a spyglass, which he turned into a telescope proper by directing it toward the heavens. Within a year he reported in a booklet that propelled him to sudden fame, *Sidereus nuncius* (The Sidereal Messenger), that the Moon has craters and mountains and valleys and seems made of earthy, rather than some kind of celestial, crystalline, or immaterial, stuff. Further, the Sun displays spots which move and which, according to Galileo's highly disputed interpretation at the time, are on, or adjacent to, the Sun's surface and rotate with it. Two decades later, on the First Day of the *Dialogo*, he devoted lengthy discussions to showing why there is no escaping the conclusion from, chiefly, these two observed phenomena that the Aristotelian distinction between the sublunary and the superlunary spheres just breaks down. Other considerations of his were not so straightforwardly empirical; for instance, an argument meant to show that there is nothing particularly attractive about a celestial realm as perfect and altogether changeless as Aristotle made it out to be.

The astronomical evidence.   Galileo's remaining arguments drawn from telescopic observations make their appearance on the Third Day, which is mostly devoted to the an-nual orbit of the Earth around the Sun. Here he adduced two other telescopic discoveries. One is that Jupiter, like the Earth, has moons (in 1609 Galileo discovered the four most lu-minous ones, which swiftly proved to be decisive to his own career; p. 417). The observation, if duly acknowledged, gets rid of the anomaly that, in the original Copernican system, the Earth is the sole planet in possession of a satellite. Furthermore, observation of the planet Venus through the telescope shows it to display phases as the Moon does. This is as it should be according to the Copernican hypothesis, whereas phases for Venus were ruled out in geo-centrism. Unless one took Venus to be self-illuminating rather than deriving its light from the Sun, its phases might well seem to make for a decisive difference.

So much for the major arguments put forth on the first three Days of the *Dialogo*. How did they work out on balance, from the point of view not only of Galileo but also of a non-committal reader at the time?

Copernicanism in the balance.   In the previous chapter I presented the pros and cons of Copernican heliocentrism as considered from the vantage point of 1543, when *De revolu-tionibus* appeared. At that time, very little appeared to speak in favor of that hypothesis in its realist interpretation, whereas a great deal spoke against it (p. 111). We examine now what had changed in the intervening ninety years.

The primary thing to have changed, so the attentive reader of the previous section will be tempted to remark, was that Johannes Kepler had turned a convoluted collection of circles moving in complicated ways without any kind of dynamic mechanism to govern the whole setup into a beautifully simple solar system in which six planets orbit the Sun in neatly ellip-tical curves with speeds varying in almost equally neat, predictable ways. Kepler made these discoveries public in 1609. He had explained them at length in 1621, in books written in rea-sonably clear Latin and available (though not too easily) in Italy as well. Kepler had eagerly responded to Galileo's overtures; in 1612 Galileo had received a letter from a patron of his, Prince Cesi, in which the elliptical shape of the planetary orbits was referred to as a matter of common knowledge;[99] yet nothing at all of this transpires in the *Dialogo*, despite the re-peated mention of views held by Kepler on other subjects. Instead, Galileo acted throughout the *Dialogo* as if the circular mess were all in Ptolemy's system and as if Copernicus, by the sole move of eliminating five major epicycles, had simplified planetary astronomy down to

188   six neat circular orbits uniformly traversed by the six planets. Thus, when Simplicio is reasonably made to ask "what anomalies are there in the Ptolemaic arrangement which are not matched by greater ones in the Copernican?" Galileo has Salviati respond: "The illnesses are in Ptolemy, and the cures for them in Copernicus".[100] This amounts to saying that the very subject of mathematical astronomy from the days of Hipparchos down to Kepler's time and beyond, the very subject to which Copernicus had devoted books II–VI of *De revolutionibus* – that is, the full range of observed irregularities in the motions of the planets – was ignored by Galileo. He simplified the problem in a manner that made his position as attractive to the lay readership he sought as it made him vulnerable to the experts, several of whom stood ready to tackle him. Not that Galileo's ignoring of Kepler's laws exposed him to attack, as the latter's theoretical work was mostly ignored anyway. The point is rather that Galileo might have availed himself at no cost whatever of a far more solid, technical-astronomical basis for his arguments precisely where the matter of relative simplicity came up. Worse still from a contemporary view was another feat, though – Galileo managed equally to ignore Kepler's onetime research director, Tycho Brahe.

True, Tycho's name, like Kepler's, comes up frequently in the astronomical passages of the *Dialogo*, but only for his observations, not for his geoheliocentric compromise system. And this is important because many astronomical arguments directed by Galileo against the Ptolemaean system – for example, the phases of Venus – failed to affect Tycho's compromise, which carries the same implication (as Kepler pointed out in his 1625 treatise *Tychonis Brahei dani hyperaspistes* [Tycho's ... Shield-Bearer]). In short, throughout the *Dialogo* Galileo set up a false dichotomy, which pervades even its very title,

> *Dialogue of Galileo Galilei, member of the Academy of the Lynx-eyed, extraordinary mathematician at the university of Pisa, and primary Philosopher and Mathematician of the Most Serene Grand Duke of Tuscany; where in course of meetings over four days the two great systems of the world, the Ptolemaean and the Copernican, are discussed by putting forward without arriving at a definite conclusion the Philosophical and Natural reasons for the one as well as for the other part ...*

Such tactics were all the worse since those expert astronomers whose support Galileo originally enjoyed, and might have continued to need most – those in the Jesuit Order – were well on their way to leaving Ptolemy's apparently indefensible bastion behind and were feeling their way toward Tycho's compromise. It is not that Galileo could possibly have avoided contradicting them on this score, yet little was to be gained by acting as if their preferred position did not even exist.

Next, in one respect the argumentative situation had actually deteriorated. No parallax had been found by Copernicus with the naked eye; no parallax had since been observed with a naked eye so exquisitely armed as it had been by Tycho; no parallax was *still* being observed through the very instrument capable of revealing so many celestial novelties – the telescope. Galileo might spend rhetorical flourishes on the idea of the sphere of the fixed stars turning around in twenty-four hours being no less odd than the assumption of a huge amount of void space between it and Saturn. But this could not conceal the fact that one inescapable inference from the Copernican hypothesis still failed to manifest itself. (Actually this particular argument could cut two ways, as there were those at the time who extended Copernicus' void space, introduced by sheer necessity, into the bold supposition of an *infinite* universe, thus turning an evident liability into something of an asset; but Galileo was not among them.)

The balance pro and con further reveals that, as Galileo was well aware, many of his best arguments, notably the ones he derived from motion or from his telescopic observations as hitherto reported, served not so much to speak positively for Copernicanism as, rather, to remove standing objections.

Was this bound to be all, then?

Or did he have conclusive proof?

In search of proof. This was the very question that was in effect put before Galileo in 1615 by Cardinal Roberto Bellarmino in his 'Letter to Foscarini'. The sheer fact that for seventy-two years no one had found this seemingly obvious question worth raising illustrates better than anything else how little occasion, prior to Galileo's advocacy of Copernicanism using *observed facts* for the first time, there had been for the uninitiated to take the idea of a moving Earth sufficiently seriously to worry about the exegetical consequences if it were adopted.

Bellarmino was the most authoritative theologian in the Catholic Church at the time. His point in the letter was that the Church had ways and means available to review, and if need be alter, interpretations of biblical passages customarily taken literally (as indeed it had done, for example, with the issue of the spherical shape of the Earth) but that it stood ready to wield these exegetical instruments only in matters securely proven. No such proof, so the cardinal ended his brief letter, had as far as he knew been forthcoming yet on the matter of the Copernican hypothesis taken as depicting reality, with the obvious implication being, Where, Signore Galilei, are your proofs?

Seventeen years later, near the end of the *Dialogo*, Sagredo pronounced three arguments out of the plethora taken up over four days to be "assai concludenti" (very conclusive).[101]

The first was the argument from the elimination of retrograde motion, which for Copernicus had been one of his arguments from harmony but which to Galileo, in his undue simplification of the Copernican system, meant the overall reduction of planetary theory to six simple circles uniformly traversed.

The second was an argument from telescopic observation never previously put forward by Galileo. The sunspots, he argued, do not just move across the surface of the Sun. Their observed trajectory, in showing the Sun's axis of rotation to be tilted against the ecliptic, can be accounted for in a satisfactory manner only in the Copernican, not in the Ptolemaean, setup. Remarkably, Galileo made Simplicio, in full command here of the doctrine of the relativity of motion, object that the phenomena present themselves no differently be they observed from an Earth at rest or in orbit around the Sun. He then went on to make Salviati turn the objection into a conclusive point in favor of Copernicanism by means of a convoluted argument, the merits of which are disputed to the present day.

Galileo's third and final argument, to himself the most conclusive of all, derives from the tides. It occupies the whole of the Fourth Day. It says that the continual alternation between ebb and flood can be explained only by the simultaneous action of the daily rotation and the annual revolution of the Earth, thus proving these conclusively. The chief difficulty is that on this account only one change of tides should manifest itself every twenty-four hours, whereas in most places there are two at less regular intervals. So Galileo had to introduce quite a few ad hoc suppositions, of a kind he knew so well how to scorn in others, to bolster his argument. There is evidence that already in the 1590s he had conceived of the idea, his dogged pursuance of which has always remained a bit of a mystery.

In any case, the *Dialogo* ends on another note than that of proof or disproof. After all, as Galileo recalled with marvelous irony in a preface 'To the discerning reader', there had appeared in Rome in 1616 "a salutary edict" which forbade one to hold or defend the Copernican hypothesis other than as a mathematical fiction, and to the zealous defense of which, in the face of frivolous accusations that it had been ill-founded, he pledged himself in the book that followed.[102] In other words, his hands were bound. So much transpires even in the very title of the book, where it is said that the debates that follow will proceed "indeterminatamente". After four busy days filled with debates from which no discerning reader could possibly have inferred anything else but that the author was wholly convinced of the reality of the two motions attributed by Copernicus to the Earth, and which only fools could deny, Galileo proceeded to keep this promise of inconclusiveness at the end of the Fourth Day. There he recalled with no less exquisite irony 'a wonderful and truly angelic doctrine' ("mirabile e veramente angelica dottrina"),[103] according to which, irrespective of what we humans may think about the causes of phenomena, God in His omnipotence might have arranged everything in quite other ways wholly beyond our understanding. Galileo managed to put this final consideration, which he knew from personal conversation to be the favorite argument of his friend, admirer, and patron-of-a-kind, Pope Urban VIII himself, into the mouth of none other than Simplicio. In this manner he secured for good the papal wrath. It had already been aroused by the very publication of the book and, thus enhanced, it exacerbated greatly the outcome, the infamous trial of 1633. Treatment of the trial and how it came about belongs to a later chapter (p. 419). Suffice it here to say that, on abjuring under

considerable pressure his Copernican conviction in public, Galileo, now sixty-nine years old, $\quad$
returned to what he had allowed to let rest ever since he had left Padua for Florence, to wit,
his new conception of motion and all that he had inferred therefrom.

The laws of fall and of projectile motion. $\quad$ Galileo gathered the major finds of his
Padua days in his last book, the *Discorsi* (Discourses and Mathematical Demonstrations
concerning Two New Sciences). It came out in 1638 in the Dutch Republic, a free haven for
those in Europe who had trouble publishing work at their home base. The three protagonists
of the *Dialogo* appear once again, though in somewhat shifted roles and with less literary
flourish – not a plea for a particular conception of the universe but the sober presentation of
two new, exact sciences was at stake here. The first of these, which is about the cause of cohesion in bodies and their resistance to separation, is treated on the first two Days (p. 315). In
the two Days that follow Galileo presents the most important treatise he had ever produced.
They are spent discussing the motion of falling and of projected bodies, with the Third Day
covering the former topic, and the Fourth the latter.

The subject treated on the Third Day is called by Galileo 'naturally accelerated motion'.
What kind of question is Galileo seeking an answer to? Sagredo begins by putting forward
the 'accidental lightness' account of free fall that Galileo had sought to elaborate long ago in
'De Motu' (p. 180). Salviati then shrugs off, not only the account itself, but the very question
to which it had been directed:

> The present does not seem to me to be an opportune time to enter into the investigation of
> the cause of the acceleration of natural motion, concerning which various philosophers have
> produced various opinions, some of them reducing this to approach to the center; others ...
> [two more examples follow]. Such fantasies, and others like them, would have to be examined
> and resolved, with little gain. For the present, it suffices our Author [i.e., Galileo] that we
> understand him to want us to investigate and demonstrate some attributes of a motion so ac-
> celerated (whatever be the cause of its acceleration) that the momenta of its speed go increas-
> ing, after its departure from rest, in that simplest ratio with which the continuation of time
> increases, which is the same as to say that in equal times, equal additions of speed are made.
> And if it shall be found that the properties that then shall have been demonstrated are verified
> in the motion of naturally falling and accelerated heavy bodies, we may deem that the defini-
> tion assumed includes that motion of heavy things, and that it is true that their acceleration
> goes increasing as the time and the duration of motion increases.[104]

This pithy statement sums up the principal conclusion drawn on the Third Day: the acceleration of falling bodies may be assumed to be uniform, and this proves true when checked
empirically. It also stresses that the conclusion has been reached by deliberately abstaining

192   from a causal search such as customarily undertaken hitherto. The point has been interpreted as a lack of interest in the causation of the uniform acceleration of fall, or even in causes
generally. In the context of Galileo's work as a whole it makes better sense to understand the
decision here presented by Salviati as a move specifically fitting this one, particular occasion.
Not until the prior question of *how* bodies fall has been settled does it become profitable
to return to the *why* question, which, owing to the discoveries meanwhile made by asking
'how?' can now be posed in a more precise manner, thus becoming susceptible to being answered in its turn. For the vague question 'Why does the speed of falling bodies increase on
being let loose?' which over time had yielded nothing but "fantasies", could now, thanks to
Galileo's discovery, be reformulated as 'Why do bodies fall with uniform acceleration?' Now
this, though hardly an easy question, is at least answerable, because it is put in a precise,
quantified, or at least quantifiable form allowing geometric treatment, measurement, and
computation. While it is certainly true that Galileo confidently left the answering of that
further question to posterity, much of the required geometric treatment was already being
performed by him on the Third and the Fourth Day. All in all he derived from one axiom
and one auxiliary postulate, and from theorems stated and proven along the way, thirty-
eight propositions with their corollaries. The final aim toward which the whole structure
tended was to establish the curve of quickest descent (wrongly identified by Galileo with the
circle). On the Fourth Day he went on to combine his axiom of uniform acceleration in free
fall with his rule of horizontal motion being retained. This allowed him to show with the
help of pertinent theorems in Apollonios' *Conics* that a ball thrown (or fired or whatever;
in any case, an object in horizontal projectile motion) partakes of both motions at the same
time and thus describes a parabolic curve (Figure 5.5).

He went on to use this result for the derivation of exact conclusions regarding the maximum range of gun shots. In short, the one novel question that Galileo had asked turned
out to yield at once a whole array of results different not only in outcome but in kind from
whatever had previously been argued verbally with respect to the phenomena of fall and of
projectile motion.

Still, it is not only this highly significant shifting of the question that made such unheard-
of mathematical treatment of free fall and projectile motion possible. What remains of the
scrap papers of Galileo's Padua years does not appear to allow a reconstruction uncontested
even in outline; still, at least some steps taken by Galileo seem to be beyond controversy.
One basic step has to do with the way he arrived at his principle of motion retained. Recall
first from p. 185 how Galileo introduced that principle in its full-fledged state in the *Dialogo*.
There he argued as follows: The smaller the tilt of an inclined plane against the horizon,
the smaller the acceleration or the slowing down of a ball thrown up or down the incline;
therefore, only the tiniest motion need be imparted to a perfectly spherical ball resting upon
a perfectly smooth plane not tilted at all but parallel with the horizon so as to make the ball

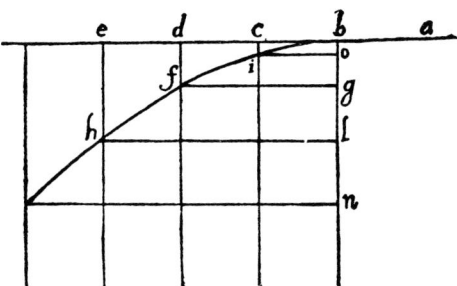

Supported on plane ab, a body moves uniformly from a to b and, on losing support, descends successively through i, f, and h.

Figure 5.5: Galileo: Parabolic trajectory of projectile motion

move horizontally forever. The general idea of horizontal motion providing a borderline case where opposite tendencies cancel out had occupied Galileo before, albeit as yet with less radical consequences. In his Pisan 'De Motu' of the late 1580s he had argued in passing that

> a body subject to no external resistance on a plane sloping no matter how little below the horizon will move down [the plane] in natural motion, without the application of any external force. ... And the same body on a plane sloping upward, no matter how little, above the horizon, does not move up [the plane] except by force. And so the conclusion remains that on the horizontal plane itself the motion of the body is neither natural nor forced. But if its motion is not forced motion, then it can be made to move by the smallest of all possible forces.[105]

In his Paduan *Mecchaniche* of c. 1593–1600 the same limited conclusion, with the same provisos about perfect sphericity of the ball, perfect smoothness of the tilted and horizontal planes, etc., is reached by means of a more extensive argument. The underlying idea of the "indifference" toward motion and rest of an object on a horizontal plane is now being made explicit.[106] There is good reason, then, to regard Galileo's apparently growing awareness of this indifference as the route his thinking took from just one fleeting paragraph in 'De Motu' toward the full-fledged conception of motion retained that he explicated at length in the *Dialogo* and applied in the *Discorsi*.

By the same token, this novel idea first entertained in Pisa and then elaborated early on in Padua of a vanishingly small force being sufficient to bring the transition from rest to motion about served Galileo effectively [to] demolish ... the barrier between [modern] statics and dynamics by introducing a bridge between mechanical [i.e., here, static] force and motive force: it could be used to show that only an infinitesimal difference existed between the force needed to maintain equilibrium and that required to initiate motion, and to argue that the two quantities should therefore be regarded, in functional definition and in magnitude, as identical.[107]

This, then, is the crucial step Galileo had to make to overcome Guidobaldo's Archimedean purism. This way he could replace the analysis of falling bodies offered in 'De Motu' in strictly Archimedean static terms with an analysis in which the analogy with the balance is given up in favor of a more flexible approach in keeping with the variability of the phenomena under consideration. And that is how Galileo came to spend a good deal of his most creative work on the problems of free fall and projectile motion during his eighteen Padua years. In the end, he succeeded in blending empirical trial and error (e.g., he established experimentally the true shape of the curved trajectory of projectiles) with conceptual analysis (e.g., his false but quickly corrected assumption that in vertical descent speed increases as the distance) and with empirical testing by often very ingenious means.

To test the outcomes was certainly necessary. It is one thing to assume on grounds of nature's simplicity that in free fall speeds grow in proportion to times elapsed, but quite another to show that such an arbitrary-looking axiom is actually obeyed by bodies actually falling.

In the polished form given to the argument on that Third Day of the *Discorsi*, the empirical test went as follows. Galileo first introduced one auxiliary postulate: whatever the angle of inclination, bodies let loose from equal heights will have equal speeds at the bottom. He could not do without the assumption, since there was as yet no way to measure time in perpendicular fall, so for empirical testing he had to slow down the descent by having it take place along an inclined plane. But an assumption it had to remain, as he could not rigorously prove it but only render it plausible. With the postulate assumed in addition to his original axiom of uniformly accelerated motion in free fall, he derived numerous theorems, one of which lent itself particularly well to empirical confirmation. This theorem states that

> if any moveable descends from rest in uniformly accelerated motion, the spaces run through in any times whatever are to each other ... as the squares of those times.[108]

Galileo claimed to have verified this property of falling bodies by means of a groove cut out of a wooden beam and glued over with vellum

> as much smoothed and cleaned as possible. In this there was made to descend a very hard bronze ball, well rounded and polished, the beam having been tilted by elevating one end of it above the horizontal plane from one to two braccia [c. 1 to 2 feet], at will.

The time of descent was then determined repeatedly by weighing the amount of water flowing meanwhile out of a tank, with hardly any difference between the various, measured
outcomes ever manifesting itself. Next, with the length of the path of descent varied in accordance with all kinds of proportions,

> by experiments repeated a full hundred times, the spaces were always found to be to one an-
> other as the squares of the times. And this for all inclinations of the plane ... with such preci-
> sion that ... these operations repeated time and again never differed by any notable amount.[109]

In short, even with this rather primitive mode of timing (Ktesibios' water clock would have
done a better job), results measured invariably matched results predicted without a fault.

It is not strange that some of the best historians of an earlier generation refused to take
such claims at face value. But a faithful reconstruction of the experiment carried out in
1961 has shown that confirmation can be reached this way; research on Galileo's scrap papers conducted in the 1970s then showed that he indeed conducted these experiments. Still,
there are basic problems concerning Galileo's experiments, the use he made of them, and
the veracity of the claims he went on to make on their behalf. These problems are all part
of the complicated knowledge structure that marked the abstract/realist-mathematical science newly wrought in their partly different, partly similar ways by Galileo Galilei and by
Johannes Kepler.

The knowledge structure of Alexandria-plus

The cumbersome yet accurate expression 'abstract/realist-mathematical science' points at
the key feature of the revolutionary mode of nature-knowledge that these two men pioneered. That key feature is as tacitly familiar to the initiate as it is baffling to the newcomer.
It rests in this: Outcomes of research conducted in this mode inform us about nature's realities, yet in an indirect, idealized sense only. Mathematical science in this new mode does
transcend everyday reality, yet not entirely so – unlike in Platonism it does not refer to a
separate realm of being. Things fall to the ground in everyday reality as they do in idealized
reality, yet in the latter they fall in a decisively different manner, which Galileo sought to
make operational by insisting time and again on polishing, on smoothing, on immaterial-
izing, as it were, the object that falls and the inclined plane along which he makes it fall. We
must distinguish here between three levels:

(1) *A basement level of everyday reality*, where loose branches or shot fowl or other irregu-
larly shaped, tangible objects fall through the air or a liquid in a fairly irregular manner
about which, beyond asking why, little can be said.

196   (2) *An upper level of idealized reality*, where perfectly spherical objects are imagined to fall through no medium at all to the ground, in a manner which on various, mathematics-governed grounds may plausibly be taken to be uniformly accelerated; one may also imagine the same process to take place along an immaterial, inclined plane so perfectly smooth as to prevent sliding, friction, and the occurrence of similar impediments altogether. Why call these circumstances impediments? After all, the air through which an object falls cannot, on the basement level of reality, be an impediment to the 'true' phenomenon of motion; rather, it is integral to it. Air friction, or the ball's sliding, etc., appear as impediments only from the point of view of this imagined upper level of idealized reality. One can *think* those impediments away, so to speak, yet how to *do* away with them in the everyday real world? This is done by introducing

(3) *a middle level* of *experimental reality*, best called the stairway of experiment, by which the upper level and the basement level are connected. Here balls and inclines cannot help being material and a medium cannot fail to present itself, yet one may to the best of one's equipment seek to offset their materiality by polishing, by smoothing, and by getting rid of the air; in short, by artificially *imitating* as well as can be done in our real-life world properties that are rigorously valid only on the upper level of reality.

By making experiment serve as a stairway between the basement and the upper level of reality, Galileo put experiment (incidentally used before only in a fact-finding context) to use in a wholly novel way. We can now better understand Galileo's presentation of this unheard-of, intermediate level of reality to a public accustomed to considering natural phenomena on the basement level only. After all, this is what the Athenian mode of nature-knowledge was all about – each philosophy of nature took empirical phenomena in their everyday sense, subsequently to catch them causally in the net of its grand scheme of things. Only those reared in Alexandria were accustomed to the idealization of natural phenomena (balance beams idealized into straight lines; consonant sounds into ratios of whole numbers, etc.). These few specialists, however, were in the habit of doing so *without* seeking to set up a link with reality, other than (in the sole case of planetary theory) to get one's predictions right while leaving the reality status of their mathematical models in the realm of useful fiction. In short, Galileo faced an audience wholly unaccustomed to the complex mingling of natural reality and mathematical idealization that marks the knowledge structure of the mode of nature-knowledge he pioneered.

It was to meet this difficulty that Galileo developed a knack for displaying precisely such real-life phenomena as come closest to representing the idealized level of 'true' motion (e.g., the ship to Aleppo; p. 184). But the same predicament also compelled him to present his experimental outcomes as if they were in rigorous conformity to outcomes reached theoretically on the idealized plane of reality. To an audience in no way accustomed to the latter and, hence, to the role of experiment as a mediator between an unknown, idealized level of reality and a solely known, everyday level of reality, every outcome other than that of perfect agreement must appear to count as a *rejection* rather than as a confirmation of preconceived theory. Today, we are taught the impossibility of experimental setups ever matching theory fully and are happy to regard measured values that come close to predicted ones (by degrees of approximation that we are taught to determine beforehand) as confirmations (for the time being) of our theories. Galileo's contemporaries naturally lacked this sort of education. To what extent Galileo was aware of the inevitable gap in exactness between the middle and the upper level of reality is hard to say; the point, however, is that he had good reasons to handle it in public as if it were nonexistent.

Which is not equivalent to saying that nothing can be asserted at all about Galileo's own conception of the relation between those three levels of reality that I have distinguished and that he was to all intents and purposes the first to explore.

Galileo's views on natural reality and its distinct layers.   First and foremost, Galileo was a realist. He did not believe in just probable knowledge, nor did he believe that our knowledge is confined to the phenomenal level of things only. Rather, the mathematical philosopher (as he insisted on calling himself; p. 417) can find out what things really are like. However, whereas God's knowledge of the world is both complete and instantaneous, our knowledge is inevitably finite and acquired through much laborious effort.

Galileo's idea of the world is that it is fundamentally mathematical: "The Book of Nature … is written in the language of mathematics."[110] In the vocabulary here adopted this expresses his conviction, not only that an upper level of 'idealized' reality may be distinguished, but also that it expresses something about the basement level of everyday reality, in that it forms the latter's essence. *There is a mathematical essence in the things themselves:* In this sense (but in this sense only) did Galileo in effect blend Plato's belief that the essence of things is in a transcendent domain of Ideal Forms modeled after the structure of mathematics yet not themselves mathematical with Aristotle's belief that the essence of things is in the things themselves.

Still, Galileo held varying views on the relation between the upper and the basement levels of reality. There is an exchange on the Second Day of the *Dialogo* where, in facing squarely this very issue, he appears to opt for the upper level as the sole kind of reality that counts:

SIMP. This [i.e., a geometrical theorem involving spheres, just proved by Salviati] proves it for abstract spheres, but not material ones. ... Material spheres are subject to many accidents to which immaterial spheres are not subjected. Why might it not be that a metallic sphere being placed upon a plane, its own weight would press down so that the plane would yield somewhat, or indeed that the sphere would be mashed at the contact? Besides such a plane

198   can hardly be perfect, if for no other reason than that matter is porous; and maybe it will be no less hard to find a sphere so perfect that it has all lines drawn from the center to the surface exactly equal point by point.

SALV. Oh, I readily grant you all these things, but they are beside the point. For when you want to show me that a material sphere does not touch a material plane in one point, you make use of a sphere that is not a sphere and of a plane that is no plane, so that, by your own statement, such things [spheres and planes] are either not to be found in the world, or if found they are spoiled upon being used for this effect. ...

SIMP. ... doubtless it is the imperfection of matter which prevents things taken concretely from corresponding to those considered in the abstract.

SALV. What do you mean, they do not correspond? Why, what you are yourself saying right now proves that they correspond exactly ... even in the abstract, an immaterial sphere which is not a perfect sphere can touch an immaterial plane which is not perfectly flat in not one point, but over a part of its surface, so that what happens in the concrete up to this point happens the same way in the abstract. It would be novel indeed if computations and ratios made in abstract numbers should not thereafter correspond to concrete gold and silver coins and merchandise. Do you know what does happen, Simplicio? Just as the calculator who wants his computations to deal with sugar, silk, and wool must discount the boxes, bales, and other packings, so the geometrical philosopher, when he wants to recognize in the concrete the effects which he has proved in the abstract, must deduct the material hindrances, and if he is able to do so, I assure you that things are in no less agreement than arithmetical computations. The errors, then, lie not in the abstractness or concreteness, not in geometry or natural philosophy, but in a calculator who does not know how to make a true accounting.

Even so, there is something odd about Salviati's final conclusion:

Hence if you had a perfect sphere and a perfect plane, even though they were material, you would have no doubt that they touched in one point; and if it is impossible to have these, then it was quite beside the purpose to say that a bronze sphere does not touch in one point.[111]

However ineluctably the conclusion follows from the premises, it also begs Simplicio's original question. For Simplicio had not denied that one can in the abstract imagine a nonperfect sphere. Rather, he had denied that one can in the concrete construct a perfect one and, hence, that results reached in the abstract can be valid at the level of concrete reality. Still, the passage brings out in a brilliant manner Galileo's ability to show how hard to grasp the procedures of his brand-new, abstract/realist-mathematical science inevitably were for those not at all used to thinking in terms of two levels of reality, one mathematical, one not, and of the nature of the connection between them. In his time, not even the Platonists thought that way – no one did. We at present, with centuries of science education behind us, can hardly fail to think that way. So on seeking to reconstruct what Galileo was doing here, I take this passage to provide the grounds underlying his claim that, in abstract/realist-mathematical science, outcomes predicted mathematically and outcomes reached experimentally can match without a fault.

Which seems tantamount to saying that, strictly speaking, to perform experiments is superfluous – mathematical reasoning alone suffices. And indeed, there are places in Galileo's work where he affirms so much. For example, in the *Discorsi* Salviati is made to say at one point that "the knowledge of one single effect acquired through its causes opens the mind to the understanding and certainty of other effects without need of recourse to experiments".[112] Here and there he even acts in accordance with this prescription. For instance, the Second Day of the *Dialogo* contains a discussion of stones falling alongside masts on sailing ships. Galileo manages within the space of just one page to make Salviati rebuke Simplicio for having failed to see for himself what happens experimentally and then to make the same Salviati declare that he, Salviati, has not bothered to carry out the experiment since, in solid possession of the mathematical rule, he already knows what will happen.

For students of Galileo's work, he is as much a pleasure to deal with as he is a pain in the neck. Very little in the sum total of his writings is just unambiguously *such* and in no way *so*. Take the present issue – even in the selfsame *Discorsi* there are passages with quite a different slant. Early on the Fourth Day the presuppositions are debated on which the deduction of the parabolic trajectory of projectiles rests. Sagredo and Simplicio adduce familiar kinds of real-life objections: The horizontal component of projectile motion is not truly rectilinear; "it is impossible to remove the impediment of the medium", etc., so that, on conclusion, "all these difficulties make it highly improbable that anything demonstrated from such fickle assumptions can ever be verified in actual experiments". But now Salviati is made to respond in a much humbler vein. Acknowledging at the outset that "all the difficulties and objections you advance are so well founded that I deem it impossible to remove them", he goes on to invoke the same way out that had traditionally been granted to Archimedes, namely, that assumptions of exact rectilinearity, etc. are "liberties [to be] pardoned" because, given the small distances involved, they are very good approximations to the real state of affairs.[113] So Salviati, still in the same humble vein, now concludes that

no firm science can be given of such events of heaviness, speed, and shape, which are variable in infinitely many ways. Hence to deal with such matters scientifically it is necessary to abstract from them. We must find and demonstrate conclusions abstracted from the impediments, in order to make use of them in practice *under those limitations that experience will teach us.*[114]

Here as elsewhere, the point of the discussion has not been to catch Galileo in the act of committing an inconsistency. Or rather, such is precisely the point, though not for reasons of cheap exposure but rather to show how much he was, inevitably, grappling with these wholly novel, unheard-of problems that his inauguration of Alexandria-plus put before him. How are the upper and the basement levels of reality related to one another? Where is the center of gravity of the experimental stairway that connects them situated? Is its weight concentrated near the upper level of mathematical abstraction (as Galileo appeared to affirm in the passages quoted first) or rather in the immediate neighborhood of the basement level of real-life concreteness?

Right at the onset of the Scientific Revolution, then, there was an as-yet-unresolved problem concerning the status of experiment in mathematical science. At its center stood the capacity of experiment to reflect real-life experience in spite of its built-in artificiality. What, meanwhile, did Kepler's partly different, partly similar exploration of abstract/realist-mathematical science reveal at the time about its knowledge structure?

Kepler and the realities of nature.   Unlike in Galileo's domain of terrestrial motion, in Kepler's chosen domain, planetary theory, making mathematical science realist required very little removal of real-life impediments, be the removal done materially or mentally. Real-life impediments do present themselves at the level of observation, due to distortions of light rays in the atmosphere and to inaccuracies in our instruments and in our reading of them. But these are impediments insofar as they serve as sources of imprecision in need of correction – they do not occur until astronomical entities (the trajectories of Sun, Moon, planets, stars) enter the terrestrial domain. That is, beyond the Earth's atmosphere the basement level and the upper level coincide; no stairway connecting them is called for. Only one step was needed for turning mathematical astronomy realist. That was to strip it of its Ptolemaean aura of wholesale or erratically partial fictitiousness and anchor it firmly in the reality of what Kepler called his celestial physics.

Where the issue of empirical confirmation is concerned, this unprecedented move had radically novel consequences for Kepler, too, in a manner not quite like and not quite unlike Galileo's case. He faced it with less ambiguity. In mathematical astronomy of the standard Alexandrian variety, if a mismatch between predicted and observed celestial events occurred, one could always adapt some component model so as to make it 'save the phenomena' after all. For a 'celestial physicist' it now suddenly became imperative to subject his abstract/realist-mathematical hypotheses to exacting, empirical tests. We know already, from Kepler's moving words on God's goodness and Tycho's treasure, how at a decisive moment he met the challenge.

The mathematization of nature.   Ever since Alexandre Koyré coined the phrase, the
achievement of Kepler and of Galileo has with deserved frequency been defined by saying
that their accomplishment inaugurated the 'mathematization of nature'. Nothing at all is
wrong with that felicitous expression, as long as we keep in mind that they did so in two
quite different ways. With Kepler the emphasis is on 'nature' – he found a science already
mathematical for many centuries and went on to show why, and how, it is really about *nature.* With Galileo (one more contrast between the two men, this time not arising from their
different personalities or approaches but from the logic of the different situations in which
they found themselves) the emphasis is rather on 'mathematization' – he dealt with some vital aspects of nature in a novel, *mathematical* way. But even this expression is not quite right,
for Galileo first had to show why free fall and 'horizontal' and projectile motion are vital aspects of nature, which they are only at the upper level of reality, hardly at the basement level.
In this sense alone was Galileo's job the more arduous of the two. Here resides the deepest
reason for why, even more so than Kepler, Galileo has customarily been taken as the central
figure in the history of how modern science came into the world. This has little to do with
the number of 'mistaken' views either of these two men held in terms of present-day science
(even though on this unhistorical criterion Galileo is decidedly the more 'modern' figure)
and much more with those inherent properties of abstract/realist-mathematical science just
outlined.

So far my analysis of the intricate knowledge structure of abstract/realist-mathematical
science, marked by that forever-confusing blend of, on the one hand, abstract derivation
and, on the other, claims made for it to represent the very realities of nature. From now on I
drop the adjective 'abstract' from that cumbersome expression. To mark off Alexandria-plus
from Alexandria I shall henceforth speak of 'realist-mathematical science'. In *both* cases we
are dealing with profoundly abstract modes of nature-knowledge, to be sure, but Alexandria
was so almost exclusively, while the abstractness of Alexandria-plus was mitigated by the
variety of reality-infused ways just surveyed.

Causes of the first transformation

So much for the difference that Kepler and Galileo made in turning Alexandria into Alexandria-plus. In previous sections I have *measured* the difference between these two modes of
nature – knowledge; the time has now come to seek to *explain* it.

Recall to that end our imagined observer filing away her imagined report on the state
of European nature-knowledge around 1600 and its probable extrapolation (p. 150). The
report's careful predictions have by now turned out to be badly misconceived, so we must
ask how extrapolation could fail so abysmally. What, if anything, caused the transforma-

202   tion – more than that, the altogether quite sudden and really unpredictable twist in a time-honored pattern, that is to say, the *revolution* – that took place instead?

What needs to be explained, then, is neither some imaginary, historically impossible creation out of nothing, nor just an ongoing sequence of events taking their smoothly next turn, but something in between: a radical *twist* such as extraordinary individuals may under propitious circumstances bring about. To remind ourselves of what exactly it was that got twisted, let us summarily retrace the Alexandrian line of development. In the mathematical mode of nature-knowledge, certain properties of certain bits and pieces of empirical reality – some of inherently geometric shape, like light rays or vibrating strings; others first given geometric shape through 'idealization', like floating bodies or planetary trajectories – were derived by simple arithmetic or by Euclidean geometry. The outcomes (e.g., specific states of submersion or the ratios for the consonances) were either kept confined to or were at once drawn back into the same domain of abstract mathematics. As a rule, then, an at least *twofold* process of geometric abstraction was being applied in the process.

This is how it was done in Alexandria and its spheres of influence (Rhodes, Syracuse). This is likewise how it was done, and extended, in Islamic civilization. This is *not* how it was done, let alone extended, in medieval Europe, but this is how it was done and once again extended in Renaissance Europe up to 1600. This is *not* how it was done by the two extraordinary individuals who gave so marked a twist to the standard pattern by drastically enhancing the reality content of mathematical science, thus reducing the manifold process of geometric abstraction in Alexandrian science to its onefold counterpart in realist-mathematical science. But this is how it *kept* being done and extended for some four to five decades beyond 1600 at the hands of all mathematical scientists but Galileo's own disciples and a few early converts. Prominent among the late practitioners in an Alexandrian vein were Thomas Harriot, Willebrord Snel, and (in one facet of his variegated scholarly personality) René Descartes. Almost simultaneously these three men cracked one tough nut in Alexandrian mathematical science: the rule governing refraction. It was well known that the rule that Ptolemy had inferred from his measurements on light rays through air and water was at best approximately valid. In Islamic civilization the sine rule of refraction was found in the 10th (4th) century by Ibn Sahl. In Europe between 1601 and c. 1628 the sine rule was found independently by Harriot, by Snel, and by Descartes. This confirms the structural likeness between the pursuit of nature-knowledge in Islamic civilization and in Renaissance Europe, but it also shows that, while in Prague and in Padua the Alexandrian pattern was being thoroughly transformed by two individuals operating on their own, elsewhere in Europe mathematical science in standard Alexandrian fashion remained quite as unscathed and quite as capable of gradual, piecemeal enrichment as ever. *Or even a little more so.* Does not the fact that the discovery was made three times in a row, and also within a narrow time span, suggest that, for all the qualitative similarity between the state of Alexandrian science in Islamic civilization and in Renaissance Europe, the latter involved a significantly larger number of practitioners and forged ahead with significantly greater intensity? If that is so, we have put ourselves in possession of at least one historical circumstance favoring (though certainly not demanding) the occurrence of the radical twist we seek to explain here. For consider from all the foregoing what our explanatory task really comes down to.

The core explanation: An inherent possibility realized. In my treatment of Kepler and Galileo I have insisted on the extent to which they fitted into the tradition of mathematical humanism. Like Regiomontanus or Copernicus or Benedetti before them, and like Stevin or Snel contemporary with them, they took on the Alexandrian corpus with a view to helping restore it further and in so doing extend it so as to fulfill more thoroughly the destiny apparently set for it by the ancients. Only in the course of their respective endeavors did they begin to conceive of ideals far beyond that ancient destiny. Nor did either man cease to regard his new ideal as well in line with his original, Alexandrian point of departure, Ptolemaean/Copernican with the one, Archimedean with the other.

In establishing all this and more in previous sections, I have now given sufficient flesh to a thesis summarily announced in my first chapter on the Greeks. The first revolutionary transformation actually accomplished around 1600 in Europe rested in the Alexandrian corpus as an as-yet-unrecognized and unrealized possibility (p. 47). It is profoundly misleading to regard the onset of the Scientific Revolution as a "cataclysmic event" bursting on the scene out of nowhere – the favorite straw man view for many historians who over recent decades have made their varied, continuist points against it.[115] It is hardly less misleading – under the lingering influence of what Pierre Duhem once freshly argued in the 1910s – to consider the event as fundamentally a continuation, one way or another, of medieval precedents. The analytically and also causally more fruitful line is to consider the event as, at bottom, the revolutionary transformation of one particular portion of the Greek body of nature-knowledge as received and further enriched by a range of practitioners in Islamic civilization and then in the Renaissance movement of mathematical humanism – the twisting, that is, of Alexandria into Alexandria-plus.

This core explanation determines the explanatory task that remains, at least insofar as it pertains to Europe's prime feat of transforming nature-knowledge (for two *other*, near-simultaneous transformations will require mostly *other* explanations). The task now before us is no more than twofold. It is to account for the twisting and to account for the plus.

Whence the twisting? Two preparatory steps require a quick recall. It follows from the nature of the twist as an inherent yet so far unrealized developmental possibility that the emergence of realist-mathematical science out of the Alexandrian corpus in early-17th-century Europe was not, on the one hand, a *fortuitous* event – it had been there all along as

204 an inherent possibility. Nor, on the other hand, was it *foreordained* – it might or it might not have happened, yet even when it might it need not. Conceivably, Archimedes' and Ptolemy's known works might today still represent the summit of individual scientific achievement, so that we would still live in the premodern world, with the astrolabe or the mechanical clock as the supreme piece of toolmaking, and death within a year of birth as the likeliest human fate by far. This is so in part because of absent necessity but also, and more interestingly, because inherent possibilities cannot be realized anytime, anywhere.

This truism means two things. It means that realization can occur only (if indeed at all) under certain historical constellations, not under others. But it means principally that the opportunity for such a feat of realization actually to occur must be able to be repeated in essentially similar fashion. That is to say, a given corpus of ideas like the one I am treating here must not be so tightly bound to a time or place as to withstand a replay of any kind anywhere or anytime else; also, the temporally and/or spatially new context must be such as to allow for some kind of sustained reception. Precisely here rests the world-historical significance of the Abbasid translation movement of the 8th (2nd) century and beyond. In adopting and creatively enriching the Greek corpus, those to embark upon it showed what no one could have known for sure previously, which is that the corpus was capable of transcending its original situatedness in the culture of Hellenism. The translation movement has rightly been said to

> demonstrate for the first time in history that scientific and philosophical thought are international, not bound to a specific language or culture. Once the Arabic culture forged by early Abbasid society historically established the universality of Greek scientific and philosophical thought, it provided the model for and facilitated the later application of this concept in ... the Latin West.[116]

Here is what, for present purposes, all this comes down to. I argued earlier that this transformation wrought in Europe around 1600 rested as a hitherto-unrealized possibility in the Alexandrian corpus and also that at the onset of the Islamic Golden Age that body of knowledge proved capable for the first time of transcending the confines of the civilization from which it sprang. Consequently, every single case of transplantation of that corpus offered new chances for the transformation actually to occur. That is why (as stated baldly on p. 72) there is no decisive, *inherent* reason why no Galileo-like figure appeared by the 11th (5th) century to cap the ongoing enrichment of the corpus by men like Ibn al-Haytham or al-Biruni. Still, those fresh chances differed in every case, and that is why causal analysis of the specific transformation we are discussing here must be done by pointing at case-specific factors less or more propitious for the realization of such an outcome.

Why medieval Europe hardly provided a favorable environment for the decisive twist to be brought about is obvious at a glance. With the Alexandrian legacy partly falling to the wayside and partly being disfigured in scholastic categories, not even enrichment, however modest, was feasible but only, at the very best, the incidental quantification of a few Aristotelian categories displayed in Oresme's work (all of which was of course far from lacking merit in its own right, yet required the novel frame of realist-mathematical science to come to any fruition; p. 307).

The true problem facing us is why the high level of enrichment of the Alexandrian corpus in *both* Islamic civilization and Renaissance Europe eventually petered out in the one case and gave rise to wholesale transformation in the other. What propitious factors emerge from sustained comparison between the two? At first sight plausible candidates are improved tools of communication and the larger number of practitioners.

It has been argued that the hugely expanded and improved communication offered by commercial print shops in Renaissance Europe was the decisive causal factor in the emergence of modern science. This monolithic thesis needs to be cut down to proper size. It is true that in one specific case the ongoing "shift from script to print" will prove indeed to have been an indispensable component in propelling the Scientific Revolution forward (p. 610).[117] But for our present transformation of revolutionary proportions the shift appears to have had little effect. One indication that print had hardly yet replaced script in fostering discovery in mathematical science is the story of the threefold finding of the sine rule. Harriot found it first, in 1601. Kepler would have picked it up without fail if only he had kept corresponding with the Englishman, who never published it, for a few more months. It was independently found again in the early 1620s by Snel, who likewise did not publish it. In the late 1620s it was found a third time by Descartes, either (in all likelihood) independently or using information picked up from one of the late Snel's former colleagues. Unlike Harriot or Snel, Descartes went ahead, in his *Dioptrique* of 1637, at long last to inform the world of a property known already from the moment Ibn Sahl buried his find in manuscript – a habit apparently not quite shaken off yet by every mathematical scientist when the early 17th century dawned. For where, in this hardly untypical story, is the print? Where, more pointedly, is the role of print in the transformation wrought by Kepler and Galileo? With neither man was it a matter of remaining abreast of the latest literature that made him ripe for his creative breakthrough, but rather a growing conviction of its relative worthlessness. The conviction went together with thorough mastery of precisely such printed classics as in the scribal culture of Islam could already be ordered to be carefully copied by a scribe for a few dirhams. It is certainly true that, in book *illustration*, print did make a decisive difference almost right from the start; but for the Alexandrian variety of nature-knowledge this counted least of all. Without the stunning pictures of the Moon's craters Galileo's *Siderius nuncius* would not have made nearly the splash it did make; but this book, however fateful for Galileo personally and for his Copernican crusade, affected his revolutionary transformation of Alexandria into Alexandria-plus in an indirect manner only.

206 With *number of practitioners* things are different. Trivially obvious as it may sound, the more intensively work in a creative spirit was being done on the Alexandrian corpus, the greater the chance that someone would hit upon the unrealized potential hidden near its core. Here Europe held an advantage over Islamic civilization simply by virtue of being next. And there is another, more intricate side to the numbers factor. The European university system, with its comparatively huge turnout of people with at least a nodding acquaintance with Greek thought, provided a deeper and also wider soil on which true mathematical talent could grow and on occasion flourish than in Islamic civilization. A significantly larger number of both secular and ecclesiastical courts (as likewise a variety of educational institutions) were in a position to absorb such university-schooled talent, thus multiplying chances for it to come to some measure of fruition. If we count from the onset of the Baghdad translation movement, by the late-8th (mid 2nd) century, to the end of the Golden Age of nature-knowledge in Islam, by the mid-11th (early 5th) century, 41 mathematical scientists were of sufficient repute to make it into the *Dictionary of Scientific Biography* in the 1970s. For Europe between 1453 (the fall of Byzantium) and 1600 (onset of the Scientific Revolution) the number is 105. This yields averages per century of *16.4* for 'early' Islamic civilization and *70.0* for Renaissance Europe. That is, numbers of mathematical scientists during the upswing stand in a ratio of *1 : 4.3*.

The ratio is of course anything but exact. Still, it is solid enough (p. 219) to provide a good impression. And it had noticeable effects. It provides the background to the heightened intensity of discovery just illustrated by means of the sine rule of refraction. It comes to the fore likewise in the remarkable procession, of some twenty years' duration, of over thirty research assistants to pass through Uraniborg at Tycho Brahe's beck and call. All these men had picked up enough quadrivial knowledge to advance, under his tutelage, toward some more intricate portions of planetary astronomy.

Not only was nature-knowledge pursued the Alexandrian way in Renaissance Europe with greater intensity than in Islamic civilization or a fortiori in ancient Alexandria itself, but a related effect was that talented men were drawn to the cultivation of mathematical science who under less propitious circumstances would have been lost entirely or would have been drawn into quite other careers. Galileo, himself the son of a 'musical humanist', smoothly came to mathematical science through the 'Urbino school' along the pathway of patronage, which dovetailed nicely with the direction the greatest of his varied gifts drove him to. In contrast, the career of Kepler, who halfway through his theological studies received an appointment as a mathematics teacher, might easily have been sidetracked into quite other channels. Still, in the European context there was nothing untypical about his career. A counterpart in Islam seems most unlikely.

Please note carefully the very limited scope of this consideration. It is not meant to rule out the possibility of someone achieving broadly what Kepler achieved over his lifetime in a context other than the European. It is meant solely to illustrate that paths toward a career in mathematical science were in Europe more numerous and more varied than before or elsewhere. And the more numerous and varied those career paths were, the greater the chance that some extraordinarily gifted individual might come forward to perceive what no one else had so far perceived.

What, then, did those two extraordinarily gifted individual perceive? I have dealt at length with the decisive innovation of their shared (albeit rather differently arrived at) *realism* in linking mathematics to the empirical world in a novel way. As Kepler phrased the point, "I take more delight in geometry expressed in physical things than in the abstract".[118] What decisively distinguishes Alexandria from Alexandria-plus, then, is this radical drawing of the basic modalities of mathematical science into empirical reality, though not to be sure the reality of our everyday experience but rather that peculiar reality of nature mathematized, which is still decisively more real than the twice or even thrice abstract nature of mathematical science in the mode of Archimedes or even Ptolemy. Whence, then, these two giant strides from Ptolemy to Kepler and from Archimedes to Galileo?

Whence the plus? In line with what has just been said, the search for propitious circumstances ought now to be focused upon the question of what, over and above their extraordinary perceptiveness, helped turn these two lone pioneers into mathematical realists in the sense defined.

Was their revolutionary transformation due to the *importation of foreign knowledge*? Arguments have repeatedly been made that this or that discovery first made in Islamic civilization and later made known in Europe had the unintended effect of enabling Europe to reap the full fruit of what its predecessor in nature-knowledge of Greek provenance had sown. Did not those Europeans who, from the 12th century onward, occupied themselves with problems of light and vision benefit to the point of indispensability from Ibn al-Haytham/Alhazen's perspectivist conception? Did not Copernicus avail himself of the Tusi couple for his radical reform of mathematical planetary theory? Does not the depiction of the constellation Perseus in Bayer's pioneering star atlas of 1603 emulate a 5th/11th century Arabic manuscript drawing?

The answer to these and numerous similar questions is undoubtedly 'yes'. But what does that mean; how far do the implications actually stretch? They remind us for sure that every subsequent recipient of the Alexandrian corpus received it in somewhat-enriched fashion. But the central feat at issue here – the mathematization of nature – is not touched thereby. This feat, after all, cannot just be decomposed into a range of incidental discoveries. Purported explanations of the revolutionary transformation from Alexandria into Alexandria-plus must come to terms with the core achievement itself. Even the Tusi couple, for all its providing Copernicus with a handy alternative to his chosen enemy, the equant point, hard-

208   ly begins to point in the direction of factors propitious for turning mathematical science realist.

If, then, no plausible resources of realism can be seen to have come from individual discoveries in Islamic civilization, let alone in the European Middle Ages, we must ask whether more plausible resources rested in the two rival modes of pursuit of nature-knowledge current in Europe by the late 16th century. Do we find propitious factors in Europe's practice-oriented, coercive empiricism and/or in the comprehensive understanding of the real world sought for in natural philosophy? Inside those two broad resources I shall now specifically explore half a dozen, partly overlapping stimuli toward realism in the sense defined, none of which had much of a counterpart in Islamic civilization. These possible stimuli are the significance to Kepler of Tycho's observations; the tradition in practical mathematics Galileo encountered in Padua; the unprecedentedly wide intellectual and social chasm obtaining in Europe between the mathematical scientist and the philosopher; Copernicus' ambiguous brand of cosmological realism; Kepler's personal vision of "the astronomer as a priest of God to the book of nature";[119] and Jesuit promotion of mixed mathematics.

I shall first consider *possible stimuli toward realism arising out of Europe's coercive empiricism*. I have insisted on the continuous existence of thick walls between mathematical humanists and practitioners involved in those varied, Europe-colored activities here jointly labeled 'coercive empiricism'. If the respective adherents took notice of each other's activities at all, they were wont to do so in a polemical vein. For instance, Kepler took almost personal offense with Ramus' outspoken preference for mathematics as practiced on the streets (p. 131) over the allegedly sterile, axiomatic structure of Euclid's 'Elements'. Still, the walls, however thick, were permeated at a few isolated spots.

One case in point is Kepler's creative usage of the data patiently accumulated by Tycho. These enabled him, among other significant feats, to perceive as no one else did the full consequences of the dissolution of the heavenly spheres that those data appeared to entail. Another is Galileo's exposure, during his Padua years, to the current application in practice of basic mathematical insights. This contributed significantly to how the pure Archimedean of the Urbino school, and even the 'Archimedean with a difference' of his Pisan days, turned into the man drastically to transform that body of knowledge (p. 180). This is not to say that no other paths to transcend the legacy might have been taken under other circumstances – a possible counterpart in Islamic civilization would surely have done it differently. It is only to say that the presence of an environment enriched by a broad, control-oriented, and accuracy-seeking empiricism may reasonably be taken to have contributed its share to transforming the Alexandrian corpus.

Next, I shall consider *possible stimuli toward realism in connection with natural philosophy*. Of course, philosophy might serve as another ready-made source of realism – an understanding, grounded in first principles, of the real world was the very thing the natural

philosopher was aiming at. But how was that source to be tapped? And what might make
it attractive for anyone located outside the realm of philosophy to do so? Here the lingering effects of the drastic curtailment of the Greek corpus during the medieval episode were
significant in many ways – never had mathematical science sunken so low in comparison.

The fall of Byzantium had given the Alexandrian corpus a chance to come into its own
and to enter the European scene other than through, at best, the backdoor. Even so, mathematical humanists had to fight an uphill struggle to regain some recognition (in terms of
both prestige and income) for their own concerns, as compared to those of the philosophers, whose unprecedented near monopoly during the medieval period had put them in
secure possession of the first institutionalized system of higher learning ever established.
Regiomontanus' and others' speeches 'in praise of mathematics' made the point quite clearly
(p. 100). They drew a sharp contrast between the certainty of knowledge gained the mathematical way and the perennial intellectual rivalry, hence less-than-full certainty, that beset
the domain of the natural philosopher. The latter point, directed against the philosophers,
had been made before, by Ptolemy (who had called natural philosophy 'guesswork'), by al-
Ghazali (who had gone on to seek an indubitably secure stance in religious mysticism), and
by the skeptics (who lacked a positive alternative). Now mathematical humanists availed
themselves of the point all over Europe. But only in rare cases did their advertising the virtues of mathematical science and its timeless verities lead to transgression of the boundaries
separating mathematics and philosophy. One of these rare cases was provided by *Copernicus*.

It may look as if throughout my account the contribution made by Copernicus to the
Scientific Revolution has been almost ludicrously downplayed. The principal reason I have
not followed custom and taken the publication of Copernicus' *De revolutionibus* in 1543 as
the onset of the Scientific Revolution is that that seminal book, as likewise the story of its
reception over the half-century that ensued, fits far too seamlessly the ongoing mode of
recovery-with-some-enrichment of the Alexandrian corpus. It is far more enlightening to
regard Copernicus as Ptolemy's last and greatest heir, who throughout books II–VI of *De
revolutionibus* carried the hoary art of 'saving the phenomena' to new heights by means of
his heliocentric, Aristarchos-inspired hypothesis. Only in retrospect can this change be seen
to have been instrumental, through the unambiguously realist interpretation of Copernicus' heliocentrism given to it by two exceptionally perceptive men, in bringing about the
upheaval in the way of doing mathematical science to which Copernicus himself remained
faithful in almost every respect. However, those two exceptionally perceptive men, whose
exceptional perceptions we are here seeking propitious factors for, did find one ready-made
source of realism in the introductory book I of *De revolutionibus*. In adducing a motley
selection of ad hoc natural-philosophical arguments in the context of a simplified version
of his system, Copernicus made the most of the realist claim laid down there that the Earth
really and truly orbits the Sun in a year and really and truly rotates around its own axis

210 every twenty-four hours (p. 109). That claim was at bottom irreconcilable with what he went on to do over the full remainder of his book. There he settled down to the real business of working in all required planetary details, which he did using time-honored fictional gadgets honed by Ptolemy. Whence, then, Copernicus' claim for the reality of heliocentrism if he did not even seek to uphold it in books II–VI of *De revolutionibus,* where he carried out his principal job as a mathematical scientist, and if he failed (as he knew he had) to produce convincing arguments for it in book I?

For an answer, recall first that Copernicus was once again following Ptolemy's example. Ptolemy had likewise put some considerations selected from a variety of philosophies of nature at the head of his *Almagest,* and he had preceded Copernicus in his felt need some-how to link mathematical science with pieces of natural reality philosophically conceived (p. 109). But however limited Copernicus' realism was, it did go a step farther than Ptolemy's bridge building. There is little in the literature to explain it. Copernicus may have shared in the upward social striving of his fellow mathematical humanists, sensing that to snatch a piece of the philosopher's gown might enhance his stance in society. Less controversial is the observation that over the period 1543–1600, when the fictional schemata of books II–VI were as widely commented on and applied by the common run of mathematical scientists as book I was ignored by them, just a dozen scholars at most were prepared to take Copernicus' realist claim at all seriously. And only two of them did so, not only without any remaining ambiguity, but also from a profound awareness that the inner tension in Copernicus' helio-centrism (with its broadly realist claim incongruously joined to a 'saving the phenomena' mode of operation) could, and ought, to be resolved, not by glossing it over, but by trans-forming it. Kepler, then, transformed Copernicus' heliocentrism-in-Ptolemaean-garb into what he called with full justice a 'New Astronomy, Based upon Causes; or, Celestial Physics'. And Galileo went on actually to shatter the Aristotelian worldview disrupted only poten-tially by Copernicus, using for the purpose the novel, mathematical conception of motion that he developed during his Padua years. Clearly, then, one factor propitious to their shared realism resided in the precedent that, in however incongruous and ambivalent a fashion, Copernicus set in book I of *De revolutionibus.*

Still, the manner in which Kepler worked components of natural philosophy into his mathematical science differed widely from how Galileo did it. Prior to the 17th century, natural philosophy in the tradition of the Greeks had been of one kind only – settled *a priori* by the positing of a few first principles as keys to the whole wide world, the logical deduction of a range of consequences thereof, and their joint illustration by means of apparently well fitting pieces of empirical, everyday reality. To the quite limited extent that mathematical scientists felt a need to enter the domain of natural philosophy, they did so in the 'oppor-tunist' way of Ptolemy or Copernicus – a way that has been revealingly characterized (and on occasion adopted) by Albert Einstein when he wrote that "the scientist ... must appear to the systematic epistemologist as a type of unscrupulous opportunist".[120] Galileo's work, too, is marked by a lack of fixed philosophical allegiance. He threw in bits and pieces of Platonic cosmology, of Aristotelian methodology, and of atomist speculation wherever it suited his specific purpose. Nor did he even begin to advertise heliocentrism as a mathematical *philosophy* until he felt driven by an urgent need to have highly placed patrons support his newfangled mathematical science (p. 419).

With Galileo, then, philosophy came *after* his newly realist mode of mathematical science, so that the former yielded none but ad hoc justifications for the realism he had come to instill in the latter. With Kepler all this was quite different. He was the first (and, but for Newton, the last) creator of a uniquely hybrid philosophy of nature. As in natural philosophy of the common run, Kepler's hybrid, too, operated on first principles – in his case, the geometric, archetypal ratios that God had at the Creation worked into His created world at various strategic spots. In it, too, a variety of consequences were drawn from that basic setup (e.g., Kepler's argument on how it is possible for humans to gain knowledge of those ratios). It, too, displays an ongoing search for well-fitting empirical evidence. But the fundamental difference with the four Greek philosophies of nature is that these were *closed* systems. Their respective first principles were constructs of the intellect alone. Instead, Kepler adopted a characteristically *open* stance in regard to the two issues basic to his own brand of philosophy: what exactly, on investigation, are those archetypal ratios going to prove to be, and where and how does God appear actually to have placed them in nature? Not *a priori* reasoning alone, but reasoning freely developed *a priori* yet both expanded and held in check *a posteriori* by empirical evidence ought to serve as the arbiter in the search for a fully satisfactory account of the world, however much that account was itself imbued with the hoary idea of cosmic harmony. It is through this (for a natural philosopher) unprecedented openness that, in putting mathematical procedures of Alexandrian origin to quite novel ends, Kepler time and again sought to anchor his hybrid mathematical science–cum–philosophy in empirical reality and refused to regard it as enduringly settled for as long as potentially countervailing empirical evidence had not been brought into line with the hybrid philosophy or the hybrid philosophy with the evidence. More often than not, it was indeed the evidence that Kepler allowed to steer the course of his investigation of celestial realities – as one student of *Kepler's Physical Astronomy* noted, "many of the logical weak points in Kepler's physics occur where observational facts seem to have forced him into positions he might not otherwise have taken."[121] In the last resort Kepler's openness accounts for the peculiar kind of realism that imbued his mathematical science. Take the openness away, and it would not have been a science at all but would have remained (for all its thoroughgoing Alexandrian origin and inspiration) just a bizarre mathematical philosophy. And that openness of his was reinforced in its turn by the peculiar manner in which Kepler had worked God into his first principles, as also by his profound respect for the phenomena of nature as God given and therefore to be accepted in all humility. That un-Greek sense of humble respect for what observation shows to be the case came to the fore most decisively in his rejection of his own 'vicarious hypothesis' on Mars in view of a seemingly tiny discrepancy with Tycho's observational data (p. 171).

One final cause of the revolutionary transformation here under scrutiny is suggested by the surface resemblance between, on the one hand, Galileo's and Kepler's efforts to produce a newly realist kind of mathematical science and, on the other, the effort at solidifying and extending 'mixed mathematics' that Clavius had embarked upon some decades earlier (p. 146). The resemblance was not lost upon the men concerned. Kepler, albeit a Protestant, was held in high regard (and was discreetly protected) by one of Clavius' disciples, Father Paul Guldin SJ. Galileo was well connected with the Jesuit 'mathematical wing' at the Collegio Romano. In 1611, within a year of the publication of *Siderius nuncius*, Clavius himself stood ready with his disciples to take the lead in the celebration of Galileo's telescopic discoveries on his triumphal tour to Rome. It was only later, for personal reasons chiefly, that the order turned against Galileo.

The surface resemblance notwithstanding, there is a vast difference between the tentative infusion of natural-philosophical tenets with a certain amount of quantitative exactness that was practiced by Clavius and his disciples and the specific knowledge structure of realist-mathematical science. That difference was soon enough to come to the fore in a widespread lack of comprehension for what Kepler had wrought (p. 303), as also in the intellectual clash Galileo's work on falling bodies was to run into soon after his death (p. 410). In 'mixed mathematics' the verities of Aristotelian natural philosophy were taken for granted, only to be given some enhanced, quantitative exactitude by means of some appropriate mathematics. In Alexandria-plus, nothing at all was taken for granted but the fundamentally mathematical structure of natural reality, to be revealed piecemeal for each of its distinct segments (planetary trajectories, projectile motion, etc.). Clavius' example may well have served to some limited extent as one stimulus among others for both Kepler and Galileo when embarking on the adventure of making mathematical science realist. To that extent the example of Jesuit mixed mathematics may count as one final, propitious factor. Even so, the uses to which realism was in the end put in either case were decisively different. The one case was marked by self-evident incorporation of some quantitative extras into the ready-made schemata of Aristotelian natural philosophy; the other, by exploration of one natural phenomenon after another never yet subjected to mathematical treatment and carried out in accordance with the partly abstract, partly realist knowledge structure of Alexandria-plus.

But is not the point just raised true of all those propitious factors now listed?

To answer the question, we must first sum up the net yield of the causal inquiry.

Summing up and some 'what-if' history.   The core explanation is and remains the
realization of the hidden potentiality that rested from the start in the Alexandrian corpus.
For what caused that realization (i.e., 'the twist') actually to occur in Renaissance Europe
rather than in Islamic civilization, we have found one propitious factor of some restricted
viability – a larger number of practitioners. Finally, for the 'plus' (i.e., the realism that Kepler and Galileo infused in thus far almost wholly abstract mathematical science) we have
for *both Kepler and Galileo* encountered viable propitious factors in the ambiguous realism
(possibly enhanced by the mathematical humanists' striving for status) of Copernicus' book
I and, perhaps, in Clavius' example; for *Kepler alone*, in Tycho's observations and in his own
open-mindedness (fed perhaps by his religion) toward his own *a priori* constructions; for
Galileo alone, in his Paduan exposure to practical concerns, especially those in a mathematical context.

   This is all, and (not counting to be sure the core explanation itself, which is of paramount
significance) it does not amount to much. It fails in particular to amount to much in view
of the apparent lack of proportion between, on the one hand, the extraordinary significance
for the course of human history of the revolutionary transformation here under scrutiny
– the onset of the subjection of natural phenomena to mathematical analysis in the frame
of an intricate structure of idealized abstraction and empirical/experimental confirmation
– and, on the other hand, a certain poverty in the range of propitious factors encountered
along the way. Somehow the harvest of propitious factors looks a bit bleak in comparison.
A certain lack of explanatory imagination may of course be responsible, even though careful scrutiny of the vast literature on the subject has revealed little by way of viable causes. It
may further be (as Pascal reminds us with reference to Cleopatra's nose) that big events do
not necessarily require equally big causes. But most of all does the relatively small size of the
explanatory harvest serve to underscore the principal causal thesis here defended: although
we can see what turned Europe into a somewhat readier place to realize potentialities hidden in the Greek corpus than were earlier recipients, these events were not bound to happen
regardless. In each case the hidden possibility might conceivably have been realized before
(in locally different fashion, to be sure) or later or not at all. Indeed, the less numerous and
the less weighty the causes that help account for Europe's greater readiness in this regard, so
much the more does this confirm that contingency forms a major term in the full historical
equation.

   Precisely this is behind the preliminary announcement (p. 72) that there is nothing *a priori* impossible about some Galileo- or Kepler-like figure making his appearance in the generation after al-Biruni and Ibn al-Haytham if external events had not robbed the upswing of
its momentum at a truly critical moment. After all, the heightened realism required to realize the 'plus' inherent in Alexandria might have come from numerous other resources than
the ones somewhat accidentally around in Renaissance Europe. For example, both Kepler

and Galileo grounded their respective varieties of mathematical realism in God's attributes; but so could an Islamic counterpart have done with his One God. Again, some hypothetical Galileo- or Kepler-like figure or figures in Islam would have arrived at by and large the same final outcome along no doubt different pathways, and numerous alternative scenarios may be thought up which have nothing *a priori* impossible about them. Just as al-Birjandi fleetingly suggested the persistence of circular motion in the context of Ottoman mathematical science, which focused upon the Golden Age of nature-knowledge in Islam, just so might someone prolonging the Golden Age itself have hit upon the idea and, with the zest of the Golden Age explorers still fully intact, have pursued it further.

A similar point may be made for Kepler's first law, of the elliptical shape of planetary trajectories. It is true that even for Mars, the planet with the most eccentric trajectory, the ellipse's deviation from circularity is so tiny as to be negligible for any accuracy of measurement attained prior to Tycho. But take the case of David Fabricius. Once an assistant to Tycho, during Kepler's first years of discovery he served Kepler as far and away his most knowledgeable correspondent. Informed of Kepler's ellipse, he rejected it out of hand as an outrageous trespassing of received ideas. But he quickly changed his mind and went ahead to adopt the ellipse after all, yet in a manner that Kepler found so repulsive (in that it seemed to make his own 'physical' reasoning superfluous) as quickly to break off their exchanges for good. In deriving for Kepler's hoped-for benefit the elliptical orbit from epicycles on uniformly moving deferents, Fabricius succeeded, not of course in persuading Kepler to desist from his 'celestial physics', but rather in demonstrating that the elliptical shape of planetary orbits is not necessarily bound up with Kepler's celestial arrangement. As Kepler found to his horror, it is compatible after all with the full panoply of Ptolemy's planetary equipment, its geocentricity included. Fabricius' construction, then, of elliptical planetary orbits in a Ptolemaean frame opens up an alternative scenario that is not historically impossible. Just conceivably, the Golden Age of Islamic nature-knowledge might have yielded a practitioner of mathematical science ready to resolve apparent trouble arising from fancy combinations of circular orbits by positing (out of the blue, rather than along Kepler's carefully reasoned route) their elliptical shape. Centuries down the line the suggestion might then have yielded for the newly founded Maraghah Observatory a program of systematic observation designed to confirm or reject that inspired hunch of our hypothetical man to cap its venerated Golden Age.

So much for historical speculation, undertaken here as elsewhere on the basis of the solid fact that in mathematical science Islamic civilization and Renaissance Europe were, as it were, stepbrothers, descending both from the classic, Alexandrian tradition and enriching it in ways that were overall quite comparable. Here as elsewhere, the point has been to underscore that human events are not one-way-only affairs. At every moment there are possibilities and impossibilities, with the possibilities split between the one possibility actually realized that very moment and the far more numerous possibilities either ignored or discarded, most often for good. We can, with the benefit of hindsight, see that the big break-through toward the mathematization of nature was wrought in late Renaissance Europe at the hands of two extraordinarily perceptive men, Kepler and Galileo, and we can invoke some historical factors that were favorable for the occurrence of that world-shaking event. By way of explaining the event that is all we can do. We cannot, in particular, say that because it *did* happen that way, it was *bound* to happen that way.

NOTES ON LITERATURE USED

Pre-1600 Copernicans. The prime authority is Robert S. Westman, whose findings are as yet spread over a number of articles. His oft-cited list of ten 16th-century Copernicans, in note 6 on p. 136 of his 'The Astronomer's Role in the Sixteenth Century: A Preliminary Study'. *History of Science* 17, 1980, pp. 105-147, comprises Thomas Digges, Thomas Harriot, Giordano Bruno, Galileo Galilei, Diego de Zuñiga, Simon Stevin, Joachim Rheticus, Michael Mästlin, Christoph Rothmann, and Johannes Kepler; probably Giovan Battista Benedetti should be included in the list as well. Westman's comprehensive monograph *The Copernican Question. Prognostication, Scepticism and Celestial Order* will be published by the University of California Press in 2011.

Kepler.

- **Writings, original and in translation**. Ongoing publication of the modern *Gesammelte Werke* ('Collected Works'), which has had many editors, starting with Max Caspar and Walter van Dyck, seems to be nearing its end: München: Beck, 1938 – . This edition comprises both his published work and what remains of his manuscripts and correspondence. English translations of three of Kepler's four astronomical main works have appeared over the past decades. MYSTERIUM COSMOGRAPHICUM. *The Secret of the Universe* (translated by A.M. Duncan, with an introduction and commentary by E.J. Aiton and a preface by I.B. Cohen: New York: Abaris, 1981). ASTRONOMIA NOVA. *New Astronomy* (translated by W.H. Donahue: Cambridge: Cambridge University Press, 1992; the way I render the book's full title in my main text is adopted unaltered (but for one word) from Donahue's). HARMONICE MUNDI. *Five Books of the Harmony of the World* (translated by E.J. Aiton, A.M. Duncan, and J.V. Field: Philadelphia: American Philosophical Society, 1997).

- **Scientific biography**. The best survey still of Kepler's life and works is Max Caspar's German biography *Johannes Kepler*. Stuttgart: Kohlhammer, 1948 (four reprints so far, with the 1995 one supplying the sources for the many quotations Caspar inserted but frustratingly omitted to give references for). C. Doris Hellman's translation *Kepler*. London: Abelard-Schuman, 1959, has in reprint (New York: Dover, 1993) been similarly enriched by Owen Gingerich and Alain Segonds.

- **Accomplishment**. Much writing on the subject has been superseded by three lucidly insightful books, one by James R. Voelkel, *The Composition of Kepler's* Astronomia nova. Princeton UP, 2001, and two by Bruce Stephenson: *Kepler's Physical Astronomy*. New York: Springer, 1987 (mainly about *Astronomia Nova*) and *The Music of the Heavens. Kepler's Harmonic Astronomy*. Princeton: Princeton University Press, 1994, with book V of *Harmonice Mundi* for its principal subject. Also instructive on the basics of Creation according to Kepler is Judith V. Field, *Kepler's Geometrical Cosmology*. Chicago: University of Chicago Press, 1988. A very good summary of Kepler's overall accomplishment in just a dozen pages is by Jim Bennett, 'Johannes Kepler: the New Astronomy'. In: R. Porter (ed.), *Man Masters Nature. Twenty-Five Centuries of Science*. New York: Braziller, 1988; pp. 51-62. Michael Dickreiter, in *Der Musiktheoretiker Johannes Kepler*. Bern/München: Francke, 1973, detailed among other things Kepler's geometric derivation of the consonances, which procedure I

went on to compare, in *Quantifying Music*, with similar treatments of consonance in the works of   217
some of Kepler's best contemporaries (I have detailed Kepler's derivation of the consonances on
pp. 19-23).

- **Tycho's compromise system.** My side remark in the main text is meant to point at uncertainties
over its true authorship revealed in a detective-like story by Owen Gingerich and Robert S. Westman, *The Wittich Connection. Conflict and Priority in Late Sixteenth-Century Cosmology*. Philadelphia: American Philosophical Society, 1988. The same topic is treated from a point of view more
favorable to Tycho in ch. 8 of Victor E. Thoren, *The Lord of Uraniborg. A Biography of Tycho Brahe.*
Cambridge: Cambridge University Press, 1990.
- **Meeting between Tycho and Kepler**. This utterly captivating story has often been told. I regard
Arthur Koestler's mode of telling it, in his problematic yet enticingly written classic *The Sleepwalkers,* as still unsurpassed in insightful psychological acumen (London: Hutchinson, 1959; often
reprinted as a Penguin paperback): Part IV 'The Watershed'; ch. 5: 'Tycho and Kepler'.

Galileo.

- **Writings, original and in translation.** Galileo's published and manuscript treatises, as well as what
remains of his correspondence, have been collected in the 20 volumes of the 'Edizione Nazion-
ale' of Galileo's *Opere* edited between 1899–1909 by Antonio Favaro and published in Firenze by
Barbèra (numerous reprints). In preparing the Galileo section I have worked chiefly from Stillman Drake's English translations of both the *Dialogo* and the *Discorsi*: Galileo Galilei (transl. S.
Drake), *Dialogue Concerning the Two Chief World Systems – Ptolemaic and Copernican.* Berkeley /
Los Angeles: University of California Press, 1953; and Galileo Galilei (transl. S. Drake, with intro
and notes), *Two New Sciences*. Madison: University of Wisconsin Press, 1974; while consulting also
S. Drake (ed., transl., intr.) *Discoveries and Opinions of Galileo*, Garden City (NY): Doubleday, 1957;
idem (ed.), *Galileo's Notes on Motion, Arranged in Probable Order of Composition and Presented in
Reduced Facsimile*. Firenze: Istituto e Museo di Storia della Scienza, 1979 ('Supplemento agli Annali
dell' Istituto e Museo di Storia della Scienza, Fascicolo 2'); I.E. Drabkin & S. Drake, *Galileo Galilei
on Motion and on Mechanics*. Madison: University of Wisconsin Press, 1960, and Galileo Galilei
(transl. Albert Van Helden, with intro, conclusion, and notes), *Sidereus Nuncius or The Sidereal
Messenger*. Chicago: University of Chicago Press, 1989. I have checked all passages quoted in the
main text against the originals, notably those in vols. 7 (*Dialogo*) and 8 (*Discorsi*) of the *Opere* (for
the *Dialogo* the outcomes proved at times rather remarkable).
- **On Galileo: general.** The literature is virtually boundless, and I am not able to acknowledge fully
even the fraction to have passed before my eyes over some four decades. For the provenance of the
variety of historians' views on Galileo summed up on p. 179, I refer to *SRHI* (best approached for
the purpose through the detailed index s.v. Galileo).
- **On Galileo: some specifics.** Stillman Drake, *Galileo: Pioneer Scientist*. Toronto: University of Toronto Press, 1990, sums up forty years of research entirely focused on Galileo, in the course of
which Drake (1910–1993) translated all of Galileo's published, and many of his unpublished, works,

investigated all his scrap papers, and went on to offer increasingly heterodox interpretations of all this material. Drake's greatest find has been his incontrovertible proof that a good deal of empirical trial and error went into Galileo's discovery of the laws of falling and projected bodies. The magnitude of Drake's knowledge of Galileo's work and life is not likely soon to be surpassed, so one would be happy to avail oneself of his final Galileo synthesis without more ado were it not for some flaws that make his results unreliable to an extent no one lacking similar expertise can assess with precision. Prominent among these flaws are Drake's at times urgent desire to present Galileo as, above all, a modern scientist making a complete break with all previous, inherently worthless *philosophizing*; a resulting, strangely flat picture with depth, tragedy, and even lifelikeness excised; an urge to find the conclusive solution to every possible 'puzzle' and remaining ambiguity in the reconstruction of Galileo's career by means of sometimes rather odd ways of reasoning or far-fetched, ad hoc interpretations; a tendency to withhold from the reader's sight just about all possibly countervailing evidence; an at times unfair, sweeping condemnation of virtually all other historians' views rendered often with an equal lack of fairness; in short, the presentation of Galileo as a faultless scientist according to present-day standards and unlike a man of the early 17th century steeped in modes of thought impossible for even the greatest genius to get rid of at one stroke. These flaws are expressed as well in some of Drake's translations, especially of the *Dialogo*. Luckily, in his translations of *Le Mecchaniche* and the *Discorsi* they are much less in evidence.

In all these respects the very opposite is E.J. DIJKSTERHUIS' picture of Galileo, as sketched in sections IV, II.C and II.G of his *The Mechanization of the World Picture*. Oxford: Oxford University Press, 1961 (Dutch original: 1950). These sections, while outdated in several important respects, are still outstanding for the very feature Drake lacked most – a nicely drawn balance between "on the one hand, [Galileo's] being rooted in a past that does not bind him the less powerfully for the vehemence with which he disowns it, and, on the other hand, his preparing a future in which his ideas are to lead to conclusions far beyond anything he could ever have foreseen" (section IV, 78; my translation). A similarly balanced view is in ch. 1, 'Galileo and the New Science of Mechanics' of R.S. WESTFALL, *Force in Newton's Physics. The Science of Dynamics in the Seventeenth Century*. London: MacDonald, 1971; p. 1-55. Particularly striking is Westfall's conclusion (pp. 41-42) "that [Galileo's conception of nature] was an impossible amalgam of incompatible elements, born of the mutually contradictory world views between which he stood poised." More than any other of the books here mentioned, this one by Westfall was consciously made to fit a historiographical tradition started by Alexandre Koyré in his *Etudes Galiléennes*. Paris (Hermann), 1939-1940 (on which seminal book I have lengthy passages in *SRHI*, 2.3.3.). W.R. SHEA, *Galileo's Intellectual Revolution*. London: MacMillan, 1972, focuses in an enlightening manner on Galileo's often neglected, 'middle period', 1610-1632.

- **The second 'conclusive argument'** in the *Dialogo*: David Topper, 'Galileo, Sunspots, and the Motions of the Earth: Redux'. *Isis* 90, 4, December 1999; pp. 757-767.
- **Galileo and Kepler.** The riddle of Galileo's neglect of Kepler's laws of planetary motion is dis-

cussed in Erwin Panofsky, *Galileo as a Critic of the Arts*. The Hague: Nijhoff, 1954, and in a review written by Alexandre Koyré in 1955 ('Attitude esthétique et pensée scientifique'. In: idem, *Etudes d'histoire de la pensée scientifique*. Paris: Gallimard, 1973 (1st ed.: Presses Universitaires de France, 1966), pp. 275-288).

Causes.

- **From script to print.** *SRHI* 5.2.9.
- **Europe's advantage over Islamic civilization by virtue of being next.** I owe this ingenious idea (ingenious in that it is so deceptively simple as easily to be overlooked) to Aydin Sayili (*SRHI*, 6.2.3.). A pupil of George Sarton, he did not direct it toward the transformation of mathematical science but toward the discovery of something Sarton deeply believed in, 'the' experimental method.
- **Number of mathematical scientists.** For counting I have used vol. 16 (*Index*) of the *DSB*, 'Lists of Scientists by Field'. Under 'Astronomy' (pp. 479-481) and 'Mathematics' (pp. 493-496) I listed everyone who appears (from years given in the principal Index, pp. 1-439; checked where needed in the *DSB* entries themselves) to have operated between 800 – 1050 and 1450 – 1600, respectively, and then eliminated double counts. The numbers may of course reflect an unconsciously Eurocentric bias, even though A.I. Sabra's presence on the Editorial Board makes that unlikely. On the other side (if indeed there is an other side) Europe suffers somewhat from mathematically adept cartographers like Ortelius being listed in the *DSB* under 'Earth Sciences', not under 'Mathematics'.

    An extensive discussion of Tycho's numerous assistants is in ch. 6, 'The Flowering of Uraniborg', of Victor E. Thoren's book cited above.
- **Exchanges in nature-knowledge and invention between China, the Islamic world, and Europe.** *SRHI*, 6.4. offers a critical view on the general run of Needham's ideas on transmission. In *The Dialogue of Civilizations in the Birth of Modern Science*. New York: Palgrave MacMillan, 2006, Arun Bala has proposed a general criterion for accepting transmission in the absence of material evidence for it that is somewhat less biased than Needham's but that still does not satisfy basic requirements of empirical scholarship. I have benefited much from an unusually balanced survey presented in Edwin J. van Kley's entry 'East and West': *Encyclopaedia of the History of Science, Technology, and Medicine in Non-Western Cultures*, compiled by Helaine Selin (Dordrecht: Kluwer, 1997).
- **Galileo's encounter with practical mathematics.** *SRHI* 5.2.1. (on Leonardo Olschki).
- **Galileo's philosophical commitments.** *SRHI* 4.4.3, 5.2.1 on Galileo's alleged Platonism and Aristotelianism. Pietro Redondi, *Galileo Heretic*. Princeton UP, 1987, has further made a case for Galileo as an atomist.
- **Social chasm between the mathematical scientist and the philosopher in the Renaissance.** This issue is part of the 'Zilsel thesis' (*SRHI* 5.2.4.), and has been examined in greater detail in Robert S. Westman's article 'The Astronomer's Role in the Sixteenth Century: A Preliminary Study' cited above, and by Mario Biagioli: 'The Social Status of Italian Mathematicians, 1450-1600'. *History of Science* 27, 1989, pp. 41-95.

220 · **Jesuit promotion of mixed mathematics.** Literature listed at the end of the previous chapter.
· **Poverty of propitious factors.** *SRHI*, ch. 7 (especially p. 498).

VI

THE SECOND TRANSFORMATION:
A KINETIC-CORPUSCULARIAN PHILOSOPHY OF NATURE

In 1618, in Breda (the Netherlands) a momentous encounter took place. Isaac Beeckman, then thirty years old, was a candle-maker, graduated theologian, and spare-time philosopher of nature. He lived in Middelburg but had come to Breda for courtship. René Descartes, twenty-two years old, was in garrison there as a soldier in the army of Prince Maurits. When the two met, they congratulated each other on their rare, shared capacity to "join physics with mathematics", an activity they then pursued together for several months.[122] They did not attach quite the same meaning to the expression, however. For the young Descartes it referred as yet to the 'mixed mathematics' in which he had been raised at the Jesuit college of La Flèche. For Beeckman the expression stood for something new, something that he had already been working on for almost a decade. At the time of their encounter, he was far advanced in his ongoing construction of a natural philosophy ('physics') of a novel and on occasion somewhat quantitative kind. He enriched the atomist doctrine with something decisively new: a conception of motion retained. Ancient atomism had postulated the existence of particles that moved through the void in ways left almost wholly unspecified. With Beeckman the way these assumed particles actually move became far more specific. For instance, Demokritos and Epikouros had explained sound in general by invoking the emission, by some source, of sound particles, which on arrival at the ear produce the sensation of sound. Beeckman now explained *consonant* sound by invoking the emission, by the vibrating string, of sound particles *of specifically different sizes and speeds* that, once arrived at the ear, produce the sensation of *consonant* sound.

Beeckman's idea of motion, independently conceived, was similar to Galileo's insofar as he, too, posited its being retained once an object has acquired it. He sought to explain a large variety of natural phenomena and effects by assuming a variety of differently sized and shaped particles to move with various velocities in various directions. In the literature this manner of explaining all and sundry natural phenomena goes by an expression that Robert Boyle coined in the 1660s, 'the mechanical philosophy'. Because already in the 17th century the term 'mechanical' carried a confusing number of meanings, I shall from here on mark its most distinctive feature by speaking rather of the natural philosophy of 'kinetic corpuscularianism'.

It was not through Beeckman directly that this partly novel philosophy was to become known to the world. He lacked the capacity and also the self-confidence to organize his dis-

222 jointed notes on a disjointed range of topics into a publishable treatise. The one time in his life when he gave it a try, in the late 1620s, Descartes, in his anxious search for priority, succeeded in nipping the effort in the bud. In this manner Descartes cleared the way for his own effort to work out and make public his own, philosophically far bolder variety of a kinetic corpuscularian conception of the world. He gave it coherent expression during the 1630s in his unpublished 'Le monde' ('The World') and then, in 1644, in his immensely influential *Principia philosophiae* ('Principles of [Natural] Philosophy').

The revolutionary transformation that we are to examine now involved more interaction among investigators than that of the Alexandrian corpus. The three pioneers, Isaac Beeckman, René Descartes, and Pierre Gassendi knew and influenced each other. Together with some more marginal figures they brought about a revolutionary transformation in natural philosophy, thus turning Athens into 'Athens-plus'. In discussing what these men wrought between c. 1610 and c. 1645, we shall consider

- the many similarities and the one truly distinctive difference between what was transformed (ancient atomism) and what came out of the transformation (kinetic corpuscularianism);
- what each of these men (but Descartes most decisively so) contributed to it;
- what consequential interactions took place between them;
- what withheld Descartes from following the path of mathematical realism taken by Galileo; and
- what caused the revolutionary transformation to occur.

Revival continued

The Renaissance scene of natural philosophy, so I have argued on p. 99, came down basically to a replay of the original, Greek performance. What renewal did take place remained confined to the Aristotelian fringe, where philosophizing was extended in the direction of mathematics ('mixed mathematics') in Clavius' vein or of natural magic (as with Fernel) (p. 146). For the rest, the printing press added speed and liveliness to the recovery of ancient texts and to exchanges between their respective followers without thereby altering the performance in kind. Nor was it altered at those rare occasions where exact measurement entered the debate. Tycho Brahe's and others' uncommonly precise observations on the comet of 1577 made it hard to maintain the standard Aristotelian position on the subject – comets are hot and dry exhalations of the Earth which catch fire in the outer sublunary sphere of fire, through which they keep moving for a while. Characteristically, most commentators sought refuge in the alternative provided of old in stoic doctrine – comets are celestial omens. The obvious move to solve a natural-philosophical conundrum was still to seek for a natural-philosophical answer.

Exchanges in natural philosophy.   By the late 16th century, then, the revitalization
and extension of Aristotelianism and the resurrection of Platonic and stoic conceptions,
together with their skeptical nemesis, had been in full swing for quite some time already.
Debates between adherents of these various philosophies kept consisting of ever inconclusive exchanges of points of view on basics and consequences thereof, shored up by appeals
to carefully chosen bits and pieces of empirical evidence. Was the world to be apprehended
through logic applied to facts observed in a frame of potentiality and actuality (Aristotelian)
or through a mathematics-inspired insight into its Ideal Forms (Platonic)? Was the world, at
bottom, continuous (stoic) or fragmented (atomist)? Or were, as the sheer rivalry between
these four purportedly true systems of the world and the very inconclusiveness of their debates strongly suggested, human beings incapable of gaining secure knowledge of these as
well as all other matters (skeptical)? Such was the stuff of ongoing exchanges in European
natural philosophy, as it had been in its Greek predecessor, with the exception that one of
these, atomism, was rather a latecomer to the European exchanges. On the one hand, the
core doctrines of atomism as embodied in Greek fragments and with greater coherence in
Lucretius' poem had already by the late 15th century gone through the well-established machinery of textual criticism followed by a printed edition. But a broader reception of these
basic texts took place in a more hesitant fashion than with any other philosophy of nature at
the time. Not until the turn of the 17th century did the revival of atomism begin in earnest.

The revival of ancient atomism.   Recall the bare fundamentals of the atomist doctrine (p. 7). Our subjective perception of appearances notwithstanding, the world is really
made up of nothing but empty space through which the ultimate constituents of matter –
unobservably small, indivisible, perfectly hard corpuscles of various shapes and sizes – are
either moving with equal speeds or temporarily coalescing into the visible bodies of our
everyday world. While entirely speculative in its overall design, the doctrine was reinforced
by an appeal to empirical phenomena like the filtering of grains of salt from seawater
through pores in a jar or our ability to pick up scents from far-away objects – what, if not
similarly shaped particles emitted by the object in question and picked up by our noses,
could account for it?

Atomists differed over the basic properties to be assigned to the atoms – in particular,
whether or not weight ought to be attributed to them as an original property besides shape,
size, and position. Demokritos held that weightless atoms move in all directions, whereas
Epikouros attributed heaviness to his atoms, which therefore fall down in a straight line.
To explain how atoms can nonetheless cohere for a while and form larger bodies, and also,
against the determinist Demokritos, to save some freedom for human will, Epikouros left
room for a tiny 'swerve' for his atoms, that is to say, an irregular, oblique motion affecting
their fall. Also, not every atomist adduced quite the same everyday examples to support or

224   defend the doctrine, the empirical evidence for which was rather meager in any case (certainly as compared to the Aristotelian conception of things and the wealth of phenomena treated in its light). Its ethical directives for the good life notwithstanding, atomism (swerve or not) seemed uncomfortably close to a determinist stance. In the absence of any obvious function for a deity of any kind it even seemed to border on atheism. In view of such ill-repute as atomism could easily slip into, its early resurrections are rather surprising, notably those in the context of *kalam* (p. 56) and of the Carolingian Renaissance in Europe (p. 81). There is nonetheless good reason to suppose that the delay between the printing of Lucretius' poem in 1473 and the actual spread of the doctrine starting almost one and a half centuries later was due in good part to the suspicion that it went against the very core of Christian belief.

When early in the 17th century the revival of atomism began in earnest, it took two different directions. Pharmacists and physicians began to suspect that what underlies the experimental analysis and reconstitution of chemicals is the mutual separation and recombination of particles – how else could stuff like copper vitriol undergo all kinds of changes and at the end emerge with its original properties intact? But more properly Athenian, wholly speculative thinkers also began in the decades around 1600 to adopt the essentials of the atomist doctrine, blended more often than not with remnants of Aristotelian ideas on the constitution of matter in its finest state (*minima naturalia*). That is also how, when still a student, Isaac Beeckman came to the doctrine he was soon to invest in revolutionary fashion with a novel conception of motion. Who, then, was Isaac Beeckman?

Beeckman and the transformation of ancient atomism

Beeckman was born in 1588 and died in 1637. But for two inconsequential exceptions, he never published a word of his varied views and findings. Rather, he was in the habit of jotting these down in a diary that he kept from 1604 onward. Only in recent decades has it become clear how consistent and also influential a thinker hid himself behind so unsystematic and so private a way of proceeding.

At first Beeckman linked his atomist views to Aristotle's doctrine of the four elements in that he posited atoms of four kinds: one for earth, one for water, one for air, one for fire. Gradually such quality-linked atoms gave way to more abstract ones, distinguished from one another in the classic manner by nothing but shape, size, and position. By 1620 Beeckman began to worry over the incompatibility he felt to exist between two apparently opposite attributes of atoms that he found equally indispensable: perfect hardness (needed to prevent them from falling apart) and perfect elasticity (required to explain rebound on impact). He sought for a way out of the dilemma by invoking an aether composed of even tinier fire particles. Not noticing that he had only shifted the difficulty from the atomic to225 the subatomic level, he never again questioned the basics of atomism.

To Beeckman, the essence of the world rested in its picturability. His broad program was to explain natural phenomena by means of material processes which, while in most cases not directly visible, can yet be represented to the mind in visual terms. Take that staple phenomenon in the magical conception of things, sympathetic resonance (p. 133). When a string is touched, a similar one at some distance away may make itself heard in unison with it. Customarily some hidden force that attracts through mutual sympathy, or nature's active eschewing of the void, was held responsible for the phenomenon. Beeckman felt nothing but scorn for explanations along such lines. In his view the vibrating string performs two feats. It cuts the ambient air into tiny particles which, on reaching our sense of hearing, are perceived by us as sound. Also, its regular to-and-fro motion causes successive condensations and rarefactions of the air which extend to the untouched string and put it in motion, too, in those cases when the number of concurring motions exceeds the number of contrary motions.

What we have here, then, is one particular variety of atomism augmented (and this is the essential point) by a Galileo-like conception of motion. Beeckman began to develop its characteristics by the early 1610s. This is what enabled him to widen the explanatory range of ancient atomism by large stretches. His favorite explanatory resources became (1) the pressure exerted by the air and (2) the tendency of bodies, once put in rectilinear or circular motion, to retain it. By these means did he come to grips with everyday phenomena that were less intricately, or not at all, handled in ancient varieties of atomist thought, like heat, combustion, cohesion, magnetism, the tides, numerous manifestations of light and sound, and the planetary motions as rendered mathematically (yet, Beeckman was sure, misconceived physically) by Kepler.

In all this Beeckman remained a natural philosopher in the classic, Athenian sense, happily instantiating one case after another of one more wholly speculative account of the constitution of the world. He admired Galileo, particularly when late in life he had a chance to read the *Dialogo*, yet his aims were quite different. So was his view of mathematics, which he did not regard as the preeminent key to the real world of natural phenomena the way both Galileo and Kepler did. "Mathematics," Beeckman wrote, serves as the "hands of physics", for the "dignity" of natural philosophy is "as much greater as the shadow is less noble than the body itself".[123] In producing an almost entirely qualitative conception of the world aimed at the elucidation of everyday phenomena in terms of the first principles of that conception, Beeckman, for all his creative extension of ancient atomist doctrine, remained true to the Athenian tradition of nature-knowledge in which he had been reared. So did two other pioneers of the kinetic-corpuscularian philosophy of nature, both French, both with an intellectual debt to Beeckman.

226   One of the two was Pierre Gassendi, a slightly younger contemporary who looked him up for a couple of days in 1629 when touring the Netherlands. Gassendi, a late humanist in many respects, started scholarly life as a skeptic, with a special knack for disproving the knowledge claims of the Aristotelians. When in the 1620s skepticism became a subject of hot debate in France, Gassendi realized that the skeptical position might all too easily be associated with an impiety he was not in reality far removed from. So he sought cover in a likewise humanist effort at recovery of the ethical views of Epikouros. His meeting with the both atomist and deeply religious Dr. Beeckman caused him to widen his enterprise – henceforth he dedicated himself to proving that atomism and Christianity are compatible after all. In his long-winded rendition of the standard ancient views he produced two principal alterations, bound to become highly significant in the next generation (p. 551). He excluded our immortal soul from the realm of atomic structure; he had God create the atoms and endow each single one of them with motion. The latter move marked a significant difference from another natural philosophy of moving particles – the one that, also in the late 1620s, René Descartes began to develop.

The Cartesian variety of kinetic corpuscularianism

The first meeting between Beeckman and Descartes was marked by their shared urge to 'join physics with mathematics'. The meaning of the phrase evolved over the next decade. For Beeckman 'physics' remained primary. But Descartes, who thought far more highly of the mathematical discipline, tried out a variety of increasingly novel approaches under that banner, until by the late 1620s he settled for good upon a world made up of corpuscles in incessant motion. As yet, late in 1618, his ambition went no farther than the one entertained by the mathematical fringe of the Jesuit Order: to render a collection of known phenomena in geometric terms. Musical theory seemed to offer chances for carrying their program forward, and this is how, by way of a New Year's gift to Beeckman, Descartes wrote a brief treatise that after his death was published as 'Compendium musicae' ('Outline of Musical Theory'). In it, he rendered Zarlino's views on consonance, measure, and composition in terms derived as much as possible from geometry without significantly altering their content thereby. For instance, he derived the consonant intervals more rigorously than usual by means of a more strictly geometric criterion: a threefold process of bisection of the monochord. The argument of 'Compendium musicae' remained unaffected by a brief passage inserted at Beeckman's suggestion about the vibrations actually made by the string. In the course of some more truly collaborative work undertaken during the preceding month and a half, the young Descartes became acquainted with the approach Beeckman had been pioneering for almost a decade in terms of ultimately corpuscular processes guided by some principle of motion. In their joint work they broke some new ground, in ways that go far to explain the remarkable fact that in the end the young mathematical prodigy was to land on the Athenian, not the Alexandrian, side of revolutionary transformation.

Around New Year's Eve Beeckman went back home; four months later Descartes resumed his grand tour across Europe. When they met again in 1628, much had changed in Descartes' conception of how to go about the pursuit of nature-knowledge. His travels completed, he was on the lookout for a suitable place, preferably in the Dutch Republic, where he could settle and devote himself to the undisturbed elaboration and subsequent dissemination of a natural philosophy meant to be entirely of his own making. This new philosophy of his, in marked contrast to all its predecessors, with their built-in inconclusiveness, ought to be just as indubitably secure as that unique carrier of epistemic certainty, the mathematical way of thinking, which therefore had to infuse it through and through.

By 1628 Descartes had persuaded himself that a radically transformed atomism was in the best position to help him accomplish such a feat. No doubt he was a sharp enough thinker to reach such a conclusion independently. Even so his lengthy conversations with Beeckman first directed his mind toward the idea, and awareness of that fact did not sit easy with him. When on his explorative trip he looked up Beeckman in the fall of 1628, their renewed exchanges and his renewed reading in the latter's diary came as a shock. For there he saw his host busily engaged in the very enterprise for which he was seeking tranquil surroundings to carry out himself – the system of a world made up of particles in motion, written up with a view to eventual publication. Descartes quickly forced a rupture with his friend and, seizing upon Beeckman's undersized ego, managed to bully the elder man back into silence. From 1629 to 1633 he worked hard on a book with the modest title 'Le monde' ('The World'), which, however, he filed away unpublished as soon as he learned of the condemnation of Galileo for the same Copernicanism that formed one cornerstone in his own argument (p. 422). A few hints then appeared in his first publication, the *Discours de la méthode* of 1637, but not until 1644, when he had hit upon a way to circumvent the issue of the Earth's motion, did he publish the full system of his natural philosophy in a book for which in the same humble spirit he chose the title *Principia philosophiae* ('The Principles of Philosophy').

Here he distinguished fundamentally between two substances in the world, *res cogitans,* or (translated literally) 'thinking stuff', and *res extensa,* or 'extended stuff'. As the grammatical form shows, the former is active, the latter passive. The former, identified with what in Aristotelian terms used to be called the 'rational soul', which in effect stands for consciousness and volition, is wholly confined to human beings. These must be regarded as extended entities to which a rational soul has been joined.

In *Principia philosophiae* Descartes showed himself concerned foremost with the workings of *res extensa*. In his view extension and matter are identical – no space without matter and no matter without space. Matter is infinitely divisible, even though for practical pur-

228   poses it suffices to distinguish between three kinds: gross matter; invisibly small particles polished over time so as to form little spheres; and a fluid-like 'aether' of extremely tiny particles of all kinds of shapes, which fill up all space between, and pores inside, both the spherical particles and those which temporarily cohere in gross matter. Matter has at Creation been fitted out by God with a quantity of motion that remains forever constant – motion is passed on all the time, yet the total amount never changes. Every particle tends to move uniformly in a straight line, yet in actuality it is forever prevented from doing so because matter is everywhere, so that every particle is surrounded by adjacent particles with a similar tendency to move in a straight line. Motion, in other words, is passed on solely by particles of matter pressing upon, or even bumping into and rebounding against, other particles or temporarily cohering heaps thereof. This endless crowding in upon one another of endless ranges of particles of diverse shapes and sizes has led over time to the formation of heaps of particles swirling around together in whirlpools of matter, some as large as the vortex that swirls planets around the Sun. While gifted with ongoing rectilinear motion, particles actually tend therefore to move in circles, or rather in broadly circular, closed curves. The qualification 'broadly' is indispensable. Never did Descartes make an effort to express in quantitative terms the tension thus set up between the rectilinear and the curved motions of his particles. And this is exemplary for the constitution of the universe as Descartes saw it – even though he had his reasons for claiming mathematical certainty for it, hardly any mathematics or even quantities entered into his almost entirely verbal explanations of its manifold properties.

From ancient atomism to the clockwork universe.   How does the Cartesian conception of the world compare with original atomist doctrine? Among the most significant similarities is this qualitative approach to phenomena. Take, for example, Lucretius' and Descartes' respective explanations of taste and of magnetic action.

Some foods and drinks taste sweet, others bitter, etc., in accordance, so Lucretius asserted, with the different shapes of the atoms these stuffs emit. When "smooth and round atoms" reach the sense organ proper to them, the mouth, they elicit sweetness. Bitterness is caused by "atoms more hooked, ... accustomed to tear open their way into our senses and to break the texture by their intrusion", whereas atoms "rightly thought to be neither smooth nor altogether hooked with curved points, but rather to have small angles a little projecting ... rather tickle our senses than hurt them [as with] tartar of wine."[124] Seventeen centuries later Descartes explained how the taste of salt and of acid liquors differs in that the particles of the former are of such grossness that, "when separated from one another and agitated by the action of saliva in the mouth, [they] enter the tender pores of the tongue in a pointed manner, without folding", whereas the smaller (and therefore pliable), sharper particles of the latter, on entering obliquely, cut into the more tender parts of the surface skin of the tongue while leaving its coarser parts intact.[125]

Nor was Lucretius at a loss to explain how it is that pieces of iron, when brought close to   229
stone of Magnesia, move toward it. Making liberal usage as well of ubiquitous 'pores' in
gross bodies through which particles may stream, he offered as principal cause that a whole
current of atoms is blown out of the stone. In thus removing the air between the stone and
the iron, it creates a void space filled up at once, first by iron atoms and then, on account of
their exceptionally close intertwinement, by the whole piece of iron itself thus moving across
the evacuated space onto the stone. Descartes, too, knew what particles may be like and
how they behave. Particles in the shape of screws are produced by the whirlpool around the
Earth. They pass through similarly screw-shaped pores of both the iron and the lodestone,
and thus drive out the air from between them so as to make them move toward each other
(Figure 6.1).

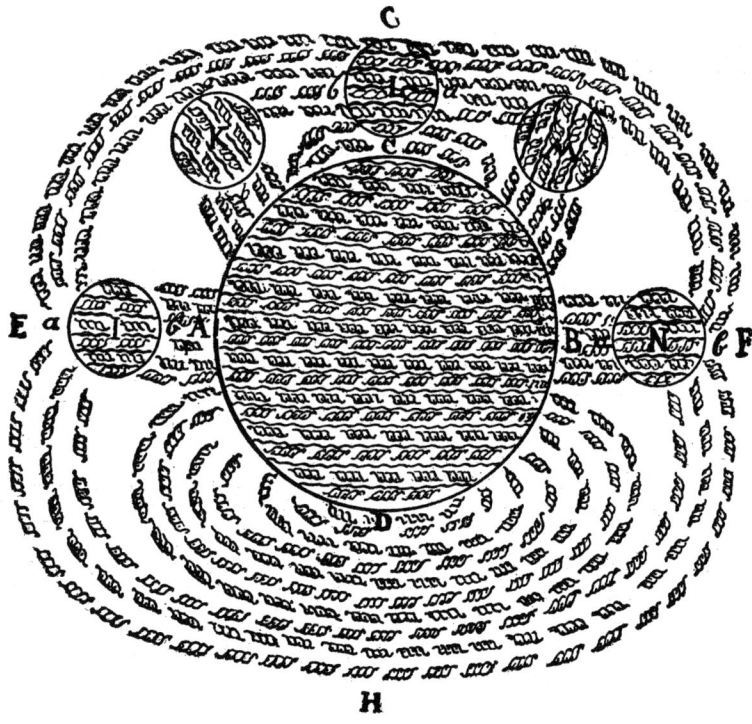

Figure 6.1: Cartesian magnetic action

230 While surely somewhat different in detail, what all such explanations have in common is an appeal to nothing else but certain invisibly small particles of certain shapes and sizes on their way through the infinity of space. No other properties than these – shape, size, and position – are assigned to the particles. Hence, no other properties of things we perceive in the macroworld have a counterpart in the microworld. The role allotted to sensory experience is therefore very limited. It presents us with a prima facie awareness of the phenomena to be explained, yet the phenomenal world is different in kind from how we perceive it and its true explanation takes place on another level entirely. Indeed, explanation is all there is to both Lucretius' and the 'kinetic' variety of corpuscularianism. As in every Athenian account of the world, the phenomena of nature are given, and what the natural philosopher adds to his noticing them is solely his account of what, deep down, they truly *are*.

These are very significant similarities, and the criteria by which Descartes, near the end of *Principia philosophiae*, claimed the deepest of possible chasms to yawn between his own approach and that of ancient atomism dwarf in comparison. To Descartes the decisive difference rested in his insistence on the infinite divisibility of matter and his consequent rejection of the void. It is certainly true that this difference was responsible for some broad features of Descartes' microworld, such as, notably, his whirlpools of matter, which necessarily lack a counterpart in the rectilinear, weight-driven, almost fully downward motion Lucretius claimed for his atoms. It is also true that during the 17th century much heated debate was devoted to the question of whether or not the void could be thought to exist (p. 410). Still, already one generation later it began to dawn upon certain particle thinkers, notably Robert Boyle, that these differences between the atomist and the Cartesian variety of kinetic corpuscularianism were much less significant than the broad mode of explanation they held in common, which Boyle went on to fuse into one single expression, 'the mechanical philosophy' (in the Prologue I have explained why I refrain from using the term: p. xxxvi).

No, the truly profound difference lies elsewhere. Even though the ancient atomists claimed their universe to be run by 'necessity', such necessity had always remained somewhat indefinite. What Descartes now presented to the world was (to use one of the great catchphrases of the age) a *clockwork* picture of the universe. The kind of necessity the universe was held to be subject to was the necessity by which a piece of automated machinery like a clock, once put together, goes on running all by itself. And how it runs, so Descartes realized much more clearly than Beeckman, is governed entirely by a set of *laws of motion*.

Now this was truly a big change. The entire history of ancient atomism had yielded just one dispute over the kind of motion atoms are subject to: in all directions or downward in straight lines with a swerve. Explanations of natural phenomena had consequently been focused on the specific kinds of particles that objects participating in some action were supposed to be made of. Attention among adherents to the corpuscular view of things now turned toward those actions, in terms of the various kinds of motions held to underlie them.

In giving them such attention, Beeckman confined himself to subjecting to such treatment whatever natural phenomenon happened to evoke his explanatory urges. Descartes, in contrast, grounded his full conception of motion in his dualism of consciousness and matter. The contrast extends beyond foundational issues to motion itself. In Beeckman's diary notes the key exemplar of his own novel conception of motion, to wit, his idea of its being retained, stands alone, by way of an isolated albeit frequently reiterated assertion that he went on to apply to a range of distinct cases. In Descartes' much more orderly thought-through conception, the same principle, expressed in a more specific manner and joined now to its corollary, a Galileo-like principle of relativity of motion, and also to a law of conservation of quantity of motion, was made to underlie in a coherent fashion the how and why of the perennial changes of position of particles that the phenomenal world ultimately reduces to. It is at this very point that Descartes' view of things touched another revolutionary conception – the one pioneered a few decades earlier by Galileo.

Poised between Athens and Alexandria.   One crucial component of Galileo's Alexandria-plus mode of nature-knowledge, then, also featured prominently in Descartes' quite differently tinged Athens-plus setup: their shared conception of motion as retained and as relative. This, however, is where the similarity ends, and the question is why. *For if the two men shared so much, then why not the remainder?* What made Descartes prefer in the end to follow Beeckman on the route to Athens-plus rather than follow a path, similar to Galileo's, toward Alexandria-plus?

The question is exacerbated by inspection of one collaborative feat that Beeckman and the young Descartes accomplished at Breda in 1618. They managed to derive the squared relation between time and distance in vertical descent from Beeckman's proposal to conceive of its cause as a force "which pulls by tiny jerks"[126] and from Descartes' acquired ability to handle what were in effect mathematical limiting procedures. At least ten years earlier, Galileo had derived the same rule by means of an argument involving motion only, not any operation of forces. He had turned the rule into the cornerstone of his paradigm case for realist-mathematical science, where motion is considered in abstraction from whatever is to be regarded as an impediment to the true, ideal phenomenon (p. 191). Beeckman and Descartes, in contrast, did not find the rule exemplary in any way; they did not think they had done more than just resolve an interesting problem among many others. In so doing they also passed by another feature of tremendous significance to Galileo: what the problem of fall meant for the possible confirmation of the Earth's daily rotation. Furthermore, Descartes' own (as distinct from Beeckman's) rendition of their joint solution was enveloped in some remnants of conceptual confusion – the very same that Galileo had been wrestling with in Padua. Descartes was surely a sharp enough thinker to overcome these in time as, at some moment between 1604 and 1610, Galileo had. But he never did; rather than pursuing

232 the issue, he dropped it for good. Beeckman, neither mathematically astute nor a very orderly thinker and by 1618 fully set in his own, unreservedly Athenian ways, was not to explore a realist variety of mathematical science at any greater length or depth. But what prevented Descartes from doing so? What led a man so deeply concerned about mathematics in more than one sense to travel ever farther away from a Galileo-like approach?

The answer rests in the various meanings that mathematics acquired for Descartes in the first decade of his career.

In these early months of 1619 already, he tried to 'join mathematics and physics' in a different manner. In this case, the 'mathematics' rested in the strictly Archimedean-style proof that Stevin had given for his 'hydrostatic paradox' (Figure 4.2, p. 105). The 'physics' naturally absent from Stevin's proof was now added by Descartes. He devised a corpuscular mechanism meant to match at the microlevel the mathematical arrangement at the macrolevel. Stevin had dealt with macroworld entities (containers of various shapes, the fluid that filled each) in the abstract, while deriving the required proof from one specific assumption, followed by a rigorously Archimedean *reductio ad absurdum*. Descartes focused instead on a causal account such that the property in need of proof followed from certain motions, or tendencies to motion, that he posited for the corpuscles that made up the container and the fluid. The approach foreshadows the procedures he was to follow in 'The World' as also in *The Principles of Philosophy*.

In 1619, then, Descartes left behind 'mixed mathematics' as cultivated in Clavius' vein in his 'Compendium musicae', so as to probe two distinct approaches to the challenge of how to arrive at a truly mathematical way of doing natural philosophy. At this early stage he stands poised between Athens-plus and Alexandria-plus – the former newly represented to him by his new friend, Beeckman, and also explored by him in his effort to rewrite Stevin's hydrostatic paradox in natural-philosophic fashion; the latter explored in unwittingly Galileo-like fashion in their joint solution to the problem of how objects fall. A decade later, Descartes no longer stood poised. By then, he had unambiguously chosen Athens-plus. Again, why?

Between 1619 and 1621 Descartes conceived in quick succession of three general methods for pursuing nature-knowledge. The first, inspired by his encounter with Beeckman, was the method of combining mathematics with physics for which the hydrostatic paradox served him as a training ground. Next came a project for "'universal mathematics', a supposedly general analytical discipline spanning mathematics and 'physico-mathematics'". This was followed in its turn by a third project, a "method commanding all rational disciplines [which pleased him so much] that on St Martin's Eve 1619 he dreamt that the project had been consecrated by God himself."[127] He laid down this third method in 'Regulae ad directionem ingenii' ('Rules for the Direction of the Mind'), which in spite of several revisions he had not yet completed when, by the late 1620s, he found the method to run into insuperable difficulties.

The difficulties flowed from an almost equally ambitious project that had as its principal concern the determination of all those lower- and higher-order curves (circles, conics, …) whose properties can be defined with complete, untainted exactness. He published the outcome, *La géométrie*, in 1637, together with two more treatises in a mathematical vein, *Les météores* and *Dioptrique*, in both of which he used his work on the conics for analyzing the rainbow and deriving the sine rule of refraction (pp. 229; 352). In all these regards, then, Descartes operated as an Alexandrian in the received sense, an innovative one to be sure, yet all the same fitting in the traditional mold. His combining Alexandria with his Athens-plus project even led him to irresolvable contradictions: his derivation of the sine law required a finite velocity of light, whereas his whirlpool account of the nature of light in 'Le monde' and in *Principia philosophiae* compelled him to make its velocity infinite. It is characteristic of the hoary separation between the Alexandrian and the 'Athenian´ approaches to the pursuit of nature-knowledge that Descartes never even attempted to reconcile his opposite assumptions with each other.

Might not adoption of a Galileo-like Alexandria-plus approach, followed no doubt by brilliant extension, have saved him from running into such incompatibilities? It very well might have, and the motives that drove Descartes back toward the Beeckman-like pursuit of Athens-plus, which, finally, he settled for by the end of the 1620s must have been quite deeply felt.

A glimpse of those motives may be caught from the reminiscences that a man who as a precocious youth was still utterly captivated by *Principia philosophiae*, Christiaan Huygens, wrote down near the end of his own life (1693):

> But Mr. Descartes, who seems to me to have been very envious of Galileo's renown, had this great urge to pass for the author of a new philosophy, as appears from his efforts and his hopes to have it taught at the universities instead of Aristotle's, or from his wish that the Society of Jesus embrace it. … In his desire to make us believe that he has found the truth (as he does everywhere in founding himself upon, and in rejoicing in, the succession and the beautiful connectedness of his expositions) he has done something at great prejudice to the progress of philosophy, as those who believe him and those who have joined his sect fancy to possess knowledge of the causes of everything, insofar as it is possible to know them.[128]

In Huygens' view as, more than anything else, a true Galilean, then, a bid for the glory that marks a leader to whom the truth has been revealed once and for all is what caused Descartes to undertake a bold grab for the whole, rather than more humbly following in Galileo's footsteps. But more than just a character trait like 'envy' is at stake here – the very features that made Descartes' mind tick resisted Galilean and Keplerian mathematical realism. Insofar as Descartes, in his approach to nature generally, was a realist at all, he definitely was a realist of a different sort.

234   *The reality of things.*   "Larvatus prodeo" ('I present myself from behind a mask') – this early piece of self-characterization offers a clue to René Descartes and how he saw himself in relation to others.[129] In his writing he was concerned all the time to establish a most self-centered kind of communication with his prospective readers. Hardly any human being is in private quite the same person as in public, yet Descartes was exceptional among 17th-century thinkers for the extent of his obsession with what the world might think of him. Indeed, his rare capacity for deception and self-deception spilled over to his entire way of thinking. However ready he was time and again to commit himself to one or another sweeping statement about the world, he always made sure to keep a backdoor open for the possibility that things might really be quite otherwise.

Take the manner in which Descartes handled matters of truth and falsity, of reality and imagination, at the very end of *Principia philosophiae*. There he maintained in quick succession the following: (1) on matters beyond our perception (such as treated in a vein of indubitable certainty for hundreds of previous pages) it suffices to explain, not so much what they are, but only what they might be like; (2) we nonetheless have moral certainty that things are that way; (3) this certainty extends even farther, in that none but some details in the preceding account of things terrestrial and celestial can be doubted at all; (4) everything said in the book is nonetheless subject to possible correction "by the wisest of men and by the church authorities".[130] Even if we discard the final phrase as a pious expression of timely prudence (p. 422), the course taken here from 'it might be so' to 'it is certainly so' back to 'it might be so', all within four successive pages, is as breathtaking as it is telling. Nor is it exceptional to find it stated, in an earlier caption of the same book, that "in order to find the true causes of all there is on Earth one must retain the hypothesis already posited, even though it [i.e., the hypothesis] is wrong."[131] But also take a statement like the following: "I would like people to think that, if what I have written ... on any ... matter that I have treated in more than 3 lines in my printed works appears to be wrong, all the rest of my Philosophy is worthless."[132] Somehow Descartes managed throughout his work to convey the impression that he stood ready to vouchsafe as incontrovertibly true even the most insignificant of his assertions and, at one and the same time, to disallow even his strongest assertion as just a noncommittal figment of the imagination.

This unbridled imagination of Descartes', then, is what stands out in his entire philosophy of nature. For every conceivable phenomenon, he was capable of thinking up some corpuscular action meant to explain it, with his treatment of the magnet being an all-time paradigm case for what a fertile imagination, once directed toward nature, may produce. Huygens, on looking back in 1693 upon his own youthful enthusiasm for *Principia philosophiae*, compared the book to "pleasing novels, that impress one as if they were true stories". But how, then, did Descartes manage to claim indubitable certainty for such fantasies? Or, to join in with the slightly malicious irony in which Huygens packed the issue, how ex-

actly "had Mr. Descartes hit upon the way to have his conjectures and fictions accepted as   235
truths"?[133]

A seemingly obvious answer to this question of how Descartes sought to establish his
ambiguous claim to indubitable truth for his kinetic-corpuscular explanations of phenomena would be: by means of empirical evidence. But this is hardly so. In the Cartesian philosophy of nature, just as in ancient atomism, experience furnishes prima facie indications
of how the world is constructed. But, so Descartes readily granted near the end of *Principia
philosophiae*, the creative step of the natural philosopher – to translate those surface indications into corpuscular mechanisms – is beyond direct, empirical confirmation. We can in no
way perceive the corpuscles posited, nor, for that matter, can we observe their shapes or their
motions. What, then, justifies our positing them in the first place?

The short answer to this question, which it took Descartes a good portion of his life
first to hit upon and then to work into a complete epistemology grounded in its turn in a
metaphysics of his own, is: 'the warrant supplied by mathematical certainty'. The question
of what that means sends us back to the early stages of Descartes' career.

The quest for certainty. Descartes' obsession with obtaining certainty in the pursuit of nature-knowledge passed through two stages: a felt need to take up the challenge of skepticism
and his subsequent desire to have the indubitable certainty proper to mathematical derivation permeate somehow the pursuit of nature-knowledge.

Paris, in the 1620s, was a hotbed of philosophical-theological contention. Even in the
midst of the intellectual heat generated by the religious wars that were laying waste ever
more regions of Europe, the Parisian atmosphere was saturated with exceptionally intense controversy, known in historiography as a 'skeptical crisis'. "Que sais-je?" ('What do I
know?') – this concise challenge from the skeptical quarter as coined by its most influential
expounder, Michel de Montaigne, now acquired an intensified meaning. It did so in disputes
between Catholics and Protestants concerned alike to prove their respective doctrines. Involved likewise were conveyers of a Germany-based, 'Rosicrucian' message issued in various
installments in the 1610s. The message contained an appeal to Christianity to reunite under
an esoteric banner of natural magic, and Descartes was under some suspicion of having
dealings with the movement during his travels across Europe. Also involved in the crisis were
so-called erudite libertines whose posture of superior worldliness seemed to hide a stance
bordering on what passed at the time for atheism (actually impiety rather than an all-out
denial of God's existence). Gassendi was close to the group, and he reneged in the end on
his risky skepticism by seeking refuge in the recovery of Epikouros' doctrine (p. 226). By the
same token he developed an epistemological stance that has been dubbed 'mitigated skepticism'. It implied that, of the various creations of the human intellect, nature-knowledge
alone can overcome the skeptical critique, but only if it renounces the claim to secure knowledge of the essence of things in favor of just probable knowledge of appearances only.

236   This was not at all the direction in which René Descartes was to seek a solution when, upon his return to an agitated Paris from his roaming in Germany and Italy, he became involved in the controversies of the day. To him, the renunciation of certainty was too high a price to pay for the one acknowledgment he was prepared, eager even, to share with the skeptics: their ever-reiterated point that no previously attained knowledge of the empirical world had been sufficiently certain. After all, securely true, proven knowledge was possible, mathematics produced it all the time, and the sole problem lay in how to carry such certainty over to the empirical realm, not incidentally but in a thorough, orderly, methodical way. How, in other words, to snatch from the jaws of skeptical doubt the victory of secure knowledge?

Descartes' master stroke was his realization that skepticism could be overcome, not by rebutting it at some arbitrary point, but by joining it all the way, by following it through every single link in its relentless chain of argument and even, where possible, by fortifying the chain further, only to find in the end a core of certainty so unshakable as to compel even the toughest skeptic to yield to it. If such a procedure worked, the benefits would be immense, for they would enable Descartes, from the core of certainty thus found, to build bridges back to both the empirical world and to God, in other words, both to prove God's existence and to create wholly from scratch an indubitably secure system of knowledge of all there is in the world.

This, then, is how Descartes set out on his path of methodical doubt carried to extremes never yet heard of. All previous knowledge has been doubtful, so he agreed with the skeptics; all appearances as conveyed to us by our unreliable senses are unreliable, so he agreed as well. More than that, each and every of our thoughts may have been conjured up by a "*malin génie*" – the celebrated 'malicious ghost' who formed Descartes' personal contribution to the skeptical armory. The only thing in all this that we can be certain of is *that* we think, and it is here that Descartes hit upon his no-less-celebrated principle of certainty, *cogito ergo sum*, 'I think, therefore I am'.[134] My thought presupposes and, hence, proves my existence; not even a malicious ghost could make me doubt this elementary given, and the task that remains is to carry the certainty thus gained back to the empirical world. What carrier is available? To Descartes' mind, only one presented itself – it resides in whatever it is in mathematics that makes its pronouncements securely true.

At first, Descartes set his hopes for the certainty to be wrested from mathematics very high. As a youth at La Flèche mathematics had made an overwhelming impression upon him, in that it is built up in orderly fashion, along a chain of infallible deductions. They are 'infallible', that is, if our thinking from one theorem to the next proceeds correctly. When does it proceed correctly? Descartes decided that it does so when our thinking is "*clair et distinct*". Here, then, in these two joint features of mathematical thought, clarity and distinctness, the warrant of certainty that he sought resided (at least until the mid-1630s, when work on *Géométrie* caused him to give up this particular hope, too). If we start from the rock bottom of certainty that resides in the '*cogito*' and proceed from there clearly and distinctly, our thought cannot go awry. Others have without exception failed in their varied quests because they set out from uncertain or even absent foundations and/or did not know how to think clearly and distinctly. In contrast, Descartes has now obtained a secure foundation, and, in full possession of the gift of clear and distinct thinking, the only feat that remains for him to accomplish is to rebuild the world from scratch in accordance with his own precepts.

How to reconstitute the world, after having doubted it away? Once again the mathematical way of thought, this time by means of a by-product of his failed method, yielded a ready-made solution. In the opening stages of his campaign against skepticism Descartes had sought to meet it halfway by conceding our capacity for true knowledge to be very limited, in that it is bound by what is defined in geometry – shape, size, and position. Now, around 1628, he transferred these properties from our way of knowing to the realm of being. We are unable truly to apprehend anything beyond shape, size, and position *because that is what the world is like* – the world itself is defined solely by shape, size, and position. By the end of the 1620s, then, Descartes had taken definitive hold of the building blocks from which, over the decade and a half that ensued, he was to construct the corpuscular world that acquired its final shape in *Principia philosophiae*.

So within ten years of standing poised between the pursuit of realist-mathematical science and his first exploration of a mathematics-inspired natural philosophy Descartes had opted for the latter for good. The answer to the question of what it was that drove him down that path, then, comes down to this. To pursue Alexandria-plus in the manner he probed first with Beeckman for the case of perpendicular fall might at first sight look attractive for addressing the real world with the certainty of mathematical derivation. Unlike the Athenian mode of pursuit of nature-knowledge, however, it left no room for the idiosyncratic, imagination-infused kind of realism uppermost in Descartes' mind. In contrast, Athens-plus provided him with a loophole to evade refutation. It also granted plenty of scope to Descartes' wholly personally motivated urge to pose as the single-handed overturner of everything ever thought previously. Finally, it left room for the monumental self-deception implicit in what Christiaan Huygens was to call, half a century later, 'Mr. Descartes' way to have his conjectures and fictions accepted as truths'.

All this, then, is how over the most creative decade of his life Descartes abandoned any further pursuit of nature-knowledge in the vein of Alexandria-plus, so as instead to carry to novel heights those hoary, defining characteristics of natural philosophy Galileo had broken with in his own search to come to grips with the realities of nature – its catch-all scope, its know-all pretensions, its less or more explicit claim that, but for the filling in of some left-over details, no riddles remain. And since such exercises in explaining the world of phenomena from first principles had been carried out before in much variety, Descartes was driven

238   to using a particularly strident manner to repudiate and, where possible, replace those rivals at all costs.

In striking the pose of a man able to rethink the world from scratch, Descartes accomplished something quite different from what he fancied he was leaving to posterity. In the pursuit of nature-knowledge his true accomplishment rested in his infusing ancient atomism with this new, mathematics-inspired, counterintuitive idea of motion. Descartes' conception of motion – notably, its tendency to persist – was not that different from Galileo's; however, in embedding it in laws of nature he made it far more general, as general, really, as could be. What Galileo had demonstrated in his *Dialogo* for the Earth moving and in the *Discorsi* for falling and projected objects was shown by Descartes to reside in a law of the persistence of motion valid all over the universe.

Causes of the second transformation

As with the slightly earlier transition from abstract mathematical science to abstract/realist-mathematical science, in the present case, too, the event is best characterized as a case of revolutionary transformation. It was a *revolution*, in that the change was radical indeed – again, in 1600 the imaginary observer was very far from foreseeing the outcome. Not since the foundation of the four Athenian schools of natural philosophy had so drastic a change befallen them; never had even one of them undergone so vast a widening of its explanatory range as now happened to the doctrine of atomism. Still, it was a *transformation* in that, as before with Alexandria-plus, this was not a creation out of nothing – rather, something already existing was given a radical twist. Not that it was quite so radical. Alexandria-plus entailed an almost wholly novel knowledge structure (p. 195), whereas with kinetic corpuscularianism the hoary Athenian structure of knowledge (a speculative grasp of the natural world as a whole, based upon first principles and shored up with pieces of empirical evidence) remained fully intact. Descartes in the first stage of his career played with the possibility of moving in the general direction of those intricate, never-ending processes of idealizing and experimenting that Galileo substituted for the ultra-abstract procedures of the Greek geometers. But in the end he wanted nothing to do with it – he preferred the wiggle room that the Athenian all-or-nothing approach left to his unbridled imagination and to his obsession with certainty rather than the piecemeal approach and the intricate to-and-fro between mathematical regularity and ever-provisional reality confirmation that has always characterized Alexandria-plus. Insofar, then, as Beeckman, Descartes, and Gassendi stuck to a preestablished structure of knowledge, their endeavor displays greater continuity with the past they built upon than is true of the previous case. Even so, a revolution it was, and the question of what made it possible requires an inquiry of its own.

The core explanation: An inherent possibility realized.   As with the revolutionary transformation of Alexandria into Alexandria-plus, just so are we facing in the present
case a familiar state of affairs – the transformation actually accomplished in Europe in the
early 17th century rested as an unrecognized and unrealized possibility in the Athenian corpus. Here, too, it is profoundly misleading to regard the onset of the Scientific Revolution as
a 'cataclysmic event' bursting on the scene out of nowhere. Once again the analytically and
causally more fruitful line is to consider the event as, at bottom, the revolutionary transformation of one particular portion of the Greek corpus; in this case, of Athens into Athens-
plus. And once again this core explanation determines the explanatory task that remains,
which is once again twofold: to account for the twisting and to account for the plus.

Whence the twisting?   In principle, two stages are involved in a twist of revolutionary
proportions: the productive stage of coming up with something new is preceded as a rule
by a critical stage of dissatisfaction with the reigning view. With Kepler and Galileo, the two
stages effectively merged into one: to decide that the customary way of doing mathematical
science suffers from an overdose of abstractness easily merges into a search for ways and
means to enhance its realism. In contrast, such dissatisfaction as might arise with the Athenian corpus could not by itself serve as a pointer toward the direction in which to seek a
remedy. Take one of the earliest notes in Beeckman's diary, dated July 1613. There he posited
his principle of motion retained as a more insightful way to explain what enables a projectile
to keep to its course than impetus (force impressed) could, since, so he summarily pointed
out, "Who can mentally conceive what that [force] is, or how it keeps the stone moving, or in
what part of the stone it resides?"[135] Impetus, conceived by Jean Buridan three centuries earlier, had ever since made perfect sense not only to those who adopted it but also to those who
preferred other accounts of projectile motion (p. 84). But in Beeckman's diary the concept is
suddenly declared, not just wrong, but incomprehensible. In published work of the period
this is quite a theme. Galileo and Descartes and Bacon and Hobbes and Boyle and Huygens
all railed at times at Aristotelian doctrine. They exposed and held up for ridicule the sheer
verbalism and the manifest lack of explanatory power and real meaning of basic categories
like 'substantial form' or like motion defined as "the actualization of a potentiality insofar as
it is potential". Evidently, by the early seventeenth century,

through changes in the habitual use of words, certain things in the natural philosophy of Aristotle had now acquired a coarsened meaning or were actually misunderstood. It may not be
easy to say why such a thing should have happened, but men unconsciously betray the fact that
a certain Aristotelian thesis simply has no meaning for them any longer.[136]

240 Why this happened is indeed not easy to say. Might it, or at least widespread dissatisfaction with Aristotle's doctrine and the search for a replacement, have happened in Islamic civilization? There are no grounds for ruling such an event out absolutely, yet if the hidden possibility were to become manifest at all, Europe was definitely the more likely place. Not only that, as with Alexandria, chances for such an event increased in proportion to the sheer number of practitioners, which in Europe was definitely larger. But also, dissatisfaction could flow into a larger variety of channels. Insofar as it arose in Islam, it led away from natural philosophy, as in al-Ghazali's case (p. 56). In Europe, where philosophy used to be cultivated in a more naturalist vein (p. 96), other outlets more readily presented themselves. Some were conventional, like taking a skeptical stance toward allegedly certain knowledge of any kind or taking refuge in some syncretist hodgepodge. A more constructive solution, again more likely to be taken in Europe than in Islamic civilization with its more spiritualist philosophical inclinations, would be to take stock of available philosophies of nature and seek there for opportunities for reform. Ever since the post-1453 recovery of three more Athenian philosophies, rivals to the Aristotelian conception of nature were available which might in principle afford chances for seizing upon constructive alternatives. What opportunities for such a 'plus' presented themselves?

Whence the plus?   The question divides into two: How is it that, among four Athenian schools of natural philosophy potentially receptive to some feat of creative transformation, atomism was the one the three pioneers unanimously turned to? And how do we account for the new conception of motion by means of which this particular philosophy of nature actually was transformed?

A general answer to the first question – whence the turn toward atomism? – is that on closer inspection there were not so many other possibilities left.

In stoic philosophy the ethical and political message had already in Roman times gained primacy over its naturalist portion. More importantly, the core idea of all changes in the natural world being brought about by varying tensions in the *pneuma* was just too specific and not malleable enough to lend itself easily to transformation or extension of any kind.

In contrast, Plato's teachings did prove fit for transformation of a kind. The first to give them a radical twist was Plotinos, whose brand of neo-Platonism, however, with its spiritualist disparagement of natural phenomena, was even farther removed from a philosophy of *nature* than Plato's. In the late 15th century, Plato's dialogues were translated and the original doctrine was reinstated due to the efforts of Marsilio Ficino. Under the banner of Hermes Trismegistos, his efforts also produced a magical worldview, which was not a natural philosophy at all and for which the usual label 'neo-Platonism' is really a misnomer (p. 133). A twist proper, not, as with Plotinos, in a spiritualist but this time in a naturalist direction, was given to Platonism around the turn of the 17th century when Kepler took the geometric

constituents of the world that Plato had sketched in *Timaios*, replaced them with their three-
dimensional counterparts, and then literally built the solar system around them. It is only because of Kepler's subsequent, revolutionary determination to subordinate his philosophy of nature to the empirical exigencies of his realist-mathematical science that this second transformation of Platonism did not take shape as a full-fledged doctrine the way this happened with the transformation of ancient atomism about a decade later (p. 210). Besides Platonism, atomist doctrine alone was sufficiently plastic in its first principles and in its scanty empirical ramifications to make such an act of transformation feasible. In the atomist picture of a world made up exclusively of particles of matter moving, a shift of emphasis from the particles themselves to the nature of their motions had rested as a potentiality from the start. Altogether novel, however, was the idea of motion retained with which it was now being infused. Where did the pioneers of kinetic corpuscularianism get it from?

An attractive answer is: from the man who first conceived of the idea, Galileo. But is it also the right answer? Gassendi for one was much impressed with the theories of motion expounded in Galileo's *Dialogo* and *Discorsi*, which he went on to defend in public. What about Descartes? Upon leafing through the *Discorsi* right after the book came out, he was so quick at vehemently denying, without any provocation, that he had ever learned anything from Galileo as to make those historians familiar with Descartes' gift for prevarication in similar cases of priority (which obsessed him) readily suspect that that is precisely why he always enveloped in silence his travels through Italy in the early 1620s, when Galileo's new conception of motion began to spread through word of mouth. Still, the suspicion, even if well warranted, is superfluous, since it is certain that Descartes' first exposure to this new idea of motion already took place in 1618 at Breda and also that Beeckman had conceived it on his own – there is no possible way Beeckman could in his isolation of the early 1610s have heard about it from Italy. How, then, to explain how Beeckman came to transform atomism by means of this novel conception of motion of his?

At this final point of the causal inquiry we appear to get stuck. Recall the passage on impetus quoted a few pages ago. In it, Beeckman coupled his pronounced inability to form a mental picture of what kind of force 'impetus' might be without more ado to an explanation of ongoing projectile motion in terms of the motion just persisting. His phrasing wrongly suggests a necessary connection, as if not many intermediary steps still separate radical dissatisfaction with Aristotelian doctrine from the move of infusing atomist doctrine with this decisively novel conception of motion. So, to understand where that novel conception came from we must abandon Beeckman's diary, where it appears as one of the first jottings he made in it, and search outside. An obvious move has been to list exhaustively the components of Beeckman's upbringing and environment – his roots and subsequent practice in the arts and crafts; his study of theology and his Calvinist allegiance overall; his introduction to Ramist logic by Rudolf Snel; his reading of Stevin's *Wisconstighe Ghedachtenissen*

242   (*Hypomnemata mathematica*); and so on. But such a listing is bound to raise the question of how, then, did other people's similar backgrounds lead to quite different outcomes and also how Galileo's different background could meanwhile have led to broadly the same revolutionary conception of motion retained.

The net conclusion is that even in radically different modes of nature-knowledge, separated from one another by high fences, acts of revolutionary transformation could on occasion lead to similar outcomes. For the most part the three transformations that together make up the onset of the Scientific Revolution require distinct causal analyses, applicable only to the case at hand; but apparently there are also some leftover elements of concurrence – pieces of a larger causal puzzle. To complete the number of pieces and get a chance to play around with them we must first explore the third revolutionary transformation and discover how far explaining it in its own terms can get us.

NOTES ON LITERATURE USED

Debate in natural philosophy. The example goes back to Tabitta van Nouhuys, *The age of two-faced Janus. The comets of 1577 and 1618 and the decline of the Aristotelian world view in the Netherlands*. Leiden: Brill, 1998.

Atomism. The most comprehensive history of atomism is Kurd Lasswitz, *Geschichte der Atomistik vom Mittelalter bis Newton*. Leipzig: Voss, 1890 (2nd ed. 1926). Its two volumes contain a mine of information for all relevant periods. Fundamental texts in ancient atomism are collected in A. Stückelberger (ed.), *Antike Atomphysik*. München: Heimeran, 1979. I further consulted Lucretius's *De rerum natura* in the edition (with English translation) of the Loeb Classical Library: Cambridge (Mass.): Harvard University Press, 1975 (4th, revised edition). E.J. Dijksterhuis, *The Mechanization of the World Picture*. Oxford (Oxford UP), 1961; section II-5, discusses atomism in early medieval thought (NB: Discussions by some medievalists of Aristotle's '*minima naturalia*' under the heading of atomism founder on the principal difference between these respective doctrines, as demonstrated by A.G.M. van Melsen, *From Atomos to Atom: The History of the Concept Atom*. Pittsburgh: Duquesne UP, 1952 (originally *Van atomos naar atoom. De geschiedenis van het begrip atoom*. Amsterdam: Meulenhoff, 1949).

Beeckman's notes (written partly in Latin, partly in Dutch) were published by C. de Waard as *Journal tenu par Isaac Beeckman de 1604 à 1634* (4 vols.). The Hague: Nijhoff, 1939-1953. Whereas several scholars (in particular, E.J. Dijksterhuis, C. de Waard, R. Hooykaas, M.J. van Lieburg, and myself) examined specific aspects of the diary, K. van Berkel devoted his book *Isaac Beeckman (1588-1637) en de mechanisering van het wereldbeeld*. Amsterdam: Rodopi, 1983 (to appear soon in an up-to-date, English version) to the diary as a whole. Most of what is asserted in the main text on Beeckman as the man to construct the first 'mechanized world picture' and on his dealings with, and impact upon, Gassendi and Descartes goes back to van Berkel's book; further to E.J. Dijksterhuis' penetrating analysis of Beeckman's and Descartes's joint derivation of the law of falling bodies in *Val en worp*. Groningen: Noordhoff, 1924, pp. 304-321 and pp. 342-348, and (for Beeckman's musical theory, his broad picture of the world, and his relations with Descartes) to my own *Quantifying Music*. Dordrecht: Reidel, 1984; pp. 116-161; 187-201.

Gassendi's 'baptism of atomism' is discussed in Margaret J. Osler, *Divine Will and the Mechanical Philosophy: Gassendi and Descartes on Contingency and Necessity in the Created World*. Cambridge University Press, 1994.

Descartes. The latest, still authoritative edition of his works is by Charles Adam & Paul Tannery, *Oeuvres* (11 vols.). Paris: Vrin, 1971-1974 (second impression of the second edition; the first edition, which I have used, appeared in 12 vols. in Paris: Cerf, 1897-1913). The literature on Descartes, in much of which the focus is on the *Meditationes* considered in isolation from Descartes's primary concern, the philosophy of nature, is quite voluminous. Books focused upon his views on nature include Daniel E. Garber, *Descartes' Metaphysical Physics*. Chicago: University of Chicago Press, 1992; William Shea, *The Magic of Numbers and Motion. The Scientific Career of René Descartes*. Canton (Mass.): Science

244 History Publications, 1991; Stephen Gaukroger, *Descartes: An Intellectual Biography*. Oxford UP, 1995, and Desmond M. Clarke, *Descartes. A Biography*. Cambridge UP, 2006. John A. Schuster, who reviewed the first three of these books in *Perspectives on Science* 1995: 3, p. 99-145, has pioneered examination of the historical development of Descartes's thinking, which he has summed up in the entry 'Descartes, René', in W. Applebaum (ed.), *Encyclopedia of the Scientific Revolution from Copernicus to Newton*. New York/London: Garland, 2000; pp. 184-188. I have further benefited from a preview, kindly granted by the author, of chapters in Schuster's book-in-the-making 'Descartes Agonistes. Physico-Mathematics, Method and Mechanism 1618-33'.

For raising the question of what made Descartes opt in the end for Athens-plus rather than for Alexandria-plus, I found inspiration in Dijksterhuis's *Val en worp*, pp. 301-303 and 343-357, and in Alexandre Koyré, *Etudes Galiléennes*. Paris (Hermann), 1939-1940; Part II: 'La loi de la chute des corps. Descartes et Galilée' (*SRHI*, 2.3.2.–3.); for solving it, in Stephen W. Gaukroger & John A. Schuster, 'The Hydrostatic Paradox and the Origins of Cartesian Dynamics'. *Studies in History and Philosophy of Science* 33, 2002; pp. 535-572, and in Daniel E. Garber, 'A Different Descartes. Descartes and the Programme for a Mathematical Physics in His Correspondence'. In: Stephen Gaukroger, John Schuster, & John Sutton (eds.), *Descartes' Natural Philosophy*. London: Routledge, 2000; pp. 113-130. The account of the skeptical crisis in Paris in the 1620s is based on R.H. Popkin, *The History of Scepticism from Erasmus to Spinoza*. Berkeley: University of California Press, 1979 (*SRHI*, section 4.4.5.); of the Rosicrucian manifestoes, to Frances A. Yates, *The Rosicrucian Enlightenment*. London: Routledge & Kegan Paul, 1972 (*SRHI*, 4.4.4.).

Loss of meaning in Aristotelian doctrine. The general point has been made by Herbert Butterfield, *The Origins of Modern Science 1300 - 1800*. London: Bell, 1957 (first edition: 1949); p. 118. Peter Dear has extended it in ch. 1 of his *The Intelligibility of Nature. How Science Makes Sense of the World*. Chicago: University of Chicago Press, 2006. Some of Descartes' criticisms are listed in Garber's book cited above, pp. 96-97 and pp. 157-158. Mario Biagioli, on p. 212 of his *Galileo, Courtier. The Practice of Science in the Culture of Absolutism*. Chicago: University of Chicago Press, 1993, mentions several more by Bacon and by Locke.

VII

THE THIRD TRANSFORMATION:
TO FIND FACTS THROUGH EXPERIMENT

The recovery and ongoing enrichment of the Greek corpus in both mathematical science and natural philosophy that took place in Renaissance Europe were accompanied by a specific coloring around its edges (p. 113). On a far larger scale than in either Islamic civilization or medieval Europe certain subjects were pursued in ways marked by the culture, leading to a specifically control-oriented mode of empiricism that reflected specifically European drives and values. During the late 15th and the 16th centuries, an urge for accurate, factual description and a search for ways and means to blend understanding with action in the sense of manual operations greatly furthered endeavors as far removed from each other as the making of maps, chemical operations, and the preparation of herbals and anatomical atlases. Here and there, now and then, with men like Leonardo da Vinci or João de Castro or Vincenzo Galilei all this even gave rise to the setting up of coherent series of experiments aimed chiefly at the discovery of phenomena as yet unknown. How did such somewhat-disjointed activities fare over the first four decades of the 17th century, at a time, that is, of Scientific Revolution, when the Greek corpus, in the margin of which all this took place, was itself subject to two distinct, near-simultaneous transformations of a very drastic kind?

Briefly stated, a revolutionary, albeit not nearly so drastic, transformation occurred here, too. The big change in this case was that, starting around 1600, such incidental 'setting up of coherent series of experiments aimed chiefly at the discovery of phenomena as yet unknown' began to become the rule rather than the exception. In due course almost all areas where accurate description and magic-infused operation reigned were affected by the experimental revolution, from animal and human dissection to museum collections; and four men, each operating in a distinct branch of learning, brought about the feat of transformation itself: Francis Bacon, William Gilbert, William Harvey, and Jean Baptiste van Helmont. The latter three pioneered the experimental way of description and of operation in unprecedentedly systematic and consequential fashion on an unprecedentedly large scale; the former proclaimed that way in the guise of a visionary program. Here is how he did that.

Bacon's vision

Francis Bacon was an English politician who in his spare time thought and wrote about how

246 to reform the whole of nature-knowledge from the ground up. During James I's reign he rose to the august job of Lord Chancellor and then fell into disgrace. He is still known chiefly for his inductivist methodology, his vision of a New Science, his advocacy of an experimental approach to nature, and his authorship of the first scientific utopia. None of these feats is attributed to him undeservedly, yet they acquire a different complexion once hindsight is dropped and they appear as outgrowths of ongoing, magic-infused ways of dealing with natural phenomena. Bacon gave these approaches a decisive twist in more than one respect. He did so in his *Novum organum* (1620; New Organon) and in several other projected parts of his incomplete 'Instauratio magna' (The Great Instauration; really his life's work).

Bacon was steeped in Paracelsian thought, however critical he was of the man, and the fairly detailed, speculative cosmology that he constructed still reeked with the smell of sophic sulphur and mercury. While believing in natural magic, he was repelled by the esoteric bent given it by many adherents, whom he called upon to come out of their closets and proclaim to the world the benefits that might come from an operative stance toward nature. Never missing a chance to sum up his views in a pithy phrase destined to live on, he assigned to those engaged in the pursuit of nature-knowledge the task of 'conquering nature by obeying her', in other words, to gain mastery of nature through an understanding of her hidden regularities. That understanding had to be gained through the carefully accurate description of pertinent instances of a rather limited number of principal subjects. The only subject he ever explored in some detail in accordance with his own prescriptions was the phenomenon of heat. Whether for heat or for any other principal subject Bacon required three lists, or 'histories', to be drawn up:

(1) one describing all those various instances in which the principal phenomenon in question does occur (e.g., heat goes together with light if coming from the sun; with force if coming from a chemical reaction; and so on);

(2) one describing all otherwise similar instances in which the principal phenomenon *fails* to manifest itself (e.g., the rays of the moon give light without heat);

(3) one comparing lists 1 and 2 with each other.

From list 3 a more generalized description may be derived. For example, we have now learned that heat cannot be reduced to light, which is not always copresent with it, whereas heat might still be reducible to motion, for which case no counterinstances appear in list 2. Such a generalized description in its turn is subjected to a similarly threefold listing at a higher level of generality, and so farther upward along the ladder of induction to the highest generalization possible in the systematic investigation of natural phenomena.

Here, too, Bacon went beyond customary coercive-empiricist practice in that he extended the accurate descriptions that these listings required beyond the almost purely observational. Borrowing from alchemical but especially from craft practice, he proclaimed the need to make manifold experiments so as to compel nature to reveal under severe yet respectful interrogation her most hidden properties. In other words, a kind of 'artificial nature' was being called up,[137] or rather, in Bacon's view the kind of artificial phenomena craftsmen inadvertently produce all the time offered the best exemplar available of how to go about the systematic, inductively guided investigation of the phenomena of nature as well.

This, then, is how Bacon desired to see the ongoing, all-too-haphazard 'hunt' after the secrets of nature, and the all-too-uncritical collection of natural facts the world over, turned into a truly methodical enterprise. It was clear to him that no one person could carry it out in full. Rather (and here again he borrowed from standard practice in the crafts), nature-knowledge ought to be pursued by way of a collaborative effort, with clear-cut assignments given to every member of the group. He lobbied in vain for the funds required to do the job, so by way of consolation perhaps he sat down toward the end of his life to imagine a civilization – which he called the 'New Atlantis' – where the job had been performed already, thus giving humanity its first glimpse of what life in a thoroughly science-imbued society might look like. For, and this lay at the heart of Bacon's vision, once nature had been conquered by the human race it was incumbent upon us to go ahead and employ the insights gained "to the effecting of all things possible",[138] as he expressed his intention in one more immortal catchphrase.

In short, all those disparate themes and ideas and notions and practices that marked the coercive-empiricist mode of nature-knowledge were now brought by Bacon to a synthesis as potent as it was innovative. He himself did not contribute much to the finding of facts the experimental way; he only claimed to know how it should be done. One pertinent example is his proposal of two hundred experiments on musical sound in a book, *Sylva sylvarum*, that appeared posthumously in the year of his death, 1626.

Bacon's proposed practice: The natural history of sound

Whether deliberately or not, Bacon's ideas on musical sound followed the empiricist trend in Italian musical theory, and carried it to extremes. In the second half of the 16th century, following Aristoxenos' recently discovered views, Vincenzo Galilei and others took the Pythagorean regularity in consonant intervals to be of subordinate significance to the irreducible autonomy of the melodic flow in a musical piece (p. 144). Bacon now went so far as flatly to deny any connection between sound and number. How then to account for the powerful effect music may exert upon us? Here the worldview which he developed as the background to his disquisitions on method stood him in good stead. Only in recent decades has his worldview been subjected to painstaking reconstruction from hints scattered over his many works. At first given to atomist conceptions, Bacon exchanged these for an aethereal, noncorpuscular yet by and large material spirit of mixed Hermetic, iatrochemical, and pos-

248   sibly stoic provenance, which in the earthly realm interacts with gross matter. Sound, as one exemplar of spirit, acts upon both its animate and its inanimate receivers in accordance with a sympathy holding all that partakes of spirit connected. Such a cosmic sympathy, then, accounts not only for resonance between musical instruments but also for the various ways in which music resonates with, and affects, the mood of the audience.

Such was the general background against which the empirical problems listed in *Sylva sylvarum* acquired for Bacon their full meaning. He took many without acknowledgment from either the ancient collection of *Problemata* ascribed to Aristotle or from Gianbattista della Porta's *Magia naturalis* of 1589. Specifically Aristotelian were Bacon's notions on, for example, the speed of sound in relation to pitch (the higher, the faster), the range of sound in relation to the time of day (farther at night), or the various media through which sound can travel (air is best, but other media will also do, provided they are porous). Bacon's chief interest in the domain of sound, however, was in keeping with his general concern for the improvement of human destiny by harnessing natural effects to human ends. The end in this particular case was the artificial 'majoration' of sound. Here della Porta provided a rich source of ideas, in the best tradition of natural magic, for devices to make sounds louder or carry farther. Speaking trumpets, 'ear spectacles', echoes, and whispering galleries all found a place in Bacon's program for a natural history of sound and how to make proper use of it. So did contemporaneous schemes for 'perspective lutes', automatic virginals, and the like. Bacon further suspected that makers of musical instruments had amassed treasures of empirical knowledge on how sound behaves under artificial conditions. From their instruction one might infer how pitch, loudness, and timbre vary with such factors as the material and the shape of viols and clarions and bells and a whole plethora of other instruments with strings and pipes and resonating bodies. Rather than looking up the artisans in person, Francis Bacon, Lord Verulam, characteristically confined himself to admonishing others to find out what could be learned in such plebeian quarters.

Sound was but one subject in an array of topics covered in *Sylva sylvarum* that its author deemed fit to provide the kind of stuff natural histories ought to be made of. For most of the rest of the 17th century, phenomena as varied as light, colors, cold, wind, or (a matter of deep concern to Bacon himself) life, illness, and death were investigated in the same manner (ch. 13). Still, the most significant contributions by means of which the experimental search for facts was to drive the Scientific Revolution forward do not reside in this, altogether rather unproductive genre of natural history. Those contributions were rather to follow the example of three near contemporaries of Bacon's whose efforts, albeit not undertaken under Bacon's aegis, nonetheless fit smoothly his vision of impending reform in the direction of consistently undertaken fact-finding experimentation.

Gilbert: Lodestone and amber treated the Baconian way

The section heading is misleading insofar as King James' Lord Chancellor heartily despised Queen Elizabeth's court physician as a disproportionately pretentious handler of not nearly enough observational data much too rashly subjected to a pet theory arrived at a priori. Nonetheless, posterity can see as Bacon apparently could not that William Gilbert's way of examining magnetic and electric phenomena, while situated indeed at some distance from the methodological letter, nonetheless was quite close to the empirical, magic-infused, and practice-oriented spirit of Bacon's slightly later prescriptions for how to restart from scratch the pursuit of knowledge of nature.

In his book of 1600, *De magnete* ('On the magnet'), Gilbert in effect turned the previously scattered, more or less random examination of two strictly circumscribed phenomena hitherto immersed in heaps of inaccuracies and folklore, namely, the enigmatic behaviors of pieces of lodestone and of amber, into two distinct, budding branches of specialized inquiry – those of magnetism and of electricity. So far the two ancient topics were held to be related only in the puzzling nature of the attractions apparently involved. Gilbert now pooled, checked, and orderly recorded much practice-gained experience gathered by others; established by means of firsthand experimentation numerous other empirical properties of electric and magnetic substances; constructed a view of the world fit to direct his empirical researches and account for the outcomes; and boldly advertised all this by tirelessly labeling it brand new. 'New physiology, demonstrated by means of several arguments as well as experiments, concerning the magnet and magnetic bodies and that grand magnet, the Earth', is how Gilbert entitled this one book of his to appear during his lifetime. Here is how he went about the investigation.

Pierre de Maricourt had established numerous properties of the magnet (p. 89), and Gilbert took over from him far more than he cared to acknowledge. Both men established that like poles repel and opposite poles attract one another. They both saw that a magnet with its north pole and its south pole, once cut into two, becomes two magnets with a north and a south pole each. They described identical procedures for determining the poles of a spherical magnet. Both experimented with little magnets fixed to pieces of wood (or, in Gilbert's case, of America-imported cork) to make them float on water. Gilbert further made liberal usage of later literature in the 'natural magic' genre. Gianbattista della Porta had found that a magnet's weight did not increase even when it had been in close contact with iron filings for months. Similarly, Girolamo Cardano had listed five respects in which the attractions of amber differ from the attractions of the lodestone: for instance, straw or pieces of paper or other kinds of stuff may be drawn by amber, whereas a magnet draws only iron. All these findings duly turned up in *De magnete*. In further seeking data on the distribution of iron ore all over the Earth and on diverse ways to work iron, Gilbert likewise drew heavily on ex-

250 tant literature (notably, on Agricola's book on mining, *De re metallica,* of 1556) and on what English mining operators told him firsthand.

Much in Gilbert's extensive treatment of magnetic declination and inclination and the varieties thereof derived from seafarers' experiences in watching the behavior of their compasses. He relied in particular on Robert Norman's booklet of 1581, *The New Attractive,* with its extensive discussion of irregularities in the north-south alignment of compass needles observed at different latitudes and longitudes. Norman also discovered and, by means of an 'inclinometer' of his own invention, investigated magnetic dip – the direction a compass needle takes in a vertical plane. All this was likewise adopted by Gilbert with more care than he took to assign proper credit to a mere craftsman. He further left in good part unacknowledged a specific contribution by Edward Wright. Originally an academic, by the 1590s Wright had turned himself into the nation's leading expert in the mathematical aspects of navigation. To Gilbert's book he contributed not only the signed preface but also some further portions only recently unmasked as not by Gilbert at all.

Still, Gilbert differed from all previous authors on the subject but Pierre de Maricourt in that, whether adopting earlier accounts of alleged facts or establishing novel ones of his own, he never took anything for granted but, whenever possible, went out of his way to check. For instance, Gilbert showed that the story about Fracastoro suspending a piece of iron in midair between the ground and a magnet was demonstrably false. Both amber and the lodestone had for a long time been associated with gems and diamonds; Gilbert sorted matters out in many patient trials and showed that the only valid connection resides in the capacity of a rubbed diamond to act the way amber does. Besides disproving fables, he also charted unknown magnetic and electrical territory by setting up many fact-finding experiments wholly his own. He armed magnets with sheet iron. He found out that the power of a magnet to influence the direction of a piece of iron extends farther than the largest distance over which that piece can be drawn to the magnet.

Further, in close analogy to the compass for magnetic action Gilbert used a small, pivoted, metallic needle (Fracastoro's *versorium*) to get a better grip on electric action. With it he established that, besides amber, many other materials – for example, sulfur, glass, and sealing wax – may also give rise to 'electric' attractions. He also found that magnets act across interposed substances (e.g., a stone or a flame) but that electrics do not. Measurement was not involved in all this – the only quantities to come up in *De magnete* arose in close connection with measured values already in common usage among mariners as reported by Norman and Wright. It is in their domain, where haven finding and the determination of latitude and longitude were paramount issues, that Gilbert's principal interest in magnetic effects was located. The directive properties of the lodestone, to an understanding of which he contributed such experimental discoveries as the variation of dip with latitude, captivated him far more than its powers of attraction. This preference had much to do with greater apparent utility; but Gilbert's theoretical views on magnetism were also involved.

Gilbert regarded the Earth as possibly in daily rotation, and perhaps in annual orbit around the Sun as well. The vindication of Copernicanism in whole or in part may have been among his motives for writing his book in the first place. On these issues, however, the preface and title are silent, whereas his principal concern is loudly announced in the title – he conceived of the Earth as one grand magnet. This is why he preferred spherical magnets (*terrellae*, 'little earths') over the stronger bar magnet for his experiments. He held the Earth to be made of iron, with only its crust contaminated by impurities. The iron has been bred in the Earth's womb by condensation of the inner Earth's excretions and exhalations. This vegetative life of iron is not incongruous in Gilbert's conception of nature, which is a thoroughly vitalist one. Nature is animated and fully lifelike; terrestrial magnetism is an expression of the world-soul, and the apparent tendency of the lodestone to 'attract' iron is a matter of voluntary union – the very opposite, therefore, of electrical attraction. Gilbert's efforts to distinguish so carefully between magnetic and electrical effects, which seemed justified by his experimental outcomes, corresponded to his insistence on a basic difference between the two. Enriching at this point the vitalist frame of his theorizing with an Aristotelian distinction, Gilbert held electricity to be caused by invisible yet material *effluvia* (outward streams) emanating from the electric to whatever was attracted by it and, hence, to stand for 'matter', whereas incorporeal, all-penetrating magnetism stood for 'form'. This magnetic form is forever replenished; it is, in fact, the primary living force in nature and serves as a model for nature's most hidden operations. As such, it can be harnessed to serve human ends – our magnetic Earth, in aligning pieces of lodestone in accordance with the North and South Poles of the universe, helps us determine both our terrestrial and our cosmic bearings. Twenty years later Francis Bacon, in his *Novum organum* of 1620, cogently presented a magic-inspired, broadly empiricist, practice-oriented program for the pursuit of nature-knowledge – how could he *fail* to see that program prefigured in the late William Gilbert's New Physiology of Magnetism?

The 'new' in the titles of both these works – the 'Organon' and the 'Physiology' – was hardly a new 'new'. Advertising novelty was by 1600 commonplace. Or so it was in the generally forward-looking mode of nature-knowledge colored the European way – heaps of books in this vein carried a bold 'new' in their titles, such as Patrizi's *Nova de universis philosophia* of 1591 and a Spanish book on American herbs translated as *Joyfull Newes Out of the Newe Found World* (1565/1577).[139] Such programmatic insistence on novelty differs tellingly from the tamer titles of those Renaissance books that were concerned instead with the recovery and gradual extension of the Greek corpus. Here Kepler's 'New Astronomy' and Galileo's 'New Sciences' marked the point where mostly backward-looking recovery was consciously transformed into decidedly forward-looking discovery. In any case, the adjective 'new' had not, at the dawn of the 17th century, ceased, in such books as those of Gilbert on the magnet and of Bacon on how properly to investigate nature, to stand for a confidently future-oriented, dynamic approach to things much more than for unadulterated originality.

Harvey: Bodily processes revised

William Harvey's discovery of the circulation of the blood, as well as his later work on the nature of reproduction, forms another case of 15th- and 16th-century coercive empiricism beginning by the turn of the 17th to condense into large-scale experimentation. Decades younger than Gilbert, Harvey, too, rose to the position of physician to England's ruler, Charles I, to whom he remained loyal throughout the civil war. Although ensconced in the English medical establishment in his maturity, Harvey had spent his student days at the University of Padua – ever since Vesalius the international center of studies in human and animal anatomy.

In 1602 William Harvey, then twenty-three years old, completed his academic studies. Half a century had passed since Vesalius published his unprecedentedly accurate, captivatingly illustrated descriptions of the human body, and in the meantime Galen's account of the body's internal functions had become ever shakier. In Galen's setup blood serves a dual role. It feeds the bodily organs; it serves as raw material for a variety of spiritual substances, the natural, vital, and animal spirits which emerge from its successive transformations. The first transformation takes place in the liver, which prepares blood out of the nutrition that the body receives. Endowed there with natural spirits, the blood flows through the veins to the various parts of the body to strengthen them with its nutritious components. One portion flows to the right ventricle of the heart, where, through the pores in its internal separation wall (the *septum*), it seeps to the left ventricle. There its natural spirits are blended with air descended from the lungs through the pulmonary vein and transform into vital spirits which spread across the body through the arteries. Finally, part of the vital spirits, on arriving in the brain, is transformed into animal spirits, which spread through the nerves to activate the muscles. In short, blood is distributed all over the body and is used up; it does not return. New blood is forever being supplied by the liver, the central organ, therefore, of the body. The venous and the arterial flows are separate; each reaches its destination in the various bodily organs independently. The heart, far from serving as the agent of the flow of blood, just sucks blood in; its dilation, not its contraction, is the motion that counts.

The chief strength of Galen's conception rested in its capacity to give a coherent account of a large variety of bodily processes. Its coherence stood it in good stead when, over the second half of the 16th century, a range of new observations proved hard to square with it. Vesalius found the septum not to be permeable at all, and a later incumbent of the Padua anatomy chair, Fabrici, discovered the presence of membranes in the veins. These might have suggested that the flow of venal blood from the heart to the organs was blocked, but to Fabrici they rather suggested the means by which the force of the flow was kept in check. Likewise, possibly adverse consequences for the Galenic system of the apparent impossibility for blood to seep through the *septum* were thwarted by Colombo's discovery of the pulmonary transit – through the pulmonary artery, blood loops from the right ventricle to the lungs, there to be supplied with air, which it carries through the pulmonary vein to the left ventricle of the heart. Thus, it remained possible after all for blood to reach the left ventricle from the right, and also for the observer to keep regarding the function of the lungs as dual: to supply the air needed to transform natural into vital spirits and to cool off the chest.

When in 1616 Harvey first expressed himself about these issues in public, he still accepted the standing account of how blood is formed and distributed. One major difference rested in a conclusion he reached after sustained vivisection of various animals, from frogs to dogs: that contraction, not dilation, is key to the operation of the heart. The heart expels blood into the arteries, with a vigor that is reflected in their pulse. It then relaxes, so as to allow the ventricles passively to be refilled with blood from the principal veins.

The crucial step Harvey made soon after was to shift attention from one individual heartbeat "to the cumulative effect of many beats in succession".[140] Even a deliberate under-estimation of the frequency of heartbeat and of the amount of blood discharged in every beat showed that more blood is expelled from the heart in contraction within half an hour than the entire body contains; where else could all that arterial blood go than back to the heart along another pathway, the veins?

In his book of 1628, *De motu cordis & sanguinis* (On the Motion of the Heart and Blood), Harvey combined two leading themes. He made an all-out attack on Galen in view of those numerous refractive discoveries made in Italy over the past decades plus his own experiments on contraction. He further set up an argument in favor of the circulation of the blood. He could not, to be sure, make it fully clinching. Just as Copernicus had declared parallax to be present indeed, yet so minute as to escape observation (p. 111), just so was Harvey driven to declaring the connection between the arterial and the venal systems in the bodily organs themselves to be too fine to yield to human sight. Unlike with parallax (not to be observed until 1830), the capillaries were to show up within decades after Harvey's death, by means of the microscope (p. 456). Also unlike Copernicus in defense of the reality of heliocentrism, Harvey had other empirical arrows to his bow. Using his own arm, he made an experiment with a tight ligature to hold up the blood flow in various ways that went far to show that, indeed, arterial blood flows to the extremities and then returns through the veins. And of course he found confirmation in Fabrici's membranes, which now appeared as valves, serving to prevent venal blood from returning to the bodily organs.

This is how Harvey pulled the rug out from under Galen's venerable account. What enabled him so decisively to go beyond other anatomical experts familiar with a range of observations (like Colombo's pulmonary transit or Fabrici's veinous membranes) that had already posed awkward questions for decades? Two conceptions were instrumental in his transcending these boundaries. One was the idea of *circulatio*. Following Aristotle, certain thinkers at Padua during Harvey's stay there took the heart, not the liver, to be the central

254 organ of the body. Expanding on Aristotle, they ascribed to it a certain circulation, not in Harvey's later, directly tangible sense, but rather by way of the heart's representing through the unceasing alternation between contraction and dilation the cyclical course of events both human and celestial. In line with such exaltation of the heart, Harvey ascribed to it all kinds of vital functions. His own discovery, to be sure, turned the heart into a pump. But Harvey saw the heart as far more than a pump, just as he held the liquid it sends on its way through the body to be a far more elevated substance than just the subordinate mediator between inanimate food and the fully animate, vital parts of the body that Galen had taken it for. Rather, blood serves the body as its true source of vitality, cyclically replenished in the heart:

> So in all likelihood it comes to pass in the body, that all the parts are nourished, cherished, and quickned with blood, which is warm, perfect, vaporous, full of spirit, and, that I may so say, alimentative: in the parts the blood is refrigerated, coagulated, and made as it were barren, from thence it returns to the heart, as to the fountain or dwelling-house of the body, to recover its perfection, and there again by naturall heat, powerfull, and vehement, it is melted, and is dispens'd again through the body from thence, being fraught with spirits, as with balsam, and that all the things do depend upon the motional pulsation of the heart: So the heart is the beginning of life, the Sun of the Microcosm, as proportionably the Sun deserves to be call'd the heart of the world, by whose virtue, and pulsation, the blood is mov'd perfected, made vegetable, and is defended from corruption, and mattering; and this familiar household-god doth his duty to the whole body, by nourishing, cherishing, and vegetating, being the foundation of life, and author of all.[141]

The same blend of vitalist thinking and experimental endeavor made its appearance in Harvey's later examination of reproduction. Here in similar fashion he held a vital principle to regulate the process of cyclical regeneration "from the fowl to the egg and from the egg back to the fowl, which gives them perpetuity".[142] Prior to Harvey, different accounts of reproduction had been given for different classes of animals – one for insects, another for viviparous beings. In Harvey's unprecedented view, reproduction takes the same basic form throughout the animal (and human) realm. "Ex ovo omnia" (everything from the egg) is how he expressed the idea on the frontispiece of *De generatione animalium* (On the Generation of Animals; 1651). Not that his idea of 'egg' was like ours: "An egg, the origin of every being, was to Harvey an homogeneous point of matter which an indwelling formative principle molds and converts into an articulated individual able to produce, as its ultimate act, an homogeneous point of matter, the primordium [first beginning] of another generation."[143] Male seed does not materially fuse with the egg, so Harvey established on patient examination of the uteri of does held in the royal parks, but rather stimulates the dormant vital principle in the egg to become active, with the egg itself providing the food needed to start growing. Opening a hen's egg three days after it was laid, he observed a point of pulsating blood, which in due time would turn into the heart of the chick. Thus did Harvey confirm the experimental way both his conception of blood as primary in life's processes and his idea of reproduction as universally operating by way of a homogeneous beginning that grows into a mature, heterogeneous being due to some formative principle activated by seed.

Harvey examined numerous other subjects besides the two to which he devoted distinct treatises, but he completed no other works. As most of his readers, adherents and opponents alike, saw at once, what he did publish was bound to change things. The circulation of the blood was clearly incompatible with Galen's account of bodily processes, even though Harvey himself, in sticking to the doctrine of the four bodily humors, was not prepared to replace Galen's account with a similarly comprehensive model of his own. Harvey's views on generation, albeit not quite so disturbing or so radical in their consequences, were bound likewise to engender dispute. The latter dispute was much aided, albeit not resolved, by a piece of equipment that Harvey more or less deliberately ignored, the microscope. The experimental usage of scientific instruments, in their infancy at best, was as yet quite limited.

Van Helmont: Paracelsianism reformed

Paracelsus' legacy was one final domain where a sustained search for facts the experimental way was being pioneered. In an array of rambling writings in the German vernacular, only a small part of which were published during his lifetime, Paracelsus arrived in the first half of the 16th century at what has been called 'a unified chemical approach to nature' (p. 132). He blended selected components of alchemical, metallurgic, and pharmaceutical knowledge with folk medicine, natural magic, and Hermetic wisdom. Out of this came an extraordinary, chemistry-infused doctrine meant to provide both a conception of all change in the world and the master key to curing humanity of its varied ailments. Consistency was not the greatest virtue of Paracelsus' collected views, however, and it is only through the efforts of a range of thinkers gripped by his thought after its author had passed away in 1541 that a distinctly Paracelsian doctrine was finally presented to the world. Its career up to the middle of the 17th century, when corpuscular conceptions began to take over, is best divided into two. By 1600 a somewhat coherent doctrine of iatrochemistry had come into being, in a more empirical vein than taught by Paracelsus himself. The doctrine was accompanied by ever more controversy, which reached its peak in the 1620s and 1630s. During that later period one prominent victim of these controversies, Jean Baptiste van Helmont, proposed in consequent isolation a number of reforms and extensions, all directed toward making the Paracelsian variety of a magic-laden worldview more specifically experimental.

256   At the first stage, sustained efforts at systematization were undertaken by the Danish court physician Peder Sørensen (Petrus Severinus) and the Prague-seated, German physician Oswald Croll(ius). Due to their efforts chiefly, the following statements sum up expurgated Paracelsianism as settled by the turn of the 17th century. The world has been created through chemical processes. Fire is the central chemical agent, whether directly (combustion) or indirectly (distillation). All chemical substances partake of the *tria prima*, the three principles. These are not to be found in a pure state; that is, they are 'sophic' substances that cannot be isolated, yet their presence is revealed by chemical analysis. This appears empirically from the three fractional products yielded by the decomposition of organic substances: a volatile fraction (corresponding to sophic mercury), an oily one (corresponding to sophic sulphur), and an ashen, solid one (corresponding to sophic salt). In one way or another, the three principles coexist with a limited number of elements, which may be one or two or three or five but never the four that Aristotle distinguished and that Galen made to correspond to the four temperaments and the four bodily humors (p. 16). The Earth has a central fire, which nourishes volcanic outbursts; in its crust base metals ripen into the higher ones, ultimately into gold. All of nature is lifelike and organic; macrocosm and microcosm are closely analogous and mutually dependent. The macrocosm/microcosm analogy can be read in detail by means of the signs that the forms of things display. In view of the chemical composition of nature generally, the analogy implies that human beings are assemblies of chemicals, too. Mutual balance attained between these chemicals in terms of the *tria prima* constitutes health; when disturbed, the balance can be restored by administering supplementary doses of chemical drugs.

    Van Helmont, who was born in 1579, at first went along with these doctrines without much ado. But by the 1630s he had turned himself into their most radical innovator. While sticking to numerous aspects of Paracelsus' work, notably his ceaseless attacks on the Aristotelian and Galenic establishments, he freely dissented from him on many specific points. Above all, van Helmont was forever on the alert to make Paracelsian assertions more specific and better anchored in experience. Rejecting as unbiblical Aristotle's four elements, he replaced them with just two, air and water. His reasons were that fire is not mentioned in Genesis and that earth can be shown to come from water. He proved the latter point by means of a well-known experiment. He planted a little willow tree in 200 pounds of dried, carefully weighed earth. For five years he took care to provide it with nothing but rainwater or distilled water, while covering the earth to keep away dust. In the end a big tree had grown, and when he dried and weighed the earth it had grown in, it still weighed "200 pounds, wanting about two ounces. Therefore [so he concluded] 164 pounds of Wood, Barks, and Roots, arose out of water onely".[144] He further argued that it is always possible, by means of chemical processes, to reduce earth (even metals) to water without loss of weight.

    While van Helmont agreed with the profound significance of the *tria prima*, he nonetheless cautioned that sometimes only one or two of the three fractions actually make their appearance (e.g., fire applied to sand yields neither ash nor oil), or even none at all. Moreover, Paracelsus had wrongly taken fire to be a neutral agent. In reality it often takes an independently transformative part in chemical reactions, for which the chemist, being in van Helmont's striking phrase a "Philosophus per Ignem" (Philosopher by Fire), ought always to be on the alert.

More significantly still, at least in theory van Helmont broke with the macrocosm/microcosm analogy, which to him signified no more than a fine metaphor devoid of real import. He did not cease to search for chemical remedies; only, in his customary vein, he strove to monitor the effects of their application more critically than was usually the case.

As with the willow tree, in some of his day-to-day experimentation van Helmont took care in unprecedentedly systematic fashion to weigh the various substances that went into his reactions, as also those that came out. Medieval alchemists had preceded him in operating with an implicit conception of specific weight (p. 88). Van Helmont combined their regular usage of the assayer's balance with the newly Paracelsian idea of chemistry as devoted to chemical analysis. In his view 'analysis' involved not only the separation of substances by various means but also their subsequent recombination. The quantitative approach to chemical reactions that went with these ideas entailed fine-grained bookkeeping. It enabled him to identify end products with ingredients or, if thus alerted to missing portions, to investigate why the original ingredient had not fully returned. To conceive of chemistry the way van Helmont practiced it "as the art of analysis and synthesis ... would provide a disciplinary identity to the field lasting well into the nineteenth century".[145]

Nothing in van Helmont's empiricist bent, critical sense, and quantitative finesse struggled against his being a thoroughgoing vitalist, profoundly convinced that metals grow in the earth, that fermentation is central to both organic and (so we would say, in a distinction van Helmont would not have acknowledged) inorganic processes, and, in short, that everything, whether animal, vegetable, or mineral, proceeds from water and an *archeus,* or life-giving seed. It is in such a vitalist context that van Helmont arrived at some of his most productive findings. One is his painstaking analysis of the neutralization of an acid by an alkali. He caught on his tongue a drop of a sparrow's gastric fluid and recognized the process at work in both animal and human digestion. He further asserted that, when a body made of a given substance is forced out of its fixed state (mostly by heating), it releases a 'wild spirit' which "is in essence a spiritual form of matter that still contains the seed (and therefore the archeus) of that substance".[146] For this wild (in the sense of uncontainable) spirit van Helmont coined the term 'gas'. It was destined to a more impressive career than 'blas' (for a certain kind of motive power) or 'magnal' (for some stuff intermediate between air and void).

Van Helmont's ceaseless search for empirical evidence, whether acquired directly or

258 through deliberate experimentation, as also his interest in putting insights gained about nature to work for human ends, were to animate the work of a range of later chemical/alchemical practitioners (p. 466). The vehicle by means of which they became acquainted with van Helmont's insights was a vast volume of collected works entitled *Ortus medicinae* ('The Dawn of Medicine') that appeared in Amsterdam in 1648, four years after he died. His own country, the Southern Netherlands, still being ruled by the king of Spain, he came into conflict with the local bulwarks of Roman Catholic orthodoxy, the Louvain Theological Faculty and the Spanish Inquisition. In 1633/34 (the same time, that is, when Galileo was being tried in Rome) his work on Paracelsian doctrine was charged with heresy in view of its being tainted with diabolic magic and its going against Aristotle's philosophy of nature. Not to be released from investigation and/or custody until two years before he died, van Helmont quietly committed records of his experiments and of his accompanying thoughts to paper without attempting to make them known during his lifetime.

Theory: Experimentation, theorizing, and background worldview

It is time to sharpen the distinction made earlier between 'natural philosophy' and 'worldview'.

No one comes to the pursuit of nature-knowledge without preconceptions – even those who deny they do so really smuggle their preconceptions into their ostensibly pure, empiricist accounts through the backdoor. And indeed, such coercively empiricist approaches as emerged in late 15th- and 16th-century Europe were invariably undertaken not only with an appeal to some ancient authority (Pliny for collectors in natural history; Hermes Trismegistos for iatrochemists, etc.) but also with some preconception in mind. It might take the guise of a well-focused, theoretical starting point like, for Vesalius, Galen's account of the structure and operation of the human body. It might also (as, notably, with conceptions of magic) take the broader guise of what I shall call 'a background worldview'.

For a proper understanding of much that follows in this book, it is imperative to keep 'natural philosophy' and 'worldview' analytically distinct (p. 9). Irrespective of contemporary parlance, in which the expression 'natural philosophy' was used in a wide variety of meanings, I reserve the term for the 'nature' portion of those four tightly structured Athenian philosophies that claim to explain with certainty the whole wide world from indubitable first principles. In contrast, by 'worldview' I denote more loosely structured sets of ideas about, or approaches to, the natural world, such as China's worldview of organic materialism. Natural philosophy in the circumscribed sense here insisted upon is what the Stoa engaged in, or Ibn Sina or Buridan or Descartes. In the European context, a background worldview is what singled out in particular Paracelsus' conception of the world as a chemical workshop. The experimentation of the four pioneers of fact-finding experimentalism, Bacon, Gilbert, Harvey, and van Helmont, involved a significantly larger degree of interaction with their respective worldviews.

The rise, with these men, of a pattern of interaction between their worldview and their experimentation is better explained with the help of one further distinction. Paradoxically, natural philosophies in the strict sense might (but of course need not) be handled *as if* they were background worldviews. That is, one could draw inspiration from ranges of, say, Aristotelian or stoic tenets without committing oneself to their strict coherence, to their comprehensive scope, to their claim to indubitable certainty. That is how Bacon could so easily shed his atomist views and replace them with a spirit-governed worldview made up of personally revised and enriched components of stoic and magical provenance. That, too, is how Harvey, for all his vitalism, drew on Aristotelian tenets for the concept of *circulatio* or for taking the heart as central to the flow of blood. All this is akin to the 'opportunist' posture that mathematical scientists took at those points where willy-nilly (Ptolemy in the *Almagest*, Copernicus) or quite deliberately (Galileo) they were dealing with the realities of nature. With the pioneers of fact-finding experimentalism, who were less agnostic than most mathematical scientists about the constitution of the world, it is better to speak of an 'eclectic' approach – 'eclectic' not only in the customary sense of picking what one likes from all available philosophies but also in the wider sense of ignoring the knowledge structure and the all-encompassing knowledge claims of philosophy *as such*. To be sure, such 'eclectic' usage of natural philosophy differs from the 'syncretism' I distinguished before, in that in the latter the dogmatic knowledge structure of natural philosophy was scrupulously observed, with only the contents of different philosophies being blended.

In what manner did Bacon's, Gilbert's, Harvey's, and van Helmont's mutually different, specifically-theoretical *and* looser worldview backgrounds affect their experimentation, and vice versa?

In Bacon's ideal view, as set forth in his 'natural history' methodology, it did so in one manner only – as unprejudiced inferences drawn solely from the sustained comparison between lists of data acquired the empirical/experimental way.

In Bacon's proposed mode of proceeding, as in those questions he listed for a natural history of sound, not much more so – his notion of sympathy through spirit only rarely touched upon such observations and experiments as he eclectically took from pseudo-Aristotle and from the natural-magic tradition. Indeed, Harvey had *some* good reason for dismissing the Lord Chancellor as too high-handed for properly pursuing knowledge of nature.

With Harvey, then, as also with Gilbert and with van Helmont a pattern of interaction began to manifest itself between three constituents: *facts*, found by means of observation or even experiment, and *inferences* drawn from those facts against the *background* of some broad worldview and/or more specific theory. To recall one example from each, this is how

260   Gilbert came to list and to insist upon differences between magnetic and electric effects (which his worldview induced him to regard as distinct), how Harvey (using the idea of *circulatio*) arrived at his conclusion of the heart acting as a pump, and how van Helmont sought to demonstrate by means of his willow tree that earth comes from water (in view of his conception that all matter arises from water enriched by vital seeds). Absent from one and still rare with the three other pioneers, such interaction patterns were soon to become both tighter and more intricate (chs. 13; 16).

Causes of the third transformation

The spectrum of change in history runs from just a smooth, predictable unfolding in the general direction of the current course of events to convulsive upheavals impossible to anticipate. The revolutionary transformation of Alexandria into Alexandria-plus (the least predictable upheaval of the three) and of Athens into Athens-plus are both situated fairly close to the latter extreme. In partial contrast, the present case of transformation, while surely an upheaval of revolutionary proportions, was not revolutionary to the same degree and was somewhat more of an unfolding than the other two. There was good reason to make our trend watcher foresee it in part (p. 150).

What we have actually watched unfolding is how a fairly incoherent range of approaches to natural phenomena, governed jointly by a control-oriented mode of empiricism, consolidated into a more systematically undertaken series of fact-finding experiments over the first three decades of the 1600s. This process of consolidation, while surely the most significant innovation from a present-day point of view, took place within the fundamentally unaltered frame of a broadly Hermeticist worldview with its numerous magical overtones and its organic conception of things generally. Activity in nature kept being regarded as the varied manifestations of a variety of forces, of attraction and repulsion, some material and some spiritual, and more or less hidden ('occult') from human understanding. Bacon's call for a general reform of knowledge directed toward the conquest of nature for the benefit of human welfare and longevity has rightly been read as a ringing declaration of a boldly new, experimental mode of nature-knowledge to be erected upon that very foundation. It can be read just as well as a timely summing up of all that seemed most dynamic and forward oriented in what the coercive-empiricist mode of nature-knowledge had achieved over the preceding century and a half.

In this one case among the three transformations, then, we have encountered, not a sudden break involving a radically novel element – *kinetic* corpuscularianism; mathematical *realism* – but a more equably flowing transition. One component already incidentally present – progressive series of experiments deliberately set up to detect hidden properties of

nature, as with Leonardo, Castro, or Vincenzo Galilei – received around 1600 a further boost
in the general direction of its constitutive elements. The degree of continuity present in this
one process of revolutionary transformation, then, is substantially greater than in the two
previous cases. Galileo and Kepler, as also (albeit to a lesser degree) Beeckman and Descartes
and Gassendi, transformed the achievement of their respective forebears more drastically
than Bacon and Gilbert and Harvey and van Helmont did.

Up to this point, then, the question of 'why' almost resolves itself – a process of consolidation undertaken in an ambience that remained by and large the same. Things get more
complicated when we ask next how it is that the turn toward fact-finding experimentalism
took place in Europe and not elsewhere. A Galileo-like figure may by some little stretch of
the historical imagination be conceived to have arisen in Islamic civilization (pp. 72; 204).
Might something similar be said of an equally hypothetical, *Bacon*-like figure?

The question sends us back to the mode of nature-knowledge out of which fact-finding
experimentalism emerged – the mode of coercive empiricism that (prefigured by Friedrich
II and by Pierre de Maricourt) arose by the mid-15th century in Europe around the edges
of the wholesale revival of Greek learning. I have argued that this coercive-empiricist mode
reflected certain specifically European events and circumstances, notably a dynamism that
gave rise to feats like the Voyages of Discovery and the emergence of interfaces between
nature-knowledge and the crafts (p. 134). Inevitably these events and circumstances were
marked in their turn by certain underlying values specific to European civilization at the
time and determined to a large extent by the predominant religion. To understand how a
mode of coercively empiricist nature-knowledge could arise in Europe, not in the world
of Islam, we need to compare these two varieties of monotheism. Specifically, we need to
compare 'this-worldly' and 'other-worldly' value orientations in Christianity and in Islam
and their development over time. To create room for discussing these matters with the dispassionate distance proper to historical scholarship, one major stumbling block must first
be removed: the idea, self-evident to some, rejected by others on the most varied grounds, of
Europe's alleged superiority.

Theory: 'Western superiority' and the lessons of global history.   The rise of recognizably modern science in the 17th century and of such modern technologies as appeared
in its wake from the early 19th century onward was instrumental in providing Europe with
two major attainments without any world-historical precedent: wealth and luxury for more
than just a tiny elite and domination over all other parts of the world. Much in this has
changed greatly in the meantime. For all the poverty still around today, wealth and luxury
have come to affect, in however unequal a measure, the living standards of ever more people
all over the globe. Also, Europe's colonial overlordship has crumbled to the vanishing point.
Nonetheless, much of the picture that Europe formed of itself once its early wealth and

262 predominance emerged in the course of the 19th century is still firmly rooted in Westerners' collective consciousness. The core of that picture was, and is, a conviction of Europe's superiority over all other civilizations. After all, to what else than its superior character traits could Europe possibly owe its unique wealth and predominance in the first place?

What exactly made Europe so superior, and by the same token what made other civilizations allegedly inferior, are questions that have received a multitude of answers. The decisive difference has been sought in everything that went with the idea of 'the white race' – the apparent pinnacle of evolution in many a Darwinist's view. It has been sought in Europe's uniquely early, capitalist mode of production, in Western pluralism versus Oriental despotism, in Europe's individualist cult of the personality versus more collectivist and conformist orientations elsewhere, in the dynamics of a newly emerging civilization as compared to the stasis of sleepy elder ones (a view taken by, e.g., Hegel and Marx), or in a combination of such viewpoints. Further, with one or more of these purported contrasts being taken as solidly established, their first stirrings have been tracked down to deep layers of European history, back all the way into the Renaissance or even the early Middle Ages.

Over the past decades a branch of historiography has come up that goes by the name of 'global history'. Its practitioners have challenged constructions of history along such lines to the very core. Not only that, with the end of the Second World War, explanations along racial pathways have found a well-deserved burial. But comparative work done by global historians has made it ever more apparent that, prior to the early 19th century, there were between advanced, premodern civilizations like China, Japan, India, the Ottoman Empire, and Europe no structural differences of such a kind as to make the great divergence ensue as a matter of course by dint of native superiority and/or preordained necessity. Surely the Old World of premodern times was marked by numerous differences. Yet differences existed between any pair of civilizations, not just between 'the West' on the one hand, and some undifferentiated 'Rest' on the other. No such polar opposition is called for. In Europe the state was not so weak, nor in 'the Orient' so strong, as to explain without more ado the great divergence of the early 19th century. Explanations involving a vast prosperity gap, or the alleged indolence of Oriental civilizations, tend to dissipate likewise under closer scrutiny.

Certainly the core message that goes with investigations along such lines serves as a major act of intellectual liberation: it delivers us from the long-standing hold of an unreflexive Eurocentrism. It is in this spirit that I have sought to examine on equal footing ways in which nature-knowledge was pursued over long stretches in civilizations other than the European one, notably those of China and the Islamic world. Likewise, in my successive efforts at explaining the three revolutionary transformations that began to take place in Europe around 1600, I have deliberately abstained from invoking the standard picture of European superiority and possible grounds for it. But this does not amount to taking *everything* in the standard picture to be false beyond hope of redemption. Even prior to the 'great divergence'

of the early 19th century, when rapidly increasing wealth and predominance began to make
an impact worldwide, Europe developed certain traits distinctly its own – for reasons amply
discussed before, Europe tended from early on to be less self-sufficient and more outgo-
ing than the other civilizations at the time (p. 134). In addressing as we are about to do the
religious values that sanctioned and fostered Europe's penchant for dynamic, extraverted
action, I am once again examining developments distinctively European, yet once again
without attaching a value judgment to them. That is, I emphatically do not take such distinctively European traits as I am about to invoke and whose effects I am about to analyze to
be of a higher or, for that matter, of a lower order than their counterparts in other advanced
civilizations.

   None of this, to be sure, is equivalent to a renunciation of all value judgments. I do not
subscribe to a posture of wholesale cultural relativism; I rather think that just about every
advanced civilization has been superior in some specific regards, inferior in certain specific
others. Europe, albeit not superior in an across-the-board sort of way, has definitely produced certain superior things. Modern science, with its firm grasp of natural reality, is vastly
superior to any alternative for it. The core values of the Enlightenment, with their insistence
on the equal value of all human beings, on settling arguments by debate rather than by
dogma or force, and on human autonomy and its free deployment in a humane society, are
in my view worthy of adoption everywhere. However, value judgments like these are beside
the point of the present investigation. In the period 1600–1640, no modern science was yet
around – it is its bare beginnings that I am seeking to elucidate. Nor was anything like the
Enlightenment already there at the time – the very emergence of the movement cannot even
be thought of without the prior advent of recognizably modern science. In short, although
everything in the distinctive features and developments that I am about to address carries
a label 'made in Europe', none of it should be construed as really meaning 'and therefore
worthier than its counterparts in other parts of the premodern world'.

Fact-finding experimentalism as a specifically European development.   The aim
for now is to understand how an approach of coercive empiricism, with its increasingly
experimental way of finding facts in nature, could come about in Europe rather than in the
Islamic world. Key to any explanation is a major difference between the two civilizations
that rests in a long-term religious development first conceptualized by Max Weber. He made
a principal distinction between '*other*-worldly asceticism' and '*inner*-worldly asceticism'. In
the great religions of the world, be it Taoism, Hinduism, Buddhism, Islam, or early Christianity, the former variety has been predominant as a rule. Quiet contemplation, inward-di-
rected spirituality (or even a mystic search for how to lose oneself in the Deity), the monastic
life, abstention from worldly goods or at least the renunciation thereof on one's deathbed,
all served as pathways toward salvation in the afterlife. This is the customary religious de-

264 velopment, albeit differentially shaded – Hinduism, together with mystical currents in the other religions (e.g., Sufism) has been the most otherworldly of them all. In a long chain of empirically well-supported argument, Weber has built up a claim for one wholly exceptional development in the direction of a much more this-worldly conception of how salvation might be attained. Feeding on certain roots in ancient Judaism, this conception and its attendant practice emerged by the early Middle Ages in Latin Christendom, as distinct from the Byzantine variety, which has remained in conformity with the standard pattern. An austere way of life, conducted without ostentation, and practical, methodical, well-regulated labor in the world directed toward the acquisition of wealth but without consuming it right away, might serve as tokens of, or even as the prime pathway toward, salvation of the soul once the body has gone the way of all flesh. As a rule in the world history of religions, then, abstention from earthly goods was directed *away from* the world; in the exceptional case of European Christianity, it was directed *toward* it.

St. Benedict's rules for the monastic life as promulgated in the 6th century provided one powerful boost in such a this-worldly direction. They embodied a quite novel idea of manual labor regarded as a form of prayer (*ora et labora*) and led within a few centuries to medieval monasteries serving as centers for land reclamation and other instances of hard physical work – a world-historically unique feat. Other boosts followed, as notably with the foundation of the Franciscan Order in the 13th century. Three centuries later a further, possibly decisive boost occurred with the Reformation. As one unintended consequence of the early Reformers' effort at restoring what they saw as a biblical conception of life and the afterlife, a particular work ethos came in due time to mark, not so much Lutheranism or High Church Anglicanism, but rather Calvinism and many Baptist sects. This is when the development of forms of asceticism within, rather than beyond, the world, which was on the rise in European Christianity from the early Middle Ages onward, reached its peak.

So much for Weber's grand thesis on Europe's distinctive religious development. He went on to claim that this long-term, uniquely Latin-Christian development in the direction of worldly labor regarded as a major pathway toward salvation gave indispensable religious incentive and sanction to the ongoing rise, in Europe alone, of specifically capitalist modes of production. In his view, to conduct a life regulated by inner-worldly asceticism fostered the routine accumulation of capital in a framework of free labor. This part of his argument, which is known as 'the Weber thesis', has been fiercely disputed ever since he first made it in 1904/5. We may leave the full thesis and the dispute aside, as they are mostly irrelevant to our concerns here, which are with the origins of modern science, not of modern capitalism. However, two portions of Weber's thesis are of great help for our purposes. One is his *specific* point about the methodic conduct of life in a Protestant setting. The other is his *general* argument about inner-worldly asceticism and how at an early stage it distinguished Latin Christendom from the other world religions.

As a matter of course, in every premodern civilization people took their primary value orientation from the predominant religion. As a consequence of the major differential development just outlined, Europe's value orientation acquired a uniquely extraverted character.
This is directly reflected in the nature and the multiplicity of the 'cultural markers' that
emerged around the edges of the Greek corpus of nature-knowledge. In Islamic civilization
those cultural markers were few and also confined mostly to issues derived from the faith in
a direct manner, as with the direction of prayer or the division of legacies (p. 62). By contrast,
in Europe such markers were vastly more numerous, to the point of turning into a mode
of nature-knowledge all by itself – coercive empiricism (p. 113). Also, the connection was
almost entirely indirect. In Europe, with the sole exception of the determination of Easter,
no other pursuit of nature-knowledge undertaken outside the Greek corpus stemmed from
obligations imposed by sacred texts. The 'third' mode of coercive empiricism reflects instead
the very extraverted, inner-worldly asceticism under scrutiny here. Under its sway, ideological room emerged for a gradual *rapprochement* between the arts and crafts, on the one hand,
and the investigation of nature, on the other, that began to make itself felt during the Renaissance and that had been lacking in antiquity and in Islamic civilization alike. Likewise, in
this context novel probings into the secrets of nature with a strongly practical streak could
come to the fore. Not even the most erudite seeker of hidden truths felt ashamed of getting
his hands dirty, and practitioners of iatrochemistry and alchemy were proud to turn the accusation of being just 'sooty empirics' into a title of honor.

To recapitulate, medieval Europe's increasing separation between church and state, the
growing autonomy of city life, medieval Europe's geographic, political, and linguistic divisions, its lack of precious metals and consequent search for them elsewhere, in short, its
lack of self-sufficiency and its inclination to rove, all contributed to turning Europe into an
ever more extraverted civilization, all the while reinforcing (but also being reinforced by)
a religious development that went in the same direction. Not only was overseas discovery
undertaken on a larger scale and sustained more consistently than elsewhere, as a comparison between Vasco da Gama and Chêng Ho has shown (p. 136). Also, worldly activity of a
kind that was at best condoned in other civilizations found religious sanction in a similarly
activist, inner-worldly conduct of life. Out of this context arose the specifically coercive turn
given to Europe's empiricist undertakings, from Leonardo's emulation of birds' flight to
Vesalius' incisive dissections.

It is characteristic in this regard that the great majority of protagonists of Europe's 15th-
and 16th-century coercive empiricism came in two specific clusters, one centered on exploration overseas, the other on Protestant loyalties. Portuguese pioneers like Garcia de Orta,
Pedro Nunes, and João de Castro, and mapmakers like Mercator and many others took a
formative part in activities connected one way or another with the Voyages of Discovery.
Among the remainder there is a definitely Protestant preponderance, not so much among

266 those concerned with the contemporaneous recovery of ancient mathematical science or of ancient philosophy (to which Catholics and Protestants contributed roughly in proportion to their representation in the population at large), but among those concerned with forms of accurate description and practice-oriented activity specifically, such as Brunfels, Bock, Fuchs, Ramus, Tycho, and Paracelsus (Protestants all) were occupied with. And when, around 1600, these empiricist activities began to consolidate into a more consistently experimental way of exploring nature's phenomena, the same still applied. Unlike with the emergence of Alexandria-plus and Athens-plus, with Bacon, Gilbert, Harvey, and van Helmont there is a definite preponderance of Protestants (the Southern Netherlander being the only Catholic) and also, with both Bacon and Gilbert, a significant background in overseas discovery and trade. In short, the empiricist/experimental approach to nature's phenomena represents, in ways that the recovery of Greek learning did not, certain developmental features uniquely Europe's own – its outward-bound dynamism and its attendant, likewise extraverted, religious values. It is for all these reasons that, unlike with Galileo, a hypothetical, Bacon-like figure could not possibly have come forward in the civilization of Islam.

The question of time. So much for explaining why it was Europe where the turn toward fact-finding experimentalism occurred. What, finally, about the *timing* of the event? Is the bald fact of history that it coincided in time with the transformation of Alexandria into Alexandria-plus and of Athens into Athens-plus no more than that – a coincidence? So far we have encountered little evidence to the contrary. A case can certainly be made for young Galileo´s presence at his father´s musical experiments. Further, in seeking to solve some major puzzle each, both Kepler and Galileo made grateful use of some of Gilbert's insights on magnets. But this is hardly tantamount to ascribing their radically novel approach to natural phenomena generally to incidental influences like these; no more is at stake here than the one propitious factor already invoked (p. 208). Nor, conversely, did the consolidation of coercive empiricism benefit in any way from the simultaneous transformation of the Alexandrian corpus that Bacon so tellingly ignored. True, there would come a time when the two principally different modes of experimentation involved in each – directed at fact-finding in the one case, at validation in the other – would almost blend, in ways prefigured in Mersenne's plodding yet Galileo-inspired experimentalism (ch. 14). It is also true that that distinction does not even hold fully in the case of Galileo himself, who carefully effaced his previous, tentatively experimental search for natural regularities confidently declared in his published work to be experimentally proven a posteriori. On the whole, however, where the timing of the third revolutionary transformation is concerned we seem reduced to the observation that this was one possible, next step to be taken in the inner dynamics that propelled the coercively empiricist mode of nature-knowledge forward.

Doubtless this is a less than fully satisfactory way to complete the effort at explaining how the Scientific Revolution got under way. Nor is the search bound to end here. In the next chapter I first list the full panoply of *specific* explanations we have now come up with for each of the three distinct, revolutionary transformations apart – the ones which gave rise to, respectively, realist-mathematical science, the natural philosophy of kinetic corpuscularianism, and fact-finding experimentalism. Then I consider whether the sheer fact of their near simultaneity does not, after all, call for some further explanation on a higher level of generality.

NOTES ON LITERATURE USED

Conceptualization. The historian to recognize in particularly cogent ways what he called 'the Baconian sciences' as a distinct cluster with a distinct role in the making of the Scientific Revolution was Thomas S. Kuhn. Fully aware that their emergence around 1600 was well prepared, he declined to treat the process of their emergence as revolutionary. I have discussed Kuhn's pertinent ideas at length in *SRHI*, 2.4.4. (where, as I now think, my various criticisms suffer from overstatement), 3.4.1., 4.4.4., 5.1.2., 5.2.6.

Bacon. My treatment is based in the main upon A.G. Debus, *Man and Nature in the Renaissance*. Cambridge: Cambridge UP, 1978, pp. 102-105; upon Mary Hesse's lemma in *DSB* 1, pp. 372-377; upon a sketch of Bacon's conception of the natural world by Graham Rees in his *Francis Bacon's Natural Philosophy: A New Source*. BSHS Monograph 5, 1984; ch. 2 especially, and upon Peter Pesic, 'Wrestling with Proteus. Francis Bacon and the "Torture" of Nature'. *Isis* 90, 1, 1999, p. 81-94. For Bacon's natural history of sound I have used Penelope M. Gouk, 'Music in Francis Bacon's Natural Philosophy'. In: M. Fattori (ed.), *Francis Bacon. Terminologia e fortuna nel XVII secolo*. Roma: Edizioni dell'Ateneo, 1984, pp. 139-154; the broader context for this article is in idem, *Music, Science and Natural Magic in Seventeenth Century England*. New Haven / London: Yale UP, 1999.

Gilbert. My account is pieced together from the following literature. E. Zilsel, 'The Origins of Gilbert's Scientific Method'. *Journal of the History of Ideas* 2, 1941, pp. 1-32 (*SRHI*, 5.2.4.). Stephen Pumfrey's entry 'Gilbert, William', in W. Applebaum (ed.), *Encyclopedia of the Scientific Revolution from Copernicus to Newton*. New York/London: Garland, 2000; pp. 266-268 (an early summing-up of his book *Latitude & the Magnetic Earth*. Duxford: Icon Books, 2002). R.S. Westfall, *The Construction of Modern Science*. Cambridge: Cambridge UP, 1971, pp. 25-28. J.L. Heilbron, *Electricity in the 17th and 18th Centuries. A Study of Early Modern Physics*. Berkeley: University of California Press, 1979, ch. 3.

Harvey. Jerome Bylebyl's entries for Harvey in *DSB* 6, pp. 150-162, and in W. Applebaum (ed.), *Encyclopedia of the Scientific Revolution from Copernicus to Newton*. New York/London: Garland, 2000; pp. 285-288. Also enlightening proved to be Roy Porter, *The Greatest Benefit to Mankind. A Medical History of Humanity from Antiquity to the Present*. London: Fontana, 1997, pp. 211-216, and R.S. Westfall's summing-up in *The Construction of Modern Science. Mechanisms and Mechanics*. Cambridge UP, 1971, pp. 86-92; pp. 97-99.

Van Helmont and Paracelsianism. Two authoritative authors are W. Pagel and A.G. Debus; my account is based upon the latter's *The Chemical Philosophy. Paracelsian Science and Medicine in the Sixteenth and Seventeenth Centuries*, 2 vols. New York: Science History Publications, 1977, with the remarks on van Helmont's quantitative experimentation based on William R. Newman & Lawrence M. Principe, *Alchemy Tried in the Fire. Starkey, Boyle, and the Fate of Helmontian Chymistry*. Chicago: University of Chicago Press, 2002, ch. 2.

Global history. Patrick O'Brien, 'Historiographical Traditions and Modern Imperatives For the Restoration of Global History'. *Journal of Global History* 1, 2006, pp. 3-39.

Weber thesis. *SRHI* 3.6.1. The fundamental notions I am prepared to adopt from Weber's famous treatise of 1904/5 'The Protestant Ethic and the Spirit of Capitalism', part of his incompleted series of studies 'The Economic Ethos of the World's Religions', are (a) his conceptual distinction between customarily 'other-worldly' and exceptionally 'this-worldly' religious orientations; (b) his idea that whether means of salvation are primarily sought in spiritual preparation for the future world ('other-worldly') or in actions in the present one ('this-worldly') contributes significantly to the regulation of believers' day-to-day conduct; (c) his argument that a this-worldly orientation, rooted in the singular development of ancient Judaism, came increasingly to mark Latin Christendom and by the late 16th century reached its most extreme consequences yet, to wit, the world-historically unique stance in life adopted by adherents of certain Protestant denominations in particular, which Weber called 'the Protestant ethic'.

Weber made these points of his serve an argument meant to show, *more specifically*, that a consequent posture of sternly methodical conduct, especially noticeable in 17th- and 18th-century Calvinists and Baptists of various stripes (not so much in Lutherans or Anglicans), came to foster and sustain the already ongoing unfolding of capitalist modes of commerce. I do not commit myself to this particular portion of Weber's argument, as my subject is the emergence of modern science, not the routine reinvestment of capital regularly acquired through free labor. Nor is my argument touched more than tangentially by Weber's corollary that, alone among ancient religions, Judaism was marked by an anti-magical strand which in later times was likewise carried to its ultimate consequences in Calvinism and the Baptist sects, thus giving rise to the 'disenchantment of the world' (*Entzauberung der Welt*). Unlike many of his critics and followers, Weber was well aware that how modern science emerged obeys the logic of an*other*, in good part independent causal concatenation of events and circumstances, which he never went on to explore. He did, however, drop a hint that, whereas modern science quite evidently came into being in Catholic and Protestant circles alike without any noticeable preponderance for either, the very notion of its possible exploitation for economic gain was indeed bound up mostly with Protestantism. The hint was picked up by the historical sociologist Robert K. Merton, in a pioneering work which I discuss on p. 596.

VIII

CONCURRENCE EXPLAINED

An explanatory overview

The core explanation, applicable to each of the three initial revolutionary transformations alike, is the realization of hitherto-unperceived potentiality. Realist-mathematical science rested as a hidden potentiality in the Alexandrian corpus; so did kinetic corpuscularianism in the Athenian corpus; so did fact-finding experimentalism in Europe's coercive-empiricist mode of nature-knowledge.

It follows from this core explanation that none of these revolutionary transformations was bound to happen regardless. In each case the hidden possibility might conceivably have been realized before (in locally different fashion, to be sure) or later or not at all. But here already explanatory differences set in. Why the third transformation came about in Europe is fairly obvious once the distinctively extraverted propensities of the civilization are taken into account; why the first and second did is much less clear. In Islamic civilization, the Greek corpus had flourished in very similar fashion to how it did again in Renaissance Europe. With two unpredictable twists (in mathematical science and in natural philosophy, respectively) resting as potentialities in both civilizations, the question becomes why what *might* have happened in both *did* happen in Europe. Hence, the search must be for causes propitious to the contingent outcome, that is, for historical circumstances and events *favoring* Europe though not *necessitating* the bringing out into the open of what was potentially there, that is, for factors furthering the occurrence of the two big twists, of Alexandria into Alexandria-plus and of Athens into Athens-plus.

Under the heading 'whence the twist?' two propitious factors proved common to both cases. The first involves the general point of cultural transplantations serving as prime stimuli for novelty. In that regard, Renaissance Europe had somewhat better chances for transforming the Greek corpus, in view of its having become the next all-round recipient. Further, the presence of a considerably larger number of practitioners in Europe enhanced chances for individuals to hit upon the potential hidden inside that legacy.

In the case of Alexandria, this latter point means that, with larger numbers, chances were enhanced for one or more individuals to come forward and perceive what no one else had so far perceived. Mathematical science need not remain almost wholly abstract; rather, it can be linked up with the real world, as specifically demonstrated for planetary trajectories (Kepler) and for free fall and projectile motion (Galileo).

272   In the case of Athens, larger numbers similarly enhanced chances for creative dissatisfaction with current conceptions to arise, yet such dissatisfaction could still flow into many distinct channels. Here the main additional finding was that, whereas alternatives had indeed come forward in both Hellenist and Islamic civilization, only in Europe did these take shape as philosophies of *nature*.

From that point onward in the two distinct causal analyses, under the heading 'whence the plus?' everything was listed that may plausibly be considered to have contributed to making ancient mathematical science *realist*, and ancient corpuscularianism *kinetic*. For mathematical science five distinct, albeit rather-ambiguous stimuli came from the two contemporary, rival approaches to nature with their built-in realism: from natural philosophy generally and from all that was going on in the coercive-empiricist mode of nature-knowledge. For natural philosophy I argued first that, of all four available, atomism was the one that lent itself best to feats of creative transformation. In considering next that the new conception of motion persisting emerged with Galileo and Beeckman independently, we found ourselves eye to eye with the issue of *simultaneity*. That same issue came up again when we asked next whether the rise of fact-finding experimentalism had anything to do with the near-simultaneous rise of experimentation in its overall quite different, Galilean mode of providing a stairway between mathematically idealized reality and everyday reality. So few and so relatively insignificant factors of a *specific* nature did we find on that score, but also *so little have the three transformations appeared so far to owe to one another*, that our causal needs are unfulfilled, and we are forced to wonder whether there was really nothing more to say on the subject.

Causal gaps identified

This leftover issue of simultaneity, then, is the point where the outer limits of my original explanatory strategy have apparently been reached. That strategy (to repeat) comes down to doing things in proper order. In historiography a plethora of putative causes have been around for decades, each purportedly covering the Scientific Revolution whole. This approach has not led anywhere definite (p. xix). For monolithic explanations of the kind hitherto advanced in the literature the event is just too complex. But it is not necessarily too complex for any causal analysis at all. That is why I have sought to steer clear of all possible explanations of the customary, all-encompassing type. Instead, I have given priority to setting up a search for *specific* explanations, which I have striven to map with some care onto the three *specifically* distinguishable portions of that truly complex event, the onset of the Scientific Revolution. Only upon completion of this search for specifics have we now arrived at the point where it makes sense to identify any causal gaps that still remain.

Gaps of two kinds have presented themselves. One rests in the apparent poverty, as compared to the world-shaking significance of the events here treated, of the range of propitious factors encountered along the way. The onset of the subjection of natural phenomena to mathematical analysis in the frame of an intricate structure of idealized abstraction and empirical/experimental confirmation; the natural philosophy of particles in motion now brought to bear in much more intricate fashion upon a far broader range of empirical phenomena; the routine setting up of whole ranges of fact-finding experiments – these are no mean feats of history, and such causes as have so far been advanced for each lack proportionate substance. True, this serves to confirm an important point (p. 204). Late Renaissance Europe was indeed in a somewhat-better position than its predecessors to bring into the open possibilities that lay hidden in the Greek corpus, yet it was not *bound* to do so – the opportunity might still have remained unnoticed or been left underexploited. Even so, the range of explanatory factors adduced so far looks slimmer than it needs to.

On another plane, too, we face a causal gap. So far I have been concerned to adduce a range of specific factors that help explain why *each* of the three major transformations occurred in Europe, not elsewhere. But how is it that *all three* occurred there and did so at almost the same time, too? Does it not defy belief to ascribe their concurrence in time and space to sheer chance?

Of course it does. So the search must now finally be directed toward possible causes of that concurrence. The effort, to be sure, is *not* tantamount to explaining the Scientific Revolution whole after all. The question that remains to be answered is *only* how it is that its three major components, each individually the outcome of a distinct chain of historical events, still made their appearance at the same time and place.

An underlying sense of values shared across the culture

The only viable answer resides in the unusually extraverted propensities of European civilization – its 'roving' qualities and its outer-directed religious orientation (p. 134). These propensities do not, to repeat, 'explain the Scientific Revolution'. They only go some way toward accounting for the concurrence of three near-simultaneous revolutionary transformations, in that they mark Europe as a comparatively suitable place to grab opportunities for realizing latent possibilities for revolutionary transformation. 'Comparatively' is a necessary qualifier, to be sure, for other civilizations, too, experienced bouts of outer-directed curiosity and open-minded exploration. The consistency and tenacity with which such strivings continued to be pursued and the extent to which such activities were valued were peculiar to Europe, however (p. 263). The premium placed upon these activities was not necessarily in the literal sense of monetary rewards received for catering to demands but in the wider

274 sense of meeting the kind of overall approval that comes from an underlying sense of values shared across the culture. Compared to other advanced civilizations at the time, this latecomer felt less bound overall to received tradition, more ready to try out something new and see just how far you could get that way. To illustrate the point, here are two examples.

Novelty at a premium. On 11 November 1572 young Tycho Brahe was struck by the apparent presence, in the constellation Cassiopeia, of a star he had never seen there (Figure 4.10b; p. 125). Unable at first to believe his own eyes, he checked with his servants and with some nearby farmers, who, blissfully ignorant of Aristotle's insistence on the immutability of the heavens, had no difficulty confirming the presence of the first nova noticed more than fleetingly in Europe. Soon the new star was observed all over the place, and a fierce, Europe-wide pamphlet war broke out over what ominous meaning the *nova* might convey. Tycho's contribution to the debate, observational proof that the phenomenon was located in the supralunar realm far outside the terrestrial atmosphere, helped undermine the credibility of the Aristotelian scheme of things and offered support to the stoic conception of the universe, which after all did allow for change in the heavens. The point of the story rests in the human inclination to pass over the unexpected and to ignore things whose possible existence is ruled out beforehand on theoretical grounds. In Europe, apparently, a new kind of openness had emerged that was capable at times of overcoming this inclination and kindling the imagination of the multitude with respect to natural phenomena not previously regarded as worthy of note.

On 25 September 1608 the States of Zeeland made a request on behalf of a spectacle maker by the name of Hans Lipperhey for permission to show to Prince Maurits in The Hague his alleged new invention, a concave and a convex lens placed at some distance from each other and mounted in a tube. This instrument enabled one to see faraway things as if from nearby. Two rivals quickly came forward with claims for priority of their own, and no patent was granted. Even so, within a year and a half the telescope made its way all over Europe. Three channels were involved, one diplomatic, one mythological, one astronomical.

In The Hague negotiations were under way that were within a year to lead to a twelve-year truce between Spain and its rebellious province, the budding Republic of the Netherlands. This fortuitous circumstance made for a quick spread of the news and, on occasion, of the instrument itself through diplomatic couriers:

by April 1609 the telescope was commercially available in Paris in a shop on the Pont Neuf, by May it was in the possession of the Count of Fuentes, the Spanish governor of Milan, and by some time that summer it appeared in cities as distant as Ansbach in Bavaria, London, Rome, Naples, Venice, and Padua.[147]

The pace of events was further enhanced by the circumstance that vast amounts of lore were already around regarding instruments allegedly capable of making remote objects visible. The legend of the lighthouse in Alexandria's harbor (the Pharos) mirroring ships from afar was notorious in the Islamic world centuries before it reached Renaissance Europe. There it led to efforts, undertaken by Gianbattista della Porta in a sphere of magical allusion, to reconstruct the lens-mirror combination that allegedly made such feats possible. Consequently, news of the novel lens-lens combination from the Netherlands did not strike everyone as news at all. Galileo, too, needed some prodding before he realized that the instrument could truly do what he felt sure its alleged forebears could not. But as soon as the realization dawned upon him, he perceived what no one (except for Thomas Harriot, simultaneously) had perceived so far: an opportunity to direct the instrument to the heavens and discover there a new world (p. 186). In less than ten weeks he both wrote *Sidereus nuncius* and had the booklet licensed and printed. Its powerful statement of the presence in the sky of objects (moon craters, planetary satellites, a star-studded galaxy) unseen by any mortal from Creation onward did indeed raise doubts with some about possible deception by the unknown, possibly magic-infused instrument that appeared to reveal them. Even so, the existence of these objects was speedily confirmed, and Galileo was catapulted to instant fame all over Europe. Less than two years after Lipperhey's invention was shown to Prince Maurits and his Spanish adversary, the tube with its two lenses had made Galileo's name a household word throughout Europe and raised its owner to the job of court mathematician and philosopher of the Grand Duke of Tuscany (p. 417). Surely novelty in connection with natural phenomena was at a premium in those days and place.

The self-confidence of the pioneers.   Not only did Europe by the late Renaissance provide a (comparatively speaking) innovation-friendly, this-worldly environment facilitating perceptive individuals to devote themselves to the investigation of nature as their prime task in life. But the most daring among them also went about their innovative pursuits with a marked sense of self-confidence.

In one sense their achievement exceeded the accomplishment of those who, way back in ancient Greece, had built entire intellectual worlds from scratch – there were now so many venerable ancient notions to be cleared away first. The greatest of Galileo's achievements has been said to reside in his capacity *not* to think certain things all his contemporaries were in the complacent habit of thinking.[148] The capacity to acquire and then sustain a conviction of having it just about right in utter opposition to views taken for granted from time immemorial requires indeed the kind of monumental self-confidence Galileo displayed throughout his life. He himself knew it well:

> it being very true that our reputation starts from ourselves, and that he who wants to be esteemed ought to have self-esteem first.[149]

276   So did Descartes:

> if from my youth onward I had been taught all the truths of which I have since sought the
> proofs ... I would never have acquired this ability and facility I think I have of finding new ones
> whenever I apply myself to seeking them.[150]

And so did Bacon:

> Myself, then, I found to be equipped, more than for other things, for the contemplation of
> truth.[151]

What did such unreserved self-confidence stem from?

Clearly, personality traits were involved – these were men with formidable egos. In part the size of their egos and the shock-proof nature of their self-confidence stemmed from their noble parentage. Even so, with lowly-born social climbers like Kepler or Beeckman the many insecurities or even shyness that both men displayed in dealing with their fellow human beings was no match at all for the quiet self-confidence with which these so much more attractive personalities carved out their respective paths through unknown territory. The latter felt insecure to the point of finding himself happiest in the role of unpaid assistant to his younger brother, who was a school principal. Just like Kepler and quite unlike either Galileo or (even more so) Descartes or Bacon, Beeckman was in the habit of acknowledging what he owed intellectually to others to the point of self-effacement. And yet his diary notes make it clear that he knew exactly what he was up to, just as Kepler knew what he was doing on his lonely path toward creating a new astronomy and finally laying bare the heavenly harmonies that God had for six thousand years kept hidden from mortal sight.

If, then, such marked confidence in the self-chosen job of radically transforming received views about nature can be seen to transcend both social rank and a merely accidental character trait, where does it in addition come from? It seems obvious that, in particular with fundamentally modest men not in the habit of feeling so self-confident overall, it could only stem from a quiet sense of values shared all over the culture.

An upswing luckily not interrupted

How far, really, did all this differ from the earlier occasions in history when the Greek corpus of nature-knowledge was given a chance to reveal its developmental potential?

Just when the Golden Age of nature-knowledge in Islam was reaching its peak in the early 11th (5th) century, waves of invasion intervened and a centuries-long period of inner-directed contraction set in for the civilization at large. This led in turn, not to wholesale cessation of the pursuit of nature-knowledge to be sure, yet to a certain sapping of the will at the exact point where radical transformation might have been the next step. In our modern world, where innovation has become routine, momentum reversed at one point does not as a rule betoken a loss for good – with some luck a restart is still possible. In the Old World zest interrupted on the grand scale was zest gone for good, at least within one given civilization. Nature-knowledge in Islamic civilization was revived at no fewer than three distinct localities (al-Andalus, Persia, the Ottoman Empire). Practitioners were enterprising enough, yet in each case the Golden Age served as their natural point of orientation. Once the original zest is gone, it cannot apparently be regained. Scholastization, petrifaction, or at best incidental bits and pieces of innovation encapsulated at once in the standard conception of things take over for good.

Three times in history did feats of cultural transplantation open opportunities for the Greek corpus of nature-knowledge to reveal its developmental possibilities. The first time when momentum was unleashed it was destroyed from the *outside,* due to the Mongol and earlier invasions (p. 65). The second time it was held up from the *inside,* due right from the start to the 'Andalusian' curtailment of the medieval upswing (p. 90). The third time, by a stroke of luck, momentum did *not* get lost. The Greek corpus was transplanted in full as it had been in the civilization of Islam, but this time no chieftains on horseback appeared to lay waste the European centers of innovative nature-knowledge at so critical a moment. It goes much too far to say that Europe owed the Scientific Revolution to *nothing but* the accidents of military history; but luck certainly enters into the full historical equation.

To what extent it does, is another question. The pioneers of the Scientific Revolution in its three varieties benefited hugely from the apparent harmony between the search for a novel understanding of natural phenomena and the underlying, outward-bound values of the civilization where the event took place – if nothing else, their self-confidence serves as a clear-cut marker. In its (as the world's spectrum of civilizations goes) unusual penchant for extraverted activity, Europe had by 1600 reached a point where the will toward innovation was furthered by an overall climate in which daring novelty stood at a certain premium. Galileo, in particular, made himself a public, widely admired figure in the very land where the cultivation of individual feats of high achievement had gone to the farthest extremes yet reached. His "rugged individualism" has been called by the economist Joseph A. Schumpeter "the individualism of the rising capitalist class".[152] Although the capitalist mode of doing business is really less relevant here than the values Schumpeter held to underlie it, he was quite right in sensing that Galileo conspicuously symbolized a peculiarly European set of increasingly ascendant values.

The achievement still at risk

Once again this line of argument must not be carried beyond its proper limits. This-worldly activity and an individualist-innovative attitude toward it stood at a certain premium at the very time when the imaginative reception of the Greek corpus at a level previously attained in Islamic civilization had once again reached its peak, and this helps us understand further how steps toward the radical transformation of that corpus could now be made as they had not been previously. But it does not follow that the *outcome* of these transformations, once it had a chance to sink in, was necessarily to look quite so satisfactory as the efforts that had given rise to it in the first place. The Medici court, eager to hire Galileo in 1610 on account of his discovery of Jupiter's satellites, could still within three years begin to raise doubts about their theological implications (p. 420). So the presence of an innovation-friendly atmosphere in which around 1600 the Greek corpus had a chance to be transformed in revolutionary ways did not imply that, if resulting modes of pursuing nature-knowledge were to find themselves presented as going against the reigning faith and its deepest values, such accusations could in the European context be lightly dismissed. In no way was their eventual triumph over an onslaught undertaken in view of alleged sacrilege and other grave defects a foregone conclusion. To the contrary, the long-term survival of three modes of nature-knowledge now radically transformed hung very much in the balance of their no doubt numerous assets and their possibly (for who could safely predict the outcome?) even weightier liabilities. What, then, did the prospects for these novel modes of nature-knowledge look like from the vantage point of the early 1640s, when their original shaping had come to an end?

NOTES ON LITERATURE USED

Tycho and the new star of 1572. Tycho mentions how he first made the discovery on the first page of his leaflet *De nova et nullius aevi memoria, a mundi exordio prius conspecta stella, quae in fine Anni superioris omnium primo apparuit*. Knudstrup, May 5, 1573.

Early spread of the telescope. Eileen Reeves, *Galileo's Glassworks. The Telescope and the Mirror*. Cambridge, Mass.: Harvard UP, 2008.

The Scientific Revolution as an effort of the will. I owe my original insight in this regard to Frances A. Yates' work (*SRHI*, sections 3.3.5. and 4.4.4.).

IX

PROSPECTS AROUND 1640

Chronology and continuities

The pioneering stage of the Scientific Revolution may indeed be considered completed by the early 1640s. It is true that, conceptually, the major breakthroughs had already taken place decades earlier, and even if they had not yet come into print, word about them had begun to spread by the 1620s. Still, the two founding documents of the – as yet – greatest moment of all saw the light of day with considerable delay, in 1638 and 1644, respectively. Schema 4 lists the pioneers' most innovative work in chronological order.

Schema 4:   Dates of the onset of the Scientific Revolution

1592–1610	Galileo in Padua at work on free fall and projectile motion
1600	Gilbert, *De magnete*
1605	Kepler determines elliptical orbit of Mars
1609	Kepler, *Astronomia nova*
1610	Galileo, *Sidereus nuncius*
1613–1634	Beeckman develops kinetic corpuscularianism in his diary
1617–1621	Kepler, *Epitome astronomiae Copernicanae*
1620	Bacon, *Novum organum*
1627	Bacon, *New Atlantis*
1628	Harvey, *De motu cordis*
1629–1633	Descartes at work on manuscript of 'Le monde'
1629–1642	Gassendi prepares restoration of Epicurean atomism
1630s–1644	van Helmont revises Paracelsianism in private
1632	Galileo, *Dialogo*
1637	Descartes, *Discours de la méthode*
1638	Galileo, *Discorsi*
1644	Descartes, *Principia philosophiae*
1648	van Helmont, *Ortus medicinae*
1649	Gassendi, *Animadversiones in decimum librum Diogenis Laertii*

Years of publication italicized; nonitalicized dates to be taken as approximate in most cases.

282 By the mid-1640s, then, unprecedentedly drastic innovation in the pursuit of nature-knowledge had not only come about but also become clearly visible. But what next? Now that revolutionary nature-knowledge had come into the world in ways hard to miss, was it also going to stay there? If so, was it going to be subject to further change? And if that, too, was the change going to be of a similarly drastic kind? Before we can even begin to answer these questions, we ought to remind ourselves of the numerous things that so far had not changed at all.

To begin with, the three older modes of nature-knowledge (two Greek, one specifically European) had not of course vanished with the mere arrival of what was meant to replace them. Was mathematical science in its classic Alexandrian mode quietly to give way to the realist claims built into Alexandria-plus? Was kinetic corpuscularianism bound to do more than add one more philosophy of nature to the foursome already in existence for so long? Was fact-finding experimentalism due to overtake without more ado the looser observational procedures by means of which research had been conducted as a rule in the mode of control-oriented empiricism?

Also, nothing much had as yet changed in the customary compartmentalization of nature-knowledge. As before, each mode was pursued by different individuals, with little interaction as yet taking place between them.

Further, by the early 1640s nature-knowledge was still being pursued as a rule in the long-standing frames of the court and the university. It continued to be communicated (if at all, since much was still being buried in manuscript) by means of lessons taught, letters exchanged, books printed, and travels undertaken.

Finally, by the 1640s Europeans' sense of ongoing dialogue with antiquity had not been broken in any essential way. The ancient authors, Greek and Roman, continued to make their presence felt as if they were still around in person; they kept occupying the background of almost every learned person's thought; they kept providing the measure against which every novelty had to be gauged. Right after making a big discovery regarding centrifugal motion, Christiaan Huygens quoted Horace ("free steps have I been the first to make into the void");[153] when his proud father wished to label him in a way everyone could grasp, he called him 'my little Archimedes' ("*mon petit Archimède*").

Amid this pattern of overall continuity in which the three novel modes of nature-knowledge were as yet enveloped, several pointers toward further drastic change could nonetheless be discerned already by the early 1640s.

One such pointer was the prophetic vision of three of the pioneers. Whether perusing Bacon's *Novum organum* or *New Atlantis* or Galileo's *Dialogo* or Descartes' *Discours de la méthode*, a contemporary reader susceptible to what the authors of these in many other ways so different books sought to convey in the first place was bound to pick up from their rousing prose an exciting message of novelty, not just piecemeal novelty but novelty at the very

foundation of things. There are new, unheard-of ways to investigate nature, so their prime
message ran, and models for how to do it are ready to be taken up and applied. Together
with this vision of exciting novelty, an almost wholly new variety of sense perception had
come into being. It had revealed that nature counts many more phenomena than our un-
armed senses can identify, such as, notably, those announced to a startled public in Galileo's
Sidereus nuncius of 1610. Also, Galileo's and Bacon's prophetic utterances suggested in terms
hard to mistake that it was time to make a clean break with the time-honored approach to
nature by means of all-encompassing, speculative reasoning on the allegedly solid basis of
indubitably secure first principles.

At bottom, the visionaries had it right. The future belonged to a kind of science encapsulated in the three novel modes of nature-knowledge that by the early 1640s had come
into the world; to approaches to nature going far beyond unaided sense perception; to ways
of reasoning not comprehensive, speculative, and overwhelmingly intellectualist but rather
piecemeal, hypothetical, and also operative as much as theoretical. Furthermore, all those
continuities so noticeable still by the 1640s were in due time to be swept away. As time went
on, Alexandria, Athens, and the 'European' mode of coercive empiricism were absorbed or
replaced by their successors; barriers between these successors broke down; new institutions
and ways of communication came into being; and the authority of the ancients turned more
and more into background noise.

We know the outcome, but the outcome is not self-explanatory. Our present-day knowledge that the visionaries had it mostly right does not serve as a sure sign that no other
outcome was possible. Rather, the question is what its possible future looks like from the
vantage point of the early 1640s.

Take the triptych in nature-knowledge as it had acquired shape by the 1640s, and compare it with the triptych in its prerevolutionary guise – by the 1590s, that is. Extrapolation of
the likely fate of the latter, carried out in view of the known fate of comparable instances of
flourishing nature-knowledge, led our imagined trend watcher to expect (a) for mathematical science in the Alexandrian way, if not imminent decay, then at best current enrichment
continuing for a while; (b) undiminished philosophical rivalry among the five Athenian
schools, and (c) some ongoing expansion of accurately descriptive, practice-directed investigations; all this naturally destined to peter out in the customary way to near extinction, followed by some later resurrection elsewhere (p. 150). Instead, a mostly new mode
of mathematical science, a mostly new philosophy of nature, and a far more focused range
of experimental investigations had by the 1640s come into the world. If, on the basis of this
triptych's novelty but also in view of its many continuities with the past, we are invited to
predict by extrapolation what will happen from the early 1640s onward, what might we expect? Do we have sufficient reason to expect what by hindsight we know to have happened,
namely, no petering out at all but the unceasing flourishing and ongoing expansion of the
scientific enterprise up to and including our own day?

284   Dynamics of the revolution, in brief

Since this time we have no precedents fit for comparison to go by, no more definite answer to the question can be given than *not necessarily so*. When discussing the downswing of nature-knowledge in Hellenist and then in Islamic civilization, I insisted that, with decline being the natural course for events to take, our modern permanence is rather the singular phenomenon in urgent need of explanation (pp. 28; 64). In Part III we shall watch such permanence in the making.

To be sure, permanence-early-in-the-making is bound to differ considerably from how permanence is safeguarded in our time. Our modern permanence rests on two props: an ever-advancing research front and the prosperity widely seen to derive from science-based technology. But we must not project the presence of these two props back upon earlier times. We may not assume a priori that whatever it was that over the later stages of the Scientific Revolution served to keep recognizably modern science going is identical with what serves today to maintain its unceasing flourishing. Rather than assuming *that*, we must rather ask *whether* and, if so, *to what extent* a science-based technology, perceived to enhance prosperity, came into being in the same move with the arrival in the world of recognizably modern science. Similarly, to what, *if any*, extent can one already for the 17th century speak of an ever-advancing research front? Finally, we must find out whether perhaps props of an altogether different nature manifested themselves to ensure that revolutionary nature-knowledge, once arrived in the world, did indeed manage to stay there.

These, then, are the fundamental questions to be answered in Part III. In addressing them, we must take into account all the time that such core components of present-day science as in Part II we have watched come into the world – a mathematical approach to the realities of nature undergirded by efforts at experimental checking; the world conceived as particles moving in law-governed ways; sustained efforts to find facts the experimental way – were split up as yet into three distinct, almost entirely separate modes of nature-knowledge.

As one consequence, we must direct our core questions at each revolutionary mode of nature-knowledge separately, *and expect different answers for each*. This is what I do in ch. 10 for how realist-mathematical science fared between Galileo and Newton; in ch. 11 for the kinetic-corpuscularian philosophy of nature in Descartes' wake; and in ch. 13 for fact-finding experimental science between Bacon et al. and the final decades of the 17th century. Each of these chapters is dedicated to detecting, and then to analyzing down to its component parts, the specific nature of such advances as were made and also what if any impact these advances had upon contemporary craftsmanship. For instance, in the next chapter I discuss a variety of ways in which findings in realist-mathematical science came about, with a view to uncovering the underlying dynamic of advance in this specific mode of nature-knowledge. I do not take for granted that this dynamic conformed to any present-day, grand-philosophi- cal conception of how science is supposed to proceed (e.g., through ongoing falsification or through successive paradigm shifts). Rather, in ch. 10 I look for regularities in how a second generation of realist-mathematical scientists actually operated in their attempts to obtain valid results, and I try to find a pattern in those regularities. And with respect to their efforts to use their science to improve or even overturn contemporary craftsmanship, I proceed in similarly empirical-historical fashion. That is, I seek to find out case by case to what if any extent elements of a science-based technology did indeed begin to emerge from such efforts in the course of the Scientific Revolution and, if not, why not. In none of these three chapters does the investigation remain confined to a succession of theories and concomitant practices. For ch. 13, in particular, where I discuss the post-Bacon fate of fact-finding experimental science, my two leading questions will turn out to be closely bound up with the rise of institutions newly emerging in the two most prominent European cities from midcentury onward, Paris and London.

As another consequence, we are bound to face the further circumstance, to be uncovered and explained in ch. 14, that by the 1660s barriers between the three modes broke down to an unprecedented extent. The pioneers of the breakdown, which began to take place in the late 1650s, were Huygens, Boyle, Hooke, and the young Newton. In their hands, kinetic corpuscularianism ceased to be pursued as a speculative philosophy. Instead, they began to use the conception of particles in incessant motion as a hypothesis fruitfully to be infused for certain specific topics of inquiry into either the mathematical-experimental or the fact-finding experimental approach to nature. In so doing, these men gave rise to two more revolutionary transformations, each co-constitutive of the Scientific Revolution. I analyze them in chs. 15 and 16. Here as in the earlier chapters, feats of discovery (whether real or retrospectively spurious) make up a good part of my account. But here, too, such discoveries do not by themselves constitute what I am after. Rather, I am concerned with what exactly it was about them that served to keep the pace of events going.

Not that the forward pace of events proceeded unhindered.

In ch. 12 we shall find that the strangeness and the sacrilege that certain key authorities believed were implied by realist-mathematical science and the natural philosophy of kinetic corpuscularianism produced a veritable crisis of legitimacy of quickly intensifying proportions. The forward pace of events was noticeably slowed down thereby during the late 1640s and the 1650s (definitely the low point of the Scientific Revolution). But we shall see in ch. 17 how remarkably quickly, before momentum had a chance to seep away for good, the downward trend was reversed. The Peace of Westphalia and the shift of Europe's center of gravity away from the Mediterranean to the Atlantic were the key events that made the reversal possible. Instrumental in how the reversal actually came about by the early 1660s was not so much the *reality* as rather the *promise* of a prosperity-enhancing, newly science-based tech-

286 nology. This promise was shot through, not only with a specific set of religious views most prominently present in England, but also with the relatively outward-bound values that had come to distinguish Europe from other civilizations and with the unprecedented amount of autonomy gained for the pursuit of nature-knowledge in London, in particular.

In ch. 18 I draw my various findings together. There I inspect the state of affairs in European nature-knowledge by 1684. I call up our familiar fictional observer and charge her with reporting on strengths and weaknesses of revolutionary science-in-the-making at that particular point in time. She identifies for us one notoriously weak spot standing in the way of much possible advance at a critical juncture. This is the multilayered confusion surrounding the concept of force. In ch. 19 I examine how, in reconceptualizing force in ways that add up to one final revolutionary transformation, Newton opened a novel pathway toward the further advance of innovative nature-knowledge.

Not that Newton's *Principia* and also his *Opticks* henceforth guaranteed the forward pace of modern science. Insofar as the survival of ever more recognizably modern science depended upon social support for it, no more than time-bound props were as yet there to sustain it. But insofar as its *inner* dynamics was concerned, the decisive step toward its staying power had been made.

NOTES ON LITERATURE USED

Chronology of the onset of the Scientific Revolution. I have gleaned the approximate dates of unpublished work from the standard biographies; for Gassendi I found Kurd Lasswitz, *Geschichte der Atomistik vom Mittelalter bis Newton*. Leipzig: Voss, 1890 (2nd ed. 1926) vol. 2, pp. 128-133, particularly helpful.

Ongoing sense of continuity with ancients. A forceful reminder of the 'cessation of dialogue' with the Greeks, which she dates from the mid-17th century onward, with a few earlier signs of it, is in Tabitta van Nouhuys, *The age of two-faced Janus. The comets of 1577 and 1618 and the decline of the Aristotelian world view in the Netherlands*. Brill, Leiden 1998 (especially pp. 20-25).

Dynamics of the Scientific Revolution. The very idea (left without follow-up in the literature so far) that a distinguishable dynamic drove the Scientific Revolution forward lies implicit at the core of two quite different books both published in 1971, by Richard S. Westfall and by Joseph Ben-David, respectively (*SRHI*, pages listed in index under 'Scientific Revolution, dynamics of').

PART III

DYNAMICS OF THE REVOLUTION

X

ACHIEVEMENTS AND LIMITATIONS OF
REALIST-MATHEMATICAL SCIENCE

What future lay in store for mathematical science in its newly realist form? Kepler resigned himself to a very long wait before his own achievement would attain its well-deserved recognition. In the introduction to what he regarded as his crowning achievement, book V of *Harmonice mundi*, he famously wrote:

> well, then, I throw the dice and write a book, no matter whether it be read by contemporaries or later generations – it may await its reader for a hundred years, as God himself waited six thousand years for a witness.[154]

Galileo, in facing the future of his own variety of realist-mathematical science, took a different position. Convinced likewise that the very structure of the world is ultimately mathematical, yet sure as well that nature's properties are not given a priori but can be unmasked only by laboriously investigating them one by one, Galileo exuded a quiet confidence in both the built-in power of realist-mathematical science and in the capacity of his successors to finish what he himself had left undone.

At the outset of the recovery of Alexandrian mathematical science in mid-15th-century Europe, Regiomontanus in his oration on the dignity of mathematics invoked as its principal asset the indubitable certainty of conclusions reached the mathematical way, as opposed to the perennial controversy philosophies of any kind are bound to engender (p. 100). Galileo extended this idea of the unique certainty of Euclidean/Archimedean/Ptolemaean science to the mathematically realist mode of nature-knowledge for which he himself was laying foundations. "The force of necessary demonstrations is full of marvel and delight; and such are mathematical [demonstrations] alone," he had Sagredo exclaim on the Fourth Day of the *Discorsi*.[155] Similarly, in the *Dialogo* he had the same spokesman claim that "trying to deal with the questions of nature without geometry is attempting to do that which is impossible to be done".[156] But he also felt the power of realist-mathematical science to stretch beyond his own efforts. Other investigators would take up the mathematization of nature at the point where he himself had perforce to leave it. On the Third Day of the *Discorsi*, in the run-up to his derivation of uniform acceleration in vertical descent, he announced that the path shall be cleared and made accessible for a very ample, very excellent science, of which these our labors will be the elements, and in the more hidden recesses of which farther-seeing minds than mine shall penetrate.[157]

That is, the advance of nature-knowledge in its realist-mathematical mode proceeds, not by the all-in-one-stroke approach customary in natural philosophy, but by an ongoing, possibly never-ending advance over time. Galileo's understanding of this was no doubt enhanced by the circumstance that in his Padua days he had begun to surround himself with a dedicated circle of gifted students – one more contrast with Kepler, who had none. Their names will crop up with some frequency in the account that follows of the first post-Galilean (often directly Galileo-inspired) manifestations of the power that proved to be inherent in realist-mathematical science. These manifestations, to be sure, were far from uniformly successful. They surely included a number of *extensions* of the range of realist-mathematical science. But early probings of its power revealed that it was subject to certain major *limitations* as well. I shall examine here both the extensions and the apparent limitations in an account dedicated solely to the question of how over the remainder of the 17th century realist-mathematical science advanced under its own steam. Later on I shall discuss three major clashes over turf contested between Galileo's disciples, on the one hand, and, on the other, those natural philosophers who, by dint of their first-principle reasoning, claimed to possess superior prior knowledge regarding issues connected with specific weight, with vertical descent, and with the void (p. 403).

The power of realist-mathematical science worked two ways. It worked toward the productive *absorption* of what was already there; it also worked toward the *creation* of almost entirely new ways *and* subjects of mathematical analysis. In the category of productive absorption I shall treat successively
- the realist transformations undergone by the five *classic Alexandrian subjects*;
- a newly gained productivity for *scholastic* concepts like impetus or graphic representation almost as soon as they found themselves exposed to efforts at mathematization; and
- mostly vain attempts to improve certain trial-and-error products of the *arts and crafts*, as with claims to determine with mathematical exactitude geographic longitude or musical temperament or the breaking point of wooden beams from which a load is suspended.

In the category of wholesale creation three further areas of incipient mathematization come up:
- the invention and usage of two mathematical *instruments* of a new type – the telescope and the pendulum clock;
- such unprecedented advances as came from drawing sustained *analogies* between one kind of motion (e.g., vertical descent) and another (e.g., the swing of a pendulum bob); and

- the further increase in power accruing to mathematical science when ratios expressed by means of straightforward Euclidean geometry gave way to the infinitesimal procedures of the *calculus.*

By way of conclusion, I shall pin down what underlies all these advances – that is, the specific nature of the *forward dynamic* that propelled Alexandria-plus between Galileo/Kepler and Newton.

The classic Alexandrian subjects absorbed

In the Alexandrian tradition five classic subjects had been subjected to mathematical treatment: solid and fluid equilibrium, consonant intervals, light rays, and planetary trajectories. In the course of the 17th century each subject was absorbed into the new realist-mathematical science. Here is how the process of absorption went.

Solid equilibrium.   The big difficulty with available texts on the equilibrium of solids concerned the precise relation between Archimedes' treatise and two others that (as became more and more apparent) derived from the Aristotelian tradition. Once the former was recognized for what it was, the adoption of Archimedes' approach and insights proceeded without further conceptual and/or philological confusion. The picture that remains upon bracketing for now the story of the confusion (on which see p. 336) is unambiguous and simple. *Qua* content, equilibrium states of solids were by the early 17th century less in need of transformation than any of the other Alexandrian subjects. What Archimedes and a range of later Alexandrians up to and including Pappos had accomplished regarding balance beams, centers of gravity, and inclined planes had been reiterated in Islamic civilization and then again in the late-16th-century 'school of Urbino'. It had been enriched with additional finds by Thabit ibn Qurrah (weighted balance beams) and by Simon Stevin ('wreath of spheres' proof of the rule for equilibrium on inclined planes). The net outcome of all this and some later, more refined mathematical theorizing pretty much equals the basics of the subject as established over the remainder of the 17th century and finally included in textbooks under the definitive name of 'statics' (p. 630).

Fluid equilibrium.   On the equilibrium of fluids, too, much of the basics had already been accomplished by Archimedes. But since the somewhat arbitrary assumptions from which he derived his theorems (notably, his 'bath' theorem) presupposed a fixed, central Earth, they were ready candidates for revision by dedicated Copernicans. This was done by Stevin, who also, in his *Beghinselen des waterwichts* (Principles of Water Equilibrium; 1586), went on to extend Archimedes' theorems farther. He enriched them in particular with the

294   hydrostatic paradox (p. 105). Galileo, in his *Discorso delle cose che stanno in su l'acqua o che in quella si muovono* (Discourse on the Things That Stay on Top of Water or That Move in It; 1612), rather turned Archimedes' approach in terms of specific gravity to his own ends in an effort to demolish the Aristotelian account of heaviness and lightness as polar opposites. He thereby caused newly realist mathematical science to put natural philosophy on the defensive for the very first time (p. 404). In 1650 young Christiaan Huygens further refined some of the basics and then, in a display of late-Alexandrian virtuosity, went on to extend still further Archimedes' investigations into how certain intricate geometric figures float in water.

At the same time, a comprehensive effort at extension and reorganization of the domain came from the discovery that a substance other than ordinary fluids, namely, air, may in equilibrium situations be treated as if it were a fluid. The pioneers of the discovery were several pupils of Galileo's, notably Evangelista Torricelli, who made it in direct connection with the issue of the void. He thus gave rise to a further clash between realist-mathematical scientists and natural philosophers over terrain covered by both albeit in radically different modes (p. 410). The protagonist of the clash became Pascal. He perceived this particular piece of contested turf to provide an ideal, public battleground for mathematical-experimental science to prove its superiority over what he felt to be the empty speculations of Aristotelians and Cartesians alike. Hence, to him the chief point of his final account of the subject, his *Traités de l'équilibre des liqueurs et de la pesanteur de la masse de l'air* ('Treatises on the Equilibrium of Fluids and on the Weight of the Mass of Air'; written in 1654 but not published until 1663) rested in this very demonstration. To later generations, Pascal's two treatises, with their careful, never overstretched analogy between the conduct of fluids and of air in equilibrium states, served as the definitive account of the subject. All the real-world phenomena established in the meantime – floating, suspension, sinking, buoyancy, hydrostatic paradox, communicating vessels, hydraulic press – now found themselves jointly derived from a comprehensive, unifying viewpoint.[158] What remained to be done was, once again, to take up the disciplines thus constituted, hydrostatics and aerostatics, and rewrite them in didactic fashion for such textbooks of Newtonian mechanics as began to appear from the early 18th century onward.

An interim distinction.   Unlike with these equilibrium problems, the three subjects that remain – musical consonances, light rays, planetary trajectories – had in earlier times been deeply affected by individual efforts at increasing their reality content by means of an infusion with pertinent pieces of natural philosophy. Ptolemy was the pioneer of such efforts in every case (p. 24). In consciously extending Ptolemy's early bridge building, Ibn al-Haytham worked a geometry of light rays into his grand optical synthesis known in Europe from the 13th century onward as Alhazen's *perspectiva*. In similar emulation, Zarlino and

Salinas wove the arithmetic of musical consonance into a wide web of Platonic-Aristotelian 295
viewpoints and polyphonic music by the 16th century. Likewise, Kepler built a grand synthesis of radically novel planetary theory, Pythagorean harmony, and Platonic conceptions of
the universe into his *Harmonice mundi* and his *Epitome* (1617–1621). *In the course of the 17th*
century, each of these three efforts at synthesis came unstuck. The main tenets of each proved
incompatible with the exigencies of realist-mathematical science. The main tenets were perceived to hinge on pointless number symbolism in the case of Zarlino's grand synthesis of
Renaissance music and to be geometrically mistaken in the case of Alhazen's *perspectiva*. In
the case of *Harmonice mundi* it was rather the exigencies of realist-mathematical science
that were felt to be too fancifully applied. Here is how each synthesis came apart in its turn.

Light rays. From antiquity onward, light has customarily been taken to propagate in
a straight line. But problems with this natural assumption arise over how, then, images are
formed. For one single point emitting or receiving light one can easily conceive how, on
rectilinear propagation to (or, as Euclid thought, from) the eye, it gives rise to a clear image
of that point – a one-to-one correspondence of the punctiform object outside and the point
image inside is easily established in this elementary case. But it was not at all obvious how
rectilinear propagation could account for cases of more complex image formation. The most
important of these cases was vision – how is it that, if every single point of an object emits
light rays in all directions, the eye can form a sharp image, not an utterly confused one? Ibn
al-Haytham's grand synthesis arose from his awareness that a Euclid-like, geometric account
of the propagation of light, an Aristotle-inspired conception of its general nature, and physiological inferences from knowledge of the anatomy of the eye are required all together. How
does the eye go about establishing a one-to-one correspondence between every single point

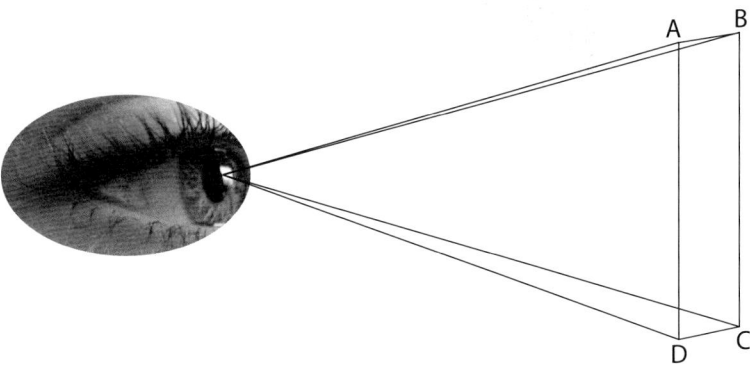

Figure 10.1: The visual pyramid

296   of the object and every single point of the image somehow created inside? His answer (p. 59) was to exclude from the process of vision every light ray that fails to fall perpendicularly upon the eye – all other rays are refracted by the fluid lens at the front of the eye and *therefore* do not contribute to vision. Hence, vision takes place through the establishment of a 'visual pyramid', with the aggregate of points of the object for its base and the center of the eye for its apex (Figure 10.1).

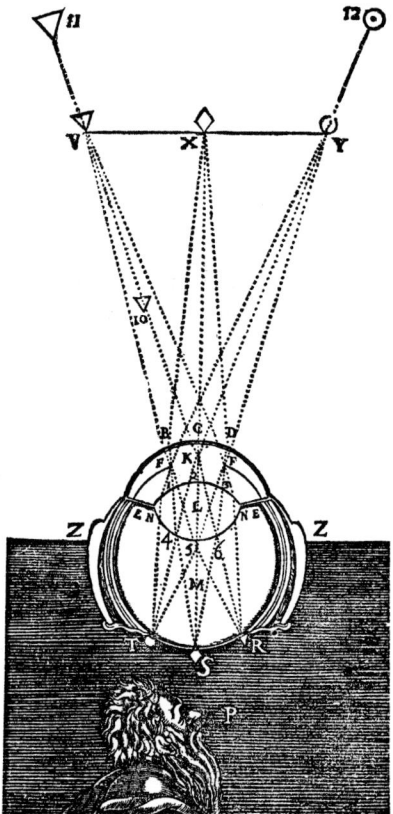

In the absence of a picture in Kepler's own work, the one drawn in 1637 in Descartes' *Dioptrique* is customarily shown instead.

A riddle remains, to be sure: How can we see an inverse image upright? Kepler did not venture an answer – what happens inside the optical nerve and the brain is a serious problem in its own right, but it has nothing to do any more with the tracing of light rays.

Figure 10.2: Formation of an image on the retina

Kepler took up the subject at the very points where Ibn al-Haytham/Alhazen and his European followers had left it. In his modestly entitled *Ad Vitellionem paralipomena* ('Additions to Witelo'; 1604) he raised the principal objection that it is most unlikely for adjacent, *almost* perpendicular rays, which refract only very little, abruptly to fail to contribute anything at all to the process of vision. Hence, each ray that meets the eye from any point of the object contributes to vision; hence, we are back at the earlier problem of why one-to-one correspondence rather than total confusion obtains. Kepler's solution was made possible by his coming into possession of a recent, more accurate description in Vesalius' vein of the anatomy of the eye than the Galenic one available to his predecessors. He joined this to an independent, far deeper analysis of the process of refraction than hitherto undertaken for the purpose. He concluded that all rays from every point of a luminous object, insofar as these reach the eye at all, are refracted in the eye's lens to a focus such that an inverse image is formed upon the retina (Figure 10.2).

What enabled Kepler to unmask as a glaring absurdity a geometric assumption that had been adopted without question for centuries was the same realism that he brought to his astronomical investigations. Indeed, it was a problem in astronomy that drew Kepler to questions of light and vision in the first place. Tycho had noted that the moon, if observed in solar eclipse by means of projection through a pinhole, appears inexplicably smaller. Kepler attacked the problem by means of an investigation of image formation through pinholes. He approached it in a manner tellingly different from how his perspectivist forebears used to go about it. Unlike them, Kepler was not prepared to sacrifice the rectilinear propagation of light, which to his realist manner of thinking could not be dispensed with. Rather, he cobbled together a model, with threads that he attached to a book playing the role of light rays (Figure 10.3).[159]

Kepler's revision cut one cornerstone away from under the perspectivist synthesis. More was to follow. For Kepler as for everyone else ready to adopt his new account of vision, the hoary Alexandrian problem of finding the rule for refraction gained a new urgency. Ptolemy, Ibn al-Haytham, and now Kepler, too, had sought in vain for the regularity obeyed by light rays in being refracted at a surface separating two media (e.g., air and water). Kepler's realism had led him to a new concern for what it is at that very surface that gives occasion for refraction to take place at all. Now the sine rule of refraction had already been found by Ibn Sahl in the 10th (4th) century in another, more abstract context (p. 59). It was found all over again by two other Alexandrian mathematical scientists among Kepler's contemporaries, Harriot and Snel. The first to share Kepler's concern with refraction due to adoption of Kepler's account of vision was one more mathematical scientist occupied with the propagation of light rays in an otherwise unKepler-like, traditionally Alexandrian vein. This was Descartes, who found the rule at about the same time as Snel did. One difference with his co-discoverers was that he went ahead to publish it (in his *Dioptrique* of 1637); another was that

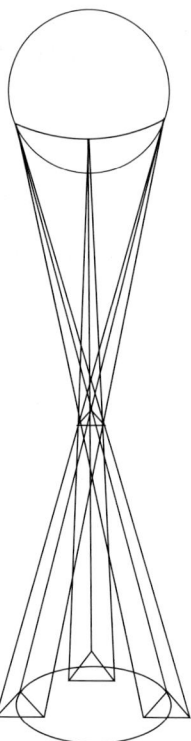

"In *Paralipomena*, [Kepler] describes how he replaced a ray of light by a thread. He took a book, attached a thread to one of its corners and guided it along the edges of a many-cornered aperture, thus tracing out the figure of the aperture. Repeating this for the other corners of the book, and many more points, he ended up with a multitude of overlapping figures that formed an image of the book. In the same way, he argued, all the points of the sun project overlapping images of the aperture. The resulting image has the shape of the sun, albeit with a blurred edge. In the projection of an eclipse, the image of the shadow of the moon is partially overlapped by the image of the sun. Consequently, the diameter of the moon seems too small."

Figure 10.3: Kepler's solution to the pinhole problem

he did not just state the rule but sought to add proof. The assumptions underlying his proof stand fully in the tradition of *perspectiva*, Kepler's revision included. But this Alexandrian mathematical scientist was also, in another compartment of his mind, a natural philosopher of the Athens-plus variety. Undiminished loyalty to Alhazen's Aristotle-infused notion of the nature of light was therefore impossible for Descartes, so he replaced it with hints at light being instead a tendency to motion that certain particles display. For all his claims to restart the investigation of nature from scratch, he stuck to tradition in this domain as in so many others, in that he kept his mathematical science separate as much as he could from his natural philosophy. There was no possible way to reconcile the finite velocity of light that he had perforce to assume in his proof with the instantaneous propagation that followed with necessity from the kinetic-corpuscularian account of light that he had first rendered in his 'Le monde' some four years prior to *Dioptrique* and that, seven years after that essay, he was to publish with little alteration in *Principia philosophiae* in 1644 (p. 233).

The discovery of the sine rule was indispensable for bringing into line for the first time the analysis of rays of all kinds, those that go on unimpeded as well as those that are reflected or refracted. It was nonetheless due to remain underexploited for as long as it remained in sole possession of mathematical scientists of the strictly Alexandrian variety. For the time being, its chief effect was the geometric solution to another long-standing riddle pursued in vain by Ptolemy, Ibn al-Haytham, Kepler, and many another first-rate student of phenomena of light and color – the rainbow. Dietrich von Freiberg, otherwise the lesser brother of Ibn al-Haytham's European disciples, had realized what is going on here – both reflection and refraction take place in every single droplet of a rain cloud between an observer and the sun (p. 82). Using a glass sphere filled with water to find out in repeatable detail how rays of light travel under such circumstances, Dietrich had arrived at an explanation of some signal properties of the rainbow quite as satisfactory in retrospect as it was ignored for centuries. Descartes was probably not aware of it when he went ahead to accomplish the same and more. Combining similarly meticulous observation of similar, water-filled glass spheres with his newly found sine rule, Descartes now managed (in his *Météores* of 1637) to account for *why*, rather than just *that*, we always see the rainbow under the same angle of 41½°, and also why the even more baffling, secondary rainbow appears only at about 51½°.

Fifteen years after Descartes' *Dioptrique* came out, Christiaan Huygens began to put the sine rule at the center of an ongoing effort already pioneered in less definitive ways in Kepler's *Dioptrice* of 1611. He aimed to determine foci and focal distances and images real or virtual, upright or inverted, enlarged or not, in all kinds of configurations of lenses such as found in telescopes, the operation of which he hoped to understand and optimize that way (p. 327). Such geometric tracing of lines representing paths of light rays, similarly pursued in the 1650s and 1660s by Isaac Barrow, was during the 18th century to yield a distinct discipline of geometrical optics that was basically as we still know it. However, it was noticed in the meantime that the sine rule fails to hold in two distinct cases. In 1665/6 Isaac Newton began to observe color fringes around prisms, not in the immediate neighborhood of a prism as Descartes had done, but in projection upon a screen at some distance away from it. To Newton the elongated spectrum thus produced suggested a different angle of

300   refraction for each colored ray. In 1668 another mathematical scientist, Erasmus Bartholinus (Bertelsen), became aware that in Iceland spar not only ordinary refraction occurs but in addition 'strange' refraction such that a perpendicularly incident ray is broken, too, whereas certain oblique ones are not. In 1672 Huygens, unconvinced by Bartholinus' effort to bring strange refraction in line with the sine law, attacked the problem along a path remarkably similar to how Newton, in his 'Lectiones opticae' of the same year, sought to establish a rule for color dispersion. Sticking as yet to the revised-perspectivist tradition laid down in Descartes' proof of the sine rule, each man sought for a new mathematical regularity to save the anomalous phenomena of varied refrangibility and of strange refraction, respectively. Not until these searches turned out to lead nowhere did they reconceptualize their respective problems in truly revolutionary manner. Each sought to blend newly refined versions of his *natural-philosophical* conception of light (emission-based with Newton, wavelike with Huygens) with his *mathematical* tracing of light rays in new, unheard-of ways. These novel ways mark a fourth revolutionary transformation, to be treated as such in ch. 15. There we shall see how the beginnings of a new branch of science, physical optics, came to replace the remnants of Ibn al-Haytham's perspectival synthesis (p. 539).

Consonant intervals.   Of the two fundamental problems of consonance, one is to define a quantitative rule that matches the demarcation between the consonances and the dissonances; the other, to explain how such a match between a phenomenon given in sense experience and the quantitative regularity thus established is at all possible. Solutions had so far rested upon the symbolic meaning ascribed to the numbers in question, as derived from successive divisions of a string struck or plucked: $1 + 2 + 3 + 4 = 10$ (the *tetraktys*) with the Pythagoreans; 6 (the *senario*) with the two great musical humanists, Zarlino and Salinas. Properties of the vibrating string itself, so far left out of account entirely, began to be touched upon briefly and in ambiguous ways in the second half of the 16th century by Benedetti and by Vincenzo Galilei. The definitive farewell to accounts of consonance founded upon the arithmetical properties and the supposed deeper meanings of numbers now took place in two stages: an abstract-geometric and a mathematical-realist stage. The quantities assigned of old to the consonances are very simple, and they were deeply ingrained in elementary education. That is how, *in this Alexandrian subject alone*, and only for as long as the quantities dealt with were the elementary ones of the first few integers, thinkers in the mold of Alexandria-plus arrived at results that overlap in good part with those attained by some pronounced 'Athenians'. By the 17th century, the former were out to infuse a long-standing, abstract subject with a larger degree of reality; the latter, to enrich a variety of available natural-philosophical accounts with motion of a kind marked, in this case alone, by known quantitative regularities.

    Between 1605 and 1619 dissatisfaction with Zarlino's *senario* account of consonance be-

gan to be articulated – the consonance-producing integers are all in the range 1–6 (the perfect number). Stevin, Kepler, and the young Descartes developed instead a specific (also each
a quite different) geometric criterion for distinguishing the consonances from the dissonances. Each also added at the outskirts of his endeavor some pointers in the general direction of the vibrations of the string whose successive divisions produce the consonances. In
the 1620s and 1630s, another mathematical scientist and another natural philosopher went
much farther in founding a theory of consonance upon properties of vibrational motion.
The former was Galileo Galilei, whose findings (published in 1638, at the end of the First
Day of the *Discorsi*) probably go back to his Padua days, 1592–1610. The other was Isaac
Beeckman, whose pertinent jottings in his diary started in 1615. Both men recognized that
pitch is uniquely determined by the frequency with which a vibrating string moves to and
fro. Possibly inspired by the stoic analogy of the propagation of sound with waves in a pond,
Galileo further adopted a pulse account of sound production. He held that musical sound is
yielded by the successive pulses (which he called 'shocks' or 'percussions') transmitted from
the vibrating string through the air to the sense of hearing. This implies that, if two different
notes are made simultaneously, pulses coincide in those cases when the intervals in question
happen to be given by ratios of the first few integers. In the case of the octave (2:1), every
second pulse yields such a coincidence; with the fifth (3:2), every sixth one (Figure 10.4),
and so on, up to and including the minor sixth (8:5). As a result, the traditional range of
consonance-yielding ratios is reinstated, but now linked solidly to the real-world parameter
of vibrational frequency.

Beeckman's derivation of, at bottom, the same account was characteristically different. He thought that the vibrating string cuts the ambient air into little globules, which, on
reaching our sense of hearing in proportion to vibrational frequency, are perceived by us as
correspondingly consonant or dissonant sound. In passing on his variety of a coincidence
account of consonance to Mersenne he took care to present it shorn of its corpuscular sub-
stratum, and it is in this decorpuscularized, more Galileo-like version that Mersenne took it

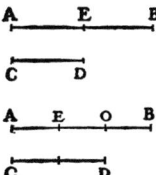

Figure 10.4: Galileo: Pulses coinciding at the octave and at the fifth

302   up and developed it in his *Harmonie universelle* (1636/7). The big difference between Galileo's concise presentation of the coincidence theory and Mersenne's lengthy pages is that the latter, just as Beeckman himself, proved better aware of the many respects in which musical experience *fails* to conform to what the theory would lead one to expect. For example, why is the minor sixth, with pulses coinciding every 40th time, still a consonance, but not the dissonant second (9:8) or (even worse for the theory) such harshly dissonant intervals (e.g., the augmented fourth or tritone, known of old as 'the devil in music') as were customarily associated with the ratios 7:4, 7:5, and 7:6? Also, as Newton was to realize, pulses can coincide only if they start at the very same moment.

To the extent that solutions to such difficulties were sought at all (Galileo just ignored them), these were of an ad hoc nature, designed above all to save an account for which no ready alternative in conformity with current modes of thought presented itself. It is, however, characteristic of the realist transformation which the mathematical subject of musical consonance had meanwhile undergone that such ad hoc solutions were no longer sought in any deep meaning of constituent numbers but rather in certain real-world phenomena of sound, some newly discovered. Thus, Beeckman found himself juggling with sympathetic resonance and with beats (p. 375). Mersenne invoked the upper partial tones that he had discovered – he distinguished, above 1 as the fundamental, harmonics 2 through 5. And Huygens, who invoked the 6th and 7th harmonics as well, sought plausible reasons for declaring intervals with the number 7 in their ratios pleasingly harmonious after all. Not until the 18th century was enlarged knowledge of upper partials and other sounding properties of vibrational motion turned into the cornerstone of a mostly new, once again mathematical account of consonance, destined to be replaced in its turn by one by Helmholtz (1863) in which the *perception* of musical sound received its full due for the first time. Unlike in any of the four other Alexandrian subjects, however, the problem of consonance first posed by the Pythagoreans is in several major regards with us still.

Planetary trajectories.   Planetary trajectories had been the very first subject to undergo revolutionary transformation – Kepler had got rid of epicycles, eccentrics, and all other fictional devices and replaced them with his 'celestial physics'. Still, the absorption of his results into the mainstream of realist-mathematical science took longer than in all the other cases just treated. What made the reception of Kepler's planetary work comparatively slow, gradual, and halting?

During Kepler's lifetime, responses to his astronomical theorizing outside the small circle of his lay admirers came down in most cases to either outright rejection or silence. This was due above all to the centrality in his work of the heliocentric doctrine, which continued to be rejected wholly or in good part by the great majority of scholars (astronomers included) until at least the 1650s. The way Kepler handled that doctrine provided further grounds. David

Fabricius was one of Kepler's most expert correspondents. When Kepler informed him that he could resolve all his quandaries by taking the orbit of Mars to be an ellipse, Fabricius suggested in response that Kepler better try one more little epicycle.[160] Another correspondent, the astronomer Peter Crüger, noted upon receipt of the *Epitome* that

> in trying to prove the Copernican hypothesis from natural-philosophical reasoning, Kepler introduces strange speculations which belong, not in the domain of astronomy, but of natural philosophy.[161]

Mästlin likewise felt that his admired pupil was trespassing the proper bounds of the astronomical discipline. To fellow experts, then, the jump was just too big, the departure from received modes of thought (elliptic vs. circular orbits; celestial physics vs. fictional models) too radical by far. Still, Kepler's fellow Copernican Galileo, no less prone than Kepler to making big jumps and taking radical departures from received modes of thought, chose in no lesser measure to ignore this very portion of Kepler's life's work (p. 187). Here an additional reason for silence-cum-rejection strongly suggests itself: the idiosyncratic synthesis, part astronomy, part astrology, part harmony, in which Kepler had enveloped his finds.

What began to break this pattern of almost unanimously felt lack of persuasiveness was not so much Kepler's reconstruction of how God had established the world, but rather the increasingly apparent, matter-of-fact superiority of Kepler's *Tabulae Rudolphinae* of 1627. This book contained more than Tycho's long-awaited treasure trove of planetary and star positions – it came adorned with clear-cut pointers to Kepler's first and second laws. More than half a century earlier the day-to-day and night-to-night superiority of the heliocentric Prutenic Tables over their geocentric, Alphonsine predecessors had brought Copernicanism numerous adherents, albeit only in the fictionalist, 'saving the appearances' mode Copernicus himself had at one and the same time rejected and employed. In similar fashion did the *Rudolphine Tables* lead numerous astronomers occupied with the details of planetary trajectories toward Kepler's planetary conception, albeit (once again) in the fictionalist, 'saving the appearances' mode Kepler himself had so unambiguously taken leave of. Particularly revealing of its superior performance (in this case, by a factor of thirty) was a transit of Mercury across the Sun. In accordance with his tables Kepler accurately predicted it for 7 November 1631 at noon. Alas, he was not able himself to rejoice in the cloudy yet in the end visible transit, as he had died one year earlier.

Adoption of Kepler's rules for the planetary orbits in a fictionalist vein not only fit in with a little contested tradition of centuries-long standing but also had the clear advantage of circumventing the problems raised by what, in the subtitle to *Astronomia nova*, Kepler had called 'celestial physics' – whether such an entity is acceptable at all, and if so, what to think about how Kepler had set it up. Some tough difficulties presented themselves even so,

304   in that over the centuries astronomers had stuck to circular orbits not only out of respect for Plato or the tradition established by him but also because under any assumption other than circularity their computations became far more troublesome. This was true in particular of the area law. Not everyone stood ready in effect to emulate Kepler's laborious determination of planetary positions at given times by calculating each and every one of the 180 segments into which he had divided half of his as yet unmasked oval. Just as there were still many semi-Copernicans, accepting, for instance, daily but not annual rotation of the Earth, just so numerous astronomers now adopted some portions of Kepler's variety of heliocentrism but not others. Many saw how their predictions of planetary positions could benefit from adopting such less radical revisions as Kepler's putting the Sun itself rather than the empty center of the Earth's orbit at the center of the universe, or attributing to the orbital planes of all planets an equal inclination with respect to the celestial equator. Some adopted elliptical orbits but made them sprout from epicycles on uniformly moving deferents. The first to try this out, to Kepler's horror, as it seemed to make his physical reasoning superfluous, was David Fabricius at a later point in their correspondence (which Kepler promptly broke off for good). Ismaël Boulliau, in his influential *Astronomia philolaica* of 1645, adopted the ellipse as such but rejected the area law as well as any consideration of celestial forces producing the ellipses. Even so he smoothed the further reception of Kepler's work, which received further boosts in England.

There, preparatory work was done by Jeremiah Horrocks in the late 1630s. Still in his teens, he managed with some help to cull from the *Epitome* what – thanks to him in the first place – we now know as Kepler's three laws, and in effect adopted them all. His own observations enabled him both to use the laws for a vastly improved lunar theory and to lend empirical support to the third law. Immediate publication of his results, with their careful focus upon the most productive elements in Kepler's work, might have sped up reception considerably; but Horrocks died in 1641 and publication did not take place until the early 1670s. By then, the state of Kepler's three laws and the perceived difficulties with them was broadly as follows.

Heliocentrism had in the meantime become not only the standard computational basis for accurate prediction but also the accepted conception of the universe in the view of both the experts and most others engaged in the creative pursuit of nature-knowledge, notably adherents of kinetic corpuscularianism (p. 396).

Considerations of *heavenly harmony* had certainly not lost their spell yet, not even over the mind of so down-to-earth a Galilean as Christiaan Huygens. His aloofness vis-à-vis Kepler's astronomical work in all respects but the ellipses that he needed for the design of a planetarium stemmed from his lack of tolerance for any fanciful cosmology that mixed up astronomy with music and astrology. This did not prevent Huygens in his own theorizing about the true sizes of the planets, for which he knew the required measurements to be lacking, from falling back upon considerations based on harmony. But whatever interest in 305 harmonic speculation might still remain, no one was prepared to go to such lengths of fancy in constructing a planetary synthesis as Kepler had. Meanwhile a general sense was growing that some synthesis of another kind was needed instead (p. 631). The question then became one of how to use Kepler's *specific* contributions to planetary theory to make such a synthesis.

One or more *forces emanating from the Sun* were now perceived by several practitioners (notably, Borelli, Hooke, and Wren) to be indispensable for accounting for the workings of the solar system. Again, Kepler's own conception of such forces had to be rejected. Horrocks had already dispensed with Kepler's idea that in its axial rotation the Sun drags the planets along while alternately attracting and repelling them magnetically so that elliptical shape and nonuniform speed result (p. 173). In the 1670s Hooke proposed instead an inertial path deflected by some agent, which might be a force and, if it was, probably diminished with the square of the distance (p. 632).

Kepler's *three laws together* had meanwhile attained straightforward textbook state in *Astronomia Carolina* (1661) by Thomas Streete, who had gained access to Horrocks' posthumous papers and made good use of them. For British practitioners in particular it was henceforth no longer necessary, if they wished to make their acquaintance with the laws, to have recourse to Kepler's rare, somewhat opaquely written, and still rather fanciful *Epitome*.

Kepler's *first law* now stood accepted, be it in fictional or (more and more often) in realist mode, by the great majority of those aware of the intricacies of planetary theory (not, to be sure, by the far larger number of natural philosophers, whether of geocentric-Aristotelian or heliocentric-Cartesian persuasion).

Kepler's *second law* in its full rigor was at first employed reluctantly if at all in view of the computing drudgery involved. It was embraced with more gusto as easier apparent equivalents and simpler approximations were worked out. But when Nicolaus Mercator in the 1660s demonstrated that the equivalents were not really equivalent and the approximations often were not close enough, the area law was reinstated – apparently, laborious computation provided the only feasible pathway toward maximally accurate results.

Kepler's *third law*, which states that the squares of the orbital periods of any two planets are as the cubes of their mean distances from the Sun, was used by Horrocks for arriving at more accurate values for those distances than attainable by means of straightforward observation and also for determining the relative sizes of the planets. By the 1670s the law found itself regularly employed for these purposes by practical astronomers aware of such procedures chiefly through Streete's textbook. Also (unknown to the world outside his student's room at Trinity College, Cambridge), in or shortly after 1666 a certain bachelor of arts named Isaac Newton compared the tendency of the Moon to recede from the Earth with gravity at the surface of the Earth. On finding the latter value to exceed the former by a fac-

306   tor of c. 4,000, he went on to feed Kepler's third law into his own rule for centrifugal motion and then to calculate that for each planet "the endeavours of receding from the Sun will be reciprocally as the squares of their distances from the Sun."[162] What 'endeavors to recede' means, and what the whole sequence of events signifies, come up for extensive discussion on pages 534; 647.

Such, then, was the state the problem of how to account for the planetary orbits had attained when, in August 1684, Edmond Halley traveled from London to Cambridge and asked the same Isaac Newton, now Lucasian professor of mathematics, what orbit would ensue from an inverse-square, center-directed force acting upon a body in inertial motion. Halley received the wholly unexpected answer, calculated some five years earlier by his host, that such an orbit had to be an ellipse. That is to say, he had *proved* it. From that seminal point onward, it took three more years for Newton to produce the *Principia*. There Kepler's laws finally found themselves fitted into a large-scale, coherent structure of realist-mathematical science, albeit without his name being so much as mentioned in connection with these laws (p. 660).

Conclusion: rates of absorption.   It took the five Alexandrian subjects some fifty years altogether to abandon their pristine, ultra-abstract character and become fully absorbed into the realm of realist-mathematical science. The process involved the definitive breakdown of two ingrained habits of thought. One was the 'saving the phenomena' posture that characterized not only the handling of Kepler's results but also such remnants of the standard Alexandrian approach as came to the fore in how Harriot and Snel and Descartes went about the geometric treatment of light rays. Even harder to break were the syntheses in which, following Ptolemy's 'bridge-building' example, light rays, consonant intervals, and planetary trajectories had been enveloped by Ibn al-Haytham, Zarlino, and Kepler, respectively. These syntheses were originally undertaken with the aim of enhancing the reality content of Alexandria. Each had to come unstuck before it could find a new home in the novel frame of realist-mathematical science. Not until this dual task had been completed could light rays and consonant intervals and planetary trajectories take their fitting place in the fast-growing range of domains of nature-knowledge properly mathematized.

As is to be expected in a process of adaptation to radical change, absorption occurred haltingly and at an uneven pace. Even so, there were no regressions. If considered whole, the process displays a relentless drive in the direction of an ever more widely embraced realist mode of mathematization – by the end of the century, all three premature syntheses had vanished from sight, and no trace of Alexandria remained. Nor is this the only case where such a relentless drive manifested itself at the time.

Scholastic concepts mathematized

Mathematization in a realist mode did not remain confined to subjects customarily treated the mathematical way already. It likewise was applied to several concepts that first emerged in a natural-philosophical and, hence, almost entirely nonmathematical context: medieval thinking about motion. The most important cases are Oresme's 'latitudes of forms' and Buridan's impetus.

In the 14th century Oresme had invented a way to represent graphically the uniform or uniformly difform or difformly difform intensity of certain 'forms', or qualities (p. 86). One form thus represented was speed, which he treated without considering the possibility that 'uniformly difform' (i.e., uniformly accelerated) motion might occur in nature. A century later, vertical descent was mentioned quite in passing as an example of a uniformly difform quality by Domingo Soto (p. 90), yet the passage remained without noticeable consequence, nor was graphic representation extended in any other direction until the early 17th century. By then, Galileo took it up. Whether or not he picked up graphic representation from the medieval tradition or reinvented it himself is immaterial to the point to be made here, which is that the technique began at once to display its latent potential in this novel intellectual frame.

Take the 'mean speed theorem' discovered by the Oxford *calculatores*, which Oresme derived in the abstract and which appears in Galileo's *Discorsi* as an early theorem of vertical descent (p. 195; also Figure 10.19 on p. 359). The rule states that, over a given period of time, a body moves an equal distance irrespective of whether its speed is uniformly accelerated or uniform at the mean value of the uniformly accelerated motion. Derivation and proof of the theorem enabled Galileo to go on and derive dozens of more complicated theorems connecting time passed, distance traversed, and speed acquired in vertical and inclined descent. What had once been the semiplayful exploration of a possibility left open in Aristotle's natural philosophy of qualitative and quantitative change was now replaced with the beginnings of a new, mathematical science of falling bodies confirmed experimentally along inclined planes. In that novel frame, graphic representation has kept lending mathematical-experimental science indispensable support ever since.

In a similar manner *impetus* was mathematized during the 17th century. Three centuries earlier, Buridan had introduced the concept by way of a solution to Aristotle's problem of what causes a projectile to move on rather than fall down at once upon being thrown (p. 84). In the Aristotelian scheme of things every motion requires a mover. Buridan replaced Aristotle's solution – the throwing hand passes its motive force on to the layer of air nearest the projectile, and so on from one layer to the next all the while the force keeps decreasing – with the idea that, instead, the throwing hand transfers its likewise-decreasing motive force to the projectile itself. Buridan dubbed this inner force, which expends itself over time, 'im-

308 petus'. For over two centuries impetus found its way into the majority of current Aristotelian accounts of projectile motion. It was also assigned a part in various causal accounts of free fall. The concept remained confined to such usage until Torricelli employed it for both underpinning and extending Galileo's almost fully kinematic, deliberately noncausal account of falling bodies (p. 191). Torricelli did this in lectures delivered in 1642. Upon his early death five years later, the text got lost, not to be found back and printed until 1715. Seizing upon the concept of impetus, and

> taking a dynamic [i.e., force-oriented] view of vertical and inclined descent, with heaviness (*gravitas*) operating as motive force, he recognized the proportionality of force and acceleration. Equally he saw that the product of a constant force and the time it operates is equal to the total momentum generated in a body falling from rest. [In further analysis along such lines] Torricelli ... recognized that the destruction of momentum is dynamically identical to its generation. ... The product of a uniform force and the time it operates equals the change of momentum.[163]

Impetus finds itself transformed here from a wholly qualitative, natural-philosophical idea into a specific, mathematically clear-cut definition of momentum by way of accelerative force considered in its operation over time.

Impetus also came to the fore in another 17th-century effort to clarify concepts of force the mathematical way. In the context of Buridan's original problem – how to repair a perceived weak spot in Aristotle's account of local motion – the attribution of a motive force, no longer to the air but to the projectile itself, had been of little consequence. Its chief benefit rested in that the air now appeared in its proper role of impeding, rather than furthering, projectile motion. But the very idea of an outside force, applied by the hand that throws the projectile, being liable to transfer to the projectile itself so as to act as an inner force for the remainder of its movement had implications which made themselves readily felt in the newly arising frame of realist-mathematical science. The tacitly assumed conception that underlay the concept of impetus is of force as a substance. There is a 'thing-likeness' in force, so that its transfer from the outside to the inside looks no more problematic than (for example) the pouring of water out of one vessel into another. That is to say, the action of motive force is not regarded as being mediated by space and time, which is our modern conception of force. Rather, it is held to result from "the identification of the action of a force with the force itself."[164] A range of apparent ambiguities in the way that Newton and Leibniz still were to handle the concept of force in a variety of mathematics-infused modalities stems from this tacitly underlying conception of the substantial, 'water-like' nature of force (p. 623; 660). It had been implied in ancient ideas of force in a broad sort of way. Scholastic impetus had given it sharper relief. It was in the new frame of realist-mathematical science that impetus served as the point of departure for the mathematization of the transfer of motive force.

Craft techniques mathematized

One further domain which early Galileans, on the example of their master, sought to subject to treatment in the newfangled vein of realist-mathematical science was the arts and crafts. So far, mathematics had impinged upon craftsmen's activities only in a few exceptional cases: the invention of linear perspective by Italian painters and geometers conjointly, the layout of fortresses, and a range of issues bound up with seafaring, notably the determination of geographic latitude, the operation of the compass, and the production of accurate maps. In other cases where an attempt was made, the benefits for practice proved illusory, as, for instance, with Stevin's effort to improve the mathematical way the construction of windmills (p. 131). There, too, then, craft practice remained confined to rules of thumb handed down from one generation to the next.

Now that, with Galileo's achievement, a new way had been found to connect mathematical science with the real world, this seemed to open a highly promising, as yet untrodden pathway toward the wholesale improvement of the arts and crafts by means of mathematics. Early Galileans confidently foresaw the rise of a radically altered kind of craftsmanship. Liberated once and for all from what they saw as its imprecise tinkering, craft practice could henceforth be elevated to an unprecedentedly novel plane of mathematical exactitude.

Their aspiration to turn the trusted procedures of the crafts upside down gave rise specifically to a dozen or so bold promises. What a few samples of practical mathematics had been able to accomplish so far was as nothing compared to what realist-mathematical science now pronounced itself capable of achieving. Galileo and his early pupils and disciples confidently peddled their ability to tame water currents, to determine geographical longitude, or to improve the accuracy of gunshot. Such promises did not fail to evoke the interest of princes and other secular authorities throughout Europe. They also had vast consequences.

It was in good part owing to the underlying vision of a new, truly science-based technology that those revolutionary modes of nature-knowledge which first arose around 1600 managed to survive beyond the generation of the pioneers. This is the 'legitimation' aspect of the aimed-for mathematization of 17th-century craftsmanship, which I shall examine in ch. 17 (p. 583). In the present section I am concerned instead with another or, rather, with a prior question. What about actual achievement? To what extent were these promises actually fulfilled in the 17th century? From about 1450 onward, incidental efforts at building interfaces between mathematics and the crafts had met with a measure of success, albeit in a highly circumscribed number of cases only. What did the newly realist mode of gaining

310 knowledge of nature the mathematical way add to those successes? In short, to what extent did the revolutionary transformation wrought by Galileo and then taken up by others affect the ongoing pace of events in the realm of the crafts?

In the inquiry that follows I refrain from any a priori assumption that Galilean mathematical science, which indeed was to make so much of a difference in the longer run, necessarily began to make that difference all at once. I shall confine myself to the period of the Scientific Revolution and approach the issue empirically by means of an item-by-item survey of the literature devoted to the various crafts in 17th-century Europe and efforts to improve them by means of mathematics. Since we need a systematic approach, expressly designed to leave all possible outcomes open, I distinguish at the outset between five modalities of possible interaction between mathematics and the crafts, as follows. We may encounter cases where craftsmanship of the customary kind prevailed during the Scientific Revolution and those where it did not. If and where customary craftsmanship did prevail (1), this might be so for the following reasons. Mathematical exactitude might hardly or not at all have been attained (1ª), or it might have been attained but have been proved irrelevant for craft practice (1ᵇ). Alternatively, the prior infusion with mathematics in the course of the Renaissance might already have been spent by the end of that period (1ᶜ), or it might just have sped up in an autonomous development (1ᵈ). The opposite to these four possibilities, then, is that in certain cases traditional craftsmanship indeed yielded to incipient, Galileo-style mathematization during the Scientific Revolution (2).

Quite a bit hinges on the outcome, to be sure. The more preponderant (2) turns out to be, the more we shall be obliged to ascribe the Europe-wide acceptance of revolutionary science to its apparent capacity to make a difference, not just by way of wide-ranging promises but in actual practice. Conversely, the more preponderant we shall find (1ª⁻ᵈ) to be, the more we shall have to seek in due time for important *non*material grounds for societal support for revolutionary nature-knowledge in the making.

Mathematization not attempted.   In terms of the full sweep of the arts and crafts the premodern world over, the inquiry is definitely situated in the margin. Absence of any crafts/mathematics interaction whatsoever has been the rule anytime anywhere. In course of the Renaissance Europe had produced the few exceptions listed above. These gave an inkling of what might be achieved and also, to the extent that they impinged on Europe's commercial enterprises, significantly enhanced the prosperity of the realm. Yet even in Europe, traditional craft practice, unaffected by mathematics, remained the rule throughout the period. The primary crafts to govern everyday life, those responsible for food, clothing, and shelter, remained in their traditional, rule-of-thumb and trial-and-error state, unaffected as yet by efforts at mathematization of any kind.

Incipient mathematization proves as yet irrelevant. The category of incipient mathematization leading as yet nowhere is historically much more intricate and instructive. In at least five distinct areas of craftsmanship, often quite sophisticated mathematical thinking was directed at the solution of practical issues of a more or less pressing nature but proved impracticable for the time being. The same fate befell similarly sophisticated thinking on a higher level of generality, with respect to the performance of laborsaving machinery of any kind. As a result, in each of these cases traditional craftsmanship prevailed for the time being as if nothing had changed at all. Here is a brief survey of all six craft areas, undertaken with a view to finding out what specific circumstances contributed to the apparent gap between mathematical scientists and the relevant craftsmen.

The first case is *musical tempering*. Calculation shows the consonances to be mutually incompatible and a tonal scale with pure thirds *and* fifths to be inherently unstable. Vincenzo Galilei and his onetime teacher Zarlino had come to verbal blows over the issue of how singers resolve their quandary – what musical scale do they adopt in practice (p. 145)? Singers sing as they go along, but keyboard players must tune their organs or harpsichords beforehand, so the same problem of what practicable scale to employ arises here in a more pressing manner. Pythagorean intonation was out, in view of its intolerably false thirds and sixths, as was just intonation for the reason just mentioned – its inevitable instability. To add one or more keys per octave proved unwieldy. The solution was most often sought in temperament – a slight mistuning of selected intervals that our hearing appears to some variable extent willing to tolerate. The most common variety was 'mean tone temperament', with eight out of twelve major thirds pure and all but one of the fifths audibly yet not too disturbingly mistuned. This worked fine with music using chromatic alteration sparingly, that is, when the five 'black keys' on the keyboard are unambiguously used for $C^\#$, E^b, $F^\#$, $G^\#$ (or alternatively A^b), and B^b exclusively. But due to major changes in musical style, by the 1630s keyboard composers like Frescobaldi began to seek effects which required an overstepping of such narrow chromatic boundaries. The battle of temperaments that ensued was won in the end (with the early-19th-century advent of the pianoforte) by the equal variety which has since become standard. It allows free modulation through all twenty-four keys, albeit at the previously well-noted cost of both much subtle tonal shading between the various keys and the lovely purity of those major thirds. As the choices that were decisively involved depended greatly on issues of style and fashion, little if anything about this outcome may count as foreordained.

Among those who joined the battle right from the start in the 1630s were several practitioners of realist-mathematical science. Mersenne, who throughout his work toned the Galilean approach down to a preference for computation and measurement (p. 515), characteristically restricted himself to calculating a large range of conceivable temperaments. More thoroughgoing Galileans aimed for solutions both more creative and more mathematically

312   elegant. Prominent among the latter was Christiaan Huygens, who discovered that the values for the mean-tone intervals turn up, with negligible differences, in a regular thirty-one-tone tuning – produced by taking, not eleven times the mean proportional between 1 and 2, as in equal temperament, but thirty times – which he calculated using logarithms. This way the strong points of mean-tone temperament are of course preserved, and in order to make easy transposition feasible Huygens invented a 'mobile keyboard' which he claimed to have been admired by 'great masters' in his Paris days. But the principal issue – how to use more than five chromatic alterations apiece – was not even touched by his proposal. Nor did any contribution from the side of mathematical scientists ever affect the ongoing battle in any way. For all the possible clumsiness of the calculations that men of musical practice came up with, in their trusted trial-and-error ways they had a better sense of the kind of workable compromise required. Huygens expressed his contempt for the author of the first 'well-tempered' tuning for keyboards to find widespread adherence, the organist Andreas Werckmeister, by calling him "devoid of erudition, and of little worth".[165] The expression serves as one valuable clue as to the nature of the 17th-century gap between mathematical scientists and craftsmen, even when the latter were learned and inventive far above the average.

The next case I will examine is *water streams*. The Venetian lagoon and the Po delta, along with the below-sea-level, western provinces of the Netherlands, were obvious sites in Europe for local administrators to apply what water-taming skills were available. In the Netherlands, draining, land reclamation, etc., remained a matter of craft practice throughout. But in Italy events took a different turn. In 1625 one of Galileo's pupils, Benedetto Castelli, was invited to give his learned opinion on how to undo a diversion of the river Reno that had been undertaken near Bologna a quarter century earlier with unfortunate consequences and to redirect it into the Po. In response Father Castelli applied his teacher's geometric approach to the problem. This led to the publication, in 1628, of a brief tract containing as its principal fruit what has since become known as 'the law of continuity'. It states that "the cross-sections of the same River discharge equal quantities of water in equal times, even though the cross-sections themselves are unequal".[166] Castelli went on to prove the rather counterintuitive theorem by means of an equally geometric argument.

Such an analysis by means of geometric idealization was accorded a fairly hostile reception by the other consultants, who worked rather in the customary, trial-and-error style. They argued that Castelli's solution, even if correct in theory, was irrelevant in practice, for two principal reasons. Castelli's model ignored numerous determinants of flood behavior that are equally if not more influential than the one that he did take into account. For instance, flow speed appears to vary with height as also with slope. Furthermore, such a Galileo-like approach (soon extended by the master himself when consulted in his turn about a Tuscan river, the Bisenzio) is too far removed from reality to be of use anyway. In-

deed, a good deal of practical expertise had over previous centuries been built up by Renaissance engineers – among their sometimes splendid, intuitive insights those that Leonardo
noted down in private stand out in retrospect. The men who by the early 17th century took
up their heritage, mostly Jesuits raised in the mode of 'mixed mathematics', were loath to
give up their accumulated expertise.

In the following decades controversies along these lines flared up time and again. Still,
behind the scenes an intricate process of mutual rapprochement took place. Practitioners
of realist-mathematical science sought to include more and more determinants of proven
practical value in their mathematical models. For instance, by the 1680s academic experts
like Montanari and Guglielmini began to take effects of fluid pressure into account. On
the other side, practitioners of mixed mathematics kept insisting all along that the proper
way to proceed rested in a cautious, preferably yet not necessarily quantitative estimation
of pertinent elements in carefully observed, everyday reality. They nonetheless began to acknowledge that, in principle, processes of mathematical idealization might contribute to a
fuller picture of the vagaries of streaming water.

Even so, definite limits were set to all their groping toward compromise. In 1692 a papal
committee was charged with adjudicating between the conflicting preferences of the cities
of Bologna and Ferrara over (once again) the Reno. Domenico Guglielmini, formerly superintendent of Bolognese water management and now the city's professor of mathematics,
represented the Galilean approach. He built a sophisticated mathematical model that was
free from many a Galilean prejudice. This did not keep the outcome of the cardinals' delib-
erations from being decided more by partisan infighting than by reasoned balancing of the
respective merits of the two distinct approaches put before them, neither of which appeared
clearly superior to the other.

Gunnery provides another illuminating case, but for partly different reasons. The issues
of how, with a given cannon, to increase precision and attain the longest possible shooting range became more pressing as artillery spread and medieval techniques for battle and
siege more and more lost their point. At the dawn of the 17th century these issues still used
to be solved as before, by means of rules of thumb established through trial and error. Nor
did the intervention of the mathematician Niccolò Tartaglia, who by the mid-16th century
advanced the guess that an elevation of 45° maximizes shooting distance, make any significant difference in practice. Once again, mathematization proper began with Galileo. On the
Fourth Day of the *Discorsi* he derived the same figure from his general analysis of projectile
motion as parabolic (p. 192) and added a detailed table expressly meant to make shooting
more precise.

It hardly turned out to have done so. Nor was a sophisticated contribution by Torricelli
of any help. A given cannon may be thought of as enveloped by a particular paraboloid of
revolution, which gives you the distance beyond which you can no longer be hit by a ball

314 shot by that cannon. Torricelli's derivation of the three-dimensional curve and its properties was a mathematical virtuoso piece, yet it also proved useless in practice. So did a treatise
written in 1683 under the title *L'art de jetter les bombes* (The Art of Throwing Cannonballs).
The author, François Blondel, was a member of the Paris Academy with a military background. The aim of his book was to render Galileo's and Torricelli's best insights into bal-
listic theory in such a way as to help gunners in the field improve upon their job. Here is a
contemporary practitioner's scathing comment, followed by a later piece of retrospective
criticism:

> [Surirey de Saint-Rémy; 1697; 'An officer's duties':] The late Mr. Blondel wrote a large treatise
> on this, claiming to give a far more secure demonstration for proper shooting than done by all
> those concerned with it previously.
>
> But it seems to me far better to stick to the example of those who exercise with cannonballs
> all the time and who find themselves at ease with their method, being certain that experience,
> with powder above all, always gets the better of even the most learned considerations.

> [De Ressons; 1716; 'Method for shooting cannon balls with success':] Even though it remains
> true that, in attaining perfection in the arts, theory joined to practice forms the high point,
> nonetheless experience has made me aware that theory has been of very little utility in the
> usage of mortars.
>
> Mr. Blondel's book described well enough for us the distance of parabolic lines, according
> to the different degrees of elevation of a quarter circle, but practice has shown that there is no
> theory at all in the effects of powder, for on committing myself to direct the mortars with all
> possible exactitude in conformity to the calculations, I have never been able to establish any
> foundation upon their principles.
>
> Still, I would not claim that, if each ball proved to be of equal weight, if for each shot one
> could give the powder the same arrangement, and if the platform were so solid as never to
> change its situation, one might not put theory to useful service.[167]

Beside the empirical variables here listed with damning irony, one further culprit was
air resistance, which Galileo and his disciples also neglected, confining themselves to the
ideal, friction-free case. To be sure, mathematical analysis advanced quickly. In the late 1660s
Christiaan Huygens expressed a suspicion that for both components of the parabolic motion this resistance varies with the speed squared. A few years later James Gregory independently rediscovered Thomas Harriot's find (to remain unpublished until 1979) that not an
upright parabola but a slightly tilted one best represents the trajectory of a projectile. And
in 1687 Newton used his newfangled calculus to arrive at sophisticated models for the trajectories of projectiles meeting resistance. Still, none of all this sufficed to close the gap with

shooting practice, or even to narrow it to any significant extent – as de Resson argued, by the
early 18th century trusted rules of thumb were still the best for an artillery man to rely on if
he wished to give his enemies a solid beating.

The *strength of materials* belongs to the neglected 'first' of the 'two sciences' introduced
in Galileo's *Discorsi*. On the Second Day he developed an argument concerning the capacity of beams to withstand being broken by a suspended load. Taking for his model a solid
prism fastened horizontally into a wall, Galileo managed by means of an ongoing process of
geometric idealization to reduce the case to the law of the lever, which in the end yielded an
expression for what weight is still sustained under what specific conditions. Inasmuch as he
did not allow for the elasticity of the beam, his results as yet lacked value in practice. Some
results (e.g., a hollow cylinder can be calculated to hold a heavier load than a solid one of
the same material and weight) made more readily understandable certain experiences well
known either in nature (e.g., with reeds and straws) or in craftsmen's actual construction
practice. That practice, however, was not to be affected by mathematical reasoning until the
theory of elasticity had moved far beyond such early attempts as Edme Mariotte undertook
later in the 17th century.

A case of more general import concerns the *theory of machines*. Leonardo had arrived
at his precocious understanding of the optimal operation of machines using an empirical
approach: avoid such needless friction and loss of effective power as occur when composite
parts are multiplied beyond necessity. What one gains one way, so he observed, is necessarily lost another (p. 117). About a century later, Galileo unwittingly took his departure from
the same fundamental insight into the nature of machine tools, so profoundly at variance
with most craftsmen's received wisdom on the subject. Instead of Leonardo's early, heuristic experimentation, Galileo characteristically went on to pioneer the other, mathematical
component of the beginnings of a theory of optimal engineering practice. In so doing he
demonstrated that all machines are jointly reducible in the end to the law of the lever:

> How could one possibly compare a fulling stocks with a corn mill, a saw mill with a blast fur-
> nace, a mine pump with a pump for supplying a mansion with water? All of them did entirely
> different things and the only common question to ask was whether each machine served its
> purpose well. The answer ... could only be normative: it was a good machine or it was not. But
> according to Galileo's arguments all machines, no matter what purposes they serve, have the
> common function of transmitting and applying 'force' or power as efficiently as possible and,
> moreover, the performance of machines can be quantified, for ideally the product of the driv-
> ing 'force' and its velocity equals the product of the load multiplied by its velocity.
>
> [Consequently] a machine croaking and groaning under its load is not, in spite of ap-
> pearances, necessarily doing the most work. ... Galileo saw that the inequality between the
> equilibrium force and the force required to set the machine in motion was not a principle of

Nature; it is merely a consequence of the imperfection of all machines. A good machine will move at a uniform, unchanging velocity when the force and the load, including residual friction, are in equilibrium. A perfect machine, free from friction, distortion and other defects, will accelerate (slowly if loaded) under an extremely small force until it reaches an infinite velocity.[168]

These were insights of (in the longer term) world-shaking significance. They intimated what could be accomplished once craft practices were drawn into the counterintuitive, all-disturbances-removed realm of mathematical science. It was one thing, however, to correct commonsense convictions in theory, quite something else to draw practical benefits from these and related insights. Not Galileo or his disciples but Mariotte in the second half of the 17th century pioneered a range of efforts, undertaken with increasing frequency over the 18th century, actually to calculate the work capacity of, notably, waterwheels and, later, engines using steam for their source of power.

Finally, we come to the determination of *geographical longitude*. How does one determine one's whereabouts on the high seas in terms of the distance to a given meridian? No practical problem believed resolvable through mathematics was considered more pressing during the 17th century, at least by the governments of Europe's seafaring nations. The hoped-for derivation of longitude from compass readings of the magnetic North Pole or known deviations therefrom proved ever more elusive as, in Gilbert's wake, knowledge of Earth magnetism increased (p. 478). Rather, mathematical scientists set their hopes on three possible solutions, each based on the clock principle. Two solutions had been suggested already in the 16th century; Galileo added one more.

In 1530 Gemma Frisius proposed to compare readily observed local noon on board ship with the reading of a portable timepiece set on departure to time at the home port meridian, with every hour of difference standing for $360:24 = 15°$ farther from that meridian. For more than a century, it was out of the question even on land to measure time within a quarter-hour per day at best, so no timepieces even came close to the far higher accuracy required for usefully determining longitude at sea. This changed with the invention of the pendulum clock. The hopes the instrument raised for the determination of longitude and the difficulties into which it ran for a century to come are treated in the next section (p. 332).

The two other solutions pursued by mathematical scientists concerned clocks in the sky. Could successive positions of the Moon against the background of the stars be used as a sky-clock, as proposed likewise in the 16th century? The solution proved intractable, due above all to difficulties encountered in both lunar theory and the empirical basis thereof. Big strides were made over the century by observers who used ever better telescopes (p. 451) and also by theorists on the move from Kepler's lunar theory to a far better one that Newton devised on the basis of universal gravitation (p. 669). Further efforts were to turn Newton's lunar theory into what by the mid-18th century the Moon's motion was to become indeed for a while – the chief rival to the marine chronometer that eventually won the race.

Likewise defeated in the end was the solution that occurred to Galileo soon after his telescopic discovery of four moons around Jupiter (his 'Medicean stars': pp. 186; 275). The motions of our one Moon are highly irregular, so why not turn the apparently much more predictable, also quickly alternating motions of Jupiter's satellites into a heavenly clock readily usable on board ship? Negotiations with the king of Spain and with the States-General of the Republic of the Netherlands faltered over what soon proved to be one major stumbling block shared by all three mathematical solutions pursued at the time. Even if astronomers could prepare the required tables such that mariners not skilled in astronomy could employ them, and even if tools became available to measure pertinent phenomena with sufficient accuracy (precision clocks; better telescopes with more minute readings), how then to perform those measurements from the deck of a ship subject to the often violent movements of the waves? For all the successes meanwhile attained with time measurement on land, this proved to be a major stumbling block until far into the 18th century.

Prior infusion with mathematics meanwhile spent.   Besides the numerous domains of craftsmanship where, *at least up to the early 18th century*, mathematics either was not applied at all or for all practical purposes kept overshooting its mark, there were also cases where, over the period of the Scientific Revolution, the application of mathematics did make (or had made already) some appreciable difference in practice.

With regard to linear perspective and fortress building, the significant stimulus each received between c. 1450 and c. 1600 from geometry was exhausted by the end of that period. From that time onward these mathematically enriched areas of craftsmanship went on very much as before, *without* being touched any further by what was meanwhile taking place in the mathematical sciences.

In *perspective*, by 1600 the ways had already parted. The sole issue of interest to *quattrocento* painters was how geometry could assist them in creating a convincing illusion of space for objects and persons depicted. This problem was to all practical purposes resolved at the time in a major collaborative effort. Within a century or so, the drawing of square-tiled floors that had been so helpful in early perspective became superfluous, so thoroughly were perspectival thinking and at least the basic rules underlying it taken up in craft practice. The practice was passed on in the usual manner by means of rules of thumb illustrated by examples likewise drawn from practice. In the opinion Christiaan Huygens aired for the benefit of his elder brother, an aspiring artist, no more than that was needed in any case: "There is so little difficulty in this science, which can be understood in one or two rules, that I don't doubt your ability to find it out all by yourself."[169]

By contrast, mathematical scientists like Benedetti, Guidobaldo, and Stevin moved be-

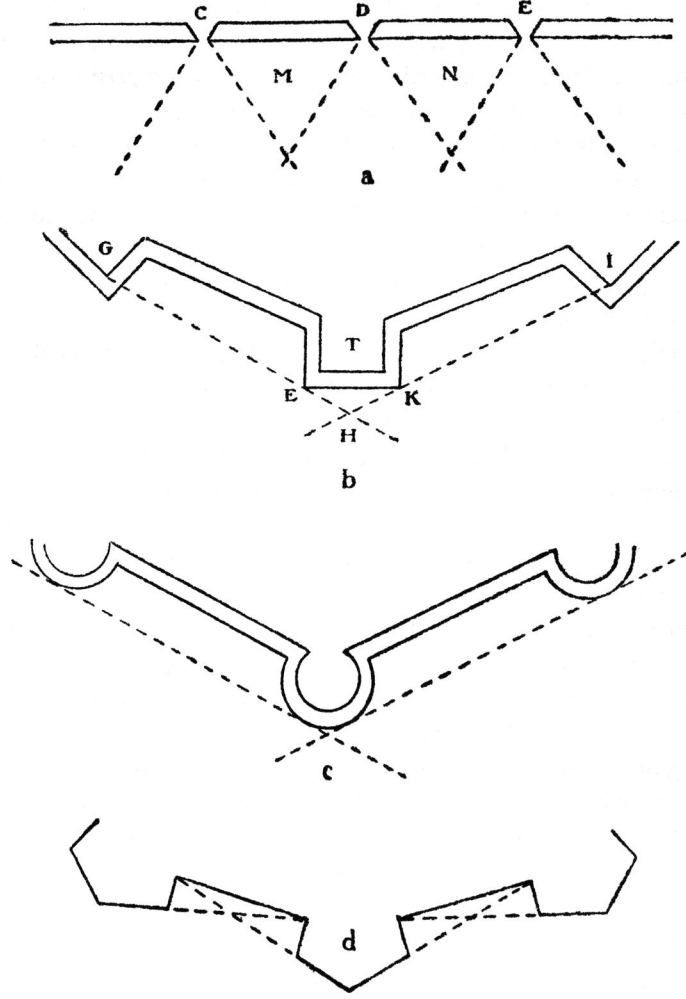

(a) (prior to the usage of gun powder:) ramparts with loopholes C, D, and E for shooting projectiles; M and N are still areas safe for attackers
(b) (early gunnery:) reduction of dead angle by means of square protruding towers
(c) further reduction of dead angle by means of circular protruding towers
(d) elimination of dead angle by means of angle bastions.

Figure 10.5: Development of bastions

yond the question of 'how to' toward a search for orderly axiomatization and for proof. Next, Girard Desargues and other gifted mathematicians expanded their work in setting up a more general investigation of projection and its geometric properties, which in the end was to turn into a field of abstract mathematics in its own right.

Sophisticated mathematical work like this was no longer of any use to practitioners. In his 1605 treatise on perspective Stevin still believed to be of great service to the artisans, whom he sought to enlighten in his wonderfully didactic manner on how to extend farther than usual the range of directions of lines up for foreshortening. What he and later theorists of perspective kept overestimating was the level of mathematical sophistication the craft was prepared to handle. Albrecht Dürer's *Unterweysung der Messung* (1525; 'Textbook of Perspective') still served painters and geometers alike. By the 17th century mathematicians' and craft manuals on the subject were worlds apart. But for some exceptional displays of perspectival virtuosity in Baroque art, the temporary alliance had come to an end.

The art of *fortress building* became a preeminent concern in the 15th century, when guns were turning into a serious threat. What drew geometry into the art appears at a glance (Figure 10.5) from four consecutive drawings in Stevin's treatise *De Stercktenbouwing* (1594; 'The building of fortresses'). Figure 10.6 displays the outcome advocated by Stevin, chief engineer at the service of one of the great military innovators of the age, Prince Maurits of Orange.

The design of such angle bastions obviously demanded some basic geometric skills. But what more was demanded once, by the early 17th century, so much of the theoretical groundwork had been firmly laid? Stevin's proposed solution shows the ideal case. In practice the design always required adaptation to local circumstances like a hill or a river or a harbor. Was there anything that geometry could usefully contribute in addition to what commonsense experience could tell the man of war in any case? Most authors of mathematical treatises on the subject acknowledged that the answer was 'no'. Army commanders all over Europe did not cease to keep military engineers in their employ, but their job became less creative:

> [B]esiegers would have to measure heights and distances accurately across a field of fire, which allowed the engineer an occupation in peacetime, although, in practice, periods of peace were as often spent in expanding and reconstructing fortifications for the next war.[170]

Just as with linear perspective, then, fortress building had become routine business; new infusions of the art of war with a healthy dosage of mathematics had to await Napoleon's large-scale reform.

Prior infusion with mathematics continues its autonomous advance. The class to be treated now covers one single case, the *determination of place on Earth*. 'Cosmography', as the subject was called at the time, included three narrowly intertwined components: navi-

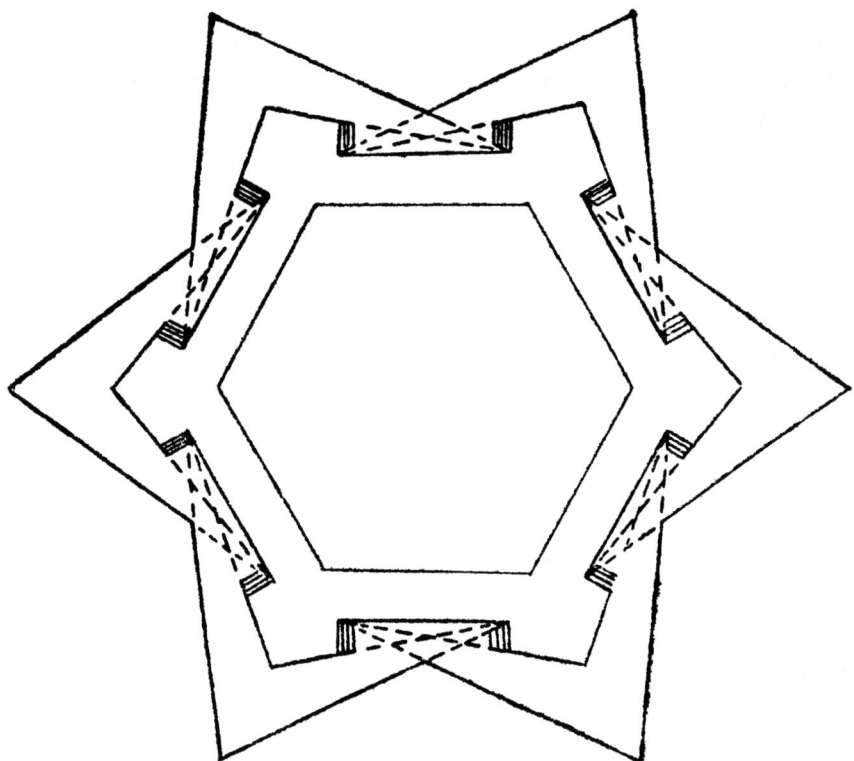

This fortress enables both free shooting and full coverage.

Figure 10.6: Stevin's ideal fortress

gation, surveying, and mapmaking. The difference with the previous class is that the original stimulus from mathematics, far from being spent by the early 17th century, kept exerting itself. The difference with the next, final class is that it did so in straightforward continuity with what had gone before, under its own steam as it were. Craftsmanship in this particular category was ever more deeply informed by mathematical treatment, yet this advance had nothing to do with the rise at the time of realist-mathematical science.

Practicable techniques for sailing out of sight of any known coastline were worked out from the mid-15th century onward. Portuguese navigators who explored the African coast and then set out for the Indies collaborated with mathematical theorists like Pedro Nunes.

The governments of Spain, England, and the Netherlands followed suit in setting up or at least fostering similar forms of cooperation between pilots and mathematicians. Not counting here the compass and the debates it produced over the true magnetic north, the principal achievements were as follows.

By the end of the 16th century the problem of determining geographical latitude had been satisfactorily resolved in practice as it had in theory. Two instruments that combined observation and measurement, the cross-staff and its more refined descendant, the back-staff, made it possible accurately to determine the altitude, not only of the Sun at noon, but also of the polestar. North of the equator, this star marks latitude if corrections of two kinds are made, which were also worked out at the time – a feat of considerable mathematical sophistication. The day-to-day or place-to-place tables ('Regiment of the Sun'; 'Regiment of the North Star') that were required to use such data while on the high seas had by then been worked out as well.

The coverage and accuracy of the maps produced to record an ever accumulating range of data kept pace with these developments (Figures 4.11; p. 127). How best to project three-dimensional shapes on a plane surface became more and more of a problem, as helmsmen in particular needed to have all lines of constant bearing represented as straight lines. Proper techniques for the purpose were devised by skilled mapmakers, notably by Gerard Mercator, albeit at the well-known price of distorting angles and shapes in ways that increase with distance from the equator.

Finally, the ongoing business of surveying on land benefited from these marine activities, so that by 1600 tools refined far beyond pole-and-rope equipment were widely available. In the 1530s Gemma Frisius invented the principle of triangulation, whereby

> bearings taken from either end of a measured base-line can be used to locate an object at an unmeasured (or unmeasurable) distance. By a series of such triangles an entire map can be constructed without further linear measurements.[171]

This is how Willebrord Snel measured with minute care the distances between his own house and three Leiden church spires, going on to triangulate his way around the Earth in his *Eratosthenes Batavus* of 1617. He used a quadrant for the purpose, which is still on view in the Museum Boerhaave at Leiden. The quadrant was one of three instruments Gemma himself, and another mathematical scientist, Leonard Digges, had adapted from seafaring practice to measurement on land. Each of these instruments, besides supplying in its own way a baseline for measurement, increased measurement accuracy in roughly the same manner as it did at sea.

So much for the accomplishments, all attained well before 1600, that mark the heroic age of the use of practical mathematics to undergird early maritime exploration and overseas

322   commerce. Comparatively little was added in the 17th century (not counting, to be sure, ongoing efforts to solve the problem of longitude). One mathematical addition concerned surveying practice; the other, the determination of latitude.

The quadrant and the two other marine instruments that Gemma Frisius and Leonard Digges had reconfigured for surveying purposes were in need of rules for how to handle them and for how to rework the raw data yielded by them. These rules, duly worked out by mathematical scientists, quickly proved to suffer from the usual defect – excessive sophistication. Craft practitioners found them too mathematically esoteric to adopt them. Instead, they opted for such easy-to-handle equipment as had meanwhile been developed specifically for the exigencies of on-land measurement. An example is the 'plane table', tellingly snubbed by Thomas Digges (Leonard's son) as "an Instrument onely for the ignorante and unlearned, that have no knowledge of Noumbers".[172] Exceptionally, however, the gap between the mathematical 'theory of practice' and craft practice itself was bridged at least in writing by two British authors, Aaron Rathborne and William Leybourne. In their successive textbooks they managed to strike a workable compromise between rule-of-thumb routine on the one hand, and mathematical virtuosity-for-the-sake-of-it on the other.

Another effort to bridge the gap was made in regard to mariners' use of maps and other measuring devices. Whereas the techniques for mapmaking remained essentially unchanged over the 17th century, their application did not. Not only did Blaeu and other renowned firms produce large to very large, sumptuously ornamented atlases and globes, but more importantly from our point of view, a tool was invented to make Mercator's projection truly serve the needs of those numerous mariners not versed in handling the trigonometric relations involved:

> The seaman needed a straightforward, instrumental means of calculation, suitable for use at sea, together with a relatively simple set of rules for applying the instrument to specific problems. Both were provided by Edmund Gunter.[173]

Gunter was a professor of astronomy at Gresham College – a London-based center for navigation-oriented studies. In the 1620s he began to adapt the recently invented 'sector' (an early kind of slide rule, to which Galileo contributed in his Padua days) to usage on board ship. Somewhat later, Gunter added logarithmic relations to the lines engraved upon the instrument. In this manner he made it possible to turn multiplication and division into addition and subtraction done by means of just a pair of dividers. Even so, over the entire period here considered many a sailor found the procedures with the 'Regiments' (instructions) and with the slide rule so cumbersome as to take refuge in traditional, typically rule-of thumb 'dead reckoning', if not as a substitute then at least by way of reassuring corrective.

What we have here, in sum, are two rare cases of help offered by mathematical scientists to craft practitioners turning out to be at least moderately successful. Only, neither case reflects in any way the great innovation that mathematical scientists in Galileo's vein promised to bring to the crafts – their wholesale revamping once they were subjected to mathematical rule and order. The two refinements just listed were nothing but modest extensions of an infusion with mathematics that had taken place long before – they owed nothing to what the revolutionary transformation of mathematical science had meanwhile brought.

Craft practice improved due to the advent of realist-mathematical science.

Class no. 2 (as opposed to 1ᵃ⁻ᵈ) finally concerns those cases where the application of idealized-mathematical insights grafted on to Galileo's revolutionary treatment of motion did induce significant changes in traditional craftsmanship. Here we find, not the plentiful examples all too often assumed on essentially ideological grounds in some of the older historiography, but a category that (to my own astonishment) turns out to be *devoid* of actual historical content. When I set out on my inquiry, I tentatively assigned several cases to the present class, only to find as I went through the literature that one case after another had to be placed elsewhere. By the end of the inquiry the present category was empty. It appears, then, that with the six cases that ended up in class 1ᵇ we have already exhausted the considerable effort spent over the entire 'century of genius' by numerous truly able thinkers in Galileo's realist-mathematical vein to be of practical service to a plethora of well-known, often-pressing practical issues, *none of which made as yet any appreciable difference*. In not one of the various crafts involved did practitioners feel that such efforts at mathematization should impel them to alter their customary ways of proceeding. How could this be so?

Conclusion: Achievement and limitation in the mathematization of craft practice.

The net outcome of the investigation, then, is that by 1700, when the Scientific Revolution came to some provisional completion (p. 722), the usual gap between craft practice and mathematical science yawned almost as widely as it had at the onset of the Revolution, about a century earlier. What exceptional crafts had already been successfully subjected to mathematical treatment over the 15th and 16th centuries either settled into routine (perspective, fortress building) or advanced under their own steam in the direction of gradually enhanced sophistication (determination of place on Earth, with measuring instruments and appended 'Regiments' serving as fairly effective mediators). In all other craft domains mathematics was either not applied at all or proved to overshoot its mark, thus in effect leaving the gap between them almost as wide as before. How is it that so much effort, always inventive, sometimes brilliant, was spent in vain?

The most basic answer is that there was scarcely reason to suspect at the outset that the gap was so wide in the first place. A marvelous new tool of, in principle, great generality had become available, the subjection of the empirical world to mathematical rule and order.

Why, then, should the empirical world of craft practice fail smoothly to fall into line? Slowly but surely in the course of the Scientific Revolution a message was brought home to mathematical theorists to the effect that the gap could not be overcome by jumping over it in one big mathematical leap but required a good deal of patient, laborious bridge building.

What bridges, then, and who had to build them, craftsmen or mathematical scientists?

This differed from case to case. Craftsmen's professional conservatism – their sensible inclination to leave things that work intact and to seek solutions for upcoming practical problems in the general direction of what had already proved workable – is sometimes quite functional, sometimes less so. It is hard to decide in advance when we are dealing with experience-based, sound judgment and when with no-less-experience-based prejudice. Werckmeister can in retrospect be seen to have done well to ignore mathematical scientists like Huygens and their preference – at least in the case of musical temperament – for elegance over practicality. By contrast, Galileo's analysis of machine efficacy, which likewise went against craftsmen's intuition and appeared likewise devoid of possible practical consequences, was in due time to alter craft practice almost beyond recognition. It is hard not to empathize with the predicament of those cardinals who, in the committee on the Reno of 1691/2, had to choose between the competing claims of Guglielmini, the haughty mathematical modeler, and of his Jesuit opponents with their down-to-earth experiential, cautiously quantifying bent. But we may also understand all too well Huygens' complaint, in a report about the performance of his clocks on the route back from the Indies, about how much their overseers had suffered "from the crew's frequent scolding and mockery of this effort to measure longitude in a new way". As Huygens rightly foresaw, the effort would eventually save the lives of countless sailors.[174]

Not everything differed from one case to the next. My survey has also yielded a number of shared features. When considered together, they allow us to draw some conclusions that are valid for all cases of craft practice subjected to efforts at Galileo-like mathematization in the 17th century.

Remarkably, the degree of urgency of the particular practical problem for which mathematics was adduced as a solution made no difference at all for the outcome. Whether perceived at the time as pressing (notably, longitude and water management) or not (in particular, efficacy of machines and strength of materials), none of these problems had even come close to resolution in practice by the century's end.

Demand, then, did not sufficiently foster resolution; but neither did craftsmen's conservative leanings necessarily thwart it. In some cases the very initiative came from craftsmen; even if not, when plausible modalities of change in a new direction were offered by mathematical scientists (in perspective, fortress building, and cosmography), craftsmen ready to go along came forward without fail. The more open-minded among them, then, could overcome their native preference to stick to what had worked well enough in the past and

embark on new adventures. What, then, prevented innovative tools, methods, and concepts
from being accepted in all other cases?

My survey has yielded four impediments to practical success for the one big mathematical leap:

(a) The weightiest impediment was a serious underestimation of the real world's messiness. For craft after craft it turned out that, besides determinants that were indeed taken into account in the mathematical model, there were others of at least equal significance, and also many second-order effects, which somehow had to be brought under mathematical rule as well.

(b) The mathematization of second-order effects required the ability to handle nonuniformly varying magnitudes, which the Euclidean doctrine of ratios was inherently incapable of. We shall find below how, late in the century, the calculus began to make it possible (p. 346).

(c) Even if confined to Euclidean means only, craftsmen's ability and/or readiness to grasp the esoteric language of mathematics were very limited. Here is a drastic expression of this sentiment at the end of the 17th century. It was voiced by the same officer, Surirey de Saint-Rémy, who was so skeptical about the value of theory for the practice of gunnery (p. 314):

> some of those [to treat of gunnery] have made it too speculative: their books are burdened with infinitely many Mathematical rules, suppositions, and reductions more suitable for refined tastes than for instructing young men, most of whom lack education and several of whom cannot, in view of the nature of their minds, apply themselves to so abstract matters, knowledge of which presupposes knowledge of the principles of geometry which they do not have or are hardly capable of acquiring.[175]

Nor were there as yet enough incentives for the drastic educational reform required to make fruitful communication between craftsmen and mathematical scientists a matter of more than just rare good luck.

(d) Such communication was further impeded by social distance. Many a mathematical scientist stemmed from the lower or higher nobility or, if not, at least had to adopt its standards and thus tended to regard the equal footing required for fruitful exchange as beneath his dignity.

Due to these four retrospectively distinguishable barriers, then, and possibly to others as well, the crafts stood to gain very little from those numerous efforts undertaken with a view to their mathematical-scientific improvement. In the reverse direction, however, mathematical science benefited greatly from the experiences thus undergone, which in most cases centered on the utterly novel practice of mathematical modeling. This novel practice was tried out for the most part in two recognizable clusters of crafts/mathematics interaction.

326   One cluster consisted of Galileo and his Italian pupils and disciples. They shared a tendency to stick very much to one-sided, mathematical idealization in the now more, now less complacent expectation (fed by their still brand-new conception of the world as inherently mathematical) that experiment would confirm it or, if not, could conveniently be reasoned away. In the second half of the 17th century, in the Parisian Académie Huygens and Mariotte began, also mathematically but in a more open-minded experimental vein, to take the world's messiness into account and to explore second-order regularities.

The removal of impediments (a) and (b), thus instructively explored by mathematical scientists in the 17th century, was to be pursued with ever-increasing zest and refinement over the next. This happened, for instance, with the mathematical taming of water streams, with inquiries into the strength of materials, and with efforts to determine the efficacy of certain machines, notably waterwheels and atmospheric engines. One major effort that in the 17th century came to naught was crowned with success in the 18th – in 1759 John Harrison solved the problem of longitude. To the extent that impediments (c) and (d) were overcome in the 18th century, this was due to men like Harrison, Smeaton, and Watt – all of them craftsmen of a thus far wholly unknown type, who combined first-rate engineering gifts with at least good working knowledge of the science required (pp. 478; 729).

Looking, not forward but once again backward from the vantage point of c. 1700, it turns out in final conclusion that the most significant change wrought by the advent of realist-mathematical science in the domain of the arts and crafts rested in a shift of initiative. Prior to the Scientific Revolution, incipient mathematization had taken place if, and only if, upcoming demand (for ways to create a painted illusion of space; for the design of fortresses protected against enemy cannon fire; for methods to project nonflat areas upon flat maps) could be satisfied by interested mathematical scientists able to rise to the occasion in craft domains where no more than Euclidean tools happened to be required. In contrast, from the Galilean revolution onward realist-mathematical science began, from the ever more richly flowing sources of its own discoveries, to tap a supply of its own. Only, the new supply failed as yet to provoke demand to match it or, if it did, paradoxically turned out not yet to be able to meet it.

These, then, are the general conclusions about achievement and limitation in the 17th-century mathematization of craft practice to be drawn from the inquiry. The question that remains in this regard is whether and, if so, to what extent these conclusions are also valid for contemporary crafts of another, revolutionarily novel kind: those as yet rare cases where mathematical science was not applied from the outside but, as it were, from the inside.

Mathematical instruments

Mathematical instruments had been around before the Scientific Revolution. Some originated in China and/or the Islamic world; others in Europe (a few with a precedent in ancient Greece). Some served in mathematics generally, for assisting calculation (e.g., abaci); others in mathematical science specifically, for assisting naked-eye measurement (e.g., cross-staffs) or for representation (e.g., planetaria) or for both (e.g., astrolabes). During the Scientific Revolution more sophisticated measuring instruments like sectors and slide rules were added to this stock. Also added were several instruments that specifically served the new modes of nature-knowledge. Most of these were instrumental in driving fact-finding experimental researches forward (p. 448), whereas two contributed greatly to the forward drive of realist-mathematical science. These were the telescope and the pendulum clock.

The way in which mathematics entered these products of craftsmanship differs significantly from the cases just treated. The pendulum clock was the very first craft product to emerge, not from artisanal quarters, but from mathematical science in the very process of turning realist. The telescope, emerging as it did from the crafts, was quickly taken up into realist-mathematical science due to the need felt by certain mathematical scientists for understanding and then improving its mode of operation and due to the kind of services it was increasingly well equipped to render to the advance of astronomy. It may not be taken for granted, therefore, that the 17th-century history of the two instruments conforms without more ado to the pattern of interaction between mathematical science and the crafts just uncovered. In the present section the question is to what, if any, extent they do so conform.

Telescope. The telescope started its career as a craft product – the spyglass invented c. 1608 by Hans Lipperhey, a Dutch maker of eyeglasses who aligned a convex lens on the object side with a concave one as an ocular and put a tube around them. The man to turn an improved version of this gadget (power of magnification raised by means of his trial-and-error grinding from c. 4 to c. 20) into the first instrument of the Scientific Revolution was Galileo (p. 274). This was due to his early awareness that the spyglass could do more than trivially assist reading from afar or less trivially identify enemy ships while still dots on the horizon. *Sidereus nuncius* is not informative about Galileo's actual understanding of how magnification actually came about in the telescope. Accounting for its powers was left to the man Galileo appealed to for an authoritative confirmation of his claims regarding the Moon's mountains, Jupiter's moons, and his other early telescopic discoveries. That man was Johannes Kepler. True to his past as Tycho's assistant, he was quite exceptionally concerned with gaining a thorough understanding of the tools of his trade. In his second optical book, *Dioptrice* of 1611, Kepler drew on his first to analyze the formation of images both in the 'Galilean' telescope and in what he showed to be a theoretically possible, alternative setup.

328   There the ocular lens is not concave but also convex, thus producing an inverted image that may be turned upright if need be by means of a third lens. Well aware that he was not in possession of an exact law of refraction (p. 298), his tracing of appropriate lines for the rays' trajectories yielded no more than approximate foci and focal distances, thus perforce leaving his account more qualitative than he wished.

   Proposals to improve telescopes from the side of mathematical science then went in two distinct directions. Both were responses to the known fact that spherical lenses do not refract incident rays to a focus, thus producing somewhat blurred images (this is called 'spherical aberration'). Descartes in 1637 proved in his *Géometrie* that ellipses and hyperbolae do refract to a focus, and in another 'essay' appended to his *Discours de la méthode*, entitled *Dioptrique* after Kepler, he encouraged craftsmen to grind lenses to such shapes. This, however, was a feat quite beyond the powers of glassmakers at the time, as his own effort in the late 1620s to make Guillaume Ferrier accomplish it already suggested. That is what, decades later, caused Huygens to stick to the spherical shape. In a range of successive treatises which he likewise entitled 'Dioptrica' but never published, he began doing in the early 1650s what Kepler would have done – to use Descartes' and Snel's sine rule of refraction for an exact, generalized derivation of pertinent properties of telescopes with spherical lenses in a variety of configurations.

   What improvement did occur before the 1660s came chiefly from the side of craftsmen acting on hints coming from users. These craftsmen, to be sure, were no longer the glassmakers who had started it all but artisans who turned the building of telescopes into a highly respected specialty of their own. Their major concern was to grind their lenses as nearly and evenly spherical as they could, which required much dexterity and trial-and-error experience but no profound dioptrical knowledge. In the course of their efforts, Kepler's alternative configuration with two or three convex lenses turned out to have two unforeseen advantages over its 'Galilean' counterpart. One advantage, noted in passing in 1630, proved in the 1640s to rest in its having a much larger field of vision. This allowed for the attainment of ever higher rates of magnification until, by the late 1660s, an as yet insuperable limit was reached of a little over 100 ×. During this period Huygens switched from the mathematical analysis of telescopes to learning how to grind them as expertly as any craft professional. This prepared him well for the first major celestial discoveries to be made since Galileo: Saturn, too, has a moon, and (as Huygens already suspected on theoretical grounds) the planet's enigmatic appendages are really a ring around it.

   By the 1660s Kepler's configuration revealed another unsuspected advantage, in that it permitted the insertion of crosshairs or similar devices to make telescopic measurement viable. From then on the 'Keplerian' telescope switched from an instrument of discovery to a measuring device finally superior to naked-eye measurement (p. 451). Even so, both theoretical and practical hurdles in the ensuing race toward ever increasing accuracy quickly made themselves apparent.

On the practical side, a dispute between the Danzig astronomer Johannes Hevelius and several British colleagues revealed how much depended on such messy details as Hevelius had painstakingly taught himself to master, notably the problem of how to keep one's instruments accurately and consistently aligned.

On the side of mathematical theory, when Huygens in the early 1660s returned to his dioptrical treatise, he set his aims higher than just elegantly derived, rigorous proof for what was already known. He now hoped to derive a configuration of lenses capable of eliminating spherical aberration in principle rather than (as the craftsmen did) just mitigating its effects by adding diaphragms and by making their lenses flatter and flatter. He arrived at a mathematically impeccable configuration in which the lenses were built and placed so as to cancel out each other's aberrations. Unfortunately, the solution soon proved quite impracticable in view of another mathematical discovery. Isaac Newton made it in 1665/6 when still a student, and six years later he made it public (p. 543; 683). In whatever way one configures one's lenses, they refract incident light, not to one focus but to as many adjacent points behind one another as there are colors. Hence, the elimination of spherical aberration does nothing to prevent much larger 'chromatic aberration' from producing a blurred image of its own. Newton concluded that refracting telescopes are probably beyond hope of further improvement. In making the light enter the tube by means of a mirror rather than an objective lens, Newton now built several reflecting telescopes – the only ones to prove viable for many decades. Among refractors, meanwhile, the race went to those craftsmen (Huygens still included, although no longer in the front rank) who could produce very long tubes or even knew how to dispense with them altogether and to those who produced the best compound eyepieces.

By the end of the century, the novel mathematical science of dioptrics, which owed its very emergence to the telescope, had advanced considerably. Little of this became known at the time. Of the two men most concerned to focus their analyses upon issues raised by the instrument, Kepler lacked the powerful mathematical tool of the sine rule and Huygens lacked the incentive to publish. Instead, two lesser books carried the day. One was Isaac Barrow's *Lectiones opticae* of 1669, which pursued the refraction of light rays in an old-fashioned, classic-Alexandrian vein of 'mathematics for the sake of mathematics'. The other book was William Molyneux' *Dioptrica nova* of 1692, which did address the users of telescopes but had little to offer beyond a none-too-rigorous compilation of previous work by others. This remained the state of affairs until, finally, in 1704 Newton's *Opticks* took refraction up as one vital aspect of the budding discipline so pithily announced in the title.

Pendulum clock.   The mechanical clock was invented in the early 14th century (p. 47). Prior to 1657, its big advantage rested more in its constancy and in its wide use (on city and church towers as well as in portable guises) than in its accuracy. Its steadily accumulating error of at least a quarter hour a day still required regular correction by the sundial or the sandglass. The problem was that neither the foliot bar (Figure 10.7) nor the wheel that replaced it in portable clocks oscillates with natural frequency.

By the late 16th century a first-rate craftsman with some extraordinary training in mathematics, Jost Bürgi, managed by means of manipulations of the going train to work his way around the problem. His solution proved far too complicated for smooth handling and was hardly adopted by others. The germ of the definitive solution – an oscillator with natural frequency – can be credited to our usual suspect, Galileo. Early in his career he found that the time it takes a pendulum freely to swing to and fro is nearly independent of the width of that swing (its amplitude). Near the end of his life he dictated to his son, Vincenzio, a very

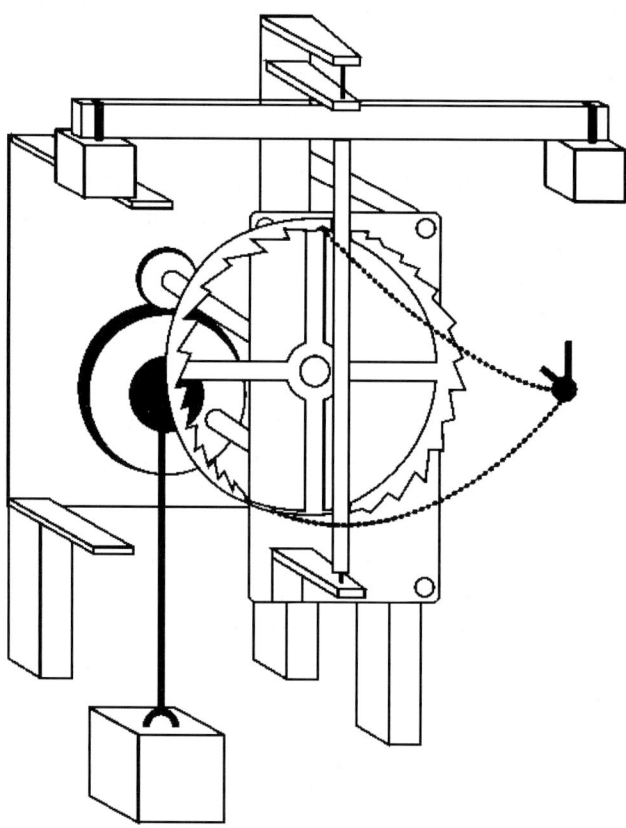

Figure 10.7: Foliot clock

sketchy design for a pendulum clock. Lack of knowledge of that unwieldy device did not prevent Christiaan Huygens from having a Hague clockmaker, Salomon Coster, build the first pendulum clock in 1657 in accordance with a detailed design of Huygens' own, which one year later he published in a brief tract, *Horologium*. Here already one finds described the solution to one pressing problem of a kind that was to crop up again and again. With accuracy enhanced by an order of magnitude (in the present case the error was reduced from minutes to seconds a day), previously insignificant sources of error suddenly demanded urgent attention. Thus, right from the start Huygens devised a mechanism for keeping the clock moving while being rewound.

A *principal* shortcoming, of which Huygens was likewise well aware, rested in the qualifier 'nearly' just appended to the amplitude independence of the pendulum's period (which is called the 'isochrony' of a pendulum). Galileo attributed its approximate nature to air resistance, but Huygens came to realize that the circle is not really the isochronous curve. In *Horologium* he sought to mitigate the consequences by forcing the pendulum into very small arcs of swing, but this increased the amount of friction and nullified the effect. In 1659, as one outcome of researches in mathematical science with quite another starting point (p. 342), he discovered that the truly isochronous curve is the cycloid (Figure 10.8).

He also found that there is a mathematically exact way to force the pendulum into a cycloidal path, namely, by causing the thread from which it hangs to run up against 'cheeks' of (once again, but not at all trivially so) cycloidal shape (Figure 10.9).

The exactly vertical mounting this requires proved hard to attain even on land, so recourse had to be had to those very small arcs of swing, which not Huygens but British clockmakers in the 1670s managed to realize without undue friction by means of an 'anchor' escapement. Thus, error was reduced to about ten seconds per day, with further reductions along such lines of trial-and-error ingenuity to follow in the next century.

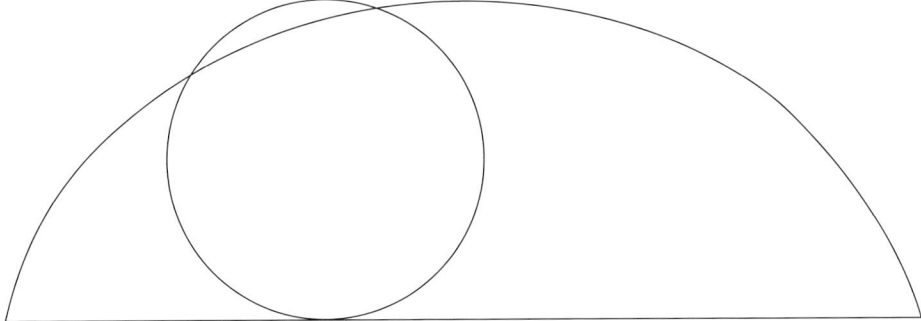

Figure 10.8: Generation of a cycloid by a point on a rolling circle

332   Right from the start Huygens was well aware that his pendulum clock made Gemma Frisius'
theoretical solution to the problem of longitude (p. 316) at one stroke practically viable. His
subsequent discoveries in isochrony raised his hopes even higher. These hopes, however,
were thwarted again and again. A cycloidal clock with cardan suspension adorns Huygens'
masterpiece, *Horologium oscillatorium*. First tried out at sea, its performance proved disappointing in practice. So did Huygens' invention (the fruit, once again, of sophisticated
mathematical deduction) of the balance spring, which brought to the portable clock what
the pendulum had brought to its standing counterpart – a natural oscillator. This device

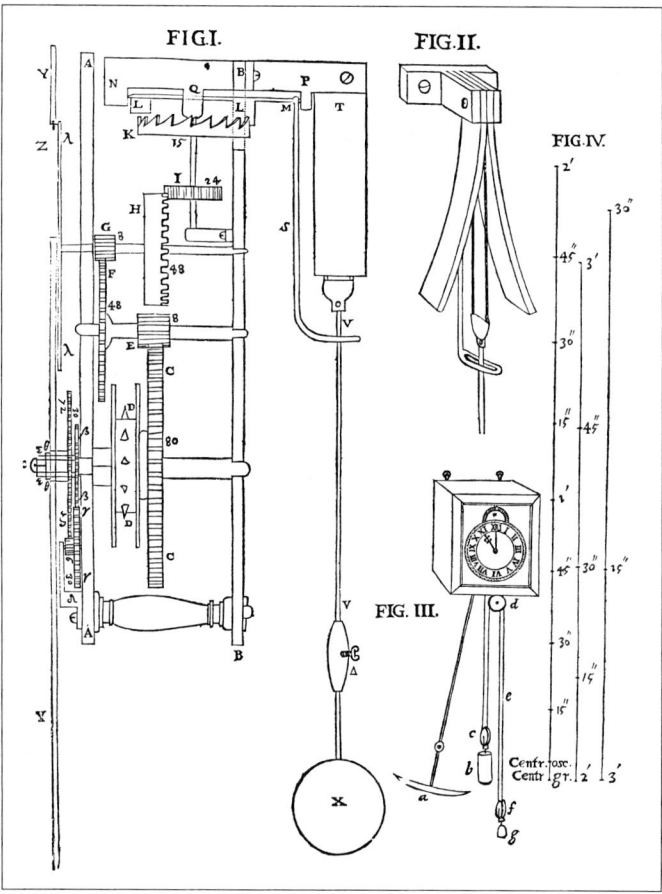

Figure 10.9: Huygens' pendulum clock with cycloidal cheeks

proved damagingly subject to as yet incorrigible temperature changes. The same, albeit to a smaller extent, proved true of pendulum clocks of partly other construction than before, for instance, on board ship in 1686/7 with that skeptical crew watching. A worrisome divergence appeared between courses sailed (as guesstimated by means of customary 'dead reckoning') and courses charted in accordance with successive readings of the clock. In retrospect we know that the divergence is due to the expansion of metal in the higher temperatures of the tropics, but for pendula Huygens never acknowledged this.

Convinced that the real world had to succumb in the end to mathematical rule and order, near the end of his life Huygens devised a range of mathematically very ingenious marine timekeepers, none of which was capable of improving things in practice. For the 'pendulum clock' line in which the solution to the problem of longitude was being sought, in rivalry to the 'Moon' and the 'moons of Jupiter' lines (p. 316), temperature compensation became the most urgent requirement. For this and for a host of related requirements, however, not ever more sophisticated mathematical science held the proper means in store but rather first-rate craftsmanship feeding on expanding empirical-scientific knowledge. It was by moving in this general direction that John Harrison was to build the marine chronometer (his celebrated 'Harrison no. 4') which on trial expeditions in 1762 and 1764 enabled ships to keep course to within less than one-third of the required maximum of thirty minutes longitude.

Conclusion: Another pattern?   To what extent did the mathematical instruments of the Scientific Revolution conform to the pattern of interaction between craftsmanship and mathematical science analyzed in the previous section?

The starting points are identical: domains of craftsmanship untouched by nature-knowledge of any kind. Eyeglasses were produced and elderly people wore them from the 14th century onward without anyone asking what optical properties enable lenses to improve eyesight. Mechanical clocks from about the same time onward woke people up all over the realm without anyone wondering what it meant to slow down the accelerated motion of a weight and make it (as nearly as trial-and-error practice could attain) uniform, then to break it up into countable units. Such lack of interest looks odd in retrospect only. The phenomena in question did not pertain to natural philosophy, as neither mathematics nor improvement were in its causal province, and as 'art' was regarded as the very opposite of nature. Nor did they pertain to mathematical science in its Alexandrian mode, for which they were too tangibly concrete. Before 1600, mathematics entered the craft domain solely on demand, be it from the state or from artisans. With the advent of Alexandria-plus the initiative shifted, not exclusively but indeed decisively so, to the side of mathematical science now turning realist. That craft practice nonetheless benefited so rarely during the 17th century was due to those four impediments: (1) the world's messiness, (2) the standard manipulation

334 of Euclidean ratios often not being good enough, (3) mathematics (whether Euclidean or more advanced) being outside the craftsmen's province, and (4) social rank standing in the way of free exchange. What about the telescope and the pendulum clock in these regards?

The world's messiness was not at issue in how to turn a spyglass into a telescope and proved only a little more so in how to join a real-world pendulum to a mechanical clock – the sheer emergence of realist-mathematical science did just about all the work necessary to make these events happen. Messiness in the guise of second-order effects then did provide an unexpectedly tough barrier with temperature compensation and nonrecoil escapements in pendulum clocks and with blurred images in refracting telescopes. Still, sometimes messiness *was* overcome in the 17th century, with the outcome alone allowing one to determine whether craftsmanship or mathematical science had offered the solution. For instance, Newton's at least partly mathematical science of color sifting appeared to offer a way (the reflecting telescope) to sidestep aberration. Further improvement of escapements, however, remained a craftsmen's affair throughout, and the proper alignment of telescopes was a messy problem characteristically underestimated in science circles yet laboriously overcome by an as yet rare bird, the craftsman-cum-mathematical-scientist (Hevelius).

As luck had it, ray tracing guided by the sine rule was all the mathematics required for understanding the telescope. Such understanding, to be sure, proved irrelevant for its practical improvement, which (with the sole exception of what led Newton to the reflector) remained a craftsmen's affair. Nor was any more complicated mathematics needed for the pendulum clock than an awareness that small arc swings approximate isochrony well enough. The very sophisticated mathematics (Euclidean to the point of infinitesimal overstretch) which enabled Huygens to derive the truly isochronous curve proved one particularly splendid case of mathematical science overshooting its mark.

All in all, common ground was easier to find for craftsmen and mathematical scientists in the cases of telescopes and pendulum clocks than in any other case of realist-mathematical science meeting craftsmanship. Often cooperation was deliberately sought and proved fruitful. Not that the domains fused – not by chance, Huygens wrote the Latin installments of his 'Dioptrica' in quite another vein than the likewise-unpublished leaflet he wrote in the vernacular for the express benefit of his fellow lens grinders under the title "Memorien aengaende het slijpen van glasen tot verrekijckers" (Notes concerning the Grinding of Glass into Telescopes, 1685). Yet the very fact that one and the same man made himself not only the foremost observational astronomer and the foremost mathematician of refracted rays of his time but also the foremost grinder of lenses points at the much more intricate crafts/mathematics interaction that was brought about by the mathematical instrument. Here the professional self-interest of the mathematical scientist was palpable – if the crafts could not deliver to his specification, then he could turn himself into a craftsman and further his own researches that way. This was true in particular of Galileo and Huygens, and also of Newton insofar as the reflector was concerned. By contrast, the mathematical analyses of light rays that Kepler and Barrow committed to writing remained far removed from the craft domain, with Barrow even lacking all interest in the instrument itself. With the pendulum clock, only one mathematical scientist in the Galilean mold entered the craft domain, Huygens. Unlike with his lens grinding, however, he did not turn himself into a clockmaker. His dealings with the practitioners ran smoothly at times, yet they were interspersed with frictions and quarrels in ways that betray his less than full understanding of the craft intricacies of clockmaking.

This divergence between the craft making of clocks and of lenses also involves impediment (d), the social barrier. For instance, when Huygens received a letter addressed by Louis XIV's chief minister Louvois to 'Mr. Huygens, mathematician', his father, secretary to two successive *stadhouders* from the Orange family, complained that Louvois "seems to take him for one of his fortification engineers. I was not aware of having craftsmen among my children."[176] Even so, in view of the incentives that the mathematical instrument uniquely offered the social barrier could indeed be overcome, albeit differentially so. A distinction apparently kept being felt over the 17th century between the craft of the lens-grinders, with whom even independently wealthy aristocrats like Huygens could exchange views and dispute points as among equals, and the clockmakers. The latter did not cease being perceived as socially inferior, and, hence, to be kept rather at arms' length, sometimes to the detriment of what might have been learned from them.

In sum, then, impediments to productive interaction between mathematical science and craft practice were overcome with greater ease with respect to the two new, mathematical-scientific instruments than in all other cases. A frequently halting yet, in net terms, fruitful train of exchanges began to manifest itself already in the period of the Scientific Revolution. Something faintly resembling our modern two-way chase between science and technology, with the one side coming up with finds that are soon picked up by the other and vice versa, shines through the events analyzed here. As in so many cases of craftsmanship subjected to efforts at mathematization, Galileo was the man with whom it all started. As in so many, Huygens was the man to give them a particularly significant boost in the direction of enhanced practicality *and* sophistication. As in so many, much that was set in motion in the 17th century came to fruition in the next, due, once again, to the emergence of a new kind of craftsmen: the scientifically expert artisan prefigured by, for instance, Hevelius.

Analogies of motion

Drawing analogies is almost a fixture of the human mind. Medieval and Renaissance literature abounded with edifying tales about apparently different things perceived to be of com-

336   parable form, where the one was taken to signify or symbolize at some deep level the other, as between walnuts and brains. Also, findings in one area of nature-knowledge were on occasion applied in another. Aristotle's doctrine of the goal-oriented motions of inanimate objects was inspired by his extended investigations into the form and function of living beings. The stoics accounted for the propagation of sound by means of an analogy with pond ripples. Early in the 17th century, the pioneers of realist-mathematical science gave the hoary idea of analogy a new, much more pointed meaning and significance. Particularly where motion was concerned, analogy often proved the best or even the only way to proceed. Here are a few telling examples.

Balance beam analogies.   Both the first and by far the most often employed analogy of motion prior to the advent of Newtonian mechanics was with the balance beam. After all, equilibrium states alone had in pre-Galilean times been subjected to mathematical treatment, so what else was there for a newly realist mathematical scientist to fall back on? But the vicissitudes of history had made the European reception and further enrichment of the concepts pertinent to such states particularly knotty. The branch of nature-knowledge in which the equilibrium of solids was being handled in the 16th and early 17th centuries had an Archimedean and an Aristotelian component. The latter came in part from the *Mechanical Problems*, with virtual speeds as the central concept, and in part from a medieval tradition instigated by Jordanus de Nemore, with virtual displacements as their (closely related) central concept. Although to the modern eye the Archimedean and the Aristotelian approaches to the subject are just incompatible, this was not the case when these texts were recovered. Throughout the 16th century Archimedes was regarded as the man who applied the principles that Aristotle had laid down. Even such Archimedean hardliners as Guidobaldo del Monte and Simon Stevin took that view. Lack of clarity over the actual historical lineages contributed its part to a time-consuming process of disentangling strands so closely interwoven. For as long as the two remained intertwined, both Archimedean weights and Aristotelian conceptions of virtual speed and/or displacement in proportion to weight came to serve as possible sources of analogy for such nonequilibrium situations as Galileo faced in at least two cases: with freely falling bodies and with the force of impact.

The transition Galileo made from *vertical descent* conceived as equilibrium restored toward vertical descent as kinematics provides an early case where not adopting an analogy but rather giving it up was the decisive step. In his early 'De motu' Galileo sought in vain to reduce the acceleration of free fall to the law of the lever (p. 180); at Padua he crossed the bridge from statics to kinematics by means of his emerging conception of indifference toward motion or rest of a perfectly spherical body resting on a perfectly horizontal plane (p. 193).

Another case of analogy employed by Galileo concerns *hammers and similar objects in impact*. On the *Sixth* Day of the *Discorsi* (incomplete and not published until 1718) the interlocutors search for a measure for the force of impact. Galileo was well aware that any hammer, however light, in being made with however small a velocity to hit a pile, can always overcome in the end the pile's resistance, however large, and drive it into the ground. His search was for a mathematical expression for the force of impact exerted by the hammer. He hoped to be able to gauge it against the effect of a weight that does nothing but rest on the pile – what weight-in-rest can, through sheer pressure, produce in the end the same effect as the hammer? To find out, Galileo imagined an analogy between the process of impact and a balance arrangement in the guise of a frictionless pulley, as shown in Figure 10. 10.[177]

In the balance situation here created, Galileo could demonstrate using elementary Archimedean theorems that, if the slope of the incline is chosen such that weights G and g are in equilibrium, the fall h of weight g causes a vertical rise H of weight G such that $H{:}h = g{:}G$. Therefore, the larger G is, the smaller is H.

The argument hinges on the dual role assigned to G, which not only represents the weight pressing *on* the pile but also represents the resistance exerted *by* it. As a result, full analogy has been attained in Galileo's view between the two situations. In the equilibrium situation, G is inversely proportional to H, as both g and h remain unchanged. In the impact situation, the weight-in-rest G is inversely proportional to the depth of the blow of the hammer, as both g and v remain constant. The significance for us of all this wasted ingenuity lies not so much in Galileo's final conclusion that the force of impact is infinite with respect to a resting weight, but rather in the assumption on which the analogy rests, which is that motion arising from impact can be used in a fruitful analogy with motion in an equilibrium situation. The presupposition could not but look natural to 17th century thinkers. When a century or so later it turned out that different dimensions are involved, the very idea came to look absurd in retrospect. In its time the analogy contributed to blocking a smooth pathway from Galilean kinematics to Newtonian dynamics (p. 623).

Pendulum analogies. Analogies drawn with the balance beam were almost bound to lead the investigator astray; the pendulum came to provide a better source of analogy. What enabled it to serve in that capacity was Galileo's discovery of its near isochrony.

Pendular motion was first used by way of analogy in Galileo's analysis of *falling bodies*. To make his axiom on free fall (speeds increase as the times do) and the theorems following from it testable on the incline, Galileo had to posit an auxiliary postulate: Irrespective of the tilt of the inclined plane against the horizon, bodies released from the same height have at the end of their descent down the incline acquired the same speed (p. 192; also Figure 10.11).

Since in the buildup to his doctrine of falling bodies, Galileo had deliberately renounced all consideration of the action of forces, he found himself unable to prove this indispensable

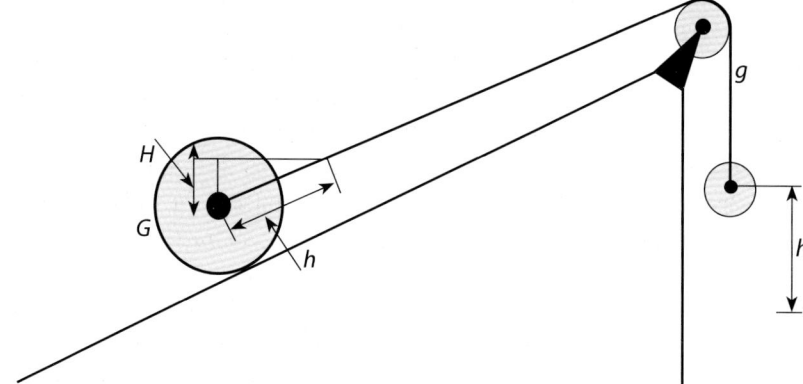

h is the fall of a weight g. In the analogy, it represents the force of impact of a body with weight g (the hammer) which in the act of falling attains velocity v. H is the vertical rise of a larger weight G. In the analogy, it represents the effect of the force of impact, that is, the displacement of a given, constant resistance over a given distance. G represents the weight pressing the pile so as to produce the same effect as attained by the force of impact exerted by the hammer, and of which the measure is being sought.

Figure 10. 10:   Galileo: Balance analogy for force of impact

auxiliary postulate. He sought recourse in an analogy which he hoped was persuasive – as it happened, rightly so. It concerns a pendulum and the assumed interruption of its free swing (Figure 10.12).

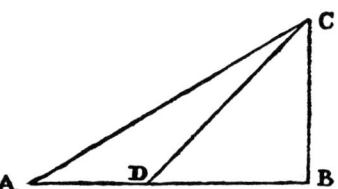

A body falling from C will have acquired the same speed at A, at D, or at B.

Figure 10.11:   Galileo's auxiliary postulate

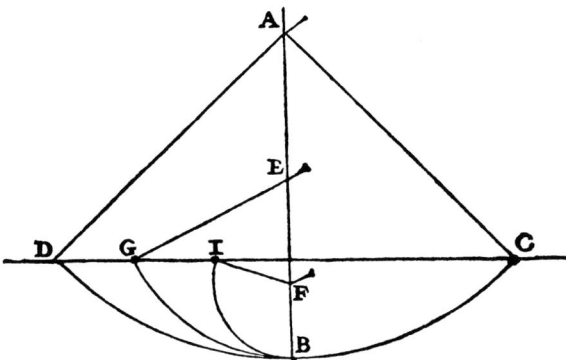

Figure 10.12: Galileo: Analogy to support the auxiliary postulate

If released from C and meeting no hindrance at AB, the bob will reach D, being enabled to do so, in Galileo's argument, by virtue of the '*momento*' acquired by descent through CB being sufficient to make the bob ascend to the same height. If released from C but at AB meeting with a nail E (or F), experiment confirms that the bob, in that it now reaches G (or I), has once again acquired sufficient *momento* to ascend to the height of the line CD from which it started. Galileo conscientiously added that descent along a curve may not be set equal to descent along straight lines – the analogy is imperfect and therefore renders the postulate plausible rather than rigorously proven.

Elsewhere in the *Discorsi* Galileo undertook the *visualization of coinciding sound pulses.* Consonance comes about in his view when sound pulses coincide often enough (p. 301). The pulses of the octave (2:1) coincide on every second 'percussion', those of the fifth (3:2) on every sixth one, and so on, up to and including the minor sixth (8:5). This not only explains why the octave is more consonant than any other interval than unison but also accounts for the peculiar nature of the fifth. This particular interval

> produces such a Titillation upon the Cartilage of the *Timpanum*, that, allaying the Sweetness
> by a Mixture of Tartness, it seems at one and the same Time to kiss and bite.

The regularity thus obeyed by the consonances allows for clarifying visualization by means of properties of the pendulum. Suspend three lead balls from threads of lengths 16 (= 4^2), 9 (= 3^2), and 4 (= 2^2) and make them swing simultaneously. Isochrony means that it does not matter how small or large the arcs are – at every fourth full swing of the longest bob the oscillations will coincide.

This Mixture of Vibrations is the same with that which being made in Strings of Instruments, presents to the Ear an *Eighth* [octave] with an intermediate *Fifth*.[178]

Strictly speaking, the experiment is impossible – earlier on in the *Discorsi* Galileo had shown himself well aware of the *inverse* squared proportionality between pendulum length and period. But no matter, the experiment added to the variety of ends for which analogies of motion were being employed right from the start. Galileo used them for at least four distinct objectives: (1) categorically to extend the scope of thinking about motion (from static to kinematic situations), (2) to find the solution to a particular problem (the measure of the force of percussion), (3) to attain plausibility in the apparent absence of proof (postulate of equal final speeds), and (4) to make more compelling through visualization a result just attained (consonance yielded by the coincidence of percussions).

The inventive power that might come from drawing analogies of motion extended beyond Galileo's pioneering efforts. Seizing upon the method, his best pupil, Torricelli, and the man to embody more than anyone else the Galilean approach beyond the 1640s, Huygens, both started from Galileo's concern with freely falling bodies to attain certain specific results that went far beyond it.

Water outflow.   In his treatise 'On the Motion of Water' ('De motu aquarum'; 1644) Torricelli derived a theorem on the speed with which water flows out of a hole drilled in a vessel at some point below the water surface ('Torricelli's law'). He took the case to be analogous with Galileo's treatment of falling bodies. Hence, he assumed

> that those water jets which flow out with violence possess at their point of outflow the same impetus which any heavy body, or one drop of that very water, would possess if it were to fall naturally from the upper surface of that water down to the orifice out of which it flows.[179]

In the Galilean context 'impetus' means a capacity, acquired in falling down, to return to the original height. From the principle here assumed of equal impetus Torricelli inferred his theorem that outflow speed is proportional to the square root of the distance between that hole and that surface. He sought to shore up the principle by some theoretical arguments drawn from what would happen in communicating vessels but then went on to grant that he had trouble confirming his principle experimentally. His derivation of the speed of horizontal outflow implies that a jet not allowed to flow out but at once redirected upward should reach the level of the upper surface (Figure 10.13).

This, however, it failed to do to such an extent as to cause Torricelli to complain in some exasperation that "the experiment itself seems in a certain sense to prove the principle, even though in a certain sense it also seems to destroy it."[180] Whence this lack of conclusiveness?

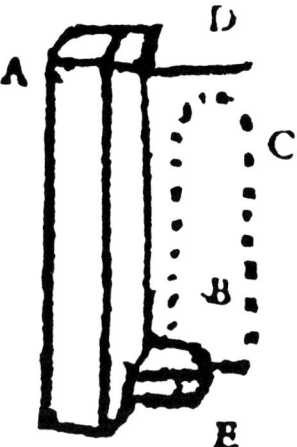

Figure 10.13:   Torricelli's experimental test of the principle underlying his law of water outflow

Torricelli's test runs revealed that optimal, though still not sufficient, conditions for the experiment were that the hole be small and the vessel be both large and always full of water. By the latter requirement he may have meant either that one may neglect the level drop as too small to count or that the water must actually be replenished over the duration of the experiment.

The issue was taken up again in the late 1660s by Huygens and two colleagues in the Paris Académie. Just as with their ongoing effort to reduce the gap between mathematical models in a Galilean vein and the application of those abstract models in craft practice, here too their search was after second-order variables. Huygens decided in the end that these resided in air resistance, in the fall of the water back onto itself, in the 'adhesion' of the water to the vessel walls, and in its mode of outflow, thus jointly accounting for those optimal conditions specified by Torricelli. But Huygens went on to voice a worry of another kind. The principle from which Torricelli derived his theorem had none but an experimental foundation (and a somewhat shaky one to boot) rather than being "demonstrated by reason".[181] In other words, Huygens no longer accepted as sufficiently persuasive the analogy with freely falling bodies originally invoked by Torricelli; in addition, he required better theoretical proof for it.

In the 1690s another Paris Academician, Pierre Varignon, went a step further. No such better proof would ever be forthcoming, he argued, since the analogy itself is false. However plausible at first sight, the resemblance with freely falling, hence, uniformly accelerated bodies is misleading – "since the water is contiguous over its entire length, the water above descends with the same speed as the water below; consequently, there is no acceleration in the

342 vessel at all."[182] Varignon then managed by means of the calculus that Leibniz had recently invented to derive Torricelli's law from another principle. This principle obeyed Varignon's own point of departure in uniform, rather than uniformly accelerated, motion. But it also satisfied Huygens' requirement of being established 'rationally' (i.e., mathematically) rather than just by means of what Varignon evidently took to be the more fallible source of demonstrative knowledge – experiment.

So here is one further usage to which analogies of motion were being put at the time. However fertile a given analogy may be for deriving certain previously unknown properties of nature, it might in the end turn out to be false and, hence, better to be discarded.

From the constant of fall to 'time unrolled'.   With what speed do bodies actually fall? If, as Galileo had sought to demonstrate in the *Discorsi*, bodies fall with uniform acceleration, what is the constant of acceleration, or (to phrase the same question in contemporary parlance) what distance does a falling body cover during the first second? Galileo was more interested in finding mathematical regularity than in empirically establishing quantitative exactitude in a given case. Regarding the constant of acceleration, too, he shrouded himself in evasion or ambiguity or pseudoprecision, with the most significant hint he gave being a reasoned proposal to use the oscillations of a pendulum for purposes of timing. The hint was taken up in the 1640s by Giambattista Riccioli and by Marin Mersenne. The former supervised the counting of pendulum swings over an entire sidereal day from a tower in Bologna manned by a line-up of fellow Jesuits (p. 409). Mersenne compared the fall of one ball with the swing of another. He held two little lead balls, one of which was suspended from a three-foot-long cord so as to need ½ second for a quarter swing in his estimate. The idea was to let go of both balls at the same instant, with the former falling to the floor and the latter swinging toward a stone wall – when, in the course of the test run, the height of fall had been adjusted such that the sounds of the balls hitting the floor and the wall came at the same moment this was taken to yield the distance of vertical fall in ½ second.

Reported outcomes varied between 12 and 15 modern feet. In 1659 Christiaan Huygens took an interest in the problem. In view of the many sources of inaccuracy that had produced so widely varying results, he quickly decided that an approach was needed more in keeping with the fundamentals of realist-mathematical science. Within days of his careful repetition of Mersenne's experiment and its frustratingly inconclusive results, it occurred to him that there is a deep analogy between the tension in a cord set up by a heavy body suspended from it and the tension arising in a cord from a body fixed to it that is swirled around. Can situations be defined in which the two tensions, or tendencies toward motion, produced by heaviness and by being swirled around respectively, counterbalance? That is, *can one species of uniformly accelerated motion be represented by another, in view of the two being mathematically equivalent?* If so, the determination of the constant of fall can be realized by other, much more theory-infused means. It quickly proved that such situations can be defined indeed. Huygens realized that a conical pendulum – that is, a pendulum made to swing, not to and fro in a straight line, but rather along a circle – provides the real-world embodiment of the counterbalancing situation he sought (Figure 10.14).

From there on, his effortless command of the whole of Greek geometry led him to a comparison of properties of circles and parabolas and of infinitesimal segments thereof that enabled him to derive a value for the constant of vertical descent which he went on to check experimentally and which is very close indeed to the value still accepted at present. Meanwhile, he established as a matter of course the mathematical rule governing the pertinent parameters in what he went on to call 'centrifugal' motion: the tension is proportional to the orbiting body's size and to the square of the velocity and is inversely proportional to the radius of the circle.

With the problem that had set him going thus resolved, Huygens returned to the theoretical underpinnings of Mersenne's original experiment. The latter had wrestled in vain with the comparison between rectilinear fall and fall along the circle arc traversed by the bob of his pendulum. Here a representation of one motion by another was already taking place – whereas Riccioli had employed the pendulum only to make time countable, Mersenne had sought to benefit from the fact that both vertical fall and the swings of a pendulum are manifestations of gravity. However, he had not known how to equate the one with the other. Huygens, after two early efforts that got him nowhere, did. Shifting his problem from the full quarter arc traversed by the bob to the infinitesimally small segment near the vertical, he "asked what ratio does the time of a very small oscillation of a pendulum have to the time of perpendicular fall through the height of the pendulum".[183]

Underneath the surface of the question lay his awareness (p. 331) that the isochrony of the pendulum posited by Galileo is no more than approximately valid, as it applies to very small arcs only, just around equilibrium position. With the theorems of Greek geometry at his fingertips, and with his acquired virtuosity in dealing with infinitesimal ratios, Huygens found something he had not originally been looking for. There is a curve which is isochronous, not only for very small arcs of swing, like the circle, but for every segment of the curve however large. After much to-ing and fro-ing of geometric ratios, Huygens found this particular curve – examined a few years previously in quite a different context by some of Huygens' fellow mathematicians – to be the cycloid (for how it is generated, see Figure 10.8 on p. 331).

This great find, if considered now in the light of Huygens' invention of the pendulum clock two years earlier, at once presented him with a new problem. That clock, after all, was not really isochronous. He had therefore fitted it out, at the top of the cord, with metal 'cheeks' bent into an empirically determined curve so as to compel the bob to describe a more nearly isochronous curve slightly above its uninhibited, circular path. But now an

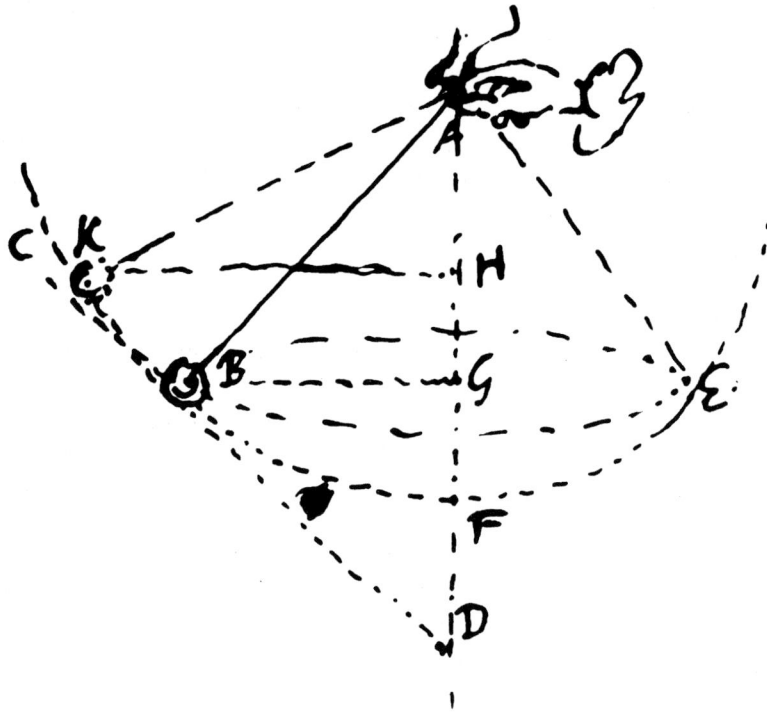

Figure 10. 14: Huygens' conical pendulum

opportunity had turned up to determine the shape to be given to the cheeks with rigorous exactness. The question to be answered was: What curve, in being 'unrolled', can generate a cycloid? Within two more weeks Huygens, using his trusted mathematical techniques plus his instant recall of properties of the cycloid picked up from the debates between his colleagues, discovered that the curve sought is, once again, a cycloid.

From here, Huygens moved in three different directions, which in due time came together in *Horologium oscillatorium*. He expanded the 'unrolling' technique he had explored in the cycloid for obtaining a comprehensive theory covering the unrolling (or, more technically, the 'evolution') of curves generally. He extended the rules for pendular motion he had found along the way from the abstract, 'geometric' pendulum toward an investigation of the real-world, materially rigid pendulum (p. 537). And he sought to exploit the newly found property of isochrony for the definitive practical solution to the problem of determining geographical longitude at sea.

The vicissitudes of Huygens' clocks have occupied us in the previous section (p. 331). But three other issues in connection with the developments just sketched require further comment: the unspoken origin of the two analogies of motion here employed by Huygens, his extending analogy to the point of mathematical equivalence, and the nature of his mathematical tools.

Huygens handled his two analogies with as yet unsurpassed surefootedness and virtuosity, yet neither was original with him. Galileo had already perceived the analogy between vertical and curvilinear fall (Mersenne had dabbled in it, too) and also that between gravity and centrifugal motion. In his campaign for heliocentrism, Galileo had on the Second Day of the *Dialogo* been compelled to face the objection that, if it were true that the Earth spins around its own axis, terrestrial objects (ourselves included) would be hurled off. Rather than examining, as Huygens was to do, under what conditions gravity and centrifugal motion *balance*, Galileo rather overshot his mark in an argument meant to show that, irrespective of the size of the Earth or its speed of revolution, no terrestrial object will ever be able to fly off the Earth at a tangent. Whatever the parameters, never will the centrifugal tendency set up by the Earth's daily rotation be large enough to exceed gravity. Huygens reasoned differently. After establishing his rule for centrifugal motion, he fed the best available value for the Earth's radius into it and concluded that only if the latter figure were 265 times larger than it is would the Earth's spin be capable of overcoming the downward effect of gravity. Another source of inspiration for Huygens' pursuit of the analogy between a body's fall and its tendency, in circular motion, to fly off at a tangent is likely to have been certain aspects of Descartes' world, notably the inclination there posited for particles to move in closed curves (pp. 228; 530).

In the course of his investigation Huygens turned analogy of motion into an even more powerful tool than it had already become in the hands of Galileo and his pupils. Uses to which they put the method included (to repeat) transference of a known mathematical property to another situation held to be comparable with it, rendering an unproven postulate plausible, visual representation, and domain extension (even though the specific analogy employed might in the end prove untenable and therefore be discarded without necessarily damaging results obtained thereby, as with Torricelli's law). What Huygens did in his wonder year 1659 went a decisive step further, in that he turned the drawing of what had so far remained somewhat loose, intuitively plausible analogies into full mathematical equivalence. *Irrespective of real-world differences between the phenomena taken as similar, mathematical properties of the one might without reservation be extended to the other.*

From the rigorous point of view of ancient geometers like Euclid or Archimedes or Apollonios, Huygens' virtuoso handling of infinitesimal ratios was formally impeccable. Still, examination of the mathematical thought and practice at work behind Huygens' elegantly rigorous theorems and proofs shows that principal limits set by the Greeks were actually being overstepped here. Huygens made his pertinent discoveries in 1659 but was not to publish

346 them until 1673. Halfway through those fourteen years, a novice mathematician less bound than Huygens to ancient rigor devised a method that was able to engage in a general manner questions of motion that were definitely out of reach of Euclidean techniques even when stretched to their limits. That mathematician was Isaac Newton; that general method, the differential and integral calculus.

From Euclidean ratios to the calculus

Among the highest attainments of Greek geometry was Archimedes' handling of the method of exhaustion (p. 10). For all its exactness it was a very cumbersome way of reducing the curved to the straight, and its rigorous strictness might have been alleviated (but actually was not) using contemporaneous, stoic ideas pointing in a qualitative manner at the idea of convergence to a limit (p. 19). I shall now compare the two conceptions involved – exhaustion and passage to the limit – by considering Figure 10.15 and appended comments.[184]

It looks as if all this is just about as close to a limiting procedure as can be. In all likelihood Archimedes adopted it himself when engaged in *finding* his result of 4/3 A. Also, the sole technical difference between his eventual, formal proof and an effective passage to the limit resides in the three italicized words – he "added the remainder". Is it not, then, a mere quibble to deny to Archimedes mastery of the basic procedure underlying the calculus?

It is not, because a vast conceptual chasm lies hidden behind so seemingly tiny a technical difference. The nature of the chasm becomes apparent from a range of polar opposites, each of which points from a somewhat different angle at what was at stake: between measuring the rectilinear and the curvilinear; between a static approach and one stressing the variable; between the finite and the infinite; between a rigid distinction between the continuous and the discrete and a readiness to blur that distinction; between rigorous proof and looser demonstration. How is it that, in each of these pairs of opposites, Greek geometers are generally to be associated with the former rather than with the latter notion?

The answer resides in two defining moments in the history of mathematical thought in ancient Greece. The first is the invention of the idea and practice of rigorous proof. The other is the discovery of such logical difficulties involved in the infinitely large and the infinitesimally small as put down so compellingly in Zeno's paradoxes (p. 19). Each hinges on the apparent irreducibility of the continuous to the discrete, and this lesson dominated mathematical proof ever since. At least in their public writings (i.e., with the property in need of proof already well in hand) Greek mathematicians confined themselves to a static, strictly geometric mode of arguing, with infinities avoided at all costs. The method of exhaustion represents a particularly ingenious way to circumvent the apparent impossibility of matching the rectilinear and the curvilinear without giving up the ideal of rigorous proof.

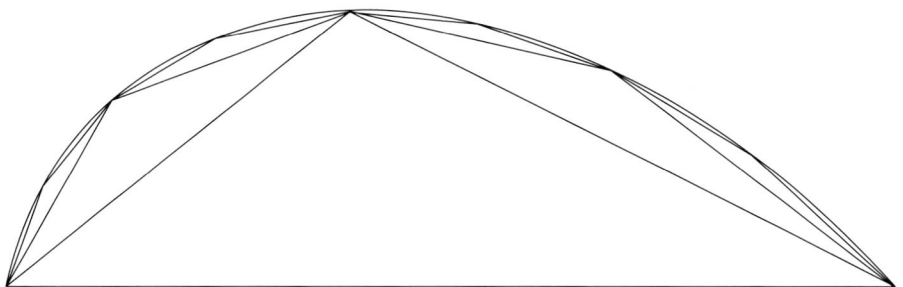

The proof starts with "inscrib[ing] within the parabolic segment a triangle of area A, having the same base and vertex as the segment. Then within each of the two smaller segments having the sides of the triangle as bases, he similarly inscribed triangles. Continuing this process, he obtained a series of polygons with an ever-greater number of sides, as illustrated ... He then demonstrated that the area of the nth such polygon was given by the series $A\left(1 + \frac{1}{4} + \frac{1}{4^2} + ... + \frac{1}{4^{n-1}}\right)$, where A is the area of the inscribed triangle having the same vertex and base as the segment. The sum to infinity of this series is 4/3 A, and it was probably from this fact that Archimedes inferred that the area of the parabolic segment was also 4/3 A.

However, he did not state the argument in this manner. Instead of finding the limit of the infinite series, he found the sum of n terms and *added the remainder*, using the equality

$$A\left(1 + \frac{1}{4} + \frac{1}{4^2} + ... + \frac{1}{4^{n-1}} + \frac{1}{3} \cdot \frac{1}{4^{n-1}}\right) = \frac{4}{3}A.$$

As the number of terms becomes greater, the series thus 'exhausts' 4/3 A only in the Greek sense that the remainder, $\frac{1}{3}\left(\frac{1}{4^{n-1}}\right)A$, can be made as small as desired. This is, of course, exactly the method of proof for the existence of a limit, but Archimedes did not so interpret the argument. He did not express the idea that there is no remainder in the limit, or that the infinite series is rigorously equal to 4/3 A. Instead, he proved, by the double reductio ad absurdum of the method of exhaustion, that the area of the parabolic segment could be neither greater nor less than 4/3 A." [my italics]

Figure 10.15: Archimedes' proof of his determination of the area of a segment of a parabola

True, in search of properties as yet unknown Archimedes felt free to take the leap from the finite to the infinite (in the 17th century Torricelli and others began to suspect this, with confirmation of their hunch eventually coming from a Byzantine manuscript rediscovered in the early 20th century). Still, a consequent, more general loosening of mathematical thought could not come from taking such incidental license but only from a readiness squarely to face the infinite, the variable, the continuous. A conceptual breakthrough of such magni-

348   tude, while not fully inconceivable, was quite unlikely to appear in the very civilization that had learned so well how to work its way around the underlying issue. Here, too, then, a feat of cultural transplantation was to offer chances for such a breakthrough – in this case, for finding pathways toward taming the infinite.

Conceptual pathways toward taming the infinite.   Islamic civilization adopted much of the Greek corpus in geometry and enriched it in various ways, most innovatively by way of a deepened understanding of foundational issues in Euclidean geometry (p. 57). But two principally new elements were added to the corpus as well: the adoption from India of the decimal place system including zero and the emergence of early algebra. Arithmetic and algebra could now come into their own as independently valid ways to solve equations and ascertain the outcomes, even though most often those outcomes were known already in Greek geometry. Still, an exceptional effort was undertaken by Umar al-Khayyami to find roots for as yet unsolved equations by recourse to the intersection of certain conic sections and translation of results back into algebraic form. The effort looks highly promising in retrospect, but it was not pursued any further. It also remained unknown in Europe. There a somewhat similar probing of ways to reason back and forth between algebraic equations and geometric figures was undertaken, and this time the effort was pursued to the end.

Two principally new developments starting in late-Renaissance Europe cleared the path toward what eventually became the calculus. One development, from inside mathematics, resided in the discovery of the principal equivalence of algebraic equation and geometric curve – the very breakthrough incidentally opened up as a possibility in Islamic civilization. I discuss it in the next section (p. 351).

The other novel development came for the largest part from outside mathematics proper. It rested in the creation of conceptual space for a decisive shift away from the Greek preference for the finite, for the nonvariable, for the measurement of straight lines, and for keeping the continuous and the discrete wide apart, in the direction of their polar opposites. This conceptual space came about as follows.

In the first place, a readiness to confront the infinite squarely rather than surreptitiously was furthered by how, from the late 16th century onward, acceptance of the heliocentric system compelled adherents (however few in number originally) to take problems of infinite space seriously. To counter the anticipated objection from unobserved parallax, Copernicus had put a very large amount of otherwise useless space between the spheres of Saturn and the fixed stars (p. 111). Among the near-dozen Europeans ready between 1543 and 1600 to adopt heliocentrism in realist fashion (p. 159), two men felt the obvious next step to reside in acceptance of infinite space. Neither Giordano Bruno nor Thomas Digges contributed to mathematics, yet their bold extension of heliocentrism was bound to attract others once, in the course of the 17th century, more and more considerations began to speak for Coperni-canism realistically conceived. One example is Pascal and his profound thoughts about the 'two infinities' between which humanity finds itself stretched. Consequently, mathematicians in 17th-century Europe operated in a climate of thought about infinity to an extent not matched elsewhere or before.

Two significant resources enhanced a readiness to stress the variable over the static. Stoic thinking in terms of dynamic continuity, resurrected by the humanist movement along with other Greek philosophies of nature (p. 99), may have contributed toward making conscious passages to the limit prevail over rigorous exhaustion methods. Another, vastly more significant line of thought in the general direction of the mathematical taming of the infinite resided in the conceiving of geometric figures, not as static entities, but rather as drawn through *motion*. Greek geometers had generated several specific curves, such as the spiral or the quadratrix, by imagining a point to be subjected to two specific, simultaneous motions (Figure 10.16).

It even seems likely that Archimedes himself managed to find the tangent to one particular spiral from considerations of motion (Figure 10.17). Here once again Archimedes hit upon an approach fraught with developmental possibilities. Just as, from a *technical* point of view, the difference between the method of exhaustion and actual integration is almost insignificant, just so is Archimedes' determination of the tangent to the spiral close to the differentiation procedure practiced from Newton onward. But once again, *conceptually* no passage to the limit is taking place here, and it is not by chance that no other example occurs in Archimedes' known works, just as he kept the laborious method of exhaustion confined to a few select cases. Despite what this approach looks like in retrospect, it did not burst out of the frame of the mathematically static. In Euclid's definition of a tangent as a line touching a circle at only one point, the determination to stick to the mathematically static was embodied in more orthodox fashion than in the mathematical-kinetic approach here probed by Archimedes. Even so, his probing took place at the outskirts of a conceptual frame not in the end transcended even by him.

Also by hindsight only can we see that a more motion-oriented climate of thought than provided by the intellectual environment of the Greeks served as a powerful incentive to tame the infinitesimal. The onset of the Scientific Revolution, in making motion central to mathematical science as well as to the natural philosophy of kinetic corpuscularianism, provided such a climate. The geometric treatment of motion could much more easily lead to what is in fact its converse – a motional treatment of geometry, so to speak – than had been likely to happen in ancient Greece, where the treatment of motion remained confined in the Alexandrian sphere to the equilibrium case and in Athens' schools to no more than a side issue (for there change as such, not local motion, had been at the center of natural-philosophical concern).

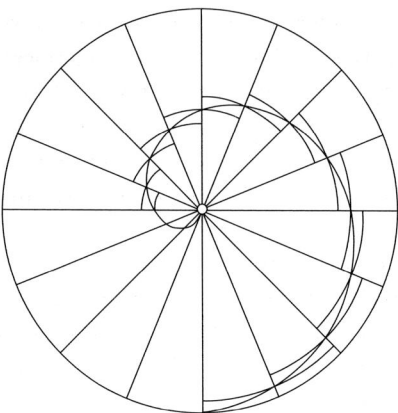

A straight line rotates uniformly around a fixed point, while a point on that line moves out uniformly.

Figure 10.16: Generation of a spiral

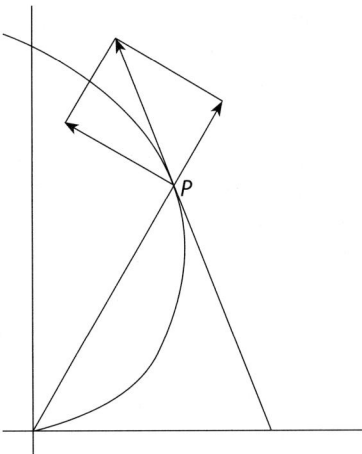

The tangent is drawn "through a determination of the instantaneous direction of motion of the point P, by which [the spiral $\rho = \alpha\theta$] is traced", according to Boyer's hypothetical reconstruction.

Figure 10.17: Archimedes' kinetic drawing of a tangent

Take, finally, the dual *reductio ad absurdum* by means of which proof is rendered in the method of exhaustion. Not only was it undermined by the various intra- and extramathematical developments just sketched, but it is also very cumbersome. The method pays for rigor in the currency of complexity; it certainly does not invite free-riding, leisurely exploration. In Europe laborsaving equipment had been on the rise already for centuries (e.g., water-powered sawmills, automated machinery for fulling), and the urge for it reached mathematics with the early 17th century invention of logarithms. Similarly, the time-consuming drudgery for the sake of sheer rigor which the method of exhaustion entails gave way to ever looser methods, up to the point that they proved not just alternatively viable but also capable of yielding fresh discoveries. Rigorous proof could be cast to the winds as it became clearer by the day that to ignore boundaries once set by the Greeks brought its own rewards. An exciting sense of discovery moving ever farther beyond the achievements of the ancients swept over Europe's 17th-century mathematicians. The vehicle to lead in the end to the recognition of curves and equations as equivalent and of geometry and algebra as, in the final analysis, interchangeable was the attempted recovery of an ancient Greek achievement known as the *analysis* of mathematical *problems*.

Curves and equations.   Analysis was the name given to a procedure for solving problems (i.e., constructions and the determination of loci) in which, upon assuming the solution to have been found already, one then works one's way backward. Around 320 CE Pappos had collected results of the procedure in one manuscript text. On making their acquaintance with Pappos' collection when by the late 16th century it appeared in print, a suspicion quickly arose that the Greek geometers had achieved many more of their results along this path than they had allowed to shine through the elegance of their formal presentations. Acting on the hunch, mathematical humanists set out to recover that lost method of analysis. As also in other mathematical domains, recovery soon gave way in its turn to genuine innovation (p. 102). The names of three discoverers, all French, stand out. Viète and Fermat, drenched in the waters of mathematical humanism, went on to break through its limits. Descartes, in contrast, was concerned to leave no doubt in his readers' minds that he was on to something quite novel and unheard of.

To solve equations of a given type is often done best by example, and this is how, in the 16th century, algebraic techniques had been developed for solving certain third- and fourth-degree equations in one unknown (much of this had a parallel in the mathematics of Islamic civilization and in any case proceeded under its influence). When, in 1591, François Viète introduced letter algebra, his aim was not so much to facilitate the solution of equations. After all, to denote parameters by a, b, c, ... often makes finding the solution more cumbersome, not easier. Rather, his aim was to invoke the help of algebra for solving problems – geometric problems like those Pappos had grappled with. Viète did this by introducing a generalized

352 method which allowed him to refrain from specifying the magnitudes his letter algebra dealt with. He called it *logistica speciosa*. This came down to the translation of a geometric problem into algebraic terms, followed by its transformation such that, upon translation back into geometry, the problem could straightforwardly be solved. Two subsequent treatises, meant to flesh out Viète's claim that his specious logistics offered the means "to leave no problem unsolved",

> culminated in the proof that all geometrical problems which can be reduced ... to fourth- or lower-degree equations, can be constructed by ruler or compass, or can be reduced to two special problems not so constructible, namely either the trisection of an angle or the determination of two mean proportionals between two line segments. He combined this new result with a classical one, namely that these two problems themselves may be reduced to a single problem, called the 'neusis problem'. The combined result was new and truly significant because at one stroke it structured almost the whole field of geometrical problem solving.

Sufficient correspondences between algebraic and geometric operations were at hand (notably, equality and proportionality) to make such to-and-fro shifting possible. What tended to get lost, was the very insistence on the principal difference between continuous magnitudes (lines) and discrete magnitudes (figures) which was so paramount in the Greek tradition. Instead, a concern arose over which lines (curved lines, that is) were admissible in the solution of problems and which not.

This concern was voiced, above all, by Descartes in his *Géométrie* of 1637. This book provided a range of techniques for solving problems by moving back and forth between geometry and algebra. Implicit in Descartes' treatment throughout is the identification of a curve and an equation – a curve can be represented by an equation, and equations can be manipulated and rewritten such that, rendered back into line segments (i.e., coordinates), their solution becomes possible. By such means Descartes likewise claimed to be able to solve "all the problems of geometry". More specifically, he employed specific categories of conic sections to extend the possible solution of equations beyond the fourth degree up to the sixth degree (in practice) or even to any degree whatever (as yet, in theory only). From this he borrowed the criteria he went on to work out for which curves were allowed in analysis and which not.

Contemporaneous with Descartes but in a mostly different spirit and in more fragmentary fashion, Pierre de Fermat stated almost explicitly what had been implicit in Viète's work and remained so in Descartes'. Equations are not only very helpful in solving problems involving curves but equation and curve are really equivalent. In Fermat's expression, the equation is the "specific property" of the curve. Another significant difference is that in Fermat's work the actual handling of infinitesimal magnitudes (e.g., in the determination of certain tangents to certain curves) does not take place nearly so deeply beneath the surface as was the case in the more covert manner in which Descartes hid effective passages to the limit from sight.

From the analysis of finites to the analysis of infinites.   Rather at odds with Descartes' overriding, methodological concerns in his *Géométrie*, and really more in keeping with the tenor of Fermat's approach, Descartes' book was seized upon by the next generation mostly for the techniques he had used and for the vistas these opened upon the solution of other classes of problems. Descartes had presented an 'analysis of finites' – his techniques showed the way toward an 'analysis of infinites'. That is to say, a new, powerful mode of operation had now become available to tackle such topics bordering on the infinitesimal as Archimedes had handled in the rigorous manner of the method of exhaustion. That new technique, then, was interchangeably to treat curves as equations and equations as curves. To find tangents to certain curves, or to 'rectify' them (i.e., to find a straight line equal in length to the arc of a curve), or to determine areas bounded by them ('quadrature') could now be attempted with the help of series expansions and other algebraic techniques more easily and on far larger scale than had hitherto seemed sensible to undertake. All such operations involved infinitesimal magnitudes, to be sure. These were handled with increasing license – the logical paradoxes involved in the twin ideas of the infinitely large and the infinitesimally small, which had so bothered the Greeks, were now felt to be less significant than the extraordinary gains in solving power offered by the new, algebraic methods. Mathematicians all over Europe engaged in the determination of tangents to both old and new but always individual curves, of their asymptotes if they had any, of their maximal and minimal values, and of areas bounded by them; they also engaged in rectification by evolution and other means, with the identification of curve and equation yielding whole arrays of curves (e.g., the cycloid and the logarithmica) never considered previously. Descartes' methodological strictures over certain categories of novel curves were cast to the winds in the excitement of discovery.

Individual style very much predominated in all this. Some mathematicians glossed very lightly over the conceptual difficulties involved in the infinite and the infinitesimal – to get valid, new results was the only thing that seemed to count. Thus, Galileo's disciple Cavalieri took a plane figure to be composed of infinitely many 'indivisible' parallel lines, which he then felt free to sum. Kepler, too, in his pioneering efforts to determine the optimal shape of wine casks geometrically, walked what he called a "bridge of continuity" without too many qualms.[188] Pascal and Huygens, in contrast, handled the infinitesimal in more purist fashion, by means of ingenious transformations of geometric figures carried out in the Greek style. One example is Huygens' 1659 discovery and proof of the isochrony of the cycloid (Figure 10.18; see also p. 343).

354 From such handling of the infinitesimal to the calculus is no longer a big step conceptually. Although Huygens remained as true as he could to the Greek *style* of proof, he fully bridged the *conceptual* chasm between the method of exhaustion and what, in other hands than his, would be turned into the calculus within six more years. One decisive feature of Huygens' thought that kept him (as a little earlier it had kept Fermat) from inventing the calculus or something close to it was their shared lack of interest in generalized procedures. A sense of enchantment with their own facility in facing, and solving, individual challenges stood in the way of an interest in generalizing their techniques such that, instead of the geometric imagination required in each individual case, just the application of a standardized algorithm would suffice to yield the solution sought in any given case. The two men thus to crown mathematicians' preceding efforts to tame the infinite, then, were not Fermat or Huygens or any of the other early- and mid-17th-century explorers of the infinite and the infinitesimal but Newton in 1664/5 and Leibniz in 1672.

Newton's fluxional calculus. On 8 November 1676 Isaac Newton wrote to John Collins, a mathematical enthusiast and news broker, that

> there is no curve line exprest by any aequation of three terms, though the unknown quantities affect one another in it, or ye indices of their dignities be surd quantities ... but I can in less than half a quarter of an hower tell whether it may be squared or what are ye simplest figures it may be compared wth, be those figures Conic sections or others. And then by a direct & short way (I dare say ye shortest ye nature of ye thing admits of for a general one) I can compare them. ... This may seem a bold assertion because it's hard to say a figure may or may not be squared or compared wth another but it's plain to me by ye fountain I draw it from.[189]

Newton hid from Collins, as he kept hiding from everybody else for at least ten more years, what that fountain was. It had begun to spout a dozen years earlier, when Isaac was a twenty-two-year-old student, within little more than a year of making his acquaintance with the mathematical discipline in the first place. His mathematical apprenticeship had not been through Archimedes (far too advanced for the Cambridge curriculum) or even through Euclid (too much enveloped, in that same curriculum, in scholastic categories no longer to this unruly youth's taste) but through his own private reading in post-Greek analysis. Descartes' *Géométrie* in the expanded, second Latin edition by Frans van Schooten Jr. (1659/1661) served as his principal entrance gate. Particularly inspiring to young Newton proved to be supplements to that book by van Schooten's pupils Huygens, Hudde, van Heuraet, and de Witt, as well as John Wallis' treatise *Arithmetica infinitorum* (1656). The latter work more than any other marked the point where Newton's advanced apprenticeship began to give way to independent discovery. Wallis, in seeking to determine sums toward which certain

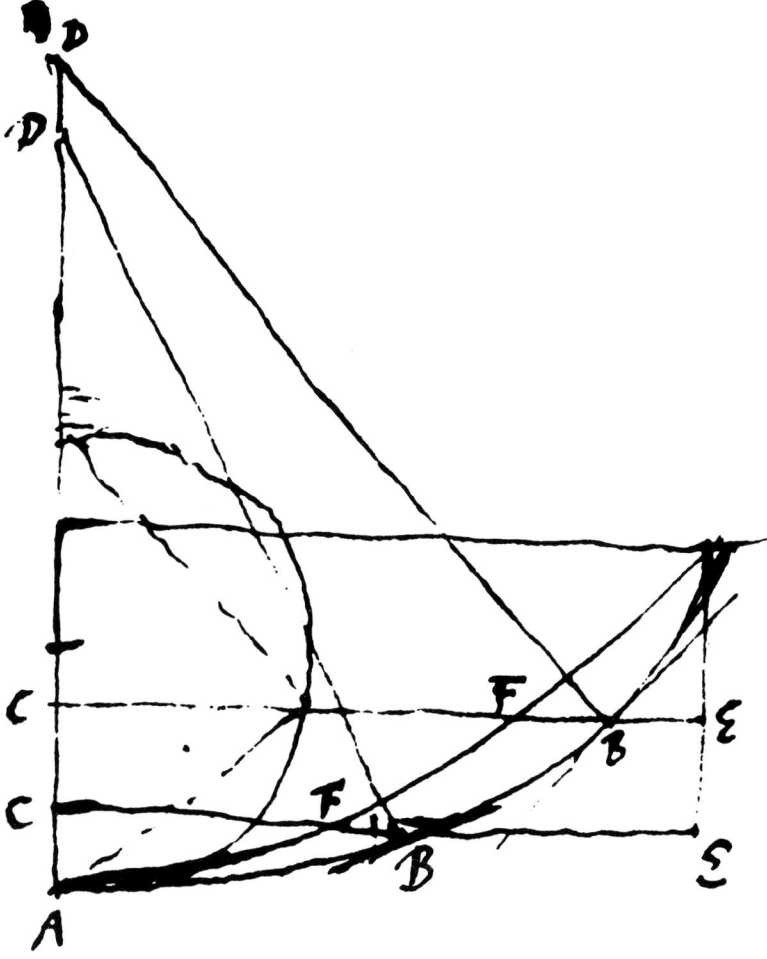

ABB represents the cycloid.

Figure 10.18: Huygens' original identification of the cycloid

infinite series approach, with a view to evaluating areas under certain specific, elementary curves, had adopted a highly informal, inductive form of attack. So did young Newton when he went on to address more complex equations with exponents of less regular fractional powers. The net outcome of his efforts so far was an expansion rule for any binomial of the form $(b+x)^{m/n}$.

356   Here already, what made Newton's approach different in kind from those preceding was his unceasing effort to generalize intervening results. So many others were after unique solutions to unique problems: for example, the way Grégoire de Saint Vincent sought to square the equilateral hyperbola or Wallis sought to sum the series of terms into which the equation for the circle expands for certain integral values. Newton, however, aimed at general methods for solving entire classes of mathematical problems, as when, next, he invaded territory previously explored at its outskirts by Hudde.

Hudde sought to facilitate the finding of tangents to given conic sections along lines of attack first developed by Descartes and Fermat. In so doing he hit upon a way to multiply the curve's equation by an arithmetic progression chosen such that unmanageable, constant terms cancel out. Such a procedure, so we realize as Hudde did not, literally gives one the derivative equation for the tangent in question, in which the slope of the curve at the tangent point finds itself expressed. Newton's essential insight, as it began to dawn upon him in the spring of 1665, was twofold. One part of it was to see what Hudde's find truly implied, once it was lifted out of the issue-by-issue frame in which its author enveloped it. In his customary way Newton tried to generalize the approach using parabolas and hyperbolas of various equations. The other part of his new insight was as follows. In the course of his effort at further development it began to occur to Newton that one may just as well *derive* the equation for the tangent from that of the original curve as one may *integrate* the latter out of the former. In other words, drawing tangents and finding quadratures are inverse procedures. In again other (not his own) words, differentiation and integration are each other's inverse.

The decisive step in exploring what the second insight implied was for Newton to begin considering curves (conic sections for starters) as traced by motion. This, to be sure, was not a novel idea. There were, of course, those incidental, Greek precedents with the spiral and the quadratrix. More importantly, Johan de Witt, in his *Elementa curvarum linearum* appended to van Schooten's edition of Descartes *Géométrie*, had recast Apollonios' theory of conic sections in modern, analytical form by treating them as drawn through motion. Newton, during his stage of self-education, had been fascinated by the approach. He now adopted it in his search for a generalized method of quadratures using Hudde's find and used it to demonstrate his new insight.

Newton's certainly not unique yet thoroughgoing awareness that to determine the slope of a curve at a given point is mathematically equivalent to the determination of nonuniform velocity at a given instant made him cast his insights in the guise of what he was forever after to call his 'fluxional calculus'. 'Fluents' stand for quantities varying in time (hence, flowing), and 'fluxions' for their rate of change. As yet, Newton felt that he had successfully avoided the logically questionable usage of infinitesimal quantities made as small as one wishes by benefiting instead from the apparently continuous flow of time in motion. Time now came to bear the (in Newton's original view) lighter burden of being considered as a regular succession of moments 'passing to the limit' (as we would say). Still, in the end he was to seek another foundation for his fluxional calculus, which he would introduce in the first publication presenting his, now twenty-two years old, still mostly secret calculus – that is, in the *Principia* of 1687 (p. 660).

Motion and the calculus: a look back and a look forward.   From the outset, the centrality which the category of motion acquired in revolutionary fashion in a Galilean/Keplerian (as also, in a different vein, in a Cartesian) context greatly stimulated the creation of conceptual space for taming the infinite. The stimulus worked two ways, back and forth between motion and mathematical technique, so as to lead to ever closer and more productive intertwinement. What Newton once called the "generation of figures by motion" had been pioneered by Archimedes at the outskirts of a framework of ancient rigor.[190] From the early 17th century onward the approach contributed greatly to those mathematical innovations out of which the calculus eventually emerged. But the new, mathematical treatment of phenomena of motion also appeared to require clarification of basic concepts, notably space, time, velocity, and acceleration. The entities denoted by these concepts, and the relations between them, at once proved susceptible to Zeno-like paradoxes, which now reemerged in unprecedentedly subtle fashion. Mathematics had meanwhile become entrenched in the real world in unprecedented ways, and this development made it so much the harder to keep ignoring those paradoxes. One way to measure the conceptual leap wrought by the emerging calculus in this regard is by comparing how, on the Third Day of the *Discorsi*, Galileo dealt with uniform and then uniformly accelerated motion and how, about half a century later, Newton and then Varignon treated the concepts involved.

Uniform motion first. When, in preparation of his treatment of falling bodies, Galileo laid down the rules for uniform motion, he did not simply set speed in uniform motion equal to the quotient of distance traversed and time elapsed. Rather, he took care to write them up as six successive proportionalities. In Galileo's time even those who were prepared to attach a satisfyingly clear meaning to a distance 6 or a time 3 were far from inferring that the former may be divided by the latter so as to yield a speed 2. Space and time are mutually heterogeneous, as apples and pears proverbially are, and they were quite literally taken to be incommensurable. That is why Galileo made sure to posit as one axiom for uniform motion that at greater speed more space will be traversed in a given time than at lesser speed. He posited five more axioms in similar fashion, such that the full collection of axioms related distance, time, and speed in uniform motion in conformity with the standing rules for homogeneity of Euclidean ratios.

None of this preparatory matter would have looked out of bounds to an ancient geometer. But then Galileo went on to establish axioms for uniformly accelerated motion in free fall and to derive theorems from them. Here one almost hears the ice crack under Galileo's

358 Euclidean skates. Take the very first theorem he derived from his axiom that velocity in vertical descent increases in proportion to time passed. The 'mean speed' theorem states that, whether a body falls from rest in uniform acceleration or uniformly at the speed attained halfway through the fall of the former, equal distances are traversed in either case. On page 307 we examined the big conceptual difference with how Oresme had handled the theorem in the abstract; we now consider in greater detail Galileo's mode of proof, which centered on a diagram reproduced in Figure 10.19.[191]

The proof is elementary insofar as it invokes the congruence of triangles EFI and GAI, from which the equality of the surfaces of triangle AEB and rectangle GABF follows at once. So far so smoothly; but of course there were difficulties involved in the proof. Galileo's care in introducing (albeit without definition) notions of 'instants of time', 'degrees (also 'momenta') of speed', and 'final speed' show that he was well aware of these difficulties. They culminate in the problem of how to arrive from that 'final speed' at those distances for which he had set out to prove equality. As he was also well aware, nothing in Euclidean geometry allowed him to make AEB represent the space traversed in uniformly accelerated motion just by *summing* those parallels representing increasing degrees of speed, as if any number of discrete parallel lines can ever add up to a surface. Therefore, he carefully spoke of the "aggregate" of these lines. But how to make the conceptual leap from speeds to distances? On having shown "the deficit of momenta in the first half of the accelerated motion (the momenta represented by the parallels in triangle AGI falling short)" to be "made up by the momenta represented by the parallels of triangle IEF", Galileo declared it to be "patent" that the theorem had thereby been proven. What we have here, then, is an ingenious attempt to squeeze into an (at first sight) impeccable handling of Euclidean ratios a line of thought fundamentally at odds with it.

Now compare all this with what the limiting procedures of the emerging calculus made possible, whether directly or by opening vistas beyond the uniform acceleration Galileo was dealing with here.

In the *Principia*, in a setup similar to Galileo's, Newton identified without more ado the surfaces of triangles drawn like those in Figure 10.19, with the distances covered by the falling body. Justification for such a procedure came from a geometric argument that he devoted to establishing the idea of instantaneous velocity. That entity, Newton knew, could be measured by the slope, at any given tangent point, of the specific curve representing the body's motion. Like Galileo, he was also well aware of those paradoxes of the infinite over which Galileo had skated in his proof. That is precisely why in the *Principia* Newton chose to present the numerous limiting procedures he handled there in a geometric vein. That way, so he felt, he could travel a sufficiently rigorous, middle road between the ancient, laborious mode of proof through exhaustion and the modern, convenient, and therefore often-used yet really illicit mode of summation of 'indivisibles'. As a consequence, the calculus-inspired

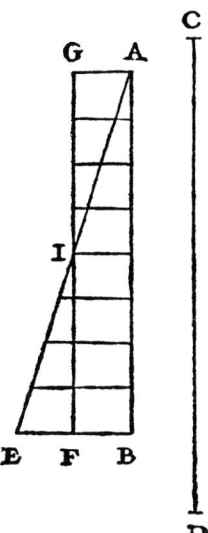

AB represents the time of fall from rest at C over distance CD; AB is divided into "instants of time". EB, drawn perpendicular to AB, represents the "maximum and final" speed attained on arrival at D; line segments parallel to EB, drawn to AE from each point on AB that marks an instant of time, represent "the increasing degrees of speed after the instant A". I is the midpoint of AE; GF is drawn through it parallel to AB.

Figure 10.19:   Galileo's proof of the mean speed theorem for falling bodies

'method of first and last ratios' that he employed throughout the *Principia* forced him to operate on a case-to-case basis rather than through the reiterated application of general algorithms.

Such algorithms, then, still lay dormant in Newton's own fluxional calculus, for which he was not to devise algebraic notation, with dotted x's and y's, until several years later. But three years *before* the *Principia* came out, another mathematician than Newton presented the calculus to the world by way of a range of algorithms, however opaquely rendered in the first instance. For Newton's explicitly 'motional' treatment of curves was not the only possible road toward the calculus. With inspiration drawn from hints from Newton, but with Pascal's work in geometry and arithmetic as his prime point of departure, Gottfried Wilhelm Leibniz reinvented in essential independence the calculus in a more algebraic, by and large motion-free context between 1672 and 1684. To Newton the apparent link between his own calculus and the problem of how to handle nonuniform motion mathematically had become ever more important. By contrast, Leibniz' primary concern was and remained with the calculus as a versatile, powerful new technique in its own right.

Along came Pierre Varignon in the 1690s to derive procedures and results of Newton's *Principia* by means of the algebraic symbols of Leibniz' calculus. As a result, these procedures and results became both easier to compute and more generally applicable. More than with Newton they also acquired the look of the self-evident. Varignon made short shrift of Galileo's failure, in his proof of the mean speed theorem, to sum without more ado his 'degrees of speed' into the triangle's surface as a straightforward representation of distance traversed. To a man steeped in the infinitesimal ways of the calculus, Galileo's scruples in trying to appear to adhere to the niceties and purist requirements of Euclid's doctrine of ratios had lost their point.

So much for the past; even more important were the vistas into the future that Varignon's work helped open up. He derived a precise expression for 'instantaneous velocity' as the first derivative of distance with respect to time ('$v = dx/dt$' in his notation). He further used Newton's advanced results about the acceleration produced by central forces for deriving a similar expression for 'force of instantaneous acceleration', as the second derivative of distance with respect to time ('$y = ddx/dt^2 = dv/dt$'). But this meant that the treatment of motion need no longer remain confined to acceleration at a uniform rate, as it had been in Huygens' *Horologium oscillatorium* no less than in Galileo's *Discorsi*. With the concepts of instantaneous velocity and instantaneous acceleration thus undergoing increasingly clear-cut definition, the mathematical science of motion ceased to remain confined to the treatment of the uniform and the just uniformly variable. As proper techniques in differentiation and integration began to be worked out, any rate of change whatever became subject to mathematical handling, in Newton's *Principia* for the first time and in Varignon's successive memoirs in more lucid and accessible fashion. Here rested the most significant achievement of the calculus to further the investigation of motion in the realist-mathematical manner inaugurated by Kepler and by Galileo.

Power and pull of realist-mathematical science

Before Kepler and Galileo, mathematical science had been a fundamentally backward-directed business, with efforts aimed at restoring the ancients' achievement to its original state of perfection. In the wake of these two pioneers, mathematical scientists operating in newly realist fashion turned from recovery interspersed with an occasional step forward toward a deliberate voyage of discovery, punctuated to be sure by the occasional glance backward. What did their efforts yield? Galileo had a prophetic sense of the power that he claimed resided in his newfangled manner of coming to grips with the realities of nature (p.

291). What exactly did the power that Galileo envisioned appear to consist of within the first half century of its being probed? Over the period, realist-mathematical science expanded in both depth and scope. What made for such rapid expansion?

For answers we compare Alexandria-plus with Alexandria in three distinct respects: what approaches were taken, what subjects were tackled, and what concepts were developed.

The approach customary before 1600 was one of abstract mathematization. Its one and only vehicle was the process of abstraction itself, carried out by means of Euclidean geometry and arithmetic. By the end of the 17th century another approach, of abstract-*realist* mathematization, had become settled practice. Its principal vehicles, all new, were validation-oriented experimentation, mathematical modeling, selective adoption, analogies of motion, and infinitesimal procedures.

Natural phenomena to give occasion to the customary, abstract mode of mathematical treatment were by 1600 still confined to solid and fluid equilibrium situations, musical consonances, light rays, and planetary trajectories. The sole products of art to which known mathematical rules were as yet being regularly applied were pictures in linear perspective, fortresses, measurement of geographical latitude, and maps. By the time of Newton's *Principia*, the full panoply of natural phenomena or products of art either mathematized for the first time or refashioned in the new, realist-mathematical manner had come to include not only planetary trajectories and falling and projected bodies (with which it all started) but also solid and fluid equilibrium situations (with air newly included), musical consonance *and* temperament, light rays, geographical latitude *and* longitude, maps, telescopes, pendula, clocks, pressing hammers, streaming and outflowing water, guns, materials at breaking point, and the mode of operation of machine tools generally.

Concepts already extant yet newly involved in this ongoing process of realist mathematization, or developed in the course of it, included virtual displacement, graphic representation, infinite magnitudes, and centrifugal motion, but also a variety of forces, such as impetus and its mode of transfer, the force of impact, and forces believed to keep the solar system together.

These lists are still far from exhausting all that happened at the time in mathematical science. In particular, the motion- and light-related products of mathematical-*corpuscularian* theorizing, as well as the revolution Newton brought about therein, still await treatment (chs. 15; 19). Even so, both the pace and the scope of advance attained by mathematical science over the 17th century were quite without precedent in the history of the pursuit of nature-knowledge. *What drove this relentless advance?*

In view of the preceding survey, four distinct driving forces jointly account for such unprecedented scope and pace. One is the versatility of Galileo's mind; another, the close intertwinement of subjects with one another; another, the problem/solution/problem chase; finally, the ongoing search for a balance between mathematical theory and experimental

362   outcome. Nor did these driving forces fail to reinforce one another. I shall consider them first in turn, then in their mutual and joint impact.

Conceivably, Galileo might have confined himself to just a first step in realist-mathematical science, with the rules he established for falling bodies and projectile motion serving as its sole exemplar. In that case, he would have left its possible expansion in both scope and depth to later generations. For all his awareness that later generations were to see farther than he did, it is still astounding how many novel approaches, subjects, and concepts taken up by those who followed in his footsteps were already explored by the pioneer himself. The lists reconsidered with a focus on Galileo alone show that newly mathematical experimentation and modeling, analogies of motion, and infinitesimal procedures were all tried out by him, and that solid and fluid equilibrium situations, musical consonance, geographic longitude, telescopes, pendula, clocks, pressing hammers, streaming water, guns, materials at breaking point, and the general operation of machine tools made a more than fleeting appearance in his written works. By the same token, among the concepts handled or treated by him in newly realist mathematical fashion were graphic representation, centrifugal motion, infinite magnitudes, and the force of impact. How could such wide probing fail to serve as a source of inspiration and fitting point of departure for younger men sensitive to the attractions of mathematics? To be sure, planetary trajectories and the tracing of light rays (the domains of Kepler's major breakthroughs) remained outside Galileo's province. It is also true that between Torricelli and Newton some new terrain wholly untrodden by Galileo was to be explored as well. Still, much of the width of the front over which realist-mathematical science marched during the remainder of the 17th century had already been covered right at its inception by this extraordinary man alone.

Not so personal driving forces were also involved. Close intertwinement is one of them. Geographic longitude provides as telling an example as any. During the 17th century the problem turned out to have close links not only (obviously) with latitude but also with the telescope, with the measurement of time, and, from there, with the pendulum and with the cycloid and the general unrolling of curves by means of infinitesimal procedures. Indeed, very few subjects of realist-mathematical science stood apart from one another the way the five Alexandrian subjects had. Consequently, discovery in one field as a rule carried implications for another, and this, too, helps account for the rapid pace as well as the wide scope this revolutionary mode of pursuit of nature-knowledge acquired almost at once.

So did veritable problem/solution/problem chases. The special way that solutions to some problem in realist-mathematical science gave rise to a further and/or deeper problem formed another significant driving force, this one contributing to quickness of pace more than scope. The three intense months that Huygens spent moving from his original problem of how to improve the determination of the constant of fall to his discovery of the evolution of curves is an exemplary case of the problem ⇨ solution ⇨ next problem ⇨ next solution dialectic here unfolding. Such sequences might mark the creative outbursts of just one individual, as here with Huygens. Or one mathematical scientist might move on from the point where a predecessor or a colleague had left off, as with Torricelli, then Huygens, then Varignon on the problem of water outflow. In either case there appeared to be open ends, obscurities demanding clarification, questions left dangling or newly arising, and (as the process became more common) the confident expectation of unknown territory lying just around the corner. The philosopher of science to mark such open-ended, ongoing sequences as a regular trait of the scientific enterprise was Kant; historians of science have noted their inception with Galileo and with those who followed the path first explored by him.

A final driving force resided in the search for a balance between mathematical theory and experimental outcome. Recall from page 199 Galileo's ambiguity over the role of experimental tests (which he himself pioneered) in the mathematical treatment of natural phenomena. On one occasion he held that "the knowledge of one single effect acquired through its causes opens the mind to the understanding and certainty of other effects without need of recourse to experiments"; on another, that "we must find and demonstrate conclusions abstracted from the impediments, in order to make use of them in practice under those limitations that experience will teach us". No doubt he meant both pronouncements sincerely when he made them. In his actual practice, his root conviction of the ultimately mathematical structure of reality drew him much closer to the former position of experiments being really superfluous for purposes other than persuasion than to the latter, humbler stance of readiness to accept experimental outcomes as guides toward how far one may actually go in abstracting from empirical impediments standing in the way of the mathematical-ideal phenomenon. But more important even than how he himself chose to proceed in the face of this central tension in the working life of the mathematical scientist was the clarity with which he laid it out right from the start. Experimental outcomes may lead one astray insofar as they may represent no more than, indeed, irrelevant 'impediments' that do no more than obscure some underlying mathematical pattern. Mathematical deduction may lead one astray in view of all that is unpredictably messy about the world of natural phenomena. There are no hard and fast rules here but only room for confidence that interaction of some kind between mathematical abstraction and its experimental testing may help one move forward, with the balance between them to be struck anew in every next case at hand. Recall Kepler's instant readiness to let go of years of drudgery leading to an at-first-sight splendid, new theory when just one countercase from the domain of empirical fact presented itself. But mark also the words of a highly successful, present-day practitioner, Steven Weinberg:

I did some of my own best work [leading to the unification of the 'weak' and the electromagnetic forces] because I had certain preconceptions about the way forces ought to work, and ignored experimental evidence to the contrary, and I did not succeed in taking the next step in this work because I was prejudiced against certain mathematical methods.[193]

364   He went on to say, with considerable understatement, "It's not an atypical story". Indeed, more or less intuition-guided, prejudice-laden processes of interaction and balance seeking between mathematics and experiment have been going on for four centuries now. The *onset* of exploration of this perennial field of tension is precisely what we have been watching in the present chapter, with ongoing efforts to correct the perceived one-sidedness of predecessors providing one more, powerful force speeding up the advance of mathematical science over the 17th century.

Take Torricelli's worries over the apparently arbitrary conditions he felt he needed to introduce to make the disappointing outcomes of an experiment match the principle he had used for deriving his rule for the speed of water outflow (p. 340). A quarter of a century later, Huygens and several colleagues at the Paris Académie, out of a felt dissatisfaction with where Torricelli had left the issue, sought to account for the latter's conditions by pointing at various possible second-order effects of an empirical nature. Even though in this manner they adjusted the experimental outcome to their own satisfaction, in the end they noted with regret that the underlying principle could still be sustained only by the outcome of an *experiment* rather than by "reason". Another quarter of a century later, Varignon, in full possession of the calculus, managed to strip the issue of almost all remaining empirical content. Replacing Torricelli's underlying and, in Varignon's reasoned view, ill-founded principle by one abstract prior assumption, Varignon could now derive the rule for the speed of water outflow by none but strictly mathematical means. In other cases (with craft experience, in particular) we have instead seen *experiment* gain the upper hand. The essential point about the driving force here addressed does not, however, rest so much in whether mathematical theory or experimental outcome came out victorious in any single case. It rather rests in the interplay itself. That interplay was to become even more dynamic once, over the 18th century, mathematical science developed to the point of engaging, in a more even-handed way than possible before, such empirical phenomena as craft practice kept providing.

Fancy reined in by systematic feedback.   Of the four agents thus driving realist-mathematical science forward, then, only one (Galileo's versatility) was bound up exclusively with a contingent and, as such, transient fact of history. In contrast, the three others (narrow subject intertwinement; problem/solution chase; mathematics/experiment interplay) had come to stay. It is here that we must seek for the enduring power of this revolutionary mode of nature-knowledge. Certainly, there were limits set to that power. None of the results here surveyed proved as yet capable of transforming craft practice; some proved partly or wholly mistaken. Even so, many more results have proven of lasting value, even if adapted later to altered conceptual contexts. But over and beyond even the most significant results attained in the wake of Galileo stand those three enduring drives toward wide scope and/or quick pace. With the last-mentioned driving force, in particular, that is, ongoing interplay

between mathematical theory and experimental practice, a quite novel, truly crucial element   365
had now been introduced into the pursuit of nature-knowledge: *the opportunity to check one's conceptions against natural reality, not incidentally but in a way built into one's general procedures.*

Mathematical science in its Alexandrian guise had not required much of a check against reality because it scarcely addressed reality (but used it only as a starting point). Empiricist modes of nature-knowledge (even if embedded in some broad background worldview) had remained confined as a rule to accurate observation, with possible empirical correction confined to the degree of accuracy. Natural philosophers alone had ventured to address empirical reality in ways transcending that reality, yet outcomes, for all the indubitable certainty claimed for these in view of the first principles from which they flowed, had always remained plausible at best. That is, one could always make a persuasive-sounding case for an assertion and then cling to it regardless. Means to rein in the perennial temptation of the human mind and its wonderful powers of imaginative discovery to move unnoticed from the imaginative to the fanciful did not present themselves. Craft practice as a rule offers feedback of a very basic kind – a bridge holds, or it comes tumbling down; a mixture stop sounds brilliant, or it jangles. But the pursuit of nature-knowledge does not spontaneously provide nature with the opportunity to demand correction in a similarly striking way. Systematic experimentation in the framework of some theoretical structure, however, comes close, and Galilean/Keplerian mathematical science is where this was discovered and worked out first. It appeared soon enough that the constraints set by nature always leave room for decisions on how in everyday practice to act upon the opportunity for correction which feedback may thus provide. One may (dogmatically) decide not to seek correction, or (sometimes foolishly, sometimes prudently) not for the time being to heed it, or (in a skeptical-positivist vein) to settle for none but the observable in despair of any human ability to go beyond what our senses tell us. Such ways out are always open to us. Here, too, there are no hard and fast rules; 'falsification' is not, and never can be, a self-regulating process left untouched by the endlessly varied ways of the human mind. But the point is that, with the onset of realist-mathematical science, nature-knowledge did acquire quickly expanding features of *falsifiability* and has kept refining those features ever since.

The pull of realist-mathematical science.   So much for the question of what it was that, right from its inception, drove realist-mathematical science so relentlessly forward and what its apparent power consisted of. But in addition to the discovery of properties and the expansion into domains as yet unheard of, realist-mathematical science also turned out to possess formidable powers of absorption. Long-existing, nonmathematical concepts and approaches were drawn into its sphere. Practitioners engaged in other kinds of activities turned to realist-mathematical science either wholesale or to significant degrees.

366   The pull of realist-mathematical science made itself felt most obviously in the certainly not smooth or easy, yet once again relentless process of absorption of the classic subjects of abstract, Alexandrian mathematical science into abstract-realist, Keplerian/Galilean mathematical science. By the end of the 17th century, equilibrium states of solids and fluids, light rays, consonant intervals, and planetary trajectories had all found their place in the new frame. Also, scholastic concepts like graphic representation or impetus were put to productive use as soon as they were touched by realist-mathematical science. In due time, one area of human concern after another found itself subjected to the special pull that processes of mathematization are in the habit of exerting – the issue is taken up in the Epilogue (p. 724).

There was a personal counterpart to these developments. How did the apparent attractions of realist-mathematical science affect the practitioners of the time?

Mathematicians first. It is telling how quickly the pool of mathematical scientists who kept operating in trusted Alexandrian fashion dried up. The generation roughly contemporaneous with Kepler and Galileo still counted quite a few, like Stevin, Snel, Harriot, and Descartes. One generation later they had already become the exception. Take the case of two mathematical scientists in the first post-Galileo generation, Torricelli and Huygens. Enchanted with mathematics as youngsters, the former set out to emulate and improve certain findings of Apollonios, the latter of Archimedes. As soon as they made their acquaintance with Galileo's work, both men switched to his realist mode of doing mathematical science and soon turned themselves into its most gifted practitioners. Most others followed suit. Not counting those numerous mathematical practitioners who all over Europe were engaged in just routine business, by midcentury only a few men, like Barrow and Bartholinus, still engaged in mathematical science the Alexandrian way, as also those mathematical astronomers who clung to a posture of 'saving the phenomena'. One generation later again, realist-mathematical science had turned into the sole mode of mathematical science still being cultivated – Alexandria had become a thing of the past.

So much for how the new realist-mathematical science affected the mathematical domain itself. But there is a circumstance that speaks even louder to the power of absorption exerted by realist-mathematical science so quickly after coming into the world. Several men engaged in *other* modes of pursuit of nature-knowledge, notably in natural philosophy, also began to be drawn into its sphere of attraction, albeit never fully so. Mersenne's case is telling. Steeped in the Aristotelian tradition of mixed mathematics, he never quite surrendered to Galileo's mode of approach. Still, the *Dialogo* and the *Discorsi* made a profound impression upon him, to the point that he quickly rendered the latter book into a shortened text in French under the title *Les nouvelles pensées de Galilée* (Galileo's New Thoughts, 1639). In his investigations into 'universal harmony' as also in his efforts to supplement Galileo's rules for vertical descent with the quantitative determination of the constant of acceleration (p. 343; 409), Mersenne arrived at a peculiar blend of plodding experimentalism and quantitative expression, reinforced at times by insights gained from his ongoing correspondence with two innovative natural philosophers, Beeckman and Descartes. Something similar is true, albeit in not quite so idiosyncratic a manner, of certain mathematics-oriented Jesuits. Reverend fathers like Riccioli, Le Cazre, and Wendelin came to treat many a Galilean issue in the manner of quantitatively handled Aristotelian commonsense reality. In so doing they built bridges of sorts with experimentation of the fact-finding variety (p. 509).

True, the pioneers of the kinetic-corpuscularian philosophy of nature, Beeckman and Descartes, deliberately remained outside, although even they adopted certain attainments of both Kepler and Galileo, going on, however, to rewrite these in their own idiom. But the third pioneer, Gassendi, went one step further. At the outskirts of his wordy expositions of his own variety of kinetic corpuscularianism there are incidental efforts to come to quantitative grips with Galileo's rules for free fall and to check by means of as accurate telescopic observation as the clouds of the day allowed him Kepler's prediction of a transit of Mercury for 7 November 1631 (p. 303; 452). Mixed forms like these testify to the tenacity of the natural-philosophical way of dealing with natural phenomena to be sure, yet more surprisingly to the power of attraction exerted by the new realist-mathematical science even upon those whose customary approach was so squarely at odds with it.

Not, to be sure, that the apparently remorseless advance of realist-mathematical science has ever been uncontroversially straightforward. From its very inception its practitioners were embroiled in heated controversies over both actual outcomes and basic points of departure. These come up for treatment in chapter 12, 'Legitimacy in the Balance' (p. 403). Here I mention those revealing clashes only to make one final point. Neither its manifest power of absorption alone nor the resistance it has never ceased to engender alone, but a blend of pull-*cum*-resistance has marked the vicissitudes of realist-mathematical science on its march through the world ever since it made its disputed entrance there.

NOTES ON LITERATURE USED

Surveys. Two comprehensive books dedicated to the 17th-century treatment of motion are Richard S. Westfall, *Force in Newton's Physics. The Science of Dynamics in the Seventeenth Century*. London: Macdonald, 1971, and Domenico Bertoloni Meli, *Thinking with Objects. The Transformation of Mechanics in the Seventeenth Century*. Baltimore: Johns Hopkins UP, 2006.

Absorption of the classic Alexandrian subjects. The very idea of the various Greek mathematical sciences being radically transformed early in the 17th century is due to Thomas S. Kuhn's wonderfully inspiring 'Mathematical versus Experimental Traditions in the Development of Physical Science'. In: idem, *The Essential Tension*. Chicago: University of Chicago Press, 1977; pp. 31-65. But for Westfall's and Bertoloni Meli's books mentioned above, the process of transformation itself has not so far been taken up in the literature. Here is a subject-by-subject listing:

- **Solid equilibrium.** I have gleaned bits and pieces from several works by E.J. Dijksterhuis: *Simon Stevin*. The Hague: Nijhoff, 1943 (with a posthumous, abridged version in English edited by R. Hooykaas and M.G.J. Minnaert, *Simon Stevin. Science in the Netherlands Around 1600*. The Hague: Nijhoff, 1970); *Archimedes*. Kopenhagen: Munksgaard, 1956; *The Mechanization of the World Picture*. Oxford UP, 1961 (originally *De mechanisering van het wereldbeeld*. Amsterdam: Meulenhoff, 1950), and a book cowritten with R.J. Forbes, *A History of Science and Technology* (2 vols.). Harmondsworth: Penguin, 1963 (originally *Overwinning door gehoorzaamheid. Geschiedenis van natuurwetenschap en techniek* (2 vols.). Zeist: Phoenix, 1961); further from Stillman Drake's introduction to a book he edited together with I.E. Drabkin as co-translator, *Mechanics in Sixteenth-Century Italy. Selections from Tartaglia, Benedetti, Guido Ubaldo, & Galileo*. Madison: University of Wisconsin Press, 1969.
- **Fluid equilibrium.** Based on pertinent passages in the works by Dijksterhuis just cited. Pascal's carefully analogical reasoning is treated in R. Hooykaas, *Fact, Faith and Fiction in the Development of Science*. Dordrecht: Kluwer, 1999; p. 346; p. 372.
- **Consonant intervals.** Drawn from my *Quantifying Music. The Science of Music at the First Stage of the Scientific Revolution, 1585-1650*. Dordrecht: Reidel, 1984. I have treated the downfall of Zarlino's grand synthesis in 'Benedetti's Views on Musical Science and Their Background in Contemporary Venetian Culture'. In: *Cultura, Scienze e Techniche nella Venezia del Cinquecento. Atti del Convegno Internazionale di Studio 'Giovan Battista Benedetti e il suo tempo'*. Venezia: Istituto Veneto di Scienze, Lettere ed Arti, 1987, pp. 301-310.
- **Light rays.** Most enlightening for my purposes proved to be D.C. Lindberg, *Theories of Vision from al-Kindi to Kepler*. Chicago: University of Chicago Press, 1976, and two works by Fokko Jan [another than E.J.] Dijksterhuis: his book *Lenses and Waves. Christiaan Huygens and the Mathematical Science of Optics in the Seventeenth Century*. Dordrecht: Kluwer, 2004, and his article 'Once Snell Breaks Down. From Geometrical to Physical Optics in the Seventeenth Century'. *Annals of Science* 61, 2004, pp. 165-185. Problems connected with shadow, half-shadow, and the *camera obscura* are

treated in Lindberg's 'Laying the Foundations of Geometrical Optics: Maurolico, Kepler, and the   369
Medieval Tradition'. In: D.C. Lindberg and G. Cantor, *The Discourse of Light from the Middle Ages to
the Enlightenment*. Los Angeles: Clark Memorial Library, 1985; pp. 1-65. For the rainbow I have used
Carl B. Boyer, *The Rainbow. From Myth to Mathematics*. New York: Sagamore Press, 1959 (reprinted
in 1987 by Princeton University Press).

•   **Planetary trajectories**. The authoritative work to consult on the whole of 17th-century astronomy
    is René Taton and Curtis Wilson (eds.), *Planetary Astronomy From the Renaissance to the Rise of
    Astrophysics. Tycho Brahe to Newton* (vol. 2, part A of: M. Hoskin (ed.), *The General History of As-
    tronomy*). Cambridge UP, 1989. Further, Arthur Koestler's *The Sleepwalkers*. London: Hutchinson,
    1959, has an instructive, brief section 'Reception of *Astronomia nova*' (pp. 351-353 in the Penguin
    paperback edition), on which subject Max Caspar's biography *Johannes Kepler*. Stuttgart: Kohl-
    hammer, 1948 (several reprints; also an English edition with reprints) is also informative. At least
    for David Fabricius as Kepler's earliest critic these accounts have been superseded by James R.
    Voelkel, *The Composition of Kepler's Astronomia nova*. Princeton UP, 2001. A helpful, short account
    of the later fate of Kepler's work is W. Applebaum, lemma 'Keplerianism' in idem (ed.), *Encyclope-
    dia of the Scientific Revolution from Copernicus to Newton*. New York/London: Garland, 2000; pp.
    346-348.

Scholastic concepts mathematized. Both *IMPETUS* and *LATITUDES OF FORMS* in the 17th century have
been treated mostly in a continuist vein not here adopted. I have supplemented what (for my purposes) useful hints remain in the literature with what R.S. Westfall, in his *Force in Newton's Physics.
The Science of Dynamics in the Seventeenth Century*. New York/London: Elsevier/MacDonald, 1971,
pp. 125-138, and J.C. Boudri, in ch. 2 'Force as Water' of his *What Was Mechanical About Mechanics.
The Concept of Force between Metaphysics and Mechanics from Newton to Lagrange*. Dordrecht: Kluwer,
2002, have found out about (respectively) the 'momentum' and 'substance' guises impetus took in a
frame of realist-mathematical science.

Craft techniques mathematized. In SHRI 5.2.1.–6 I went over the still influential, ideologically charged
controversies that raged between c. 1930 and c. 1960 regarding the general issue of science and the
crafts in the 17th century, while in 3.4.3 I listed fruitful insights sprinkled through some more-recent
literature on the topic. I have imposed my fivefold categorization as presented in the main text upon
data found in the following articles and books, in particular: R.S. Westfall, 'The Background to the
Mathematization of Nature'. In: J.Z. Buchwald & I.B. Cohen (eds.), *Isaac Newton's Natural Philosophy*.
Cambridge (MA): MIT Press, 2001 (pp. 321-339); D.S.L. Cardwell, *Turning Points in Western Technology.
A Study of Technology, Science and History*. New York: Neale Watson, 1972; idem, *The Fontana History
of Technology*. London: Fontana Press, 1994; E.J. Dijksterhuis, *Simon Stevin* (esp. ch. 9, corresponding
to ch. 7 in the abridged version in English); Klaas van Berkel, *Isaac Beeckman (1588-1637) en de mecha-
nisering van het wereldbeeld*. Amsterdam: Rodopi, 1983 (soon to appear in an English version); Kirsti
Andersen, 'Stevin's Theory of Perspective: The Origin of a Dutch Academic Approach to Perspective'.
Tractrix 2, 1990; pp. 25-62; Cesare Maffioli, *Out of Galileo. The Science of Waters 1628–1718*. Rotterdam:

370 Erasmus Publishing, 1994; L. Bagrow, *History of Cartography*. London: Watts, 1964; James A. Bennett, *The Divided Circle. A History of Instruments for Astronomy, Navigation and Surveying*. Oxford: Phaidon/Christie's, 1987; Michel Blay, 'Le développement de la balistique et la pratique du jet des bombes en France à la mort de Colbert'. In: L. Godard de Donville (ed.), *De la mort de Colbert à la révocation de l'édit de Nantes: un monde nouveau?* Centre Méridional de Rencontres sur le XVIIe siècle, 1984; pp. 33-51, as well as on pertinent lemmata in Ivor Grattan-Guinness (ed.), *Companion Encyclopedia of the History and Philosophy of the Mathematical Sciences*. London: Routledge, 1994 (2 vols.), and in Wilbur Applebaum (ed.), *Encyclopedia of the Scientific Revolution from Copernicus to Newton*. New York/London: Garland, 2000.

Mathematical instruments. In *SRHI* 3.4.2. I have discussed Daumas' and Van Helden's views on the general subject of scientific instruments. For all its one-sidedness, Alexandre Koyré's article 'Du monde de l'"à-peu-près" à l'univers de la précision'; in: idem, *Études d'histoire de la pensée philosophique*. Paris: Armand Colin, 1961 (on pp. 341-361 of the reprint published in 1971 by Editions Gallimard) remains a source of inspiration.

- **Telescope**. Albert Van Helden's lemmata 'telescope' and 'telescopic astronomy' in Applebaum's *Encyclopedia* mentioned above, and chapters 2 and 3 of Fokko Jan Dijksterhuis' *Lenses and Waves* likewise mentioned above.

- **Pendulum clock**. In the vast literature (well summarized in Denys Vaughan's lemma 'horology' in the same *Encyclopedia*), David S. Landes, *Revolution in Time. Clocks and the Making of the Modern World*. Cambridge (Mass.): Harvard UP, 1983, and Joella G. Yoder, *Unrolling Time. Christiaan Huygens and the Mathematization of Nature*. Cambridge UP, 1988, stand out. Also enlightening is Jan Hendrik Leopold, 'Christiaan Huygens and His Instrument Makers'. In: Henk J.M. Bos et al. (eds.), *Studies on Christiaan Huygens*. Lisse: Swets & Zeitlinger, 1980, pp. 221-233.

Analogies of motion. Cognitive scientists give much attention to (among other heuristic devices) analogical reasoning in modern science, with 'cognitive historians of science' mining the past for extensive, historically responsible case studies like Nancy J. Nersessian's pioneering *Faraday to Einstein. Constructing Meaning in Scientific Theories*. Dordrecht: Kluwer, 1984, or her paper 'Kuhn, Conceptual Change, and Cognitive Science'. In: T. Nickles (ed.), *Thomas Kuhn*. Cambridge UP, 2002, pp. 178-211, where the entire approach is explained. Meanwhile, the principal novelty of analogies of motion seems still to await acknowledgment in the historiography of the Scientific Revolution. In broaching the theme, for Galileo on IMPACT and on PENDULA I have taken data and specific analyses chiefly from section 2.3.1. of Boudri's *What Was Mechanical About Mechanics* and from ch. 1 of Westfall's *Force in Newton's Physics*; for Galileo on CONSONANCE from my own *Quantifying Music*, p. 91; for WATER OUTFLOW from Michel Blay, *Les raisons de l'infini. Du monde clos à l'univers mathématique*. Paris: Gallimard, 1993, pp. 114-123 (translated as *Reasoning With the Infinite. From the Closed World to the Universe of Mathematics*. University of Chicago Press, 1998), and idem, *La naissance de la mécanique analytique. La science du mouvement au tournant des XVIIe et XVIIIe siècles*. Paris: Presses universitaires de France, 1992, pp. 332-371, and for CURVES UNROLLING from Joella G. Yoder's *Unrolling Time*.

From Euclidean ratios to the calculus. The comparative element in this section is due to an effort at nonpresentist reading of chapters 2–5 of Carl B. Boyer, *The History of the Calculus and Its Conceptual Development*. New York: Dover, 1959 (originally published in 1939 as *The Concepts of the Calculus. A Critical and Historical Discussion of the Derivative and the Integral*). For the European development I owe much to what I have picked up from both Henk J.M. Bos' patient oral instruction and a range of his papers: 'Tradition and Modernity in Early Modern Mathematics: Viète, Descartes and Fermat'. In: C. Goldstein et al. (eds.), *L'Europe mathématique*. Paris: Maison des sciences de l'homme, 1996; pp. 183-204; 'On the Interpretation of Exactness'. In: J. Czermak (ed.), *Philosophy of Mathematics*, vol. 1. Vienna: Hölder, 1993; pp. 23-44; 'The Structure of Descartes' Géométrie'. In: G. Belgioioso et al. (eds.), *Descartes: il Metodo e i Saggi*. Rome: Istituto della Enciclopedia Italiana, 1990; pp. 349-369; 'Descartes en het begin van de analytische meetkunde'. In: H.J.M. Bos et al., *Vacantiecursus 1989. Wiskunde in de Gouden Eeuw*. Amsterdam: Mathematisch Centrum, 1989; pp. 79-97 (with much in these papers being preparatory to Bos' later *Redefining Geometrical Exactness: Descartes' Transformation of the Early Modern Concept of Construction*. New York: Springer, 2001), and 'Introduction' to R.J. Blackwell (trans.), *Christiaan Huygens' The Pendulum Clock*. Ames: Iowa State University Press, 1986; pp. xi-xxix. I have further used with profit: Kirsti Andersen, 'Precalculus, 1635-1665'. In: Ivor Grattan-Guinness (ed.), *Companion Encyclopedia of the History and Philosophy of the Mathematical Sciences*, vol. 1. London: Routledge, 1994; p. 292-307; chs. 2 and 4 of Peter M. Engelfriet, *Euclid in China. The Genesis of the First Translation of Euclid's Elements in 1607 and its Reception up to 1723*. Leiden: Brill, 1998 (with many thanks to the author for his incisive comments on an early draft of the present section); section 2.2. 'Quantity and quality' of J. Christiaan Boudri's *What Was Mechanical About Mechanics*; Joella G. Yoder, *Unrolling Time*; four books by Michel Blay: *La naissance de la mécanique analytique*; *Les raisons de l'infini*. Paris: Gallimard, 1993; *Les 'Principia' de Newton*. Presses universitaires de France, 1995; *La naissance de la science classique au XVIIe siècle*. Paris: Nathan Université, 1999; finally, Richard S. Westfall, *Never at Rest. A Biography of Isaac Newton*. Cambridge: Cambridge UP, 1980, Derek T. Whiteside, 'Sources and Strengths of Newton's Early Mathematical Thought'. *The Texas Quarterly* 1967, 10, 3, pp. 69-85, and Niccolò Guicciardini, *Isaac Newton on Mathematical Certainty and Method*. Cambridge (Mass.): MIT Press, 2009.

Power and pull of realist-mathematical science. Salient ideas in this section have been inspired by Karl R. Popper's broad idea of falsifiability; further by ideas of Steven Weinberg (notably his *Dreams of a Final Theory. The Scientist's Search for the Ultimate Laws of Nature*. New York: Vintage, 1992, and *Facing Up. Science and Its Cultural Adversaries*. Cambridge: Harvard UP, 2001), and of Rob Wentholt, in many writings and conversations on the indispensability of feedback for human beings generally and on what makes it so hard to find it and so easy to get around it. The reference to Kant (which I owe to section 1.3., 'Kant's Principle of Question Propagation', of Nicholas Rescher, *The Limits of Science*. Pittsburgh UP, 1999 (2nd ed.)) is to section 57 of his *Prolegomena*.

XI

ACHIEVEMENTS AND LIMITATIONS OF KINETIC CORPUSCULARIANISM

What, from the perspective of René Descartes on completing his *Principia philosophiae* and seeing the treatise through the press in 1644, was still left to be done in nature-knowledge?

In terms of *content*, very little, really. The author of 'Le monde' and *Principia philosophiae* shared to the full the dream of earlier founders of a school of philosophy, at one stroke to produce the definitive account of the world by positing a reasoned set of first principles and showing how the world's phenomena could truly be made sense of in their light alone. Descartes was surely willing to allow that he had not covered all phenomena but only the most important ones, so that numerous leftover details had still to be filled in. He was also willing to allow that certain subordinate phenomena situated near the bottom of his top-down chains of orderly, clear, and distinct reasoning might in consistency with the system as a whole turn out to be *either* this *or* that. For such second-order dilemmas empirical inquiry alone enabled one to determine how God had actually resolved them. Still, as he argued at the very spot in his *Discours de la méthode* where he conceded these points,

> making my mind pass once again over all objects which at any time have presented themselves to my senses, I venture to say that I have never come across any thing which I could not explain adequately enough by means of the Principles I had found.[194]

So the task he left to posterity, while surely colossal in scope, concerned only the fixing of details in a system that Descartes had already worked out.

In terms of *adoption by others*, however, a great deal still had to be done. True, Descartes himself had taken great pains to promote acceptance. He had tried out a variety of stratagems of persuasion in presenting his views in a variety of written formats and also by actively, albeit as yet none-too-successfully, seeking converts in social positions well placed to spread the word widely. Among them were Elisabeth, eldest daughter of the deposed queen of Bohemia, and Christina, not-yet-abdicated queen of Sweden. Other hoped-for converts included the Society of Jesus, several philosophy professors at the universities of Utrecht and Leiden, and his friend Mersenne at the center of a vast network of correspondents. As Descartes saw it, the task of gaining adherents was both smooth and arduous. Smooth, in that hearing is believing – "the foundations of my Natural Philosophy ... are, almost all of them, so evident that one need only hear them to believe them." Arduous, in that no one, however bright, seemed capable of getting it just right:

374   although I have often explained some opinions of mine to people with a very good mind and who, while I was talking to them, seemed to understand them quite distinctly, I have noted nonetheless that, when they reproduced them, almost always they had changed them in such a way that I could no longer avow them as mine.[195]

Our job, then, is to find out to what extent Descartes proved to be right in his dual expectation that kinetic corpuscularianism, in the way he had given it orderly shape and solid foundation, was indeed *the* definitive and by and large complete conception of nature, and that, as such, it was suitable for adoption by all. In the present chapter I shall tackle these two questions only insofar as they pertain to kinetic corpuscularianism as a speculative natural philosophy aimed at a grasp of the whole. I shall leave for treatment in chapters 15 and 16 the issue – of outstanding historical significance in the longer term – of what kinetic corpuscularianism accomplished once it was turned by some highly perceptive Galileans and some highly perceptive Baconians from speculative dogma into a ready-made source of hypotheses. And I leave for treatment in chapter 18 the question of what happened to those philosophies of nature (Aristotelianism and its four classic rivals) that the doctrine of particles in motion was meant to replace. In the present chapter I discuss in successive sections

- reasons for the speedy as well as widespread *adoption* the doctrine attained indeed in the second half of the 17th century;
- a variety of *modifications* that kinetic corpuscularianism underwent as a philosophy of nature; and
- the one lasting achievement of the doctrine as such, to wit, its fostering the acceptance of *heliocentrism*.

By way of a conclusion, to be drawn in both parallel and contrast to the previous chapter, I shall consider the kind of power and the kind of pull exerted by the natural philosophy of kinetic corpuscularianism.

As a reminder of the kind of thing the doctrine offered and also by way of an introduction to these three main topics, I shall now first address the partly similar, partly different ways in which two of the founders, Beeckman and Descartes, dealt with a range of phenomena pertaining to musical sound. The aim is to give the reader a feel for how the natural philosophy of kinetic corpuscularianism worked in everyday explanatory practice. That way we enhance our grip on a whole range of issues pertinent to the development of the kinetic-corpuscular conception of the world, notably, its explanatory range and power; what constraints were imposed in practice upon its seemingly boundless plasticity; its concern with the relation between body and mind; its handling of the phenomenal world as just given; and finally, the sort of reasoning and the sort of empirical evidence that counted as arguments pro or con in any given case.

Musical sound as moving corpuscles: An explanatory sample

Beeckman followed the ancient atomists in taking sound to be made up of a succession of particles emitted by the sounding body. On how this comes to pass he was far more specific than they had been. He assumed that the vibrating string cuts the ambient air into tiny globules that are set in motion by the force of the vibrating string. As for pipes, be they wind instruments or human throats, sharp edges near the hole where wind enters cut the globules, and the resulting increase in inside air pressure ejects them. He further associated loudness with the amount of globules or, alternatively, with the density of their aggregate. Pitch he attributed to their speeds or alternatively to their sizes – a string twice as short as an otherwise equal one cuts off globules twice as rapidly and/or twice as many small ones. Consonance comes about in the following manner. In moving to and fro, the string at midway cuts off its maximum number of globules. At the points of return from 'to' to 'fro' and vice versa, which provide moments of 'intermediate rest', no globules are cut off at all. Hence, at each given length of the string each note is marked not only by a definite pitch but also by a sound-silence pattern of its own. In musical intervals, then, the more often such momentary 'sounds' coincide with 'sounds' and momentary 'silences' with 'silences' (as they do all the time in unison; every second time in the octave, etc.), the more consonant the interval is. From this follows a table of degrees of consonance; for example, the fifth (2:3) is more consonant than the major third (4:5). The table, however, could not be adopted without more ado but required correction in view of a range of apparent anomalies.

In strict counterpoint the fourth (4:3) is taken by musicians to be a more problematic consonance than, notably, the major third (5:4). Beeckman sought to repair this anomaly by means of the following assumption. As a given interval is produced, the 'sounds' and 'silences' can be clearly distinguished at first, but soon they languish a little and get bisected thereby. Consequently, the mental representation of the major third 'degenerates' into 5:2, whereas that of the fourth degenerates into 8:3. So the major third affects our hearing as more consonant after all.

Further, from a numerical point of view, intervals with 7, which do not even occur in the scale that musicians ordinarily employ, would take precedence over certain accepted consonances, notably over the minor sixth (5:8), which had already plagued Zarlino (p. 145). To solve the difficulty, Beeckman expanded his original account of sympathetic resonance – the phenomenon that, when a string is plucked, a similar one nearby may spontaneously sound a note as well. Resonance at the unison, known since antiquity, was a staple of Renaissance accounts in terms of 'hidden sympathies' (p. 133). Its occurrence at other intervals began to be noticed as well. Beeckman always regarded his entire philosophy of moving particles as the best way to get rid of sympathies and other occult forces. In his rival account of sympathetic resonance, vibrating strings, in cutting globules off the surrounding air, cause the

376   air to be rarefied and condensed in quick alternation, thus affecting the untouched string nearby so much as to put it in similar motion as well. He then showed through a detailed, almost wavelike analysis of their respective vibrations that the resulting sound/silence patterns are such that resonance at 5:1 undergoes much more reinforcement from the strings' motions to and fro and is therefore much stronger (as he believed to be the case) than at either 3:1 or 7:1. As one further consequence, the straightforward 1:1, 2:1, 3:2, 4:3, . . . , 8:5 table of degrees of consonance must be revised accordingly and thus brought back in line with musicians' experience.

There were further difficulties of a more principal nature. The very distinction between consonance and dissonance gets blurred by Beeckman's account in terms of sound/silence patterns. Why should the coincidence every 40th time of a 'sound' with a 'sound' and a 'silence' with a 'silence', such as marks the minor sixth (8:5), still be perceived as consonant, but no longer coincidence every 72nd time, as with the major second 9:8? His account of sympathetic resonance allowed Beeckman to come up with a solution here, too: beyond these very markedly coinciding intervals with 5 in their ratios, sound/silence coincidences become too rare to produce a consonant effect.

Further, tiny, at first unnoticeable irregularities in the production of consonant intervals may, on constant reiteration, make themselves heard to the increasing detriment of the consonant effect. At first Beeckman made the point only in theory. He soon found it to be musically relevant when an organist of his acquaintance made him aware of the phenomenon of beating, which he went on to explain along these particular lines. He concluded that beats are coincidences getting out of sync, innocuously so if occurring, say, every 50th cycle, but causing dissonance if taking place much more often.

When Mersenne informed Beeckman of his discovery of a range of harmonics, Beeckman knew at once how to give the new phenomenon a fitting place in his own conception of musical sound, answering promptly that

> a string, by its tremor dispersing the air, breaks it into nearly equal globules; however, as all parts of the string tremble equally frequently indeed, but not equally fast, and as some particles of air are perhaps more fragile than others, and as the thickness of the string is not everywhere exactly the same, it happens that a certain amount of those globules is broken into two, three, four, etc. parts. Those that are broken in two represent to the ear the *octave*, because in the same time it is affected by a twofold sting,[196]

and similarly so for the higher partials.

The 'sting' just mentioned refers to Beeckman's account of the perception of musical sound. Globules ruffle the eardrum the way a pair of sticks ruffles a military drum. As the trembling of the drumhead is transmitted through the air inside the drum to the bottom

membrane, so the ruffles on the eardrum are passed on in the same order through the air in
the middle ear (and also through the three ossicles co-vibrating with it) to the oval window.
At that point (which is as far as Beeckman's advanced yet not quite up-to-date knowledge of
the anatomy of the ear went), the auditory nerve takes over. It passes the vibrations on to the
brain by means of very fine, material 'spirits' moving either through the nerves or alongside,
the way water may stream both through a conduit and alongside it. Pitch differences are
duly reproduced by these spirits. They contract or dilate in accordance with the fineness or
coarseness of the original globules, with particularly fine globules causing the spirits of the
brain (the center of our sensation) to be stung likewise. Generally speaking, whether musical sound affects us sweetly or not depends on whether or not the globules "correspond to
the pores of the brain, or of the members, or of the collection of spirits."[197] Note that in this
account the original problem of the peculiar sweetness of the consonant intervals, as well as
their underlying, numerical ratios, has been lost sight of.

Descartes was acquainted for the most part with Beeckman's private notes on the subject. Here is how he dealt with these same issues. He did not go along with Beeckman's
globular conception, which he called "ridiculous" without more ado.[198] Rather, he regarded
sound as a succession of vibrations of the air, with their frequency determining pitch and the
amount of agitated air their loudness. The propagation of sound through the air takes place
by way of wavelets. Descartes showed how the successive rarefactions and condensations of
the air brought about by the vibrations of any object give rise to these wavelets. This enabled
him to account for consonance by means of the same 'coincidences of strokes' that Galileo
later put forward in his *Discorsi* (p. 301) and to which Beeckman's account, once shorn of its
globules, reduces as well.

Descartes dealt with the array of anomalies highlighted by the table of degrees of consonance in rather a different way than Beeckman. He offered a broadly similar correction for
the fourth, but he saw no need to make the table match the musician's experience – degrees
of ratios are fixed, but degrees of 'sweetness' depend entirely on individual taste. To the
distinction between consonance and dissonance as such, however, he did attach objective
validity. He derived it from an argument first made in 'Compendium musicae', to wit, that 2,
3, and 5 are the only consonance-generating numbers. The octave (with 2 in its ratio) comes
from the first bisection of a string; the fifth and fourth (with 3) from the next one, and the
thirds and sixths (with 5) from one more, final bisection. Further bisections do not count
any more – the obviously dissonant intervals with 7 stem from such a further bisection,
which, in producing dissonance, proves unable to generate consonances. By means of the
same argument 9:8 was dispensed with as well.

To Beeckman's account of beating there is no counterpart in Descartes' semipublic statements about musical sound. But Descartes did seek to explain harmonics. Upon likewise
learning of their existence from Mersenne, he ascribed them likewise to certain irregularities

he posited to occur in the vibrating string. He further sought to account for the perception of sound. The difference with Beeckman's explanation is not nearly so large as their distinct points of departure – in air globules and trembling air, respectively – might lead one to suspect. Descartes similarly invoked the nerve ends (which, unaware of the oval window, he thought situated right beyond the third ossicle). But in his vibrational conception these are touched by a trembling of the air passed on and kept in vibration from the eardrum to the ossicles. Then again, as with Beeckman, this is where the auditory nerve takes over. But with Descartes no spirits run up hollow conduits to reach the pores of the brain. In his view the trembling movement touches one of the filaments that together fill up the nerve. This causes it to be pulled and thereby to open pores in the brain so as to release animal spirits of a likewise material nature, and this release is what makes us undergo the sensation of hearing. As with Beeckman, the sweetness of the consonances and the numerical order of their ratios have been lost sight of in the course of his account.

Possible checks upon explanatory fancy.   A sustained comparison between Beeckman's and Descartes' ranges of explanations of musical sound yields numerous insights into the everyday practice of explaining the world by means of nothing but corpuscles in incessant motion. Indeed, *explanation* is their be-all-and-end-all. Neither a Bacon-like, patient description of phenomena as yet unknown, nor a Galileo-like, mathematical analysis is going on here. Such mathematical features as are already given in the subject are being made some use of, but not extended in any way (it is in this Pythagorean presence of numerical regularity alone that the subject of musical sound is not fully representative of kinetic corpuscularianism and the qualitative accounts that philosophers were in the habit of rendering). The range of known phenomena is passed in review, and when experimentally inclined practitioners like Mersenne bring new phenomena to a philosopher's notice these are duly taken up in the explanatory mechanism as well. Yet no active search for phenomena as yet unheard of is being undertaken, and possible empirical counterevidence comes up only so as to be brought in line with the main explanation already rendered. In none of this is there any significant difference between the procedures of kinetic corpuscularianism and of the four classic, Athenian philosophies of nature.

What *is* different, though, is the sheer range of phenomena now covered. Ancient atomism had dealt only in the broadest and most summary fashion with the production and propagation of sound, regarded as the emission of particles of various shapes, from smooth to rough. With the focus not so much on the particles themselves any more as on their movements, the explanatory range has vastly increased. It has increased in terms of not only the number of phenomena up for explanation and the level of detail reached but also the variety of possible explanations. However restricted the primary entities adduced for the explanation may look at first sight – the motions of invisible particles differing in nothing

but size, shape, and configuration – the freedom left by these first principles actually appears
to have no limit, and the plasticity of the doctrine is really endless. Sound may consistently
be held to be either emitted air globules or air tremblings propagated like waves. Sensation
may consistently be due either to material spirits streaming up hollow nerves or to fine nerve
filaments being pulled down. By their very nature, these particles and their movements are
beyond observation. *So what check is there upon such explanations?*

In the natural philosophy of kinetic corpuscularianism, checks of four kinds appear to
be explicitly or implicitly acknowledged. These are (1) foundational certainty, (2) consistency, (3) analogies with the macroworld, and (4) a human faculty that with due caution
may best be called by its modern name of 'physical intuition'.

Foundational certainty could take a more or a less elaborate guise. It could take the elementary guise of Beeckman's a priori criterion for an acceptable explanation of natural
phenomena, to wit, that it be *picturable*, in the sense of it being to him inconceivable "how
something immaterial can move something material."[199] Foundational certainty could in
addition take the stricter guise of those securely grounded first principles *cum* secure deduction that marked from the outset the Cartesian variety of kinetic corpuscularianism
(p. 235). Its very strictness required a neat distinction to be made in everyday explanatory
practice between what the philosopher felt to be objectively certain and what not. That way
the uncertain tends to be identified with the subjective, as in Descartes' distinction between
how degrees of consonance come about and how each of us may judge their comparative
sweetness.

Consistency did work as a possible check on the foundational level, but not beyond. As we
have just seen Descartes argue in theory and Beeckman act in practice, at the level of phenomena one might, consistent with one's first principles, change one's mind about explanatory details, for example, whether pitch depends on globule size or globule speed. Another
question is whether, in everyday explanatory practice, consistency with one's first principles
might not prove to be more apparent than real. Stoics and Aristotelians had over the centuries pestered their atomist rivals with three conundrums: how to account for activity in
nature, for the inner coherence of things, and for sensation. The only way out left to corpuscularians of any stripe – attribution of further movements or invocation of a particularly
subtle matter – could easily appear to paper over rather than resolve their quandary. A case
in point is the particulate 'spirits' both Beeckman and Descartes invoked to help explain the
sensation of sound. These were meant to materialize the more ethereal spirits found of old
in Galen's doctrine of the operations of the body. Yet the very usage of spirits, with their
inevitably 'spiritual' overtones, points at limitations set to the doctrine which some in the
second generation were to notice and to exploit (p. 552).

The demand of picturability, joined to the invisibility of corpuscles and their movements, left *analogies* with the macroworld as the only empirical check upon any given expla-

380   nation. For example, Beeckman sustained his globule account of musical sound by means of an analogy taken from the sound one may produce by gently rubbing the rim of a water-filled glass with a moist finger. Beeckman had observed how the sound is accompanied by regular jumps of water droplets, easily explicable by assuming that the vibrating glass cuts little drops off the water (hence, the 'boiling' one watches) as well as off the air (hence, the sound one hears). In this use of analogy, the new focus on motion enhanced and intensified a practice already standard in ancient atomism. A certain preference for analogies drawn from the domain of the arts and crafts is further noticeable, especially where the operations of the human body are concerned (e.g., Beeckman's drum ruffles or conduit-like nerves, or Descartes' nerve threads). Both men were familiar with craft practice. For years Beeckman earned his living by making candles and installing or repairing water conduits. Descartes for a while employed a glassmaker in the vain hope of getting the hyperbolic lenses ground which he had theoretically shown to refract light rays to a focus (p. 328), and his sense of excitement aroused by the domain of the arts and crafts is palpable throughout his writings. Something more than just experience or interest is at stake here, though. The analogy so frequently drawn between craft products and mechanisms of particles in motion was meant to express an underlying commonality (next section, p. 387).

One final check is *'physical intuition'*. It is more intricate than the others, and its historically responsible handling requires considerable circumspection. In a positive sense it was at work in Beeckman's perceptive account of beating, which betrays an unmistakable sensitivity to vibrational cycles. Descartes applied it negatively in his curt yet well-deserved dismissal of Beeckman's globule conception of sound as just 'ridiculous'. In fully modern times, 'physical intuition' is attributed as a rule to seasoned practitioners with a special, piercing gift for grasping how in some given case things actually do, or conversely could not possibly, work. We may not without more ado transplant the notion to earlier times, certainly not to the first half of the 17th century. By then the very term 'physical' found itself in an early process of change, still meaning 'natural-philosophical' yet in a slightly more modern sense than had been true of any of the ancient philosophies of nature, atomism included. That is to say, Beeckman's and Descartes' defining move of focusing on the particles' motions made it possible for explanatory mechanisms to get so intricate and so detailed as now to bring this human faculty of 'physical intuition' into play for the first time on a more than incidental scale. We must take care to differentiate here, to be sure. On the one hand, there is our *modern* appreciation, due to which we can hardly avoid being impressed with Beeckman's account of beating (even if partly mistaken) or with Descartes' account of wave formation through successive rarefactions and condensations. On the other hand, there is the kind of implicit intuition that *at the time* found expression, more often than not, in such curt qualifications of somebody else's favorite mechanisms of particles in motion as 'ridiculous' or 'just a fantasy', out of an apparent awareness that certain otherwise well-conceivable mechanisms of particles in motion just could not possibly work. Take a phenomenon for which two rival explanations were put forward, each well in line with the three other checks. That is, each was dutifully conceived in terms of movements of particles alone, internally consistent at least in outline, and, finally, well provided with fitting analogies drawn from the macroworld. In such a case (as with Beeckman's and Descartes' rival explanations of musical sound) 'physical intuition' could and did act as a fourth check, independent of the others. On making one's first acquaintance with kinetic-corpuscularian explanations even in the hands of its gifted founders, one may well get the impression that there was no check on possible absurdity at all; yet on occasion practitioners knew how to restrain themselves to some extent.

From particle movements to human sensation.   One further feature shared by all explanations that ran along the lines just discussed is the radical dichotomy originally posited by Demokritos between the apparent world of macrophenomena and the true world of invisible microphenomena. Analogy, as noticed, may be invoked as a *bridge* between them so as to assist our thought about them. But there is one place where the two worlds actually meet – the human brain. They do so in the sense that some corpuscular mechanism at the micro-level makes itself seen or heard or felt in the macro-world as a sensation that is radically different in kind. Beeckman's claims regarding how this comes about were relatively modest. The soul, he thought, is definitely not reducible to moving corpuscles, and the task of accounting in material terms for its interactions with the body, albeit not a priori hopeless, is nonetheless daunting. Even so, at least "this foundation [ought to be] posited that [e.g.] the sounds of the consonances actually cause, through motion, something inside us to change place".[200] In kinetic corpuscularianism the ambition is not or not necessarily to explain the soul out of existence. Rather, it is to explain in terms of the sheer movements of particles bodily phenomena and events previously ascribed to the action of the soul or soul-like substances. To push our immortal soul back as far as possible, special attention is given to the fine structure and the functioning of the body in straightforward connection with our sensations and reactions to them. This attention found expression in a rather up-to-date awareness of ongoing innovation in these medical domains. With Descartes it also gave rise to one of the few sustained empirical pursuits of this otherwise prototypical armchair philosopher: his investigations of animal carcasses. Whether a soul reduced (as it was by Descartes) to the sheer self-reflexive capacity of human beings still proved a theologically acceptable soul is a question to be addressed in the next chapter (p. 422). At issue now is the widespread adoption of kinetic corpuscularianism soon after publication of *Principia philosophiae* and how to account for its remarkably quick spread.

Rapid adoption

How wide and how quick was that spread, really?

Among those in the second half of the 17th century actively committed one way or another to the renewal of nature-knowledge, a very significant number, probably the great majority, adopted a view of the world as made up of particles in motion. This might be in the guise of a natural philosophy or in a looser vein, and by way of adherence to Descartes' surely prevalent variety or to varieties of somebody else's or one's own making. A few among the innovators sought to stem the flood they perceptively saw coming. So did Gilles Personne de Roberval, who made a point of contradicting everything Descartes had ever upheld. So did a young acquaintance of Descartes', Blaise Pascal, who already during Descartes' lifetime expressed himself in memorable one-liners like "all things of such a nature, the existence of which does not manifest itself to any of our senses, are as difficult to believe as they are easy to invent" (pp. 415; 517).[201] Also, there were numerous Galileans and Baconians who, in confining their work to just mathematical treatment or just experiment-guided description of some chosen segment of nature, remained noncommittal on the subject. But of those scholars with confidence in the attainment of some causal understanding of what underlies the surface appearance of natural phenomena, just about everyone actively engaged in the ongoing search for a new understanding at that level did so, from the 1650s onward, in terms of particles in motion.

There were, further, many scholars who were expected, not to come up with ideas of their own, but to pass received doctrine in natural philosophy on to the next generation. These were university professors chiefly but also numerous innovation-prone yet institutionally constrained Jesuits. Among these men the reigning doctrine of Aristotelianism gave way with remarkable speed to some, most often Cartesian variety of kinetic corpuscularianism. However, Descartes had been one forcefully outspoken, yet not of course the only, man to challenge the scholastic conception of the world. So had Bacon and Galileo, each on behalf of his own conception of nature-knowledge. The demise of Aristotelianism should not be attributed therefore to the Cartesian assault alone. Even so, and whatever the prime culprit, there is no mistaking the speedy loss of its prominence. In 1661 Isaac Newton, as a freshman at Cambridge (admittedly a late adapter even among universities), was routinely assigned an uncompromisingly Aristotelian work written in the 1620s as a textbook in natural philosophy and then exceptionally went ahead to find out about Cartesian and Gassendist particle movements all by himself. Compare this with the situation some sixty years later, by the 1720s, when Newton's definitely post-Cartesian, mature teachings began to be disseminated in textbook form in the universities of Britain and the Continent. By then, Cartesian teachings constituted the doctrine up for replacement in its turn. By 1660 an academic natural philosopher could still ignore Cartesian conceptions without looking oddly backward; by 1700, this was no longer so.

Then, as in later episodes of major philosophical shift, the didactic vehicle of the textbook $\qquad$ 383
was instrumental. Descartes deliberately wrote *Principia philosophiae* in such a way as to
make it look like a standard scholastic work and thus easier to be accepted by academics.
Even so, further adaptation to the habits of the schools proved necessary and was indeed undertaken. Particularly widespread were a succession of Latin textbooks by Johann Clauberg
from 1664 on and two authoritative French ones based on lecture series and soon translated
into Latin, Jacques Rohault's *Traité de physique* (1671; 'Treatise of natural philosophy') and
Pierre-Sylvain Régis' more comprehensive *Système de philosophie* of 1690. All these works
were marked by close adherence to Descartes' variety of kinetic corpuscularianism, yet with
such discoveries as had meanwhile been made duly accounted for in terms of motions of
particles, and also with the underlying conceptual apparatus handled more in conformity
with the strict logical distinctions and subdivisions customary in the presentation of Aristotle's philosophy of nature.

Whence the shift?   Whence, then, this major shift of allegiance?

To begin with, similar major shifts had occurred in philosophy before. The Stoa in late-
republican and early-imperial Roman times; Platonism in late-imperial Roman times and
to some extent in eastern Islam; Aristotelianism in western Islam and then in medieval and
Renaissance Europe – each had over a significant segment of its lifetime had the upper hand.
In Europe between c. 1300 and c. 1450 the latter doctrine had even enjoyed a unique near
monopoly coupled to a unique institutional stronghold. But this had not prevented the subsequent comeback of all its Athenian rivals, with which, by means of a deliberate movement
of revitalization, it was made to enter into renewed competition. In principle, then, there is
little out of the ordinary in one more turn being given to Fortuna's wheel at some point in
the thus far cyclical history of philosophical predominance in the Greek tradition – now at
long last the turn was to atomism or at least a variety thereof.

Still, just as special reasons have been adduced by historians for why it was *this* particular
philosophy that at *this* particular juncture gained widespread adherence at the detriment
of its rivals, just so we may well inquire what attracted so many European scholars in the
second half of the 17th century to the partly novel philosophy of kinetic corpuscularianism.

For starters, here is what Christiaan Huygens, an early adherent to Descartes' teachings,
meanwhile turned stern and perceptive critic, had to say on the subject:

> The novelty in the figures of his particles and whirlpools make[s] it a great pleasure. When I
> read this book of *Principles* for the first time it seemed to me that everything in the world ran
> as well as it could, and when I found some difficulty in it I believed that it was my fault for not
> grasping well what he meant there. I was only 15 or 16 years old. [202]

384 And indeed, still today, on perusing *Principia philosophiae* and fancying ourselves dispossessed of our Newtonian habits, we can well understand how this book could have overwhelmed a reader, certainly an impressionable youth – only thinkers born old have never experienced the temptations of the bold grab for the whole here fondly recalled by the sixty-four-year-old Huygens.

Still, Descartes was hardly the only salesman marketing bold grabs for the whole. I shall leave aside for a while the 'novelty' that Huygens appealed to, as it could and did serve as a liability as well as an asset. Scholars might have more specific reasons for shifting allegiance on a scale and at a pace not witnessed for many centuries.

In ancient times, against the radical dichotomy posited by Demokritos and his atomist followers between the world as we perceive it at the surface of things and the world as it truly is deep down, Aristotle had upheld the basic reliability of our senses. In the Aristotelian conception, a thing *is* what it looks like, and the philosopher's task is to go ahead and assign that thing its proper place in the full system of knowledge. By contrast, in the corpuscularian conception a thing is *not* what it looks like but something else, really. What debate the basic difference between these two distinct conceptions had lent itself to had inevitably taken place on intellectual grounds alone. But now, from the second decade of the 17th century onward, the latter conception began to receive empirical support from the sudden appearance on the scene of the telescope and the microscope. Both at once revealed entire worlds of sense beyond those of everyday perception. Those willing to accept as genuine what these new tools seemed to show were compelled at a minimum to draw the conclusion that our senses are subject to much larger limitations than previously suspected. That is, confidence in their prima facie reliability and capacity to cover the whole of reality could not fail to be undermined by these new tools.

Although Aristotelianism thus found one of its major props unexpectedly undercut, it was by no means inevitable that atomism (whether or not in its 'plus' variety) would stand to benefit. Platonism and (albeit to a lesser extent) the Stoa upheld, each in its own way, a basic distinction between a surface world of sense and a radically different world lurking behind it. Why, then, did not either of these two philosophies subsequently regain the upper hand?

Note here that each owed its previous rise to prominence over its rivals to changes in political climate and concomitant shifts in ethical needs, with predominance in natural philosophy following in the wake of meanwhile established political/ethical doctrine. In characteristic contrast, the first major turn of Fortuna's philosophical wheel to take place in Europe was toward naturalism and novelty (p. 240). How the philosophy addressed *nature* (preferably in an *innovatory* frame) now uniquely became a decisive criterion, though subject always to ultimate theological approval. Here Platonism and the Stoa, with their spiritual overtones and their quite broadly sketched philosophies of nature, had relatively little to offer – not by chance, their major attractions kept being felt to reside in the political/ethical

sphere. With atomism things were different. While fitted out by Epikouros with an ethical
doctrine, its outspoken materialism had caused it as a rule to be perceived as a philosophy of
nature above all. By the early 17th century what had once been a liability had turned into an
asset, particularly in view of a related development occasioned by the microscope.

The telescope showed things above the Moon to be remarkably similar to those below,
thus helping undermine Aristotle's own dichotomy, the one between Heaven and Earth.
Meanwhile, the microscope revealed a world of ever more stunning *un*likeness to the one
of human measure. It thus tended to confirm empirically Demokritos' dichotomy between
things as perceived at the surface and as understood deep down. The microscope had been
around about as long as the telescope, the optical properties of which it largely shared. However, the microscope took much longer to come into its own as an instrument of sustained
observation – a feat not accomplished until the 1660s, when a breakdown of barriers between modes of nature-knowledge provided the decisive stimulus (p. 454).[203] Significant for
now is that the microscope increased confidence in corpuscularianism by fostering an op-
timistic belief that its powers of enlargement and resolution might enable practitioners to
make those invisible corpuscles so far only posited visible after all. And even if we discount
such ultimate hopes, the countless irregularities and unevennesses that the microscope did
reveal in what the naked eye perceives as smooth and even formed an argument of some
strength in favor of a world made up of discrete particles rather than of homogeneous substances. This is how, for example, in 1654 an early practitioner, Walter Charleton, a Helmontian turned Gassendist, could sum up a book section thus: "The Microscope of great use, in
the discernment of the minute particles of Bodies: and so advantageous to our Conjecture,
of the exility of Atoms."[204]

This particular consideration drawn from the microscope points indiscriminately at atomism in its received, ancient mode and in its 17th-century 'plus' mode of kinetic corpuscularianism. The former was in due time to make quite a comeback among Enlightenment
philosophers, yet it was in the latter mode that corpuscularianism gained the upper hand
over the second half of the 17th century. Specific attractions of the kinetic mode of corpuscularianism were (1) an appeal to think for oneself framed in such a way as to appear both
challenging and safe, (2) an infinite clockwork universe governed by laws of nature, and (3)
incorporation of the latest astronomical finds into a broad, new picture of the universe. I
shall review these attractions one by one.

To Descartes' principal message, that it was time to rethink the world from scratch, there
were two almost opposite sides. On the one hand, the message carried with it a forceful appeal to every reader to renounce guidance by someone else's thinking and take a fresh look
at things. This had been Galileo's message in the *Dialogo*, too. Yet unlike with Galileo, Descartes' call 'think for yourself' was accompanied by a reassuring 'I have already done it for
you'. This was a tempting combination – to go ahead and think for oneself is as seductive as

386   it is scary. Here was an invitation to take a walk through excitingly unknown territory with the help of a roadmap to prevent one from getting lost along the way. Descartes' message simultaneously addressed itself to quite different audiences. To those sharing the dogmatic posture that had been part and parcel of Athenian philosophical doctrine, kinetic corpuscularianism was offered as the new dogma to follow, which is how many an adherent was indeed to take it. To others it offered a chance to take seriously Descartes' rousing appeal for once in one's life-time to doubt everything one has ever been led to believe, but without running the risk of falling into the abyss of everlasting uncertainty. The skeptical abyss was skirted, yet abandoned in time for a more constructive pursuit.

That pursuit, moreover, culminated in the positing of nothing less than an infinite clockwork universe governed by immutable laws of nature. There is a certain majesty in the universe conjured up by Descartes. Infinitely many worlds like our own were spread over an infinity of space, running in a predictable manner according to a few definite laws of motion governing all of nature. The general idea had of course been around since Demokritos, yet those Cartesian laws of motion as worked into the fine structure of phenomena and events anywhere on Earth or in the heavens made the idea appear so much the more tangibly realistic. Its inherent attractions were further enhanced by the ready availability in the European imagination of a very special metaphor – the image of the universe running the way a clock does. After all, the mechanical clock could serve as a powerful symbol in view of both the inexorable lawlikeness with which the hands appear to move on their own accord and the radical dichotomy between its surface (those hands) and its interior (its intricate gearing). Starting with Dante and his exquisitely detailed comparisons in *Paradiso*, the metaphorical use of the mechanical clock became a literary topos traceable down the centuries. Soon after arriving on the scholarly scene of Descartes' world, the image found itself transplanted and adopted on the grand scale by the increasingly numerous adherents of that world, thereby revealing wherein one of its attractions lay. Descartes himself confined usage of the metaphor to a frequently reiterated analogy with the kind of automaton to which he reduced all of animal, and a good deal of human, action. The multiplication of clockwork similes increased with the popularity of Cartesian doctrine, and sank with it as well. In Britain, where the metaphor dwindled with the triumph of Newtonianism, it subsided far earlier than on the Continent.

In its heyday, the image came to be invoked for a variety of purposes. Robert Boyle was wont to invoke the most intricate clock of his time, the one on a wall of Strassburg cathedral, as a fitting metaphor for a universe of moving particles. In an edifying vein, the priest Nicolas Malebranche used the clock for arguing that so intricate a design presupposes a Maker (this is the famous 'argument from design', destined for unceasing repetition over subsequent centuries). In a polemical vein, the future bishop Simon Patrick came up with a witty story about a fraudulent clockmaker-repairman dumbfounding his clients with

Aristotelian lingo about whether or not the hammer is the 'informant' of the clock and if so, whether it be *forma informans* or *forma assistens*, until finally an unruly client well acquaint-ed with modern thought opens the defunct clock and points at a wornout wheel-tooth. Fi-nally, in a vein both polemical and constructive, the microscopist Henry Power argued that

> the old Dogmatists and Notional Speculators, that onely gaz'd at the visible effects and last Resultances of things, understood no more of Nature, than a rude Countrey-fellow does of the Internal Fabrick of a Watch, that onely sees the Index and Horary Circle, and perchance hears the Clock and Alarum strike in it: But he that will give a satisfactory Account of those *Phaenomena*, must be an Artificer indeed, and one well skill'd in the Wheelwork and Internal Contrivance of such Anatomical Engines.[205]

Here, then, resides the profound reason for why the only possible empirical check upon kinetic-corpuscularian explanation was so often found, not just in macro-world analogies that could hardly be avoided, but in analogies taken where possible from the domain of the arts and crafts.

So much for the clockwork analogy and its many-sided attractions. But the pull of Des-cartes' universe is still to be considered from one more point of view. Not only was it attrac-tively lawlike, but it also appeared as the fitting repository for a range of new views and new discoveries about the heavens which between Copernicus' *De revolutionibus* and Galileo's *Dialogo* had contributed to the stock of arguments in favor of heliocentrism. Views and discoveries that did not fit current conceptions of the world – the idea of the Earth revolv-ing; the occasional appearance of new stars; the heavenly, not terrestrial, abode of comets; the apparent absence of solid spheres; the Earth-like nature of the Moon; Jupiter's moons; Venus' phases; etc. – seemed to fit smoothly into Descartes' philosophy of nature, with its infinite space; its particulate aether; its infinitely numerous stars, each surrounded by a plan-etary system of its own and its planets surrounded by moons; its comets moving from one solar system into another, etc. There were bound to be many who stood ready to accept, even if only in broad outline, Galileo's views and discoveries on matters astronomical while, un-like Galileo, feeling themselves in need of a unifying grand scheme of things in which to fit them. For such people Descartes' philosophy of nature could serve as the perfect framework by means of which to get rid of all anomalies. To be sure, the reverse applied as well, in that the Cartesian conception of the universe served as the intellectual framework par excellence in which, over the second half of the 17th century, heliocentrism achieved broad acceptance (p. 394).

There is good reason to take the list of reasons just given for a major shift in philosophi-cal predominance seriously I hope, yet not too seriously. Without wishing to detract from the relative validity of all these considerations, it is time to point out how elusive historical phenomena of the kind here addressed are as a rule. How do intellectual fashions come; how do they go? How is it that, time and again, vast masses suddenly come to believe things more often than not unthinkable before (only to abandon belief in droves at some later time with apparent ease)? What sorts of ideas are capable of inducing a bandwagon effect on the truly grand scale? What is it that sets apart, in this regard, publications like *Kommunistisches Manifest*, *Le contrat social*, and *Principia philosophiae*? Christiaan Huygens, in answering that question for the latter book, came up in his customary offhand manner with a very good formula by means of which both to capture the problem and leave its ultimate elusiveness intact. Recall first his remark that 'Mr. Descartes had hit upon the way to have his conjectures and fictions accepted as truths' (p. 235). It is already quite perceptive to see that there *is* such a way in the first place. Even more penetrating is how, with characteristically mocking admiration, Huygens went on to speak of Descartes' gift for "fabricating . . . this entire new system, and for giving it such a turn of truth-likeness as to make infinitely many people satisfied with it and pleased with it."[206] 'A turn of truth-likeness' indeed! From time immemorial the human mind has opened itself to the temptations of total systems, of thought-constructions going for the whole, and, so Huygens is implying here, Descartes was among those rare authors with a very special knack for presenting their own world to the reader as inescapably valid – for as long as it lasts, to be sure.

Modifications of the doctrine

No necessity was felt at the time to adopt the philosophy of kinetic corpuscularianism on a 'take it or leave it' basis. The doctrine was plastic enough to allow for modification in a variety of ways, and clearly the large amount of freedom the doctrine entailed to modify it contributed its own part to the pace and scale of its adoption.

Modification could be of three principal kinds (which, with any given author, might well appear in any sort of combination): syncretist construction of intermediate forms, revision of core components, and efforts to constrain the seemingly endless plasticity of the doctrine.

Intermediate forms. By the second half of the 17th century scholarly inclinations toward syncretism were as strong as ever, and it appeared soon enough that particles in motion lent themselves well to that hoary philosophers' gambit. Two varieties presented themselves. In one, Descartes' subtly particulate aether got shot through with Stoic pneuma and a Platonic world-soul so as to yield an 'aethereal spirit' that will occupy us in chapter 16 (p. 552). In the other variety, corpuscularian views were combined with components of Aristotelianism held to be still tenable or even indispensable. Here are two examples. One concerns a genre that, often going under the name of *philosophia novantiqua* ('new-old philosophy'), was adopted over the century at many an arts faculty. The other example illustrates how the Society of Jesus sought to remain on what it took to be the forefront of the new learning without thereby transgressing the boundaries of its own religious orthodoxy.

In 1688 Wolferd Senguerd published a textbook under the title *Philosophia naturalis*. He was one of the first Leiden professors to enliven his teaching of the subject by means of some experiments, notably with air pumps. In the book he derived his conception of a universe filled with corpuscles from such Aristotelian first principles as 'matter & form' and 'substance & accident'. He conceived of motion itself as an 'accident' inherent in matter, where it appears in three distinct modes: *vis movendi* (a force in the moving body), *impetus* (impressed by that force upon the moving body), and *translatio* (the effect of *impetus*). By means of these and arrays of similar distinctions Senguerd went on to detail how the movements of corpuscles, which, Descartes-like, he recognized as the cause of all change in the world, actually work their various effects.

> Senguerd apparently belongs to those scholars who seek indeed to modernize the concepts of the Aristotelian school but actually bring about a retreat from the great pioneers of a mechanical [i.e., kinetic-corpuscular] explanation of nature. . . . This becomes particularly clear in Senguerd's doctrine of motion, which suffers in an essential way from motion being conceived as an inherent accident of a body so as to be subjected to the logical definitions of substance and accident.[207]

Syncretism of another kind appears with Father Claude François Milliet Dechales SJ. In a lengthy overview first published in 1674, he carried his eclectic approach farther down to the level of specific phenomena:

> Those who reject qualities and introduce everywhere their corpuscles and subtle spirits believe to have resolved all difficulties when by way of illustration of their opinion they adduce one or another comparison or analogy. But those who shrink back from corpuscles seem mistaken in their views, too.[208]

In cases where the analogies invoked did not seem arbitrary but rather appeared to reinforce one another, Dechales showed himself willing to adopt a 'modern' explanation by means of corpuscular effluvia. In a case like the magnet, however, he found Descartes' explanation ingenious yet wanting. Descartes had made too many prior assumptions; also, no analogy was invoked with other natural effects. Hence, for magnetism Dechales fell back upon invoking a substantial propagation of forces in iron or in a medium of some kind.

The passage is important for two more reasons. In facing the question of what is and what is not acceptable in the kinetic-corpuscularian mode of handling sundry phenomena of nature, Dechales employs here the criterion of analogy as a check upon the doctrine (p. 379). Further, the end product as put down in Dechales' vast overview represents one possible variety in an array of similarly Aristotelian/corpuscularian hodgepodges wrought by Jesuits at the time. These were the product of the order's leading policy. It flowed, on the one hand, from an acute awareness of all that was leading at the time to the ongoing weakening of untainted scholastic doctrine, coupled to institutional pressures to look maximally up-to-date. But the order remained committed, on the other hand, to retaining both a fixed Earth and such principles as seemed indispensable for accounting in a reasoned manner for, notably, the Eucharist and the immortality of the soul. Nor did the Jesuits stand alone in their theological concerns. Both Roman Catholic and Protestant circles concerned to accommodate such portions of kinetic corpuscularianism as seemed fruitful and/or timely strove to get rid of, or at least to modify, certain teachings at the very core of the doctrine.

Revision of core components. Such a revision of core components could, and did, take many forms. If guided by considerations of faith in the first place, scholars who did not wish to go along with theological rejection of the doctrine wholesale might seek refuge in two possible gambits. They might opt for a repudiation or partial adaptation of the metaphysics that Descartes had superimposed upon what he had originally conceived as a philosophy of nature alone. Or they might confine rejection or revision to such natural-philosophical teachings as inevitably impinged upon specific religious tenets. For Roman Catholics not content with revealed truth alone the Eucharist was a tough nut to crack. The substance of bread and wine can on Aristotelian principles change into Christ's body and blood as their accidental forms remain the same, but how to account for what is not for nothing called 'transsubstantiation' in corpuscularian terms is, quite literally, another matter. Further, for Roman Catholics and Protestants alike, Descartes' varied pronouncements about the relation between body and soul could easily raise questions about the immortality of the latter in Cartesian (not so much, as we shall see, in Gassendist) thought. This came to the fore during a prolonged controversy between Descartes' early opponent Gijsbert Voet (pronounced 'Voot') and Descartes' early adherent Henrick de Roy (Regius). The incompatibility became apparent between Descartes' orthodox pronouncements in his metaphysical *Meditationes* of 1641 and his natural-philosophical concern to stretch as far as he could the explanation of human feats as functions of the body, so as in the end to reserve for the soul none but thought conscious of itself. This time Descartes was still able to quell what he perceived as imprudent insubordination by publicly casting his erstwhile favorite disciple into the outer darkness of his high disapproval.

Primarily theological motives were not alone in inducing revision of core components of the doctrine. For example, de Roy was also the man to give a significant twist to Descartes' idea of rest. Descartes posited a resistance in bodies to being moved from rest, with the

amount of resistance dependent on size alone. De Roy saw that this left the doctrine wide
open to the charge voiced of old by Aristotelians and Stoics alike that sheer rest of constituent particles can never account for the cohesion of things in the macro-world. By extending
Descartes' merely passive resistance into a positive force acting as a curb on the body's being
moved, de Roy could put the doctrine in a better position to counter the objection. In the
1670s Father Ignace Pardies SJ took the same line as de Roy. He even deviated from Descartes
so far as to declare motion and rest to be equivalent principles.

There was further the question of the finite or infinite divisibility of matter. Gassendi
and Descartes were already divided on this issue. Over the decades it came up again and
again, with a noticeable tendency for otherwise mostly Cartesian conceptions to revert to
atoms in the void or at any rate in voidlike space.

There were finally those philosophers who, on the one hand, felt the need to undergird
the ongoing renewal of nature-knowledge with some kinetic-corpuscularian conception of
things while, on the other hand, worrying at a profound level about the logical tenability of
any such conception. Beeckman already expressed concern about two equally necessary yet
mutually contradictory assumptions for atoms: perfect hardness *and* perfect elasticity (p.
224). Huygens, too, experienced some concern over the issue for a while. Like Beeckman,
he sought to resolve the antinomy by invoking particles at a subatomic level. An exchange
on the subject with his erstwhile pupil Leibniz in the early 1690s is revealing. Huygens, ever
the philosophically opportunist mathematical scientist in Ptolemy's and Galileo's tradition,
readily settled for *both* properties, even if incompatible, in view of the apparent fertility of
the kinetic-corpuscularian hypothesis. In contrast, a stern metaphysician like Leibniz allowed himself to be led by these and other concerns to a revision at the very heart of the
doctrine (p. 628). Leibniz's fundamental concern was over the classic issue (broached already
by Aristotle) of how the discrete particles of the micro-world could together constitute a
macro-world of coherence and continuity. And that is how, amid the new thinking about
the infinite and the infinitesimal that in the second half of the 17th century was to issue in
the invention, in Leibniz's own hands among others, of the calculus, the hoary paradoxes of
the infinite acquired a new urgency in the domain of natural philosophy, too.

Restriction of degrees of freedom. If all of the above might perchance not suffice yet
to persuade the reader of the almost unlimited flexibility of the natural philosophy of kinetic corpuscularianism in terms of both its content and the purposes to which it could be
put, please consider for a moment a certain textbook by a certain professor, Heinrich May.
Guided by mostly Aristotelian leanings, he was also well educated in Cartesian and many
another philosophy of nature, whether ancient or modern, and was further possessed by
the gift of the muddleheaded for surface reconciliation of the truly irreconcilable. Published
in 1688 (one year after Newton's *Principia*, that is), the book made good on all that its title
promised:

Synopsis, written for the usage of studious youth, of enriched old-new natural philosophy according to the principles of Demokritos, most ancient philosopher, as reconstituted by Gassendi, Bacon, Boyle, de Rodon, Digby, and other modern authors, and as proven by means of various experiments.[209]

In this textbook the elasticity of bodies (to give just one example) found itself explained as tiny to-and-fro movements of their constituent corpuscles sustained by an inner principle of substantial form so as to ensure return to rest. The work is definitely an extreme example of speculative eclectic-corpuscularianism run rampant – the weirdest one to be gleaned from Lasswitz's near-exhaustive overview of corpuscular thought in the 17th century. The point is that by this time the congenital fancifulness of the kinetic-corpuscularian philosophy of nature had come to strike not only an early, principled opponent like Pascal but also many a scholar in general sympathy with it. The ultimate deathblow to kinetic corpuscularianism as a natural philosophy was not to be delivered until far into the 18th century, with Newtonianism as the principal agent. Yet in the 1660s already some of the better thinkers on natural phenomena began to seek ways and means to enhance what possible checks the doctrine had carried with it from within, so to speak – those criteria of foundational certainty, supraphenomenal consistency, analogy, and 'physical intuition' we have come to distinguish in the above (p. 379). Dechales, for instance, provides a case of acceptance of certain kinetic-corpuscularian mechanisms and rejection of certain others on the basis of one of those 'inner' criteria (analogy) while casting the others into the winds of his syncretism. Other, more penetrating thinkers, moved by a similar urge to sift the wheat from the chaff, sought for further checks outside the doctrine itself. They hoped to rein in the near-infinite freedom the doctrine left to a fertile imagination by means of three distinct tools: the microscope, experiment, and mathematics.

With the *microscope* the hope to find by its means direct empirical evidence for the existence of corpuscles acted as one motive among others for adoption of the doctrine (p. 385). That hope also acted as a spur, so as to usher in the brief flourishing period of 17th-century microscopy. Its very brevity (what started as a movement with great ambitions in the early 1660s was by and large over by the late 1680s) was not only due to a variety of practical shortcomings and disappointments of a technical nature (p. 454) but was due in no lesser measure to a failure actually to observe these corpuscles. Globules galore (Antonie van Leeuwenhoek, in particular, found them everywhere), yet even at the highest attainable degree of optical resolution matter still looked thoroughly organized, not at all dissolved into the kind of corpuscular mechanisms that the doctrine posited. No additional check, then, appeared in the end to come from this particular domain of empirical research.

Could it from any other? The expression 'proven by various experiments' just encountered in May's title and in Senguerd's work, and the pride with which Rohault in his popular

textbook appealed to arrays of confirming experiments carried out with a fine set of instruments, may make us suspect the presence of a lively practice of *experimentation* by natural
philosophers of a kinetic-corpuscularian bent. To some limited extent such a practice existed indeed, yet it meant less than seems apparent. Such productive interaction between
kinetic corpuscularianism and Baconian experimentation as by the early 1660s began to
flourish in Britain was definitely not about the doctrine taken as speculative dogma but
rather as heuristic hypothesis (p. 554). Inside the standard-dogmatic tradition to which Rohault, Senguerd, May, and so many others belonged, the role of 'experiment' was much more
restricted than the liberal usage of the expression suggests. The manner in which Beeckman
and Descartes accounted for musical sound as moving corpuscles is telling here.

Kinetic-corpuscularian philosophers tended to regard the world of phenomena as by
and large given (e.g., sympathetic resonance, or the problematic status of the fourth in current counterpoint). Yet both Beeckman and Descartes were prepared to bring under the
broad umbrella of their explanatory principles certain new phenomena which came to their
knowledge passively (as with higher partials) or on occasion actively, as with the structure of
the inner ear. Further, both in the range of phenomena covered and in the amount of detail
furnished, the new doctrine definitely surpassed its Athenian predecessor. Even so, in this
unadulterated, natural-philosophical context the expression 'experiment' stood for the same
kind of treatment of empirical phenomena as practiced of old. This is true even in a case like
that of Rohault. He took pains to subject, for example, magnets to various instrument-aided
manipulations to illustrate Descartes' pertinent account. Illustration was his whole point,
not 'proof' in any stricter sense. Experiment in his treatment stands neither for the open-
minded, relatively preconception-free search for phenomena yet unknown of the Baconian
experimentalist nor for the validation-oriented checking procedure of his Galilean counterpart. Although here as everywhere else in history borderlines are permeable and some
ambiguous mixed cases might perhaps be traced, the long and short of the argument on this
point is that no more than with the pioneers do we find among later kinetic-corpuscularian
philosophers an empirical approach of such a kind as to act as an effective check.

Just as for experiment to be turned into an effective check it proved in the course of the
century necessary to move from philosophical dogma to scientific hypothesis, so it was with
mathematics. The Galilean approach, if set against a background of hypothesized corpuscles
in some hypothesized mode of motion, was to prove so fertile as to yield some of the best
products of the new science (p. 521). Nonetheless, the vicissitudes of history present us here
with the very thing we failed to find with our previous, experimental check: an instructive
borderline case. Giovanni Alfonso Borelli, while embracing the doctrine of corpuscles in
motion in the good old Athenian way as speculative dogma not as heuristic hypothesis,
uniquely sought to constrain it by means of mathematics. In his *De motionibus naturalibus
a gravitate pendentibus* (1670; 'On natural motions dependent on heaviness') he was concerned

to show that all events in the world of body are up for explanation by the force of heaviness alone, if only one attributes fitting shapes to the particles of bodies, so that the force of heaviness, directed toward the center of the Earth, compels them according to mechanical laws [i.e., those of statics] to move in other directions, too, depending on circumstance. He therefore regards the corpuscles as machines, which are moved by the force of heaviness yet determined as to direction by their construction. Since the only driving force is thus the pressure of bodies upon others and themselves, he conceives of the totality of movements on Earth as a problem in *hydrostatics*.[210]

This broad conception determined how Borelli went on to detail his world. He judged certain geometric figures, like prisms or hollow cones, for present or absent capacity to pile up, to fill up all of space, or to enable fluids actually to flow. He further took into account the (in his handling) essentially Alexandrian, mathematical science of fluid equilibrium. Both the geometric figures and the rules of fluid equilibrium served Borelli as constraints upon the corpuscular mechanisms that he designed for, for example, the freezing of water or the solution of certain chemical substances in fluids. The outcome may strike us as the height of fanciful folly: for instance, the down-like shaped 'little machines' by which he held his octahedrical water particles to be surrounded in order to account for the phenomena of capillary action. But the point is that his effort was nonetheless governed from start to finish by ongoing reasoning meant to put a check on fancy and sift the few possible from the many impossible micromechanisms.

The exceptional case of Borelli notwithstanding, the upshot of the investigation into searches undertaken over the second half of the 17th century by certain believers in some form of kinetic corpuscularianism for additional checks upon unbounded fancy is that, beyond those 'inner four', such checks were just not to be had, not in the microscope, not in experiment, not in mathematics. Or rather, they were to be had indeed, but only if the doctrine was first robbed of its hoary, Athenian framework, then radically reset in a new frame of hypothetical science through equally unprecedented, thoroughgoing infusion with either of the two available rival modes of pursuing nature-knowledge, Galilean or Baconian. With the advent of realist-mathematical science and of fact-finding experimental science, then, speculative natural philosophy *as such*, almost coincident with its apparent heyday in the second half of the 17th century, began its long, halting, never quite completed way out (p. 603).

Whirlpools for a revolving Earth

The idea of the Earth as a planet revolving annually around the Sun and every 24 hours around its own axis first came up as just that, an idea, with Aristarchos. Not worked out in any detail, its evident absurdity in the face of elementary observation left it no chance of acceptance, certainly not among naturally realist philosophers but also not among mathematical astronomers. This was even more true since it proved possible to use the commonsense conception of a fixed Earth as the starting point for so satisfactory a mathematically detailed model as eventually presented in Ptolemy's *Almagest*. The idea returned as the crystallization point around which Copernicus sought to restore mathematical astronomy to pristine purity. His realist claim was expressly *denied* in his (i.e., Osiander's anonymous) preface, emphatically *affirmed* for his simplified model, and actually *ruled out* by the full model in its full complexity. This ambiguity made it possible for Copernican heliocentrism to be received over its first half-century with similar ambiguity. Many put it in the service of improved calculation while just ignoring the realist claim. Nine accepted that claim only in partial and/ or subdued ways. Just two combined unhesitating acceptance with a search for ways to get rid of the many ambiguities perceived to inhere in it from its surface down to its very core. Kepler's concern yielded in the end his 'celestial physics', among the barriers to more than incidental acceptance of which its very heliocentrism scored highly at first (p. 303). Galileo's concern led to the effort he undertook in his *Dialogo* to arrange as attractively as he could all that might plausibly be argued in favor of broad acceptance of heliocentrism, not in much mathematical detail but as a conception of the state of things beyond the Earth overall.

By midcentury, with the *Dialogo* meanwhile published in Latin translation (1635), the state of the debate was roughly as follows. Among those with some knowledge of mathematical astronomy, Ptolemy's uncompromisingly geocentric conception was slowly yet surely on its way out, chiefly because it ruled out a celestial phenomenon – the phases of Venus – that had been seen in Galileo's telescope. The choice left was

- *either* to remain noncommittal (the posture adopted by many mathematical astronomers out of the time-honored 'saving the phenomena' habit, but also by Pascal in 1647 still, in line with the strict criteria for proof in empirical science he found it necessary to uphold against Cartesian fancifulness),
- *or* to move to
  - *either* Tycho's compromise system, be it wholly or in part (the policy line of the Jesuit Order from the ban of 1616 onward, but also adopted by many others who were not so constrained),
  - *or* Copernicanism in one or another (rarely as yet Kepler's) variety.

Once on the move it is surely easier to move on than when starting from rest; even so, more was needed for broad acceptance of full heliocentrism than a debate governed by technical issues like those phases of Venus, the localization of the sunspots, or the significance of Jupiter's moons. This was especially true for natural philosophers, who as a rule cared little for such apparent niceties but, if they nevertheless found them important, could still accommo-

396 date some of them. Notably, the spectacular, hard-to-ignore appearance of comets could in 1577 but also still in 1618 be brought to order by shifting allegiance from Aristotelian to stoic philosophy. After all, the Stoa, while likewise geocentric, regarded comets as supralunary (also highly portentous) phenomena. Yet it became increasingly hard to continue accommodating the ever-growing list of refractory celestial phenomena that way, and here is where Cartesian cosmology came in.

Descartes had conceived his 'World' in part under the inspiration of a Copernicanism that he adopted in a broadly qualitative vein. He arranged the book to cover not only the central tenets of the doctrine but also those empirical phenomena that had in the meantime appeared to sustain them. Instrumental in this regard were the whirlpools of particulate matter which follow from his laws of motion and which he made to serve as the principal carriers of displacement of matter all over the universe (p. 228). He offered a broad setting for a broadly Copernican view of the world made up not of one (as with Copernicus himself) but of infinitely many solar systems (as with Bruno and Digges and several of their followers). But he managed in addition to account by means of intricate vortical mechanisms for all major astronomical phenomena, be they known of old or discovered in recent decades. Thus, he explained the motion of planetary satellites by supposing that Saturn, Jupiter, and the Earth are each at the center of a whirlpool of its own. Each whirlpool then carries along its several moons (Jupiter) or its one moon (the Earth, but also Saturn, as at this time its apparent appendages were usually taken to be a satellite). These planet/moon systems, Descartes thought, might either have come into being by prior formation as such (which he held most likely for Saturn and Jupiter) or by the satellite having been caught by the grand whirlpool of the Sun, as probably with our own Moon. Regarding the Earth/Moon system he put to good advantage the rough correlation perceived by many (albeit denied by Galileo) between the tides and the Moon by positing a slightly eccentric position for the Earth in its own whirlpool, such that the whirling matter exerts a little more pressure upon the surface of the Earth when at certain positions vis-à-vis the Moon than at others. Further, the difficulty created in 1633 for an obedient Roman Catholic still to uphold a Copernican view of things, which had caused Descartes to lay his budding 'World' aside, was resolved in *Principia philosophiae* by positing that near-Earth-centered whirlpool. In carrying the planet in its dual motion along, it allowed him to keep speaking in apparent good faith of an Earth at rest (namely, with respect to its own whirlpool).

In such a way, then, Descartes' philosophy of nature came to provide the obvious medium for plausibly spreading the word of heliocentrism. An investigation of numerous students' disputations, by means of which professors used to make their own views known at the time, but also additional evidence taken from contemporary poetry and from illustrations in atlases has recently shown for the case of the Dutch Republic that debate once carried on mostly in terms of mathematical-astronomical arguments shifted in the midcentury

to being discussed within the domain of natural philosophy. Descartes' variety of kinetic
corpuscularianism came to serve as the principal carrier of the heliocentric message beyond
the circles of the experts and in the learned world at large. Not that everyone jumped on
board. Most professors cautiously balanced their Aristotelian, Cartesian, and possibly other
leanings in some variety of new-old philosophy, and more than a few, like Senguerd, did see
the weaknesses of Tycho's setup yet rejected Copernicanism on biblical grounds. Nonetheless, through the academy the notion of a moving Earth as a proposition worthy at the very
least of being taken seriously spread among the educated elite of regents, lawyers, physicians,
and clergy.

By the end of the 17th century numerous believers in a fixed Earth were still around. Still,
the battle for a moving Earth had been won insofar as heliocentrism had lost its remaining
tinges of the utterly paradoxical (infinite space, which still upset Pascal's famous interlocu-
tor, included) and had turned into something close to the standard view. Scholarly life in
the Netherlands was exceptional in the absence of a central court and, hence, of a vigorous
scholarly culture outside the universities. Even so there is no obvious reason to suppose
that elsewhere in Europe other vehicles than Descartes' popular whirlpool-world presented
themselves to carry heliocentrism to such broad acceptance as, by century's end, it had come
to enjoy in the Europe-wide 'republic of letters'.

Another sort of power; another sort of pull

In the 1610s kinetic corpuscularianism began to take shape in Beeckman's diary. In the 1640s
the doctrine began to spread over the width and breadth of Europe, mostly in the thoroughgoing variety developed by Descartes. Already by the 1660s to 1670s it had become the predominant philosophy of nature, with every Athenian rival put on the defensive. Besides the
doctrine in unadulterated form, a large variety of mixed forms made their appearance, rang-
ing from local modification to wholesale, syncretic compromise. Even so, few known natural
phenomena escaped an explanation in terms of corpuscular motions.

The power of explanations undertaken along such lines was quite limited, certainly if
compared to the vast achievement attained in realist-mathematical science. There, Galileo's versatility, a narrow subject intertwinement, an ongoing problem/solution chase, and a
fruitful interplay between mathematics and experiment jointly made for a relentless expansion in both width and depth. In the natural philosophy of kinetic corpuscularianism, other
than the furtherance of broad acceptance of a general (not mathematically exact) conception of heliocentrism, there was hardly any lasting achievement at all. Although its better
practitioners proved quite capable of *accommodating* ongoing discovery, they were unable
to *produce* it. Not one of the great designers of a kinetic-corpuscularian conception of the

398   world can be credited with even one major discovery.[211] (Descartes' discovery of the sine rule of refraction and of the making of the rainbow was due to his prowess as a mathematical scientist in the Alexandrian mold, not to his natural philosophy).

The outcome need not cause surprise – the sheer variety of conceivable movements of diverse particles was just too large and too prone to fancy to make it possible to pinpoint anything definite. Four checks on fancy were developed and applied right from the start: foundational certainty, consistency, analogy, and 'physical intuition'. These were all vastly different from the kind of check that mathematical scientists sought to wrest from the opportunities which experimental validation appeared to offer. Even if applied jointly, these four restraints proved hardly capable of curbing flights of fancy; hence the search, exemplified by Borelli, for additional checks taken from elsewhere (in his case, from geometric form). But this search was in vain, too. Only if wholly divested of its speculative knowledge structure and hypothetically redesigned as the underlying ontology of something else, be it in mathematical or in fact-finding experimental science, was kinetic corpuscularianism in a position to bring about enduring novelty.

None of this stood in the way of the vast pull that it exerted in the original guise of a speculative philosophy of moving particles. That pull resided in the ambivalent liberation of thought the doctrine offered, in its ability to account for astronomical novelty, in the attractions of a clockwork universe, and in the 'turn of truth-likeness' that Huygens identified. The attraction proved so strong as to pull on board not only youngsters like Huygens, or Jesuits and other Aristotelians not about to miss the latest fashion, but even a few scholars whose entire intellectual makeup predisposed them to quite different conceptions. Just as men like Gassendi and Le Cazre proved unable wholly to withstand the pull of realist-mathematical science, so foreign to their own natural-philosophical habits of thought, so were two men with a mind-set similarly out of sync with the times, Kenelm Digby and Henry More, drawn almost willy-nilly inside the sphere of kinetic corpuscularianism. The thinking of both men was steeped in Renaissance sympathies and magical manipulation; their preoccupation was with the mental capacities of human beings; yet neither felt capable of treating issues of mind and spirit other than in the language of moving particles with which the French pioneers acquainted them in the 1640s and 1650s.

Take Digby's sustained interest in the kind of tricks our memory may play upon us – to mix things up, give them a false turn, or forget them altogether. Digby described memory as a vaporous fluid in which the particles of our sense impressions swim once they are carried there from the outside via our brain by likewise-material, vital spirits. Retrieval from memory by the imagination is achieved by messenger particles which from the imagination that sends them on their way have received a shape similar to the particles of the specific sensation sought. We are remembering things all the time, so there is an ongoing stream of particles from memory to imagination and back again. Particles not recalled for a long time

have become out of reach for the messenger particles, which can no longer be given ade-
quate shape and thus get lost (forgetting), unless they happen to be carried along in another stream (mix-up). Retrieved memory particles, their stay in the imagination completed, may upon return to memory carry with them some imagination particles, which get stored like-wise, so as henceforth to act as if they were genuine memory particles (false memory).

Or take Digby's distinction between passionate people, inclined to concentrate upon themselves to the point of paying little attention to the outside world, and more outer-directed ones out of touch with their own feelings. In the former case the nerves that run from the heart (the seat of our passions) to the imagination are filled to capacity with par-ticles, which, on arrival in the imagination, push out the particles from our sensations; in the latter, the 'sensation' particles gain full entry to our imagination in the absence of those passion-carrying particles from the heart.

Or take Digby's case histories of what he called 'the contagion of the imagination'. The phenomenon of persons engaging in all sorts of compulsively imitative behavior is not so much due to possession by some evil spirit in Digby's view. Rather, it is to be attributed to particles aligned to the other's original behavior to such an extent as to get a chance to stream through the otherwise overflowing nerves.

What is going on here behind the scenes? The point is that Digby allowed his subtle understanding of certain peculiar features of the human mind to be squeezed into the pro-foundly reductive categories of Cartesian mind-thought, even though his psychological acu-men was more in tune with multilayered Renaissance views of human conduct than with the one-dimensional association psychology which particle-thought was inducing at the time in, notably, Thomas Hobbes and John Locke.

Hobbes, in *Leviathan* (1651) and *De corpore* (On Body; 1655), summarily scrapped Des-cartes' *res cogitans* from the world's furniture and declared the soul to be just a name for those specific movements of corpuscles in the brain that, by means of associations of vari-ous kinds, produce what we experience as our thoughts. Whereas Digby sought to couch a much subtler conception of our thinking processes in such reductionist terms, Henry More instead rejected reductionist moves like Hobbes' out of hand. What makes More's case per-tinent is the extent to which, in combating such views, he adopted the very weapons of his adversaries.

A fling of untainted kinetic-corpuscularian proselytism behind him, by the 1660s More turned himself into the foremost of the so-called Cambridge Platonists, who were concerned to save from the corpuscularian onslaught at least some claim to an irreducible autonomy of the human mind. As he came to see it, in the Cartesian view of things *res cogitans* all too visibly fulfilled the role of a stopgap for whatever in human thought and action had as yet eluded an explanation in terms of moving particles. All this came down, so he felt, not only to a risky flirt with atheism but also to an appalling reduction of the very features that make

400 us distinctly human. The crux of More's advocacy of some notion of 'spirit' rests in his utter inability to vindicate it in other than the very kinetic-corpuscularian terms employed by Descartes and Hobbes themselves. It is a sure sign indeed of the impending victory of an idea when those who oppose it do not know any longer how to oppose it other than on a battlefield demarcated wholly by the adversary's choosing.

One final component of the pull exerted by kinetic corpuscularianism rested in the ease with which the doctrine allowed its adherents to dabble in novelty. When following in Descartes' footsteps, albeit not in a strict manner, as necessarily literal truth, but somewhat more freely, one found oneself engaged in a game easy to join. As distinct from particle motions and their convenient picturability, the cultivation of realist-mathematical science required a capacity for alternating abstract *and* realist thought *and* practice of a very special kind. Galileo's conception of things, however much open in principle to inspection by all, was bound to remain highly esoteric, effectively open only to those few with a gift for penetrating its profound and also complicated knowledge structure to the very bottom. As opposed to this, Cartesian particle-thought and its numerous varieties offered a game everybody with a modicum of learning could play – to think up particles of whatever shape, size, and specific movements seemed best to fit the occasion. As Lasswitz, the great historian of particle-thought, remarked more than a century ago, "its treatment seemed so easy that anybody could take it upon himself to penetrate by its means the inside of nature."[212] In this regard the triumphal procession that carried the philosophy of moving corpuscles from its pioneers to the first culmination point of the Scientific Revolution and beyond was something of a surrogate affair. Its greatest effective merit was to acquaint the scholarly world with the ongoing renewal of nature-knowledge by way of a poor man's exercise in fake-mathematical nature-knowledge.

NOTES ON LITERATURE USED

Musical sound. This section draws on my *Quantifying Music*, chs. 4 and 5.

Adoption. A mine of information is Paul Mouy, *Le développement de la physique Cartésienne 1646-1712*. Paris: Vrin, 1934. The points made about the microscope are due to Catherine Wilson, *The Invisible World. Early Modern Philosophy and the Invention of the Microscope*. Princeton UP, 1995, especially chs. 2 and 7. The usage, over the 17th century, of the image of the world as a clock has been examined at length in Otto Mayr, *Authority, Liberty and Automatic Machinery in Early Modern Europe*. Baltimore: Johns Hopkins UP, 1986; ch. 3 in particular. The suitability of Descartes' universe for accommodating astronomical novelty has been forcefully argued by Rienk Vermij in *The Calvinist Copernicans. The Reception of the New Astronomy in the Dutch Republic, 1575–1750*. Amsterdam: Koninklijke Nederlandse Academie van Wetenschappen, 2002; pp. 139-142 in particular.

Modifications. Most data here used are taken from Kurd Lasswitz, *Geschichte der Atomistik vom Mittelalter bis Newton*. Leipzig: Voss, 1890 (2nd ed. 1926); vol. 2. On Descartes and de Roy I owe basic insights to Catherine Wilson, 'Descartes and the Corporeal Mind. Some Implications of the Regius Affair'. In: Stephen Gaukroger, John Schuster, & John Sutton (eds.), *Descartes' Natural Philosophy*. London: Routledge, 2000; pp. 659-679. Rohault's instruments are discussed in Mouy's book cited above and in Trevor McClaughlin, 'Descartes, Experiments, and a First Generation Cartesian, Jacques Rohault', on pp. 330-346 of the same book.

Furtherance of heliocentrism. The argument goes back mostly to the book by Vermij mentioned above, and to Eric Aiton, 'The Cartesian Vortex Theory'. In: René Taton and Curtis Wilson (eds.), *Planetary Astronomy From the Renaissance to the Rise of Astrophysics. Tycho Brahe to Newton* (vol. 2, part A of: M. Hoskin (ed.), *The General History of Astronomy*). Cambridge: University of Cambridge Press, 1989); pp. 207-221.

Power and pull. The idea of someone (like Digby) being born in the wrong century occurs in Paul F.H. Lauxtermann, *Schopenhauer's Broken World-View. Colours and Ethics Between Kant and Goethe*. Dordrecht: Kluwer Academic Publishers, 2000, pp. 4-5 (Schopenhauer and Stendhal are the examples). The passage on Digby goes back to a doctoral dissertation by J. Marius Engelbrecht, *De onttovering van de waanzin. Wetenschap, het bovennatuurlijke en de opkomst van een psychologisch mensbeeld* (to appear in 2011); the one about More, to E.A. Burtt, *The Metaphysical Foundations of Modern Physical Science. A Historical and Critical Essay*. London: Routledge & Kegan Paul, 1972 (reprint of 2nd edition of 1932; 1st edition: 1924); pp. 135-150 (ch. 5, sections C-E).

XII

LEGITIMACY IN THE BALANCE

This perhaps a little too playful one-liner invites a more-sustained comparison between those two modes of pursuing nature-knowledge, realist-mathematical science and the natural philosophy of kinetic corpuscularianism. Prominent in the comparison is what principally divided the two, to wit, the *strangeness* of the former in view of the latter, as manifest in three successive clashes between realist-mathematical science and natural philosophy of any kind. Prominent in the comparison is likewise what they held in common – the *strangeness* of an all-out violation of common sense and the *sacrilege* that, in view of many theological sensibilities at the time, was implied in both. All this together will prove to have yielded by midcentury a crisis of legitimacy of unprecedented proportions in Christian Europe.

Strangeness: Three successive clashes

Perforce we return for a while to the first half of the 17th century. Pioneers and early practitioners of realist-mathematical science, in their campaign to take up into realist-mathematical science both the classic Alexandrian subjects and many subjects never before handled mathematically, invaded three domains of nature-knowledge already occupied by natural philosophers. The resulting clashes tell us much about how the differences between the two modes of nature-knowledge were perceived at the time. The first clash, over light/heavy in connection with the problem of the conditions under which objects in a fluid float or sink, was between Galileo and a number of Pisan Aristotelians. It took place in 1611/12, by way of (almost literally) a duel, fought out first by word of mouth and then pursued in acidly polemical writing. The second clash, over the validity of Galileo's treatment of falling bodies, pitted Aristotelians of a mostly quantifying bent but also natural philosophers of the kinetic-corpuscular variety against pupils and disciples of Galileo during the late 1630s and 1640s. The third clash was over the void – whether the space created around 1644 by one of those disciples, Torricelli, in a glass tube above a column of mercury was really or only apparently void, and why such a space appears at all. These two intertwined issues were exploited by Pascal in the late 1640s as a suitable occasion to vindicate the superiority of validation-oriented experiment in the face of an assortment of *a priori* convictions upheld in various ways by Aristotelian and by kinetic-corpuscular philosophers of nature.

404 *The first clash: Light and heavy.* Archimedes, in his 'On Floating Bodies', had from one axiom derived abstract-mathematical proof that objects placed in a fluid sink or float as their weights are larger or smaller than that of the fluid, in close connection with the weight of the volume of fluid displaced thereby. These theorems had soon given occasion to the incidental determination of what has since come to be called 'specific weight'. As distinct from this, Aristotle's conception of the natural motion of objects – wholly ignored by Archimedes – implied that whether they sink or float in a fluid depends on the mix of earthy (heavy), watery (less heavy), fiery (light), and airy (less light) stuff of which any real body is made up. If the light elements prevail in the mix, it floats; if the heavy, it sinks. Further, in a brief passage at the end of his 'On the Heavens' Aristotle had specifically argued that the shape of an object, while not as a rule determining whether it floats or sinks, does determine the speed with which it sinks. In some cases, as with "flat objects of iron or lead", shape even prevents it from sinking altogether, "whereas others smaller and less heavy, if they are round or elongated – a needle for instance – sink".[213]

These quite distinct views had ever since their original conception coexisted peacefully, in the Hellenist world as in the Islamic world and in Europe. During the European Middle Ages the Aristotelian near monopoly even enabled Buridan in his account of what happens in a bathtub on top of Notre Dame (p. 83) to ignore Archimedes' account (probably due to ignorance of its very existence). In the state of restored rivalry that from c. 1450 onward had given occasion to the revitalization of Aristotelian doctrine, summary rejection sometimes found a place in a textbook, for example, Buonamici's observation (1591) that an earthenware vessel, while heavier, does stay afloat in water.

This state of peaceful coexistence changed all at once, and decisively so, with the arrival on the philosophical scene of Galileo. In 1610 he was appointed as, not just court *mathematician* but, at his very express desire, court *philosopher* as well. Within a year Galileo found himself embroiled in an acrimonious dispute with an array of Aristotelian philosophers from the University of Pisa over this very issue – what makes objects immersed in a fluid float or sink?

The conceptual tools for the head-on confrontation that followed had been latent in the Alexandrian approach to floating bodies for eighteen centuries. They now emerged from the flimsiest of occasions. Proponents on both sides were quite right in perceiving the clash to be about principal differences between two rival modes of nature-knowledge, each internally consistent and each at bottom irreconcilable with the other.

The question that sparked the confrontation was why ice floats on water. Galileo contended that water is rarefied by cold, and thus, ice is lighter than liquid water. Several Aristotelians, unable to square this response with their standard account of the four elements, argued that ice, while clearly heavier than water, remains on top due to its relatively flat, thin shape. Within days of the deadlocked, oral exchange that ensued, another Pisan Aristotelian,

Lodovico delle Colombe, countered Galileo's appeal to the Archimedean account of floating   405
and sinking by proposing a decisive experiment that involved putting a thin slice of ebony
on the water surface. Would it sink, the way an ebony sphere does, or would it float? Well, it
floats.

Delle Colombe's challenge was so forceful because it appeared to meet Galileo on his
own preferred ground – the capacity of experiment to confirm or disprove a mathematical
prediction made about some segment of the real world. Unable, however, in the absence of
our modern knowledge of surface tension, to face it head on, Galileo became enmeshed in
an intellectual duel that captivated aristocratic onlookers as it still does us over how the rules
for the duel ought properly to be set.

As Galileo saw it, it was essential to focus, not on this unfortunate floating but rather on
what happens when the piece of ebony is first wetted by submersion and then released – in
that case it stays submerged. With delle Colombe's experiment thus cut down to proper size,
Galileo performed the task that he felt to remain in his subsequent *Discorso* of 1612 (p. 294).
He first set right Buonamici's objection of the earthenware vessel, for the floating of which
one must consider not the vessel alone but the vessel together with the air contained in it. He
went on to argue that the exceptionally floating piece of dry ebony, which on close inspection appears to lie just a trifle deeper than the surrounding water surface, must likewise be
considered together with the air just above it.

His opponents were surely not the only ones to regard this argument as far-fetched and
unconvincing. To them it was obvious that, since ebony sometimes floats, sometimes sinks,
specific weight has nothing to do with the issue. It was by no means the short passage in Aristotle's 'On the Heavens' alone that made them stick to the point. If Archimedes and Galileo
were right, the polar opposition between heavy and light vanishes and with it the whole idea
of natural motion toward the center of the universe for heavy objects, away from it for light
ones, as all motion upward then becomes forced motion, and no longer natural motion.
With so basic a distinction as the one between natural and forced motion thus thoroughly
blurred, what else could still be maintained? At a point like this the very strength of natural
philosophy generally, and of Aristotle's comparatively detailed one in particular – the consistency and interconnectedness of all its component parts – turned into no less formidable
a weakness. Critical Aristotelians like Buridan had still been able to repair local breaches like
the standard account of projectile motion by means of the novel, yet well-fitting concept
of impetus. But now the doctrine was facing objections stemming from a quite different
conception of how to pursue knowledge of nature. If the breach were not plugged at the
very spot, so these skilled logicians realized, their entire natural-philosophical fortress was
in danger of tumbling down. What a piece of good luck, then, that the world of phenomena
proved to be on their side – Aristotelian doctrine saved by a chip of ebony!

As no less skilled epistemologists, these Aristotelians also knew exactly what led Galileo astray: his "delirious" delusion that mathematics can teach us anything at all about the real world – the domain proper of the philosopher of nature. In the words of one of the four Aristotelians to counter Galileo's *Discorso* in writing, Vincenzo di Grazia, it was

> necessary to show how far from the truth are those who want to demonstrate natural phenomena through mathematical reasoning. ... The natural philosopher studies natural phenomena whose essence entails movement, while, instead, the subject matter of mathematics does not comprehend movement.

> I would like Signor Galileo to adopt some philosophical propriety, because, although he adorns himself with that title, he does not behave accordingly.[214]

Not only was di Grazia reaffirming here the confinement of mathematics to the subcategory of quantitative change at best, but he also reminded his opponent how a philosopher ought properly to behave – natural-philosophical argument involved shoring up one's first principles with empirical phenomena fit to serve as evidence. In the present case empirical evidence came out on the side of the philosophers, even if they were prepared to go along with most of the rules for the duel as set by Galileo, that is, with attention focused on wetted pieces of ebony on the bottom of the water tank, not dry on top of it.

No doubt Galileo was in somewhat dire straits. The complicated role of experiment in his realist-mathematical science as the stairway connecting the upper level of idealized reality with the basement level of everyday reality virtually excludes a perfect match between effect predicted mathematically and effect observed experimentally (pp. 196). It was already hard enough to get it across as such to natural philosophers, as we shall see with the next clash. But now Galileo was compelled to explain why the clear-cut outcome of a clear-cut experiment seemingly of the very kind advocated by himself ought really to be ignored. In explaining it away, he did not confine himself to his ad hoc reasoning but used the occasion to point at another principal difference between his realist-mathematical approach and that of his opponents – how causation ought properly to run. The idea defended by them that the mix of heavy and light in any given object determines whether it floats or sinks, so he argued, is operationally worthless. After all, the only way to determine that mix in practice is circular: if an object appears to sink, this means that (barring shape) its heavy (earthy/watery) components must be predominant, and if it floats, the light (airy/fiery) ones must be. Realist-mathematical science, by contrast, does not suffer from such a causal roundabout. Floating or sinking, through the immediate cause of lesser or greater specific weight, can be used to determine, by means of comparative measurement, the underlying, relative densities. To his opponents it could not but appear as if Galileo was reversing here (as likewise with his telescopic discoveries) the proper roles of sense and reason in a particularly perverse manner. Once again di Grazia perceptively pinpointed the difference:

concerning the things that fall within the domain of the senses and that we see all the time, he wants to demonstrate them mathematically. Instead, concerning those things that cannot (or only full of imperfections) be grasped through the senses, he insists on explaining them through the senses. [215]

So once again Galileo was engaged in making a perplexing point – a point really presupposing prior understanding of the knowledge structure of realist-mathematical science, in which, indeed, experience has quite a different part to play than in natural philosophy. So in the end it was only his sharply honed rhetorical skills, joined to the lucky circumstance that delle Colombe and his league stood too low on the status ladder of the Tuscan court to require being addressed on an equal level by the court mathematician-*cum*-philosopher (we just saw how galling his opponents found the appointment), that helped Galileo out.

To the historian this first clash, for all its convolutions and the element of slapstick involved, reveals at least three important things.

Galileo had started his career in the traditionally Alexandrian mold. He had first proved his mettle in 1586 with his 'Bilancetta' – a short treatise aimed at showing by means of the theorems of 'On Floating Bodies' how Archimedes might have determined an alloy of gold and silver (p. 11). He remained an untainted Alexandrian for almost a decade more, but during his years in Padua he turned himself into the first 'Alexandrian-plus'. That is when he worked out his conception of a mathematical science of the real world (p. 185). Now, within a year of his appearance on the public scene, he managed to squeeze out of an incidental question of why ice floats on water almost the full, devastating real-world implications of the Archimedean conception of light and heavy. At long last, after eighteen centuries of latency, these implications suddenly became manifest in this first clash with those who had so far monopolized deep knowledge of the real world of nature – the philosophers. That is how, almost as soon as mathematical science turned realist, peaceful coexistence gave way to open clash.

This first clash closed off a long history of mutual indifference in that, with mathematical science turning realist, reality itself *ipso facto* turned into common and (in view of the readily apparent chasm between the two approaches) disputed ground. But the clash also pointed at the future, in that it highlighted the apparent anomaly of Galileo's self-gained professional status as a 'mathematical philosopher'. Deep trouble was clearly in the making.

Part of that trouble was bound to rest in the so far unresolved contrast between experiment set, as with the ebony chip, in a context of common sense and everyday experience and experiment set in the counterintuitively idealizing context of realist-mathematical science. Natural philosophers had known all along how to dispute with other natural philosophers. Now that the customary rules were given a very odd twist in a dispute with a self-styled 'philosopher' of a hitherto nonexistent, really illegitimate kind, they willy-nilly became engaged in an effort to find out and help settle what the new rules were. For that effort, a later clash gave ample opportunity.

The second clash: Falling bodies.   Both the previous clash and the next, on the void, were regarded by their protagonists as exemplary, so they were concerned to make their opposing views reach wide audiences. The clash to be treated now was more in the nature of a prolonged controversy aimed at getting a tough problem right. The bone of contention was Galileo's laws of falling bodies. Here is an inventory of grounds for disputing them.

Among scholars concerned with falling bodies at all, probably the most numerous were authors of university textbooks who (out of sheer ignorance mostly) maintained deep silence about Galileo's laws. These men kept handling pertinent phenomena in more or less standard Aristotelian, hence, qualitative fashion, thus treating motion as one subclass of change bound as such to return to rest unless maintained by some outside agent.

Then there were those who (while operating with a conception of motion retained that was broadly identical to Galileo's) raised the objection that one just cannot deal with the speed of falling bodies in isolation from all other phenomena. Descartes made the point when Galileo's *Discorsi* came out: "It is impossible to say anything good and solid regarding velocity without a true explication of what heaviness is, and altogether the entire system of the world." Galileo had failed to derive his views on vertical descent "in orderly fashion", that is, "without considering the primary causes of nature", so that, in just seeking "reasons for some specific effects", "he has built without foundation".[216] Only Athenian possession of a prior conception of the whole of nature allows one to deduce the true nature of heaviness. In his own *Principia philosophiae* Descartes was to treat perpendicular fall in a wholly qualitative vein, considered in its full complexity as governed by motion in whirlpools of matter.

There were further those who, while sufficiently attracted by Galileo's approach to join him on a quantitative pathway, were not prepared to take quite the same track. This was true of Gassendi, who was inclined to adopt Galileo's rules but concerned chiefly with assigning them their fitting place in his own philosophy of atoms in motion. Further, Jesuits like Le Cazre or Wendelin or Riccioli but also other contemporaries (notably Baliani and Mersenne) took on the *Discorsi* using the Aristotelian categories of 'mixed mathematics' (p. 146). Three issues with falling bodies, in particular, occupied investigators of such a bent. Do bodies fall indeed with uniform acceleration? If so, do they obey Galileo's mathematical rule for it? And what, in that case, is their actual rate of fall?

Several authors held that the very conception of fall as uniformly accelerated was mistaken because in reality, owing to the resistance of the air, the motion of a falling body quickly tends toward uniformity. That is, they missed Galileo's frequent admonitions (in the *Discorsi* especially) that his handling of the matter was an idealized abstraction. This constitutive feature of the knowledge structure of realist-mathematical science, with its central yet ut-

terly confusing claim to represent reality yet to do so in a thoroughly idealized manner, was
bound to create profound intellectual difficulties throughout the century, as in fact and for
quite the same reasons it still does in elementary science education today.

Then there was Galileo's theorem of the squared proportionality between distance covered and time passed (directly consequent upon his 'mean speed' theorem; p. 358). It has as
a corollary that with every successive interval of time the distance increases as the odd numbers: 1 in the first time interval, 3 in the second, 5 in the third, and so on. By the late 1640s,
Galileo's vacillating admirer, Marin Mersenne, tapped the boundless knowledge gained from
both his omnivorous reading and his network of correspondents all over Europe and listed
all available rivals to the odd-number rule. He came up with three. In vertical fall the distances over successive moments of time were held by Father Honoré Fabri SJ to proceed as
one by one (1, 2, 3, 4, ...); by Father Pierre le Cazre SJ in geometric progression (1, 2, 4, 8, ...);
and by Ismaël Boulliau according to some complicated sine rule. Lengthy polemics over the
merits of these various proposals naturally ensued.

Finally, there were problems of an empirically quantitative nature. The general primacy
of quantity over quality is only one aspect of realist-mathematical science and hardly its
deepest feature. But the mode of operation with quantities that it often entails in practice –
to carry out precision measurements – is easier to grasp and pursue than other, more hidden
aspects like the niceties of mathematical modeling. On this very point of actual quantitative
determination, then, Galileo's pronouncements had been elusive at best. Others did a better
job of measuring by making extended use of Galileo's proposal to employ the isochronous
swings of a pendulum by way of analogy. Giambattista Riccioli in Italy and Marin Mersenne
in France proved particularly dogged in their pursuit of quantitative precision experimentally attained.

Riccioli, with assistance rendered by dozens of colleagues from the Society of Jesus,
turned himself into a master of the faithful, collaborative counting of pendulum swings
from one star transit to the next, one sidereal day later; of the delicate push every next time
the pendulum swing looked in danger of dampening out, and of the coordinated dropping
of balls of various compositions from the top of Bologna's Torre degli Asinelli. In the absence of correspondingly sophisticated theorizing, much of the precision thus gained and
faithfully recorded really came down to pointless pseudoprecision; nonetheless, Riccioli's
ultimate figure for the constant of acceleration – 15 modern feet – proved useful as a first
approximation of the first natural constant ever established.

Mersenne's less spectacular labors on the same issue suffered from a similar lack of theoretical panache. Starting out with a ball in each hand, he dropped one and simultaneously
released the other so as to make it swing toward a stone wall (p. 343). He arrived at a value of
12 feet for the constant of acceleration, which, however, he immediately began to doubt for
both theoretical and practical reasons.

410   The two principal issues on which Mersenne conducted his plodding and inconclusive investigations up to the end of the 1640s – the correct relation between distance and time in vertical fall and the quantitative determination of its rate of acceleration – were brought to quick resolution by a precocious youngster, the second son of Mersenne's friend Constantijn Huygens. In 1646 Christiaan coolly pointed out to Mersenne that, of the whole list of 'possible' laws of fall he had so meticulously assembled, all but one were mathematically impossible. Only Galileo's odd-number rule satisfies the principal requirement, overlooked by Mersenne, that the law holds irrespective of the unit of time chosen. Eleven years later Huygens repeated Mersenne's experiment with the two balls and attained similarly inconclusive results. He at once drew this issue, too, back into its proper domain of mathematical-realist abstraction. He set up a profound, theory-laden analogy with a conical pendulum, which yielded at once the locally exact determination of the constant of acceleration and, within three more, highly creative months, the isochrony of the cycloid and proof thereof (p. 342).

In conclusion, three distinct parties were engaged in the controversies of the 1630s and 1640s over the merits and further pursuit of Galileo's new conception of motion as exemplified in his treatment of falling bodies. One party was probing the power of realist-mathematical science newly created from Alexandrian roots. The other two were both conceiving of nature in customarily catch-all Athenian fashion, but otherwise they went different ways. The kinetic corpuscularians shared with the Galileans this novel idea of motion. The mixed-mathematicians *seemed* to share with the Galileans an urge to treat things the mathematical way, while actually thinking and operating all the time in terms of commonsense reality approached (if and where deemed appropriate) the quantitative way.

The third clash: Void space.   Prior to the 1630s the question of whether void spaces exist or not was an issue in natural philosophy. That is, it was a subject of speculation. Empirical evidence was surely adduced pro and con, yet convictions were established on a priori grounds. To atomists the existence of large, void spaces was a fundamental component of their atoms-in-the-void conception of the world. Stoics, with their universe-filling *pneuma*, denied it. So did the followers of Aristotle, who had proved the logical impossibility of the void in his campaign against atomism. Nothing in these debates stood in the way of, or was in any way influenced by, the unwitting creation of void spaces whenever artisans tried to make a suction pump operate over a height of more than c. 10 meters – at this height, the column of water 'breaks' (in contemporary parlance; p. 130). A connection between theoretical debate and unwitting creation was made by Galileo during his frequent visits to the Venetian Arsenal, as he recalled on page 1 of the *Discorsi*. Practitioners took the inability of suction pumps to raise water any higher as one of those facts of life one just has to accept, ascribing it at best to accidental, material shortcomings. In contrast, Galileo was the first theoretician to begin to ask questions on the subject and to seek for less ad hoc answers. He realized that the void was somehow involved.

Soon after his death in 1642, his disciples experimentally pursued the matter further. Torricelli was drenched in the waters of Archimedean thought turned realist, and to him the phenomenon suggested the presence of a balance between, on the one hand, the column of water inside the pump tube resting upon the surface water at ground level and, on the other, a 'column' (not as such visible) of atmospheric air equally weighing upon that surface. That is, at breaking point the column of water weighs just as much as the corresponding column of air. Hence, beyond the breaking point no air is left above the column of water; hence, a void space has inadvertently been created. But if this is so, Torricelli reasoned, the phenomenon can be made both clearly visible and liable to further examination by using, not water (for almost nowhere was it yet possible to have 10-meter-long glass tubes blown), but the liquid with the heaviest specific weight known, mercury. And indeed, in the 'barometer' thus produced, the mercury column appeared to 'break' at a much lower point than a water column did, namely, at a height of about 76 centimeters (c. 2 ½ feet; see Figure 13.5 on p. 493), in neat correspondence to the 1:14 proportion of specific weights as established by application of Archimedes' bathtub rule.

News of the experiment quickly spread. Repeating it proved easier, however, than determining what it signified. Two major questions were involved: Is the space above the mercury column (Torricelli's space) truly 'void'? And is Torricelli's balance explanation valid indeed? On neither of these two distinct yet obviously intertwined issues was it easy to remain noncommittal, as a good deal was at stake for students of nature of many different stripes.

Aristotelians were *a priori* committed to denial of both contentions. Obviously they could not go along with recognition of the actual existence of void space. But neither could they accept the inference, already drawn by Galileo from Archimedes' conception of buoyancy and now apparently reinforced by Torricelli's balance explanation, that their polar opposition between 'heavy' and 'light' had to be given up in favor of one continuous scale of diminishing heaviness. Even less could they adopt the further inference now being drawn that we really find ourselves living 'at the bottom of a sea of air'.[217]

Atomists (whether of the traditional or of Gassendi's kinetic variety) of course went along with the void, on likewise *a priori* grounds. In the balance explanation they held no particular stake.

Kinetic corpuscularians of the Cartesian variety, who asserted space on *a priori* grounds to be coextensive with matter, were bound to reject the void. But the same Cartesians, reinforced by Descartes' inclination to reduce specific cases of motion to the general case of a hydrostatic balance (p. 294), were only too happy to incorporate into their system Torricelli's explanation of his barometric experiment.

Meanwhile, practitioners of realist-mathematical science like Galileo's own pupils were of course the very ones originally to conceive of both the experiment and the explanation of its outcome. To them, neither the outcome nor its explanation was a matter of commitment

412 on *a priori* grounds. Rather, it was an instance of the Galilean mode of acquiring knowledge of a specific, attention-catching phenomenon of nature through ongoing interplay between *a priori* mathematical prediction and *a posteriori* experimental checking.

It is unknown to what, if any, extent the pioneer, Torricelli, felt tempted to rise to the occasion that presented itself here. Certainly, in post-trial Italy, and with atomism and its risky theological implications possibly involved, he was on shaky ground, and in any case he died too soon to make a big splash out of what at the surface was just one disputable issue of nature-knowledge among others. Instead, some four years after the original discovery Blaise Pascal became the man to clinch the dual argument for Torricelli's space being void and due to a two-column balance. With the barometer, so Pascal saw, adherents to the new, mathematical-experimental mode of pursuing nature-knowledge had been handed on a platter an opportunity to display in spectacular manner the capacity of experiment to serve as an arbiter in settling properties of nature. Actually, for him the stakes were higher still. The issue gave him an opportunity to combat at one and the same stroke the speculative approach to nature *and* the reason-guided approach to Christian faith that he saw embodied (albeit in different ways and on different grounds) in Aristotelian as also in Cartesian doctrine – his two intellectual archenemies for these very reasons.

Pascal's principal moves were to shift the issue from the speculative to the experimentally testable and to proceed with the greatest possible caution so as never to infer more than the evidence strictly allowed. Thus, he took care for a long while to speak, not of the void, but of the 'apparent void'. And in the beginning he even granted his opponents their celebrated *fuga vacui* (i.e., the alleged tendency of nature to intervene as soon as a void threatens to get established), while adding at once that, with the mercury column, that tendency was apparently overcome at a height of 76 centimeters. It was pointless, so he further insisted, to uphold that nothing whatever or, conversely, that some hitherto unknown, insensible stuff was present in Torricelli's space. At issue was rather whether or not that space contained any matter of a known or at least readily knowable (i.e., sensibly detectable) kind. Thus, in his first short publication on the subject (dated 8 October 1647) he listed all the suggestions that had been made as to what kind of stuff, in filling Torricelli's space, might be holding the column down, to wit, (1) air introduced from the outside through the tube's pores; (2) air from the pores of the mercury; (3) air from pores between the tube and the mercury; (4) a tiny bubble of air accidentally introduced when the experimenter's thumb failed completely to close off the tube and then rarefied (of which bubble Pascal grimly remarked that "some would maintain that it can rarefy to the point of filling the whole world, rather than admit the void");[218] (5) a tiny portion of vaporized mercury drawn in both from the tube's wall and by the force of the liquid mercury; (6) the mercury's spirits; (7) a subtler kind of air (of unmistakably Cartesian origin) pervading the air outside and introduced by way of the tube's pores and tending through mutual attraction to return to the outside; and (8) some unknown stuff not detectable by any of the senses.

As it happened, the man to come to the rescue of a filled Torricellian space, Etienne Noël, was an ideal opponent for Pascal. He was a syncretist Aristotelian/Cartesian. He was a Jesuit and, as such, the very embodiment of reason-guided as distinct from revelation-guided theology. He was, in short, almost a living caricature of all that Pascal found most reprehensible in the imagination-prone, speculative, know-all mode of acquiring knowledge of nature. Pascal had already demonstrated in spectacular fashion that suggestion (6) just fails to hold. He had set up two large tubes filled with water and wine, respectively, and then invited spectators to tell in advance which would stand higher, more 'spirituous' yet specifically lighter wine or water – they thought water, but the wine did. He had also undertaken (or, at least, succinctly described in his short tract) an array of auxiliary experiments partly of his own design, all meant to show that those other plausible-looking suggestions of what might still be imagined to fill Torricelli's space ran counter to what he kept finding – irrespective of the shape of the tube or of its tilt, etc., the column of mercury keeps 'breaking' at c. 76 centimeters. Noël's rushed rebuttal provided for Pascal a field day (29 October 1647) for squeezing out of these experimental outcomes of his all he could by way of a disquisition on elementary scientific reasoning. Together with Galileo's *Dialogo*, Pascal's 'Letter to Noël' is the most accessible text marking, in his dazzling, crystal-clear style replete with memorable one-liners and rhetorical flourishes yet also transcending these due to the very nature of his message, what makes speculative natural philosophy effectively barren and the mathematical-experimental approach liable to lead in the end to truly conclusive outcomes.

No doubt Pascal overdid it on both counts. Not only can we see in retrospect that he failed to perceive the possibility (perceived indeed by some of the next generation, notably Huygens and Newton) of replacing closed systems of speculative thought about particles in motion, not with a flat refusal to consider their 'machinery' at all, but with specific, open-minded hypotheses to be drawn up about them. He further laid down impossibly strict rules for falsification. Also, at a slightly later date he made an effort to clinch the matter of the balance explanation by means of an experiment that was fit indeed to get the still hesitant on board but that he sought to sell as if it necessarily settled the matter once and for all. He considered that, if indeed a balance between the columns of air and of mercury is in play, decisive proof can be found by diminishing the height (and, hence, the weight) of the former and then seeing whether the latter diminishes accordingly. In Paris it was of course hard to alter deliberately what we now call atmospheric pressure. Pascal's brother-in-law, who lived in Clermont near the Puy-de-Dôme in southern France, was in a better position to do just that. Walking up its slopes on 19 September 1648, he could not help observing what Blaise in his extensive instructions had led him to expect – that the mercury level in the barometer he carried with him sank as he ascended.

Pascal's sense of definitive proof notwithstanding, the experiment still left it possible to argue consistently that, with the *fuga vacui* having to withstand increasing rarefaction of the

414   air outside, so much the less of it remains for withstanding rarefaction inside, thus inducing the very drop of the mercury level actually observed in the tube. In empirical science
unambiguous proof is just not to be had. The experiment made a great impression in any
case – together with Newton's 'crucial experiment' on color sifting (1665/72; p. 685) it is easily
the most celebrated genuine experiment of the entire Scientific Revolution. In 1691 a French
Jesuit, Father Gabriel Daniel, was to observe that "not one little philosopher of nature is left
who does not know inside out the story of Mr. Pascal's experiment".[219] Pascal himself went
on to expand systematically Torricelli's explanation into his twin *Treatises on the Equilibrium
of Fluids and on the Weight of the Mass of Air* of 1654/63 (p. 294).

Pascal certainly achieved a thing or two in helping students of nature overcome the
widely perceived strangeness of experiment-in-mathematical-context as a (if carefully executed) valid mode of gaining nature-knowledge. It nevertheless remains true that (with
the sole exception of the experiment on the Puy-de-Dôme) Pascal, just like Galileo, left
descriptions of his experiments so brief and formal as to give rise to doubt over whether he
had truly performed them all. In his case, these doubts were voiced early on by Boyle, who
advocated quite a different ideal of experimental science (p. 491). Further, as Boyle equally
began to insist from the mid-1660s onward, not so much the weight as, rather, the pressure
of the air is the relevant variable. Nor had the battle between speculative and ultimately
empirical modes of pursuing nature-knowledge been won for good. It never has – the history of their mutual relations has not been one of complete elimination but has remained
one of shifting balance, albeit shifting very much in one direction (p. 603). Even the debate
over the void itself had to be done all over again within twenty years (p. 484). Finally, during
the 18th century the composite nature of air was discovered, and theoretical and empirical
distinctions between gases and vapors began to be made. Then it appeared that Torricelli's
space, while devoid indeed of any substance of the kind known at the time of the Scientific
Revolution, is filled with extremely rare, saturated mercury vapor engaged in an ongoing
balancing act with the liquid below, in some superficial likeness therefore to solution (5).

Nothing in all these contemporary overstatements and subsequent refinements detracts
from the core message Pascal perceived the issue of the void to be well-suited to drive home.
For arriving at valid insights into the operations of nature, one is well advised to undertake,
not *a priori* speculation about the whole wide world, but the deliberate setting up, guided by
the peculiarities of the case at hand, of a careful interplay between *a priori* reasoning and *a
posteriori* experimental confirmation.

Conclusion.   Ever since Athens and Alexandria had come into being, the lack of almost
any overlapping territory had enabled these distinct modes of pursuing nature-knowledge
to coexist peacefully. But as soon as, due to the rise of Alexandria-plus, natural reality itself
turned into the claimed sole territory of *both*, Athens (whether 'plus' or not) and Alexan-

dria-plus were bound to clash. Thanks to Galileo's foresight and success in ascertaining philosophers' rights they even did so within a year of his coming out of his neo-Alexandrian closet. In the course of the clashes that thus ensued the Aristotelians found themselves increasingly thrown back upon their own resources – their master had never been challenged by inferences about light and heavy, mathematical rules for falling bodies, or experiments and precision measurements with mercury. In this respect the kinetic corpuscularians found themselves in a somewhat more favorable situation, being less on the defensive due partly to overlapping views (on motion and hydrostatic balance) and partly to greater flexibility in incorporating new finds into their grand scheme of things.

The head-on collision between natural philosophy and realist-mathematical science manifested itself right from the start in the aura of *strangeness* that surrounded realist-mathematical science in the perception of natural philosophers of whatever bent and that was articulated so ably by di Grazia and Descartes. Two quite distinct knowledge structures were involved. I have discussed both at length (pp. 195; 231; 238) and so it suffices to highlight them one final time by means of a brief passage in Pascal's response to Noël. Among the latter's arguments against Torricelli's space being void was that light passes through it. Since it cannot do so without a carrier, something inside the tube must be there to carry it. This was by no means Noël's weakest point. The customary way of countering it in natural-philosophical debate would have been for Pascal to argue that, despite what Noël thought, the nature of light is such as very well to allow its passage through the void. The striking thing about Pascal's rejoinder is that he deliberately *abstained* from such a counterclaim altogether. Instead, he appealed to our present ignorance, perhaps to be repaired in future: We do not know as yet what the nature of light is – Noël does not and I do not – hence, we cannot as yet derive any argument either pro or con the void from that as yet unknown nature of light. It has remained a classic locus for the deliberately fragmentary, modestly step-by-step, future-oriented mode of proceeding characteristic of mathematical-experimental science and distinct from the all-encompassing claims of natural philosophy, be these laid down in books entitled, to Pascal's horror, in Aristotelian fashion *De omni scibili* ('On all that can be known') or *Principia philosophiae* the Cartesian way.[220]

Strangeness: Against common sense

So much about the strangeness of realist-mathematical science *as a mode of gaining knowledge about nature*. It is only natural that what struck other people than the practitioners themselves as so strange about its unprecedented ways of proceeding was brought to the fore and was most clearly articulated in clashes with its direct rivals, the philosophers of nature. But its aura of strangeness extended far beyond, really to everyone located outside its esoteric-looking precincts. Nor has it ceased to do so – it has remained a mode of gaining knowledge of nature that requires ever more specialist in-depth training, including the conscious *un*learning of modes of thought and ways of approach that come much more naturally to us. Why more natural to us? Because they are more in conformity with common sense and everyday experience. From his examination of children's cognitive abilities, Jean Piaget, the early-20th-century psychologist, almost identified their approach to natural phenomena with Aristotle's. One need not go all the way with him to see that one of the attractions of Aristotelianism resided in how consistently it succeeded in offering a thoroughly reasoned account of what common sense teaches us and everyday experience shows us.

So, realist-mathematical science and scholastic natural philosophy, 'Galileo' and 'Aristotle', were opposed both in the counterintuitive *approach of* and in the counterintuitive *content of results attained in* the former. Between realist-mathematical science and the natural philosophy of kinetic corpuscularianism, between 'Galileo' and 'Descartes', things were not so simple. On the *inside*, their distinct knowledge structures made for clashes that were hardly different from, and as a rule coextensive with, clashes with Aristotelians and, on occasion, natural philosophers of still other bents. But what Galileo's and Descartes' teachings held in common, and what was most striking about them from the *outside*, was that both appeared to go against common sense and everyday experience. In other words, both upheld a counterintuitive conception of the world.

Kinetic corpuscularianism, which after all dealt with the real world, did so in a very clear-cut, explicit manner – by positing a chasm between our surface world of perceived phenomena and the underlying world of particles in motion (pp. 230; 381). The atomist philosophy of nature, to be sure, had posited such a chasm all along. But the new focus on how particles actually move according to inexorable laws of nature caused the kinetic corpuscularians to handle the world in so much more intricate detail as to make this radical dichotomy between a surface world of sense and a subvisible world of moving particles so much the more conspicuous and ineluctable – hence the ready-made, widespread image of the kinetic-corpuscular world as a clock (p. 386). But in what, if any, manner did such an image fit realist-mathematical science? In being programmatically confined to the piecemeal solution of single issues (however dynamically on the move from one to the other), realist-mathematical science, quite unlike kinetic corpuscularianism, did not have on offer any large-scale conception of the world. Or did it?

In two respects it did, one constitutive, the other secondary. Constitutive was the conviction that the right way to apprehend phenomena of sense was not indeed through the senses but through the mathematical regularities held to underlie them. The world of sense is incorporated in realist-mathematical science as a rule by way of artificial experiment and then comes into its own in the secondary role of possible corrective. This core conviction might take diverse guises – with Kepler, as an ontology of harmonic archetypes; with Galileo, as a view of the world as expressed in mathematical symbols; with most later adherents, as a variety of more instrumental, not so ontologically committed views. A counterintuitive dichotomy between the everyday phenomenon and its theoretical, mathematically ideal counterpart marked it in any case. This is not, to be sure, the same dichotomy as the one that characterized kinetic corpuscularianism. Looked at from the outside, however, the two dichotomies might easily be taken as roughly coextensive. Such a view was only reinforced by Descartes' insistence that his world had been derived along the 'mathematical' pathway of clarity and distinctness and that its particles were constituted by the likewise 'mathematical' properties of figure, shape, and configuration alone.

Secondary yet quite deliberate was a feature that the two modes of nature-knowledge had literally (not just from afar) in common and that was particularly suitable to being turned into an exemplary case, a symbol even. This was the narrow link forged with heliocentrism taken as real. Not only was this doctrine as counterintuitive as any then around, but it was so in a manner more readily grasped than the knowledge structure of realist-mathematical science ever could be. Besides, obviously, the inherent interest of realist Copernicanism to anyone who had ever gazed in some wonder at the night sky, it was this very feature that made Galileo opt for it as the best vehicle available for making his mathematical realism conquer the world. Due partly to institutional and partly to character differences, he was bent on that conquest in a way that Kepler never was. Kepler was content to remain the court mathematician, even though he was sure that the mathematical astronomer really ought to turn himself into a 'celestial physicist'. He did not care whether his colleagues might need a hundred years to find that out for themselves. Galileo, by contrast, was thoroughly exasperated by the impossibility, to which his inferior status as a mathematics professor doomed him in Padua, of gaining a hearing from the philosophers. That is how he came to use his telescopic discovery of Jupiter's satellites as the ideal stepping-stone for two twin feats. As a gift for naming these satellites the 'Medicean stars', he received from his new patron, Cosimo II de'Medici, the title of court mathematician *and* philosopher. This way, he compelled the philosophers to listen to what he, the first exemplar of so utterly paradoxical a social role as that of 'mathematical philosopher', had to tell them. Further, by way of a soft sell of his counterintuitive, mathematical science of the real world, he chose for his preferred terrain the vindication of Copernicus' heliocentric setup as unambiguously real. He thereby provoked a clash, not only with the philosophers as he foresaw and intended all along, but, unforeseen and unintended, with the ruling Church authorities as well.

Sacrilege: Three successive clashes

Look again at Schema 4, 'Dates of the onset of the Scientific Revolution' (p. 281), and con-

418   sider the nonitalicized dates, which denote delay (or, in one case, absence) of publication of the most constitutive works of the Scientific Revolution in its pioneering stage. In the cases of the two Protestants, Kepler and Beeckman, the delay or absence had none but personal reasons – trouble made by Tycho's heirs in the one case and a lack of capacity to set up an orderly argument in the other. But in all other cases the reason for the delay resided either in strangeness of the kind discussed above or in sacrilege or in a combination of the two.

Gassendi's anticipation of the serious risk his work ran of being perceived as sacrilegious and then prohibited held him back for two decades (in other words, he censored himself). Van Helmont's actual persecution by the religious authorities of the Spanish Netherlands held him back for the rest of his life.

In the cases of Galileo and Descartes the situation was more complex. Galileo's reticence over his revolutionary Alexandria-plus work at Padua while publicly teaching cosmology from Sacrobosco's *Sphaera* was due to his awareness that the very notion of 'mathematical philosopher' must look extremely odd to natural philosophers of the common run. As a mere mathematics professor he knew well that these colleagues were in no mood or had no need to give his pathbreaking work a hearing even if he tried to get it. His position at the court of the Medici plus his decision to make Copernicanism serve as his showcase 'mathematical philosophy' backfired in the end, leading to the Roman ban on the doctrine pronounced in 1616 on theological grounds and then, in 1633, to the condemnation of Galileo himself for having trespassed it. It was news of this very condemnation that moved Descartes at once to delay publication of his own Athens-plus natural philosophy. It did not appear until he had published a metaphysics that was meant to furnish him with a theologically safe haven but that proved powerless to prevent an outburst on grounds of impiety at one university in the country where he lived, the Netherlands. Publication also had to await his finding a way to reconcile his Copernicanism with a nominally fixed Earth. Posthumous publication of his unfinished works and his letters then provoked a third clash with the authorities, this time in France.

I shall now discuss *and* compare these three major clashes, between Galileo and the papacy; between Descartes and a leading, orthodox-Protestant clergyman, and between Descartes' French followers and several clerical agencies as well as the French king himself. The comparison proceeds as follows. I first provide succinct accounts of the principal acts, arguments, and motives of the parties involved in each case separately. Next I examine the nature-knowledge-related, the faith-related, the institutional, and the political forces and constraints behind these clashes. A definite pattern emerges from the comparison – the perceived and actual consequences of these forces and constraints together indicated a big crisis of legitimacy for Alexandria-plus and Athens-plus combined.

Galileo vs. the papacy.   There is no more heavily charged topic in the entire history
of science than the clash between Galileo and his Church. It would be pointless to claim
sovereign objectivity for my own position. As far as the bias of sentiment goes, it is marked
by (1) intense admiration for Galileo's achievement and his spirited defense thereof, (2) lack
of a personal stake in doctrinal faith of any kind, and (3) an Enlightenment sense of indignation over most positions adopted and measures taken in the dispute by Church officials
– the indignation mitigated to be sure by due effort at historical understanding of definitely
pre-Enlightenment times. These sentiments notwithstanding, my reading of the documentary evidence moves me to go along with one particular view shared on the whole with
the 'Church party' in the ongoing historical controversy. I am unable to swallow Galileo's
portrayal of himself as seeking in vain to hold Roman authorities back from a condemnation of heliocentrism to which they tended for reasons quite their own. Rather, Galileo
actively sought the Church's endorsement of the doctrine, thus unleashing the whole affair
and tragically attaining in the end the very opposite of what he had striven for.

The principal motive for Galileo's active stance rested in his desire to use his newly gained
court platform for launching his unambiguously realist Copernicanism as his showcase for
his newly realist conception of a mathematical science of natural phenomena. That is, he
wished to set 'mathematical philosophy' up as the only valid philosophy of nature. What
were the attractions for him of setting out on such a campaign?

The outcome of his clash with the philosophers over light and heavy is instructive. The
controversy had been set off by the innocent question of why ice floats on water (p. 404).
Questions of this kind – asked on a prince's or courtier's whim; taken from everyday experience; inviting an answer more or less on the spot – formed the lifeblood of intellectual
sparring bouts at court. At a premium (metaphorically but also literally so) at Italian courts,
in particular, was the quick-witted rebuttal, preferably in a duel-like setting fit for princely
entertainment. However well Galileo's rhetorical skills were up to games of that sort, they
could also land him in considerable difficulty as delle Colombe's ebony chip revealed. Nor
did such a setting, which demanded a precipitous commitment that one could hardly renege
upon in the sequel, provide the ideal platform for laboriously setting forth the intricacies
of his 'mathematical philosophy'. Indeed, its built-in discrepancy between everyday and idealized levels of reality renders issues like why ice floats on water (or how objects fall) far
more complex than meets the eye. The setting fitted the less as Galileo himself, by means of
his telescopic discoveries and how he had gone out of his way to peddle them, had set the
standard of the age for a new intellectual commodity with courtly entertainment value –
the production of sheer novelty, the more spectacular the better. How to avoid the trap in
future? Once the controversy of 1611/12 was mercifully behind him, Galileo changed tactics.
For a vehicle for his mathematical philosophy vastly more effective on all these counts, he
now chose the realist Copernicanism he had already begun to commit himself to in public

420 anyway. A 'mathematical philosophy' not esoteric-looking but suitable for pulling on board the mathematically unskilled; not riskily dependent on questions asked on a whim but solid in and of itself (or so he thought); and also a ready-made source of, *and* vessel for, such novelties as might come along (e.g., sunspots, Venus' phases) – small wonder that, as soon as an opportunity presented itself to tackle those worried about the heliocentric message, Galileo leaped at it. The occasion was the indirect, semiofficial voicing, by the Grand Duke's mother on 12 December 1613, of the family's concern over their emblematic Medicean stars. Were these not fake products of the instrument after all? And was not the heliocentrism Galileo held them to underwrite contrary to scripture?

Within a week of learning about these queries Galileo penned a deliberately semipublic response ('Letter to Castelli'). Within two more years he expanded it into his famous 'Letter to Grand Duchess Christina'. Standard fare in these letters was Galileo's appeal to the exegetical principle of 'accommodation' – the Bible has not been meant as a textbook of astronomy, and those who wrote God's word down have adapted themselves to common speech (p. 112). Not strictly new yet in his time decidedly unusual was his express worry (undoubtedly a heart-felt motive) that the Church, as guardian of eternal truth, would be ill-advised to commit itself to a view ('the Earth stands still') which might prove wrong, the way St. Augustine had argued in his time that Bible passages suggesting a flat Earth ought not to be taken literally in view of the availability of sufficient evidence for its spherical shape. From this Galileo inferred the fully novel claim that making exegetical decisions on subjects like these with proper caution required the expertise of the mathematical philosopher. He did not, to be sure, claim for the expert a right to do the reinterpretation; instead, he expressed his confidence that theologians would know how to pull that feat off once the experts had informed them of the true nature of some state of affairs in nature with biblical implications. Still, over the letter hovers the mathematical philosopher setting himself up as ultimate exegetical arbiter. A case in point was how in his letter Galileo handled the biblical miracle related in Joshua 10:12–13. In voicing her doubts, the Grand Duchess had appealed to the story of how, in order to help Joshua complete the slaughter of his adversaries, God made the Sun stop for a while ('Sun, stand thou still'). On the accustomed view, the story highlights a moving Sun as the nonmiraculous, natural state of things. Galileo now took it upon himself in all due humility to reinterpret it as confirming (mind you, on a *literal* yet, according to Galileo, closer reading and more thorough understanding) just the opposite.

The true historical riddle about the ban on Copernicanism that ensued within three more years does not rest in the charges of heresy some men of the lower Tuscan clergy, on learning of these views of Galileo's, filed and then attested to in Rome. It rather rests in the following sequence. By late 1615 Galileo descended upon Rome to clinch the case set forth in his two letters by the force of his spoken word. Both the Tuscan ambassador and Cardinal Bellarmino, who knew their pope but also Galileo's fervor in making his cases, sought in

vain to warn him off. In spite of (or, much more likely, adversely spurred on by) that very
fervor, the pope charged a high Inquisition committee with preparing a decision. The true
historical riddle, then, rests in how that committee, and the pope himself, came to the ban
and had it imposed on Galileo. Particularly disputed is the role of Rome's foremost theologian at the time, Cardinal Bellarmino. Although not on the formal committee, he used to
act as an adviser to the Inquisition. Did he contribute to the ban, or was he overruled by the
pope in seeking to prevent it? We shall never know; what we do know is that neither the ban
itself nor its remarkably clumsy phrasing can easily be squared with a succinct letter that
Bellarmino had written one year previously for Galileo's benefit.

The letter laid down for the first time an authoritative, clear-cut position on Bible exegesis in specific connection with the Copernican doctrine. It was also the last time – one
year later the ban prohibited all further discussion. The cardinal was a high-placed member
of the Society of Jesus and regularly in touch with his brethren at the Collegio Romano, its
astronomers included. In addressing the astronomical issues involved, he knew what he was
talking about. He left his own, partly stoic, partly biblical views on the constitution of the
heavens out of account. Rather, he was concerned to argue that reinterpretation of Bible
passages in ways that deviate from the tradition of the Church fathers is a very serious business and ought not to be undertaken lightly. He therefore confined possible reinterpretation
as non-literal of Bible passages that had hitherto been taken literally to cases both real (not
just 'saving the appearances', that is) *and actually proven*. The cardinal ended his letter by
pointedly observing that so far he had not seen any proof. He left it unspecified what might
actually count as proof, but even so the challenge was obvious. If Galileo did not come forward with proof, no reinterpretation of the Bible and no Church endorsement of terrestrial
motion were in the offing.

How best to meet the challenge? That is, how to deal with Bellarmino's requirement of
proof? It is telling for Galileo's activist posture in the matter that he spurned the obvious
way out – to exploit what common ground there was and seek for precedent. Common
ground existed after all in St. Augustine's accommodation gambit, which Galileo appealed to
in his letter to the Grand Duchess and which Bellarmino, too, endorsed as authoritative. The
conditions for nonliteral, accommodation-inspired interpretation of Bible passages that
Bellarmino invoked were the same ones that had many centuries before been allowed by the
Church on the issue of the spherical shape of the Earth. Why not take this precedent as a
starting point for finding out what, to Bellarmino, would count as proof? Galileo had made
the point in his own letter; so had Kepler in the preface to his *Astronomia nova* of 1609. Why
not, therefore, take Bellarmino's letter as really asking for nothing more than that, to make
nonliteral interpretation a viable proposal, motion rather than rest of the Earth be sustained
by reasoned argument to at least the same degree of decisiveness as once adduced in favor
of its being round, not flat?

422   But this is not how Galileo chose to respond to Bellarmino's challenge. At first he confined himself to sketching possible rebuttals of Bellarmino's points by way of demonstrating the various weaknesses of the arguments of those who opposed terrestrial motion. Only during his stay in Rome did he change tactics. Now he directed himself, not to finding out empirically what would count as proof in view of precedent, but rather to writing down what he fancied to be sufficient proof: his explanation of the tides through the Earth's dual motion (p. 190). He submitted this text, *not to Bellarmino first,* but straight to the pope. And this is how, against all his intentions (yet well-warned-off) he provoked the very ban that now intervened to bind his hands.

To what extent did it bind him? This became the central issue on three successive occasions. It came up in 1624, when Galileo believed from interviews with the new pope, who admired him, that he had gained license to resume his original campaign. Eight years later it arose anew over what hide-and-seek games and double entendres were required to get the *imprimatur* for the book resulting from his campaign, the *Dialogo.* In 1633 it was the decisive issue at the trial provoked by the Inquisition's acute concern and then the pope's fury over that book. From a theological point of view the trial brought nothing new. Its net effect was to create a Europe-wide scandal by highlighting the ban of 1616 and by the inquisitorial proceedings (culminating in enforced abjuration) brought to bear on Galileo by way of punishment for trespassing it. From the point of view of what Galileo aimed to accomplish with his *Dialogo,* the ban pronounced upon the book did not prevent its appearance in Latin translation within two years of the ban and its spread over the Inquisition-free regions of Europe. There it succeeded well in getting across in an accessible manner basic elements of Galileo's 'mathematical philosophy' by way of a defense of the Copernican doctrine taken as real.

Descartes versus Voet.   One side effect of the trial, as news of it spread, was to enhance greatly the innate caution of the then thirty-seven-year-old Descartes. He at once filed away his 'Monde'. Four years later, his first publication (*Discours de la méthode* and appended essays, which he at first left anonymous so as to test the waters further) left his ideas on the constitution of the heavens largely in the dark. The 1630s were the years when, his methodological concerns behind him, he settled down in the Netherlands to give final shape to what he regarded as his principal feat as a scholar, his philosophy of nature (p. 227). His manuscript 'The world' contained accounts of how the universe, created by God out of particles obeying the laws of motion He had equipped them with, has acquired its present shape through the sheer operation of these laws. He left unfinished a chapter that, when published posthumously in 1662, was separately entitled 'L'homme' ('Man'). Here he set forth how, in the human body, actions of particles in motion bring about most acts customarily ascribed to the agency of the soul. Comments received about what little he revealed in *Discours de la méthode* persuaded him of the prudence of a prior excursion into metaphysics and theology.

In his *Meditationes de prima philosophia* (1641, 'Meditations on first philosophy'), Descartes set out by going skeptic all the way, only to reverse and ground certainty regained in the '*cogito*' (p. 236), and then to go on to prove God's existence and the soul's immortality.

Tellingly, the argument leading to the second proof is hard to square, if at all, with what both before and later was his chief concern about mind and body: to materialize the explanation of mental functions as far as he could conceive of bodily mechanisms fit to account for them. As a result, only self-reflection and intellectual creativity (in both of which Descartes felt he excelled) remained as unreservedly mental capacities distinguishing the human machine from its animal counterpart. All this was quite unlike the teaching of the schools, which distinguished three souls: a 'vegetative', an 'animal', and a multitasking 'rational soul' that distinguishes us from the beasts. However, it was also quite unlike the tack Descartes took in his 'Meditations'. There the soul appears as independent of the body, such that it might go on existing apart from it. That is, the 'push-back-the-soul' program of 'Man' is replaced here with even imagination and perception being oddly taken as incorporeal. And there is another telling difference. As a rule, Descartes displayed little interest in how mind and body may be thought to interact. Only in the 'Meditations' is this lack of interest replaced with an account that culminates in his famous localization of the soul in the pituitary gland.

The historical point of this little exercise in Cartesian double-think rests in the first two men to notice it from opposite points of view and in how Descartes allowed himself to get embroiled in the clash between them. One was Henrick de Roy (Regius), the other Gijsbert Voet(ius) (p. 390). The latter clergyman was the chief spokesman of the influential movement of 'Nadere reformatie' ('Extended Reformation') aimed at setting up what, if successful, would have amounted to a theocracy in the Netherlands: a ban on all religious doctrines but the Calvinist and clergymen having the final say in all state decisions that counted. He was also a theology professor and, in 1641, rector of the recently founded University of Utrecht, where de Roy, under the approving eye of his master, used his chair in medicine to spread the Cartesian word among his students. Already under Voet's suspicion for that reason, de Roy decided to needle his rector further by allowing several students to defend theses that concerned doctrines stemming from Descartes' manuscript chapter 'Man'. Descartes was delighted with so bright a disciple and with this apparent opportunity to strengthen his foothold at a university. He therefore sustained the theses, until the row that promptly ensued caused him to dissociate himself from de Roy in private for his lack of caution in irresponsibly going beyond Descartes' 'meditational' views, and later in public, too.

The row, then, came about through a number of countertheses inspired (if not written) by Voet. One of these went against Descartes' (for Voet knew very well who stood behind de Roy) true idea of the soul, another against Galileo's and Kepler's heliocentrism. But the thrust of the whole appeared most clearly from a third counterthesis:

424   the essence and existence of [substantial] forms is being denied. ... Once this dangerous axiom is granted, the vanity, skepticism, and excessive license of the human mind shall lead it down the slope of adversely arguing that there is no rational soul, no generation and conception of man in the mother's womb, no wind, no light, no Trinity, no Incarnation, no original sin, no miracles, no prophecies, no awakening of a sense of God in the human mind and will, no regeneration of man through God's grace, no demonic action inside the body of man or around his mind, etc.[221]

In short, take the core of Aristotelianism away and you render the world incomprehensible and rob Christian doctrine of all its basic tenets.

De Roy's immediate counterblast of February 1642 gave Voet the desired opportunity to convene the professors' assembly and have it lodge formal complaint with the city council charged with overseeing the university. The move led, not indeed to the dismissal sought by Voet, but to an order to de Roy strictly to confine himself to his own field of medicine and no longer meddle in philosophy. It was not of course so easy to get a hold on Descartes himself. He, however, upon counsel received from theologically moderate acquaintances high up in the city's power structure (who had their own scores to settle with Voet), obliged his adversary by going on the counterattack. This centered on two successive efforts (with a Voetian counterblast in between) to demonstrate that Voet's learning was a sham. Rather than properly and methodically deriving it from solid first principles, he had cribbed his learning from a hodgepodge of secondhand resources, sustained by empty syllogisms. Descartes' confidence that the city council would hasten to rid itself of a Voet thus exposed was shaken when not Voet but he was summoned to present himself, which summons he felt free to ignore in favor of one more written rehearsal of his charges. The whole issue quickly degenerated into a dispute over whether or not Voet himself was the author of that anti-Descartes counterblast signed by Voet's Groningen colleague (he was). It came to an end with the condemnation of Descartes' actions and with a reinforcement of the formal order to have none but Aristotelian philosophy taught at the University of Utrecht that was so effective (in spite of a brief, more liberal interval) that even as late as 1742 its faculty of philosophy was proud to hire a full-blooded Aristotelian. In the meantime Descartes' worries deepened over the apparent capacity of a party like Voet's to make charges of impiety stick. That is how, a few years after the clash with Voet, he responded to de Roy's outspoken intention of publishing a book on the foundations of natural philosophy with a formal, remarkably effective disavowal of the views de Roy had originally borrowed (with full credit rendered) from Descartes' own, still-unpublished 'Man'.

French Cartesians versus king and clergy.   These views, as well as those expressed in 'The World' and in an array of letters, began to be divulged in 1657, seven years after Des-

cartes' death, by his disciple Claude Clerselier as part of a deliberate campaign to spread the
word in France. This set off one more clash over aspects of the new philosophy. On 20 November 1663, on the initiative of Father Honoré Fabri, SJ, Descartes' "philosophical works"
(whatever that meant exactly) were in Rome placed on the Index of Forbidden Books, with
the *Meditationes*, the two attacks on Voet (the staunch Calvinist), and several discussions
of soul and body listed explicitly.[222] In France, the first prohibition concerned an eloge to
be held at the reburial in Paris of Descartes' corpse; the next, the appointment of Cartesian
philosophy professors. In 1671 apparent lack of obedience provoked King Louis XIV into
having the archbishop of Paris formally summon Clerselier and his son-in-law, the Cartesian lecturer Jacques Rohault, and address them thus:

> The king has learned that certain opinions previously censured by the faculty of theology and
> forbidden by the Parlement [of Paris] to be taught or published are now spreading not only at
> the university but also elsewhere in this city and in some others in the realm, be it by strangers
> or by people from the inside. He wishes to prevent the circulation of this opinion, which might
> bring some confusion to the explication of our mysteries [of the faith]. Moved by his ordinary
> zeal and piety, he has ordered me to tell you his intentions. The king exhorts you, Sirs, to act
> such as to have no one at the universities teach any doctrine other than the one which is taken
> up in the rules and statutes of the university [of Paris] and to have no one put anything of it in
> his theses. He leaves it to your prudence and sage conduct to take the paths necessary for this.[223]

While the Parlement of Paris (the city's court of law) deliberated over a proposal to re-
assert its 1624 prohibition here referred to (which at the time had been directed against the
skeptical onslaught) against teaching any but Aristotelian doctrine at the university, several
rows broke out at provincial universities and their preparatory colleges over enforcement
of the king's order. Even at Paris the order found itself reissued several times over the decades, which of course implies recurrent lack of compliance. After all, outside the universities Descartes' teachings made steady headway. They even became quite the fashion in the
ladies-driven 'salons' that at this very time began to serve as a vehicle for gracious intellectual
novelty. They were attractively phrased in Fontenelle's widely read, both serious and wittily Cartesian *Entretiens sur la pluralité des mondes* (1686; 'Conversations on the plurality
of worlds') and mercilessly caricatured in Molière's comedy *Les femmes savantes* ('Learned
women') of 1672. Still, the king's ban on Cartesianism held to some extent. In 1666 already,
when he had a 'Royal Academy of Sciences' set up, he made sure to have it closed to all
pronounced followers of Descartes (Rohault included) and to have its sessions rule out discussion of issues with philosophical, theological, or political implications (p. 575). Also, the
lectures of one such follower, Régis, to which all of polite Paris flocked, were interrupted at
the archbishop's order. It took his all-encompassing textbook of Cartesian philosophy fully

426   ten years to pass the censor in 1691, under the sole remaining condition that Descartes' name be deleted from the title – within a year, it popped up in an Amsterdam reprint.

All this took place amid a flood of tracts pro and con Cartesianism which had not even begun to abate by the end of the century, and which clarifies to some extent what was really at stake in this prolonged clash. The issues most heatedly debated were the Eucharist and the immortality of the soul – those 'mysteries' of the faith that were hinted at in the king's order. Hidden behind them was a maze of power struggles of almost Byzantine intricacy.

Some of Descartes' French followers unintentionally fired the debates by casting the prudence their master had sought to maintain to the winds, by making public his not always so prudent, private utterances, and by reinforcing the few cases (notably, the Eucharist) where Descartes had gone beyond prudence in his own published writing. Views on the Eucharist and on the soul held rightly or wrongly Cartesian found themselves attacked with particularly fervent zeal by Jesuit theologians. The unceasing debates reflected struggles for authority on at least four levels: the absolutist claims and measures by means of which Louis XIV sought to shore up his reign, liberties still left to the Protestants (with several Cartesians among them), restrictions sought to the Roman hold on French Catholicism, and the position taken by several distinct religious orders on the central theological issue in France at the time, the efficacy of God's grace. These acrimonious disputes were further exacerbated by mutual "hopes of training young people in the true faith before they were exposed to the errors of other religious traditions".[224] It is the interlocking of these multilayered struggles over state power, church power, and the minds of the young that goes some way toward explaining why, at culmination point, the king and the archbishop involved themselves in this final clash witnessed by 17th-century Europe over new modes of philosophizing widely perceived as both strange and sacrilegious.

Strangeness and sacrilege: A looming crisis of legitimacy

One remarkable thing about the three clashes just surveyed is that, when it came to formal condemnation, each tended to get bogged down in apparent trivialities. In what precise respects had Galileo with his *Dialogo* trespassed the ban of 1616? Had Voet, as Descartes alleged, personally written the anti-Descartes treatise promulgated under another name? Would deletion of Descartes' name from the title alone suffice for Régis to get his comprehensive overview finally published? The observation makes us wonder what was really going on behind these elaborate smoke screens and what their apparent necessity implies about the political/theological constellation obtaining in 17th-century Europe.

'Theology' (its contemporary intertwinement with philosophy included) *plus* 'politics' (contemporary institutional arrangements included) indeed forms the proper unit of analy-

sis if we wish to grasp what these clashes signified for the chances of survival in 17th-century   427
Europe of the emerging modes of nature-knowledge thus coming under attack. Were the
clashes just isolated incidents, leaving everyone not directly implicated pretty much free
to pursue nature-knowledge as he pleased? Or did they form the tip of an iceberg of suspicion? If the latter, what did the fears of the suspicious truly come down to? And how far did
their powers of banning, condemnation, and physical control, in short, of stifling the whole
movement, extend? As before, a comparison between civilizations proves instructive.

There were prolonged periods when nature-knowledge of Greek provenance could
flourish in Islamic civilization, by and large undisturbed by the occasional voicing of clerical
suspicion. Even so, efforts undertaken by the mutazilites to fortify its ongoing cultivation
by means of an underlying theological structure came to naught (p. 93). No ideological
harmony was attained to cement the pursuit of secular knowledge. The balance between
the 'foreign sciences' and the 'Arabic sciences' of Quran, *hadith*, and *sharia* could shift back
and forth from just a division of labor to a lower stance in the hierarchy or even to a sharp
conflict between them, without there being a generally accepted framework of thought to
regulate the relationship. In medieval Europe such a theological structure did come about,
when Thomas Aquinas fortified Augustine's idea of philosophy as a possible handmaiden
to theology by means of his doctrine of God's absolute and ordained powers (p. 79). When,
by the fifteenth century, Greek learning was revived wholesale, the harmony thus attained
was not called into renewed question. With the partial exception of atomism, tenets from
the other Greek claimants to deep knowledge of the world were likewise held to be compatible with sincere Christianity, whereas mathematical science scarcely held implications for
the constitution of the world anyway. Religious trouble over philosophic issues remained
confined to a concern over skepticism. This culminated in the Paris outburst of the 1620s
(p. 235), which affected the philosophy of nature only in a very indirect manner. Where the
pursuit of nature-knowledge as an outflow of philosophy was concerned, then, harmony
prevailed as a rule.

What, then, came to disturb it? Or rather (since the answer has become abundantly clear
in the account of the three major clashes just given), what was felt to be so disturbing about
these new modes of pursuing knowledge of nature that were unmistakably intent on replacing all rival modes in spite of their long standing?

Four highly controversial basic issues present themselves from the foregoing account.
Two are quite specific: maintenance of the Bible as God's word and the tenability of long-
standing philosophical sustenance of Christian verities. Two others are more general but
also more diffuse sources of unease and suspicion: the picture of the world evoked by these
radical innovators and the very freedom they claimed for everyone to do his own thinking
about the world and God's and our place in it.

These four issues came to the fore, to be sure, in a European civilization saturated with

428 Christianity to an extent hard to recapture nowadays. The charge of atheism was hurled with depressing regularity at presumed or actual theological opponents. Really denoting 'impiety in thought and/or demeanor' rather than a genuine lack of belief in the existence of God, atheism was actually none but a theoretical counterposition thought up by clever philosopher-theologians like Thomas Aquinas but not in all likelihood adopted seriously until the first decades of the 18th century. The qualifier 'in all likelihood' cannot be dispensed with, as church-influenced state censorship was maintained (to varying degrees of strictness, to be sure) all over Europe. The apparent sincerity of professions of belief can always be dismissed therefore as sheer lip service. It is certain that for just about every educated European, even if deeply immersed in the veneration of antiquity, the basics of the Christian faith provided the central category against which to gauge every new utterance that he (or, very rarely still, she) encountered. These basics were the Bible, an array of established doctrines of faith, and a conception of how the world is broadly constituted – not each of these on its own but, as the outcome of a centuries-long development, intricately intertwined.

God's word in dispute: An inevitable clash?   The issue, then, of the Bible as God's word had by the early 17th century gained a new urgency due to the Reformation starting in 1517 and to the Counter-Reformation provoked thereby. For Roman Catholics, the dispute over how to interpret the Bible and who ought to have the final say over that interpretation was resolved by the Council of Trent in 1546. The rules adopted there left those of St. Augustine in place but with more emphasis on the accumulated pastoral tradition and, above all, with a ringing reassertion of Rome's final say in these matters. These rules were not particularly suited (nor, obviously, had they been intended) for dealing with the exegetical problems a realist conception of heliocentrism was bound to produce. Whether that conception was also bound to provoke the outright clash that did ensue is another question, though – the difference between Galileo's and Bellarmino's express views on the issue, *both* of whom took the Trent rules as their point of departure, was rather too subtle for that. If, under the same rules, the protagonists had proceeded in only slightly different ways, a solution along the lines of the sufficiently comparable issue of the Earth's spherical shape might quite conceivably have been reached. At issue, after all, was not the crude literalism of many in the lower clergy but those much more sophisticated Trent rules. What provoked the clash more than anything else was Galileo's not quite so crudely articulated yet (to more than one cardinal, notably Bellarmino) unmistakable suggestion that the mathematical philosopher ought really to have the final say and in the feverish pressure Galileo put behind his personal engagement in the issue. Whence that pressure, and whence that claim?

The pressure first. No doubt it had something to do with the inclination of the exceedingly bright among us toward impatience with what they may just a trifle too easily mistake for sheer slow-wittedness. More importantly, it had to do with court patronage or (to be more precise) with its shortcomings or (to be even more precise) with its shortcomings from the specific viewpoint of a 'mathematical philosopher'. Patronage extended by some secular or clerical prince or nobleman had from the Ptolemies in Alexandria onward been the mainstay in society of the mathematical scientist (unlike with the philosopher, whose perceived ability to explain the world gave him a far wider audience); take away the patronage, and mathematical science ceased to be cultivated. But patronage *was* 'taken away' all the time – any death or deposition of any patron might redirect his successor's *largesse* or cause it to dry up entirely, so it was only a multiplicity of courts that over otherwise favorable periods could keep mathematical science from falling to the wayside. Further, patronage involved an ongoing exchange of gifts, and inability of the client to keep his side of the bargain might spell ruin for him. This was the given state of affairs that every client had to make the best of. For instance, Kepler's *Astronomia nova* originated in a promise to his emperor, hastily made in a desperate attempt to save his position at court. The court of the Grand Duke of Tuscany was no exception. Exceptional was rather the new kind of gift Galileo was (on the precedent of the Medicean stars) expected to bring – ongoing celestial novelty, which embroiled him time and again in cumbersome, ill-fated polemics over priority, particularly with Jesuit astronomers. More exceptional still was his claim to the unheard-of status of 'mathematical philosopher'. The very strangeness of that claim is what propelled Galileo toward turning his realist Copernicanism into his showcase (p. 417). It also propelled him toward a search for a safer haven for it than any ordinary patron could possibly have to offer.

Where to find that safe haven? In all of Europe there resided only one patron who, not through personal but through institutional immortality so to speak, might warrant perennial legitimacy for Galileo's realist-mathematical science as embodied in his realist Copernicanism, in that that patron had the power to enshrine it forever in the very doctrine he was by divine appointment and apostolic succession called upon to guard. That hoped-for patron held court no more than 175 miles away from that of the Grand Duke; it was, of course, the pope himself, or rather, through him, the papacy itself. That is why Galileo threw himself into his campaign, casting all prudence counseled to him aside; that is also why he threw himself into it all over again when the windfall of the election of an admirer of his to the papacy appeared to give him a second chance. From this viewpoint, therefore, it was the native *strangeness* of realist-mathematical science, with the consequent difficulty of gaining legitimacy for it so acutely perceived by Galileo, that led in the end to his clash with the Church authorities.

What appeared to make it *sacrilegious* as well was Galileo's subdued yet unmistakable claim to the final say. A solution of sorts, such as to take the heat off, might have been reached over the exegetical issue at hand; only, and this is crucial, even if reached, that solution was bound to be temporary. At this point a personal statement is once again called for. The late evolutionary theorist Stephen Jay Gould has defended with his customary pun-

430 gency the view that science (modern, post–Scientific Revolution science, that is) and religion constitute two distinct 'magisteria', each with concerns and rules of its own which, if handled wisely, need not overlap (let alone clash) in any way. By contrast, I share the view recently defended with his customary pungency by the physicist Steven Weinberg that the history of religion – in the only historically realistic sense of belief in a Deity's personal concern for the universe and its human inhabitants – has ever since the Scientific Revolution been a history of clash in the sense of a halting yet unceasing retreat before what modern science has successively demonstrated to be the case; consequently, that *pace* Gould ever since the Scientific Revolution got under way there has been a good deal of inevitably disputed territory. Not that modern science can validly be held to imply a worldview of its own (as I shall insist in the Epilogue; p. 734) but that modern science appears increasingly unlikely to be compatible with any tenable view of the world as governed by divine care, is the assertion here endorsed. Now why bring these rightly controversial issues up here? Because it follows that, indeed, the question of *who* has the final say over *what* ought to yield is ineluctable: Is it the clergy, or is it the 'mathematical philosopher' (or, in modern parlance, the scientist)? It is in no way accidental that in each of the three clashes, whether over biblical exegesis or over reasoned proof for Christian doctrine or over divine concern at large and ways to cast doubt upon it, not just an exchange of theological views at variance with each other took place but, inseparably from it, a struggle over authority. And this struggle could not but reflect the general structure of authority in 17th-century Europe, and crucially depend on it for the outcome. More specifically, that structure determined how far the weapons of censorship and ban and condemnation could actually reach.

The structure of European authority. The political state of the western European subcontinent is best characterized as one of mitigated divisiveness. Europe was not a universal (in the sense of civilization-wide) empire such as had been established in China at the end of its Warring States period or under the Ottomans. The sole remaining territorial entity in Europe with some claim to universality, the 'Holy Roman Empire of the German Nation', was at the very time of the first two clashes being rent apart over the very remnants of that hoary claim. More than that, the acute state of (in the end, thirty years') war over Germany was spreading all over Europe. The war exacerbated running conflicts between states that were themselves unstable to the point of civil war in that each was split over loyalties to either Protestantism in a variety of denominations or a Roman Catholic center powerfully supported in its determination to undo the big religious divide. Not anarchy as a rule but a heterogeneity of political centers each with its own internal powers and jurisdictions and external claims and means to make them stick marked Europe at the time. The divisiveness was mitigated, in part by the at times threatening presence of a common enemy, the Ottoman Empire but also, more importantly, by the absence as yet of a war of all against all. As

the war over Germany went on and on, however, it threatened more and more to degenerate 431
into precisely that. Not until the Peace of Westphalia was concluded in 1648 in a package deal
that settled just about all outstanding conflicts did Europe offer itself a chance, which over
the 1650s and 1660s it grabbed, to settle down to more constructive pursuits. How, then, did
this broad, shifting state of mitigated divisiveness affect events after the sudden entrance
upon the scene of strangely new modes of philosophizing that claimed an authority of their
own?

Storms in teacups?   For all their mutual differences, the three clashes share some remarkable characteristics.

   To begin with, the established authorities sought as much as they could to handle the issues in concretely tangible ways. This appears most clearly from the negotiations over apparent trivialities that marked so much of the respective 'endgames' in each case. These reflected
Europe's peculiar separation between church and state – the two could, and frequently did,
ally, yet to lay down the law and maintain it were jealously guarded, secular prerogatives. In
the two clashes involving Descartes' philosophy, the Utrecht city council and the French king
called upon by Voet and by the Sorbonne theologians, respectively, to make their case for
them had either to set it up in terms of legally ascertainable charges or to enter into negotiations over degrees of prohibition. And even in the one clash where church and state power
were in the same hands, not only the established procedures of the Inquisition but also due
consideration taken of the fact that a highly placed subject of another sovereign Italian state
was on trial turned what had started as a problem of the true meaning of God's word into
the legal issue of whether a formal ban had been trespassed.

   Further, ulterior motives were invariably involved, as with Galileo's urgent desire to
break out of the legitimation deadlock to which his newfangled 'mathematical philosophy'
appeared to doom him. There was Descartes' posture of caution-guided proselytizing that
sprang from his effort to have teaching of his own philosophy of nature replace that of
Aristotle. There was the French Cartesians' stake in the power struggles under Louis XIV.
But so it was on the other side, too. In Galileo versus the papacy, policies of the Jesuit and
the Dominican Orders that were at cross-purposes helped determine the outcome in 1616
and 1633, as did the pressure under which Pope Urban VIII found himself in 1633 over the
failure of his anti-Habsburg policies in the Thirty Years' War. In Descartes versus Voet, the
latter's concern over the former was exacerbated by remembrance of the civil war the budding Dutch Republic had come perilously close to some decades earlier due to events that
had started as an academic dispute in theology (in this case, over predestination). Decisive,
further, for Descartes' defeat were his miscalculations over the unusually restricted powers
of the Utrecht city council and over the true nature of the Dutch regents' resistance to theocracy and of their well-known toleration policies, which proved not so much Erasmian as,

432 rather, pragmatically aimed at the maintenance and restoration of peace and quiet. Equally decisive were the many stakes involved in Louis XIV's efforts to ban Cartesian teaching, all of which came down in the end to his persistent, dual effort at centralizing his reign over France and at expanding it into domination over all of Europe.

Also telling is that no one punished for and/or put under prohibition of promulgation of his views had to pay for them with his life or even with captivity. The worst thing that happened to Galileo personally was a formal threat of torture that was really empty in view of his already-expressed willingness (extracted by the Inquisition through sheer verbal pressure) to recant. (This was not pleasant at all, to be sure, yet worse had happened to others put as he was under 'vehement suspicion of heresy'.) The worst thing that happened to Descartes personally (due in part to the intervention of the French ambassador with the Dutch *stadhouder*) was condemnation at Utrecht of his anti-Voet publications, with his stay elsewhere in the country disturbed by nothing but his own annoyance over that condemnation. And while dismissal from their university jobs befell some of his later French followers, no more serious consequences ensued for any of them either.

Finally, no banned book remained unpublished in the long term, or even in the short term of the merely two years required for Latin translation and subsequent publication of the *Dialogo* in Strasbourg, a Protestant city. Galileo's second masterpiece, the *Discorsi*, was on completion smuggled out of Italy and published in Leiden. Descartes' works, although effectively banned at times from being taught at the University of Utrecht, spread nonetheless, first to Leiden and then (after he died) elsewhere. Nor did the posthumous placement on the Index of an ill-defined selection prevent them from becoming the talk of the town wherever in France fashionable salons sprang up.

It may seem from this list of shared characteristics as if, for all the uproar they caused, these clashes were really no more than storms in teacups, in that no protagonist was seriously damaged personally and in that what was forbidden at one place in Europe quickly popped up elsewhere. Such a conclusion, while true as far as it goes, nonetheless seriously underestimates their impact and their wider implications. A second look at Voet's professed difficulties with Descartes' teachings as defended by de Roy is most instructive in this regard. Recall in particular that Voet saw Descartes' rejection of substantial forms (i.e., the very core of Aristotelianism) as tantamount to the ruin of all of philosophy and all of Christianity (p. 324). How could this skilled theologian draw so drastic a conclusion?

Theology and philosophy intertwined.   For an answer, it is important to realize first that the significantly enhanced ease with which the pursuit of nature-knowledge was accepted in Christian Europe compared to the Islamic world was bought at a price. Philosophy entered into close association with theology to an extent without counterpart in Islam, where the mutazilites' striving after a reasoned account of the faith had never gained the upper hand. That there was a price to be paid for such close intertwinement of theology and philosophy had indeed been perceived by some, with their unwillingness to pay it culminating in the condemnation by the bishop of Paris in 1277 of a whole range of Aristotelian tenets (p. 80). But most others went happily along. The point is not so much that from then on until well into the 18th century most theologians tended to accept without much questioning the bulk of what used to pass for Aristotle's teachings. However tight even after the mid-15th-century revival of the full Greek legacy the intertwinement remained, it never prevented anyone from taking other positions on numerous issues, as with Bellarmino and his biblical-stoic notion of the heavens. The point is rather that the way theologians habitually spoke about their faith became saturated with the idea of *proof*.

What made adducing proof for God's revealed verities so attractive? Why did not revelation alone suffice? The reason is that reasoned argument provided a powerful weapon in the conversion of educated nonbelievers. This had been so in Islam, too, where one major motive for having those Greek treatises translated into Arabic rested in intellectual sustenance for the conversion of the learned Persians who had to be recruited for manning the Abbasid bureaucracy in Baghdad. In Christendom the need for proof had been even more urgent in view of the comparatively large number of doctrines that, by the 4th century, had come to define the Christian variety of monotheism. Even the admirably condensed version of the faith settled upon in the Nicene Creed takes 162 words to cover Creation, prophecy, baptism, the Incarnation, the Trinity, Ascension, Judgment Day, and many other tenets, in short, just about all those doctrines Voet's mind's eye already saw evaporate together with the source of their philosophical proof. There is some irony in this acute concern of his, as the pioneers of the Reformation, to which Voet gave his ultra-orthodox allegiance, had originally striven for a renewed grounding of their faith in God's word alone, only to have successors like Voet find out all over again the difficulty of conversion to the whole doctrinal set in the absence of reasoned proof for it. More than that, the very struggle between Reformation and Counter-Reformation that had ensued in the meantime tended mightily to enhance already present inclinations toward doctrinal hairsplitting. For example, not until the Council of Trent had the doctrine of the Eucharist been so closely, albeit still a trifle ambiguously, bound up with its philosophical explanation. Struggles over niceties of the faith did not of course involve philosophical issues alone. The troubles, for instance, that Kepler ran into with an array of Lutheran pastors over his deviant views on the sacrament and the omnipresence of Christ's body had nothing to do with his astronomical work. Rather, it is the *conjunction* of theology with philosophy and with politics that carried a potential for blowing up, with claims to natural philosophy of a strangely new kind providing the fuse. And since, by its very nature, natural philosophy was part and parcel of philosophy as such, by the first half of the 17th century the established linkage with theology proved ready-made to explode the full compound in any situation where the built-in element of contested authority was exacerbated by

434 some specific power constellation or unpredictable shift therein; all this in a highly volatile setting of states on or over the brink of war both within themselves and against each other.

So much for how the clashes could come about and run the courses they did. But before determining their joint impact upon the pursuit of new nature-knowledge Europe-wide, we must consider that it was not the claim to a new philosophy alone (the 'denial of substantial forms') that Voet and many who thought like him regarded as a lethal threat to their faith. There was a whole list of more specific teachings that were liable to upset those who thought like him. Some of these teachings were identical, to be sure, with what rendered the philosophy of kinetic corpuscularianism so *attractive* to other, definitely un-Voet-like minds (p. 383). On this list of teachings attractive to some, pernicious to others, figure, notably, the issues of the Earth as a planet, of man versus beast, of freedom to doubt established verities, and of a clockwork picture of the universe.

Four pernicious teachings.   What problems arose from *removing the Earth from the center*? The theological issue overtly at stake in the dispute over realist heliocentrism resided in biblical exegesis. This indeed is how both Galileo and his opponents saw it in the first place. Still, the way in which Descartes went on to elaborate possible implications carried with it a whole range of questions hard to answer in any standard manner, particularly so if beyond sheer revelation of the mysteries of the faith a certain amount of proof were required. All such questions centered on the no longer central place of the Earth. Why would God have chosen to incarnate on this one planet? How to think of a caring God if His concern is spread over the infinity of the universe? True, questions like these did not find much of a place in contemporary writings. Even so it is hard to imagine that they utterly failed to be raised by many a thinking mind. Surely the vehemence with which the Copernican issue kept being pursued over the century cannot solely have been on account of the exegetical issue. After all, for those not bound by the Roman ban the solution lay so ready to hand in the accommodation gambit. The fierceness of the ongoing debate suggests some anguish over these deeper issues as well. Also, to Voet and no doubt many like him the names of Kepler and Galileo raised the same sort of associations as Descartes' did – all utterances on the subject were indiscriminately thrown into the melting pot of Voet's vision of a very bad future in store for anything that, to him and his ilk, could still be called the Christian faith.

There was further the issue of *man versus beast*. A few pages back I took the liberty of applying George Orwell's unfriendly expression 'double-think' to the two really incompatible ways in which Descartes dealt with the issue of what distinguishes humans from animals. In his unpublished 'Man' and later manuscripts, he followed a 'push-back-the-soul' program unconcerned with defining the soul's relation to the human machine, whereas in his *Meditationes* he pointedly asserted the soul's autonomy as a warrant for its immortality. To call this maneuver double-think may be apt in some ways, but it is also unfeeling in that Descartes

had good grounds for doing some prudent double-thinking – all the trouble that his follow-   435
ers experienced after his own demise is there to prove it. But the main point for now is that,
prudence or not, *either* view of Descartes' on soul and body could easily be seen as leading
down a slippery slope of obliterating the fundamental dichotomy between human and animal beings that was so much a part of 17th-century Christian faith. Even under the softer,
'meditational' view only one more, hardly far-fetched step – to scrap *res cogitans* from the
world's furniture altogether in view of its already minimized link with infinitely extended
matter – sufficed to turn *both* humans *and* animals into sheer *res extensa*, in other words,
to effect a wholesale reduction of human beings to machines, and of the human mind to
an epiphenomenon of bodily processes. Descartes himself is likely to have been sincere in
shrinking back from this very step, and even in his pronounced astonishment that anyone
could attribute it to him as his own view or as a possible consequence thereof. Voet, in contrast, as many a like-minded theologian after him,

> objected, with an astute anticipation of later developments, ... that, if the 'souls' of animals
> were reducible to mechanical causes, it was only a matter of time before human minds could
> be equally well explained in terms of stimuli, animal spirits, and brain functions.[225]

Nor was the questionable dichotomy between man and beast the only case where it was easy
to infer from Descartes' expressed views consequences he claimed to have gone out of his way
to counter. This concerns notably the issue of *freedom to doubt*. There was this ambiguity in
Descartes' appeal to think for oneself: a call at least once in our lifetime to doubt everything
ever taught us, together with a new dogmatic philosophy ready-made to fill the resulting
void (p. 385). There is another way to phrase the issue. By means of his process of methodical
doubt carried even beyond the customary skeptical armory but then wrecked on the *cogito*,
Descartes claimed to reinstate indubitably certain knowledge with a vengeance. Against this,
opponents like Voet saw that nothing but the path leading to the *cogito* need now be called
into doubt in its turn to have the skeptical enemy reinstated more triumphantly than ever.
And let us not, in thus implying the presence in Europe of numerous 'opponents like Voet',
forget that that arch-*Roman* agency, the Congregation of the Index, in banning Descartes'
philosophical works had taken care to list specifically the two assaults of this fellow *Catholic*
on the very man out to establish a *Calvinist* theocracy! Just as in the case of human and animal we may with only a little exaggeration say that conservative theologians of Voet's bent,
in spotting Descartes, saw, as it were, La Mettrie coming, just so but even more properly
may we say that they likewise saw Spinoza coming. Indeed, just as Hegel's followers were to
split themselves up between a 'right' and a 'left' wing, just so may we distinguish between a
'right' wing of orthodox Cartesians like Clerselier, which soon enough began to bear all the
marks of a sect, and a 'left' wing which, under the broad inspiration of Descartes' work, set

436 out on a path of independence from established authority leading to Spinoza's sustained critique of the Bible or to Poullain de la Barre's insistence that women, in being entitled to an equal share in *res cogitans*, have minds of their own. In short, Cartesianism was perceived by some perspicacious adversaries (and theologians had centuries of training behind them in sniffing the scents of heresy) to lead to something that has rightly or wrongly been called in retrospect a 'proto-Enlightenment' but that may in any case be regarded as one stream among others destined in the end to flow into that broad river. It was out of a clear-headed sense of what was at stake that, when the Sorbonne Theological Faculty proposed to the Paris Parlement reconfirmation of its ban on nonscholastic philosophy, the satirist Boileau produced at once a mock decree starting with the following consideration:

> Whereas has been seen by the Court a request presented by the regents, Masters of Arts, doctors, and professors of the University, both in their own name and as tutors and defenders of the doctrine of Master Aristotle, ancient Professor Royal in Greek at the college of the Lyceum, and preceptor of the late king of quarrelsome memory, Alexander called the Great, acquisitor of Asia, Europe, Africa, and other places, which [request] is of the content that for some years an unknown dame called Reason would have undertaken to enter through force the schools of said University and, so as to attain that effect, would have prepared herself, with the aid of certain fractious persons taking the fractious names of Cartesians, of new philosophers, of circulators, and of Gassendists, all people on the loose, to have Aristotle, ancient and peaceful possessor of said schools, expelled from them.[226]

As if the three issues just discussed were not quite sufficient to give food for thought to anyone taking his faith seriously, there was also the *clockwork sort of universe* called up by the natural philosophy of kinetic corpuscularianism and (by possible extension) by realist-mathematical science as well. It was this idea of a universe run according to inexorable law that upset, perhaps even more than its opponents, many of its most expert adherents. Devout Christians themselves, large numbers of them sought in public not only to justify in the face of charges of 'atheism' their enduring commitment to the pursuit of nature-knowledge but to put their own conscience at peace as well. This is how the 17th century turned into the age of what in a felicitous phrase has been called 'secular theology' in the sense of a spate of publications on theological issues written by laymen engaged in the ongoing renewal of nature-knowledge. Galileo's letter to the Grand Duchess is dedicated wholly to exegetical issues. Descartes made a special effort in his *Meditationes* to prove God and the immortality of the soul from principles in line with those of his natural philosophy. Pascal in his unfinished *Pensées* made a rare attempt at apologia of revealed Christian doctrine alone, undertaken in this mathematical/experimental scientist's acute awareness that reasoned proof of many centuries' standing was in the very process of backfiring so as to render "the God of Abra-

ham, Isaac, and Jacob" in dire need of vindication against "the God of the philosophers".[227]   437
Most other practitioners and/or convinced adherents of the new ways of pursuing nature-
knowledge were to take another line. Rather than forsaking proof for God as Pascal did, they
sought it in nature – in its laws as originally laid down by Him, to be sure, but also in its
intricate design.

The menace to religious sensibilities that the clockwork universe appeared to carry with
it could be quelled to some extent in this way. But not all of it could, or certainly not so easily – where, in such a universe, was a place for divine concern and for human purpose?

Two outstanding historians of precisely these issues have perceptively summed up what
it was that robbed *both* devout Scientific Revolutionaries *and* their most acute adversaries
of their sleep. In phrasing the core concern that emerges from leafing through thousands of
folio pages, E.A. Burtt focused on the depressing view of the human spirit that seemed to be
implied in the clockwork universe: that is, of

> man [a]s but the puny and local spectator, nay irrelevant product of an infinite self-moving
> engine, which existed eternally before him and will be eternally after him, enshrining the rigor
> of mathematical relationships while banishing into impotence all ideal imaginations; an en-
> gine which consists of raw masses wandering to no purpose in an undiscoverable time and
> space, and is in general wholly devoid of any qualities that might spell satisfaction for the
> major interests of human nature, save solely the central aim of the mathematical physicist

to go ahead and discover what makes the engine run.[228] R.S. Westfall focused on the despair
to which such an image led among those most responsible for its further spread:

> Despite the natural piety of the virtuosi [the British adherents to the new science of the 17th
> century whose works Westfall examined], the skepticism of the Enlightenment was already
> present in embryo among them. To be sure, their piety kept it in check, but they were unable
> fully to banish it. What else can explain the countless dissertations on natural religion, each
> proving conclusively that the fundamentals of Christianity are rationally sound? They wrote
> to refute atheism, but where were the atheists? The virtuosi nourished the atheists within their
> own minds. Atheism was the vague feeling of uncertainty which their studies had raised, not
> uncertainty of their own convictions so much as uncertainty of the ultimate conclusions that
> might lie hidden in the principles of natural science. With wonderful certainty and assurance
> each virtuoso proved the existence of God from the creation; yet repeated too often, the as-
> surance acquired an odor of insecurity. With Newton the insecurity was growing toward open
> fright. ... Following the birth of modern science the age of unshaken faith was lost to western
> man.[229]

438   With the stakes set so high, the ongoing flourishing of nature-knowledge of such a kind as to appear to entail such consequences was not, to put it at its most mildly, a foregone conclusion. Was an outcome as befell the pursuit of nature-knowledge in Islam, i.e., loss of legitimacy leading to a sapping of the will and to ensuing, steep decline what ought to be expected for Europe likewise, if we consider this from the viewpoint of, roughly, the 1650s?

Loss of legitimacy.   Of course, circumstances were not nearly identical. In Islam a turning away from secular concerns due to waves of foreign invasion is what came to sap the will. In Europe, the strangeness of novel modes of nature-knowledge, and their proneness to being perceived and condemned as sacrilegious, might amidst the acute threat of war of all states against all others lead to such an outcome. In the civilization of Islam, perception of sacrilege had not so much involved philosophical proof for the faith, which in less doctrine-bound Islam had never been that prominent. Rather, it involved the reading of its sacred writ, for which (since it had been dictated at one stroke to one prophet by one messenger of God) Augustine-like rules for interpretation as nonliteral had not been worked out. More than in Christendom, in Islam it was the clergy's sense of the superfluousness of the pursuit of any nature-knowledge in the face of revealed truth that contributed so much to stifling it in adverse times. But this means no more than that the paths to ruin would have been somewhat different in the two civilizations; the net effect of a loss of momentum leading to decay might well have been the same.

To some extent it was indeed. It would be a serious mistake to take the effects of the three clashes lightly. I shall consider them from two points of view: of the territories and of the practitioners involved.

For all the absence of capital punishment or captivity, and for all the remaining possibilities for publication elsewhere, little innovative pursuit of nature-knowledge remained after 1633 in countries where the Inquisition held sway. The picture is not quite so bleak as it has sometimes been rendered. In Italy, thus far at the forefront of developments in nature-knowledge, some realist-mathematical science remained (notably for purposes of river taming), as did some bits and pieces of Cartesianism. There, as well as in Spain, Portugal, and the Spanish Netherlands, the walls proved at times somewhat permeable in spite of Inquisition guardianship. For any coherent pursuit of innovation, however, after 1633 these countries could just be counted out. Italy quickly lost its centuries-long cultural supremacy, though for other reasons also; the others never had any. Leaving out of account both geographically and culturally marginal territory to the north and east of Germany, and taking events in Britain as definitely a story apart (to which we shall of course return, as in the end these proved nearly decisive for the whole of Europe), on the Continent we are left with Austria, Germany, France, and the Dutch Republic. With both Austria and Germany quickly pushed into the cultural margin as well due to the devastations of the Thirty Years' War, by the 1640s

the future of innovative nature-knowledge had come to depend in good measure on the
Dutch Republic and France – hence the supralocal significance of the two later clashes. But
with respect to the first clash as well. The 1633 ban led not only to an Italy where Galileo's
disciples no longer dared pursue the possibly contentious issue of the void but also to a good
deal of self-censorship all over the Continent, affecting Descartes, Gassendi, van Helmont,
and many another here left unmentioned. *Precisely this effect might smoothly yet surely have
led to a loss of momentum such as might then have become the first step in a process of decay,
petrifaction, and ultimate extinction.* Indeed, other manifestations of semivoluntary self-
restraint followed upon the Voet affair – outright banishment from the University of Utrecht
and, more importantly, a cautious syncretism holding sway elsewhere in Dutch academia.
By the early 1660s still, in the Netherlands as a whole the prospects for truly innovative
nature-knowledge would have seemed fairly bleak, due in part to aftereffects of the clash but
also to the absence of a court and to the presence of a national culture already settled in its
unusually narrow-minded utilitarianism. So on this side of the Channel what was to come
of the formidable impulses of, notably, Galileo and Descartes depended to a large extent on
what happened in France.

A survey of the post-Galileo development of realist-mathematical science from the point
of view of its personnel leads to the same conclusion of a precarious balance attained in
France especially. Numerous approaches were taken, and subjects of the most varied kinds,
from artillery to musical temperament to post-Kepler planetary theory, were examined by a
remarkably small number of practitioners (ch. 10). Apart from the odd Englishman (Horrocks, Newton) and a few scattered individuals like Boulliau, these practitioners were ag-
gregated in only two clusters: Galileo's Italian disciples and a few men, such as Huygens and
Mariotte, who met regularly in the French Academy. *Not part of this scene at all were the
Jesuits,* who stuck mostly to a cautiously quantifying Aristotelianism enriched by some no
less cautious thinking in terms of particles. Some limited measurement and calculation as,
for instance, practiced by Castelli's opponents over the future of the river Reno might within
decades, the original sense of strangeness overcome, have yielded to the Galilean approach.
The ban of 1616, exacerbated by the condemnation of 1633, nipped any such development
in the bud, to the serious detriment of the number of realist-mathematical scientists of the
second generation, who, as noted, remained confined almost entirely to a few staunch yet
highly constrained Italian pupils and disciples and to those members of the Paris Academy.
Descartes' self-pronounced followers were kept out of that Academy on express royal order.
Who, then, did the king fill the institution with instead?

Here a revealing picture emerges. In Paris during the late 1640s and 1650s informal groups
with an interest in furthering innovative nature-knowledge sprang into existence. Little is
known of the proceedings of the scholars and patrons who met regularly in the 'Académie
Montmor', named after a wealthy patron, or who assembled in other private houses. The

440 main reason is that nothing was published and no records of any kind were kept, in view, precisely, of the risks involved – a glaring case of collective self-censorship. Numerous competent mathematical scientists were involved in the oral exchanges these men felt compelled to confine themselves to, such as Pierre Petit and Gilles Personne de Roberval.

All this changed in the early 1660s. By then a new king, Louis XIV, decided that for the foundation of his own, state-funded Académie he would do well to tap the ranks of these men while keeping out those philosophically and therefore politically dangerous dogmatists against whom in due time he managed to mobilize his own archbishop. Now the preparatory work done in these earlier informal groups could finally come to full fruition. Louis' exclusion of the dogmatic Cartesians was to have far-reaching consequences for the culture and for the fate of the country at large (with those who were to educate the 18th-century 'philosophes' thus being driven into opposition to the king's regime), but on the shorter term it was all to the good for ongoing scientific innovation.

All this, however, already anticipates the extraordinary story of how, against all possible expectations, by the late 1650s the looming crisis of legitimacy was actually overcome. For that story to be told (ch. 17), a working knowledge is required of several further developments in revolutionary nature-knowledge Europe-wide. Prominent among these developments is how fact-finding experimentation fared once the pioneers, Bacon, Gilbert, Harvey, and van Helmont, had left the scene.

Three clashes between realist-mathematical science and natural philosophy.

- **Floating bodies.** Data taken, and many a piece of analysis adopted, from ch. 3, 'Anatomy of a Court Dispute', of Mario Biagioli, *Galileo Courtier. The Practice of Science in the Culture of Absolutism*. University of Chicago Press, 1993. I do not go along with Biagioli's relativist view of this clash between truly irreconcilable conceptions as being characteristic of the process of science as such (in line with Kuhn's 'paradigm' conceptions; see Biagioli's ch. 4). I see the clash rather as marking the historically unique event of modern science emerging in inevitable collision with the standard natural-philosophical mode of pursuit.

- **Falling bodies.** I have used ch. 5 of E.J. Dijksterhuis' *Val en worp*. Groningen: Noordhoff, 1924, with conclusions checked against Carla Rita Palmerino, 'Infinite Degrees of Speed. Marin Mersenne and the Debate Over Galileo's Law of Free Fall'. *Early Science and Medicine* 4, 4, November 1999, pp. 269-328 (a recent collection of papers on the same subject is idem & J.M.M.H. Thijssen (eds.), *The Reception of the Galilean Science of Motion in Seventeenth-Century Europe*. Dordrecht: Kluwer, 2004).

- **Void.** Among the numerous accounts I still find the one by E.J. Dijksterhuis outstanding: sections IV: 261-277 of his *The Mechanization of the World Picture*. Oxford: Oxford University Press, 1961 (Dutch original: 1950). I have also made use of two recent, succinct renditions with some novel elements added: chs. 4 & 5 of Cesare S. Maffioli, *Out of Galileo. The Science of Waters 1628-1718*. Rotterdam: Erasmus Publishing, 1994, and ch. 7, 'Pascal's Void, Natural Philosophers, and Mathematical Experience' of Peter Dear, *Discipline and Experience. The Mathematical Way in the Scientific Revolution*. University of Chicago Press, 1995. With gratitude I remember from my student days an inspiring seminar by R. Hooykaas dedicated in part to examining Pascal's 'Letter to Noël' and the lasting impression that text made upon me. Hooykaas wrote two wonderful pieces about him, 'Pascal: His Science and His Religion'. *Tractrix* 1, 1989, pp. 115-139 (authorized translation, by myself, of an article first published in Dutch in 1939), and ch. 11, 'The Thinking Reed', of his book *Fact, Faith and Fiction in the Development of Science*. Dordrecht: Kluwer, 1999.

Three clashes between counterintuitive nature-knowledge and theological authorities. In *SRHI*, 5.1. I dealt with issues of 'science and religion', both in general and with regard to a range of causal theses.

- **Galileo vs. the papacy.** Maurice A. Finocchiaro put together English translations of a number of basic documents pertaining to the relations between Galileo and the Church that culminated in the trial of 1633: *The Galileo Affair. A Documentary History*. Berkeley / Los Angeles: University of California Press, 1989. Several further significant documents, preceded by enlightening comment, are in Richard J. Blackwell, *Galileo, Bellarmino, and the Bible. Including a Translation of Foscarini's Letter on the Motion of the Earth*. Notre Dame (Indiana): University of Notre Dame Press, 1991, and in idem, *Behind the Scnes at Galileo's Trial. Including the First English Translation of Melchior Inchofer's Tractatus Syllepticus*. Notre Dame (Indiana): University of Notre Dame Press, 2006. Also interesting are two books by William R. Shea and Mariano Artigas: *Galileo in Rome. The Rise and Fall of a Troublesome Genius*. Oxford UP, 2003, and *Galileo Observed. Science and the Politics of Belief*. Sagamore Beach: Science History Publications, 2006. Informative on astronomical views among the Jesuits and their Tychonian drift by the early 1610s is James M. Lattis, *Between Copernicus and Galileo. Christopher Clavius and the Collapse of Ptolemaic Cosmology*. Chicago: University of Chicago Press, 1994. In Rivka Feldhay, *Galileo and the Church. Political Inquisition or Critical Dialogue?* Cambridge UP, 1995, the shifting divisions inside the Church are disclosed. Galileo's exegetical views as set forth in his 'Letter to the Grand Duchess' have been analyzed in painstaking detail by Ernan McMullin in 'Galileo on Science and Scripture'. In: P. Machamer (ed.), *The Cambridge Companion to Galileo*. Cambridge UP, 1998; pp. 271-347.

Most literature on the trial and the events that led to it is rather intensely partisan (be it granted, though, that the affair was so deeply polarized as to render a wholly nonpartisan analysis quite unfeasible). The basic divergence running through serious historiography on the subject separates those who take Galileo for the active party from those who take him for the mostly passive party in 1613–1616. My own position on this divergence has been influenced by a book with a bad scholarly reputation that it deserves only in part, Arthur Koestler's *Sleepwalkers*. I reject the principal thrust of his argument, that at bottom 'science' and 'faith' are one and their 17th-century 'split' a needless mistake. Even so I still find his historical account convincing overall. One actually need only replace what Koestler regards as Galileo's personality defects with the logic of patronage dynamics uncovered by Biagioli in his book mentioned above to see how neatly both accounts fit into (and then begin to reinforce) one another. But the decisive issue concerning the decree of 1616 is whether it was the culmination point of Lorini's and Caccini's denunciations or whether (with these dismissed higher up) it rather resulted from the pope's adverse intervention provoked by Galileo's insistence. On this crucial point I have yet to see convincing counterevidence to Koestler's coherent, well-supported treatment leading to the latter conclusion.

- **Descartes vs. Voet.** Klaas van Berkel, 'Descartes in debat met Voetius. De mislukte introductie van het Cartesianisme aan de Utrechtse universiteit (1639–1645)'. *Tijdschrift voor Geschiedenis van de Geneeskunde, Natuurwetenschap, Wiskunde en Techniek* 7, 1, 1984, pp. 4-18 (summed up in K. van Berkel, A. Van Helden, L. Palm (eds.), *A History of Science in The Netherlands*. Leiden: Brill, 1999, pp. 47-49). Catherine Wilson, 'Descartes and the Corporeal Mind. Some Implications of the Regius Affair'. In: Stephen Gaukroger, John Schuster, & John Sutton (eds.), *Descartes' Natural Philosophy*. London: Routledge, 2000; pp. 659-679. Theo Verbeek, 'Crisis te Utrecht: 1641-1642'. In: Willem Koops, Leen Dorsman & Theo Verbeek (eds.), *Née Cartésienne / Cartesiaansch Gebooren. Descartes en de Utrechtse Academie 1636-2005*. Assen: Van Gorcum, 2005.

- **Cartesians vs. French authorities.** Account pieced together from pertinent pages in Francisque Bouillier, *Histoire de la philosophie Cartésienne*. Paris, 1868 (3rd ed. in 2 vols.; facs. reprint: Genève: Slatkine Reprints, 1970); Paul Mouy, *Le développement de la physique Cartésienne 1646-1712*. Paris: Vrin, 1934; A.C. Kors, *Atheism in France 1650–1729. Vol. 1: The Orthodox Sources of Disbelief*.

Princeton UP, 1990, and ch. 1, 'The Religious and Political Context' of Desmond M. Clarke, *Occult* 443
Powers and Hypotheses. Cartesian Natural Philosophy Under Louis XIV. Oxford UP, 1989 (a splendid
summary of the theological/philosophical debates and the multilayered power struggles behind
them).

Crisis of legitimacy. The first to sense the mid-17th-century occurrence of such a crisis was Joseph
Ben-David (*SRHI*, 5.3.). The two modern views on science and religion cited in the text are in
Stephen Jay Gould, *Rocks of Ages. Science and Religion in the Fullness of Life.* New York: Random
House, 1999, where he defended what he called NOMA ('Non-Overlapping Magisteria'), and Ste-
ven Weinberg, *Dreams of a Final Theory. The Scientist's Search for the Ultimate Laws of Nature.* New
York: Pantheon, 1992; ch. 11 'What About God?'

XIII

ACHIEVEMENTS AND LIMITATIONS OF
FACT-FINDING EXPERIMENTALISM

In 1686 the 'Royal Society, for the improvement of naturall knowledge by Experiment', received from one of its fellows, Isaac Newton, news that the manuscript of his 'Principia' was ready for the press. Alas, the society felt compelled to bow out of its commitment to have the book published at its own cost. Its funds had been depleted by the recent publication of a lavishly illustrated *Historia piscium* ('History of Fishes') by the late fellow Francis Willoughby. In the end the society's clerk to see the *Principia* through the press, Edmond Halley, got the printing costs reimbursed, not in pounds sterling but in fish, or rather, in tomes filled with accounts of a large, thoughtfully categorized assortment of fishes. The story encapsulates in a nutshell the relative importance attached at the time to the predominantly mathematical and the fact-finding modes of nature-knowledge in Britain and (in somewhat lesser measure, given the Continental popularity of a third rival, Cartesianism) elsewhere in Europe as well.

> Had anyone in the circle close to the Royal Society been asked, in the years before 1700, where the future of science lay, they would almost certainly have identified as its focus the compilations of data and systematic taxonomics that absorbed the interest of Sloane and his fellow physicians and natural historians. Even during the period of Newton's presidency [1703–1727], ... , the physics, astronomy and mathematics we associate with the birth of modern science today was a minor, specialist interest, regarded with a certain distaste because of the personal quarrels about priority and intellectual ownership these more abstract domains of inquiry seemed frequently to provoke.[230]

Were these various domains of inquiry really so different? In examining the achievements and limitations of fact-finding experimentalism, I shall define my leading questions, as much as the subject allows, in parallel with our previous inquiries into achievements and limitations in the wake of Kepler/Galileo and of Beeckman/Descartes/Gassendi (chs. 10 and 11). As in those inquiries, two issues demand particular attention. To what, if any, extent did experimenters succeed in their ongoing efforts to revolutionize the crafts? And in what manner did they seek to check the validity of facts found through experiment and conclusions inferred therefrom? But practitioners of fact-finding experimental science also faced a problem of their own that plagued neither mathematical scientists nor natural philosophers

446 – the vast number of facts found the experimental way. How to impose some measure of order upon masses of purported facts that might well run into the tens of thousands? The difficulty had already begun to plague the more loosely observational approaches of men like Garcia de Orta and Aldrovandi. It became vastly more acute with the advent of experiment when used not as an incidental but as a routine tool for finding facts.

Hence, the imposition of order, the reform of contemporary craftsmanship, and the search for genuine facts and valid conclusions are the three basic issues to be pursued throughout the present chapter.

Facts collected and categorized

Not one domain of nature-knowledge of the coercively empiricist kind that had arisen during the late 15th and the 16th centuries remained wholly unaffected over the next century by the more pointed variety that came out of its revolutionary transformation – sustained ranges of experiments habitually directed at the discovery of new phenomena of nature. What Leonardo, Castro, and Vincenzo Galilei probed incidentally in the 16th century (ch. 4), and what Bacon, Gilbert, Harvey, and van Helmont preached and/or practiced on larger scale and in more systematic fashion between c. 1600 and 1640 (ch. 7), subsequently altered even the most purely descriptive activities.

A case in point are those 'natural histories' of which Bacon had recommended the assiduous preparation in accordance with his general method of inductive ascension. Not that the numerous methodological niceties prescribed in *Novum organum* made it very often into the natural histories actually compiled in droves over the remainder of the century by mostly British authors. In many cases these remained unfinished due to the apparent impossibility of finding a natural stopping place for the collection of facts pertinent to the topic under consideration. More literally Baconian about them was that, as a rule, the facts collected derived no longer from unaided observation alone but also, if not more so, from the deliberate setting up of artificial situations suitable for producing phenomena hidden by nature from our unaided senses. One prolific author of natural histories of such a mixed observational *and* experimental kind was Robert Boyle, whose numerous works include *General History of the Air*, *The History of Fluidity and Firmness*, and *Experimental History of Waters*. A telling example, taken from his *Experiments and Considerations Touching Colours* of 1664, is his observation that the yellow solution of a tropical wood known as 'nephritic wood' loses its blue opalescence when acid is added but regains it when an alkali is added next (so that it can be used to discriminate between them).

Musea of natural history underwent similar changes. The leading ideal of putting the whole world in a showcase remained the same (p. 130). So did the aura of natural magic that

enveloped them. What did change was that certain musea became centers of experimentation of a kind. What kind is best illustrated by the curator of a large museum connected to
the Collegio Romano in Rome, Father Athanasius Kircher, SJ. Descartes curtly dismissed
him as "plus charlatan que sçavant" (more of a charlatan than of a scholar), and posterity
has concurred in his judgment.[231] In his own time Kircher outshone everybody else in reputation and in the amount of funding he managed to accumulate. Here is how he sought to
sustain his idea

> that fossils were inorganic, produced by the mysterious lapidifying [= stone-producing] juices
> that coursed through the veins of the earth [by means of experiments meant] to prove the ex-
> istence of the *vis spermatica*, the universal generative principle responsible for the appearance
> of all natural creations. To this end, he calcified urine, produced samples of *arbor metallica*,
> performed 'crystallogenesis', and mixed chemicals to create a marbling effect in stone, all ex-
> perimental 'proofs' of nature's plastic virtues.[232]

Collections surged as the Voyages of Discovery gave way to routine overseas trade, and
strange plants and beasts were routinely sent along with silver and spices. How, then, not to
drown in an unprecedented flood of new data, beings, and objects? The writing of natural
histories and of museum catalogs tended with depressing regularity to be abandoned halfway. If these data, beings, and objects were to be put to any discernible end at all, they had
somehow to be given some measure of order. How indeed?

Leonardo, in examining in excruciating detail how water falls in a pool, had faced essentially the same problem. Scholastic concepts like impetus had served him as tools for
ordering the endless variety of assiduously observed phenomena into distinct patterns (Figure 4.5; p. 116). Other organizing principles and/or carriers thereof came along in due time.
Devices introduced in medieval times to make books readily accessible (e.g., pagination,
section headings, running heads, indices) proved in the 17th century of great help to prevent
object overflow. Encyclopedic works, organized alphabetically or by subject, proved helpful
in gaining a grip upon new data and new views. Classification schemes were the usual tool
to arrange plants and beasts in systematic fashion. In the course of the 17th century available
schemes well-nigh exploded, particularly for plant taxonomy. The Dutch and British overseas companies transported to their respective capitals vast amounts of tropical vegetation
or at least pictures and descriptions thereof, such as those prepared on the spot in loving,
painstaking detail by Rumphius on the Moluccan Islands and by van Rheede tot Drakestein
on the coast of Malabar. The practice raised difficulties of classification far beyond the capacities of the ancient schemes. Two late-17th-century botanists, John Ray and Joseph Pitton
de Tournefort, met the demand for amplified classification schemes based in part on such
novel principles as seemed suitable for deciding, for instance, whether some tropical plant

448 was indeed an unknown species or just a larger variety of a known one. For animals visible with the naked eye Aristotle's classification scheme remained of sufficient service for the time being.

In chiefly description-oriented domains like these, experimentation assisted in the finding of new phenomena but not in imposing some measure of order upon them. But if applied in more rigorous ways, experiments did serve as a tool for order. Four different types of instruments assisted in making the search for new facts proceed in an orderly manner: musical instruments, telescopes, microscopes, and air pumps. Further, five distinct phenomena gave occasion for imposing a measure of fact-finding experimental rigor: magnetic, electric, chymical, medical, and those pertaining to light and color. I shall treat them in that order.

Instrument-driven fact-finding

The four instrument types differ both in date of invention and in the time it took for each to be turned into a tool for the sustained finding of facts. Musical instruments have of course been around from time immemorial; what turned them into scientific instruments was the advent, in a mostly Baconian setting, of research into sound, for which they supplied numerous data. The telescope started life in 1608 as a craft product (p. 274). Within a year Galileo turned it into an instrument at the service of his budding, realist-mathematical science, but by the 1660s it became an instrument for sustained, high-accuracy measurement. Its derivative in an optical sense, the microscope, also emerged c. 1610 and also began to yield masses of data by the 1660s. The air pump, finally, was first constructed in or around 1649 as one material consequence of the discovery of Torricellian space less than a decade earlier. In the late 1650s it was redesigned for a sustained program of experimental research. For each of these instruments I shall examine the kind of facts yielded by their frequently experimental usage, to what extent and how these facts were made to interact with theory, and the limitations to which the use of these instruments proved subject.

Musical instruments. In his *Sylva sylvarum* Francis Bacon sought to encourage the creation of an empirical but also useful science of sound by listing two hundred experiments on the subject, many taken from books on natural magic (p. 248). He appealed to future investigators to go to the artisans and find out from their expertise how properties like pitch or loudness vary with material, shape, and dimensions of musical instruments. In the 1640s an early Baconian, Edmund Chilmead, adopted the 'natural history' mode of research to the letter and sat down to correct factual mistakes Bacon had made on the subject. He also pointed out that in the meantime Bacon's wishes had in effect been fulfilled by Marin Mersenne. In seven 'Livres des instrumens' ('Books on instruments') of his *Harmonie universelle* of 1636/7 Mersenne collected, experimentally supplemented, supplied with some measure of order, and where possible enveloped in some moderately quantitative theorizing such trade rules as makers of the most varied musical instruments, from viols to bagpipes to drums, had shared with him. As Mersenne found out just too late for use in his own work, a survey of similar scope had been published in German in 1619 by the composer Michael Praetorius, and it is instructive to compare how data which Praetorius in his *Syntagma musicum* employed for his fellow musicians alone were directed by Mersenne toward the advance of nature-knowledge in the first place. In his book on the church organ Mersenne carefully noted the proportions between length and width given to pipes of various kinds by organ makers. He established experimentally the mistake of those who believed that, irrespective of whether one halves a pipe's length or its width, the higher octave is yielded thereby. He also wondered how an increase in wind pressure affects pitch:

> Here the makers may help philosophy by listing pipes that rise only by a semitone, or a third, a fourth, a fifth, etc., for it will be easier to find the reason for that when one knows the properties of the pipes that are the cause of the difference between the notes.[233]

Sudden flashes of striking insight punctuate *Harmonie universelle*, otherwise filled with hundreds of pages of plodding vacillation. Among Mersenne's better attainments are his painstaking measurement of the speed of sound, of the upper and lower limits of audible sound, and of the number of vibrations a string makes to produce a given note (in other words, of absolute pitch). He also discovered a range of harmonics never before distinguished in a musical note (p. 376), investigated properties of natural tones and beats, and found by assiduous experimentation how the vibrational frequency of a string and, therefore, its pitch depend on its tension, its length, and its thickness (soon labeled together 'Mersenne's law').

None of these accounts is infused with magic of any kind – on religious grounds mostly since Mersenne feared magic to the point of obsession. Another big book aimed at covering the whole of musical theory and practice, Kircher's *Musurgia universalis* of 1650, did display the extent to which phenomena of sound had been enveloped in the kind of natural magic on which Bacon, too, had drawn, all the while seeking to reform the subject. Kircher described in his book numerous experiments of his own, yet they acquired their significance for the advance of experimental knowledge through Part 2, 'Acustica', of *Magia universalis*, which he had his pupil, Father Gaspar Schott SJ, compose from mostly Kircher's own material. Published in 1657–59, Schott's book delved at length into such topics as analogies between sound and light, the echo (for the understanding of which Schott compiled a 'history of echoes'), and how naturally and artificially produced sounds are related. All this made a profound impression upon numerous investigators in Britain. The experiments on sound there undertaken or at least contemplated go back to either Bacon's program or Schott or both.

450 Among those actually undertaken, most were carried out in the weekly sessions of the Royal Society, in a fitful manner dependent chiefly on whether or not the ruling president was himself a music lover. Where possible, the implications of these experiments for practical applications or for general issues of to-and-fro motion were discussed. Other experiments were executed in private by the society's 'curator of experiments', Robert Hooke (p. 498). In one session of the Royal Society Hooke exhibited test models for an 'otacousticon', or hearing aid. In another,

> following a reading of one of Bacon's experiments of the *Sylva* in July 1680, 'of the motion of bodies upon pressure', Hooke and Wren demonstrated that water in a glass set into motion using a viol bow would produce different wave patterns on its surface which corresponded to the pitch of the sound produced at the same time.[234]

Further, the Royal Society received report of the rediscovery, by two musicians, of what Chilmead had found half a century earlier: harmonics, such as first heard distinctly by Mersenne, arise from the string vibrating in several modes at the same time. Paper riders placed on a vibrating string were not thrown off at certain points, which were later called 'nodes'. By the end of the century Francis Robartes, in a contribution to the *Philosophical Transactions*, revealed the connection between these nodes and the multiple vibration of a string, on the one hand, and the range of natural tones of a trumpet, on the other.

Among experiments contemplated rather than actually undertaken, a 'General Scheme' devised by Hooke around 1668 to bring a measure of coherence to experiments at those weekly sessions displays once again the pervasive influence of Bacon and, through both him and Schott, of the natural magic tradition. Hooke followed Schott in subdividing the subject into natural and artificial sound. He recommended two 'Histories' to be drawn up through observation and experiment: a 'History of sounds, musical and harmonious', for the natural part, and for the artificial part a 'History of trades' meant to include the makers of musical instruments. Hooke classified them in accordance with the material (wood, tin, copper, etc.) with which they worked and with the guilds to which they belonged (p. 481).

Although by the end of the century a huge amount of information on properties of sound had been assembled at diverse places in a variety of contexts, a distinct discipline of acoustics had not yet come into being thereby. This was due in part to the aura of natural magic that still surrounded it. Experimental scientists tended to disavow the magical connotation without this making much of a difference in their experimental practice; also, the notion of magical effects was not conducive to digging deeply into possibly underlying mechanisms. More importantly, the one unifying theme holding the various parts of *musica* together previously, the problem of consonance, had been central indeed to the process of turning the *mathematical* science of music realist but it was not suitable as a unifying theme for its fact-finding, instrument-driven counterpart. Moreover, the problem of consonance itself had been part of a larger unity: the hoary idea of the whole of music (theory as well as practice) expressing the harmony of the world. During the 17th century this was gradually replaced with a conception of music as an aesthetic phenomenon in and of itself. Music was turning from a reflection of cosmic harmony toward being aimed solely at the arousal of human affects. The once close bond between music making and the cosmos had been broken; 'the sky had been untuned',[235] and for a long time no unifying theme was to present itself to the investigation of phenomena of sound. Contributions remained scattered over a variety of budding disciplines, until in the 1850s Helmholtz brought them all together.

Telescopic observation and measurement.   By the 1660s insertion of crosshairs in telescopes built according to Kepler's configuration made it possible for the telescope to be converted from an aid to realist-mathematical science into an instrument capable of accurate measurement (p. 329). How accurate, really? Huygens' hope to get rid of blurred images in refracting telescopes by means of mathematical, dioptrical theory was squashed by Newton's discovery of chromatic aberration. Newton came up with an alternative, the reflector, but for decades instrument builders were unable even to match its performance, let alone surpass it. Other hurdles in the ongoing race for ever higher accuracy proved a little easier to overcome. Much practical ingenuity went into the construction of very long tubes (or devices enabling users to omit them altogether) and into the optimization of compound eyepieces. Aligning the instruments in a consistent manner was also attempted. The following comparison highlights the results of all those efforts. For Ptolemy, c. 10 minutes of arc marked the limit of accuracy. Tycho reduced this to 2 minutes of arc – the well-known reason why Kepler regarded a divergence from Tycho's observations four times that large as lethal to his own pet theory on Mars' orbit (p. 170). By the end of the 17th century the limit of accuracy had dwindled to 3–4 *seconds* for the telescope itself and to c. 20 for quadrants and other measuring arcs that were fitted out with telescopic sights. The road from there to still further improvement was to run through the achromatic configuration of lenses attained by the middle of the 18th century and through ever better ways to correct for a factor that Kepler had first recognized as significant: the refraction that light rays undergo in the atmosphere.

   Enhanced accuracy led to the improvement of measurements in a great variety of cases. For instance, in his observatory at Danzig/Gdansk Johannes Hevelius prepared unprecedentedly accurate maps of the Moon, wrote a 'history of comets', and cataloged by position more than 1,500 stars and 56 constellations. Similarly, at Greenwich Observatory the first Astronomer Royal, John Flamsteed, spent a lifetime compiling tables of planetary, lunar, and stellar positions in accordance with the newest criteria for accuracy, leading in the end to *Historia coelestis Brittanica,* which was to dominate the field for much of the 18th century.

452   Further, Gian Domenico (Jean Dominique) Cassini prepared the accurate maps of France that King Louix XIV needed for the wars of conquest he had in mind. In the 1610s, Snel had used one sole quadrant to measure, through triangulation, the circumference of the Earth. Half a century later, Cassini could oversee a team of skilled surveyors equipped with quadrants and other measuring arcs enhanced with telescopic sights, with his own tables for Jupiter's satellites, with a range of precautions to keep procedures maximally standardized, and with leveling devices improved for the purpose by fellow Academicians Picard and Huygens. A more accurate value for the Earth's circumference was one by-product of Cassini's enterprise.

Measurements like these were carried out for the sake of accuracy almost as a value in itself. Others had research consequences that reached beyond their own specialty. Examples are the explosion of Ptolemy's measure of the universe, the discovery of the velocity of light, and the discovery of the irregular shape of the Earth.

Ptolemy was the first person to make a serious attempt to measure the universe. Combining "philosophical tenets, geometric demonstrations with spuriously accurate parameters, planetary theories, and naked-eye estimates", he put together a seemingly plausible, grand schema of the order of the planets and of their distances and sizes (p. 25). He established the distance between the Earth and the fixed stars at 20,000 times the radius of the Earth. Numerical details aside, the schema was challenged between Ptolemy and the advent of the Scientific Revolution only twice. Al-Urdi recast certain specific sizes and distances but also made the interior planets change place, with Venus rather than Mercury nearest the Sun (p. 68). Copernicus' heliocentric setup also had consequences for the schema. It notably entailed a necessary order of the planets rather than one requiring auxiliary assumptions (p. 110). Further, Copernicus' failure to observe star parallax caused him to posit a vast amount of empty space between Saturn and the fixed stars, which in its turn entailed some shrinkage of the distance between Saturn and the Sun.

Suspicion of deeper trouble with the grand schema arose with Kepler. His researches on atmospheric refraction made him aware of the consequent lack of reliability of the sizes of planets and stars, be they observed with aided or unaided eye. He further inferred from other researches of his that at least one of the two theoretically brilliant, geometric methods on which Ptolemy had relied for establishing an absolute value for the Sun–Earth distance did not work adequately in practice. The two conclusions taken together left Kepler facing a near void. Guided by his persistent idea of a solar system governed by the Platonic solids, he filled it with harmonic speculation, hopefully kept in check by such empirical data as would be yielded by the transit of Mercury that he calculated to occur on 7 November 1631. His death one year earlier prevented him from watching it himself (pp. 177; 303). Even so, the transit came with a big surprise. When the clouds over Paris finally broke, Gassendi saw the Sun duly project its image through his telescope onto a screen. On its surface he noticed

a tiny black spot. So deeply ingrained was the standard measure of the universe, with its attendant idea of the size of the planets, that he did not at first identify it as Mercury, whose
predicted transit he had after all set out to observe, but took it for a sunspot.

Three decades later the telescope came into its own as a measuring device, so that the
crucial parameter on which the determination of all other distances depended, solar distance, could be established with greatly enhanced accuracy. Only now did it become apparent that Ptolemy's grand schema had been mistaken not just by a factor of 2 or so but by an
order of magnitude. By the end of the 17th century, Flamsteed in Greenwich and Cassini in
Paris settled on a value for solar distance that was no longer Ptolemy's 1,210 Earth radii but
rather c. 20,000. Also by the end of the century it was widely agreed that stellar parallax, the
observational absence of which in Copernicus' time had set the astronomers' expansion of
the universe in motion, had still not been observed (p. 111). A striking explanation of the
failure to observe it came readily forward. In the *Philosophical Transactions* of 1694 it was
commented "that Light takes up more time in Travelling from the Stars to us, than we in
making a West India Voyage (which is ordinarily performed in six Weeks)."[236] The measure
of the universe, no longer conveniently expressed in Earth radii, was now well on its way
to the light-years that modern astronomers use as their unit for star distances in our own
galaxy.

By 1694, then, light was known to take time for its travels. The train of events that led
to this discovery started when Jean Picard, astronomer at the Paris Académie, measured all
over again latitude and longitude at the remnants of Uraniborg observatory. His findings
came down to a consistent discrepancy of 15' between Tycho's values and those established
by means of the up-to-date, state-of-the-art tables for eclipses of Jupiter's satellites that Cassini had compiled in hopes of solving the problem of longitude (p. 317). Earlier, at the University of Bologna, Cassini had successfully challenged Huygens' position as Europe's best
telescopic observer, and in 1669 he was lured to Paris by the offer of the highest salary paid
by the French treasury to any of its academicians, Huygens' alone excepted. Once Cassini
satisfied himself that the 15' discrepancy was not an accident, he began to suspect that both
Aristotelians and Cartesians had it wrong in taking light to cross the universe with infinite
speed. He soon withdrew the suggestion for good, only to see it adopted by Ole Rømer, one
of his assistants. Rømer found his own prediction of a 10' delay in the eclipse of Jupiter's
moon Io confirmed by his telescope on 9 November 1676. He had made the prediction in
view of his idea that there is an appreciable difference in the time it takes for light to reach
the Earth from a planet's orbit at apogee and at perigee, and that this ought to be taken into
account. The inference was important, not only for the accuracy of telescopic measurement
itself but also for a lecture that Rømer's colleague, Huygens, gave three years later to explain
to his fellow Academicians the nature of light – the finite speed of light was one indispensable component of his explanation (p. 541).

454   Telescopic measurement also altered current conceptions of the shape of the Earth. The determination of solar distance had involved a voyage to the tropics aimed at measuring the parallax of the planet Mars. At Cayenne, in 1672/3, Cassini's envoy, Jean Richer, stumbled upon an unexpected by-product – a first intimation of the inconstant length of the seconds pendulum. Not only did the find undermine the pendulum clock as a viable solution to the problem of longitude (p. 316), the data underlying it were also taken by some as an indication that the Earth might not, as hitherto assumed without questioning, be perfectly spherical. Newton's 'System of the World' as outlined in his *Principia* entailed an Earth bulging at the equator and flattened at the poles, so he cited Richer's data as apparent confirmation of his System (p. 667). In the 1730s this issue was to figure as one decisive difference between the Newtonian and the Cartesian systems of the world, with the latter being taken to entail a flattened equator and bulging poles. Two French expeditions were sent off in opposite directions. One was sent to the tropics again, whereas the decisive expedition, to the Arctic regions of Lapland, resolved the dispute in Newton's favor.

Microscopy.   When first invented in c. 1610, the microscope was enveloped in an aura of natural magic and employed only incidentally, such as for watching insects take elephantine proportions and then (turning lifelikeness into larger-than-lifelikeness) picturing them at full-page size. Not until the 1660s, when the walls between fact-finding experimentalism and the kinetic-corpuscularian philosophy of nature began to break down (p. 509), did the microscope become an instrument of experimentally sustained research. As such, it flourished for about a quarter of a century, until a range of limitations to which the instrument and its uses and achievements proved as yet subject put an end to its use. In that period five investigators stood out. Using expert techniques that each developed on his own, they managed to squeeze out of the instrument masses of data directed in good part at certain specific research objectives. Of these men, four operated in the context of the Royal Society: Hooke and Grew as fellows and officers; Malpighi and van Leeuwenhoek as fellows in steady correspondence with the society. Only Swammerdam sought and found encouragement and exchange elsewhere, chiefly in his own country. Although four of them shared an institutional affiliation and two pairs of them were in close physical proximity, mutual exchanges were scanty.

Robert Hooke was alone among these 'great five' to devote just one short bout of research to peering through the microscope. He reported what he saw in 1665, in *Micrographia: or some Physiological Descriptions of Minute Bodies made by magnifying glasses, with Observations and Inquiries thereupon.* He meant the book to serve, primarily, as a sustained plea on Baconian lines for using instruments in research (p. 247), secondly, as a vehicle for his broad view of how the world is constituted (p. 557), and, only thirdly, for the evocation of a previously invisible world of minute objects (e.g., 'Of Fiery Sparks') and beings (e.g., 'Of a Book-

worm'), conveyed even more by the book's wonderful plates than its detailed descriptions. 455
As a rule, he was more given to detailing their surface features than their internal structure.

Of the sixty 'Observations' Hooke put in quasi-geometric order, three were destined to later notoriety. One is his 'Observation 18, Of the Pores of Cork, and other Bodies'. He called these pores 'cells' and felt sure that they served as passages for sap transportation. Another is a phenomenon we now know as 'Newton's rings', after the man who was to make proper sense of them (p. 688). Hooke, the first to find them worthy of scrutiny, simply dubbed 'Fantastical Colours' what he saw in Muscovy glass (mica). Finally, removed far indeed from his microscopical researches, Hooke added near the end of his book a digression, larded with experiments, 'Of the Inflection of the Rays of Light in the Air', that is, differences in the refraction of light as the air through which it passes is more or less dense. Again, Newton was the man to get more out of Hooke's loose observations (p. 705).

Micrographia remained Hooke's only sustained effort at microscopical research. At the point where he left off, Nehemiah Grew took over. Taking Hooke's observations on the cellular structure of plants as a cue, Grew studied plants exclusively, only to quit research after a decade and resume his career as a physician. An assiduous worker in the 'natural history' tradition, he withstood the temptation to let himself be overwhelmed by the sheer masses of available detail. Instead, he focused on several issues in plant life: the shape and disposition of their parts, their nutrients, and the stuffs (e.g., sap) they are made of. He gave special attention to the chemical composition of these stuffs and to the various ways in which they move through the plant (e.g., by pressure or by capillary action). He subscribed to a corpuscular conception of matter, as all but one of his colleagues did, yet corpuscles entered his various explanations of plant structure much less than his chemical analyses did.

On the very day that Grew presented his book on plant anatomy to the Royal Society, an essay on the same subject reached the meeting from faraway Bologna. Marcello Malpighi was not alone in pursuing the dissection of human and animal body parts (muscles, liver, brain, ear) after Vesalius' manner, with a view to laying the indispensable groundwork for an understanding of their modes of operation. But he was the only one to pursue such researches down to the level of the subvisible. His primary tools for making sense of what he observed both macro- and microscopically were experiment and comparison.

His experimental data in animal physiology were to a large extent the result of the anatomical preparation of the objects rather than truly physiological experiments. Noting the distribution of colored fluid in the kidney, as it was injected through the artery, vein, or ureter, Malpighi drew conclusions about the connections between the blood vessels, the spherical bodies (glomeruli) which he regarded as glands, and the renal tubuli. Only infrequently did he complement his anatomical observations with physiological experiments involving vivisection. For instance, he endeavored to demonstrate conclusively the connections between the various structures of the kidney by ligaturing the ureter and renal vein in a dog and sometime later dissecting the excised kidney, without positive results, however.[237]

Malpighi's sustained comparison between corresponding parts in different living beings was guided by his assumption that what one finds to perfection in one being may be encountered in an as yet imperfect state in another. The assumption led him to modes of reasoning by analogy that he employed in particular at those increasingly frequent occasions when, in spite of his careful preparation techniques and other precautions for optimal visibility, his instrument let him down. Even so, Malpighi was the deepest digger of the five great 17th-century microscopists, who attained the most sophisticated interaction between facts observed and theoretical reasoning applied to them or derived from them. In publications on the silkworm, the silk moth, and the chicken in embryonic state and also in plant anatomy he arrived at results of lasting value. A particularly fine example of such experiment/theory interaction was his carefully comparative efforts to find out how secretion in glands actually takes place.

Antoni van Leeuwenhoek operated neither with a similar capacity for sophisticated theorizing nor with any research program at all. Rather, over a span of half a century he devoted his attention to just about everything that could be stuck on the object nails of his single-lens microscopes, the best of which attained the highest resolving power prior to the advent of the achromatic microscope. By their means he was able to confirm the presence, at those places in the body where arteries carry the blood and veins transport it back, of the capillary vessels stipulated by Harvey as a prerequisite for his circulatory conception (p. 253). Van Leeuwenhoek's haphazard choice of subjects and generally 'barbarous' reasoning (in Swammerdam's considered opinion) fostered chance discovery of the wholly unsuspected. Pepper water left alone for a few hours appeared to be chockfull of 'little animals' (infusoria). Other 'little animals' populated semen of other men's and his own making (spermatozoa). The discoveries had an impact upon two running debates, one over contagion, the other over reproduction (p. 474). Van Leeuwenhoek wrote up his observations in a factual, 'natural history' sort of style suitable for publication in the journal of the Royal Society, the *Philosophical Transactions*. At times he felt free to attach an attempt at explanation to his assiduous descriptions. At first he tended to reduce phenomena wholesale to globules of ever more minute sizes. Later he postulated the existence of mini-minifibers/vessels, alleged to underlie, beneath the limit of visibility, the minifibers or minivessels that he could distinguish through his microscope.

Unlike omnivorous van Leeuwenhoek, Jan Swammerdam took a particular interest in one sort of living beings, insects, and even more particularly in insect metamorphosis. He strayed less than the other four from the narrow path of exclusive reliance on the testimony of the senses, since we are bound by them in any case:

Oh GOD, thy Works are inscrutable, and all we know or can know of them is nothing but the
dead shadows of the shadows of the shadows of thy adorable and inscrutable works; before
which all the minds of men, however ingenious they may be, must become dull and confess
their dumb ignorance.[238]

At first, Swammerdam left more room for theory in experimental practice, as with an argument set forth in his doctoral dissertation that in respiration air is not *drawn* into the
lungs due to a partial vacuum but is *pushed* inside by the expansion of the chest. A growing
epistemic pessimism, fed by religious views of increasing bigotry, led him to regard such
researches as too imbued with conclusions going beyond the sense data from which he had
drawn them. So he turned to insects as a research territory that cried out for setting right
empirically the fables that false reason had kept accumulating over the centuries. These
fables came in three classes: spontaneous generation, lack of inner structure, and sudden
change of appearance. Definitive rejection, under carefully controlled circumstances, of the
first misconception was relatively easy. The description of insect anatomy (see Figure 13.1
for an instructive example) and the demonstration of the large extent to which the imago
of the butterfly already lies waiting in the caterpillar for its final unfolding were true feats of
microscopic prowess. What underlied them was, not indeed the kind of reasoning he had
come to find so shallow, but a broad preconception of God-given natural order.

There were others beside these 'great five' who contributed to the flourishing of microscopic observation. Among those who entertained hopes that the microscope would make
atoms visible (p. 385), Henry Power was the one to go ahead and turn that hope into the
guiding thread of extensive microscopical researches that he published without illustration
one year prior to the book destined to eclipse his for good, Hooke's *Micrographia*. Or take
Christiaan Huygens, who, in his customary vein of 'what others do, I can do better', emulated
and, indeed, improved upon van Leeuwenhoek's efforts to bring some order to a bewilder-
ing multiplicity of infusoria.[239]

Indeed, in their sometimes desperate search for pattern and order in a brand-new world
of uncertain likeness or unlikeness to our own, 17th-century microscopists were thrown
back upon their own resources. Some used corpuscles for an all-purpose kind of explanation, as with Grew and (in early years) van Leeuwenhoek; others spurned them. What all five
had in common was a capacity to ask further questions and let their researches be guided by
such answers as they found along the way. To foster the finding of such answers, the comparative method Malpighi developed more deliberately than the others proved particularly
fruitful. Malpighi's inquiries into the structure and operation of glands are a case in point.
The way he directed his observations and drew inferences from them was guided all the time
by comparing glands and their operations not only in a variety of living beings but also with
macroscopic sieves and filters. In this manner he produced a kind of three-way race in which

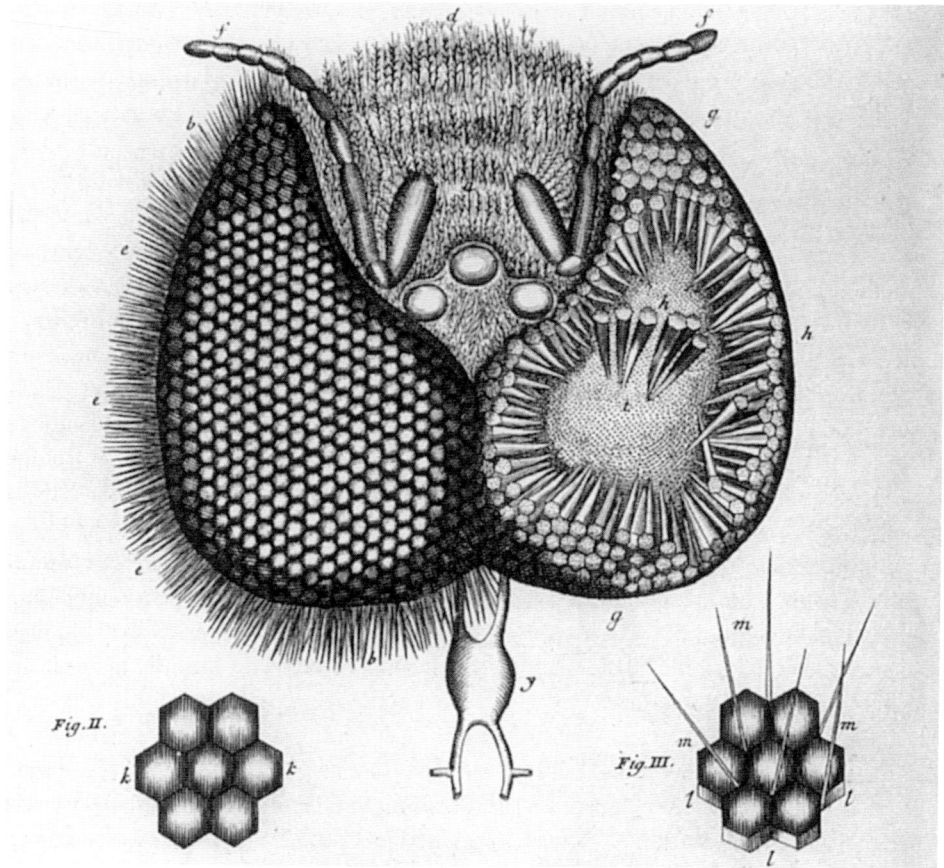

Comparison with Figures 4.8-9 (p. 120) gives an idea of both continuity in accurate description and the advances enabled by instrument-aided dissection.

Figure 13.1: Swammerdam: compound eye of the bee

every component reinforced every other. Even so, Malpighi himself was better aware than his colleagues of the limitations that the instrument and its handling did not as yet allow him to overstep.

When contrasting the high hopes set by the instrument in the 1660s with the massive flight from it taken less than three decades later, these limitations must be taken into account. How, in an age of sunshine and candles, properly to illuminate the object watched?

Where to place it? How to bring out internal structure? How to coordinate eye and hand, to ensure sufficient likeness for one's drawings? Ingenious solutions were found, for instance, to make structures visible by injecting tissue with wax or colored ink. Yet all the gains meanwhile attained in magnification and resolving power could not fully offset the struggle with tough problems like these. Worst of all, the world of the very small proved ever more subtly complex as microscopists dug more deeply into it. Organic life itself appeared to defy hoped-for chances to resolve running disputes on several larger issues by means of the microscope. One concerned an ongoing controversy between those to uphold, with Harvey, 'epigenesis' (something indispensable must be added to the homogeneous 'egg' for a heterogeneous body to grow out of it) and adherents of 'preformation' (in generation, a heterogeneous mini-being is incited by some outside agent to grow into an adult). Harvey's conviction that there is a universal unit of reproduction, which he generally referred to as eggs (p. 254), acquired a more concrete meaning when van Leeuwenhoek's fellow townsman Reinier de Graaf found definite traces of eggs in women and other females giving birth to live young. No more than five years later, van Leeuwenhoek stumbled on spermatozoa. Although these two discoveries, as well as Malpighi's microscopic refinement of Harvey's experiments with chick embryos, did alter the terms of the debate and also made it more concrete, none was capable of settling the dispute for good. The apparent lack of capacity of the microscope and attendant techniques in the state these had attained by the late 17th century to act as final arbiter in debates like this contributed greatly to the lack of follow-up that was to befall any innovative usage of the instrument until the early 19th century.

Air pump. If the microscope came into the world with a whimper, the air pump did so with a bang. The loudest bang of all sounded in Magdenburg. The mayor, Otto von Guericke, made two teams of eight horses each pull apart the halves of a brass sphere first clamped together by having the pump draw the interior space void. The air pump that he invented worked as the water pump did but now was adapted to the recent discovery of Torricellian space (p. 411). The range of noisy experiments Guericke undertook with his air pump were meant above all to show, and to measure, the extraordinary forces that appear to come from the pressure the atmosphere exerts upon us. In making the routine production of void spaces manageable, the air pump opened vast opportunities for experimental discovery. No less than three fact-finding programs were set up in the late 1650s and 1660s to exploit the new instrument, in Germany, in France, and in Britain.

Rather than publishing the design of his pump and his experiments with it right away, Guericke had them pre-published in two books by Gaspar Schott SJ, one in 1657, the other in 1664. The former work was read soon after its appearance by Robert Boyle. He was an Anglo-Irish nobleman who by the late 1640s had undertaken a sustained course of factfinding experimentation carried out in a home laboratory set up for the purpose. Recognizing at

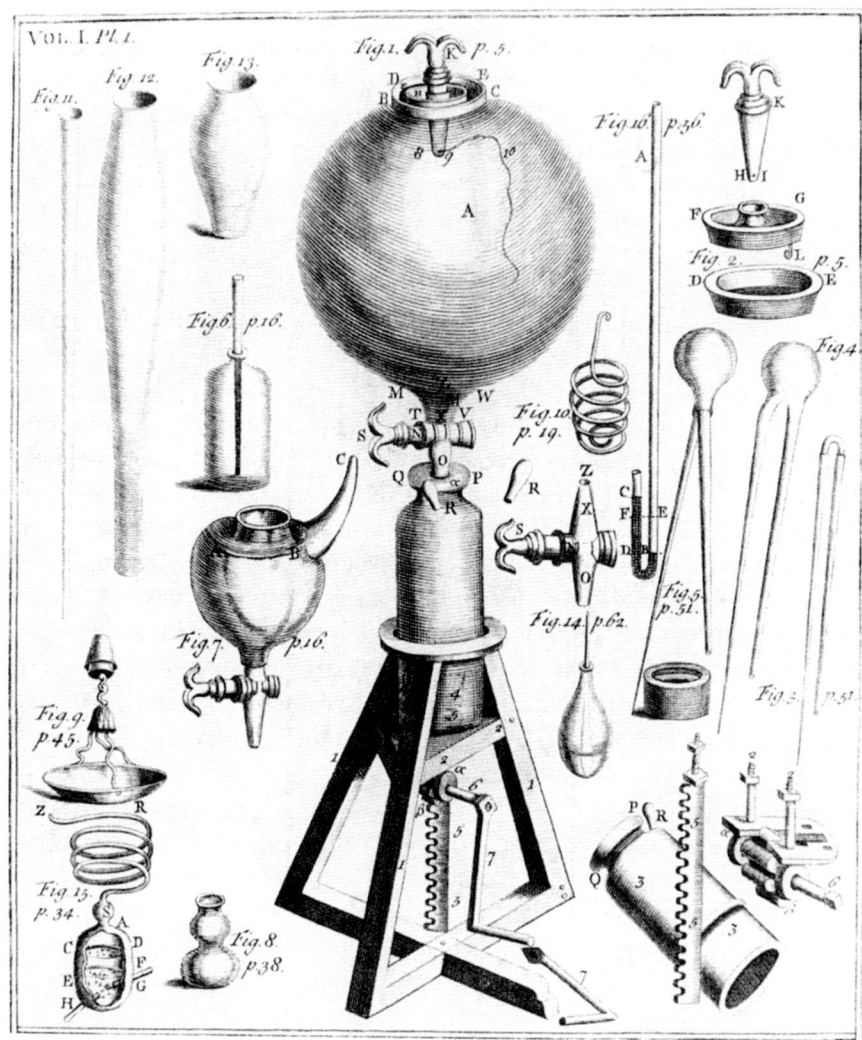

A = glass receiver (volume: c. 30 l). B-C = aperture (for putting things inside). R = valve. S = stop-cock.
3 = hollow, brass cylinderT (length: c. 35 cm; internal diameter: c. 8 cm). 4 = wooden piston, made to fit tightly into the cylinder, and topped with leather. 5 = iron rack. 7 = pinion.

Figure 13.2: *Hooke's & Boyle's first air pump*

once what wonderful opportunities the air pump described by Schott offered, he and his assistant at the time, Robert Hooke, made ingenious improvements to the device (Figure 13. 2).

Huygens rightly thought that he could improve on the Boyle/Hooke design. So (rightly or wrongly) did the pioneers, von Guericke & Schott. So, not to be outdone (but also to counter an acute outside critic, Hobbes), did Boyle & Hooke themselves. Experimental programs with successive versions of the engine were set up by the two Germans on their own (not, that is, in the context of the Society of Jesus, to which the Lutheran mayor did not of course belong); by Boyle & Hooke in the context of the Royal Society, and by Huygens partly on his own, partly in the context of the Académie. In ongoing three-way rivalry these men investigated the behavior of phenomena, objects, and beings placed inside the engine, and the properties of the air pumped out of it.

With some phenomena, objects, and beings it was readily predictable how they would behave if put inside the receiver and then wholly or partly robbed of air; with others it was anybody's guess. Examples of the former include the unhindered transmission of light, which Noël a decade earlier had regarded as a telling argument against the void (p. 415); the fading away of sound, which had always been regarded as in need of a substantive carrier; the simultaneous descent of a feather and a metal ball released from the same height, just as Galileo had claimed would happen; the expiration of animals of whatever species, size, or sex, which few seriously expected to survive. Less predictable was how Hooke himself would fare when, at the 23 March 1671 session of the Royal Society, he climbed into a large vessel subsequently evacuated by about one-fourth part (after about fifteen minutes he emerged in fine shape, reporting nothing but some pain in his ears), or how electrics would behave, or magnets. Quite unpredictable were the effects of putting a ruby inside and heating it up with focused light from outside (upon return from the void, it promptly fell to ashes) or of putting a lump of butter inside and placing a hot iron cap over the vessel (the glass got hot, but the butter did not melt). Wholly unexpected, finally, was a phenomenon Huygens first observed upon placing in the void a smaller void. He did Torricelli's experiment all over again in an evacuated receiver and carefully removed tiny remaining air bubbles. He confidently expected the mercury column to descend to a height level with the mercury dish or nearly so but found it not to do so at all, or at best upon gentle rapping only.

Among conclusions reached the experimental way about the air pumped out of the receiver was its indispensability not only for respiration but also for combustion and the growth of plants. Its chemical nature was examined, although its actual composition of a variety of gases was not to be suspected, and then readily discovered, until almost a century later. Boyle made much of the 'spring' of the air. At the Royal Society session of 1 February 1664 it was shown that air, upon being compressed, springs back – an experiment that King Charles II "mightily laughed at".[240] The density of air in proportion to water was measured in this context. Also, as a by-product of a dispute with Father Linus SJ, over other, more

462   Aristotelian interpretations of the 'spring' of the air, Boyle varied over a monotonous range of values its volume and pressure, thus confirming his suspicion of inverse proportionality between them. This was a rare case of a quantitative rule established in the context of fact-finding experimentation. Just like Mersenne's listing of variables affecting pitch, it was not attained through mathematics but by laborious measurement (Figure 13.3).

    In these experiments with air and the void three distinct components can be discerned, which were blended in a rather irregular variety of ways. Primary was the assiduous search for new facts, sometimes guided by theory, as with sound, but more often than not proceeding in more haphazard ways, as with rubies and butter. Certain specific theories or suspicions were confirmed, as with Boyle's law, or, in some other cases, contradicted. Also, at times explanations were derived, usually in an ad hoc manner, either from theories already extant or from some background worldview. An example of the latter is how the capricious behavior of mercury that remained suspended in Huygens' void-in-the-void experiment was dealt with. Huygens thought that the adhesion of mercury to the glass vessel was due to pressure of aether particles, whereas Newton thought some attractive force did the pressing.

A TABLE of the Condensation of the AIR.

A	A	B	C	D	E
48	12	00		$29\frac{2}{16}$	$29\frac{2}{16}$
46	$11\frac{1}{2}$	$01\frac{7}{16}$		$30\frac{9}{16}$	$30\frac{6}{16}$
44	11	$02\frac{11}{16}$		$31\frac{15}{16}$	$31\frac{12}{16}$
42	$10\frac{1}{2}$	$04\frac{6}{16}$		$33\frac{8}{16}$	$33\frac{1}{7}$
40	10	$06\frac{3}{16}$		$35\frac{5}{16}$	35
38	$9\frac{1}{2}$	$07\frac{14}{16}$		37	$36\frac{11}{19}$
36	9	$10\frac{2}{16}$		$39\frac{5}{16}$	$38\frac{1}{4}$
34	$8\frac{1}{2}$	$12\frac{8}{16}$		$41\frac{10}{16}$	$41\frac{2}{17}$
32	8	$15\frac{1}{16}$		$44\frac{3}{16}$	$43\frac{11}{16}$
30	$7\frac{1}{2}$	$17\frac{15}{16}$		$47\frac{1}{16}$	$46\frac{2}{5}$
28	7	$21\frac{3}{16}$		$50\frac{5}{16}$	50
26	$6\frac{1}{2}$	$25\frac{3}{16}$		$54\frac{5}{16}$	$53\frac{10}{11}$
24	6	$29\frac{11}{16}$		$58\frac{13}{16}$	$58\frac{2}{3}$
23	$5\frac{3}{4}$	$32\frac{3}{16}$		$61\frac{5}{16}$	$60\frac{18}{23}$
22	$5\frac{1}{2}$	$34\frac{15}{16}$		$64\frac{1}{16}$	$63\frac{6}{11}$
21	$5\frac{1}{4}$	$37\frac{15}{16}$		$67\frac{1}{16}$	$66\frac{4}{7}$
20	5	$41\frac{9}{16}$		$70\frac{11}{16}$	70
19	$4\frac{3}{4}$	45		$74\frac{2}{16}$	$73\frac{11}{19}$
18	$4\frac{1}{2}$	$48\frac{12}{16}$		$77\frac{14}{16}$	$77\frac{2}{3}$
17	$4\frac{1}{4}$	$53\frac{11}{16}$		$82\frac{12}{16}$	$82\frac{4}{17}$
16	4	$58\frac{2}{16}$		$87\frac{14}{16}$	87
15	$3\frac{3}{4}$	$63\frac{15}{16}$		$93\frac{1}{16}$	$93\frac{1}{5}$
14	$3\frac{1}{2}$	$71\frac{5}{16}$		$100\frac{7}{16}$	$99\frac{2}{7}$
13	$3\frac{1}{4}$	$78\frac{11}{16}$		$107\frac{13}{16}$	$107\frac{7}{13}$
12	3	$88\frac{7}{16}$		$117\frac{9}{16}$	$116\frac{4}{8}$

Column C (vertical): Added to $29\frac{2}{16}$ makes

AA   The number of equal spaces in the shorter leg, containing the same parcel of air, differently expanded.

B   The height of the mercurial cylinder, in the longer leg, that compress'd the air into those dimensions.

C   The height of a mercurial cylinder, that balanced the pressure of the atmosphere.

D   The aggregate of the two last columns B and C, exhibiting the pressure sustain'd by the included air.

E   What that pressure should be, supposing it in reciprocal proportion to the expansion.

Figure 13.3: Derivation of Boyle's law

Subject-driven fact-finding

Instruments did not provide the sole means to keep in check the propensity of fact-finding experimentalists to find *nothing but* facts – an inclination so abundantly given free rein in the collection of objects and in the drawing-up of 'natural histories'. Careful selection of subjects could render the same service. Between them, Gilbert, van Helmont, and Harvey had already performed the selection with great acuity. We shall now examine how each of their chosen subjects – magnetism, electricity, chemistry & alchemy, body & mind – was pursued by men keen to take up the challenge involved in their pioneering work.

Magnetism.　　The separation of magnetic and electric phenomena as distinct branches of inquiry was due to Gilbert. A magnet attracts iron pieces as rubbed amber attracts chaff or paper snippets or a metallic needle on a pivot. Even so, Gilbert insisted on their principal difference on theoretical grounds, and his experimental establishment of no fewer than eight differences in their respective behaviors (with only one retrospectively mistaken) confirmed it (p. 251). Since their principal difference remained undisputed until 1820, when Ørsted demonstrated experimentally that they are closely connected after all, it necessarily governs treatment here.

Gilbert did not restrict himself to the repetition and widening of a range of experiments on the behavior of magnets but had fitted these into a larger frame. He conceived of magnetism as a cosmic, life-giving force, of the Earth as a magnet, and of phenomena in it and on it as due chiefly to that force. By the end of the century, his experimental results on magnetic attraction were in as healthy a shape as they were to remain to the present day, but the cosmic frame had been given up. His magnetic worldview fell victim to the Europe-wide vogue for kinetic corpuscularianism, whether or not Descartes' specific account of magnetic action (p. 229) was adopted with it, as indeed it often was. It fell victim likewise to experiments undertaken by Jesuits who were out to undermine the heliocentrism that Gilbert's extraterrestrial views were taken to support. They used the experimentally demonstrated, consistently east-west alignment of Gilbert's *terrellae* (little earths) to show that whatever properties magnets on Earth display cannot obey any other force than that of Earth magnetism.

All that remained, then, of Gilbert's grandiose vision was an either corpuscles- or spirit-infested and, as such, still quite puzzling but otherwise somewhat pedestrian science of magnetism shorn of its vitalism and its cosmic scope alike.

Electricity.　　What, really, is electric action? Gilbert was convinced that electricity, which requires friction to manifest itself, involves contact action. This idea was of paramount importance to him because it removed the subject from his beloved realm of immaterial, cos-

464   mic magnetism. It was adopted by all but three of some twenty-five men engaged on and off during the 17th century in electric researches. Gilbert further thought that the action is due to an invisibly thin, watery *effluvium* (outward stream) which envelops the attracted object the way water tends to envelop wet twigs. Niccolò Cabeo, Francesco Lana Terzi, and most other Jesuits who dedicated themselves to the subject replaced this with another effluvial mechanism involving not moisture but air as the principal agent. They thought that the rubbing of the electric opens its pores so as to enable subtle matter to stream out, thus causing the surrounding air to rarefy and then, in regaining its original density, to drive light objects toward the electric. The possibility, denied by Gilbert, of electric repulsion thus became a natural part of the process, as matter in backward motion. It found experimental confirmation in how attracted objects appeared to spread over the electric and also in their trajectories as examined by Lana Terzi. In Britain some tried to salvage what they could of Gilbert's watery outstream, thus producing as the dominant account of the period a gluey sort of *effluvium* best visualized as tiny, viscous threads.

One Jesuit dissented, Fabri (the man to criticize Galileo's rule of falling bodies and to induce the Holy Office to put Descartes' philosophical works on the Index). In a range of experiments set up to support the British account he found that, against what both Gilbert and Cabeo had insisted, electric action does share with magnetism the property of being mutual. Meanwhile, in Britain Boyle either confirmed or independently rediscovered mutuality and then joined Fabri in showing, against Cabeo, that no air is needed for electric action. Fabri's effort, undertaken in the collective setting of the Accademia del Cimento at Florence, required some complicated maneuvers to get an electric inserted in Torricellian space. The outcome duly weakened Cabeo's air account; still, it inevitably left room to counter that the vacuum had remained incomplete. Boyle, who produced his void space by the more advanced means of his air pump, came to the same conclusion, which was open to the same objection. An enigmatic luminosity appeared as an unforeseen by-product in the Italians' mercury tube. Since it failed to manifest in the air pump, it also failed to gain the attention it was amply to receive in the next century – the age of elegantly wigged lecturers displaying catchy light effects from charges in evacuated tubes.

Other experiments, too, produced unforeseen effects. Guericke, one rare upholder of a nonmaterial agent of electricity, used for his electric a sulfur globe molded by way of a small-scale replica of the Earth and its mineral composition, which he held responsible for its many and variegated powers. Despite the ever more detailed prescriptions he kindly supplied on repeated request to Leibniz and others who asked for them, no one proved capable of replicating the range of experiments undertaken with it, which led him to conclude against Cabeo that electricity conduces. By the 1690s Huygens, who was in no need of a mini-Earth and thus felt free to replace the sulfur globe by the more regular amber, managed to obtain the same effect. Helped along by the Descartes-inspired notion then current of whirlpools specifically suitable for electric action, Huygens arrived in the end at four insights which together form the high point of 17th-century electrical science, even though his failure to publish a word of his findings gave others, some forty years later, a chance to discover them all over again. He found that attraction, conduction, and repulsion take place in natural succession, that moisture furthers conduction, that two objects electrified by a third repel each other, and that the apparent connection between electricity and luminosity is not just accidental.

In sum, whereas many of Gilbert's specific tenets had by the end of the century been left behind, "it was always *De magnete* that defined the subject of electricity, identified its problems, and provided the framework for their resolution".[241] Advances made since Gilbert had been both limited and fitful. What identifiable forward pressure the advance did display was governed in part by nature's whims and in part by how experiment, theory, and background worldview were made to interact. The various theories invoked were rarely drawn straightforwardly from some natural-philosophical orthodoxy, which did not have ready-made solutions on offer in any case. Rather, they formed a mixture of ad hoc conceptions thought up against an eclectically handled Aristotelian or corpuscularian or (in a rare case) stoic background. Experiments meant to test such theories led as a rule to highly ambiguous or even mutually inconsistent outcomes, and also at times to the puzzling appearance of phenomena unknown and unsuspected (in retrospect, their most productive service). John Heilbron wittily observed that

> the malevolence of inanimate objects is nowhere better instanced than in the phenomena of frictional electricity. Their apparent caprice consistently frustrated the efforts of early theoreticians trying to reduce them to rule. Consider the effect of moisture on the surfaces of insulators and in the air surrounding them. The early electricians realized that contact with water enervated an otherwise vigorous electric, like amber, but they did not fully recognize the effect of humidity. On a sultry summer day, or in the presence of a sizable perspiring audience, experiments that had often succeeded might suddenly and inexplicably fail; while the operator himself, sweating at his task, helped to dissipate the charges he intended to collect.[242]

It might be expected that the combination of phenomenal whimsy and ad hoc theorizing that marked 17th-century research in electricity would characterize in even larger measure the experimental handling, in van Helmont's wake, of chemical/alchemical phenomena. But that was not so. In the very domain of the transmutation of metals where, in retrospect, no substantive advance was possible at all, a range of experimental procedures was set up that were more methodically fastidious and directed in larger measure toward single-handed empirical refutation than in any other domain of fact-finding experimental pursuit, barring only the treatment of light & color in Newton's hands (p. 681).

466 *Chemistry and alchemy.*   The historically responsible examination of the theory and practice of 17th-century chemistry & alchemy has been in spectacular flux over past decades – no other pursuit of nature-knowledge at the time has come in for quite so drastic revision of the very basics. No more than for earlier periods does it make sense for the second half of the 17th century to make a sharp distinction between 'chemistry' and 'alchemy', let alone between supposedly 'rational' chemistry and its alleged forebear, 'mystical' alchemy. Those numerous practitioners who took metals to be transmutable enveloped their researches in a larger-scale effort at the separation and reconstitution of substances that they carried out by means of experimental procedures most often indistinguishable between 'chemical' and 'alchemical'. So, in any case, did the greatest experimental chemists-*cum*-alchemists of the period, Starkey, Boyle, and Newton.

Both Starkey and Boyle took their point of departure in van Helmont's posthumous works (p. 258). With Starkey this was quite literally the case. Time and again he set out on an investigation by first writing out a list of observations drawn from the literature, most often van Helmont's 1648 *Ortus medicinae*. This was only the start of a standardized, iterative procedure that Starkey himself called his "conjectural process":

> His record keeping was surprisingly orderly, methodical, and formalized. Starkey developed and deployed a consistent methodology of data collection and interpretation from both textual and observational sources and used it to direct the course of his laboratory activities. The interplay between theory and practice is clearly apparent throughout, particularly in his habit of drawing conjectural processes from theoretical principles and past experience and then submitting them to the judgment of practical trials in the fire.[243]

Starkey's notebooks show him at work on a range of long-term projects, most of these subservient to the final goal of preparing the Philosophers' Stone. One such project was to produce each of the solid metals (gold, silver, iron, copper, tin, lead) out of antimony. Another was the preparation of a universal solvent that went under the name of the 'alkahest' or, by way of an easier alternative, the volatilization of alkalies that Starkey found described by van Helmont in customarily allusive language which first needed deciphering. To Starkey the description in the alchemical literature of a certain process or substance implied proof of its existence. The problem was how to produce it, and that is where his 'conjectural process' came in. It became his wont thoroughly to decipher and examine the text from which he started; to boil the rules given there down to their essentials; to think up, with a view to current alchemical doctrine, and then try out whole ranges of practicable variations of the original process in view of its apparent failure in his laboratory to stand the test of fire. Thus, when the volatilization of alkalies kept eluding Starkey, at a certain point he returned to van Helmont's *Ortus* and to reconsideration of the nature of alkalies, which then suggested to him an alternative pathway, which he went on in his usual manner to expose to experimental trial. Those experimental trials, finally, were pervaded by book-keeping in the quantitative vein pioneered by van Helmont and pursued by Starkey with unprecedented tenacity. To give an example requires elucidation of too many esoteric concepts current in 17th-century alchemy; suffice it to say that Starkey engaged time and again in fine-grained gravimetric analyses of a kind that, until recently, were held to be the decisive feat in the late 18th century that opened the gateway to Lavoisier's 'chemical revolution'.

Who was George Starkey? Until recently he was an obscure figure, born in 1628 in Bermuda and living in New England until, in 1650, he moved to London, where he died in the plague of 1665. Until recently also, he was no more than suspected of being the person who hid behind the pseudonym Eirenaeus Philalethes (peaceful lover of truth), author of alchemical treatises remarkable chiefly for their even more than customarily luxuriant imagery. Not only has the identification recently been put beyond doubt, but Starkey's laboratory notebooks have also come to light, revealing those remarkably tight, down-to-earth working procedures that underlie all his publications, Philalethes' extravagant treatises definitely included. These treatises testify to the fact that not all long-term projects Starkey undertook ended in self-confessed failure. Philalethes' treatises describe in veiled language (readily translatable, however, in the sober records of his notebooks and even reproducible in part in a modern chemical laboratory) how he arrived at what he confidently announced to be successes. In an incidental case and on a small scale only, he even managed to market those purported successes.

It was under Starkey's direct influence that, in the 1650s, a young nobleman given to the writing of moralizing tracts, Robert Boyle, was converted to a lifelong pursuit of chemical/alchemical interests. Boyle's pertinent investigations, unlike Starkey's, were never undertaken for their own sake alone. He brought the innovative insight to chemistry that here lay the best chance for confirming the general truth that the whole of nature is built up from what he used to call the two 'catholick [i.e., universal] principles' of matter and motion. Boyle embraced the doctrine of kinetic corpuscularianism as a general source of hypotheses capable of giving some badly needed direction to Baconian experimentation (p. 554). He therefore pursued his chemical researches in a more traditionally qualitative vein than either van Helmont or Starkey – for Boyle it sufficed as a rule to demonstrate that a certain chemical reaction could be understood in terms of particle motion. This is also why Boyle kept his alchemical beliefs and pursuits (about which he remained silent in public) more separate from his chemical work generally than was true of both Starkey and another, younger adept, Isaac Newton.

To Newton the pursuit of chemistry & alchemy served, not as with Boyle to *confirm* experimentally a conception of nature as being ultimately reducible to a variety of motions of variously sized and shaped particles, but rather as the very antidote to that decades-old

468 conception. What the insights Newton gained from a sustained effort to transmute metals did for his successive conceptions of the natural world will occupy us later (p. 643). Here I shall deal solely with the experimental effort itself and how Newton carried it out.

In or around 1669 he immersed himself in alchemical theory and practice, not to give up the practice until the late 1690s. His belief in the possibility of metallic transmutation remained with him until his death in 1727, when all around him such belief went into steep and steady decline.

Newton's alchemical efforts look much like Starkey's. The quantitative rigor that marked so much of Starkey's alchemical experimentation was, if possible, carried to even greater heights by Newton, who was likewise in the habit of setting up certain long-term projects to which he clung with tenacity. Most substances central to both men's laboratory efforts were also the same, notably 'sal ammoniac' (NH_4Cl) and antimony in various guises.

More than Starkey Newton focused on a substance known among adepts as 'the net'. It derived its name in part from the reactants that went into its production: copper (symbolized by Venus) and antimony obtained from the reduction of the ore in which it occurs by means of iron (hence, Mars). Since according to Greek myth Venus and Mars were caught in adultery by Vulcan's net, the name to give to the alchemical substance was obvious. It was reinforced by the netlike appearance displayed by the compound thus produced. To Newton, the net was even more than a powerful alchemical substance; in due time it was to sum up his understanding of the constitution of matter per se (p. 708).

But the greatest difference between these two highly advanced practitioners rested in how they handled the literature. In a treatise 'Index chemicus' which he kept expanding, Newton not only surveyed all the literature he could lay hands on (through his apparent membership in a secret network of fellow adepts), but he also systematized the thousands upon thousands of data there encountered under appropriate headings, ending up with a vast, comparative survey of the entire field such as no one before had even contemplated. From his turning and returning to this formidable resource resulted ongoing interplay between alchemical theory and experimentation that differed to be sure from Starkey's practice but was no less punctilious – Newton's hallmark in his optical pursuits as well (p. 690).

Boyle and Newton were among the great chemists/alchemists of the period, but this does not imply that their accomplishment fitted in with the mainstream of chemical pursuit at the time, which was not, as with them, concerned primarily with the general constitution of matter and of the overall structure of the natural world. The mainstream of chemists, professionally occupied as pharmacists or physicians, were in their everyday undertakings occupied rather with bringing some order to their laboratory experiments by focusing on what in these reactions altered and what appeared unalterably to remain the same. Here the current conception of 'element' was of no help at all – neither the Aristotelian foursome nor the three principles of the iatrochemists served as substantive limits of chemical analysis but rather served as carriers of certain qualities, for example, sophic sulphur as that which underlies combustibility.

As a consequence, chemical practitioners felt themselves irresistibly drawn (some willy, others nilly) toward some particulate idea of matter, be it in Aristotle's tradition of *minima naturalia* or in one or another variety of atomism or in some idiosyncratic blend of these two, strictly incompatible conceptions. The philosophic rigor that obtained in the works of the scholastics or in Lucretius' poem was foreign to these practitioners. Their aim was rather some pragmatic conceptualization of what chemical reactions suggested, which was two things in particular: (1) even if stuffs on being combined display new properties, as indeed they often do, their subsequent reconstitution revives their original properties as if nothing has happened in the meantime; (2) numerous stuffs can be decomposed into their apparent constituents by means of fire or distillation or whatever tools available to the chemist, yet some appear to withstand decomposition by any means.

The first-mentioned regularity had given rise of old to the problem of whether genuine blending or rather a temporary lying side by side of stuffs otherwise unaltered is at stake. Philosophically the issue cannot be decided (if the latter, whence the new properties displayed in the combination? but if the former, whence the unscathed reconstitution?) but only papered over the verbal way. Van Helmont was compelled to take a stance in view of his own novel practice of analysis followed by reconstitution. He opted for a particle view only in the case of such reactions as take place at the surface of things, while for what happens deep down he rather invoked his vitalist 'seeds' as impregnating the fundamental stuff, water, with shape and with life. Over the century, the common run of practicing chemists felt free to take a pragmatic stance in which issue (2) was taken up, too. Slowly but surely they moved in the general direction of what Lavoisier was to turn into the defining characteristic of a chemical element – the apparent limit of chemical analysis. Substances that proved enduring, then, began in the 17th century to be connected loosely to a particulate conception; that is, atoms began to be associated, in practice more than in theory, not with matter in general as in kinetic corpuscularianism, but rather with those stuffs that kept returning in reconstitution or even refused stubbornly to let themselves be decomposed. The case is similar to the earlier usage-in-practice of specific weight. Not theoretically careful (and, in this case, ultimately sterile) conceptualization is what was taking place in these quarters, but rather a half-deliberate, half-implicit move in one direction (specific atoms recombining with others into molecular compounds yet persisting as such in every reaction) rather than another. Boyle held matter to be infinitely plastic, and he explicitly denied the very possibility of any limit of analysis. Even so, in his everyday practice he could not do without an implicit conception of durable substances composed of a finite number of chemical elements. Countless chemical tests occur in his lengthy treatises; "implicit in Boyle's use of [them] was a radically new idea of a chemical substance as one which answers to a series of chemical identification tests".[244]

470 To sum up, the chemical/alchemical literature of the period abounds in experimental accounts of reactions, most of them genuine. The principal tools by means of which to carry them out were those furnaces and glass instruments (notably, distillation vessels) that had been used for centuries in Europe as also earlier in Islamic civilization. In the wake of van Helmont four new major tools were developed to bring some order to rapidly multiplying experimentation: the iterative procedure, deliberately opened to empirical refutation, that bears Starkey's hallmark; such precision measurement as van Helmont pioneered and Starkey and Newton carried to the limits of what contemporary laboratory practice could support; the exhaustive examination of the literature by way of orderly categorization followed by sustained comparison that was carried out by Newton; finally, some loosely atomist theorizing in the general direction of a conception of stable chemical elements, such as went on more implicitly than explicitly among mainstream 17th-century chemical practitioners.

Body and mind. Early opponents realized even better than early adherents did that the consequences of Harvey's discovery of the circulation of the blood went far beyond the mere technicality it may look like at first sight. In the long-accepted conception, particularly in the extensive way it had been specified by Galen, all kinds of stuff (blood and the other humors; spirits; vapors; waste) were taken to flow freely through the body, some on their own, some together, some in opposite directions. In one specific vein two contrary streams were even supposed to run at the same time – allegedly, blood and fresh air flow through that vein from the lungs to the left ventricle *and also*, in the opposite direction, soot and vapor discharged by the heart. Circulation of the blood ruled out for good any such conception. Harvey had been led to his discovery in part by an increasing awareness that, just as liquids and the vessels through which they flow cannot possibly behave that way in artificial, closed systems like tubes and pumps, neither can they in living beings. Still, it is a big step to move from an awareness of such apparent constrictions toward conceiving a new synthesis. The implications were so radical that they

could not be taken in at once. The Galenic conception had always been that the animal spirits move through the nerves, the vital spirits through the arteries, and the natural spirits through the veins. How can this be maintained if the arteries pass over into the veins, through connections visible only by means of the microscope? Do the vital spirits suddenly change into natural spirits? Taken literally, the Galenic concepts could not be fitted into the new model. How could thick, slag-like black bile still flow freely through the body and delay processes and cloud the brain if the vessels proper to it left room only for blood in closed circulation? For melancholy to reach the brain, ought it to flow through the veins, up against circulation? Precisely that was no longer held to be possible. What could remain of humoral pathology if the humors lost a good deal of the freedom previously assigned them?[245]

All these hoary doctrines, then, had to be rethought from scratch, and this is indeed what, 471
in part the philosophic but in larger part the experimental way, happened to them in the
second half of the 17th century. Much of the rethinking provoked by Harvey's discovery took
place under the general aegis of Descartes' adoption of it. Being Descartes, he adopted it very
much in his own way, governed by his insistence that the body should be regarded as a machine, no more than loosely if at all connected to its immortal soul. All the while dropping
Harvey's vitalism and in the process reverting in many ways to Galenic conceptions, now,
however, rephrased in the vocabulary of machines, Descartes cloaked circulation in a material cloth that was to stick. During the 17th century no one conception of the human body
replaced Galen's; but the overall trend to reinterpret specific bodily parts and their structure
and functions diminished the spiritual element in favor of explanations along more narrowly material lines.

One significant reinterpretation concerned the true function of the lungs in connection
with blood. Inspired by Harvey's discovery, the physician Richard Lower became preoccu-
pied with the difference between the bright hue of arterial blood and the darker, less fresh
appearance of venous blood. Originally following his teacher, Thomas Willis, in ascribing
this to effervescence taking place inside the heart, Lower changed his views due to a range
of experiments on respiration he conducted with Hooke. The latter had devised a technique
for regulating a dog's breathing by means of an arrangement with bellows while keeping
its lungs motionless. Together they managed to bypass the lungs entirely and to have blood
flow "directly from the pulmonary artery to the aorta through an air-free channel".[246] This
way it became clear that the change of color occurs, not at the heart but at the place where
the blood takes air in. Lower found this confirmed by the venous color remaining when he
blocked the trachea. He went on to ascribe the change of hue to a 'nitrous spirit' in the air
which was to give rise to further experiments meant to disclose the chemical agent responsible for both respiration and its apparent relative, combustion.

Another significant reinterpretation concerned the relation between mind and body
and consequences thereof for conceptions of madness. Demonic possession apart, mental
disorders were no longer routinely attributed to specific disturbances of specific organs or
humors, for example, hysteria to the uterus being on the move through the female body or
alternatively to an overdose of black bile. Increasingly, disturbances occurring in the brain
and the nervous system due to underlying, subtle (often particulate) displacements or to
rearrangements taking place therein were held responsible. For instance, Thomas Willis rejected the standard explanation of hysteria by comparing the state of the uterus of women
thus diagnosed to its state in healthy women, without finding any difference. He ascribed
hysteria instead to the blood having become soiled by certain ferments. These he took to
arise from an excess of 'seed' turned inward rather than being properly expelled in a regular
menstrual cycle. The ferments, in obstructing the narrow blood channels in the brain, thus
disturb its proper operation, which in turn affects the patient's behavior.

472   Willis' preference for ascribing disturbances in the brain and nervous system to chemical agents like those ferments was not shared by all physicians who took up the challenge
implied in the demise of Galen's conception of the body. For instance, his contemporary,
Thomas Sydenham, rejected all such theorizing and relied exclusively on his vast clinical
experience. Still, what they held in common was a pragmatic reappraisal of what in human
beings may count as bodily and what as mental processes. In earlier times, human beings
had been viewed as indissolubly body/mind. In the new, experimental approach the search
was for bodily mechanisms that were capable in concretely tangible and *therefore* (in this
'therefore' rests the principal shift) plausible ways of causing certain mental effects:

> Since the body as a material entity was robbed more and more of emotional and spiritual
> characteristics, a new entity was bound to find expression where mental processes could take
> place. Sydenham speaks of 'mental emotions'. It is as if he wants to make it clear to the reader
> what he is not addressing – the passions, the emotions that surge up from the body, from the
> spleen, the heart, the blood. From now on he is addressing the nervous system, the brain, the
> animal spirits. Emotions, too, are mental states. This differentiation of domains betokens a
> fundamental shift.[247]

Experimentation, then, went ahead in tandem with such newly directed attributions. With
Willis this took shape as a detailed inquiry into the anatomy of the brain and nerves, both
in humans and in animals. Sydenham rather filled natural histories with a comparison-
drenched listing of patients' symptoms and their attendant circumstances. These men were
physicians; they were also, *grosso modo*, Baconians. Small wonder, therefore, that they meant
their explanations of bodily and mental illnesses not only to increase *knowledge* but also to
increase *power* – in their case, the power to heal. They were hardly the only experimenters
to harbor such ambitions.

Craft techniques improved through experimental science

If *mathematical* scientists, following Galileo, assiduously sought over the 17th century to
improve craft practice the way this had been adumbrated in perspective, fortress building,
star shooting, and mapmaking, it would be strange indeed if *experimental* scientists, following Bacon, had failed to act likewise. Mathematical science as such was far removed from
the crafts – generally too far to make instant success possible (p. 325). But the improvement of human life through nature-knowledge itself improved the empirical way had been
a staple of pioneers (e.g., Leonardo, Paracelsus, Agrippa) of the coercively empiricist mode
of nature-knowledge. Bacon turned the practice into a substantial portion of his program.

Even before him, empiricist investigations had drawn upon craft knowledge as, for example,
with Paracelsus and his doctrine of restoring health by means of mineral cures. In terms of
precedent, programmatic announcement, and inherent proximity, then, chances for 17th-
century efforts to improve craft practice through experimental science looked even more
promising than in the case of mathematical science. As before, I shall focus here, not yet
on those promises themselves (these come up for treatment in ch. 17), but on improvement
actually attained under their aegis by means of experimental science.

Naturally, the very mode of attack of the experimental scientist displayed more of the
trial-and-error approach that marked contemporary craftsmanship than was the case with
the mathematical scientist. Still, in the inquiry that follows ongoing trial-and-error improvement should not in and of itself be taken for an achievement due to the fresh infusion
with nature-knowledge of a new kind. What we are after is, rather, experimental science *as
such* making a demonstrable difference for craft practice in the course of the 17th century. In
what domains of craftsmanship were efforts in that direction undertaken, and what, if any,
improvement did these actually lead to in practice?

To answer these questions, a distinction must be made first. There were domains where
the improvement of craft practice was sought as a significant by-product of ongoing, experimental science. There were also domains where efforts to devise both viable and scientifically informed craft techniques were the whole purpose. To the former category belong
all but one of the four instruments and all but one of the five subjects which served as the
principal vehicles for fact-finding experimentalism. In the latter category falls a large-scale
scheme, adopted and sponsored by the Royal Society, to produce a multivolume 'History of
Trades'. Here is what came of these varied efforts.

Although in the 17th century *musical instruments* yielded much knowledge, however disparate, about certain properties of sound, by the end of the century the craftsmen were still
far ahead of the scientists. For instance, the discovery of harmonics and of the multiple vibrations that produce them neither did nor could affect the time-honored rule-of-ear manner in which organ builders kept unwittingly employing them in mixture stops (p. 130). The
time for the experimental science of sound to pay its dues to music making was not to arrive
until, by the mid-19th century, Helmholtz came along (whether or not to label the changes
that then ensued 'improvement' is a matter of taste, to be sure).

What about the hope Bacon had entertained to achieve, and then to employ for useful
ends, the artificial 'majoration' of sound that was a staple of the natural magic tradition?
Known means to make sound louder and/or carry farther were speaking trumpets, 'ear spectacles', echoes, and whispering galleries (p. 248). Kircher and certain fellows of the Royal
Society sought to enhance understanding of how such phenomena or devices worked, but
their efforts remained scattered and almost entirely inconsequential. For instance, hopes
that an increased understanding of sound attained through experiment would yield prin-

474   ciples sufficient for optimizing the shape of the 'otacousticon' (a device for making the deaf hear better, "bell-shaped at the receiving end . . . tapering towards the other end, with a small piece to fit the ear")[248] were not fulfilled. A little more successful were efforts to adapt the optimal shape for 'speaking trumpets' (megaphones of sorts) to ideas concerning the reflection of sound in a tube. These led, not only to experiments with correspondingly designed instruments, but also, in the end, to some being actually purchased by the British navy. Finally, Bacon's vision entailed the future construction (confidently envisaged in 1684 by Narcissus Marsh) of "Microphones . . . contriv'd after that manner, that they shall render the most minute Sound in nature distinctly audible, by Magnifying it to an unconceivable loudness".[249] At the time nothing came of the gadget; yet the present state of a civilized world replete with city blasters of meanwhile all-too-conceivable loudness defies even the staunchest Baconian's imagination.

The way the *telescope* was improved over the century was a matter of trial-and-error craftsmanship alternated with vain hopes to employ for the purpose recent findings in the geometry of light rays (p. 327). Fact-finding science benefited greatly from the improvement of the telescope, but not the other way round; nor did, or could, telescopic observation be pressed into service for any craft whatever. Not until a later, industrializing age were artisans other than the makers of astronomical instruments to benefit from such practices first explored in telescopic observation as standardization and systematic correction of errors.

In the case of the *microscope*, improvement of the instrument and a range of attendant techniques was as yet the sole concern of the microscopists themselves, who really were their own craftsmen. Unlike with the telescope, they saw chances for practical use in the hotly debated question of the origins of contagious diseases like, notably, the plague. Starting a line of argument destined to continue far into the next century, Kircher promoted the idea that tiny 'worms' that he even claimed to have watched through his microscope are responsible for spreading epidemics.

> Since the time of Hippocrates, 'bad air' had been recognized as a cause of disease; the theory of animate contagion, which here makes its appearance as a theory of 'vermiculation', required that the bad agent in the air be understood as living and endeavoring to persist in its being, rather than as dead, foul, or noxious. Despite some ambiguity, this seems to be Kircher's picture and also that of Frederick Slare, who, in the *Philosophical Transactions* for 1683, relayed a report of a "blue mist" that seemed to propagate a cattle epidemic in Switzerland; he suggested that it was "worth considering whether this infection is not carried on by some volatile insect, that is able to make only . . . short flights" and expressed regret that "Mr Leeuwenhoeck" had not been present at the dissections of the sick animals.[250]

Confirmation, through ongoing microscopic investigation, of the presence, if not literally of Kircher's worms in the air and in patients' blood, then at least of subvisible animals in human and animal flesh and in just about everything else readily came forward. Still, rival explanations were also around, such as unmediated divine punishment, fermentation, and the specific arrangement of the corpuscles of toxic stuffs. What kept the idea of animate contagion afloat yet prevented it from carrying the day against its rivals was that it explained some puzzling phenomena about epidemics well but not others. It did explain the propensity of contagious illnesses to multiply themselves and also their typical transmission pattern of moving from one city quarter to the next, which was readily comparable to an agricultural crop under attack by swarms of insects. But the failure of many people to catch the disease on the spot is hard to square with the assumption that these tiny animals are present all around us – here divine retribution seemed the more likely explanation. Also significant was the contemporary lack of conceptual furniture needed to make sense of a view of life as governed by "universal parasitism".[251] In short, microscopy was still far from a capacity to further the effective combating of contagion.

That it was not even up to contributing to anybody's health at all was the principal reproach directed against Malpighi by a longtime opponent in the Bologna faculty, Girolamo Sbaraglia, in 1689. In a classic case of experience-based versus clinical medicine, Sbaraglia argued that Malpighi's sophisticated comparative researches into the anatomy of the brain, the lung, the glands, etc., amounted, not to medicine as behooved a professor in that faculty, but to philosophy, since not one ailment had been made one whit more tractable thereby. Driven to distraction by this thinly veiled attempt to rob him not just of self-esteem and occupational legitimacy but of his entire living, Malpighi in the end responded by listing all the uses of microscopy he could possibly think of. Our natural sympathy with the predicament of a great man facing a bright, nasty scribbler ought not to prevent us from spotting in his lengthy list, among numerous claims for better *understanding* of a plethora of ailments (which Sbaraglia had not disputed), just two cases where better insight had led to better *treatment*:

> Treatments of kidney disease have changed as a result of the appreciation of the kidney's fine anatomy; plenty of water rather than astringents and refrigerants is the new recommendation. The frightening oily substance on top of urine that seems to be the very substance of the patient's body has been shown by the microscope to be a collection of sand particles, with a reassuringly less drastic treatment called for.[252]

In other words, near the final decade of the 17th century, when the most expert microscopist of his day, fully conversant with the voluminous literature, was challenged in a high-stakes duel to rack his brain over how the art of healing had actually benefited from a lifetime

476   of top-notch research, improved handling of two kidney-and-bladder ailments was all he could honestly come up with.

In the case of the *air pump* alone, it was not just a good deal of trial-and-error craftsmanship that went into making the instrument more airtight and more suitable for experimentation. Certain findings and practices from experimental science also contributed to these ends. Over the decade and a half before the instrument became a regular trade product, from the late 1650s to the early 1670s, improvement took place by way of that three-way race between von Guericke and Schott, Boyle and Hooke, and Huygens, with posterity agreed over giving the palm to the last. One trial-and-error technique was the rounded shape given to the container to help prevent explosion – the air pump was the first instrument to give practitioners occasion for putting themselves at sometimes deliberate risk. More important were a host of measures to prevent leakage, for instance, wholesale immersion in water. Experimental science itself helped determine the efficacy of such measures. Two ways were found to measure the quality of the void attained. One was the void-in-the-void experiment, at least if it was not marred by adhesion of the liquid to the tube's wall (p. 461) The other expedient was to take an inflated bladder, close it, then put it inside an evacuated receiver and see how long it maintained its overpressure. The importance of keeping one's equipment impeccably clean was also one by-product.

This, then, is how the artisanal improvement of the instrument benefited from experimental science thus being turned, as it were, onto itself. But what if experimental science was turned onto other things – more specifically, onto the forces apparently inherent in the pressure the atmosphere exerts upon us? Could a void be created with so much ease and in so regular succession as to enable the operator to produce what we now call a working stroke? The question came to Huygens while reading the book that von Guericke finally brought himself to publish under the title *Experimenta nova (ut vocantur) Magdeburgica de vacuo spatio* ('New (so-called) Magdeburghian experiments on void space'; 1672). As always, Huygens was inspired by somebody else's idea that he thought he could improve upon. In this case, he sought to move from spectacular demonstration of the forces involved to harnessing them to useful ends by designing a gunpowder engine (Figure 13.4).[253] He meant it to produce a working stroke

> such that it can be applied to mount large stones for buildings; to erect obelisks; to raise water for fountains; to operate flour mills, in places that are not suitable, or lack enough room, for the services of horses. And this motor has the advantage that there are no maintenance costs during the time it is not being employed.[254]

Practical difficulties were involved in replenishing the gunpowder and making it explode regularly. Also, even if the engine had actually left the design table, it would not have worked

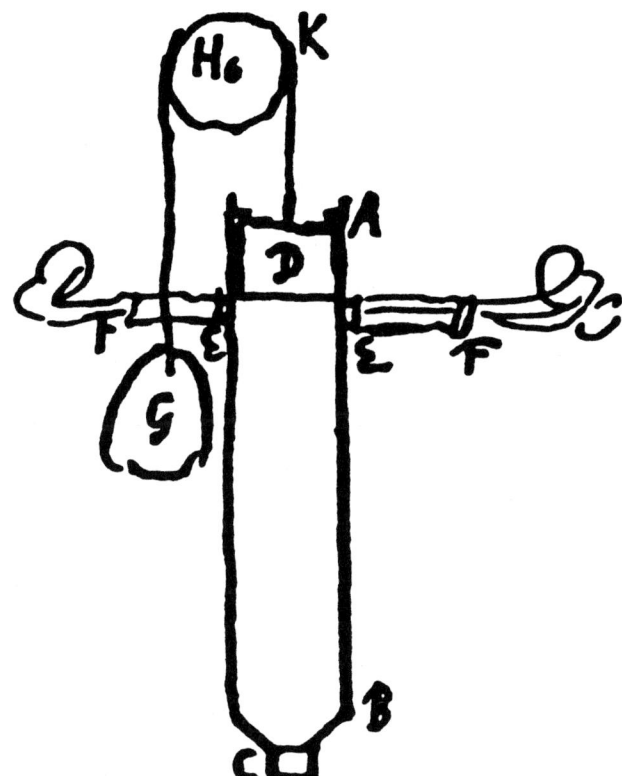

"Huygens used the effect of exploding gunpowder in the following way: the gunpowder, together with a burning wick, is placed in the container C. ...This container is then used to close the cylinder AB. A violent reaction takes place and the air in the cylinder is driven out through the wetted leather tubes EF. The cylinder AB is now empty, the tubes EF being flattened against the cylinder wall, the atmospheric pressure pushes down the piston D and weight G is lifted."

Figure 13.4: Huygens' design for a gunpowder engine

the way its inventor thought it would. All this might well have left the idea ineffectual for good were it not for Huygens' assistant in Paris, Denis Papin. He proposed replacing the successive explosions of gunpowder with a regular cycle composed of (1) bringing water on the bottom of a cylinder to boil, so as to (2) raise the piston, followed by (3) putting out the fire underneath, leading to (4) rapid condensation of the steam in the cylinder, inducing (5) descent of the piston into the void thus created. With the design-*cum*-explanation

478 Papin published in 1690, the *theory* had been completed of how to harness atmospheric pressure through the conversion of fuel into a working stroke by means of the condensation of steam. The decisive step of turning the design from 'theoretical' practice into *practicable* practice was quite something else, however. A working engine was with great ingenuity built in 1705 and then brilliantly made self-acting in 1712 by an ironmonger in Dartford (Kent), Thomas Newcomen – the first specimen of a brand-new kind of artisan, namely, a scientifically informed one (p. 326). By means of his 'engine for raising water (with a power made) by fire' (to be transformed half a century later by James Watt into a universally applicable source of power) the craft practice of deep mining began early in the 18th century to be significantly improved. In Britain chiefly but also in certain mining regions of Germany, abundant groundwater was soaked up with relative ease at affordable, profit-maintaining costs.

Gilbert's interest in *magnetism* had been aroused principally in view of the hopes that he and a circle of London scholars and navigators entertained. Would local irregularities in the variation of the compass (i.e., in its deviation from true north) make it possible to convert a given compass reading into local longitude? Such hopes began to be squashed in the early 1620s. One of these scholars, Edmund Gunter, measured a substantial deviance of nearly 6°. Might it be due to the magnetic pole being adrift? We know the phenomenon as 'secular variation', and it is incompatible with Gilbert's account of fundamentally constant magnetic variation made irregular only by local density gradients of the Earth's iron core. An array of efforts at replication of Gunter's measurements ensued:

> Marr might have been hesitant to accept the implication of incompetence for his own failure to reproduce Gunter's finding. As well, Gunter, as a professor at Gresham College, was considered to be a respected authority, making it less plausible to cast aspersions. Once again there were difficulties, the measurement was described as 'less than five degrees' and attributed to instrumental error. Refraction [i.e., possible measuring error due to light rays being broken in the atmosphere] was taken into account and logarithms were used extensively. New equipment was gathered and there was great concern over uncontrolled variables. Once again, Gunter's own needle was used as well as two 12 inch needles. The variation was measured in front of six witnesses and found to be 4 degrees 10' East. At this point Gellibrand accepted the notion of secular variation.[255]

More and more others were to join Henry Gellibrand (Gunter's successor) in so doing, and the net effect was that schemes to find longitude by means of readings of the compass, albeit proliferating over the century, ceased to be taken seriously in view of its three 'mathematical' rivals, which were directed at time measurement by means of a pendulum clock, of lunar distances, and of eclipses of Jupiter's satellites, respectively (p. 316). This, then, is how the subject of magnetism (meanwhile robbed as well of Gilbert's vitalist interpretations) came

by century's end to look so flattened out compared to the both theoretical and practical   479
excitement that Gilbert's *De magnete* had generated at its publication.

Unlike magnetism, which only lost its promise, the experimental science of *electricity* had never held any. In the state attained by the end of the 17th century, no occasion had yet arisen for even thinking of any craft practice that might be improved through electrical knowledge. Not even the staunchest Baconian ...

Products informed by *chemical* knowledge of nature had been around from Paracelsus' time onward in the guise of mineral cures (p. 133). The practice continued unabated through the 17th century and was also extended with new recipes. It was punctuated by fierce disputes over questions such as whether preparations based on antimony were safe and beneficial to patients' health (notably, as emetics and as purgatives) or were dangerous poisons in the first place. Antimony, which may have helped poison Isaac Newton to the point of a months' long mental breakdown in 1693, is a case apart. Whether other cures in Paracelsus' tradition really improved anything at all is, of course, hard to decide; so much is certain that many were believed at the time by many a healer and by many a patient to do so.

Manufacture of chemical products generally varied from pure craftsmanship untainted by any science to ambitious schemes of science-based, chemical craftsmanship. Johann Rudolf Glauber displays the full spectrum in instructive ways. He was a craftsman devoid of university training yet more than commonly learned and, as such, given to the publication of books. He was in the habit of projecting huge vistas of chemical warfare and of improving agriculture by means of chemical processes, such as the preparation of concentrates or the production of artificial manure out of wood juices. While directed toward the restoration of his native Germany from the ravages of the Thirty Years' War, Glauber's proposals for an applied, experimental science of chemistry actually evoked more interest in England. What Glauber was really good at was the invention of new chemical equipment and the improvement of certain chemical operations useful in the separation of precious metals from their ores and in the production of linen and vinegar. Through his descriptions of such operations shines a broadly Paracelsian idea of chemical transformation, but this does not necessarily imply that the former derived from the latter. More generally speaking, the improvement at the time of a range of production processes by means of more refined distillation techniques speaks rather to craft practice advancing itself by trial and error than to experimental science actually being put to the service of improving craftsmanship.

Most practitioners of the secret art of *alchemy* cared more for an enhanced understanding of nature's processes than for sudden wealth. Even so, the net effect for goldsmiths, coin makers, physicians, and pharmacists alike of the finding of the Philosophers' Stone would have been, not so much such improvement of their respective crafts by means of experimental science as was aimed for in all other cases, but rather to drive them out of business altogether. As it turned out instead, it was not their services that became superfluous at one

480   stroke; rather, the alchemical dream itself began in the second quarter of the 18th century to fade away.

What, finally, about opportunities for *prolonging life and healing the bodily and/or mentally afflicted* by means of newly experimental practices? To the extent that microscopical research might have contributed to curing anybody, we have seen how little Malpighi was able to come up with in his dispute with Sbaraglia. Likewise of little consequence were the discovery of blood circulation and such attendant shifts in the conception of what is bodily, what mental, that followed in its wake. Various attempts were made to make healing more scientific. One example with uncertain outcome is efforts to inject foodstuffs or medicinal drugs directly into the bloodstream; another, experiments with the transfusion of blood that Lower and others undertook for the benefit of the Royal Society, with the distinct aim of curing patients. More consequential was what happened to contemporary thinking about the possibility of demonic possession. Two prolonged debates on the possibility of demonic possession fought out around 1600 and in the 1660s–1680s, respectively, display a noticeable shift that was marked above all by new ways of thinking and operating that fact-finding experimental science had meanwhile put on the agenda. In 1711 the literary author Joseph Addison remarked, "I believe in general that there is, and has been such a thing as Witchcraft; but at the same time, can give no Credit to any particular Instance of it."[256] His inability to give credit derived in a straightforward manner from the application of standards of scientific argument (verification by reliable witnesses, circumstantial reporting, and the like) that Boyle and other contemporary believers in witchcraft had declared indispensable for underwriting their own belief.

So, at the level of the intellectual elite ideas about healing altered considerably due to discoveries in fact-finding experimental medicine. Also, in the longer term there were practical consequences of vast significance, notably the eventual demise of witchcraft trials. What percolated down to the level of everyday healing in the shorter term is another matter:

> Overall . . . it is a moot point whether bedside medicine was growing wiser or safer. The sixteenth and seventeenth centuries brought no revolution in medical services or treatments. There remained a mosaic of practitioners – learned physicians, herbalists, wise-women, astrologers, uroscopists, empirics, apothecaries, barber-surgeons and specialists like tooth-drawers and lithotomists. Moneyed patients shopped around, while the poor were restricted to the aid of family, neighbours, and priests.[257]

Nor did research findings, however consequential, do the doctor any good – Harvey's medical practice was said at the time to have gone into decline the very moment that his book on circulation came out. In short, what fact-finding experimentation accomplished for actual healing practice was of revolutionary significance in the longer, but scarcely of any significance in the shorter term.

So much for cases where the improvement of craftsmanship was sought as a by-product of
experimental science. But there was also one case where the crafts came first, a collective
effort under the fittingly Baconian name of *histories of trades*. This was a scheme for the exhaustive listing of artisans' diverse practices. It had a dual purpose, to learn from them and,
conversely, to make these practices benefit from the findings and/or procedures of experimental science. Advocated already during the Interregnum by Samuel Hartlib, the scheme
was spurred on by a catalog compiled by Christopher Merrett and by a reasoned taxonomy
prepared by Hooke. In terms of publications to come out of it, the scheme was rather successful for a while. Two such histories filled entire books, whereas many others were turned
either into parts of books or into articles printed in the *Philosophical Transactions*. Others
did not make it beyond the society's registers.

Examples of production processes that fellows found worth exploring in detail are glass-
making, the spinning, weaving, and dyeing of textiles, and the problem of how to use coal
rather than charcoal for burning. The last was a matter of acute concern in England, with its
ongoing deforestation, and was famously to be resolved the trial-and-error way by Abraham
Darby in 1709. Several trade procedures thus described led to discussion of their rationale
in terms of experimental science, as with musical instruments. Of the reverse process which
concerns us here, however, no trace is to be found. The decline into which the project had
entered by the early 1680s has been ascribed by its modern historian to

> the inadequacies of seventeenth-century science and the limitations of the Baconian program,
> the preference of members of the society to remain in London, their inability to do cooperative
> work, the differing interests of new members, the reluctance of artisans to reveal trade secrets,
> and class distinctions between the artisans and society members. . . . The goal of improving
> manufactures took more time; it was not until the nineteenth century that college-educated
> engineers used science-based methods to provide consumer goods.[258]

Shipwrecked on Gulliver's island?   Has our voyage through the waters of the experimentally scientific improvement of 17th-century craft practice landed us in the end on
Balnibarbi island? Have we found ourselves instructed in its capital Lagado, together with
captain Lemuel Gulliver, in the applied science of the "extraction of sunbeams out of cucumbers" and similar exploits by means of which Jonathan Swift has immortalized (through
dexterous selection more than outright caricature, really) schemes all too often entertained
in the early Royal Society?[259] Or has something more substantial been salvaged along the
way?

To find out, I shall consider the net outcome of the survey in ongoing comparison with
the pattern uncovered in the case of mathematical science. Efforts at craft improvement by
mathematical science that were undertaken before the onset of the Scientific Revolution

482 had already accomplished all they could; in only one case, mapmaking for navigation, did the effort keep advancing under its own steam. If undertaken later in the 17th century, these efforts generally overshot their mark, with the partial exception of the two instruments, the telescope and the pendulum clock. What stood in the way of the move from science revolutionized the mathematical way to science-based craftsmanship being accomplished in one big leap came down to (a) underestimation of the messiness of the world, (b) mathematical techniques unable as yet to capture that messiness, (c) craftsmen's difficulties with grasping the mathematics, and (d) social distance. Insofar as newly realist mathematical science was concerned, then, the period of the Scientific Revolution was one of exploring the gap, which began to be closed piece by piece in the 18th century and all across the board in the next.

The survey of the literature produced above has now made it clear that broadly the same pattern governed contemporary efforts at improving craft practice through experimental science. One difference is that with experimental science it has not always been as possible to speak so unambiguously of 'improvement' as we could in the case of mathematical science and the crafts – many a mineral cure looks to us nowadays as no improvement at all but was accepted as such by numerous healers and patients. Also, there were a few insignificant exceptions to the general rule of no improvement, notably, better prescriptions for two kidney-and-bladder ailments and a few speaking trumpets ordered by the British navy. Yet the general rule is, indeed, no different from the case of mathematical science and the crafts – once again the common run of efforts overshot their mark by far.

The obstacles, though not fully identical, were quite alike, too, with the exception of obstacle (b), which of course had scarcely a counterpart in experimental science. True, many experimental scientists were far better aware than their mathematical colleagues of how messy the world is – we shall find soon enough in what ingenious ways they sought to get a handle on the apparent messiness, nay, occasional caprice of nature. But those handles were subtly *different* from how the craftsman was used to resolving troubles stemming from nature's messiness. The world-historically first case where (not counting scientific instruments) the gap was effectively closed is telling here. No experimental scientist could possibly have turned Papin's design into an effective engine, let alone made it self-acting – none but a first-rate ironmonger could have coupled the various parts and movements the way Newcomen did within fifteen years of Papin's design. But no ordinary craftsman (even if a first-rate one) could have, either. A certain familiarity with outcomes and principles of experimental science (in easier reach for craftsmen than mathematical science was, to be sure) was necessarily required. Once again, then, obstacles of the 17th century began to be overcome in the 18th. In the case of experimental science these rested primarily in the wide educational and social gap between scientists and craftsmen. As such, they were easier to overcome, even though the fate of the histories of trades was in good part sealed by these very obstacles.

Once again, too, with the *instruments* of experimental science matters were somewhat

different, as they were with mathematical science. Their invention was a crafts affair in all
cases but one (the air pump). To draw them inside experimental science was, likewise and
obviously so, up to those who practiced it. Here the telescope was unique for starting its career as a mathematical instrument, only to be shunted into the realm of fact-finding experimentation when ongoing improvement made it capable of sustained measurement. Contributions to improvement were likewise a mixed affair. Effective improvement was almost
exclusively of the trial-and-error variety; the only science-derived exceptions are Newton's
reflector and Huygens' gauges for the quality of the void in the receiver of an air pump. Yet
here, too, the practitioners of experimental science themselves did most of the trial-and-
error improving, so great was their interest in the optimal performance of their microscopes
and their air pumps. In the end, though, as with the telescope and the pendulum clock, their
own concern with the fabrication and improvement of their air pumps and microscopes
was ceded to specialized craftsmen. For instance, in 1675 the Leiden foundry of the van
Musschenbroek family began the production and ongoing improvement of air pumps and
at least on the Continent nearly monopolized their manufacture and sale for the next half a
century.

In final conclusion, then, Jonathan Swift had it both right and wrong. The Baconian
promise of the improvement of human destiny through the scientific improvement of craft
practice had for the largest part by far remained empty – as empty, indeed, as cucumbers
are of extractable sunbeams. Yet Swift's radical pessimism over human character and destiny
prevented him from seeing how realistic that promise nonetheless was in the somewhat longer term. Nor need that term actually be longer than the first significant act of science-based
craftsmanship, which manifested itself in Dartford fourteen years *before* Lemuel Gulliver set
sail for Lilliput in 1726.

Problems with facts and how to ascertain them

So far I have acted as if there is nothing problematic about 'facts', be they found painstakingly or with ease. Yet neither the boundary between facts and spurious facts nor the boundary
between facts and theories is all that clear-cut or immediately obvious, let alone (as it were)
God-given. With facts and theories found the *experimental* way this is even more true. Our
senses may deceive us in a variety of ways patiently detailed by the skeptics, yet the sustained
doubt that Sextus Empiricus and his ilk cast upon the reliability of our sensory input was as
nothing compared to how plausibly the trustworthiness of experimental outcomes could be
undermined. Experiments emphatically partook of the realm of the artificial, where magic
and conjuring tricks held sway and where (at the very least in the view of most followers of
Aristotle) natural phenomena no longer come to spontaneous manifestation as in unaided

484   observation but rather present themselves in ways disfigured from top to bottom by the very experiments or instruments invoked to inform us about nature's hidden properties.

Experimentation and its discontents.   Galileo in his time was criticized by two natural philosophers of quite different persuasions – di Grazia and Descartes – who both had a very sharp eye for the empirical weaknesses and the (from the perspective of their own Athenian ways) perverse ways of arguing of a mathematical scientist improperly and impossibly out to grasp reality. In the 1660s Robert Boyle was challenged by another natural philosopher, likewise committed to an Athenian approach and likewise acute in his discernment of what made facts claimed to be fixed once and for all through experiments less than self-evidently factual and theories ostensibly derived from them less than self-evidently true. That natural philosopher was Thomas Hobbes. Although he subscribed to various philosophies of nature in succession (p. 550), these display two constant factors: he always explained nature through matter-in-motion, and his indubitably certain first principles excluded the possibility of void space. His strictures therefore have a familiar look. For instance, Hobbes' claim that one cannot even begin to understand what goes on in an air pump "unless the nature of the air is known first" is quite similar to Descartes' point against Galileo that "it is impossible to say anything good and solid regarding velocity without a true explication of what heaviness is". Or take Hobbes' remark, right at the start of his 1661 *Dialogus physicus*, "that they [i.e., experimentalists in weekly Royal Society session] may meet and confer in study and make as many experiments as they like, yet unless they use my principles they will advance nothing".[260] But there are also some instructive differences.

One concerns tone of voice. Di Grazia and his Pisan colleagues in the 1610s, attacking on behalf of a centuries-old mainstream, were still in a position to treat Galileo as a maverick upstart. Descartes in the 1630s criticized Galileo as a misguided rival. Hobbes, in the 1660s taking on Boyle, fought what quite apparently he feared to be a rearguard battle to save from impending ruin that hoary vision of natural philosophy as the sole path to proper knowledge. Other illuminating differences concern rather the kind of science being subjected to natural-philosophical attack: real-world-mathematical with Galileo and fact-finding-experimental with Boyle. What, then, did this highly perceptive outsider-critic find so objectionable about fact-finding experimentalism? Hobbes' criticisms of Boyle's 1660 *New Experiments Physico-Mechanical, touching the Spring of the Air* came down to the claim that the space inside the receiver, which cannot possibly be void on *a priori* grounds, may in actual fact be shown not to be void indeed in view of

(a) the entire setup, which given its then brand-new design was bound to leak like a sieve (in response, Boyle rejoined the race for improvement of the air pump);

(b) the very outcomes of Boyle's experiments, which, *even* if their description be accepted as a faithful rendition of what actually took place (hard enough for Hobbes to concede, as he felt deliberately kept out),

- *either* allow of alternative, plenist interpretations (e.g., Hobbes preferred to attribute
the deaths of animals placed in the receiver to compressed air making the blood in
their lungs cease to stream),
- *or* even, in one case, actually confirm, against Boyle's own efforts to explain it away,
that the space inside the receiver is full, not void.

The latter point referred to an experiment of Boyle's with smoothly polished, flat pieces of
marble. Their propensity to stick to one another unless forcefully drawn apart had from
antiquity onward given occasion to natural philosophers of both plenist and vacuist persuasions to find their own persuasion confirmed. Boyle thought that, with the spring of the air
eliminated, the lower piece would fall down and thus had to explain why it did not. Like
delle Colombe with his ebony chip and Galileo's far-fetched account of its behavior (p. 425),
Hobbes now held in his hands a nice piece of experimental evidence with which to embar-
rass Boyle, and he went on to exploit it for all it was worth. For what Boyle could adduce
to defuse it was, likewise, not a satisfactory counterexplanation (not yet available because
the phenomenon falls under a body of theory quite beyond the horizon of the period) but,
rather, *the full knowledge structure of fact-finding experimental science.*

Indeed, in perceptively putting the air pump at the center of his strictures Hobbes willy-
nilly helped his chosen adversary, and the community behind him, the Royal Society, to
make the air pump central to their laying down of rules of experimental science that are in
good part with us still. A number of men at the core of the Royal Society (notably, Boyle,
Hooke, and Oldenburg) came to practice and, insofar as their practice was taken as exemplary, to prescribe rules aimed at making 'matters of fact' as factual as they possibly could.

Four specific values have been discerned to underlie such rules. These come down to
irrelevance of personal background, knowledge communally shared, no personal stake in
one's research outcomes, and a refusal to stop short at what society at large may hold sacred.
These values have widely and rightly been read as an enlightening précis of the modern
scientific ethos, such as it has grown indeed out of experimental practices first systematically explored by Boyle *cum suis*. But one may not without much qualification ascribe those
values to these men of the second half of the 17th century. The question is rather what rules
for fact-finding experimental science they actually explored, tried out, and debated at the
time, in Britain chiefly but also here and there on the Continent. The following stand out:
(1) ways and means to avoid bias, particularly in inferring conclusions from observations;
(2) concern for replication of results; (3) elimination of (or correction for) observational
errors and material impurities; (4) invocation of trustworthy witnesses; (5) publication of
outcomes and procedures, by means of styles of reporting designed optimally to persuade
readers of the reliability of one's experimental outcomes.

486   *Avoidance of bias.* Robert Hooke was particularly concerned with listing and discussing ways and means for the practicing experimental scientist to avoid bias. In the preface to *Micrographia* he took his starting point in the 'idols' Bacon had analyzed in his *Novum organum* as a propaedeutic to his methodology of gradual ascension from particular to general by means of the comparative listing of facts in natural histories. We desperately need such preparation, so Bacon argued in that urgent prose of his:

> Do you really judge, with all the approaches and entrances to all minds beset and blocked by the most obscure idols – idols very much fixed upon and burned into them – that any undistorted and polished surfaces remain for the true and original rays of things?[261]

> For the Human Mind . . . is so far from similarity to a plane, equal, and clear mirror (which receives and reflects the rays of things without distortion) as to be rather like some enchanted mirror, full of superstitions and ghosts.[262]

The idols that in Bacon's view beset the enchanted mirrors of our minds come in four classes. The idols of the *tribe* are due to our nature, possessed as we are by an inclination to jump to conclusions, to overestimate the value of arguments that speak for our own views, and, in short, to transcend such limits as are properly set to reasoning. The idols of the *cave* are due to our own persons – to our education and our idiosyncrasies. The idols of the *marketplace* come from improper use of language, for example, the delusion that for every name there must be a thing that corresponds to it. Finally, the idols of the *theater* are due to the scenery on which one's thought plays out its role, that is, to one's philosophy, where appearance and reality are inextricably intertwined.

    In his preface, then, Hooke listed numerous ways to redress the idols, with an emphasis upon the fourth, which he found the most urgent. In line with his subject, the microscope and what it reveals, he pointed at the significance of instruments generally, not only for extending our human and, as such, quite limited sensual equipment but also for helping correct our memory and for disciplining our thought. Our intellect is tempted to seek a safe haven in either well-established minutiae, which, however, "confine and streighten it too much", or in a never-ending widening of our knowledge, due to which "we should render it weak and uncertain". Between these opposite dangers we must sail a middle course. Our intellect should not set itself up "as a Tyrant" but "as a lawful Master". Well aware of how our senses may deceive us, the intellect is nonetheless well advised not to shrug off prematurely the evidence adduced by them but diligently to search it out and weigh it critically. What we need is a perpetual cycle:

the Understanding . . . is to begin with the Hands and Eyes, and to proceed on through the Memory, to be continued by the Reason; nor is it to stop there, but to come about to the Hands and Eyes again, and so, by a continual passage round from one Faculty to another, it is to be maintained in life and strength, as much as the body of man is by the circulation of the blood.[263]

To what extent Hooke, so quick in the main text of the *Micrographia* to come up with one superficially attractive hypothesis after another, managed to stick to his own rules is another matter – he was neither the first nor the last methodologist to seek rescue from the idols of his own cave.

Replication.   When did practitioners begin to check the reliability of experiments using replication? The deliberate avoidance of stray results by means of replication, practiced as early as the 1530s by João de Castro, already looks like a routine procedure in Gellibrand's efforts a century later to decide on the empirical reality of the secular variation of the compass. Replication soon advanced to the status of a standard requirement in the motto adopted by the Accademia del Cimento, *'provando e riprovando'* (i.e., 'by testing and retesting).[264] Not that it always proved easy or obvious to achieve replication. Recall Kircher's account of fossils (p. 447), and enjoy the story of its prehistory and immediate sequel:

> Unable to reproduce many of the experiments in Della Porta's *Natural Magic*, Kircher labeled his predecessor a charlatan, impious as well as experimentally unsound. Instead, he presented himself as a Catholic magus who could discern truth from falsehood and reason from unreason by experimentally testing everything that entered the Roman College museum. He had 'seen' the processes that spontaneously formed fossils. What further proof was needed?
>
> When his *Subterranean World* appeared, naturalists from Steno to the members of the Royal Society eagerly read it and attempted to reproduce the experiments Kircher had performed. In 1665, Oldenburg reported, not without some disappointment, to Boyle that the "very first Experiment singled out by us out of Kircher" had failed, "and that tis likely the next will doe so too." Kircher's fossil experiments, like numerous other entries in his *Subterranean World*, possessed the remarkable ability of succeeding in Jesuit colleges and other Catholic centers of learning but nowhere else. Bound to his own *a priori* assumptions about the shape of the natural world, Kircher's experiments confirmed the tenets of Aristotelian natural history with the techniques of the new philosophy.[265]

In certain other cases where replication proved hard or even impossible to achieve there were good grounds in retrospect for failure to replicate what were genuine experimental outcomes. Take the trouble many scholars had replicating von Guericke's trials with his idiosyncratically sulfur sphere:

Guericke recognized the problem and tried to reduce it by distributing globes and giving lessons in their manufacture. Nonetheless success depended on the skill and even on the chemical composition of the operator. As Guericke advised Leibniz, people with soft, moist hands – like his professorial son, and probably most academics – invariably failed. He traced his own aptitude to his former employment as a mechanic, but even his tough hide was not infallible.[266]

The concept of 'physical intuition', introduced with due caution when dealing with natural philosophy (p. 380), applies here, too, albeit in the different sense that experimenters learned to sniff out those colleagues, like von Guericke, who could generally be taken seriously even if on occasion wrong and those, like Kircher, whose reports had better be taken as a rule with more than the usual grains of skeptical salt.

Still, amid all this elusiveness there was one possible source of failed replication which might provoke a sustained hunt to track it down and get rid of it.

Elimination of (or correction for) errors and impurities. Observational error and material impurities might be identified *after* the fact as causes of failed replication, even if it often required much ingenuity to make the connection. But experimentalists could also learn systematically to anticipate them. This was prefigured by João de Castro, who carefully distinguished between two possible sources of error: human fallibility and instrumental design and operation (p. 140). Care in this regard remained confined in the 17th century to a few well-circumscribed areas. Among astronomers, Cassini stood out for his concern to have Picard sent to Uraniborg, to determine atmospheric refraction at every angle so as to produce tables to take it regularly into account, and to have results in his grand survey of France reduced to a common standard. Similarly, microscopists taught themselves the hard way to be on the lookout all the time for spurious phenomena. For instance, a microscopical image may nearly always be read in a variety of ways, which repositioning can help sort out. Hooke eloquently sketched the kind of difference this could make:

> The Eyes of a Fly in one kind of light appear almost like a Lattice, drill'd through with abundance of small holes. . . . In the Sunshine they look like a Surface cover'd with golden Nails, in another posture, like a Surface cover'd with Pyramids; in another with Cones; and in other postures of quite other shapes.[267]

Further, scratches on the glass, dust, or the sheer color of an object (which may vanish under large magnification) may eliminate from sight genuine structures or alternatively produce spurious ones like Kircher's ubiquitous worms notorious already in his lifetime. Meanwhile, amid an intensifying effort to get rid of identifiable sources of error and impurity, Boyle was the man to bring to all his experiments a uniquely systematic concern for cleanliness, well

aware as he was what impurities could do to reactions inside his tubes and to events inside   489
the receiver of his air pump.

Obviously, errors and impurities can only be identified with an idea in mind of what is
right and what is pure, so it is easy to be retrospectively wise and blame these men for all they
failed to eliminate or to compensate for. For instance, chemists quite unwittingly missed
the many ways in which a variety of gases affected their reactions. It makes more historical
sense to notice the rise, as fact-finding experimentalism began to get established as a routine
mode of nature-knowledge, of ways and means to identify one big, possible source of unruly
outcomes and bring it under some measure of control.

Invocation of trustworthy witnesses.   The number of people actually present at an
experiment so as to be able to watch what happened was necessarily limited. Barring certain
educational setups (anatomical theaters and the like), the rooms where experiments were
performed were as a rule both private and fairly small, so that few people beyond the experimenter (and his technicians if he had any) could watch for themselves. The presence of
nonparticipants could serve a number of purposes. Incidental, courtly entertainment was
one; the chance to have reliable people around who might later and/or elsewhere be invoked
as witnesses, another. This was important in view of the problem of how to convince those
absent (but also, on occasion, the experimenter himself) of the reality of what the experimenter had been watching.

This was hardly a new problem. Already in the 1530s Castro routinely sought replication
and witnesses wherever he could get them, be it a lowly caulker or an even lowlier native.
The later advent of routine experimentation followed by blow-by-blow exchanges between
experimenters made the confirmation of reported facts even more urgent. Experimenters
in the 17th century were far choosier as a rule over who might count as reliable witnesses
than Castro had been. Social distance between the nobility (Castro ended his short life as
viceroy of India) and other strata of the population, while quite significant in the early 16th
century as always and everywhere else in premodern times, had nonetheless widened in
the meantime. To a nobleman in class-ridden Britain like Boyle, son of the Earl of Cork,
the testimony of an aboriginal or measurement by a caulker would have lacked the worth
required, as it actually proved to do in the case of divers. Boyle went to some length to get
divers' personal accounts of how, if at all, the pressure of the water affected them – a matter
of some importance in view of the analogous question (raised by opponents of the weight of
air) of whether an 'ocean of air' can indeed press upon us without our noticing. Other than
in the times of Buridan, who already knew for sure (p. 83), the question was now regarded
as capable of being settled the empirical way. Even so Boyle did not hesitate to point at the
inevitable bias of divers due to their material interest, to their simple-mindedness, and to
their lack of sensitivity. These fixtures of their social status made it impossible to accept their

490   testimony without more ado, particularly if it conflicted with Boyle's own expectation of how they would be affected.

Social status, then, was the obvious, prime criterion for reliable witnessing. If attainable, the invocation of a corporate, royally or papally underwritten witness was best. More often than not, one's own institutional affiliation with the Society of Jesus, with the Accademia del Cimento, with the Académie Royale des Sciences, or with the Royal Society (or, as earlier in Harvey's case, with the College of Physicians and with King Charles I himself) was sought to safeguard the reliability of one's experimental outcomes. But there was a snag. An entire art developed of associating the ruling authorities with one's research without actually having them commit to its conclusions – governing heads were concerned to keep themselves aloof at all times from involvement in disputes over the true constitution of the natural world (p. 577).

Even apart from an appeal to institutions, experimental researchers made sure to have a list ready of suitably elevated people whose testimony could be invoked if necessary. Their names were regularly listed in the publications, by means of which the absent were turned into witnesses of a kind (the practice has been aptly called 'virtual witnessing').

Publication.   Publication was of course nothing new. Still, in the case of experimental science there was a novel aspect to it: the tinge of secrecy first had to be erased. Neither mathematical science nor natural philosophy was much tainted with esotericism in the literal sense. But the coercively empiricist mode of nature-knowledge was from Renaissance times onward associated closely with natural magic and its secretive in-group habits like initiation and purification, as well as with artisans and their trade secrets. In tune with the decisive twist that Francis Bacon had given to coercively empiricist practices in so many other regards, he had advocated in his fiery manner the disclosure of these practices in view of their unique capacity, once made truly methodical, to lead the human race to understanding, improvement, and, indeed, a redemption of sorts. To attain such lofty purposes, publication was of course indispensable. Or rather, that is what Bacon's high-minded phrases proclaimed. In the utopian reality of his *New Atlantis* it was instead up to the state to keep a lid on the knowledge that it sent its experimenters out to procure, be it by ingenuity or by stealth.

This is how, when the public character of experimental science was officially proclaimed from the early 1660s onward in a state so relatively little centralized or interfering as England, it nonetheless came veiled in ambiguity. As Hobbes in his perceptive outsiderism saw so well, it might still involve exclusion of those who, like himself, refused to go along with the budding rules. Many experiments even took place in well-equipped rooms of houses privately owned; the expression 'laboratory' was as yet used but rarely for them. So, the chief respect in which fact-finding experimental science went public was by means, indeed, of the press.

To persuade their readers of the factual veracity of the accounts they were reading, authors applied a number of rules. These were regularly practiced in two of the three principal centers at the time, France and Britain; the Jesuits, for reasons of their own, did it quite differently. But it was Henry Oldenburg who, as secretary to the Royal Society and as private owner of its principal outlet, the *Philosophical Transactions*, was more concerned than anybody else with laying down the rules for how to turn readers of experimental accounts into 'virtual witnesses'.

Above all, one's way of reporting ought to be *circumstantial*. Telling here is the sustained critique to which Boyle subjected Pascal's accounts of his experiments on the void. Briefly (not one of Boyle's many virtues, by the way), Pascal, ever the mathematical scientist, had most often written them up in a vein of 'doing this leads to that happening, just as in my theory it should'. Only the account of the experiment on the Puy-de-Dôme had all the required trappings of circumstantiality and naming of highly placed witnesses. In contrast, Boyle felt that one ought rather to report by way of 'under the local circumstances that obtained, which I now proceed to list, I watched this happen, then that, and so did Lord X and Sir Y. And oh, yes, may I suggest the following way to account for it all?' Indeed, the way Pascal wrote up his experiments left all kinds of doubt behind over whether he had actually performed them or only noted down what he felt sure would happen. Among the experiments on display in Figure 13.5 there is one in particular that cannot have taken place even remotely the way here depicted, as Boyle did not fail to point out.

Another style of reporting, then, had to be followed. There is no succinct way to exemplify the circumstantial, so here is a piece by a first-rate reporter that shows up at the same time as another significant feature of fact-finding experimental science: failed replication due to phenomenal whimsy. The author is Isaac Newton; his 7 December 1675 report to the Royal Society is followed by a modern comment:

> I have sometimes laid upon a table a round peice of Glasse about 2 inches broad Sett in a brass ring, so that the glass might be about 1/8 or 1/6 of an inch from the table, & the Air between them inclosed on all sides by the ring, after the manner as if I had whelmed a little Sive upon the Table. And then rubbing a pretty while the Glass briskly with some ruff and rakeing stuffe, till some very little Fragments of very thin paper, laid on the Table under the glasse, began to be attracted and move nimbly to and fro: after I had done rubbing the Glass, the papers would continue a pretty while in various motions, sometimes leaping up to the Glass & resting there a while, then leaping downe & resting there, then leaping up, & perhaps downe & up againe, & this sometimes in lines seeming perpendicular to the Table, Sometimes in oblique ones, Sometimes also they would leap up in one Arch & downe in another, divers times together, without Sensible resting between; Somtimes Skip in a bow from one part of the Glasse

to another without touching the table, & Sometimes hang by a corner & turn often about very nimbly as if they had been carried about in the midst of a whirlwind, & be otherwise variously moved, every paper with a divers motion.

There were two novelties in Newton's message; glass replaced the usual generator, amber, and the paper bits flew to an untouched, not to a rubbed, surface. The intrigued Society undertook to repeat the demonstration; they succeeded once or twice and then failed altogether. Newton sent further instructions: the glass should be new, the paper bits triangles, the rubber rough, and the rubbing like the motion of a glass grinder. If the Fellows still could not budge the bits, they were to knock upon the glass with their finger tips; failing that, they should await Newton's next visit to London. They eventually succeeded by rubbing with stiff hog's bristles.

The story has two points. First, it shows how little even informed savants, indeed an entire Royal Society, knew about electricity in the late seventeenth century. Their collective ignorance may serve as a benchmark for evaluating work one might have dismissed as trivial. The second point is that the regularities underlying electrostatic phenomena are very well hidden. Newton's editor takes him to task for failing to find the principle of motion of the paper bits. To have discerned the laws of plus and minus electricity, of superposed central forces, of insulators and conductors, in the whirling and skipping of the bits, would have been a feat more remarkable than the invention of the gravitational theory.[268]

Aside from circumstantiality, a *matter-of-fact style* of scientific reporting, sober and down-to-earth, came to replace such rhetorical devices as had in antiquity been elaborated into a veritable 'art of persuasion'. Eloquence, erudite allusions, artfully composed arguments, carefully wrought sentences, elaborate metaphors – in short, all the customary orator's or writer's verbal equipment – ought to be shunned in order to convey a sense of the veracity of experimental events described without embellishment. However, just as the very notion that 'things' were now taking the place of 'words' did not cease the requirement for expression in words, so the avoidance of eloquence, metaphor, etc., was more a part of the rhetoric of style prescriptions than the reality (as a glance at issues of the *Philosophical Transactions* and the *Journal des sçavans* makes sufficiently clear). It is nonetheless true that, in these journals as in the books on experimental science written at the time, a more business-like, less florid, and, above all, less human-centered style of writing obtains than was customary before. In present-day scientific journals no sign of humanity beyond the authors' names has been left; the second half of the 17th century is when, not only in style of writing but likewise of illustrating, the human actor began his slow departure from center stage.

Except, that is, at the head of one's article, just before or after the title, or wherever in the journal or on the title page the author's *name* was recorded. With all the remainder of the researcher's, as well everybody else's, humanity beginning to be systematically squeezed out of science reporting, his name as the symbol of his personal prowess as a discoverer carried

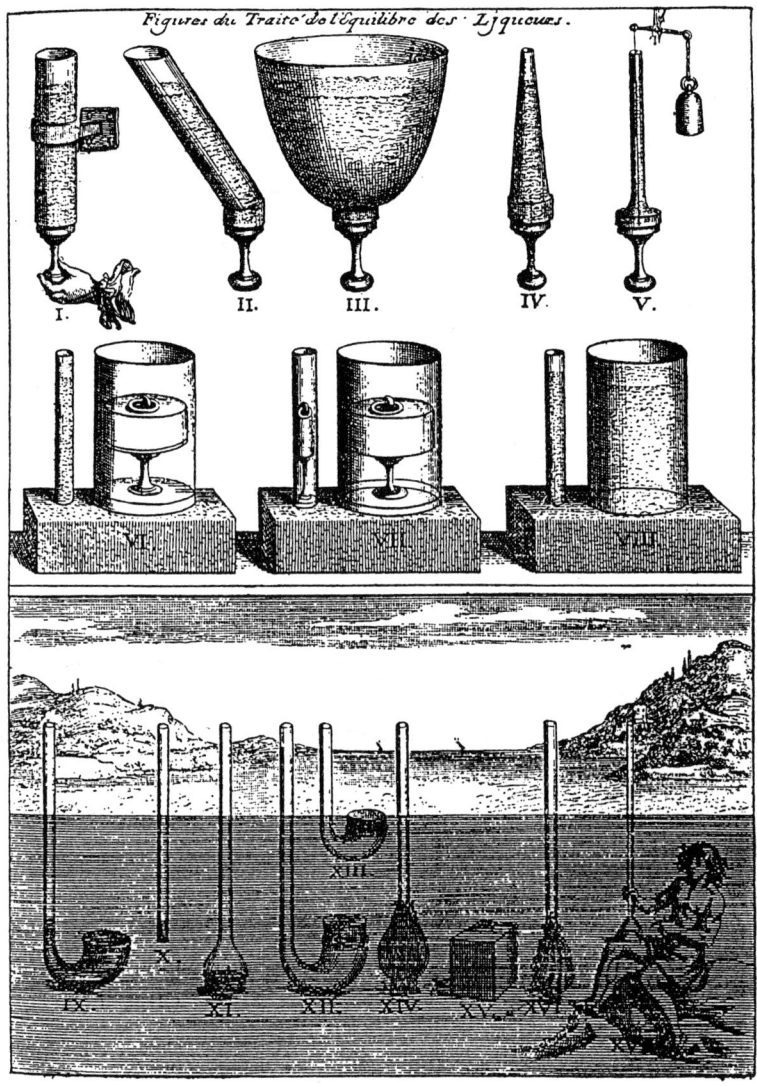

Figure 13.5: Pascal: experiments on the void

more and more weight. But how to ensure that the bearer of that name was truly the discoverer, that is, that he had not been preceded by somebody else in his claim to earthly immortality? As identifiable lines of experimental research began to get established, opportunities for unwittingly simultaneous discovery multiplied.

494 Opportunities for *wittingly* simultaneous (i.e., pseudosimultaneous) discovery, that is, for plagiarism, had of course been there all along. Tartaglia and Cardano, or Tycho and Ursus, had in earlier times fought notoriously venomous conflicts over that very accusation. As the 17th century progressed, such conflicts increasingly shaded over into more regular priority disputes (with suspicions of shadily acquired prior knowledge of the original idea hovering over some of the worst, as with Huygens versus Hooke over the balance spring or with Newton versus Leibniz over the calculus). How, then, to ascertain priority? To cover the time period between the original discovery and its appearance in print, use was sometimes made of anagrams or ciphers, devices beloved by, for example, Galileo and Huygens. With the launching of the *Philosophical Transactions*, Henry Oldenburg replaced such tantalizingly meaningless collections of letters with a rule that required registering the date of receipt along with the article received. The rule has not prevented thousands of priority disputes from arising since then, but at least it established a clear-cut criterion.

The fudge factor.   Even if all these rules were carefully maintained, the boundary between fact and spurious fact had still not been established once and for all. It never has (which is *not* tantamount to inferring that it does not exist). All these rules have proven effective over the centuries, systematic error correction perhaps most of all, and yet they could not but leave ample room for what has been called 'the fudge factor'. Here is an example showing how the ablest scientist of the age could spuriously reason his way toward an empirical fact that he rightly suspected to be anything but spurious. This is Newton's concern with the speed of sound. In book II of his *Principia* he derived it from basic principles of his dynamics. He then went on to check his result of 968 feet per second. He did so by means of an experimental setup refined from Mersenne's method using the echo and a pendulum, thus establishing a lower limit of 920 and an upper limit of 1,085. All was sunshine, therefore, until in the 1710s Derham and Sauveur independently came up with substantially higher measured values. Newton rightly decided that Derham's value of 1,142 could not be ignored. So, in preparing his second edition he sat down to disfigure his own analysis with two empirically undeterminable parameters brought in for the sole purpose of raising the calculated value to exactly the one measured by Derham. Not until the end of the century was Laplace to show that to bring Newton's original analysis in line with Derham's measurement one must take into account the heat that sound waves engender, a feat that was out of analytical reach until, by the mid-18th century, the concepts of temperature and of specific heat began to be distinguished.

Pooling of efforts

Very few individuals engaged in fact-finding experimental researches were not attached, one way or another, to some formal establishment. Academies and societies, observatories, and

laboratories, in Rome, Florence, Paris, and London, were where, in the second half of the   495
17th century, most ongoing, fact-finding experimental research was pooled and/or collectively conducted. What did they contribute to driving experimental science forward?

Rome.   Rome is the only city among these four where no scientific society was established. Nor was there any need for one – Rome is where the Jesuit Order had its center. The
order was a highly centralized institution, designed as a nonmilitary army whose purpose
was to bring or return the world's inhabitants to the Roman Catholic faith. Their conversion policy was from the top down, aimed at gaining sympathetic access to the laity through
the venue of education. An astute awareness that the knowledge to be disseminated ought
to be up-to-date is what, from Clavius' time onward, drew ever more Jesuits into the active
pursuit of nature-knowledge. Their prize figure was Athanasius Kircher – the father with the
generous funding, the far-flung network of Jesuit informants, the boundless curiosity and
erudition, and the mitigated gullibility one could find embodied in his Roman Museum. All
this also made him the ideal figure to present in endless, experimentally reinforced detail,
in more than a dozen lavishly illustrated books, what comes closest to the worldview of the
order at the time. That worldview was a hodgepodge of Aristotelian notions overlaid here
and there with numbers and with conceptions of moving particles and of natural magic, all
held together by a vast network of hidden correspondences and secret family resemblances
between the countless facts so assiduously assembled.

Permitted as a rule to publish under their own names and also to express somewhat different views on specific issues where no party line seemed called for (e.g., Fabri's dissenting
voice over Cabeo's explanation of magnetism), Jesuits were nonetheless constrained severely
in their pursuit of nature-knowledge. Even if handled at times with pragmatic leniency,
these constraints applied both in realist-mathematical science (due to the Copernican connection) and in natural philosophy (in view of its veritable interpenetration with Roman
dogma). In fact-finding experimental science the constraints of religious orthodoxy could
do least harm to the Jesuits' declared desire to be seen standing at the forefront. That is why
fact-finding experimental science became the mode of nature-knowledge where the majority were active and where they accomplished their best work. Still, there are domains of
human endeavor where permanently present ulterior motives doom even the highly gifted
to mediocrity. In art as in science, a naive sense of wonder, an urgent desire to explore uncharted territory, a readiness to pursue a clue regardless of where it may lead, are jointly
indispensable for arriving at the very forefront where the Jesuits, to whom truth was either
known beforehand or up for political negotiation, fancied themselves to be collectively situated.

Institutionally equipped with a scientific society of a kind (the sum total of the science-
minded in the Society of Jesus) and with a laboratory of a kind (Kircher's museum), the

496   Roman Church could also boast an observatory of a kind or, rather (in due time), four of them. In 1655 San Petronio Cathedral in Bologna was turned into a large, solar observatory by having one roof pierced at a suitable point. Suitable for what? For making the Sun enter its vast interior such that a line could be drawn on the pavement that connects the images of the Sun thrown there at noon from one summer solstice till the next. At the height of 76 feet where the Bolognese professor of astronomy, Cassini, had his hole drilled, the procedure yielded a line rightly calculated in advance to be 200 feet long (just grazing two pillars, it was engraved in metal in the beautiful marble pavement and is still there for visitors to see while listening, if you are lucky, to San Petronio's two uniquely early church organs being played at the same time). The arrangement made it possible for arrays of measurements of Sun-related parameters to be made with a degree of precision the telescope was not to attain until about a century later. This is how Cassini could come to Paris with a full set of corrections for atmospheric refraction. Heaps of astronomical fine-tuning followed, some of considerable value for the furtherance of the case for heliocentrism, whether in Kepler's variety or not.

Florence.   Florence did not yet boast such an observatory, but only, for a short period between 1657 and 1667, a society fitted out with a laboratory for the practice of experimental science. Called the Accademia del Cimento (Academy of Experiments), it operated in the Palazzo Pitti under the patronage and strict supervision of Grand Duke Ferdinando II de'Medici and his scientifically involved brother Leopoldo. It came to an end when the best men left and Leopoldo was appointed cardinal. The constraints of patronage in a setting seriously tainted by Galileo's condemnation, even though decades earlier, limited the number of central topics pursued. They also made for highly circumspect proceedings and one corporate publication only. The *Saggi di naturali esperienze* (Samples of Natural Experiments) came out in the final year of the Accademia. The book was meant less for reading than as a display of munificent patronage. Indeed, in relating a select number of experiments without any effort to interpret them, the book gave only a poor indication of what those active in the academy were up to. The best members were Borelli and Galileo's late disciple and first biographer, Vincenzo Viviani; they did their best work apart from the academy and, at any rate, communicated their findings to their private correspondents with far less circumspection. Most conspicuous in the group work accomplished in the Accademia were efforts at replication of experimental discoveries already made elsewhere (e.g., on the void) and experiments of their own on phenomena somehow involving heat, with chemical equipment, barometers, and thermometers with as yet nonstandardized scales among their most advanced tools.

Paris. On declaring the end of his minority in 1661, King Louis XIV began deliberately to   497
turn himself into the patron to beat all patrons. Within half a decade Paris became the seat
of a range of state-designed institutions for experimental science on an unprecedented scale.
Carefully insulated from philosophical squabbles; no less carefully manned by a highly paid,
government-selected elite of practitioners whether native or lured away from San Petronio
Solar Observatory (Cassini) or from private lodgings in The Hague (Huygens); chosen by
and overseen by the chief minister (first Colbert, then Louvois, then Pontchartrain) and
supervised on a day-to-day basis by a government official who acted as secretary; fitted out
with a binding charter and charged right from the start with government priorities, such as
mapping the realm, finding longitude, using mathematics to improve gunnery, or assisting
with the layout of the gardens of newly built Versailles – an Académie Royale des Sciences
came into being with close links to a laboratory to beat all laboratories, a Jardin des Plantes
(set up already by Louis XIII) to beat all botanic gardens, and an Observatoire to beat all
observatories except the faraway one with the hole in the roof.

To speak without qualification of a 'research program' would suggest greater substantive
coherence than was actually attained, yet several lines of research can be discerned. Huygens
and Mariotte repeated results of Galileo's Italian disciples and sought to improve on them in
a somewhat enhanced experimental vein (ch. 10). Cartesian conceptions regarding motion,
heaviness, and light were discreetly put to the test without their connection with Descartes'
views being discussed in public. But the principal day-to-day business of the members assembled in the Académie and its likewise state-financed appendages was collection, experimentation, and measurement. This is the context in which, for instance, the expedition to
Cayenne was undertaken and in which "chemists and doctors at the Académie . . . distilled
fruits and vegetables, whole and in their parts, trying to correlate known nutritive and medical properties with the chemical components found in plants or soils".[269]

At the outset all this work bore a stamp of anonymous collectivity little different from the
Cimento. Soon however a measure of individuality was regained, and most Academicians
(who until 1700 never numbered fewer than twenty or more than thirty-four) published
most of their work under their own names in the *Journal des sçavans* founded in 1665 or in
a magnificent series of books that began to appear in the 1670s. Efforts to obtain more than
incidental collaboration from elsewhere in Europe were scarce; the entire, nation-serving
setup was incompatible with the role that Henry Oldenburg carved out for himself when he
was appointed the first secretary of the Royal Society of London.

London. Shortly after the Restoration of the Stuarts in 1660, private initiative brought
the Royal Society together. In line with the different role of the English monarchy (compared to France), Charles II's concern remained confined to a very liberal, formal charter
and to one personal visit. Through a council presided over by an interested outsider of noble

498   or at least gentlemanly standing, the society governed itself. Over the first forty years of its existence, the council elected from among candidates proposed by fellows a total of 479 British fellows:

> 16% were courtiers, politicians, or diplomats; 16% were medical practitioners; 15% were gentlemen of independent means; 14% were members of the aristocracy; 12% were scholars or writers; 8% were divines; 7% were merchants or tradesmen; 4% were lawyers; 4% were civil servants or serving members of the armed forces; and 3% are unclassifiable.[270]

Surely so varied a composition made for somewhat unruly proceedings. Two officials were indispensable for the early flourishing of the society. Oldenburg set rules of scholarly conduct, reached out to the scholarly world outside England proper, and fostered publication of experimental results. The curator of experiments, Hooke, made sure to have a suitable subject of investigation prepared for the weekly sessions. Although the practical usefulness of experimental science was a permanent concern, this found expression, not as a rule in state-sponsored projects as done in France, but rather in grand schemes like the two histories of trades. A bit of an exception is the establishment, on Jonas Moore's initiative, of the Royal Observatory at Greenwich. Moore, who had advanced from mathematician-*cum*-surveyor to the king's superintendent of navy armaments, had the building paid for by his office and the equipment (telescopes, clocks) out of his own pocket, in hopes of having his appointee, Flamsteed, come up with improved astronomical data for solving the problem of longitude.

The generally private nature of the Royal Society extended to the places where its experiments were being conducted and talked about. Important as the weekly sessions were for incoming correspondence to be read and formally registered (e.g., the decades-long succession of van Leeuwenhoek's letters), creative laboratory work usually took place in the private homes of its fellows, in London or in the countryside. Most of the England-based experiments examined in the present chapter came from Boyle's and Hooke's successive home 'laboratories'. The various London coffeehouses where Hooke could be found when not at home served as the effective centers where news was exchanged and the outcomes of experiments were discussed. This made up to some extent for the lack of persistence that experimenters as a rule displayed in their work undertaken under the auspices of the society. What little coherence there was derived in part from a shared sense of Baconian inspiration but mostly from Hooke's moderately successful efforts to have subjects tackled in a certain order rather than in an order wholly governed by the whim of the moment, and to seek some depth in how these were successively investigated.

Finally, Oldenburg was instrumental in making the Royal Society a center of exchanges all over Europe. The network of correspondents set up by the Society of Jesus remained confined to the fathers themselves. The elite academies of Florence and Paris operated as

one body each, without much outside interference other than their academicians' numerous
private exchanges. But the Royal Society made a point from the start of engaging the collaboration of everyone who proved to have interesting experimental news to share. That is why, when half a decade after his invention of the reflecting telescope an unknown Cambridge professor felt unable any longer to withhold from the world the full measure of his exciting discoveries, Newton informed Oldenburg of his invention (an account of which Oldenburg published forthwith in his *Philosophical Transactions*), soon followed by Newton's communication of his discovery of color sifting, which lay behind his invention. That, too, is why between 1660 and 1700 Huygens, Malpighi, van Leeuwenhoek, and sixty-nine others on the Continent were elected fellows and either made personal visits (as Huygens did three times) or at least engaged in more or less regular correspondence with its officials.

Power and pull of fact-finding experimentalism

What chances for achievement resided in fact-finding experimentalism, once Bacon and his fellow pioneers had launched this mode of nature-knowledge into the world by way of an act of relatively smooth, yet revolutionary transformation; what constraints did it as yet prove subject to? I consider these questions first in terms of properties of phenomena handled and of ways chosen to approach them. I shall then make a sustained comparison with achievements and limitations in realist-mathematical science. Finally, I examine the apparent attractions of fact-finding experimentation.

Properties of phenomena investigated the experimental way.   Fact-finding experimental science, however serious in its aims and execution, was also a source of fun, certainly when compared to stern and forbidding mathematical science. Fun of two kinds, to be sure: fun in retrospect and contemporary fun. Whence, for starters, the retrospective fun?

The answer resides in nature's apparent whimsy. In the above I have quoted several hilariously deadpan passages expressive of the retrospectively humorous nature of the wide gap between certain phenomena and their underlying rationale as revealed by science at much later times. Electricity due to rubbing is the subject that stands out for displaying the inevitably vain struggle of earnestly dedicated experimenters with the wildly capricious behavior of their objects of research. Even though the practitioners of realist-mathematical science were also in for the occasional surprise, fact-finding experimental science is where the "whimsical tricks of nature" truly hold sway.[271] Time and again did nature's caprice make for bafflement for which there was as yet no way out. For Newton to make sense of his jumping paper snippets would have involved a conceptual leap still far beyond his immense achievement in discovering universal gravitation (p. 492). Here were definite limits set to whatever fact-finding experimentalism could achieve. But apparent caprice offered chances as well – it also served as a freely flowing source of finds not sought for, whether luminosity enigmatically co-appearing with electricity in a mercury tube, tiny animals in male semen, or indications that light takes time to travel.

Further, the ever more apparent circumstance that natural phenomena are subject to impurities and to systematic error served likewise as both an asset and a liability. Accuracy could be enhanced, as notably in telescopic measurement, once a source of systematic error had been identified and means had been established to correct for it. Advance of a more qualitative kind, as in chemical reactions but also with the air pump, could be made once a suspicion of the presence of removable impurities arose. But error and impurity presuppose some prior conception of what is regular and pure. This is why many a large-scale source of error or impurity inevitably remained undetected for a long time to come, thus placing all sorts of unsuspected constraints upon the advance of fact-finding experimentalism.

Material obstacles likewise acted as a spur in some domains and as an as yet insurmountable impediment in others. Microscopy benefited greatly from the challenge posed by bad light, scratches, or lack of contrast – within less than thirty years practitioners managed through sheer ingenuity to squeeze out of the instrument just about all it could reveal for more than a century to come. In contrast, the hoped-for improvement of craft practice by means of experimental science foundered in good part on obstacles of a material kind. Scientists, even if accustomed to getting their hands dirty, as most experimentalists assuredly were, inevitably lacked the constructive dexterity indispensable for bridging the gap that continued to yawn between 'theoretical' and 'practicable' practice.

Phenomenal exhaustion, finally, is what stood in the way of further achievement in the sole case of magnetism, where nearly all relevant phenomena up for experimental discovery had been discovered already. Except for lines of force alone, all the basic magnetic phenomena still taught in elementary physics courses today are listed in Gilbert's *De magnete* of 1600. Until Ørsted uncovered the definitive link with electricity in 1820, there was quite literally almost nothing left to be detected about magnetism.

Properties of fact-finding research.   Achievement and limitation in fact-finding experimental science were due not only to properties of the phenomena themselves but also to the way in which they were investigated. Take the ready-made source of contemporary fun that rested in the omnivorousness of the fact-finding appetite. Nothing seemed too humble, too far-fetched, or too dangerous for the true collector or experimentalist to focus his attention on. This and the related Baconian schemes were what King Charles II and, decades later, Jonathan Swift and many other commentators as well found so laughable about the Royal Society and its loudly proclaimed experimentalism. Almost from the start the society

was made an object of ridicule. In spite of various apologias, that is what it remained, as with
'Sir Nicholas Gimcrack' impersonating a crackpot Boyle in Thomas Shadwell's 's comedy of
1676, *The Virtuoso*.

More important, surely, than such not quite harmless fun-poking is all the genuine
achievement that came from experimentalists' omnivorous appetite. Their readiness to put
all and sundry under their microscopes or in their reaction tubes or in the receivers of their
air pumps, just to find out what would happen or what it would look like, yielded many
finds, due as a rule to the very whimsy of nature that experimenters had quickly learned
to open themselves up to by becoming so omnivorous in the first place. A precondition for
achievement of such a kind was the imposition of some order upon the multitude of facts
thus brought to light. Fact-finding experimentalists were bound to find ways and means to
keep themselves from drowning in what otherwise would have remained an ocean of facts.
Catalogs, encyclopedias, and taxonomies were of some help, but the sustained usage of instruments and a judicious choice of subject proved to be the two prime vehicles for order.
What these vehicles made possible in their turn was the very blend of experiment, theory,
and background worldview that had already made its occasional appearance with Gilbert,
Harvey, and van Helmont (p. 258). Such a happy blend became more common, as with
Malpighi's researches into glands, Lana Terzi's survey of electricity, and Robartes' account of
harmonics. In an as yet rare case, theory and fact were even bound together in a quantitative
rule, as with Mersenne's list of variables jointly determining pitch and with Boyle's law of
the inverse proportionality between the volume and pressure of the air in a closed vessel. In
other ways, too, quantitative precision came to mark certain experimental pursuits, such as
chemistry and alchemy. Here Starkey and Newton went to the very limit of possible precision in consistently measuring and comparing the weights of substances that went into, and
then came out of, their reactions.

Forward dynamics compared.   A comparison of the various ways in which achievement was fostered or restrained in the pursuit of nature-knowledge the fact-finding experimental way and in the forward dynamics already identified in realist-mathematical science yields some common features to be sure, but mostly a range of far more significant
differences. Realist-mathematical science advanced at a remarkably quick pace and wide
scope, due to narrow subject intertwinement, problem/solution chases, and ongoing interplay between mathematical theory and experimental validation (p. 363). Little in this has
a counterpart in 17th-century fact-finding experimental science. What pressing forward
took place there came from the often tenacious determination, and the ingenuity required
for the task, to make the best of an ongoing confrontation with nature's apparent whimsy.
Moving forward cautiously and with ever more regularized circumspection in one specific
line of research is what appeared to work best for most of the experimentalists. Techniques

of experimentation are a bit of an exception in this regard, but on the whole advances in one line of research neither elicited, nor contributed to, advances elsewhere the way this happened with, for example, the problem of longitude and its manifold extensions into other contemporary subjects of mathematical-scientific investigation. Surely advance in the experimental understanding of sound could do nothing for advance in microscopy, nor the other way round. And the relentless problem/solution chase exemplified by Huygens' prolonged bout of creativity at its peak had only the faintest of counterparts in fact-finding experimentalism. There a chance discovery might be pursued for a while without, as a rule, leading to more than either ad hoc explanation or bald registration as just stray fact. There neither was nor could be an ever-widening front over which to advance. Instead, there was a range of somewhat haphazard, incidental advances, in lines of research either already extant (as in electricity, magnetism, chemistry & alchemy, and healing) or brand-new (with the instruments), yet each and every one of them distinct. Even in the most dynamic domain, chemistry & alchemy, advance came down as a rule to either self-deception (transmutation of metals; mineral cures) or (as with the concept of chemical element) to a partly deliberate, partly adventitious sort of general drift.

Only in the unfolding dialectic between mathematical theory and validating experiment was there some substantial common ground with how fact-finding experiment was related to theory. True, theories of the latter kind lacked the rigor that went with the former; instead, the latter derived as a rule from some background worldview, or were designed ad hoc, or both. Still, unexpected experimental outcomes might cause surprise in realist-mathematical science as they so often did in fact-finding experimental science. The enduring challenge to realist-mathematical science rests in the problem of how to handle apparent or only seeming failure of experimental confirmation – mathematics/experiment interaction serves as a perennial source of falsifiability, *not* as a self-regulating process of falsification. This is due to the very whimsy of nature that enters realist-mathematical science only through the back-door of its intricate knowledge structure, but that in the 17th and also in the 18th century provided the very substance of fact-finding experimentalism. In providing that substance, nature's whimsy ruled out for fact-finding experimentalism the kind of falsifiability that marked Keplerian/Galilean science from the start. Too easily could apparent lack of confirmation be attributed to sheer caprice; too hard was it oftentimes to infer, from what for inscrutable reasons went wrong, in what direction to proceed.

In short, intricate feedback mechanisms that, prior to the Scientific Revolution, were lacking altogether and that by 1600 began to mark the advances of realist-mathematical science kept eluding, not only natural philosophy as a matter of course, but also fact-finding experimentalism. Natural philosophers did not as a rule take the absence of feedback as a liability at all, so assured were the common run of them of the infallibility of their first principles and of the few 'internal' checks upon these. But just as a few men looked for additional checks from outside natural philosophy itself (which Borelli sought in geometrical form), so were several outstanding experimentalists deeply concerned with how few checks upon the outcomes of their experiments they appeared to possess. At bottom they were left with checks of two kinds only: testimony by reliable others of the reality of the fact as such and whatever coherence was yielded by the tenuous blend of background worldview, theory, and experimental outcome that served as their prime engine of advance.

That background worldview, then, as the best of them saw, could be rendered more rigorous and consistent if given tighter shape as one coherent conception. For this, the obvious candidate was kinetic corpuscularianism. However, the natural philosophy of particles in motion could be given such needed shape only if the doctrine were not taken as the kind of vague background that allowed, for example, many a microscopist just to glue his ubiquitous corpuscles ('globules') to his efforts to explain the tiny vessels or fibers that he discerned. Nor should particle-thinking be taken any longer in the dogmatic sense of the Athenian structure of nature-knowledge. Rather, it should be taken in a new, *hypothetical* sense. The three men to perceive this possibility and to go ahead to elaborate it into a consistent approach were Boyle, Hooke, and young Newton. The 'Baconian Brew' to come out of their effort will occupy us in chapter 16.

What remains to be considered here is the sheer attraction exerted by fact-finding experimentalism, that is, its pull.

The pull of fact-finding experimentalism.

In the era of coercive empiricism (c. 1450–c. 1600) sheer unaided observation was still the rule, and progressive series of experiments were the rare exception. Fact-finding experimentation began to come into its own due to the four pioneers between c. 1600 and c. 1640. From then on, the balance kept shifting in the direction of the wholesale triumph of experimentation, to the point where even the preparation of natural histories proceeded in part by way of experiment. How did this triumph come about? Just as within two generations Alexandria fully gave way to Alexandria-plus, and Athens-plus went a long way toward replacing Athens before the 17th century was over, so did the revolutionary change in the third mode of nature-knowledge manage to stick. Why was that so? What attracted practitioners? What was it that, inside the ongoing movement of coercive empiricism, led practitioners to follow the pioneers in making sustained experimentation central to it? Three distinct answers present themselves.

The example set by the pioneers surely counted for much. Bacon's direct influence was tangible in the production of natural histories. All work done in magnetism and electricity proceeded from Gilbert's *De magnete*, whereas Harvey's experimental outcomes in the circulation of the blood and in generation dominated a good deal of subsequent experimental work concerned with the theory and practice of healing. In slightly less predominant ways had van Helmont set the problems that kept occupying chemical practitioners – scratch the

504 corpuscular surface and more often than not you are back at how he conceptualized the field. In short, the pioneers paved the way for the major subjects of fact-finding experimental research.

Still, the entire survey goes to show that instruments and societies (with their attendant observatories and laboratories) were the decisively new factors in the historical process of making sustained ranges of experimentation, not just an added component, but the primary one in the coercively empiricist mode of nature-knowledge. The bulk of fact-finding experimental research took place with the aid of instruments and found, if not its literal center, then at least its point of orientation in Rome, Florence, Paris, or London. True, von Guericke, in spite of his deliberate association with Father Schott SJ, was a case apart. So, but collectively, were most chemists/alchemists. To the extent that these men were alchemists they operated in a secret network all their own. And of course the pharmacists among them kept being organized in their own craft guilds. The vast majority of experimental researchers, however, were connected to institutionalized research centers, whether at the very core, like Hooke, or as a foreign correspondent, like Malpighi.

Further, experimentation not only offered badly needed ground for philosophically neutral pursuit of nature-knowledge, but it also found itself at the core of a rising, Baconian Ideology. The impact of these two, closely related developments extended far beyond fact-finding experimental science, in that the Continent-wide crisis of legitimacy analyzed in the previous chapter here found its resolution. I shall examine their joint impact in chapter 17 (p. 565).

The pull of fact-finding experimental science was not confined to those given to the finding of facts in the first place. With Alexandria-plus and with Athens-plus we came across similar cases. Experimentally inclined Mersenne and typically Athenian thinkers like Gassendi and hosts of Jesuits were pulled irresistibly in the general direction of a realist-mathematical science inherently foreign to their modes of approach, so as to check Galileo's rules of falling bodies and the transit of Mercury predicted by Kepler (p. 366). Just so were Digby and More unable to resist the lure of kinetic-corpuscularian modes of thought quite at variance with their deepest personal concerns (Renaissance-magical with Digby, religious-spiritual with More; p. 398). In just the same manner two men made their appearance in the present chapter whose presence looks quite anomalous. Both were mathematical scientists in the first place, and the knowledge structure of realist-mathematical science fitted them like a glove. Still, fact-finding experimentalism attracted both Huygens and Newton sufficiently to make them turn to it, the elder out of his desire to outdo all of science-minded Europe in every single case that met his eye; the younger out of a habit acquired right after making his first acquaintance with the ongoing Scientific Revolution. The consequent problem of how to integrate all this became, for Newton, a much more pressing concern than for Huygens (p. 539; 675). Not only did these two men become participants in fact-finding experimentalism,

but due to their unique gifts they made themselves masters of it wherever they touched it.   505
Huygens did so in electrical research, in microscopy, and with the air pump. Newton's outstanding gifts as an experimenter manifested themselves primarily in the domains of light
& color (p. 684) and of chemistry & alchemy. To these domains of experimental research he
managed to bring a quantitative rigor so far (but for the amazing Starkey) foreign to it.

Both Huygens and Newton, then, spread themselves all across the board in their pursuit
of anything that caught their fancy in nature-knowledge in any mode whatsoever. To act in
such a way was entirely without precedent. What made it possible for the old barriers between the various modes of nature-knowledge to break down, for the first time in history, in
Europe in the second half of the 17th century?

General features of the episode. The books that come closest to surveys of 17th-century experimental science after Bacon are Steven Shapin, *The Scientific Revolution*. Chicago: University of Chicago Press, 1996, and Lisa Jardine, *Ingenious Pursuits. Building the Scientific Revolution*. London: Little & Brown, 1999.

For almost all topics that follow, I have consulted pertinent entries (not as a rule further detailed below) in Wilbur Applebaum (ed.), *Encyclopedia of the Scientific Revolution from Copernicus to Newton*. New York/London: Garland, 2000, and/or in the *DSB*.

Facts collected and categorized. Lorraine Daston, 'The Factual Sensibility'. *Isis* 79, 298; September 1988, pp. 452-467 (essay review of three pertinent books). Paula Findlen, *Possessing Nature. Museums, Collecting, and Scientific Culture in Early Modern Italy*. Berkeley / Los Angeles: University of California Press, 1994. Harold J. Cook, *Matters of Exchange. Commerce, Medicine, and Science in the Dutch Golden Age*. New Haven & London: Yale University Press, 2007 (I have reviewed this book and its far-flung claims in *Endeavour* 32, 1; 2007). Boyle's experiment on nephritic wood is mentioned on p. 381 of M. Boas-Hall's lemma 'Boyle' in *DSB* 2, pp. 377-382.

Instrument-driven fact-finding. MUSICAL INSTRUMENTS: Penelope M. Gouk, *Music, Science and Natural Magic in Seventeenth-Century England*. New Haven/London: Yale UP, 1999 (on Mersenne and on post-17th-century developments, also my own *Quantifying Music*; I further consulted vol. 2, 'De Organographia' (Wolffenbüttel, 1619) of Praetorius' *Syntagma Musicum*). TELESCOPIC OBSERVATION AND MEASUREMENT: Albert Van Helden, *Measuring the Universe*. Chicago: University of Chicago Press, 1985, and (specifically on Flamsteed and on the surveying of France) Lisa Jardine's book mentioned above. MICROSCOPY: Catherine Wilson, *The Invisible World. Early Modern Philosophy and the Invention of the Microscope*. Princeton UP, 1995, and Marian Fournier, *The Fabric of Life. Microscopy in the Seventeenth Century*. Baltimore/London: Johns Hopkins UP, 1996. AIR PUMP: Steven Shapin & Simon Schaffer, *Leviathan and the Air-Pump. Hobbes, Boyle, and the Experimental Life*. Princeton UP, 1985, both for the early instruments and for fundamental problems with early experimentation by their means; M.J. Sparnaay, *Adventures in Vacuums*. Amsterdam: North Holland/Elsevier, 1992, for a survey of these various experiments, and Anne C. van Helden, 'The Age of the Air-pump'. *Tractrix* 3, 1991, pp. 149-172, for a survey of changes in the instrument up to the 1740s.

Subject-driven fact-finding. MAGNETISM: Stephen Pumfrey, entry 'Magnetism' in Applebaum's *Encyclopedia*; p. 385-387. ELECTRICITY: John L. Heilbron, *Electricity in the 17th and 18th Centuries. A Study of Early Modern Physics*. Berkeley/Los Angeles: University of California Press, 1979. CHEMISTRY & ALCHEMY: William R. Newman & Lawrence M. Principe, *Alchemy Tried in the Fire. Starkey, Boyle, and the Fate of Helmontian Chemistry*. Chicago: University of Chicago Press, 2002. William H. Brock, *The Norton History of Chemistry*. New York: Norton, 1992; ch. 2: 'The Sceptical Chymist', pp. 41-86. Richard S. Westfall, *Never at Rest. A Biography of Isaac Newton*. Cambridge UP, 1980; pp. 290-301; 357-371; 524-531. BODY & MIND: Roy Porter, ch. 9, 'The New Science', of his *The Greatest Benefit to Mankind. A*

Medical History of Humanity from Antiquity to the Present. London: HarperCollins, 1997; pp. 200-244;   507
J. Marius Engelbrecht, *De onttovering van de waanzin. Wetenschap, het bovennatuurlijke en de opkomst van een psychologisch mensbeeld* (a doctoral dissertation to appear in 2011; the title means 'Madness Disenchanted').

Craft techniques and experimental science. Many of the works cited above; further, for MAGNETISM: Katherine Neal's lecture 'On the 'Attractions' and 'Uses' of Gilbert's Magnetic Philosophy: Gilbert and the Practical Mathematicians' delivered at Oberwolfach on 6 January 2003. CHEMISTRY, Lawrence M. Principe, entry 'Chemistry' in Applebaum's *Encyclopedia*, pp. 139-142. HISTORIES OF TRADES: Kathleen H. Ochs' entry in Applebaum's *Encyclopedia*; also Marie Boas-Hall, 'Oldenburg, the *Philosophical Transactions*, and Technology'. In: J.G. Burke (ed.), *The Uses of Science in the Age of Newton*. Berkeley: University of California Press, 1983; pp. 21-47.

Problems with facts. The classic book to tackle head on such principal problems as first arose in experimental science of the 1660s is the one by Shapin & Schaffer mentioned above. As with Biagioli's later book mentioned in the Notes to the previous chapter, the authors found it expedient to do so on behalf of a radically social-constructivist conception of science that is neither required by their historically enlightening wish to problematize experimental science nor really consistent with their own apparent view of present-day science as very well capable of explaining in a nonarbitrary way what really went on in Boyle's air pump. Such numerous data and interpretations as remain of value once one subtracts their relativism from their account have gone far to inform the present section (more on their book in *SRHI* 3.2.4, 3.4.1., 3.5.1. (where I have also discussed Merton's four values), 3.6.1., 3.7.). Specifically on the problem of RELIABILITY, I have taken the example of the divers from ch. 6 'Knowing about People and Knowing about Things: A Moral History of Scientific Credibility' of Steven Shapin, *A Social History of Truth. Civility and Science in Seventeenth-Century England*. Chicago: University of Chicago Press, 1994; a succinct account of the entire problem is ch. 7, 'Experiment: How to Learn Things about Nature in the Seventeenth Century' of Peter Dear, *Revolutionizing the Sciences. European Knowledge and Its Ambitions, 1500-1700*. Houndmills: Palgrave, 2001. Also important on the nature of experiment in 17th century mathematical and experimental science is Peter Dear, *Discipline & Experience. The Mathematical Way in the Scientific Revolution*. Chicago & London: University of Chicago Press, 1995. I have further consulted William Eamon's various publications on PUBLICATION, most extensively so in his *Science and the Secrets of Nature. Books of Secrets in Medieval and Early Modern Culture*. Princeton, NJ: Princeton University Press, 1994. The origins of rules for securing priority have been discussed by R.K. Merton & H. Zuckerman in a 1971 paper 'Institutionalized Patterns of Evaluation in Science', reprinted in R.K. Merton (N.W. Storer, ed.), *The Sociology of Science. Theoretical and Empirical Investigations*. Chicago: University of Chicago Press, 1973; pp. 460-496. For Newton's successive accounts of the speed of sound I have followed R.S. Westfall, *Never at Rest. A Biography of Isaac Newton*. Cambridge UP, 1980; pp. 455-456 and 734-736.

Pooling of efforts. Martha Ornstein [Bronfenbrenner], *The Rôle of Scientific Societies in the Seventeenth Century*. Chicago: University of Chicago Press, 1928 (first ed. 1913; *SRHI* 3.5.2.). ROME: Paula

508 Findlen's and John L. Heilbron's books mentioned above; also John L. Heilbron's delightful *The Sun in the Church. Cathedrals as Solar Observatories*. Cambridge (Mass.): Harvard UP, 1999. FLORENCE: W.E. Knowles Middleton, *The Experimenters. A Study of the Accademia del Cimento*. Baltimore: Johns Hopkins Press, 1971. PARIS: Alice Stroup, *A Company of Scientists. Botany, Patronage, and Community at the Seventeenth-Century Parisian Royal Academy of Sciences*. Berkeley: University of California Press, 1990, as also her lemma 'Académie' in Applebaum's *Encyclopedia*, and Albert Van Helden's and Lisa Jardine's books mentioned above. LONDON: the book by Shapin & Schaffer mentioned above; also Steven Shapin, 'The House of Experiment in Seventeenth-Century England'. *Isis* 79, 298; September 1988, pp. 373-404; Michael Hunter, *The Royal Society and its Fellows 1660–1700. The Morphology of an Early Scientific Institution*. Oxford: BSHS Monographs (vol. 4), 1994 (2nd ed.; originally 1982), and Marie Boas-Hall's lemma 'Royal Society' in Applebaum's *Encyclopedia*.

Power and pull. In addition to the authors listed under the corresponding paragraph in my Notes to ch. 10, I mention here as a source of inspiration the general tenor of R. Hooykaas' *Fact, Faith and Fiction in the Development of Science*. Dordrecht: Kluwer, 1999, as expressed most cogently in the Introduction.

XIV

NATURE-KNOWLEDGE DECOMPARTMENTALIZED

Indeed, high barriers separated the various modes of nature-knowledge from the Greeks onward (p. 18). So different were the intellectual and social worlds of practitioners of mathematical science, of natural philosophy, and of more empiricist approaches that interaction between them was virtually ruled out, even in the rare case where one man was engaged in two modes at a time, as with Ibn Sina or Castro or Descartes. Consequently, the surveys just completed of achievement and limitation between c. 1640 and c. 1700 in all three modes of nature-knowledge in their newly transformed states reveal pronounced differences among them. Even so, connections have also here and there come to light. Some of these betoken nothing but sheer adaptation by natural philosophers of discoveries made in either realist-mathematical or fact-finding experimental science, or 'opportunist' or 'eclectic' uses made of some tenet in natural philosophy. But more genuinely innovative linkages were also at times forged between one mode and another. Here is a list, compiled from chapters 10, 11, and 13 and placed in chronological order.

- In the 1630s and 1640s scholars raised in the tradition of mixed mathematics like Mersenne and Riccioli felt inspired by Galileo's work to conduct ranges of fact-finding experiments directed at measurement in close connection with his laws of falling bodies (pp. 343; 409).
- From the late 1640s onward Kircher began to weave a Jesuit synthesis of nature-knowledge from numerous disparate strands (p. 495).
- By the 1650s heliocentric common ground between realist-mathematical science and the natural philosophy of kinetic corpuscularianism led to the promotion of heliocentrism in academic circles (p. 396).
- In the mid- to late 1650s men like Charleton and Power hoped to make corpuscles distinctly visible. This furthered an outburst of microscopical research in the early 1660s, in the course of which Grew, van Leeuwenhoek, and others adorned carefully observed microscopic phenomena with an arbitrary layer of corpuscles (pp. 385; 454; 503).
- In the early 1660s Galileo's assertion of equal descent in free fall was confirmed empirically in the receiver of Boyle's air pump (p. 461).
- In the early 1660s the telescope was turned from a tool of mathematical science into a tool of fact-finding measurement. As one consequence, the 'grand schema' of the measure of the universe underwent drastic revision (p. 451).

510

- From its start in the mid-1660s onward, the Académie joined the repetition and empirical correction of mathematical work pioneered by Galileo and his Italian disciples to its efforts at fact-finding experimental research (ch. 10). It also undertook a somewhat covert testing of Cartesian conceptions (p. 497).
- In 1670 Borelli invoked geometrical form and the mathematics of fluid equilibrium as 'external' checks upon kinetic corpuscularianism (p. 393).
- In 1676 Rømer discovered a way to decide empirically what so far had been a natural-philosophical controversy over whether the velocity of light is finite or infinite (p. 453).
- In the 1680s a prolonged controversy on the Reno between two mathematical scientists and their Jesuit, 'mixed-mathematics' opponents was alleviated by some cautious *rapprochement* (p. 312).
- In the early 1690s Huygens' electrical experiments were stimulated by a Cartesian conception of whirlpools (p. 464).

The list is surely far from exhaustive. After all, the aim of the present book is not to narrate the Scientific Revolution in all its colorful detail but rather to analyze its structure down to thought-provoking and/or telling detail only. It is further true that the pioneers themselves had initiated some incidental crossovers: both Kepler and Galileo made some use of Gilbert's account on the magnet; Galileo benefitted from his father's experimental work on pitch; Descartes practiced dissection.

Still, the list strongly suggests that at some point the barriers began, if not to fall down, then at least to become a good deal more permeable and also *that that point is to be situated c. 1655–1660*. That is when incidental crossovers began to become markedly more frequent. But that is also when the perception of ground shared between two modes ceased to give rise to acrimonious clashes, as in the early 1610s (light/heavy), the late 1630s–1640s (falling bodies), the late 1640s (the void), and the early 1660s (the void all over again, but now fought as a rearguard action). Instead, starting by those early 1660s previously contested, shared turf began more and more to be dealt with in a spirit of constructive reconciliation.

This, then, is the historical phenomenon we are here concerned with. As historical phenomena go, it is relative, not absolute. Surely the majority of innovative work in nature-knowledge kept being done in one mode only. Also, after 1700, mutual separation once again became the rule for a long time to come. Not until the early decades of the 19th century were major fusions undertaken once again, but they were now far more numerous, on a far vaster scale, and also permanent, even though divisions between, notably, theorists and experimenters are still readily observable today in many a discipline. For all such qualifications, the decompartmentalization between three distinct modes of nature-knowledge that began to manifest itself a little after the mid-17th century is remarkable, not for a completeness unattainable anyway, but rather for significantly disrupting for some four decades a pattern that had with only the rarest of exceptions characterized the pursuit of nature-knowledge in the tradition of the ancient Greeks.

What conditions made it possible for barriers of such long standing to break down? What research vehicles were available to give body to current efforts at reconciliation? And what revolutionary consequences did such efforts entail in their turn? These are the three questions to which answers will now be sought.

511

Whence the breakdown of barriers?

A reminder is first needed of how the barriers originally came into being, and also of what sustained them over the centuries during which nature-knowledge was pursued in Islamic civilization, medieval Europe, and Renaissance Europe.

As the shorthand expressions Alexandria for mathematical science and Athens for natural philosophy indicate, geography is what originally held the two apart. The creative melting pot of pre-Socratic Greece contained mathematical and speculative components, which subsequently came apart and solidified in ways that were distinct not only geographically but also institutionally (pp. 16; 18). Mathematical science was to a very large extent bound up with courtly support, whereas natural philosophy was indissolubly connected to teaching. From one civilization to another there might be additional occupational roles available for philosophers (such as orator and/or politician in Greece and Rome, or physician in Islam). Teaching might take different guises, with individual master/pupil relations becoming embedded in Islam in theology-centered *madrasa*s and in Europe in theologically orthodox universities with more of a place for natural philosophy. Yet the basic pattern is quite alike in each. Nor does the unique emergence in Renaissance Europe of a third, widespread, coercive-empiricist mode of nature-knowledge make a difference in this respect, in that practitioners were as a rule connected to court life as mathematical scientists were. Here, too, additional professional space might be available, as in the military or on board ship for practice-oriented mathematicians, in medicine for chemists, in overseas voyages for botanists, or in the crafts for those few to come to the pursuit of nature-knowledge without academic training, yet the broad picture hardly changes thereby. *Nor did it alter significantly during the first stage of the Scientific Revolution.*

It is tempting at this point to go ahead and attribute the breakdown of barriers without more ado to some drastic, mid-17th-century change in the institutional pattern thus sketched. But the temptation must be withstood: We must ask first whether the pattern, which quite evidently explains the presence of barriers to some extent, can *fully* explain it. Well, it cannot. Recall here those striking cases of individuals involved in two modes at a time: the thinking of, say, 'philosophical' Ibn Sina, or Thomas Harriot, was not affected by that of 'mathematical' Ibn Sina, or Harriot, just as Castro's observational/experimental practice hardly affected his scholastic textbook, nor (the other way round) his scholastic

512 disquisitions his empirical researches. In such cases there can of course be no question of institutional separation. But there is an ingrained intellectual habit, established to be sure by the originally no more than geographic distance between Athens and Alexandria and encouraged by the exclusive knowledge structure of natural philosophy, yet soon attaining and then maintaining a life of its own. So, the net picture is of an ingrained absence of curiosity about what might be picked up from a mode of nature-knowledge other than one's own, with that absence of curiosity greatly fostered by institutional barriers but not solely determined by them. To find out, then, what conditions made possible the mid-17th-century breakdown in its twin aspects – suddenly increasing ease of crossover and a more conciliatory attitude toward contested turf – we had better cast about *not only* for geographic and societal-institutional causes but for more directly nature-knowledge-related ones as well.

We look at geography first. However instrumental the geography of the Hellenist world was in bringing separation about, it is hard to see how Europe's geography, which after all did not change over the period, can have affected its breakdown. It is true that the population of the European heartland was spread over a much smaller area than that of either the Hellenist or the Islamic world. Even so, if the somewhat easier and also swifter communication made possible thereby failed to assist the breakdown of barriers as indeed it persistently failed to do from the very start in the 12th century onward, how could it possibly have begun to do so around 1655?

The speed and ease of communication, however, are not dependent on geography alone but primarily on techniques. Travel, letters, and books were readily available means to set up and maintain communication-among-equals – did any of these change in relevant ways over the second half of the century?

Travel surely affected, if not the onset of the Scientific Revolution as such, then at least specific insights of, notably, Descartes, Gassendi, Harvey, and van Helmont. Yet, if two men born at the same time in the same nation like Boyle and Hobbes could return home from their formative 'grand tours' to Italy with radically different attitudes on the very point that concerns us here, then clearly the adoption of a broadly conciliatory attitude (as with Boyle) or a particularly narrow variety of the one-mode-only posture (as with Hobbes) had little if anything to do with the travel habits of the period.

The increase in the number of books on matters of nature-knowledge being printed over the second half of the century kept pace, by and large, with the increase in the European book trade. Also, just as before, numerous significant discoveries or even entire treatises were kept hidden by their authors in private manuscripts – clearly no new pattern is discernible here, either.

Nonetheless, one additional vehicle for communication of one's finds became available from 1665 onward. It grew out of exchanges by letter. Scholarly correspondence in the second half of the century has with justice been called "a miracle of almost instantaneous transmission of information and ideas."[272] Even so, letters exchanged over its *first* half fill just about equally fat tomes, as with the seventeen modern tomes filled to the brim with letters by and to Marin Mersenne, who, from the mid-1620s until his death in 1648, conducted a Europe-wide correspondence network in the quiet of his Parisian convent. Kircher and Oldenburg fulfilled a similar role in a period that covers the critical years c. 1655–1660. Oldenburg alone found his sole calling in facilitating the work of others, whereas both Mersenne and Kircher were much given to crossover in their own researches, and it may well be that their early overstepping of barriers stands in some causal relation to their assiduous letter writing. The act of setting up such a network may well have opened the mind to cross-over, just as (the other way round) a mind-set already present may well have spurred their interest in exchanges by letter. In any case, the presence of these early networks does help account for precisely the two pre-1655 cases of barrier crossing listed above.

The third of the correspondence networks, Oldenburg's, led to something wholly novel: the first instance of the one technically new mode of communication to make its appearance in the century. Scientific journals emerged very near the critical period of 1655-1660. The first appeared in England, but France followed suit almost at once, and then Germany within two more decades. Does their appearance, and the quick pace at which reports of new finds began to fill their pages, show the scientific journal to have been a major vehicle for the breakdown of barriers? Indications for such a role for the journal *as such* are small. The *Philosophical Transactions* especially, and in somewhat smaller measure the *Journal des sçavans* as well, were much given to reporting, in the circumstantial style held proper for them, almost nothing but 'histories'-styled observation and experimentation of the fact-finding kind.

Truly instrumental for the breakdown of barriers was the institutional innovation to which the journal owed its rise and ongoing flourishing. That innovation was, of course, the foundation of two major scientific societies in 1660 and 1665, respectively. Formally, the institutional state of European nature-knowledge was not changed by the emergence of the English and French societies, in that both were under court patronage and could not have existed without it, so that realist-mathematical science as well as fact-finding experimental science remained cultivated in a courtly context above all. But, effectively, a major change was occasioned by their foundation. In the English case the change was even a dual one.

The one institutional change is that in England royal patronage provided the Royal Society with a name and a charter, which conferred on it the indispensable seal of official acceptance while leaving it free of obligations. The absence of prescriptions for the direction of research accompanied the absence of royal funding (luckily, France-like funding proved dispensable, albeit at times barely so). Key, however, were the far-going liberties of self-regulation and freedom from outside censorship that the charter entailed. Room emerged for a larger and, above all, collectively enjoyed amount of unfettered, leisurely exploration than

514 ever before in the history of pursuit of nature-knowledge. That free room could of course be used for all kinds of silly pursuits – it surely was, as Swift and his ilk were quick to point out. But it could also be directed toward better things.

What these better things might be could now more easily than ever before flow from the other major change entailed by the foundation of scientific societies (in this case, the English and the French alike). This was increased communication through regular or even intensive, weekly or even daily, personal contact, be it in formal session or in joint experimentation or in leisurely conversation over a cup of London coffee or during a sunlit stroll through the Jardin des Plantes. Examples are Halley's famous 1684 visit with Newton (pp. 637; 648), the manner in which Huygens' attention was drawn by fellow Academician Picard (upon his return from Uraniborg) to the recent discovery, by Erasmus Bartholinus in Denmark, of 'strange refraction' in Iceland spar, or how Huygens learned of the unpublished wave account by means of which a Jesuit close to the Académie, Ignace Gaston Pardies, hoped to explain the propagation of light (Huygens soon found a way, still known to every physicist as 'Huygens' principle', to improve on that explanation; p. 541).

What all this means, then, is that early on in the 1660s institutional opportunities were created that lent themselves to unprecedentedly free and easy exploration in England and, both there and in France, to enhanced communication of a more personally informal kind than even the regular exchange of letters could bring about.

To have opportunities is one thing; what use one makes of them is another. I shall consider first the 'reconciliation' aspect of the breakdown of barriers, then its crossover aspect.

The mood of reconciliation that by the 1660s so visibly replaced viciously heated clashes over the very fundamentals of the pursuit of nature-knowledge was one side effect of what may be called the 'spirit of Westphalia'. For its determination of the fate of Europe in the direction of making the best rather than the worst of its political state of mitigated divisiveness (the roots of which go back to the early Middle Ages), there is no overestimating the significance of the Peace of Westphalia, which was concluded in 1648. By then, the European state 'system' had come perilously close to a war of all against all. The numerous treaties that together formed the Peace of Westphalia served as a package deal by means of which Europe gave itself an opportunity to let its energies run henceforth in more constructive channels (p. 431). These included 'nation building' in the sense of domestic pacification, some stability-enhancing arrangement for religious divisions, and various measures for enhancing prosperity. These benefits extended, with only a little delay, to Britain. Upon the death of Cromwell and the readily apparent ineptitude of his son, rather than the sort of conflict-ridden chaos that might well have ensued in pre-Westphalian times, we find the remarkably smooth succession and the even more remarkably unrevengeful reign of the very son of a king beheaded only eleven years earlier (note in passing that the next royal decapitation, of Louis XVI in 1793 France, was to remain a national trauma for at least half a century). Indeed, the overall climate in Europe in the 1660s looks more different in many ways from that in the 1640s than that in the 1640s from the 1540s – a lid had been lifted, poisonous vapors had dissipated, and a fresher, less tense atmosphere now reigned.

All this is more or less standard political history. The additional point to make about this remarkably sudden shift from near-ubiquitous devastation, produced above all by an insistence on stamping out religious division, to the allowance of some give-and-take is that it looks as if it affected the pursuit of nature-knowledge, too – not so much in the kinds of discoveries made as, rather, in a perhaps not too deliberate decision that, if and when conflicting views were at stake, it might be more fruitful to inquire what these had in common than what kept them divided. The list of cases of crossover readily shows where these commonalities could be found. They reduce to two basic entities.

Quantities and corpuscles

I shall examine first the role of quantities in the cross-over between modes of nature-knowledge, then the chances for cross-over that resided in corpuscles.

With Alexandria as also, at first, with Alexandria-plus, the expression of mathematical relations (equality, proportionality, congruence, ...) was predominant, and quantities hardly counted. For instance, Galileo went to great pains to establish the relations that obtain between parameters in vertical descent but spent just a few perfunctory paragraphs on the actual rate of acceleration. Scholars reared in mixed mathematics like Riccioli and Mersenne quickly seized upon what they regarded as Galileo's omission and spent vast amounts of time on determining that rate (p. 409). In so doing they prefigured the prominence that efforts at measurement attained from the 1660s onward.

Leading in the effort were, of course, the astronomers. Alone among the five distinct branches of Alexandrian mathematical science, planetary theory had given occasion for measuring more than an incidental feat like Eratosthenes' determination of the Earth's circumference. Here alone, mathematical relations could not be drawn from the scantiest of quantitative data acquired through observation the way this could be done with light rays or levers or floating bodies or musical intervals. Mathematical relations in astronomy were in need of quantitative data acquired through observation to be fed back into them if the model built upon them was to have any predictive value at all. In the ongoing construction and revision of mathematical relations (epicycles, eccentrics, equants) fit to accommodate observational data, Ptolemy and his Arabic and Renaissance European successors, Copernicus included, gave primacy to the former over the latter. This remained so until Tycho Brahe came along. For the first time mathematical relations were not the center of attention; rather, Brahe granted priority to the empirical job of collecting data of the highest possible

516   accuracy. Truly intensive crossover between mathematical relations and measured quantities began to take place when, in the early 1660s, the invention of crosshairs enabled the telescope itself to cross over from mathematical science (with Galileo and then Huygens as the chief practitioners of telescopic observation) to that particular variety of fact-finding experimental science where not experiment but measurement provided the pathway toward facts of the highest attainable accuracy (which is when Cassini, Hevelius, and Flamsteed took over; p. 451).

By that time, the 1660s, mathematics-inspired measurement also came into its own in other empiricist branches of scholarship. For instance, currents efforts at resolving the problem of longitude appeared to be affected by measurements like Richer's on the inconstant length of the seconds pendulum (p. 454), and alchemy was pulled by Starkey and Newton out of its recipe-like procedures to unprecedented levels of quantitative precision. An awareness of the capacity of mathematical science to elucidate natural phenomena even informed the way in which the first two quantitative laws came into being: Mersenne's on pitch and Boyle's on air pressure. Not a mathematical relation posited *a priori* but careful measurement of variables experimentally found to be pertinent led to the discovery of these laws (p. 462; Figure 13. 3.). Here the breakdown of barriers between modes of nature-knowledge enabled fact-finding experimentation, upon passing over a bridge of measurement as it were, to shade into the finding of quantitative law.

In both different and similar fashion did the breakdown enable *corpuscles* and their various motions to extend their role beyond that of providing food for speculation in the strict framework of Athens-plus. The list of cases of crossover drawn up above shows how corpuscles came to aid understanding in a variety of ways, from the superficial labeling of otherwise seemingly indecipherable, microscopic structures to the experimental checking of certain Cartesian mechanisms. But they were also useful for wider purposes.

Recall at this point the search, discussed at length in previous chapters on post-1640 achievements and limitations, for ways and means to curb the apparent ease with which all kinds of explanations might be thought up. Some of the most perceptive of second-generation practitioners became deeply concerned over the scarcity and relative insignificance of checks upon arbitrariness available outside the realm of mathematical science. In the natural philosophy of kinetic corpuscularianism there were those four 'internal' checks (foundational certainty and consistency, analogy, and 'intuition'), which Borelli uniquely sought to supplement with a fifth, 'external' one – geometric form. In fact-finding experimentalism a tenuous blend of background worldview, theory, and experimental outcome acted as really the only check upon conclusions meant to go beyond the sheer establishment of facts. An acute sense of what little all such checks amounted to, derived from his lively awareness of the far more productive checks at work for the mathematical scientist, induced Isaac Newton, in some desperation, to speak of "natural Philosophy, where there is no end of fansying".[273]

But there *was* an end to it, or at least the beginning of an end. Corpuscles were central to the
twofold way in which this came about in the mid-1650s to early 1680s, and young Newton
himself was uniquely present, paramount even, in both these ways.

Revolutionary fusion in the making

Motions of corpuscles brought about, not just incidental crossover nor just incidental, conciliatory rapprochement, but revolutionary fusion in a range of specific subject areas. One
fusion was between realist-mathematical science and kinetic corpuscularianism; the other,
between fact-finding experimental science and kinetic corpuscularianism. In the process,
however, *and only as far as the process went*, kinetic corpuscularianism was itself transformed,
in the sense of being provided, as a prerequisite to fusion, with a knowledge structure other
than the hoary, Athenian one.

Unprecedented as fusion between distinct modes of nature-knowledge was, we may consider the superficially coherent worldview of the Jesuits, cobbled together from disparate
elements in natural philosophy, natural magic, and experiment, with stray figures standing in for mathematical relation, and with the whole held together by a network of hidden
correspondences and put down in writing by Kircher between c. 1650 and 1680 (p. 495), as
a *travesty* of how elements of the three modes might be combined. Here, however, the very
absence of any checks whatsoever was the distinctive feature (not only in retrospect, to be
sure; not one competent, contemporary practitioner outside the sphere of Jesuit activity
took their worldview at all seriously).

An awareness that the natural philosophy of corpuscles in incessant motion suffered
lethally from a built-in overdose of arbitrariness arose earlier than did ways and means to
repair the defect from the ground up. The habitual presentation of corpuscles in motion by
way of speculative, empirically untestable dogma is precisely what Pascal, the mathematical
experimentalist, already came to perceive as its native shortcoming in the 1640s. In his *Pensées* (written c. 1656–1662; published posthumously in 1670) he neatly defined what he was
prepared to accept of kinetic corpuscularianism and what not:

> Descartes. – One must say at large: 'These things happen through their shapes and their mo-
> tions'; for that is true. But to say which ones, and to construct the machinery – that is ridicu-
> lous, because it is useless and uncertain and awkward.[274]

Earlier, in his Letter to Noël, Pascal made an effort to curtail Descartes' known and confidently anticipated excesses of the imagination by pronouncing the following rule:

518 ... whenever, so as to find the cause of several known phenomena, one posits a hypothesis, this hypothesis can be of three kinds.

For sometimes a manifest absurdity follows from its negation, and then the hypothesis is true and constant; or a manifest absurdity follows from its affirmation, and then the hypothesis is held to be false; and when one has not yet been able to draw absurdity from either its negation or its affirmation, the hypothesis remains in doubt; so that, in order to make a hypothesis evident, it does not suffice for all phenomena to follow from it, whereas instead, if something ensues contrary to just one of these phenomena, this suffices to ascertain its falsity.[275]

If applied indeed, this methodological rule of immediate falsification was well-suited to kill off at one stroke all of kinetic corpuscularianism in any variety whatever. Only, in its evident origin in the structure of Euclidean geometry it was also well-suited to kill off nearly all future development of any science of the natural world whatsoever. Two subsequent decades of ongoing, mathematical scientists' experience with the messiness of the world, joined to the onset by midcentury of searches for cross-over and reconciliation, gave rise to a less impossibly strict, more creative and productive way out of Cartesian excess. This way out rested in the use of hypotheses indeed, yet in such a way as to get rid of unbridled speculation *without* in the same move getting rid of empirical science altogether. In the next chapter we shall find out in what manner invisibly corpuscular motions began to serve as a ready-made source of hypotheses meant to reconcile and thereby revolutionize previously incompatible approaches to phenomena of motion and of light. And in the chapter that follows next we shall examine how a similarly hypothetical use of corpuscles in motion meant to reinforce experimental outcomes was aimed at alleviating in likewise revolutionary manner the scarcity of checks in the realm of fact-finding experimentalism.

What we shall find most of all is how the best and most enduring researches of the century were either carried out or prepared in the frame of these two revolutionary transformations.

NOTES ON LITERATURE USED

I owe the very idea that during the 17th century Europeans began to put together what the Greeks had kept apart to a penetrating passage on pp. ix-x of Samuel Sambursky, *The Physical World of Late Antiquity*. London: Routledge, 1963. Further, the organization of R.S. Westfall's *The Construction of Modern Science. Mechanisms and Mechanics*. Cambridge UP, 1971 (discussed at length in *SRHI* 2.4.5, final subsection especially), gave me a chance to see the vast significance of Sambursky's point.

The profound significance of the Peace of Westphalia for Europe's subsequent fate is brought home with particular force in ch. 3 'From Universality to Equilibrium: Richelieu, William of Orange, and Pitt' of Henry A. Kissinger, *Diplomacy*. New York: Simon & Schuster, 1994, and in Theodore K. Rabb, *The Struggle for Stability in Early Modern Europe*. Oxford UP, 1975 (*SRHI*, 3.6.1.).

XV

THE FOURTH TRANSFORMATION:
CORPUSCULAR MOTION GEOMETRIZED

If it were possible to mark what was distinctively novel about the Scientific Revolution in just one word, that word would be 'motion'. Never at the center of any piece of mathematical analysis in abstract, Alexandrian fashion, rather neglected in the atomist natural philosophy of corpuscles moving through the void, categorized as just one among four subclasses of change in the Aristotelian philosophy of nature, motion acquired a radically novel emphasis as a fundamental unit of analysis and/or explanation by the early 17th century. By then, specific movements began to be invoked with a view to accounting for specific phenomena, as with Galileo's laws of falling bodies or Kepler's laws of planetary trajectories or Beeckman's account of sympathetic resonance. Also, motion began to be analyzed as a philosophical category in its own right – a subject on which Descartes was as vocal as Galileo was silent. But between the specific and the categorical levels of dealing with motion, an intermediate level of analysis emerged as well. On that level the challenge was to define characteristics shared by whole ranges of actual movements. I introduce for it the term 'motion considered generically'. It is situated halfway the level of principles, which it helped fill with more concrete meaning, and the level of phenomena, where it served both as abstraction from, and as exemplar for, specific cases.

For the period of the Scientific Revolution four *genera* can be distinguished: (1) motion in impact (central to Descartes' philosophy of moving particles), (2) motion retained (central to Galileo's realist-mathematical science), (3) rotational motion, and (4) to-and-fro motion. I shall now review for each *genus* what conceptual treatment it had received at the hands of the pioneers in realist-mathematical science and in the philosophy of kinetic corpuscularianism. I shall also examine how, from the mid-1650s onward, Christiaan Huygens attained for each *genus* of motion a kind of fusion of the Galilean and the Cartesian approach to it. I shall then consider the both similar and characteristically different way in which, in a brief outburst between 1665 and 1668, young Isaac Newton accomplished essentially the same feat, and more. What came out of their efforts was, not yet the genesis of rational mechanics to be sure, but decisive steps in its general direction.

Not only motion but light rays, too, stood to benefit when the customarily geometric treatment thereof began to be shot through with hypotheses about particle motion. Earlier, I abandoned discussion of optical issues at the point where Huygens and Newton abandoned their respective, initially mathematical attacks on the two distinct cases where light rays fail

522   to obey Snel's sine rule of refraction (p. 300). Here I resume their respective quests from the mid-1670s onward. What came out of these has with justice been called the genesis of physical optics.

Motion, four principal ways

The scientifically knowledgeable reader is well advised, for proper understanding of what follows, to bracket all vestiges of Newtonian mechanics in her or his head. Please drop any idea of force as defined by its distinctive capacity to alter a body's motion in magnitude and in direction. Also do not take for granted the kinematics that then remains. Not only was what Newton was in due time to enunciate in the *Principia* as the second law of motion (p. 652; 660) the product of a more mature thinker than the youngster who forms our subject for now. Even more importantly, Descartes' *genera* of motion, far from being conceived along Newtonian or even along Galilean lines, derive instead from a quite different conceptual frame, which has fittingly been dubbed Descartes' 'water world'.

The *former* portion of this caveat applies in particular to *genus* no. 2, now known as the principle of inertia. Attaining canonical form in Newton's first law of motion, it has ever since been bound up inextricably with his second law and the rational mechanics derived from it. Its *latter* portion rather concerns nos. 1 and 3, impact and rotational motion. Pertinent to all four *genera* is the profound conceptual confusion bound to result from their simultaneous presence in two quite distinct modes of nature-knowledge, Alexandria-plus and Athens-plus, respectively. The confusion became ineluctable as the barriers between them began to break down. Motion found itself retained in both, yet in different ways. Rotational motion made its appearance in both, yet in different ways, too. The motion of bodies in impact, albeit a feature of kinetic corpuscularianism alone (notably, its Cartesian variety), was treated there in a way different from the Galilean approach to motion. Or let us rather replace usage of the bland term 'different ways' with awareness of an innate *tension* running from top to bottom between mathematical and kinetic-corpuscularian conceptions of motion. The modern historian first to signalize the tension was Richard S. Westfall. He placed it at the core of the conceptual unraveling that goes on in his *Force in Newton's Physics. The Science of Dynamics in the Seventeenth Century* (1971). His analyses shall serve us here as our main guide. The modern practitioner first to notice the tension was Christiaan Huygens. He did so in 1652, within seven or eight years of making his youthful acquaintance with Descartes' *Principia philosophiae*. The *genus* of motion for which he first signalized it was impact.

Impact.   For an understanding of Huygens' difficulties with Descartes' account, we need
to examine the latter in greater detail than we have so far. The centrality of impact in Descartes' world flowed from his identification of matter and extension (p. 228). Since wherever
a particle moves it encounters other particles, things happen in the Cartesian world solely
through the particles' perennial crowding upon one another, that is, through rebound.

In *Principia philosophiae* Descartes examined rebound for the sole case of perfectly hard
bodies in central collision. He brought four variables to bear on his analysis: size, surface
touched, speed, and 'determination' (a troublesome concept, here to be understood as an
aspect of motion meant to serve as a measure for direction). The principle governing rebound is that the quantity of motion does not change before and after impact; that is, for the
pair of bodies in impact the product of size and speed remains the same. This is presented
as sufficient to yield nine rules under three distinct conditions: (a) the two bodies move
in opposite directions; (b) one body moves, the other is at rest; (c) one body overtakes the
other. The rules that ensue look baffling to the modern reader; their shared feature is that
Descartes derived them by means of a tacit analogy. Starting with Galileo, when mathematical scientists wished to analyze some as yet unknown case of movement, whether 'static' or
'dynamic' in our modern sense, only one general source of analogy presented itself from
the outset, namely, motion in equilibrium cases (p. 336). That is how, for instance, Galileo
himself invoked the law of the lever when seeking to measure the force of impact. Inevitably,
Descartes found himself in the same position; inevitably, his own handling of analogies of
motion was quite different – not mathematical in an idealized-real sense, but quantitative in
a sense meant to be pertinent to the microworld. The analogy with the lever led Descartes to
conceptualize impact in terms of a contest between opposite forces. These contests are laden
with discontinuities. As with a balance, "the difference between equilibrium and disequi-
librium may be almost infinitesimal, but the difference in outcome between winning and
losing a conflict can be very great".[276]

Considered from within the frame of Descartes' philosophy of nature there is nothing
odd, incoherent, or arbitrary about these rules. Nor, inside that frame, is one entitled to
object that a casual look thrown upon the billiard table suffices to see how wrong almost all
these rules are (rule 4 egregiously so, with its bald assertion that a moving body that collides
with a larger one at rest rebounds with the same speed as before, whereas the larger one, even
if only the tiniest bit larger, remains at rest). Descartes himself granted at once that, since
there is no perfect hardness in the world, and since two bodies never collide in isolation
from other bodies that surround them, "it often happens that experience may seem at first
to conflict with the rules I have just explicated".[277]

Considered, however, from inside another frame, that of Galilean mathematical science,
difficulties with these rules readily present themselves. And it was in reconsidering them thus
that, in 1652, to his surprise and dismay, Christiaan Huygens found them all (with the sole

524 exception of rule 1, which trivially states that when equal bodies collide with equal speeds they rebound with their speeds unaltered) to be dead wrong. It was a momentous discovery. On the one hand, from about his fifteenth year onward he had been a faithful Cartesian (p. 383). On the other hand, along the Alexandrian path taken by him as van Schooten's student he had within a few years of flexing his mathematical muscles come to abandon his efforts at beating Archimedes on his own terrain (p. 367) in favor of the kind of mathematical science of the real world that he found embodied in Galileo's work. But from that newly adopted, Galilean point of view, so he discovered in 1652, an issue at the heart of Descartes' philosophy of nature looks thoroughly mishandled.

Huygens was not much given to self-reflection, and what he found so wrong about Descartes' rules of impact is to be surmised from features of his immediately ensuing effort to derive better ones.

Little more than a guess is that (just as his pupil Leibniz was later to set forth with philosophical rigor) he may have come to feel ill at ease with the sudden jumps the rules display – how could the tiniest change in size or speed alter the entire outcome? Kepler, the mathematical realist, had found occasion to reject Alhazen's account of vision as soon as he came to realize that it hinged on the – to Kepler – absurdity of light rays failing to contribute to vision if deviating by merely the smallest amount from perpendicular incidence (p. 297). Huygens may in similar fashion have come to realize this for Cartesian impact.

More easily reconstructible is the significance, for Huygens, of the empirical aspect of Descartes' rules of impact. For in what sense can one say that these rules are 'wrong'? In Descartes' conception the warrant for their correctness rested, as always with him, in the clarity and distinctness of the chain of reasoning through which they are connected to first principles themselves erected upon the rock-bottom certainty of the *cogito* (p. 236). That they look odd in the macro-world is entirely inconsequential – there are no perfectly hard bodies in the macro-world, so no empirical test is possible. What, then, enabled Huygens to take their apparent failure to describe the conduct of billiard balls as indicative, nonetheless, of their being false? In 1656 he completed a treatise 'De motu corporum ex percussione' ('On the movement of bodies in impact') that remained unpublished during his lifetime. In early drafts Huygens made an educational effort for the benefit of his Cartesian readers. He dwelt at some length on the proper way (which of course is the Galilean way) of looking at empirical checks. True, there are no perfectly hard bodies in the macro-world, but there are bodies at least *approximating* perfect hardness. What one may ask of rules of impact is that they approximate the actual conduct of macro-world bodies and also that they do so the more closely the harder these bodies are. If, rather than approximating actual conduct on the billiard table, they go flat against it (as with Descartes' rule 4 in particularly glaring fashion, but with all others except rule 1 as well), this may be taken as a sound sign of their being false.

Quite unambiguously the case, finally, is that Huygens came to realize the incompatibility of
Descartes' rules of impact with a basic principle of motion first enunciated by Galileo: the
relativity of motion and rest as captivatingly illustrated in the *Dialogo* by means of that ship
on its way from Venice to Aleppo (p. 184). Since Descartes upheld the relativity of motion
and rest, too, it was even possible for Huygens to regard Descartes' failure to bring it to bear
on his analysis of impact as a sign of inner inconsistency of the latter's philosophy of nature
at a critical point. The inconsistency is most glaring with Descartes' rules 4 and 5. If one takes
the relativity of motion and rest seriously, the analysis of what happens when a small body
hits a larger one at rest ought to be symmetrical with what happens when a large body hits a
smaller one at rest, since the two cases are identical but for a shift in the frame of reference.
Instead, with Descartes the two cases come out quite differently (in rule 5, after rebound the
two bodies jointly move in the direction of the larger body with the same speed that it had
before impact).

Expedient shifting of frames of reference, then, so as to make each case identical to one
already treated is how Huygens went about the derivation of better rules. He, too, thought
that a boat would come in handy (Figure 15.1).[278]

"Every impact occurs on a boat coasting smoothly along a Dutch canal, its speed adjusted to the
experiment in question, and every impact is observed by two men, one on the boat and one on the
shore. Both observe the same event. Nothing more than adding and subtracting uniform velocities
is required to transpose one and the same event from one frame of reference to the other."

Figure 15.1: Huygens' 'boat' image of relativity of motion in impact

526   More difficult was to find proper axioms from which to derive the new rules. Huygens was already far advanced on his way of transferring to Galileo-style mathematical science what remained of his loyalty to Descartes if a conflict turned up. Within a year he gave up any attempt to derive his rules from the action of some 'force of collision'. Instead, he settled for principles that fitted better the state that Galilean thinking about motion had attained at the time. As a result, he came to treat the impact of bodies of equal size by grounding it in Galileo's principle of motion retained, as rectified by Torricelli (see below under (2)). Further, using an argument by Torricelli on the common center of gravity of two bodies connected over a pulley, Huygens went on to extend both it and his principle of rectilinear motion retained toward the axiomatic statement that "the center of gravity of bodies taken together continues always with a uniform motion in the same direction and is not disturbed by any impact of the bodies."[279] The more complicated case of impact between bodies of unequal size appeared to him to be reducible always to the case that the sizes of the two bodies are inversely as their speeds, which is when, upon rebound, each maintains its original speed. The axiom he in the end invoked for grounding that basic case was a further extension of Torricelli's principle; in a modern rendition:

> Imagine that the speeds before impact, inversely proportional to the bodies, were generated by falls through distances which, from Galileo's kinematics, are proportional to the squares of the speeds. In rebound, assume any possible pair of speeds whatever, other than the original ones, and it can be demonstrated that the common centre of gravity of the two bodies placed at the heights to which those speeds could raise them, would be higher than their common centre of gravity when they are in the positions from which they were imagined initially to fall. Unless the bodies rebound with their original speeds, a perpetual increase in motion is possible, and a perpetual increase in motion is absurd.[280]

At the point now reached, Huygens had rid himself of the last vestige of dynamic action to account for the collision of bodies. In impact, bodies are not brought to rest first and then have their motion restored due to a force of collision. Rather, their motion goes on indefinitely but changes direction only, and perfect hardness, not perfect elasticity, is their proper attribute.

   In the course of his investigation, as a corollary to his handling of the relativity of motion, Huygens found fault even with Descartes' starting point, the principle of conservation of quantity of motion. For the quantity 'size times velocity' may certainly change in impact; size times velocity *squared* is the quantity really to be preserved in impact, so he wrote without any foreboding yet of what the future held in store for a product which, to Huygens, remained otherwise devoid of meaning.

   In the treatise to come out of all this, Huygens took care to present his argument in such a way as to lure a Cartesian reader into nodding agreement as far into the treatise as pos- sible, so as to entice him to go the rest of the way as well. For the author himself the didactic device served as a marker that his departure from certain Cartesian tenets involved less than full rupture. Rupture of a kind it certainly was. To regard the matter as settled by means of local repair (the way Buridan had long ago repaired Aristotle's system locally by replacing the air's motive force with impetus; p. 84) clearly would not do. A philosophy of nature more emphatically than any earlier one presented to the world as indubitably secure and as fully consistent both with itself and with phenomena at large had now, in the light of Galilean mathematical science, proven to be both inconsistent and empirically wanting at one of its key points.

The depth of the impression the find made upon Huygens' mind at the time is still apparent in the comments on Descartes he penned in 1693 (p. 233). From the mid-1650s onward, kinetic corpuscularianism ceased to command Huygens' allegiance insofar as it was conceived as a system of natural philosophy. From the mid-1650s onward, he was to operate in a manner first figured out over the half-decade (1652-1656) spent on and off on the problem of impact – to use it heuristically, as a possible source of specific problems or of specific hypotheses, in those cases where sheer mathematical treatment appeared to him not to suffice on its own. This truly revolutionary move gave him the freedom to invoke Gassendi-like atoms in the void or some Descartes-like whirlpool mechanism, whichever conception suited him best at any given time. Still, however momentous such a transformation of the standard way of handling a philosophy of nature was (recall that Pascal still saw no alternative for outright rejection), it remains true that at no time throughout his life did Huygens abandon the broad conception of matter in motion as the only category suitable for arriving at some causal understanding of the phenomena of nature.

The first to put the problem of impact as arisen inside kinetic corpuscularianism on a mathematical-experimental footing, Huygens nonetheless left his 1656 treatise on the subject unpublished, except for a bald statement of results sent more than a decade later to the *Journal des sçavans* and to the *Philosophical Transactions* when he learned that in the meantime others, too, were successfully dealing with the problem. One of these men was John Wallis, a mathematician unconcerned with empirical science but for one foray into problems of motion, in the course of which he published rules for both perfectly elastic and perfectly nonelastic collisions. Another was Christopher Wren, a highly versatile practitioner who soon after publishing rules of impact abandoned the pursuit of nature-knowledge and turned fully to architecture. Some of Wren's rules were equivalent to Huygens', others supplemented them. Further, in 1673 Mariotte published a treatise on impact in which he attempted (as was his wont) to derive results attained earlier by a fellow Academician from some ill-defined, dynamic action. Meanwhile, in 1666 an obscure bachelor of arts at Cambridge University, Isaac Newton, was driven by his dual allegiance to mathematical exactness

528 and to a broadly kinetic-corpuscularian view of things to replace Descartes' treatment of impact with a mathematically consistent, empirically adequate analysis which he, too, left unpublished. The analysis ran along a path characteristically different from Huygens', in that Newton operated with a force of collision like the one so quickly abandoned by Huygens. Since Newton's pre-*Principia* concerns with *genera* of motion were tied together in ways not to be found with Huygens, and also underwent some drastic development of their own, I shall discuss them as one whole near the end of the present section.

Motion retained.   Falsified all the time in our everyday, commonsense reality, valid in mentally conceived, mathematically idealized reality, at least approximating validity in the imitated-idealized reality of, say, an ivory ball rolling over a marble floor, the radically novel, deeply counterintuitive idea that an object in uniform motion keeps moving unless impeded came into the world as true of *circular* motion. Not only the manner in which Galileo derived his principle of motion retained but also his enduring conviction that our world is finite made him stick to its strict validity for motion parallel to the horizon only, that is, for objects moving in circles, which motion he called 'natural' (p. 184). And when, on the Fourth Day of the *Discorsi*, he explained that the movement of a cannonball shot straight ahead is composed of horizontal, uniform motion retained and vertical motion uniformly accelerated, he made sure to caution that the horizontal component no more than approximates a straight line, with the approximation being permissible only in view of the short distances involved. Still, passages like these helped Galileo's disciples get rid of the inner contradictions of Galileo's mature doctrine of motion as expressed, in particular, in his conception of *both* uniform, horizontal motion *and* uniformly accelerated, vertical motion as 'natural'. The way out, for Torricelli especially, was to take the principle of motion retained as valid for only uniform motion in a straight line.

Precisely this variety of the principle appears to figure as the first and second 'laws of nature' in Descartes' *Principia philosophiae*. Descartes first adopted the principle from Beeckman, who applied it indiscriminately to rectilinear and to circular movements without regarding one case as privileged over the other in any way. But Descartes was unambiguous (at least at first sight) in pronouncing the principle as valid for only rectilinear motion:

> Each and every thing, insofar as it is simple and undivided, always remains, insofar as it can, in the same state, nor is it ever changed except by external causes. ... And therefore we must conclude that whatever moves, always moves insofar as it can.

> Each and every part of matter, regarded by itself, never tends to continue moving in any curved lines, but only along straight lines.[281]

Yet on second sight there was a snag, apparent from the innocuous expression 'regarded by itself'. For Descartes' world is filled up completely; no particle of matter can ever be regarded by itself; no particle of matter ever gets a chance actually to move straight ahead. At every instant of its tending to do so it encounters others possessed by the same tendency. Hosts of contiguous particles deflect one another all the time from a rectilinear path, thus yielding, in the reality of nature as Descartes conceived it, those unceasing collisions of his (and also the movements in closed curves we examine next).

As a result of all this, what we have in the 1610s to early 1650s are two distinct principles of uniform motion retained among mathematical scientists (circular albeit incidentally approximating rectilinearity with Galileo; rectilinear with Torricelli) and two more among kinetic corpuscularians (rectilinear as well as circular with Beeckman; rectilinear in principle yet actually constrained to become by and large circular with Descartes). To add further to the complexity, the third pioneer of kinetic corpuscularianism, Gassendi, who conceived of each atom as gifted with a capacity for motion of its own to carry it through the void, thus was in a position to adopt the principle, as indeed he sometimes did, the same way Torricelli had.

What, then, was done with this confusing legacy by thinkers reared not in just one of these modes of nature-knowledge, as with Torricelli, but in both, as, notably, with Huygens and with Newton? When, in his *Horologium oscillatorium* of 1673, Huygens needed axioms for an improved derivation of Galileo's laws of falling bodies, he adopted as a matter of course and without any indication of finding it problematic Galileo's principle as rectified by Torricelli. Indeed, in taking uniform, rectilinear motion or rest of the common center of gravity of bodies as foundational for his analysis of impact, he placed the principle of motion retained in a wider framework that was able to make it look less paradoxical. Indeed, his wholehearted espousal of its corollary, the principle of *relativity* of motion that both Galileo and Descartes had made so much of, was there to reinforce it further. Not only had Huygens directed the relativity of motion (i.e., the interchangeability of frames of reference) against Descartes' failure to employ it in his rules of impact, but in stray notes never adding up to a coherent treatise Huygens was in due time to think the relativity of motion through to the radical point of denying privileged status to any frame of reference anywhere in the universe whatsoever. The obvious thing for him to do, then, was to stick to Descartes' variety of the principle of motion retained while, unlike Descartes, taking it as actually governing actual movements in both the macro- and the micro-world. Or, to phrase the same point another way, whatever statements Huygens was to find shocking in Newton's *Principia* in 1687 (and be assured already that there were some), the latter's first law of motion was not among them. Curiously, one individual to find that first law shocking indeed was its very author, for reasons to be clarified once we have also discussed Huygens' analysis of the next *genus*, rotational motion.

530   *Rotational motion.*   Imagine a stone tied to a rope and whirled around; imagine the knot being severed and the stone flying off at a tangent. Galileo's interest in the phenomenon (not of much interest previously) had arisen from his urgent need to demonstrate that no such flying off is possible for beings or objects on an Earth in daily rotation. Descartes' principal concern with such a motion was rather its ubiquity in the micro-world. In the Cartesian universe particles are driven, through ongoing collision with adjacent particles, to move in closed curves which assemble in whirlpools of matter. Every particle, in thus being whirled around, is constrained thereby to move in a broadly circular path, yet the very constraint sets up a tendency to escape and recede in a straight line along the tangent.

This, then, is the dual legacy Huygens was facing in the fall of 1659 when, in his search to improve on Mersenne's and Riccioli's earlier efforts to determine the distance traversed by a body over the first second of its fall, he began to pursue in rigorous fashion the profound analogy between fall and rotational motion once he perceived the mathematical equivalence between them. The perceived equivalence enabled him to arrive at (in modern notation) the formula $F = mv^2/r$ for the 'centrifugal' tension set up in the cord. He went on to extend the entire setup to three dimensions by imagining a pendulum moving, not to and fro as ordinary pendula do, but conically (Figure 10.14; p. 344). Here is a summary of his achievement compared with both Galileo's and Descartes' (and also Mersenne's) in the realm of motion in constrained circles:

> The first attempts to deal with the problem were the verbal, qualitative approaches of Descartes and Galileo. Although Galileo provided a semiquantitative solution, it was Mersenne who attempted a numerical solution. Finally, Huygens provided the mathematical theory, quantifying and structuring the work of his predecessors. Certainly, Huygens' work went far beyond any of the three previous discussions, and however much he might have gleaned from his precursors in the way of general conceptualizations of the problem, he felt no indebtedness. ... Having associated gravity and centrifugal force, he proceeded very methodically from the two-dimensional geometric solution relating the two forces, to the three-dimensional description of physical reality in which both weight and centrifugal force act on a rotating object ... , to the ultimate choice of the conical pendulum, then to the mathematical analysis of that model culminating in the requirement that the pendulum maintain the same height regardless of the angle that it makes with the axis, and finally to the physical entity that embodied the foregoing theory.[282]

Huygens' entire treatment went to show that Galileo's very idea of uniformly horizontal (i.e., circular) motion being retained was not only, unlike its rectilinear variety, wrong but conceptually mistaken from the bottom up. Motion along a circular path is not really uniform at all – it involves ongoing change of direction, hence, acceleration. The question, however,

of how that acceleration comes about in either case was one that, in line with his refusal to
have truck with any 'force of collision', Huygens studiously avoided (pp. 526; 625).

Huygens and young Newton compared: Impact, motion retained, and rotation once more.

Less than a decade later, the young Isaac Newton, in dealing with the same interrelated *genera* of motion as Huygens had, arrived independently at quite similar results without sharing Huygens' need to avoid forces. Huygens had sought to resolve the inherent tension between the Galilean and the Cartesian ways of dealing with motion between 1652 and 1659 without as yet publishing a word about his researches. Between 1665 and 1668 Newton, an obscure bachelor of arts at Cambridge University who operated entirely on his own, faced the same tension. A comparison between them reveals such similarities as may be expected from two in many ways like-minded men of genius facing the same intellectual legacy. It also reveals some striking differences. Key to these is the even greater genius of the younger man, which is expressed in the amount of interconnectedness he perceived between these *genera* of motion from the outset, in the evolving conceptual apparatus with which he attacked them, and, above all, in the even wider vistas he came to perceive as he went along.

By late 1663 Newton took leave forever of the Aristotelianism fed him as an undergraduate and discovered on his own initiative all the innovative work that sixty years of Scientific Revolution had meanwhile produced. In one research domain after another, his assiduous reading notes gave way within months to the shaping of coherent views of his own. This happened first when he became acquainted with several mathematicians' ongoing pursuit of the analysis of infinites and single-handedly turned this into the calculus (1665/6; p. 355). So it went likewise with his critical reading notes on the subject of motion.

Huygens had made his acquaintance with the subject through *Principia philosophiae* alone, at the age of fifteen or sixteen. Two decades later, the twenty-one-year-old Newton, with the barriers between modes of nature-knowledge breaking down all around him, read that book and a plethora of rivals both inside and outside natural philosophy. That is, he consumed *Principia philosophiae* as one item in a varied diet containing work by Gassendi, Galileo (the *Dialogo*), Boyle, and many others. The experience enabled him from the outset to treat the categories of any philosophy of nature, not as settled dogma, but as a source of significant problems and of possibly fruitful hypotheses. In other words, a revolutionary approach to natural philosophy for which Huygens had to make a personal breakthrough looked like a given to Newton right away. In 1664 he composed a range of brief essays which blended notes taken from his reading with highly critical questions raised thereby. He penned these essays in the 'Waste Book' – his own name for a volume still full of blank paper that his late stepfather had meant to fill with edifying quotations. In the 'Waste Book', then, we already find the dual stance taken with respect to Descartes' philosophy of nature which Huygens was fully to embrace for the rest of his life but Newton for only two more decades

– whereas it is true that natural phenomena are to be explained through matter in motion alone, the question of how exactly has to be reconsidered all over again.

For motion, Newton's independent reconsideration began on 20 January 1665, under the heading 'Of Reflections', that is, collisions, which Newton followed Descartes in regarding as the central *genus* of motion. Ostensibly true to Descartes, too, he adopted the principle of uniform, rectilinear motion retained. Unlike either Descartes or Huygens, however, he went on to adopt that principle as one axiom serving as a starting point for his subsequent effort to redo Descartes' analysis of impact. He did not waste a word on the empirical and logical weaknesses of that analysis; he just without further ado went his own way. But he needed more axioms, and one that he drew up involved the concept of force. This was a profoundly problematic concept (p. 621, section 'The force knot'). The invocation of 'forces' was deeply suspect to anyone committed to a conception of the world as composed of nothing but particles of matter in motion, where force is not exerted upon anything from the outside but appears only as moving particles pressing one another out of the way. That is how Descartes had come to speak of 'the force of a body's motion', measured by speed and by size, the product of which he had called its 'quantity of motion'. But Newton introduced another, unheard-of idea of force on which to ground his analysis of impact:

> The unique position of the *Waste Book* in the history of dynamics derives from its recognition that a dynamics built on the principle of inertia demands a concept of force different from the prevailing one. He realised that the 'force of a body's motion' can be seen from another perspective. In impact, the force of one body's motion functions in relation to the second body as the 'external cause' mentioned in axioms 1 and 2 as the sole means that can alter its state of motion or rest. *Newton made the perspective of the second body his primary one.* Thus he was the first man fully to comprehend the implication of inertia for dynamics, that the prime necessity of an operative dynamics was a conceptual unit to measure the 'external cause' of changes of motion. 'Of Reflections' set out to convert the available idea of force to that use.[283]

Not, to be sure, that the road to the *Principia* now lay open. Rather than perceiving at once the incompatibility of the principle of uniform motion retained with the notion of a constant force required to sustain it, Newton sought to reconcile his new conception of force with the standard one. For two more decades his new idea of force remained confined to collisions in a world of moving particles. In that world, then, the changes of motion occasioned by impact could now be handled using the tendency of particles to retain their motion, plus this new insight that "tis knowne by the light of nature y^t equall forces shall effect an equall change in equall body".[284] Generation and destruction of motion, as well as gains and losses in quantity of motion, could thus be seen to be governed by corresponding amounts of force; as a result, "if ye body *bace* acquire ye motion q by ye force d & ye body f ye motion p by

ye force *g*, yn *d:q::g:p*."^{285} From there Newton arrived at the principle that two bodies, whether
moving independently or in rebound,

> have equal motions in relation to their common centre of gravity, and that the centre of grav-
> ity of two bodies in uniform motion is also in uniform motion (or at rest), both when the two
> bodies are in the same plane and when they are in different planes.^{286}

Unbeknownst to Newton, Huygens had already arrived at that principle in his own effort
to come to grips with impact. One subordinate difference is that Huygens used it to derive
a set of specific rules that Newton, the general result safely in hand, found it unnecessary to
bother with in anything like so systematic a fashion. A more significant difference, fraught
with promise, rested in Newton's new conception of force, however much confined as yet
to a quantitative but otherwise meaningless measure for the case of impact alone. A final
difference is that, whereas Huygens came to investigate rotational motion at a later time in a
different context, Newton went ahead almost at once to analyze it with the help of his results
on impact.

Here his initial insight was that, once a body has covered half of its circular path, the
direction of its motion has been reversed in a way not of course identical with, yet usefully
analogous to, reversal due to impact. For example, if one visualizes a ball rolling inside a
hollow cylinder, it is easy to regard its circular path as arising from the constraint that results
from an infinite succession of pressures of the ball upon successive pieces of the cylindrical shell successively pressing back. The dynamic analysis that followed upon the original
insight became entangled in conceptual trouble that is easy for us to recognize yet hard fully
to capture for Newton, who in one effort to deal with it hit upon the idea of invoking the
conical pendulum (cf. Figure 10.14, p. 344) as a means to balance the opposed forces of gravity and of centrifugal 'endeavor'. In spite of the difficulties, he arrived in the end at the same
proportionalities for centrifugal motion that Huygens had derived six years earlier, by means
of an argument based, unlike Huygens', on collision forces.

Nor did the similarities with Huygens' derivation, which had been focused on the motions themselves rather than on such questionable forces as might be assumed to produce
them, end here. Newton, too, saw at once that he was now in possession of the means to give
needed quantitative precision to Galileo's argument that a revolving Earth does not throw
off its passengers. In the course of the effort he combined his proportions for centrifugal
force with the setup embodied in the conical pendulum, so as to find, as one by-product of
his investigations, an exact value for the very magnitude that had sent Huygens on his quest
originally: the constant of acceleration for falling bodies.

The possibility of another by-product occurred to him as well – this time one that never
occurred to Huygens. The purported and, in all likelihood, historically genuine occasion for

534   its occurring to young Newton is very well known. The sudden revelation came to him in the garden of his mother's manor, to which he had retired while Cambridge University was closed due to an outbreak of the plague. He realized that whatever causes an apple to fall from a tree may not only mathematically but also dynamically be equivalent to what causes the Moon to orbit the Earth or the planets to orbit the Sun. The way, however, in which Newton, still steeped in the mode of thought that went with Descartes' whirlpool world, conceptualized their orbiting was as an endeavor to recede from the center of the orbit held in check by something, whatever it is, that presses back the way the imaginary cylindrical shell presses back upon the ball moving inside it.

What to think of the nature of whatever it is that presses back? Newton left the question open – "he was a young man; he had time to think of [the idea] as matters of great moment require."[287] But to a man in possession not only of a measure for centrifugal force but also of a setup balancing it with what Newton had begun to call 'the force of gravity', it was apparent that the entity that did the pressing lent itself to quantitative determination. This he proceeded to try out next. In this sequel paper, right after refining his attainments in the earlier paper on centrifugal motion, he found the Moon's endeavor to recede from its orbit to be 4,000 times as small as the force of gravity at the surface of the Earth. Further, Kepler's third law supplied the means to compare the various planets' endeavors to recede. True, Kepler's rule that the squares of the periods any two planets need for one revolution are as the cubes of their mean distances from the Sun applies to elliptic, not circular, orbits. Even so Newton felt free to feed the third law into the proportionalities for centrifugal motion that he had just derived. His comparison showed the planets' endeavors to diminish with the squares of the planets' respective distances from the Sun (p. 306). Whereas *conceptually* Newton still had to travel a long road, notably from Cartesian endeavors due to the incessant crowding of particles upon one another to Newtonian forces of attraction, *quantitatively* the result is identical to what Newton was to publish two decades later in his *Principia*. As yet, not even the quantitative result could satisfy him, since the outcome for the Earth–Moon comparison should have been 3,600:1 rather than 4,000:1 – Newton had adopted too low a value for the circumference of the Earth. Until he was moved to redo the correlation and feed a more accurate value into it, he needed to keep thinking of some additional whirlpool mechanism to make up for the difference. All this is how, half a century later, Newton could recall the episode thus:

> [In 1666] I began to think of gravity extending to ye orb of the Moon & (having found out how to estimate the force with w^{ch} [a] globe revolving within a sphere presses the surface of the sphere) from Keplers rule of the periodical times of the Planets being in sesquialterate proportion of their distances from the center of their Orbs, I deduced that the forces wch keep the Planets in their Orbs must [be] reciprocally as the squares of their distances from the centers

about wch they revolve: & thereby compared the force requisite to keep the Moon in her Orb with the force of gravity at the surface of the earth, & found them answer pretty nearly.[288]

In this 'pretty nearly' resided one reason why Newton left the issue pending in 1666. His examination of various *genera* of motion had not yet come to an end, however. At about the same time he sought to sum up his attainment so far in a paper in which he extended his treatment of impact to the case of perfectly elastic bodies in rotation. More importantly, the paper displays an inclination to renege upon the principle of uniformly rectilinear motion retained. The inclination was triggered by Newton's beginning to perceive conceptual friction between its apparent corollary, the principle of relativity of motion, and his growing conviction that rotation was to be associated with absolute motion. Around 1668 the tension thus threatening to disrupt the apparent coherence of Newton's analyses of impact, of rotational motion, and of uniform rectilinear motion retained came to a head in the last paper Newton devoted to generic motion prior to Edmond Halley's visit in 1684. What brought this rupture about was a full-scale condemnation of the fundamentals of Descartes' philosophy of nature.

> Even in the privacy of his study, he worked himself into a passionate fury against the philosopher who, scarcely five years earlier, had introduced him to a new world of thought. The gravamen of the charge was atheism. By his separation of body and spirit, Descartes denied the dependence of the material world on God. The ultimate cause of atheism, Newton asserted, is "this notion of bodies having, as it were, a complete, absolute and independent reality in themselves ..." To refute that notion he had to attack the Cartesian equation of matter and extension, and in consequence Descartes's relativistic conceptions of place and motion. Lest his disagreement be overlooked, he described the latter as "absurd ... confused and incongruous with reason."[289]

Relativity of motion, Newton now argued, made it impossible ever to decide whether anything at all is in motion or at rest (the very conclusion, that is, to be drawn by Huygens without apparent qualms). Where was Jupiter a year ago? With the particles of its vortex meanwhile dispersed, and the fixed stars in the background similarly floating in a sea of particles, not even God could tell where the planet had been, since whatever might be called its 'place' had vanished in the meantime. In his urge to preserve absolute motion, Newton not only replaced for good his broad adherence to the Cartesian variety of kinetic corpuscularianism with broad adherence to its Gassendist counterpart, with atoms moving through the void rather than whirlpools of matter serving as extension itself. Still, what absolute motion Newton did claim here was as yet for rotation alone. As a consequence, uniform rectilinear motion lost its defining characteristics; force took over, and the tendency of a body to perse-

536   vere in its state of uniform rectilinear motion was ascribed anew (as Newton had done in his very first reading notes) to an essentially pre-Galilean force inherently present in a body to make it persevere in its motion. When, in 1684, Newton was led to rethink problems of motion from the bottom up, he had to fight his way toward adoption of the principle of motion retained all over again (p. 653).

It is now time to look back. Huygens and Newton, guided by an allegiance to particles in motion as basic to our thinking about the world and to Galileo-style analysis as indispensable for how to do the thinking, had set out to reconcile the tension between the conceptions of motion resulting from their dual, mode-overstepping allegiance.

Huygens' success in achieving such a reconciliation was due to many factors: his sure-footedness and mastery of mathematical techniques; his steadfast renunciation of any other concept of force than the one due to a body's motion; his way of treating issues one by one, as they presented themselves; his carefully consistent handling of what occasional links he found between them. The resulting, harmonious resolution of the original tension was in no way hampered by a religiosity which, for all his apparent conformity to his family's Calvinism, turned ever more coolly reason-bound or even stoic, impersonal, and, as such, inconsequential for his thought about nature's properties.

Newton's success in achieving such a reconciliation was likewise due to many factors. What he and Huygens shared was this sure-footedness and mastery of mathematical techniques. But that is where the similarity ends. Newton embraced concepts of force from the outset. He attacked the various *genera* of motion in their full, intimate connectedness at a fundamental level, the ready perception of which helped him in addition to invade territory as yet untrodden. But in achieving what he did achieve he found as he went along that, with the original tension resolved, tensions of a wholly novel kind manifested themselves instead, only to be exacerbated by the demands his fiery religiosity placed upon him. The new tensions stemmed from the conceptual difficulties inherent in a variety of ideas of force – as generated by motion, as sustaining motion, as generating change of quantity of motion, or as generating accelerated motion – concepts as yet distinguished by Newton intuitively at best and jointly suffering as yet from mathematical intractability in all but incidental cases. By 1668 Newton was far indeed from sorting them out, let alone from fully perceiving the capacity for mathematization inherent in one of them; in the process he even felt compelled to renege on one of Galileo's basic insights. He had brilliantly given mathematical expression to key components of some hoped-for, full-fledged science of motion, on a par, at the very least, with results that Huygens was reluctant to publish as well. Even so, Newton had got stuck, and he knew it. Three successive events helped him get unstuck: a letter from Hooke in 1679, a visit from Halley in 1684, and his own growing revolt against the limitations of a conception of the world as made up exclusively of particles in motion (p. 640).

In the meantime one more *genus* of motion was being subjected to mathematical treatment against a background of particle-thinking in pre-*Principia* days: to-and-fro motion.

To-and-fro motion.   In Alexandrian, abstract-mathematical science, vibration was   537
acknowledged from Pythagoras onward as generative of the consonant intervals but had
at once been abstracted away from the string that does the vibrating. In Athenian natural
philosophy, to-and-fro motion came to the fore in the wave analogy by means of which the
stoics sought to explain the propagation of sound and in the incessant vibrations Lucretius
had assigned to atoms when not freely moving through the void but temporarily cohering
in larger bodies. In the kinetic-corpuscularian systems of the 17th century, where coherence was usually explained by other means, periodic phenomena had no obvious place and
were therefore ignored except on occasion by way of analogy. Instead, the very twisting of
Alexandria into Alexandria-plus owed a good deal to to-and-fro motion as executed by the
pendulum, the discovery of the near isochrony of which served Galileo to confirm his turn
toward a realist-mathematical science. Galileo's work on the pendulum at once evoked the
interest of Mersenne. His growing desire to construct a science of musical sound as the
centerpiece to an understanding of the harmony of the world caused him to examine vibrational motion, of the string primarily but, in view of the apparent analogy between the
two, of the pendulum as well. Mersenne's enduring adherence to the conceptual structure
embodied in 'mixed mathematics' left him powerless to pursue the analogy any further, or
even to establish any more properties of the pendulum than those discovered by Galileo.
Once again Huygens was the man who knew how to do it better.

Huygens' discovery of the functional dependencies on which centrifugal motion rests
had multiple consequences: his invention of the pendulum clock (p. 331), his discovery of
the isochrony of the cycloid (p. 345), and his idea of using the conical pendulum to balance free fall and rotational motion (just discussed). But he also investigated properties of
the pendulum itself. So much of Huygens' mathematical thinking, albeit precalculus, was
drenched in considerations of the infinitesimal that it was easy for him to perceive that a
conical pendulum, with the circle it traces out made ever smaller, in the limiting case turns
into an ordinary to-and-fro pendulum. From the properties of the conical pendulum just
established by him, joined to Galileo's finding that the period of an ordinary pendulum varies as the square of its length, Huygens could derive as a matter of course what comes down
to the modern formula for the period of an ordinary pendulum.

More challenging was the ensuing problem of how to extend his result to the pendulum
as it manifests itself in reality, as a metal bar suspended from a cord. Here Huygens reached
the outer limits of what can still be attained by means of the conceptual apparatus at his
disposal. Two components were essential to his analysis. One was virtuoso usage made once
more of the principle that a system of bodies may be regarded as concentrated at their common center of gravity. The other was his idea of imagining the bar to dissolve into an array
of adjacent particles, the velocity of which he could then determine by means of his rules
for impact (Figure 15.2). One practical result yielded by his usage of background kinetic-

538   corpuscularian thinking in a frame of realist-mathematical science was that his pendulum clock could be adjusted by placing a small weight, not at its center of oscillation, but either above (for acceleration) or below (for retardation).[290]

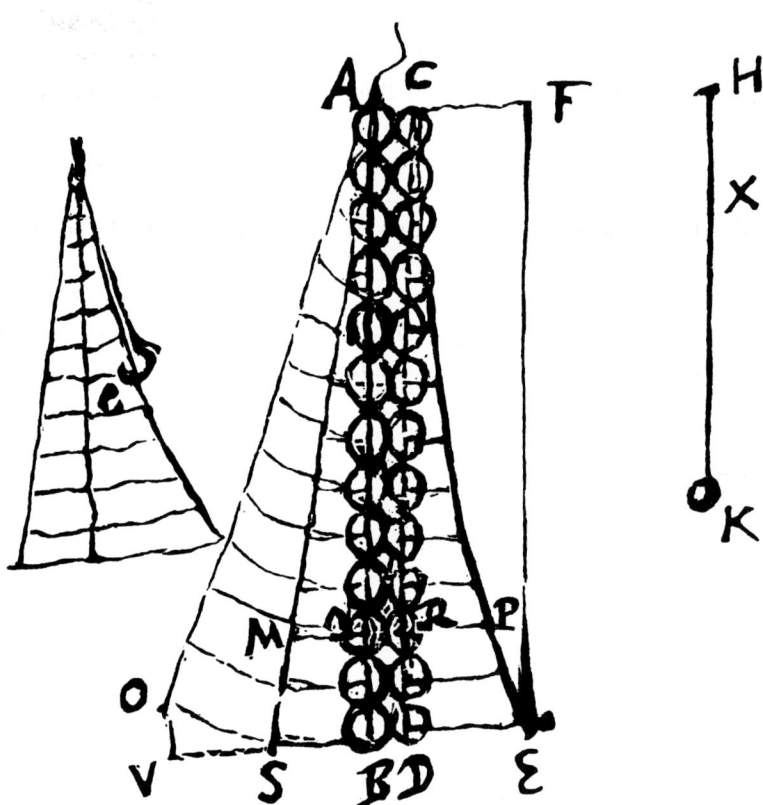

"The line of balls AB represents a solid bar that has swung from the position AO. The bar is imagined to disintegrate into its component parts when it is in the vertical position, and the line of balls CD represents the separate parts. Each part is then imagined to be deflected so as to rise straight up. The straight line AS charts the height from which each part of the bar descended. The curved line CE charts the heights to which the parts can rise when they are separated from each other. Since the center of gravity of the parts after they are separated from each other cannot rise higher than the center of gravity of the bar at its original height, the triangular area ABS must equal the curvilinear area CDE."

Figure 15.2: Huygens' analysis of the compound pendulum

Besides the pendulum, two other manifestations of to-and-fro motion received attention between Galileo and the mature Newton, to wit, waves and the vibrating string. Unlike with the pendulum, both had to await the fundamental transformation of thought about motion attained in Newton's *Principia* for even the beginnings of their resolution (p. 664).

Conclusion: A unified treatment of motion?   Not only for to-and-fro motion but for the treatment of motion as such was the *Principia* to serve as the watershed. Even if all the work discussed in the present section had been made public at the time, it would still not have come close to a coherent whole. Huygens' various treatises lacked a unifying principle (to which both his working habits and his idea of what science is about were opposed); Newton had, for the time being, become stuck in a vain search for it. The attainment of a unified, mathematical science of motion, which was not even an ideal for Huygens, was an ideal that had failed from young Newton's point of view. Not until a fresh start convinced him otherwise did the mathematical treatment of motion remain confined to *genera* that were indeed interrelated but basically stood on their own.

Anomalous refraction revisited

The sine rule of refraction discovered by Snel and Descartes had by the 1640s turned into the new centerpiece of optical thought. Within twenty years of its discovery, the rule appeared to be less than universally applicable – it breaks down with prisms and also with Iceland spar (p. 300). Newton recognized the challenge in the former case (discovered by himself), and Huygens in the latter (discovered by Bartholinus). Unaware once again of what the other was doing, both Huygens and Newton first sought to meet the challenge through purely mathematical expedients which, in resting upon the proof Descartes had given of the sine rule in his *Dioptrique*, left the basic structure of the standing, perspectivist account of light and its phenomena intact. But why did they approach their respective cases of anomalous refraction in so deliberately geometric, hence noncausal, a manner? The reason was dissatisfaction with the explanation of light Descartes had given in *Principia philosophiae*. It rested upon the instantaneous propagation throughout the solar vortex of pressures arising from the incessant collisions of its constituent particles. Descartes had kept the explanation out of his proof of the sine law as much as he possibly could, quite in keeping with the as yet unbreached barrier between mathematical science and natural philosophy, yet exacerbated in this particular case by the incompatibility of his proof (which hinged upon a finite speed of light) with his explanation (which required it to be infinite). Thus, with light, too, there was a profound tension between the mathematical and the natural-philosophical approaches which, once again, Huygens and Newton were the two men in the next generation to sense and to seek to resolve.

540   The pattern is familiar – they did so in ways similar in basic approach yet different in just about everything else. Each man quickly perceived his original, purely geometric ray tracing to be, at best, a refined restatement of the problem rather than a solution to it, and also to fail to address the tension between it and his own idea of the nature of light. But here already the characteristic difference comes to the fore between the man who sought or forged his tools one by one, as the need arose, and the man who engaged from the outset in the boldest of grabs for the whole of nature. Newton's commitment to a broad conception of the nature of light – closely linked in his view with the structure of matter as such – flowed at once from his first involvement in the literature about it, as recorded in his 'Waste Book'. In contrast, once Huygens had moved beyond his youthful adulation for *Principia philosophiae* he did not embrace any other such conception instead until he was served one on a platter at roughly the same time he began to feel a need for some plausible account of the nature of light.

So much for the broad pattern. I shall now examine how each man managed to create out of the optical quandary into which anomalous refraction had plunged him the beginnings of what has come to be known since as the science of physical optics.

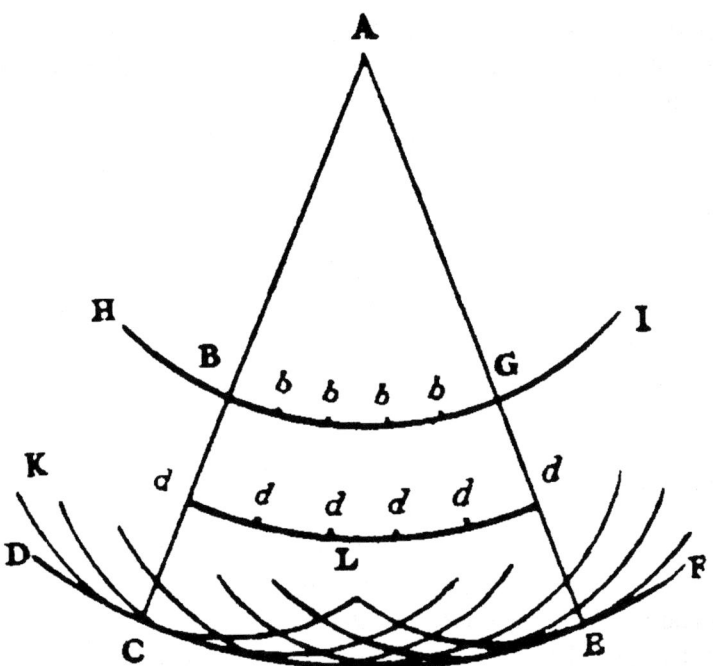

Figure 15.3: Huygens' principle of wave propagation

Huygens and refraction in Iceland spar.   Prior to 1672, Huygens' sole interest in op-   541
tical issues was focused upon the tracing of rays in the configurations of lenses in telescopes
(p. 328). At that time he regarded efforts to explain the nature of light as good only "for
satisfying that curiosity of the mind which loves to know the reason of everything".[291] In
1672 he designed a plan to turn those contents of his folder 'Dioptrica' that might still be
of use into chapters of a treatise fit for publication. Authors of optical treatises were in the
habit of prefixing them with some noncommittal account of the nature of light, and that is
how Huygens – unconvinced by Descartes' account, yet sure that a viable one had to involve
particles in motion – came to adopt the explanation Father Ignace Pardies had told him
about in Paris. Pardies thought that light consists of waves propagated through a particulate medium disturbed by the violent motion of a luminous object (e.g., a candle). Key to
this explanation was that the light rays are perpendicular to the waves. Within five years
of his original, geometry-governed effort of 1672 to come to grips with 'strange refraction'
Huygens became aware that all his ray tracing and fiddling with angles of refraction, however much of an improvement perhaps over Bartholinus', had done nothing to reconcile the
anomalous deviation undergone by a ray that falls perpendicularly upon Iceland spar with
the core of Pardies' wave account of light.

His interest had been rekindled when he learned of Ole Rømer's empirical confirmation
that the speed of light is as finite as Huygens had assumed it to be all along. Within days of
the return of his interest a specific mechanism of corpuscles in wavelike motion occurred
to him. He perceived it to be capable of explaining what light is and how the sine rule and
other basic properties follow from it. But he also saw in it a suitable means to account for
how strange refraction comes about. Huygens' key idea was that one need only assume in
addition to Pardies' basic scheme that each and every particle of the medium successively
reached by the wave front begins at once to serve as the center of another wave. A host of
wavelets is produced such that all taken together form a wave front perpendicular to the
direction of propagation (Figure 15.3).

In 1679, in successive sessions of the Académie, and in 1690, in his *Traité de la lumière*
(Treatise on light), Huygens set forth in all required detail how his principle could explain
(1) the rectilinear propagation of light, (2) reflection, (3) ordinary refraction, and (4) strange
refraction in Iceland spar. For no. 1 his rules of impact came in handy. No. 2 was easy. With
no. 3 the sine rule was explained in a straightforward manner rather than just stated (as with
Barrow) or explained ambiguously (as with Descartes) or too broadly (as with Pardies). In
no. 4, the high point of the treatise of course, strange refraction was reduced to the formation of ellipsoidal waves arising from differences in the velocity of propagation (Figure 15.4).

Not only did the finding of a viable and mathematizable specific mechanism tie moving
corpuscles and mathematical derivation much more closely together than ever before in
matters optical, but the entire setup also proved capable of holding its own in the face of two

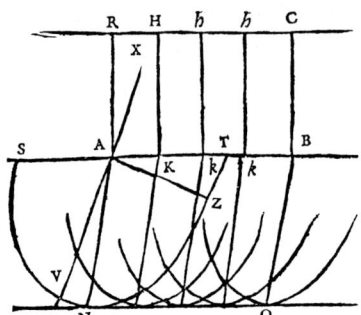

Figure 15.4: Huygens' derivation of strange refraction in Iceland spar

critical experiments. At Académie sessions in 1679 Rømer, who was married to Bartholinus' daughter, objected that Huygens' way of dealing with what happens in Iceland spar was no improvement over his father-in-law's. In response, Huygens did what he had not done before. He measured its angles of refraction with enhanced precision and found these to contradict Bartholinus' analysis after all. Only now did he manage to crown his own effort by cleaving a crystal in a special way and finding that his own theory had successfully predicted the angle of refraction at the edges thus created.

Still, not all was sunshine. What more than a century later was to prove the crucially fertile element in Huygens' theory – his principle of pooled wavelets – had arisen amid a conceptual structure unable to benefit much from it. Huygens was well aware that colors remained outside the explanatory range of his theory. He did very expressly conceive of his waves, not as periodic, but as irregular, which is why some historians insist on calling his theory a 'pulse' theory of light. He was not really able to explain why his irregular wavelets actually require pooling to become visible (the early 19th-century discovery of extinction through interference being what, by hindsight, the theory needed most to make it truly viable). Nor was he able effectively to counter Newton's main objection, which arose from the more profound examination to which Newton had in the mid-1680s subjected the propagation of motion through waves: If light is indeed a wavelike phenomenon, it should bend behind an obstacle the way sound does; in other words, Huygens' account is incompatible with the most elementary property of light, its rectilinear propagation.

For all its deficiencies, whether real or apparent, *Traité de la lumière* not only demolished what remnants of the perspectivist tradition in optics were still around but also brought together in one new, coherent picture modes of analysis that, ever since the attacks upon that tradition had begun, had been pursued in isolation from each other. Once again a geometrization of corpuscular motion had been brought about, by the dual means of taking

such motion out of its natural-philosophical frame and devising for it instead a specific   543
mechanism that could be handled mathematically. Remarkably, what pooled wavelets thus
could do for one of the two cases of anomalous refraction, no conceivable mechanism of
corpuscles in motion appeared capable of doing for the other. Or so at least it seemed to its
exacting discoverer, to whom we return now.

Newton and color sifting.   In a lecture for (probably absent) students given some four
years after his original discovery that white light is really a blend of colors which a prism sifts
from each other and projects as an elongated color spectrum upon a screen at some distance
away, Newton phrased his stance on Descartes' sine rule of refraction thus:

> what he asserted about every ray without distinction, as if the refraction of all of
> them were quite alike, we assert only about the individual kinds of them, in positing
> that the sines of refraction of equally refrangible rays are as the sines of incidence.[292]

If, then, the sine rule holds only for rays of the same kind, the question arises of how to
account for the actual variation in refrangibility from one color to another. Each color is
identified by a specific index of refraction (ratio of sine of incidence to sine of refraction),
and the question is how these are related, and why.

As Huygens did in about the same year, 1670, Newton began by assuming that a purely
geometric rule for his own case of anomalous refraction would be all he needed. And just
as with Huygens, the dispersion rule he came up with was an ingenious extension of Descartes' proof for the sine rule, which was drenched in perspectivist presuppositions about
the nature of light. Also like Huygens, he had grave doubts about Descartes' 'instantaneous
propagation of pressures' explanation of light. What, then, could justify the geometric rule
in the absence of a satisfactory, underlying mechanism for it? Like Huygens, soon enough
Newton, too, realized the need for an account that was not only mathematically sound but
also in harmony with his own views on the nature of light. For Huygens, in view of his
decades-long indifference to the nature of light, it was not a self-evident move to cling to
his recent adoption of Pardies' wave explanation to the point of feeling compelled to make
it encompass so exotic a phenomenon as strange refraction in Iceland spar. Unlike Huygens,
Newton had already in his 'Waste Book' laid down a preference for an account of light he was
never to abandon – to wit, for light conceived as a stream of emitted particles. The outcome,
however, was quite the same: In sticking to his account of light, he still had a problem with
anomalous refraction. He did not record at what time after his lectures or in what particular
manner he became aware that another approach was needed to give him a mathematical
dispersion rule grounded in some specific mechanism of particles in motion grounded in
its turn in his emission account. Nor did he record when he hit upon that mechanism. We

544   only know it from its first appearance, which was in his *Principia* of 1687. By that time Newton had ceased conceiving of such mechanisms exclusively in terms of particles in motion
and had also become very circumspect about making pronouncements (at least in public)
on the ultimate constituents of our world (p. 673). This is how he came to couch his new
derivation-cum-explanation of the sine rule in a mechanism involving an attractive force
and to present it as a general proposition associated, in an added *scholium* (gloss), with the
problem in optics – refraction of light of any kind – for which it might serve as the solution.
Here is an account of the mechanism and its unhappy fate:

> Newton's derivation of the sine law . . . comes down to an analysis of the motion of tiny par-
> ticles accelerated by a force acting perpendicularly to the boundary between two media of
> different density. . . . Newton could show that [that force] produces a fixed acceleration inde-
> pendent of the angle of incidence.
>
> In this understanding of refraction the degree of refrangibility can depend either upon
> the size of this force or on the velocity of the particles. As the force is determined by the me-
> dium, the different refrangibility of colours can only be caused by the different velocities of
> the particles in rays of different colours. This is exactly what the 'Cartesian' dispersion law of
> the [1670] *Optical Lectures* would imply, and this cannot have suited Newton very well. The
> immutability of colours was the cornerstone of his optics and, dynamically speaking, velocity
> is an unlikely agent of immutability. By the time Newton completed *Opticks*, he had indeed
> dropped his 'Cartesian' dispersion law. This may well have been brought about by the advance-
> ment of his understanding of the physics of refraction. . . . In *Opticks*, Newton provided an
> alternative dispersion law with doubtful empirical evidence and whose 'proper foundations' he
> never elaborated – not even in private.[293]

So in his second attempt to come to terms with his own case of anomalous refraction, Newton, in a manner broadly similar to Huygens', hit upon a specific mechanism of particles in
motion (albeit due now to a force acting upon them) that could make a mathematical derivation of the full sine rule possible. Only, unlike with Huygens and his *Traité de la lumière*,
the mechanism in question failed to satisfy its own author. On giving up his second attempt
at such a dispersion law, Newton had to settle for what he definitely felt to be less. Right
from the start around 1666, his chain of optical discoveries (color sifting, various periodic
phenomena) had involved elements of mathematical science, of experimental science, and
of underlying mechanisms derived from his 'emission' conception of light. How precisely
to strike a balance between these three elements had as yet remained undecided. Now that
the solid establishment of a satisfactory dispersion law had to be given up as an illusion
sometime between 1687 and 1694 (which is when, in completing *Opticks*, Newton had to
settle for that badly founded 'alternative dispersion law' of his) he felt compelled to fix the

balance for good – henceforth, he was to emphasize the experimental aspect of his optics.   545
The decision, thus induced by a technical difficulty, was to have momentous consequences.
It was to enable posterity – notably in Britain – to celebrate Newton's achievement as the
ultimate triumph of Baconian experimentalism. And that is how Newton's name could incongruously be made to serve as a figurehead for that specifically British movement toward
a science-based technology that by the late 18th century was to make large-scale industrialization possible for the first time in history.

NOTES ON LITERATURE USED

Motion. The argument here unfolded owes most to Richard S. Westfall, *Force in Newton's Physics. The Science of Dynamics in the Seventeenth Century*. London: MacDonald, 1971 (*SRHI*, section 2.4.5.). Westfall usefully summarized the book in the two final chapters of his *The Construction of Modern Science*. Cambridge UP, 1971, and in several lemmata that he contributed to Wilbur Applebaum (ed.), *Encyclopedia of the Scientific Revolution from Copernicus to Newton*. New York/London: Garland, 2000.

- **Descartes' collision rules**. Peter McLaughlin, 'Force, Determination and Impact'. In: S. Gaukroger, J. Schuster, J. Sutton (eds.), *Descartes' Natural Philosophy*. London: Routledge, 2000; pp. 81-112; also Daniel E. Garber, chapter 8 ('Motion and Its Laws: Part 2, the Law of Impact') of his *Descartes' Metaphysical Physics*. Chicago: University of Chicago Press, 1992, pp. 231-262.

- **Huygens**. IMPACT: Alan Gabbey, 'Huygens and Mechanics'. In: H.J.M. Bos *et al.* (eds), *Studies on Christiaan Huygens*. Lisse: Swets & Zeitlinger, 1980; pp. 166-199, and Christopher B. Burch, 'Christiaan Huygens: The Development of a Scientific Research Program in the Foundations of Mechanics' (unpublished doctoral dissertation, University of Pittsburgh, 1981). RELATIVITY OF MOTION: E.J. Dijksterhuis, *The Mechanization of the World Picture*. Oxford UP, 1961, sections 147–151. ROTATIONAL AND PENDULAR MOTION: Joella G. Yoder, *Unrolling Time. Christiaan Huygens and the Mathematization of Nature*. Cambridge UP, 1988. RELIGION: R. Hooykaas, *Experientia ac Ratione: Huygens tussen Descartes en Newton*. Leiden: Museum Boerhaave, 1979; section II (pp. 24-36).

- **Newton**. In addition to Westfall's work just specified I have used pertinent passages in his *Never at Rest. A Biography of Isaac Newton*. Cambridge UP, 1980. In *Force...* as also in *Never...* the revolutionary reconceptualization that led Newton to reject both aether conceptions and views of motion consonant with them is dated in or around 1679. Betty Jo Teeter Dobbs, in *The Janus Faces of Genius. The Role of Alchemy in Newton's Thought*. Cambridge UP, 1991, proposed shifting this date to 1684, on the very eve, that is, of the years of discovery that went into the *Principia*. I go along with part of this, yet not with her much-contested redating, from c. 1668 to 1684, of Newton's stridently anti-Cartesian paper 'De gravitatione & aequipondio fluidorum'.

No running account exists of 17th-century, pre-*Principia* conceptions of TO-AND-FRO MOTION, with the partial exception of C.A. Truesdell's introductions to two volumes of Leonhard Euler's *Opera Omnia* (2–13 and 2–11/2). Hobbes' frequent analogies with stretched strings are discussed in Jamie C. Kassler, *Inner Music. Hobbes, Hooke, and North on Internal Character*. London: Athlone, 1995; so are Mersenne's comparisons between the pendulum and the vibrating string in ch. 6 'Mechanics, Music, and Harmony' of Peter Dear, *Mersenne and the Learning of the Schools*. Ithaca (NY): Cornell UP, 1988.

Anomalous refraction. The argument here unfolded owes almost everything to Fokko Jan Dijksterhuis' article 'Once Snell Breaks Down. From Geometrical to Physical Optics in the Seventeenth Century'. *Annals of Science* 61, 2004, pp. 165-185, which goes back to his deliberately comparative *Lenses and Waves. Christiaan Huygens and the Mathematical Science of Optics in the Seventeenth Century*. Dordrecht: Kluwer, 2004. (His discussion of Newton's effort to treat color sifting mathematically rests upon pioneering work by Alan E. Shapiro in 'Newton's achromatic dispersion law'. *Archive for History of Exact Sciences* 21, 1979, pp. 91-128).

XVI

THE FIFTH TRANSFORMATION: THE BACONIAN BREW

Other than with the attempted infusion of *mathematical*-experimental science with corpuscles in motion, which was almost exclusively an effort of just two individuals, the effort undertaken at about the same time to infuse *fact-finding* experimental science with them took place against a recognizably national background. Fact-finding experimental science was practiced in Britain as well as on the Continent (the Society of Jesus, the Paris Académie, a variety of individual practitioners). In partial contrast, the 'Baconian Brew' mixed in the 1660s out of Bacon-inspired experimentation, particles in motion, and a broadly 'spiritual' dose of active principles was an exclusively British affair. It came about by way of cross-cultural transplantation. Compared to earlier transplantations (Greek→Arabic, Arabic→Latin, and Greek→Latin), where truly distinct civilizations were involved, it was a decidedly minor one. Still, as so often when a set of ideas developed in one place is suddenly dropped in another, it produced genuine novelty. The Baconian Brew served to make the experimental search for natural phenomena and causal accounts of them in terms of corpuscular movements reinforce one another by turning the former less haphazard and the latter less arbitrary. *In the process, the nature of these causal accounts changed.* Not only were matter and motion taken as (in Boyle's expression) 'catholick principles' to be handled as the occasion required rather than as rigid dogma of the strictly Cartesian or strictly Gassendist variety. But also, in the Baconian Brew kinetic-corpuscularian explanations and the component of natural magic that remained part and parcel of much 17th-century experimentation began to interact in such a way as to make 'spirit' more material and matter more 'spiritual'.

This, then, is what the mixing of the Baconian Brew in the 1660s came down to. In the following account of how it came to be mixed, I shall examine successively how active principles found a place in several contemporary, British philosophies of nature, how prominent that place was, and what achievement the Brew proved capable of in the hands of the three men to embody it to the full, Robert Boyle, Robert Hooke, and young Isaac Newton.

Kinetic corpuscularianism crosses the Channel

The event to set things in motion was the wave of Royalist emigration that was one outcome of the civil war over the king's rights and privileges as upheld by Charles I and his followers. Royalist figures central to the adoption of the (by the mid-1640s) still almost exclusively Parisian doctrine of kinetic corpuscularianism were William Cavendish, Duke of Newcastle, his brother, Charles, his wife, Margaret, and his employee as a tutor, Thomas Hobbes.

Of these four, Hobbes had dabbled in natural philosophy before. In 1631, when forty-three years old, he wrote down, and kept to himself, an outline of his views on nature, which were broadly similar to the 'gross matter and material spirit' account of Francis Bacon. A typical thinker in the 'systematize what you have securely deduced from indubitable first principles of nature' style, Hobbes soon thereafter took the one first-principle he was never henceforth to swerve from in his various philosophies of nature from reading Galileo's *Dialogo* in 1634 and perhaps from personal acquaintance he may have had occasion to make of its author in the same year. 'Motion', even more than matter in its varied guises, was from then on to form the be-all-and-end-all of Hobbes' successive conceptions of nature. The next installment took shape in instructive conversation with his patron, who unlike most patrons had a genuine interest in the pursuit of nature-knowledge. In 1644 William Cavendish led the Royalist army to defeat against the Parliamentary forces at Marston Moor and soon thereafter fled the country and settled in Paris with his entire household. How prominent the French thinkers met by the Newcastle Circle were, how lively an exchange ensued, and how lasting an impression the episode made upon the members of the quickly expanding circle appear from how one of them, the polymath William Petty, in a dedicatory letter of 1674 was to look back upon events:

> Your Grace doth not onely love the search of Truth, but did Encourage Me 30 years ago as to Enquiries of this kind. For about that time in Paris, Mersennus, Gassendy, Mr. Hobs, Monsieur Des Cartes, Monsieur Roberval, Monsieur Mydorge and other famous men, all frequenting and caressed by your Grace and your memorable brother Sir Charles Cavendish, did countenance and influence my studies as well by their Conversation as their Publick Lectures and Writings.[294]

The Newcastle Circle had arrived in Paris by 1644 and began to disband and drift back to England in 1649. It was during those five years that the kinetic-corpuscularian doctrine, hitherto prepared in private writing and by word of mouth, began to take on public form. Descartes, with whom Hobbes had engaged in fruitless polemics during the 1630s, now sat down (in the Netherlands, but with occasional visits to Paris and, in any case, in ceaseless contact with Mersenne) to commit his natural philosophy previously meant for 'Le monde' to publishable writing in *Principia philosophiae*. Even before this book came out in 1644 Hobbes had availed himself of a copy. It pleased him much less than the views of his by now close friend Gassendi, whose first public exposition of the atomist doctrine appeared in 1649. Other members of the circle had other preferences; Charles Cavendish, for example, who was a mathematician of independent competence, valued both the Cartesian and the Gassendist versions of kinetic corpuscularianism, thereby prefiguring a specifically English, broad-minded inclination to benefit from what the varieties of kinetic corpuscularianism held in common rather than to harp on their differences.

Out of half a decade of exposure and collaboration came almost nothing on the French side (open-minded Mersenne alone appears to have felt that the British presence might further his own inquiries) but quite a good deal on English soil once the members of the Newcastle Circle had returned.

Kenelm Digby, the first member of the circle to feel inspired by his stay in Paris to produce an all-encompassing system of natural philosophy by means of corpuscles in motion, was too much of a maverick (p. 398) to affect subsequent events much. More significantly, Thomas Hobbes used the occasion of his stay to alter his indubitably certain views on nature one more time and, upon his return to Britain, make them public, first in abbreviated form in his book on politics that brought him undying fame, *Leviathan* (1651), and four years later in a book entitled *De corpore* ('On Body'). In 1653 and 1656 Margaret Cavendish joined him, with poetry and prose devoted to expounding an atomist version of kinetic corpuscularianism of her own making. At about the same time, Hobbes' disciple and Margaret Cavendish's correspondent, the Helmontian physician Walter Charleton, undertook an effort to counter the threat of atheism that, in his and many others' view, was implied in the thoughts and works of these two authors.

Hobbes and Cavendish used what they had picked up from Descartes and Gassendi to take a radically materialist stance that, in its apparent denial of the immortality of the soul, could easily be labeled 'atheist'. In contrast, Charleton saw that he could make Gassendi's brand of atomism, about which he had in all likelihood been informed by Hobbes and Cavendish themselves, serve the same end toward which Gassendi had directed it: the promulgation of a Christianized kinetic corpuscularianism. He accomplished this in a 1654 book truthfully entitled *Physiologia Epicuro-Gassendo-Charltoniana: A Fabrick of Science Natural upon the Hypothesis of Atoms, Founded by Epicurus, Repaired by Petrus Gassendus, Augmented by Walter Charleton*.

This widely read book accomplished in Britain what Gassendi's and, much more powerfully so, Descartes' works accomplished on the Continent: the speedy acceptance of a general view of nature as made up of particles in motion. Beyond intellectual transfer it also inspired a significant enrichment. Kinetic corpuscularianism, soon after its promulgation in Charleton's book, came to be sprinkled through experimental science – that is, through experimental science of the fact-finding variety in which it used to be cultivated on British soil. In that very act it also caused certain potentialities inherent all along in Gassendi's variety of kinetic corpuscularianism to come into the open, in that in Britain (quite unlike on the Continent) subtle matter began to look less and less easily distinguishable from the very entity kinetic corpuscularianism seemed so radically opposed to – that is, from 'spirit'.

Spirit and active principles in kinetic corpuscularianism

In natural philosophy there had always been a tension between its claim of comprehensiveness, due to which the totality of empirical phenomena was held to be explicable in accordance with its first principles, and the readily apparent circumstance that some phenomena fitted those first principles nicely (e.g., the propagation of sound in stoicism) but others (e.g., projectile motion in Aristotle's account) only in very forced ways. So it was with kinetic corpuscularianism. Native to it was a crusade against 'occult forces'. The fight against them certainly yielded accounts that looked readily convincing (e.g., sympathetic resonance). But other 'occult' phenomena did not lend themselves so readily to being incorporated in a picture made up of sheer matter in motion. How to deal with them? Beeckman was the most radical of the three pioneers. Well acquainted with the natural magic literature, yet removed from it mentally by an astonishingly large distance, he neither engaged in extended polemics nor felt any need for compromise, be it open or hidden. Whether the causal mechanisms he tirelessly sought concerned magnetism or the cohesion of matter or whatever other topic more customarily accounted for in terms of forces or powers or attractions, Beeckman never made a concession. Either he moved quietly from one possible mechanism of matter in motion toward another that satisfied him better or, if enduringly dissatisfied, he gave up such a search altogether rather than ever overstepping the boundaries he had taken care to draw around his private philosophy of nature. This was not the case for Descartes, who

> in the discussion of many phenomena, of fire, for example, . . . summoned the subtle matter onto the stage, like a *deus ex machina*, to play the role of an active principle. Of course, Descartes always implied that its 'agitation' resulted from mechanical necessity. In fact, he never analysed in detail how it was caused . . . but rather dragged it in by brute force whenever a phenomenon appeared to reveal a level of activity not present before. In such passages, his subtle matter can only be seen as a mechanist rendition of an Hermeticist active principle.[295]

Such covert overstepping the self-set boundaries of kinetic corpuscularianism assumed larger proportions with Gassendi. An eclectic, rather than a systematic, thinker, Gassendi felt free at times to explain, for example, cold by the presence of 'frigorific' particles. He also took inspiration from thinkers like Paracelsus' disciple Petrus Severinus to introduce vitalist notions through the backdoor of his overtly material accounts. Most significantly of all, Gassendi 'baptized' ancient atomism in such a way as to make it possible to assign considerable activity to particles of matter. When the world was made, God had endowed every single atom created by Him with a motion of its own. The Cartesian transfer of a certain quantity of motion from one inherently passive particle to another through impact was not, therefore, the sole means available to account for phenomena, and, hence, Gassendism offered better opportunities for explaining activity in nature than any other regular variety of kinetic corpuscularianism, Thomas Hobbes' and Margaret Cavendish's included. Not that Gassendi, himself, much availed himself of such opportunities, which in his own case definitely resided at the outskirts of his system of nature. Yet in the wake of his British apostle Walter Charleton, whose allegiance had significantly shifted from van Helmont to Gassendi, British philosophers of nature were driven more and more to exploit the explanatory leeway thus gained to the utmost, and even beyond.

Take the philosophy of nature William Petty expounded in 1674 to the Royal Society. Taking his cues from Charleton as well as from Gilbert, Petty boldly declared atoms to be tiny magnets:

> all Atoms have, like a Magnet, two motions, one of Gravity whereby it tendeth towards the Center of the Earth, and the other Verticity, by which it tendeth towards the Earths-Poles, and whereby Magnets joyn to each other by their Opposite Poles.[296]

With Descartes, magnetic attraction and repulsion were phenomena to be rendered in terms of specific motions of specifically shaped particles (p. 229). Here, the other way round, they are active principles by means of which certain motions in the world ought to be explained. In similar fashion, Matthew Hale posited an

> entity I call Vis or Virtue activa, superadded to Matter, and giving immediately those motions to it, that are specifically appropriate to that Vis or Virtue activa, and without which, Matter would be stupid, dull, unactive, and always at rest in it self unless accidentally moved ab extrinseco [from the outside].[297]

Nor did Hale hesitate to extend such active virtue to life itself. Or rather, the other way round, even the very lowest among those hierarchically ordered virtues of his tended to make matter lifelike. The lengthy works of Petty, Hale, and numerous other British thinkers who in the 1660s through 1680s felt called upon to explain nature in broad brush strokes are filled with examples like these.

At the same time concepts of 'spirit' came to fulfill a dual role in British natural philosophy. On the one hand, spirit was called in by those who, like Henry More in the first place, were upset by Cartesian dualism and its relentless, effective subordination of *res cogitans* to *res extensa*, all the while expressing their opposition in terms of matter in motion (p. 399). On the other hand, 'spirit' or 'aethereal spirit' or similar locutions became catchphrases, on a par with 'virtue activa' and similar active principles, to account for that which it was felt sheer matter in motion could not sufficiently account for.

What was going on here, really? The point is that three distinct concepts, each originating

554   in an ancient philosophy of nature, were thoroughly blended in Britain in the second half of the 17th century. One of these concepts was the *anima mundi*, or world-soul, of Platonic provenance – immaterial and, as souls go, all-pervasive. Another was stoic *pneuma* – all-pervasive, continuous, material in its airy/fiery nature yet noncorpuscular and with many immaterial overtones. The third concept, ultimately going back to ancient atomism yet given a novel twist in kinetic corpuscularianism, was subtle matter, particulate and wholly material and no whit less pervasive than the other two. In British philosophies of nature like those of Petty and Hale, then, the three notions were effectively made to merge. The fragmentation of discrete particles, the continuous tension of the pneuma, and the animated activity of the world soul all came together in one novel, versatile concept. It was variously denoted by any of the terms previously employed for one or another of its constituent parts, be it 'subtle matter' or 'aether' or 'spirit' or 'aethereal spirit'. Distinct concepts created separately in ancient Greece and kept apart in Islamic civilization and in Renaissance Europe came together in Great Britain from the 1660s onward.

The Baconian Brew

The thrust of the inchoate and underarticulated philosophy of nature thus emerging was no less speculative than any of its three constituents. Even so it had a wider potential. In the hands of men who, unlike Petty or Hale, cared for more than just a speculative conception of how the world is broadly constituted, it might be turned into a powerful resource for fertilizing empirical research. Three men regarded the merger, to which they also contributed, as an opportunity to put a regular check upon the haphazard ways in which fact-finding experimentation most often proceeded and to give it some of the coherence they perceived it to be badly in need of (p. 516). Robert Boyle exploited that opportunity in his chemical theory and practice; Robert Hooke, in his conception of sympathy through the harmonic vibration of the particles of natural objects; and the young Isaac Newton, in his 'Hypothesis of Light'.

Boyle's chemical theory and practice.   Robert Boyle's experiments with the air pump mark him as a practitioner of great ingenuity and meticulous care and a thinker of great caution in arriving at any definite conclusions (p. 459). But there was still more to him. In various ways he sought to link his unceasing experimentation to a form of adherence to kinetic corpuscularianism not so much derived dogmatically from first principles as adapted hypothetically to what he took his experimental setups to require. An activating 'spirit' was not altogether absent from his theorizing, yet Boyle's overriding concern was to fertilize the finding of facts with a healthy dose of corpuscles in motion. Near the end of his life he succinctly summed up how he had found experimentation and natural philosophy to reinforce one another:

Of the Use of Experiments to Speculative Philosophy

1. To supply and rectify our senses
2. To suggest Hypotheses both more general and particular
3. To illustrate Explications
4. To determine doubts
5. To confirm truths
6. To confute errors
7. To hint luciferous [i.e., 'enlightening', a very Baconian term] inquiries and experiments and contribute to the making them skilfully

Of the Use of Speculative Philosophy to Experiments

1. To devise philosophical experiments which depend only, and mainly, upon Principles, notions, and Ratiocinations
2. To devise instruments both mechanical and others to make inquiries and tryalls with
3. To vary and otherwise to improve known experiments
4. To help a man to make estimates of what is physically possible and practicable
5. To foretell the events of untried experiments
6. To ascertain the limits and causes of doubtful and seemingly indefinite experiments
7. To determine accurately the circumstances, and proportions, as weight, measures and duration etc. of experiments[298]

Boyle's preferred domain for trying out these fourteen distinct ways of connecting experiment and speculative philosophy was chemistry. Very early on in his career he already spoke of "experimental philosophy with its key, chemistry".[299] After all, "the object of [a naturalist's] studies ought to be matter,"[300] and chemistry, in a view original with him, provided the best way for an adherent of kinetic corpuscularianism to watch matter in action and manipulate it.

While it would be an overstatement, it would not be a gross caricature to describe Boyle as a van Helmont stripped of his life-giving seeds (p. 257) and equipped instead with corpuscles in motion. Much of the sustained critique to which Boyle, in his *Sceptical Chymist* of 1661, subjected Aristotelian elements and Paracelsian principles was just a more systematic and cogent rendition of points made previously by van Helmont. Similarly, Boyle adopted and carried to extremes van Helmont's insight that fire, far from being a neutral agent in chemical reactions, often has an independent role in what changes appear to be wrought.

Unlike van Helmont, though, Boyle had no time for a view of nature as wholly lifelike. Instead, he took specific chemical reactions to be explicable in terms of the following basic scheme. Matter, Boyle believed, is ultimately made up of solid, virtually indivisible corpuscles or *minima naturalia*, which may assemble in primary concretions, or *prima mixta* (first mixtures), of the most varied kinds. If in chemical reactions the primary concretions persist,

556 as (so Boyle acknowledged) they may well do, nothing has taken place there but either a re-arrangement or a further union of parts. Yet more often than not in chemical reactions, the primary concretions do fall apart into their constituents, those original *minima naturalia*, which then recombine into quite different primary concretions with quite different proper-ties. Notwithstanding the relative stability of the *prima mixta*, matter is utterly malleable in the end, and transmutation rather than rearrangement or addition is the rule in chemistry:

> since bodies, having but one common matter, can be differenced but by accidents, which seem all of them to be the effects and consequents of local motion, I see not why it should be absurd to think, that (at least among inanimate bodies) by the intervention of some very small addi-tion or subtraction of matter, (which yet in most cases will scarce be needed,) and of an orderly series of alterations, disposing by degrees the matter to be transmuted, almost of any thing, may at length be made any thing

Boyle proceeded at once to illustrate his radical assertion by means of an appeal to ordinary, macro-world experience:

> as, though out of a wedge of gold one cannot immediately make a ring, yet by either wiredraw-ing that wedge by degrees, or by melting it, and casting a little of it into a mould, that thing may be easily effected.

He followed this with an experimentally established example of how such transmutations can actually be effected, rounded off with a consideration meant to exclude the possibility that not transmutation but only effective addition has taken place:

> And so though water cannot immediately be transmuted into oil and much less into fire; yet if you nourish certain plants with water alone, (as I have done) till they have assimilated a great quantity of water into their own nature, you may, by committing this transmuted water ... to distillation in convenient glasses, obtain, besides other things, a true oil, and a black combus-tible coal, (and consequently fire;) both of which may be so copious, as to leave no just cause to suspect, that they could be any thing near afforded by any little spirituous parts, which may be presumed to have been communicated by that part of the vegetable, that is first put into the water, to that far greater part of it, which was committed to distillation.[301]

So much for the experimentally produced, theoretically explained, orderly transmutation of water into oil and fire by means of distillation. In this extended passage we have in a nutshell both Boyle's personal brand of kinetic corpuscularianism and the kind of link he was in the habit of establishing with his experimental practice. In a dynamic interplay between his hypothesis on the configurations of particles and his experimentation, Boyle took it upon himself to "demonstrat[e] that any substance eligible for the title of element (because obtained by resolution and not further resolvable) could be transmuted or corrupted to some other apparently elementary body or bodies."[302] Even though, as with gold, the unmistakably elementary behavior of a given substance compelled him to attribute its reactions to the sheer rearrangement of its *prima mixta*, Boyle never gave up the search for ways and means to make those concretions dissolve into ultimate corpuscles so as to recombine into something else.

Such recombining, then, is what chemistry ultimately was about in Boyle's dissenting view, and fire was the best tool to carry it out:

> When bits of carbon dropped upon hot saltpeter cause a flame, Boyle views the carbon as an agent (almost a catalyst) which enhances the activity of the fire and permits a separation impossible for fire alone. Only the matter of the saltpeter is conceived to enter the reaction materially; the carbon, like the fire, is viewed as an instrumental agent. . . . [In short:] To Boyle the chemist is a micro-mechanic – an artificer who fabricates in the microscopic realm as the mechanic does in the macroscopic.[303]

In all this Boyle strayed from the mainstream in contemporary chemical research, which was increasingly committed in practice even if not always in theory to the establishment of those very elements or enduring substances of which Boyle denied the possible existence (p. 469). And there is another point to make. Just like Huygens, Boyle remained true over his lifetime to the basic categories of kinetic corpuscularianism all the while handling it, not as a closed system deduced from first principles, but rather as a fertile source of hypotheses. In systematically connecting assumed motions of particles in a specific manner to empirical research of a well-circumscribed kind, he demonstrated along a route quite different from the mathematical highway followed by one on the Continent and by one in Cambridge that it was possible to make core components of natural philosophy serve the dynamics of scientific advance.

Hooke's spiritual/corpuscularian congruities experimentally confirmed.

Robert Hooke was a man of clever ideas and ingenious devices rather than of sustained theories expounded in lengthy treatises or of solid machinery tenaciously built and rebuilt till the last nail has been driven in. On occasion, as with his personal version of a compound microscope or in *Micrographia* or in his rule for elastic force, he managed to make certain products of his head and hands formally public. Yet on the whole the sudden flash of insight was much more his forte than the single-minded pursuit by which cleverness is sometimes transformed into lasting achievement.

558   In no lesser measure than Boyle, whom he assisted in setting up his air pump and with whom he collaborated in the experiments undertaken with that machine (p. 459), did Hooke believe in solidifying experimental practice by means of an account of the world in terms of motions of corpuscles, and vice versa. Our five senses, so he adroitly wrote, suffice only for the perception and examination of a very limited number of phenomena among all those the world contains; hence, what we need for examining the remainder is a sixth sense. In experiment, we now have such a sixth sense. True, not even it can help us perceive directly "what the inward texture and Constitution of [bodies] is, and what the internal Motions, Powers and Energies are".[304] Yet experiment is indispensable for us "from Sensibles to argue the Similitude of the nature of Causes that are wholly insensible".[305]

Such an appeal to what Newton was soon to call "the Analogy of Nature"[306] had been heard before – it belonged to the stock-in-trade of the pioneers of kinetic corpuscularianism. To them, analogies with phenomena in the macro-world served as a check – the sole empirical one, in addition to the consistency and deductive security of their first-principle reasoning (p. 379). To Hooke, equipped with the fact-finding experimentalist's lack of trust in *a priori* deductions, his own appeal to the 'Similitude of the nature of Causes' acquired greater significance. He assured his audiences that experimentally achieved imitations of corpuscular mechanisms could serve as "Standards or Touchstones, by which I try all my former Notions, whether they hold out in weight, and measure, and touch, &c."[307] He even went so far as to take the former to *prove* the latter, as in his defense of a certain aether theory he held for some time:

> though the supposition even of the Aether, may seem to be a Chimera and groundless; yet had I now time, I could by many very sensible and undeniable experiments, prove the existence and reality thereof.[308]

Somehow, the time required to keep the promise seemed to elude him forever. Nonetheless, even if the epistemic power Hooke ascribed to his experiments was far overdone, the subtlety and intricacy of the corpuscular mechanisms he imagined without pause fruitfully joined his extraordinary experimental ingenuity in creating a research program of remarkable scope and interest. The best example is Hooke's conception of the vibratory motion of all particles in nature. On the one hand, he used it to explain light, gravity, cohesion, the mixing or nonmixing of fluids, and so on. On the other hand, it greatly stimulated his experimental researches into pendula, strings, sympathetic resonance, the power of music, and many more manifestations of what we now call simple harmonic motion.

All bodies whatsoever, Hooke argued in close adherence to one passage in Lucretius' atomist poem, are made up of particles in incessant vibration. If the vibrations are in congruity (i.e., in harmony) with one another, the bodies in question cohere, or mix well; if incongruous, or dissonant, they shun one another. The macro-world model for congruous action in the micro-world, then, is sympathetic resonance:

> those [particles] that are of a like bigness, and figure, and matter, will hold, or dance together, and those which are of a differing kind will be thrust or shov'd out from between them; for particles that are all similar, will, like so many equal musical strings equally stretcht, vibrate together in a kind of Harmony or unison; whereas others that are dissimilar, upon what account soever, unless the disproportion be otherwise counter-ballanc'd, will, like so many strings out of tune to those unisons, though they have the same agitating pulse, yet make quite differing kinds of vibrations and repercussions, so that . . . they cross and jar against each other, and consequently cannot agree together.[309]

So the three properties of 'bigness, figure, matter' jointly determine the congruity or incongruity of corpuscular vibrations. Hooke regarded these as the micro-world counterparts to the thickness, the tension, and the length of a string in the macro-world. Mersenne had ascertained these three properties jointly to determine pitch, and Hooke made a point of confirming this conclusion by experiments of his own. Many acoustic experiments that Hooke undertook on behalf of the Royal Society, however scattered and inconsequential they might seem, derive an underlying coherence and rationale from his micro-world model. Examples are the apparent equality between the sag displayed by a heavy string and the length of a pendulum with the same period, the observation of the ways flour moves within a vibrating glass, or the attempted production of consonant sound by means of teethed brass wheels. Hooke's experimental and theoretical work on pendula in general, as well as his frequent comparisons between sound and light, were inspired likewise by the unique significance he attached to vibrational motion.

This significance was enhanced further by Hooke's conviction that, besides cohesion and mixing, light and gravity, too, may be explained on the basis of his micro-world model. While the job he assigned to several aethers differs somewhat in his different writings (sometimes the aether itself vibrates, sometimes it only transmits vibrations), in all cases the vibration of particles is what does the trick. For instance, the compounded vibrating motion of the Earth, in being passed on to the universal aether, makes the latter vibrate "every way in Orbem [throughout the universe], from and towards the Center, in Lines radiating from the same".[310] This produces the phenomena of gravity (although Hooke remained conspicuously silent about its uniform rate of acceleration). Similarly, light is a very quick and short vibrative motion, and color the weakening and confusion thereof (p. 681).

To Hooke the supposition of such tiny vibrational motions was more than just a speculation. The omnipresence of one pertinent macroworld phenomenon, heat, goes so far as to prove it:

Now that the parts of all bodies, though never so solid, do yet vibrate, I think we need go no further for proof, then that all bodies have some degrees of heat in them, and that there has not been yet found any thing perfectly cold: Nor can I believe indeed that there is any such thing in Nature, as a body whose particles are at rest, or lazy and unactive in the great Theatre of the World, it being quite contrary to the grand Oeconomy of the Universe.[311]

The closing remark, with its invocation of the activity of matter, leads us to the ultimate causes Hooke held to be operative in the world. Here the spiritual and the material appear inextricably blended. On the one hand, he did not scruple to define his notion of 'congruity' as "a Tenaceous and Attractive power" or even as an "attractive virtue".[312] On the other hand, he reduced the entire range of phenomena that fell under the concept of congruity to a specific kind of motion of particles of matter: their unceasing vibrations. Yet this seemingly orthodox reduction to matter in motion was saturated with ambiguity in its turn. Matter and motion are the two fundamental principles of nature, to be sure. The former, so Hooke wrote without blushing, is the "Female or Mother Principle . . . [which is] without Life. . . . [It is] a Power in it self wholly unactive, until it be, as it were, impregnated by the second Principle", that is, by motion, which, Hooke went on to grant, may well be identified with "Spiritus".[313]

The ambiguity extends beyond sheer terminology – it is clearly deliberate. No simple matter/spirit dichotomy, Hooke believed, can do justice to the unending variety of phenomena the world presents us with; only a conception of the world as a gradual progression from "body without any form" up to "the highest form of a bruite Animals Soul" can do that. Hence, there is a continuous, upward range of natural processes which our inquiries must cover: "Fluidity, Orbiculation, Fixation, Angulization, or Crystallization, Germination, or Ebullition, Vegetation, Plantanimation, Animation, Sensation, Imagination".[314]

Here, as so often in his work, Hooke harks back beyond Descartes and Gassendi to certain well-known Hermeticist or at least magic-tinged conceptions; in this case, the continuity between spirit and matter. However, in blending these, as his forebears had not, with his genuine adherence to kinetic corpuscularianism and with his lifelong concern to give processes like 'crystallization' and 'vegetation' a much more concrete, experimentally reinforceable meaning, Hooke produced genuine novelty – the kind of novelty embodied in the Baconian Brew. And that is how fellows of the Royal Society, on attending its 9 December 1675 session, could not, at least if they were familiar with the effective making of that Brew by Hooke and Boyle from the early 1660s onward, be utterly surprised by the contents of a paper by another author read to them that day under the title 'An Hypothesis Explaining the Properties of Light'.

Newton's 'Hypothesis of Light'.   The paper, unpublished during Newton's lifetime, sums up eleven years of assiduous thought about the constitution of nature. It was the crowning achievement of the Baconian Brew – the pinnacle of what could be accomplished using its governing principles.

In the paper Newton provided the closest, most solid link yet attained between the corpuscularian hypothesis and outcomes of experimental practice. His central explanatory principle was an all-pervasive aether, composed of swiftly vibrating particles of subtle matter, with its density varying such that "it stands at greater degree of rarity in [the] pores [of Naturall bodyes] then in the free aethereall Spaces, & at so much a greater degree of rarity as the pores of the body are Smaller."[315] He further supposed that light is neither itself that aether nor its vibrational motion but something else that impinges upon these (namely, in Newton's personal opinion here deliberately left open, corpuscles emitted by lucid bodies; p. 683). These starting points suffice to account for all known phenomena of light. These included not only ordinary reflection and refraction but also such Newtonian finds (or Newtonian extensions of Hooke's finds) as colors with their different refrangibilities, partial refraction and total reflection, Newton's rings, and so on. For example, a ray of light moving through a region of unequally dense aether will be acted upon such that the ray "receivs a continuall impulse or ply from that side to recede towards the rarer [aether], & so is accelerated if it move that way, or retarded if the contrary." Hence, the ray follows a path that is curved in accordance with the local density gradient, and this is precisely what, in Hooke's account in *Micrographia*, can be observed "in water, whose lower parts were made gradually more salt & so more dense then the upper."[316]

Other, more involved cases follow, but the pattern has been set: An intricate, microworld mechanism of particles in motion is invoked to account in a picturable manner for a carefully observed, experiment-induced phenomenon in the macro-world. Newton further sought to render his basic supposition of a vibrating aether of varying density plausible by means of a sustained analogy with the air. The "aethereall Medium [has] much the same constitution with air, but far rarer, subtiler & more strongly Elastic."[317] It also vibrates, only much more swiftly and more minutely. It enters into the pores of solid substances as air does and stands similarly rarer in small pores, which it enters with less ease, than in wider ones. Another, very powerful argument in favor of the existence of an aether is the experimentally established fact that, whether a pendulum is made to swing in the open air or in an evacuated receiver, it will dampen out in almost equal time. Hence, the air is virtually irrelevant to the resistance offered to the pendulum's swings, which therefore must come from the other stuff that pervades every pore of the pendulum – the aether.

Newton's aether had even more attributes than those he invoked to explain the properties of light. "Perhaps the whole frame of Nature may be nothing but aether condensed by a fermental principle," he wrote, but then he struck out the last six words as unfit for public consumption and replaced them with

[... may be nothing but] various Contextures of some certaine aethereall Spirits or vapours condens'd as it were by precipitation, much after the manner that vapours are condensed into water or exhalations into grosser Substances, though not so easily condensible; and after condensation wrought into various formes, at first by the immediate hand of the Creator, and ever since by the power of Nature. . . . Thus perhaps may all things be originated from aether.[318]

He went on to outline how such ongoing processes of condensation and rarefaction of the aether might help explain phenomena like electrical attraction, gravity, capillary action, sensory perception, and animal motion. Gravity, for instance, might be explained thus:

the vast body of the Earth, wch may be every where to the very center in perpetuall working, may continually condense so much of this Spirit as to cause it from above to descend with great celerity for a supply. In wch descent it may beare downe with it the bodyes it pervades with force proportionall to the superficies of all their parts it acts upon. . . . For nature is a perpetuall circulatory worker.[319]

Newton further addressed the problem of why certain fluids do not mix well, or how it is that, for example, "water & Oyle pervades Wood & Stone, wch Quicksilver does not; and Quicksilver, Mettalls, wch water & Oyle doe not." This was the very type of phenomenon for the benefit of which Hooke had devised his principles of congruity and incongruity. Newton adopted these principles under another name, invoking "some secret principle of unsociableness" to account for what, despite sufficient subtlety of material composition, nonetheless prevented mutual penetration.[320] And just as with Hooke, Newton shrouded himself in ambiguity over whether he believed there was a mechanism of corpuscles in one or another kind of motion still underlying so secret a principle.

At that time, only a fly on the wall of Newton's Cambridge chambers could have known how to resolve the ambiguity apparent in the 'Hypothesis of Light'. Not only had Newton, like Hooke, been driven to the outer limits of what was still compatible with the principles of kinetic corpuscularianism even in the spirit-infested guise it had taken in the Baconian Brew; unknown to anyone else Newton's alchemical theory and practice, begun in about 1669, had already sent him beyond those limits. As the deleted phrase "the whole frame of Nature may be nothing but aether condensed by a fermental principle" suggests, and as Newton's private paper 'On the Vegetation of Metals' abundantly makes clear, by 1669 Newton was already persuaded that "mechanicall coalitions or seperations of particles" were not enough to explain activity in nature. Rather, "wee must have recourse to som further cause".[321] It was not until a decade and a half later, in 1684/5, that Newton, further inspired by his ongoing work in alchemy, was to settle his mind for good upon what that further cause was, deciding that not even the subtlest of aethers in whatever conceivable motion could help him out. It was then that the developments outlined in this and in the previous chapter – on the advance made possible by the blending of a hypothetically handled doctrine of kinetic corpuscularianism with realist-mathematical idealization and fact-finding experimentation – began to come together in the Newtonian synthesis.

Reception of kinetic corpuscularianism in England. My account has been pieced together from chs. 4–8 of R.H. Kargon, *Atomism In England From Hariot to Newton*. Oxford: Clarendon Press, 1966.

'Active principles' and their role in British science and natural philosophy of the second half of the 17th century have been explored by John Henry in his 'Occult Qualities and the Experimental Philosophy: Active Principles in Pre-Newtonian Matter Theory'. *History of Science* 24, 1986, pp. 335-381. E.A. Burtt's observations on the role of 'spirit' in the Scientific Revolution are in chapters 5 and 6 of his *The Metaphysical Foundations of Modern Physical Science*. London: Routledge & Kegan Paul, 1924 (revised edition: 1932); *SRHI*, 2.3.4.

Boyle. My treatment leans heavily on T.S. Kuhn's masterly article 'Robert Boyle and Structural Chemistry in the Seventeenth Century'. *Isis* 43, 1952, pp. 12-36, amalgamated with what A.G. Debus, in the sections on van Helmont and Boyle of his *The Chemical Philosophy. Paracelsian Science and Medicine in the Sixteenth and Seventeenth Centuries*, 2 vols. New York: Science History Publications, 1977, has to say on 'The "Helmontian" Robert Boyle'. Also useful were Rose-Mary Sargent, *The Diffident Naturalist. Robert Boyle and the Philosophy of Experiment*. Chicago: University of Chicago Press, 1995, and William R. Newman & Lawrence M. Principe, *Alchemy Tried in the Fire. Starkey, Boyle, and the Fate of Helmontian Chymistry*. Chicago: University of Chicago Press, 2002. An up-to-date effort to amend a range of recent revisions of Boyle's scientific persona and assemble the outcome in a coherent picture is Michael Hunter, *Robert Boyle (1627–1691). Scrupulosity and Science*. Woodbridge: Boydell, 2000.

Hooke. My treatment blends insights from Penelope M. Gouk, 'The Role of Acoustics and Music Theory in the Scientific Work of Robert Hooke'. *Annals of Science* 37, 1980, pp. 573-605 (with the broader context for this article being given in idem, *Music, Science and Natural Magic in Seventeenth Century England*. New Haven / London: Yale UP, 1999, ch. 6), with those of John Henry, 'Robert Hooke, the Incongruous Mechanist'. In: M. Hunter & S. Schaffer (eds.), *Robert Hooke. New Studies*. Woodbridge: Boydell Press, 1989; pp. 149-180. In an attempt at refutation of Henry's chapter marked more by the spirit of a lawsuit than by an effort at historical empathy, Mark E. Ehrlich in 'Mechanism and Activity in the Scientific Revolution: The Case of Robert Hooke'. *Annals of Science* 52, 1995, pp. 127–151 fails to capture the matter/spirit ambiguity displayed not only by Hooke but by several of his contemporaries as well, notably by Newton.

Newton. My treatment of the 'Hypothesis of Light' owes most to R.S. Westfall's in his *Force in Newton's Physics. The Science of Dynamics in the Seventeenth Century*. London: MacDonald, 1971; pp. 363-369, and in his *Never at Rest. A Biography of Isaac Newton*. Cambridge: Cambridge UP, 1980; pp. 269-272 and 304-308.

XVII

LEGITIMACY OF A NEW KIND

The Newtonian synthesis, the crowning achievement of the Scientific Revolution, was the product of one further revolutionary transformation, which started in August 1684. Just like Newton himself in 1669, we are almost but not yet quite ready for it. One element that drove the revolutionary movement forward from the time of the pioneers to the early 18th century is still missing from the account: the resolution of the crisis of legitimacy that over the late 1640s and the 1650s came perilously close to laying waste to the work of the pioneers (ch. 12). The achievement attained in the wake of the pioneers in all three modes of revolutionized nature-knowledge, both separately and together, makes it crystal clear *that* the crisis was overcome but not *what* had made that feat possible. To find that out is the burden of the present chapter.

At the heart of the crisis lay the 'strangeness' and the 'sacrilege' readily associated with newly realist mathematical science and with the natural philosophy of newly kinetic corpuscularianism. 'Strangeness' in the sense of an all-out violation of common sense and 'sacrilege' in a triple sense: (1) an unforeseen sharpening of current controversies over the meaning of God's Word, together with the speedy disruption of the centuries-old bond between Christian theology and philosophic proof, with grave consequences for how to conceive of the Eucharist, of the immortality of the soul, of the distinction between humans and animals, of the place of the Earth in the universe, and of what limits were set to unfettered thought; (2) the strong component in all this of contested authority, in the context of a highly unstable European state 'system' that, to all appearances, was moving from ever-deepening division to all-out war; and (3) the profound concern experienced by theologians and practitioners alike over how to think of God's purpose for human beings in a vast cosmos run in accordance with inexorable, impersonal laws.

All this did not uniquely cause but certainly contributed greatly to a precarious situation that continuing radical innovation had to overcome in the middle of the seventeenth century. In the large majority of those European territories where the pursuit of nature-knowledge had been on the upsurge for some two centuries already, cautious compromising quickly gained the upper hand. In southern Europe and in the Spanish Netherlands this was due to the Inquisition chiefly; in the Dutch Republic, to a local blend of theological controversy, political quietism, and narrow-minded utilitarianism. Also, already by the 1620s Germany and Austria had begun to drop out of the quest for innovative nature-knowledge,

566   with thirty years of warfare of the most destructive kind ahead. On the Continent, then, it appeared by midcentury to depend on France alone whether or not the originally Europe-wide will to radical innovation of nature-knowledge was to evade the fate of a gradual loss of momentum, in ways roughly similar to (albeit in circumstances quite different from) how such a fate had come to strike Islamic civilization in the 11th (5th) century. The clash over Descartes' intellectual legacy, for which the basis was laid in 1657 when Clerselier began to publish it, certainly seemed to point in that very direction.

Of course, in sketching the situation so bleakly, the presence at the time of a third movement of radical innovation in nature-knowledge, the one embodied in fact-finding experimentalism, has deliberately been left out. To add it to the overall picture in its midcentury guise, however, does little to change it. The one pioneer on the Continent, van Helmont, was enmeshed in theological controversy and persecution to such an extent that he refrained from seeking publication for his principal treatise during his lifetime. In England, amid a civil war that absorbed many passions and energies, a handful of practitioners continued Gilbert's and Harvey's pioneering work a little, as well as bits and pieces of Bacon's program. More than this need not be added to the picture to round it off.

In sum, then, the 1650s became the low point of the Scientific Revolution, as sheer counting of revolutionary achievement tends to confirm – between 1600 and 1700 no time period scores lower on a listing of major publications and discoveries. With the pioneers passed away and their accomplishments under heavy attack on the Continent, our watchful observer, whom we have left noncommittal about a decade earlier (p. 283), is likely to have sensed that the Golden Age of European nature-knowledge, its unexpected climax in a brilliant episode of revolutionary innovation notwithstanding, might well be drawing to a close.

How, then, to explain that it was not, and that no more than one decade later already the picture began to look quite different?

The edges off controversy, and a shift of center Europe-wide

The broad setting first. It is not only in nature-knowledge that the picture began to look quite different. It changed all over Europe in many other respects as well. The Peace of Westphalia, soon followed by the British Restoration, marked and then furthered a mood of reconciliation (p. 514). The edge was taken off many a standing controversy, thus inducing a rapid change in overall climate which extended to the domain of nature-knowledge as well. Theodore K. Rabb has even argued that in the decades immediately following the Peace and the Restoration, for Europe's princes to succeed in restoring stability on the large scale, the new modes of nature-knowledge were not so much passive recipients as, rather, primary tools. This radical idea of Europe's return to broad stability being due in large part to revo-

lutionary science exceeds what the evidence still can bear. But Rabb's essential point is not
affected thereby: "the quick and decisive triumph of this handful of scientists" who led the
Scientific Revolution during its second stage must surely be counted as "one of the most
amazing episodes in European history."[322] What, then, have we meanwhile found out about
this 'handful' in terms of their territorial distribution and their institutional affiliation?
Comparison with the situation in the first half of the century reveals some drastic changes.

First and most important, the 1640s through 1660s is when, fairly swiftly as such historical processes go, the center of European gravity shifted from states and regions bordering the
Mediterranean Sea to those bordering the Atlantic Ocean, with Paris and London emerging
as the principal centers of ongoing pursuit of revolutionary nature-knowledge. The shift
concerned much more than the pursuit of nature-knowledge, to be sure. Culturally, over
many centuries events in Britain, northern France, northern Germany, and the as yet undivided Netherlands had been secondary to what happened in Italy. Politically, these territories had at times been drawn into ongoing conflicts centered really on Italy, Austria, and
southern Germany, their role confined mostly to serving as theaters for sideshows. Commercially, Britain and the budding Dutch Republic had made significant inroads on Spanish/
Portuguese hegemony overseas, but as yet no more than that. All this began to alter around
midcentury. Spain and the Italian city-states quickly lost what advantages they had; by the
1660s no longer the power and the ambitions of the Habsburgs but those of the Bourbon
Louis XIV came to occupy center stage in European politics, and Britain, while at bottom
remaining aloof from Continental affairs for a few more decades, in the meantime turned
itself into the only successful rival to France's bid for commercial supremacy.

There is an evident parallel here with the geographic shift to the northwest that is so
noticeable over the same period in the revolutionary pursuit of nature-knowledge. Nor were
these just parallel developments – the large-scale movement away from the Mediterranean
was, if anything, exacerbated by the role of the Inquisition in the events that led up to the
crisis of legitimacy of the new modes of pursuing nature-knowledge. Conversely, the centrality that Paris and London had gained by the 1660s set in train various developments that,
whether deliberately or inadvertently, helped to resolve the crisis.

Of these developments, the one to affect the pursuit of nature-knowledge in the most
decisive fashion was, of course, the near-contemporaneous foundation, each under its king's
patronage, of the Royal Society in London and the Académie Royale des Sciences in Paris.
Not only did these bodies serve as focal points for the pursuit of nature-knowledge in their
own countries, but they also came to serve as centers of attraction for individual practitioners elsewhere in Europe – not all individual practitioners, to be sure, but the more creative
ones especially. Numerous men engaged in routine work (like mathematical practitioners in
the employ of city councils) or satisfied with current modes of cautious compromising (like
the Jesuit fathers) found sufficient institutional backing in their religious order (with the

Jesuits) or at their home bases in Italy, the Dutch Republic, or slowly recovering Germany. But among those engaged in innovative research the majority found institutional backing abroad, in the two new centers of pursuit of nature-knowledge. Rather than feeling compelled to withdraw into partial or wholesale self-censorship as many a pioneer (Descartes, Gassendi, van Helmont) had in the 1640s and early 1650s, men like Cassini, Huygens, Leibniz, Malpighi, and van Leeuwenhoek found themselves from the early 1660s on drawn into the sphere of activity of either the Académie or the Royal Society or (in a rare case) both. Indeed, one would be hard put to find anyone at all among the major post-1650s innovators *not* to benefit in important ways from the presence and the activities of these two institutional centers.

Clearly, then, in Europe's two most densely populated cities legitimacy for the pursuit of revolutionary nature-knowledge had been regained. The pace of discovery, after its temporary slowdown in the 1650s, was resumed. Cases of self-censorship became rare and comparatively unimportant. Royal support in the two central territories (through large-scale funding in the one case; through a charter generous with liberties in the other) indicated that, at least at these two principal courts of Europe, a careful weighing of the pros and cons of the new approaches to nature-knowledge had led to a net positive outcome. There is no escaping the question of how these reversals could come about.

Europe backing off from all-out war in a direction of stability attained through acts of some conciliatory compromise; its center of gravity shifted from the Mediterranean toward France and Britain, with institutional support for the innovative pursuit of nature-knowledge established in their capital cities through royal patronage – this is the broad setting inside which we can begin to answer that question. In casting about for answers we shall not tread wholly unfamiliar territory. What we have examined in previous chapters *at the inside* of revolutionary nature-knowledge must now be reconsidered *from the outside*. What from the inside appears as *feats* (accomplishments, controversies) takes shape from the outside as *perceptions* – efforts to mold or at least influence those outsider perceptions definitely included. The intangibles we shall deal with, then, are *aspirations, projections, expectations, promises*, and (decisively so) the *values* underlying them.

Strangeness mitigated

Nothing could, of course, alter the fundamental dichotomy between the world as we perceive it and the underlying mathematical and/or corpuscular reality of the world as proclaimed in realist-mathematical science and in kinetic corpuscularianism, respectively. Nor could the gap be conjured away between the level at which 'natural nature' presents itself spontaneously and the level at which 'artificial nature' comes to the fore in the experimental

mode of inquiring into nature's properties. If the Scientific Revolution carried one message, it was that what valid knowledge can be obtained about the natural world lies hidden and must therefore be laboriously sought for at levels deep underneath its surface. The
chasm thus opened up between the hoary 'common sense' and the newly counterintuitive
approaches to the natural world could not be closed; but in the second half of the century,
with the breakdown of barriers and with the quickly increasing visibility of fact-finding
experimental science, opportunities to mitigate it became available. Percolation of highly
theoretical, mathematical science down to the level of practical measurement offered one
such chance; the professional production of, and the spectacle offered by, the instruments
of the Scientific Revolution, another; sustained efforts at educating a wider audience in the
knowledge structure of fact-finding experimental science, a third.

Measured quantities.   More than in other civilizations did the routine handling of
measured quantities stretch in Europe beyond the intellectual elite to larger segments of the
population. This was due in part to the emergence during the Renaissance of a few interfaces
between mathematics and practical pursuits like linear perspective and mapping (p. 130).
But already in earlier, medieval times, so the historian Alfred W. Crosby has argued, did an
unusually vast variety of subjects begin to be structured in quantitative ways. Even prior to
the organization of pictures by means of perspective and cross section and of geographical
space by means of spherical projection upon maps divided into length/width grids, time
was structured by means of equal-length hours marked by the mechanical clock, music by
means of notes on staves to fix pitch and measure, and commerce by means of double-
entry bookkeeping. None of this involved particularly sophisticated mathematics, and the
"stargazers, surveyors, gaugers of tapistry, wine and bodies in general, further mint masters
and all merchants" to whom in 1585 Simon Stevin dedicated his book on decimal fractions
had little to do with the turn from Alexandria to Alexandria-plus soon to be executed and
debated by far more sophisticated mathematical scientists than the audience addressed by
Stevin. Mathematical relations as expressed in general rules and laws were far removed as yet
from actual measurement or the handling of quantities (p. 517). But as the Scientific Revolution advanced to its second stage, the sharp edges of the distinction began to wear off. As the
barriers between mathematical and fact-finding experimental science began to break down,
mathematical relations began with increasing frequency to be watered down to the level of
precision measurement, in Mersenne's work first but after midcentury in astronomy and in
several other research areas as well (p. 509). And this is how achievements of the Scientific
Revolution began to percolate down to the level of the average city calculator or high school
math teacher. Realist-mathematical science lost some of its strangeness, and chances were
enhanced for the Revolution's unhampered unfolding. The everyday practice of relatively
low-level measurement helped to make look less odd what had by 1600 started out as an

570   exercise in highly sophisticated mathematical relation but was by the early 1660s acquiring clearly visible marks of the more familiar, day-to-day handling of numbers and figures and other symbolic ways to render quantitatively divisible chunks of reality. However arcane the mathematical analysis of reality at its most sophisticated might remain, its increasing linkage with acts of measurement helped make it appear less as something beyond the ken or concern of an outsider and more as the sophisticated apex of a pyramid of run-of-the-mill quantitative exercises.

The instrument trade.   Five major instrument types were invented during the Scientific Revolution: in chronological order, telescope, microscope, barometer, pendulum clock, and air pump. Before the century was over, their production had moved from the scientists themselves to specialized professionals. A new market opened up, far beyond the already centuries-old, mostly navigation-directed trade in mathematical instruments, so that by the end of the century the entire instrument trade found itself restructured and, in the process, greatly expanded. As individual and, on occasion, institutional consumers were supplied with as yet tailor-made specimens of spectacular instruments like microscopes and air pumps, the number of people not themselves engaged in revolutionary science yet broadly familiar with its ways increased considerably.

The van Musschenbroek workshop in Leiden is a case in point. Settled there in the late 16th century as refugees from the Southern Netherlands, successive van Musschenbroeks were occupied as brass founders making lamps or chandeliers until in the 1660s Samuel van Musschenbroek began to add scientific instruments to his trade. He made some microscopes for Huygens and for Swammerdam. He built an air pump for the Leiden professor Burchard de Volder, who had obtained funds for a *theatrum physicum* to illustrate his Cartesian teaching with experiments at the same time as Rohault began doing so in Paris. Under Samuel's younger brother, Johan, the construction and sale of scientific and medical instruments to a variety of courts and universities on the Continent proved sufficiently profitable to replace the family's original trade within decades. In a prolonged correspondence in the mid-1690s with Professor Dorstenius at the University of Marburg Johan even took upon himself the role of instructor, patiently explaining to his scholarly client what sort of instruments were most suitable to set up a local collection of demonstration apparatus and how to handle them.

All this was to attain much larger proportions decades later, when university after university followed suit, when science became a fashionable subject of polite conversation, and when itinerant lecturers spread instrument-aided, experimental entertainment all over Europe. Yet in the second half of the 17th century instruments were already helping to decrease the strangeness of the new science. Britain's demand for instruments, supplied by London, was quickly expanding, while Leiden supplied what market the Continent yet provided. The

powerful effect exerted by the microscope and the air pump upon the imagination also contributed to making the public more comfortable with the new science. What a thrill, then as
now, to peruse the larger-than-life illustrations of Hooke's *Micrographia* or to learn through
van Leeuwenhoek of the tiny animals in the reader's own semen! What a spectacle to watch
Hooke spend a quarter of an hour in a partly evacuated air pump, to climb out in good
spirits but for some pain in his ears! Demonstrations like these provoked some ridicule to be
sure, yet the sense of awe they could not fail to evoke contributed to enhancing the science-
mindedness of a rather larger segment of the population than the practitioners alone.

Education in experimental procedures.   Over the 1660s Boyle, Hooke, and Oldenburg undertook a sustained effort to establish rules for fact-finding experimental science (p.
485). Out of their search came rules for the avoidance of bias, for replication, for correction
of error and impurity, for trustworthy witnessing, and for matter-of-fact reporting. The
search was not only meant to enhance the capacity of experimental research to yield reliable, nonspurious facts but involved an educational effort as well. The rules were directed at
lay audiences, too, so as to familiarize outsiders with the underlying knowledge structure of
fact-finding experimental science. That way they sought to mitigate the apparent strangeness of their experimental procedures and results. Why else did Boyle take the trouble to sit
down for a lengthy rejoinder to Hobbes? Unlike Galileo's Pisan adversaries half a century
earlier, Hobbes was no rival to Boyle, wealthy son of the Earl of Cork, in any personal or
status-related sense. True, fear of Hobbes' possible influence on Charles II, who kept his
onetime tutor around at court and used to address him with joking affection as 'the bear',
may have played some role. More importantly, Boyle was concerned to explain to a wider audience than just fellow practitioners why fact-finding experimental science was superior to
the traditional mode of nature-knowledge advocated by Hobbes, and what that superiority
consisted in. Likewise, the lengthy preface that Hooke penned for *Micrographia* was meant
unmistakably to serve not only methodological but also educational purposes.

Nor were practitioners like Boyle or Hooke the only ones patiently and laboriously to expound the virtues of fact-finding experimental science. In the Royal Society what was almost
an industry arose which was dedicated to that very purpose. In officially sponsored works,
nonpractitioners like Thomas Sprat and John Glanvill were not so much concerned, the way
Boyle and Hooke were, to make plausible from the inside that experiment, for all its apparent artificiality, could provide reliable information about events and states of affairs that
were really there in nature. Rather, they argued from the outside that experiment offered the
best possible terrain for a kind of scholarly activity far removed from any risk of sacrilege
or of principally insoluble and, hence, war-inducing conflict. This line of argument, which
found expression one particular way in Britain and several others on the Continent, had
many ramifications, to be examined now.

Sacrilege insulated

Three distinct strategies were deliberately set up to keep possible sacrilege in check. One such strategy, pursued by the Jesuit Order from the late 1630s onward, was to construe a syncretist worldview infused with up-to-date nature-knowledge from which every possibly offensive element had been removed. Another, which the order also pursued but which found more pointed expression in the scientific societies of Florence, Paris, and London, was to exploit the perceived philosophic neutrality of experiment. A third strategy, which pertained to these societies likewise, was the consistent maintenance of princely distance.

Jesuit syncretism. The syncretist worldview of the Jesuits, built up from the late 1630s onward as a way to give needed coherence to their claim of standing at the forefront of learning, was attained by first filtering out all that was felt to be sacrilegious in the new modes of nature-knowledge and then throwing all that remained into a big melting pot. From the perspective of the order, the veritable hodge-podge that came out of the pot (p. 495) provided the world with all the best that faith, reason, sense experience, and 'profane authority' (the Aristotelian and natural magic traditions) had in harmonic union brought forth. The construction of this syncretist worldview being part and parcel of its striving for universal conversion to the Church of Rome, the order spared no effort, and spent vast sums, to get the message across. As gifted educators the Jesuits were well aware of the capacity a picture possesses to bind a range of abstract conceptions together in one captivating, memorable image. If need be, years were taken to design, fund, and execute frontispieces for books of special programmatic value. One example is Riccioli's *Almagestum novum* of 1651, with the triumph of his semi-Tychonic planetary system over both Ptolemy's and Copernicus' made unmistakably visible. Or take the full set of books in which, over the decades, Athanasius Kircher brought the Jesuit worldview to detailed expression. Even today a picture like the one shown in Figure 17.1 has something extraordinarily gripping about it in its effort to convey that, over and beyond its subject matter – light in all its manifestations – a vastly intriguing world of meaning is here being opened up.

The conversion strategy of the order was designed to operate from the top down, and a full understanding of the title page required a reading knowledge of Latin, so clearly a select public was being aimed at, in the expectation that the splendid education system of the order would help the message percolate down to other layers of the population. A measure of success was certainly attained – between Galileo's trial and the onset of Newton's fame, it was Kircher's name rather than anybody else's that far beyond the confines of Roman Catholicism symbolized the ongoing search for nature-knowledge of a new, exciting kind.

Figure 17.1: Title page of Athanasius Kircher, Ars magna lucis et umbrae (Rome, 1646)

574   *The philosophic neutrality of experiment.*   With such constraints as the order imposed upon itself, revolutionary novelty was bound to remain out of sight. What the best Jesuits accomplished even so, came from their focus on fact-finding experimentation (ch. 13). Why did they choose this focus? The big advantage of experiment was that it could readily be perceived as philosophically neutral. Experimentation was farther removed from possible sacrilege than any other existing form of scholarship. The Jesuits, although perhaps the first to become aware of this, were certainly not the only ones. For the neutrality of experiment formed the very condition under which, in Florence, the Accademia del Cimento could operate at all, whereas in Paris it was at the core of the king's policy of admission to the Académie, and in London it made up a substantial portion of ongoing justification of what the Royal Society was doing.

Throughout the Scientific Revolution (considered over its full width and its entire time frame) experiment was steadily on the rise between Galileo and Newton – as the century progressed, there was a veritable *drift toward experiment.* Early on, in spite of the experimental slant that Galileo deliberately gave to the *Dialogo,* both he himself and his Italian pupils and disciples in the 1640s took mathematical theory as definitely prior to the messy distortions to which experimental validation is liable (p. 363). With the pioneers of kinetic corpuscularianism, experiment came in at best by way of allowing one to decide the factual state of affairs at a low level in the epistemic chain leading from securely laid first principles down to sheer sense perception (pp. 373; 393). Only in budding experimental science of the fact-finding kind did experiment occupy center stage from the very start. By contrast, in the second half of the century mathematical scientists at the Académie set out to reconsider the validity of the Galilean achievement in a more thoroughly experimental vein (ch. 10). The academic and salon-centered teaching of Descartes' philosophy of nature began to be enlivened by experimental demonstration at the hands of men like Rohault and de Volder. Fact-finding experimentation replaced mixed mathematics as the central concern of those Jesuits occupied with nature-knowledge. And the question is, of course, what caused this Europe-wide drift.

The short answer is, indeed, that experiment was widely perceived as well suited, ideal even, to circumvent such theological minefields as had manifested themselves to such ill effect in the 1630s and 1640s and had gone on to lay waste to so much intellectual terrain on the Continent and were still occasionally exploding in France.

Prerequisite, however, to the possible perception of experiment as philosophically neutral was to separate it from any body of ideas with proven or at least possible sacrilegious implications. The Jesuits did this by weaving fact-finding experimentation into a body of ideas itself purged of any possibly sacrilegious implication. By contrast, in the three major scientific societies of the second half of the century such separation came about by means of efforts, if not effectively to cut experiment off from any body of ideas whatsoever, then at least to present it thus to the world outside.

In the Accademia del Cimento, situated of course in Inquisition-dominated territory, the separation between experiment and underlying conceptual thought was executed in the most radical fashion. The records that were kept of its sessions show that the substantive context in which its experiments were performed was clear to its members and actually served as the frame in which these acquired what coherence they possessed. Even so, all public expression of the accomplishment of the Cimento was carefully shorn of any wider conceptual context, most thoroughly so in its one collective publication, the *Saggi* (p. 496).

In the Académie the separation, albeit not quite so drastic on the level of ideas, went even deeper inasmuch as the selection of members was in large measure governed by it. In France a remarkable situation came to obtain. On the one hand, numerous men who were associated with the new modes of nature-knowledge found themselves constantly enmeshed in strident theological dispute and subjected at times to a measure of persecution endorsed by the highest authority of the realm. On the other hand, numerous men associated like-wise with the new modes of nature-knowledge found themselves pampered, in the company of like-minded colleagues, and given high salaries and first-rate research tools by the very same highest authority of the realm. At the very same time that the clash between Descartes' faithful followers and a variety of opponents raged unabated over the 1660s–1720s (p. 424), a careful selection of mathematically and experimentally inclined scientists was assembled in the Académie and given unprecedentedly free rein by the king's prime minister to do innovative research. The criteria that the king applied for inclusion in or exclusion from his Académie can easily be deduced from his decisions, which betray considerable judgment and consistency over time – with experimentation at a premium, dogmatic philosophy had to be kept out. Which of these criteria was primary appears with crystal clarity from the case of Jacques Rohault, a man certainly no less gifted than hired Académiciens like Auzout and Petit. To his lifelong regret, Rohault's well-advertised and well-attended demonstrations of Cartesian verities by means of a fine toolkit of partly self-made instruments did not save him from exclusion from the Académie – quite rightly had the king, or his advisers, sniffed out in Rohault the very 'indubitable first-principle' mode of thought that such well-eligible men as their prize horses Huygens and Cassini but also Auzout and Petit were conspicuously free from. It was in full consistency with this policy line that the members' meetings, too, were kept meticulously free from discussing philosophy or religion or politics, focusing instead on something that thus came to look more and more as an entity in and by itself – science.

This, then, is how the kind of nature-knowledge pursued at the Académie came to be stamped by the political setting in which that institution was founded and then guided on its course. It fostered experimentation primarily in its fact-finding mode (p. 497). In less outspoken fashion, somewhat underneath the surface of activities, experiment also acquired enhanced prominence in the program effectively set up by Huygens, Mariotte, and some other mathematically gifted Académiciens to reconsider the Galilean achievement, as well

576   as in a discreet testing of empirically disputable points in Descartes' natural philosophy of particles in motion. Regarding Huygens in particular, all this means that the personal, revolutionary breakthrough he had made in the 1650s toward a hypothetical approach to kinetic corpuscularianism (p. 527) found unexpected, institutional support from the mid-1660s onward. The new theory of light he conceived and experimentally corroborated in his Paris years was the greatest find to be furthered thereby.

In the Royal Society, finally, these two salient objectives – to cut experiment off from wider bodies of ideas, so as to unleash its capacity for keeping sacrilege and consequent conflict at bay – did not so much serve as the tacit kernel of science policy at court, of which after all there was little. Rather, in conformity with the lack of funding and the attendant freedom from regulation or control left this body by its king, these efforts found public expression in specific ways of presenting results, as well as in a number of books aimed at promoting the case of fact-finding experimentation carried out the British way. Presentation was designed to accomplish the former objective, promotion the latter. For the former, Francis Bacon's name was invoked; for the latter, the sake of peace.

There was a tension between the everyday practice of fact-finding experimental science and the manner in which, in Britain especially, it was often presented. Bacon's prescriptions came in handy to present as one of the virtues of experimental science the refusal of its practitioners to enter the fray of fanciful speculators, carefully sticking instead to the observed facts alone. For instance, Boyle affected studied neutrality over the causal question of how the air comes to possess the 'spring' to which he devoted so many of his most ingenious experiments (p. 461). More often than not, the reality of experimental investigation was rather different, with the best practitioners explaining newly found facts by devising specific theories with some background worldview in mind (p. 501), or even some specific mechanism of moving corpuscles (e.g., in Boyle's chemical work). It is true that with some practitioners, Boyle himself in the first place, the affectation to present just solid facts, not fanciful theories, went so far as at times to feed back into, and thereby severely to curtail, some of their own researches, notably those set up by way of 'natural histories' of this, that, and the other (p. 446).

Such refusal, whether purported or real, to adduce causes for experimentally ascertained facts, then, was what practitioners themselves contributed to making their practice appear innocuous. The concomitant presentation of experiment as immune to possible sacrilege was put into the carefully supervised hands of two clergymen, Sprat and Glanvill, who were both fellows of the nonpracticing kind. In a sense, the title of one of Glanvill's books, *The Vanity of Dogmatizing*, says it all. In a range of works commissioned to explain and extol the activities and concerns of the Royal Society, Sprat and Glanvill took the closely related line that, since philosophy bred conflict, in the Royal Society it was shunned in favor of setting free the powers of experiment for peaceful resolution of disputes. For instance, Sprat wrote

that the contemplation of nature by experimental means "gives us room to differ, without 577
animosity; and permits us, to raise contrary imaginations upon it, without any danger of
a Civil War".[323] Such sentiments fitted in with the Church of England's acute distrust of the
very idea (still alive on the Continent) of reasoned proof for religious verities. For could not
any argument be bent at any time to serve any case, so that in the end the current dispute
was exacerbated rather than resolved? And that was why fellows deliberately abstained in
their sessions from addressing religious and political issues, so as to keep the expression
of differing views confined to those that either could be resolved at once by the outcome
of experiments or, failing that, held the promise of peaceful resolution by more conclusive
experimentation in the future.

Princely distance.   Such deliberate abstention actually worked two ways. All these measures taken and arguments put forward in favor of philosophically neutral experiment in the
context of the three major scientific societies of the second half of the 17th century found
their natural counterpart in one further feat: the careful maintenance of princely distance. It
had always been a point of court etiquette for the prince to receive gifts deemed fit to add to
his luster in such a way as to keep His Highness aloof from the specific contents of that gift.
The counterpart to this courtly rule on the side of client authors was that in their dedicatory
prefaces to works thus receiving princely sanction they had to walk a fine line. Their job was
to establish a link between work and dedicatee without implicating him in the advocacy of
anything in particular. Common practices like these acquired enhanced significance in the
case of the three major scientific societies of the second half of the 17th century and extended to their very setup. In putting themselves at the institutional center of the Cimento and
of the Académie, respectively, the de' Medici brothers and Louis XIV enforced upon their
clients anonymous procedures, whether of a very strict kind as in Florence or somewhat
looser as in Paris (p. 496). These things being a matter of well-known basic court etiquette,
in neither society was it found necessary to record them in any special, written dispensation.

In Britain meanwhile, the same felt need for princely distance worked out quite differently. Charles II deliberately chose to keep substantive distance from the Royal Society by
maintaining institutional distance as well. Neither running things personally, as Leopoldo
de' Medici did, nor through his highest official, as Louis XIV did, Charles rather left everyday
proceedings to a members-elected Council under an officially approved president equipped,
during meetings, with a royal mace as the sole symbol of the king's patronage. Such an unusual amount of institutional autonomy did require a royal charter to bring the intended
message home. Although, naturally, the arrangement did less to enhance Charles' glory as
a patron than the arrangement in France did for his royal colleague there, it provided a
characteristically different but, nevertheless, quite effective way to bind a large group of
potentially disruptive thinkers without running the risk of undue identification with views
defended by them.

Utility sanctioned in the Baconian Ideology

> ... all knowledge is to be limited by religion, and to be referred to use and action.[324]
>
> Francis Bacon

The upshot of the above is that from the early 1660s onward strangeness was mitigated and sacrilege was kept at bay to such an extent as to be felt to be manageable overall, at least in those quarters that were quickly becoming the decisive ones. That is, sufficient safeguards seemed in place to handle possibly adverse effects. But why bother in the first place? Why take a positive interest and act upon it in ways going far beyond anything ever displayed by any institution given to a measure of support for the pursuit of nature-knowledge? What brought those who gave their names and a good deal more to its collectively organized and, as such, unprecedented pursuit in Rome, Florence, Paris, and London to do so? For let us not underestimate to what lengths these governing bodies were prepared to go, in terms either of effort spent and funding provided or of freedom chartered. It has been calculated for the most sumptuous case of them all that over the first thirty-three years of the Académie (1666–1699) Louis XIV's total expenditure on infrastructure, research, and salaries amounted to more than two million livres – a sum sufficient at the time for running one of the wealthier monasteries. The vast sums spent by the Jesuit Order on the pursuit of nature-knowledge, which also included much activity overseas, are probably impossible to separate from expenses in other domains. Records of the costs incurred by the de' Medici brothers for the Cimento have not been preserved, but it is certain that the brothers paid for all the equipment used in its experimental sessions and also that Leopoldo was actively engaged in these. Finally, even in the English case, the king's surveyor-general of the Ordnance Office, Sir Jonas Moore, FRS, undertook as part of his responsibility for army and navy supplies to talk his king into hiring John Flamsteed, for £100 a year, as the first astronomer royal. Greenwich Observatory was likewise built on Moore's initiative and paid for by his office (£500). The annual bill for assistance and maintenance also went to the office, whereas the initial purchase of first-rate telescopes and precision clocks was charged to Moore's personal account.

So, once again, whence that readiness, on the part of royalty and highly placed officials, to exert themselves so strenuously and/or open their purses for such considerable sums? What did they expect in return for such monetary and other efforts?

In one case, the answer rests for the largest part in *personal interest*. As with William and Charles Cavendish (p. 549), Ferdinando II and Leopoldo de' Medici were among those rare 17th-century patrons of innovative nature-knowledge with a genuine interest and increasing expertise in more than its surface manifestations. Not so Charles II, whose one visit to the

Royal Society prompted him to laugh so mightily, or Louis XIV, whose one reluctant, utterly ceremonial visit to the Académie took Colbert fully fifteen years to squeeze out of his master. (Note here that Louis was certainly capable of taking an expert interest in activities he patronized, as when, personally joining Lully and two others on a royal selection committee considering thirty-five candidates, his own choice for one of four *sous-maîtres de chapelle*, Michel-Richard Delalande, not only proved wholly successful but is also the one still wholeheartedly endorsed by modern historians of French Baroque music.)

Another motive that applies in only one case was the fostering of universal *conversion;* about this leading concern of the Jesuit Order I have said enough.

Further, ever since Galileo's telescopic observations had made such a splash throughout Europe, *novelty* had become an attractant, as did duel-like disputes over how to interpret new finds or new tenets. Disputes of such a nature certainly kept their courtly entertainment value over the second half of the 17th century and beyond, yet it is hard to assess how important a motive this actually was in the case of the major societies.

More important surely was that evergreen among secular patrons' motives, the *glory* of the ruling house. This aim was particularly strongly and effectively pursued by Louis XIV. With the increasing identification, all over Europe, of royal houses with the countries ruled by them, it also acquired a touch of national pride not previously much in evidence. However, it is not self-explanatory that, in addition to such more customary objects of patronage as music, poetry, theater, ballet, and painting, the Sun King's radiance extended to revolutionary nature-knowledge as well. What need could Huygens or Cassini fulfill that Lully, Molière, Racine, Poussin, and other luminaries of the 'Âge de Louis XIV' could not?

One valuable clue to that question is provided by a programmatic document that Huygens drew up in the summer of 1666. Its expressed purpose was to name subjects fit to be investigated in the newly founded Académie, and Colbert indicated his approval, item by item, with a curt *"bon"* (OK). Huygens proposed (1) experiments on the void and the determination of the density of the air; (2) examination of the force of gunpowder and (3) steam; (4) examination of the force and velocity of wind, to be used in navigation and for machine tools; and (5) examination of the transfer of motion in impact. Note here that the density of the air had already been determined, notably by Huygens himself, and that he did not omit to mention his own primacy in deriving the true rules of impact. More importantly, he admonished his new colleagues jointly to operate "by and large in accordance with Verulam's design",[325] that is, Francis Bacon's design, a man respected indeed by Huygens for his methodical approach to things, yet hardly the personal patron saint of the predominantly Galilean but also Descartes-inspired scientist who drew up the list. So what was really going on here?

The answer, in one word, centers on *utility*. Even though three topics out of five on Huygens' list did occupy him at one time or another (items 3 and 4 did so at several removes at

580   best), quite evidently the selection here made by this son of a seasoned diplomat reflects not so much his own research priorities as rather those imputed to his patron, with the five official '*bons*' confirming him to be right on target.

What practically useful things, then, did Louis XIV and Charles II, and also Colbert and Moore and their ilk, expect revolutionary science to deliver?

In the first place was a possibly decisive contribution to warfare. The primary motive for hiring Cassini was not the reflected glory of telescopic observation of new celestial objects. He was hired primarily so he could mastermind the detailed mapping of France to strengthen the king's military grip on a realm felt to be in dire need of greater homogeneity and also so he could exploit such opportunities as the Earth's Moon or Jupiter's moons offered for resolution of the problem of longitude. Huygens was not hired because of the justified expectation of adding luster to the Crown or the not quite justified expectation of effective leadership over his fellow Académiciens. He was hired because of his expertise with precision clocks and what that knowledge could contribute to the resolution of the problem of longitude. Hiring Flamsteed, who came with the top-notch equipment of Greenwich Observatory, served quite the same aim – all this and more came to be "funded by the military".[326]

Both parties involved were well aware of the military connection. This was neatly illustrated in mathematics textbooks at the time. By means of a careful selection of emblems for warfare and for mathematics, as Figure 17.2 shows in particularly elementary fashion, numerous authors meant to stamp upon the mind of the reader the close connection between them.

Military usage was not the sole material benefit sought from science. Overseas trade (more often than not, a euphemism for colonial robbery) was widely regarded as a vital source of wealth and so was, in a view upheld and acted upon by Colbert, the intensification of domestic production. To the scientific furtherance of the latter, chemical research seemed especially suited; to that of the former, mathematical science had offered assistance for a long time already, in ways the frontispiece of a mathematical textbook written by none other than Jonas Moore in his earlier days was meant to remind the reader (Figure 17.3).

When taken together, these two illustrations – with gunnery figuring in one and mapping and navigation in the other – raise an awkward question. Or rather, that question arises as soon as we recall that mapping and navigation were among the few domains where the promise to improve trial-and-error craftsmanship through mathematical science was fulfilled in pre–Scientific Revolution times already, whereas gunnery was among those far more numerous branches where just about all the practice-oriented effort exerted on the side of revolutionary science had failed as yet to improve craft practice in any tangible way (pp. 323; 481). Why, then, did hardheaded practical men in government, accustomed to considering official actions in terms of net costs and net benefits, stubbornly keep seeking practical utility in a branch of activity which (its risks, which after all looked manageable, aside) promised so much and delivered so little?

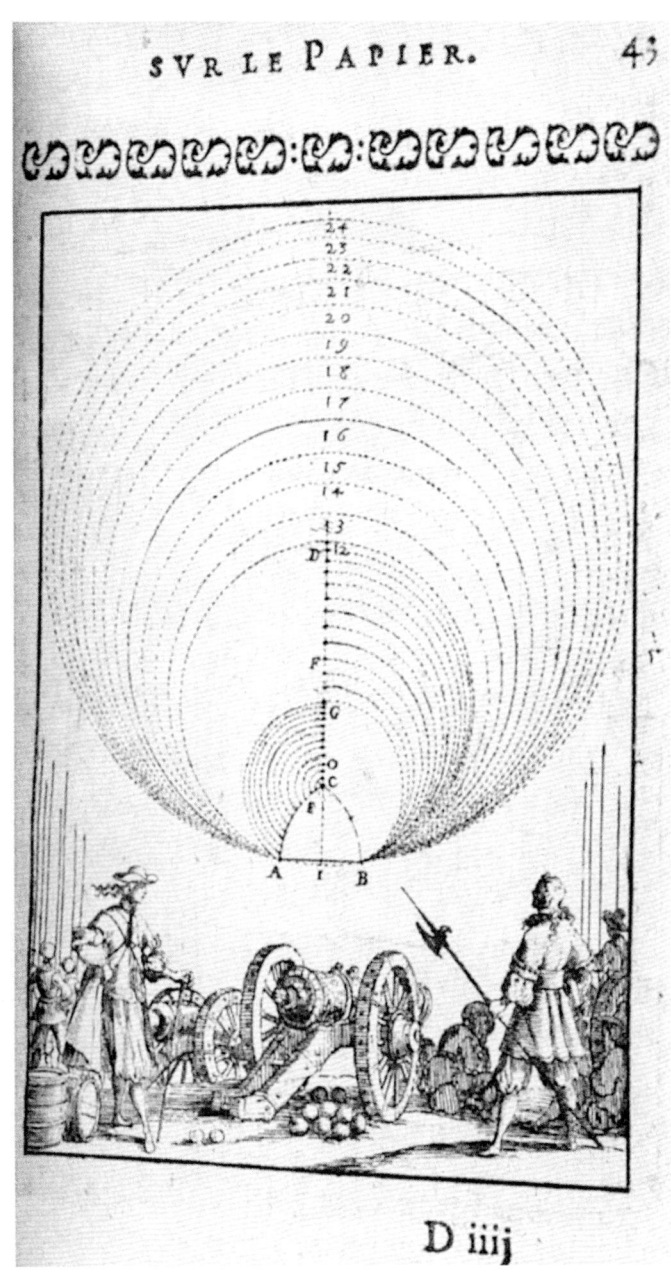

Figure 17.2: Illustration in Sébastien Leclerc, Pratique de la geometrie (Paris, 1669)

582

Figure 17.3: Frontispiece of Jonas Moore, A New Systeme of Mathematicks (London, 1681)

This, to be sure, is rather an overstated way to phrase the question.

First, we ought not to forget the other motives just listed. For instance, although Cassini did not indeed solve the problem of longitude, he did adduce pleasing novelty and did add to His Majesty's glory by his discovery of, notably, four satellites of Saturn in addition to Huygens' 1655 discovery of Titan.

But also, not quite so little was delivered. Thanks to mathematical science of pre-Scientific Revolution type, navigation had benefited greatly from the determination of latitude on board ship; likewise, in the 16th century all over Europe and in its increasing number of colonies maps had been drawn and fortresses erected in ways vastly superior to their older counterparts. Imagine taking these advances in navigation, mapmaking, and fortifications away. Europe's economy would not of course grind to a halt (agriculture being the natural mainstay of every traditional society), but it would find itself at great risk of losing what might be called its surplus income. So, the economic significance (perceived as well as real) of mathematical science was considerable. But the point is that to this significant level of science-improved craftsmanship almost nothing was added in the course of, let alone due to, the Scientific Revolution. So the real question is *not* why practice-oriented mathematicians were in demand all over Europe, turning the nature-knowledge segment of the 'market' where patron and client found each other into more of a sellers' than a buyers' market.[327] The real question is why no sharp boundary was drawn at the time between, on the one hand, the relatively low-level, routine mathematical science required to maintain and/or marginally improve mapping, navigation, and fortification practices and, on the other, all those much more sophisticated yet overall vain efforts drastically to improve craftsmanship through either up-to-date mathematical science or up-to-date fact-finding experimental science.

To that real question, two answers present themselves, one on the level of *practical* expectation, the other on the level of *ideological* expectation.

The gap between the promise and the reality of the improvement of craftsmanship through truly innovative science was filled in part by the circumstance that the promise on the side of science at its most sophisticated was vast indeed, and well worth some patient, even if expensive, waiting. Take, as one example, gunnery and the promise of enhanced range and accuracy involved in Blondel's Académie-sponsored, mathematics-infused treatise (p. 314). It was one thing for Blondel's critics to point out its practical shortcomings, another for his sponsors to cease believing that these could in time be overcome. A readiness to wait applies on still vastly larger scale to the problem of longitude, where we do possess a quantitative measure for the significance that was attached at the time to its solution in the awards offered by the nations most directly involved. By 1600 the king of Spain promised as a reward "a perpetual pension of six thousand ducats, a life annuity of two thousand ducats, and a cash prize of two thousand more". In 1600 the Estates-General of the Dutch Republic offered 5,000 florins plus an annuity of 1,000 pounds; the British Parliament in 1713/14 set the sum at £20,000 (more than 5 million modern dollars).[328]

584   The gap between the promise and the reality of the scientific improvement of craftsmanship was also filled by the circumstance that it was not widely noticed at the time (nor, truth be told, at any time later) *but rather became from early on a subject of ideological sentiment and contention.* That is to say, all over Europe, but in Britain especially, numerous outsiders came to believe in the capacity of the new science to provide opportunities for changing things for the better on grounds not taken from the bald facts of the matter alone (which, if added together, would up to the middle of the 18th century have pointed chiefly in the direction of disbelief) but from ideology. From what ideology? From what may be called the Baconian Ideology.

Nature and rise of the Baconian Ideology.   The Baconian Ideology is best defined as a leap of faith in the power of science. I use 'faith' here in a dual sense – as confidence in what practitioners of the new science (in its fact-finding experimental mode especially) could do to improve human destiny and as a Christian (more specifically, a Protestant) conviction that in doing so they were fulfilling a divine calling. The message went farther than the mere claim that innovative science did not represent the kind of lethal threat to re-ligion so often associated with it. Rather, innovative science was claimed to be in conformity with the core messages of Christianity, and even to help fulfill certain Christian aims. Thus, Sprat in his apologia for the Royal Society went to great lengths to advertise experimental science as beneficial in the material uses to be drawn from its practical findings and as truly Christian in its 'works' (an arch-Baconian expression much in vogue at the time). A list of responses Sprat made to anticipated objections is particularly revealing. The toughest nut for him to crack was the objection that sustained experimentation stood in the way of 'mor-tification', that is, of a way of life made holy by concentrating on the next world through abstention from worldly desires and from the things of this world in general (i.e., in Max Weber's terminology, salvation attained through 'other-worldly asceticism'; p. 264).

> For they tell us, that we cannot conquer, and despise the World while we study it so much; that we cannot have sufficient leisure to reflect on another life, while we are so taken up about the Curiosities of this . . . ; and that it is in vain to strive after the Purity, and Holiness of our minds, while we suffer them to spend so much time, on the labors of our Senses.

To this Sprat responded that, in accordance with the wide variety of dispositions of people, there is not just one but rather a large and quite varied array of Christian virtues, and that those not disposed toward flight from the world may find compensation in experimental science, which "makes us serviceable to the World". Indeed, Jesus, who used to withdraw from the company of others when bent on resolving an inner conflict, was in the habit of using for his conversions "some visible good Work, in the sight of the Multitude". Indeed, the very acts of mortification have their counterparts in experimental science – for example, spiritual repentance in the regular correction of errors and spiritual humility in a constant readiness, by the experimental scientist, to "misdoubt the best of his own thoughts". Indeed, "the true and unfain'd Mortification is not at all inconsistent with mens consulting of their happiness in this world, or being emploi'd about earthly affairs." Indeed, that is precisely how Christ's own apostles were employed. Indeed, if Christianity were to require attention to "heavenly things" alone, "what Traffic, what Commerce, what Government, what secular Employment could be allowed?" Indeed,

> seing the Law of Reason intends the happiness and security of mankind in this life; and the Christian religion pursues the same ends, both in this and a future life; they are so far from being opposite one to another, that Religion may properly be styl'd the best and the noblest part, the perfection and the crown of the Law of Nature.[329]

Please note that this is the future Anglican bishop Sprat speaking, in an apologia written by order of the Council of the Royal Society under careful supervision of the most authoritative fellow among its numerous nonpractitioners, the future Anglican bishop John Wilkins. What Sprat does here is to claim religious sanction for a life of worldly service and activity as exemplified in experimental science. To perform 'Works' in experimental research is as good a way to attain salvation as the resignation from earthly goods and pleasures customarily advocated – 'inner-worldly' and 'outer-worldly' asceticism as neatly contrasted as any Weberian might wish. Not only is experimental science, or really any science that obeys 'the Law of Reason', compatible with Christianity, but it even provides its fitting complement. It is almost as if we hear Francis Bacon affirm, for the 'works' to come out of the 'great instauration' so ardently advocated by him, humankind's unique opportunity already in the here and now to rise above its fallen state.

The turn from Bacon's exhortations serving as one-man prophecy toward their coming to serve as a veritable ideology for an array of practitioners and science-minded nonpractitioners alike is to be dated, not to the 1660s when Sprat and Glanvill and others wrote their like-minded pleas, but one to two decades earlier, in the Interregnum that was marked first by civil war and then by the dictatorship of the Cromwells and that ended in the Stuart Restoration of 1660. It was over these unruly decades, when official censorship broke down and all kinds of more or less half-baked ideas and projects had a chance to gain a hearing, that aspirations of the most varied kinds – millennarian, utopian, educational, commercial, colonial – came to be projected upon a mode of nature-knowledge associated in its turn with Bacon's name and writings, his scientific utopia *The New Atlantis* more than any other. Central to the creation, in the 1640s and 1650s, of a Baconian Ideology adorned with many radically reformist tinges were two Protestants from central Europe who had fled the Thirty

Years' War, Jan Amos Koménsky (Comenius) and Samuel Hartlib. Three native Englishmen, John Wilkins (Sprat's supervisor), John Dury, and John Evelyn, were concerned with gently guiding this radical body of ideas into more moderate channels at the time and then, from 1660 onward, with overseeing its continued flourishing in officially approved ways. In terms of religious denomination, in the former period the trend was chiefly set by adherents to some form of Puritanism (whether Calvinists or members of a plethora of baptist sects); in the latter, by men given to a latitudinarian (i.e., broad-minded and relatively tolerant) conception of the teachings of the Church of England.

Distinctive of Comenius was ardent advocacy of educational reform leading to the spread of some sort of all-wisdom ('pansophia') to all men of goodwill, so as to change their hearts in the direction of Protestant unity and European peace. Distinctive of Hartlib was a profound belief in the dissemination of knowledge as a prerequisite to the ardently ex-pected onset of the Millennium (the thousand years of messianic justice envisioned to reign between Christ's Second Coming and the Day of Judgment). To that end, Hartlib occupied himself with Bacon's written legacy, turned himself and some secretaries (Henry Oldenburg among them) into a clearinghouse for information and exchange, strove incessantly to ob-tain patents for inventors, and sought for experimental ways and means to improve agricul-ture. Neither Comenius nor Hartlib was an active practitioner of nature-knowledge in any regular sense; nor were Wilkins or Dury or Evelyn. But all these men came to serve as a *trait d'union* between, on the one hand, those who were and, on the other hand, numerous men who, in a variety of professional occupations in commerce, navigation, etc., strove to climb in society in ways by and large closed off in commonwealths with a more rigidly hierarchical social structure like Spain, France, and the principalities of Germany and Italy.

In all the variety of purposes (sometimes cross-purposes) to which Bacon's name was attached both then and in subsequent decades, a common denominator can be identified:

> with differences of emphasis and application, Bacon's appropriators claimed to be in favor of some or all of the following: *negotium* (employment, activity) rather than *otium* (leisure); an experimental, natural-historical, and broadly inductive approach to the natural sciences; the institutionalization of science and of the means of gathering, collating, and communicating knowledge; planned, cooperative research; rational 'utilitaria' and technological solutions to social problems (not least if there was money in it).[330]

The institution that came to comprise all this and more was, indeed, the Royal Society. Within months of the Restoration, Wilkins and others who several years before had gath-ered to talk experimental science on a regular basis in Wadham College, Oxford, drafted the founding act. In it, they sought to preserve, over and beyond such desiderata as just listed, at least two major features of what the preceding period had wrought. One was a membership composed of both practitioners actively engaged in experimental research and a surrounding circle of men – government officials, gentlemen of leisure, merchants, tradesmen, etc. – sympathetic to such aims. The other feature to be preserved was a broadly conciliatory view of things, whether this concerned the proper pursuit of nature-knowledge or theological issues. Significant examples of the former are Boyle's readiness to focus on what Descartes' and Gassendi's teachings held in common (the 'catholick principles' of matter and motion) rather than on the doctrinal niceties that kept them apart on the Continent; and the very mixing of the Baconian Brew, with its liberal blend of aethereal 'spirits' originating in no less than three distinct philosophies, Platonic, stoic, and kinetic-corpuscularian (p. 553). A striking example of the latter would be Sprat's generous insistence (cited above) that Christianity does not impose just one duty upon its adherents but that Christian duties are as varied as people's dispositions are.

Such a remarkably nonexclusive, conciliatory approach as here expressed by Sprat marked the Church of England from its very emergence out of the refusal of papal permission for a royal divorce (Henry VIII from his first wife, so he could marry Anne Boleyn) rather than out of ever-sharpening controversy over church practice and dogma. As such, it stood in stark contrast to the fight over dogmatic niceties that had for so long marked the Continent and actually marked it still, even though with the Peace of Westphalia the most pernicious edges were taken off such fights. Nor was this the only major difference. By way of comparison between the state of things in British and in French society, we shall now survey the main attainments and societal/religious ramifications of revolutionary nature-knowledge considered as a significant resource of practical utility in the decades following the Restoration of the Stuarts in 1660 and Louis XIV's decision in 1661 to declare his minority over and take the reins of France into his own hands.

Britain and France compared.   The crucial attainment was that "in Northern Europe ... science eventually became a central element in an emerging conception of progress".[331]

An increasing sense of distance from the ancients provides a clear marker of such a conception of progress from the 1660s onward. In the 1640s, for all the revolutionary novelty in nature-knowledge that had been attained and made public, the educated still felt at home in the world of the ancients, and new developments were routinely judged by criteria once set by Greek and Roman authors (p. 282). By the end of the 17th century this was no longer so. Or rather, whether or not this was how it still ought to be had become a major subject of contention, fought out at much the same time and with much heat in France and in Britain – 'Querelle des anciens et des modernes' and 'Quarrel of the Ancients and the Moderns'. Had Europeans by the second half of the 17th century managed to surpass the achievement of their ancient forebears, or had they not? In both prolonged debates the attainments of revolutionary nature-knowledge added fuel to the argument, notably so in a book by Glanvill

588   entitled *Plus Ultra* in deliberate allusion to Bacon (1668). The book had for its "main theme
. . . that the experimental philosophy, and the Royal Society in particular, have accomplished
more to advance useful knowledge in a few years 'than all the Philosophers of the Notional
way, since Aristotle opened his Shop in Greece'".[332] Lines written in the same year by the poet
John Dryden show to what extent a sense of liberation from the shackles of the past due to
the new developments in nature-knowledge was already reaching individuals even farther
removed from its practice than Glanvill:

> Is it not evident, in these last hundred years (when the Study of Philosophy has been the busi-
> ness of all the Virtuosi in Christendome) that almost a new Nature has been reveal'd to us? that
> more errours of the School have been detected, more useful Experiments in Philosophy have
> been made, more Noble Secrets in Opticks, Medicine, Anatomy, Astronomy, discover'd, than
> in all those credulous and doting Ages from Aristotle to us? so true it is that nothing spreads
> more fast than Science, when rightly and generally cultivated.[333]

In this regard, then, events in France and in Britain moved along parallel lines. But in just
about all other ways in which the new science came to be perceived as an engine and/or a
symbol for progress, things were different in France than in Britain. Whereas in France the
'engine' perception was primary, in Britain it was the 'symbol' perception, that is, what we
have called here the 'Baconian Ideology'.

   In France, the royal government rigorously cut the cultivation of revolutionary nature-knowledge into two. It neatly separated its 'engine' aspects (focused on whatever in
mathematical-experimental and in fact-finding experimental science might prove useful for
warfare or production) from its 'worldview' aspects. In taking the former up in its budding, bureaucratic machinery while carefully watching and, at times, persecuting the latter, it
overlooked nothing but the full measure of ambiguity built into Descartes' philosophy. Adherents like Rohault or Régis opted for the writ of the founder; others, in the spirit of free-
thinking Spinoza or of Poullain de la Barre, the early feminist, eagerly grabbed the messages
of supremacy of Reason and liberation from constraints upon thought that were equally
contained in Descartes' message (pp. 385; 435). In indiscriminately excluding all Cartesian
natural philosophers from the Académie, the royal government was surely effective in its effort to benefit from utilitarian science while keeping in check the sacrilege it saw involved in
the dogmatic letter of Cartesian philosophy. The unintended long-term effect, however, was
for its 'liberation' component to be developed outside the sphere of the reigning authorities,
so as to be directed ever more against them. In due time the new science, too, was to find
itself enveloped in the Enlightenment ideology of the 18th century 'philosophes' that was to
grow out of all this. But that was a matter for the future. For now, over the last third of the
17th century, such nature-knowledge as was sponsored by the state came without ideologi-

cal overtones, and the 'engine' aspect prevailed. The appeal to Bacon made by Huygens in his 1666 memoir for Colbert, however diplomatically astute at the surface, did not so much stem from a profound sense of where the French scene was heading but rather reflected an insufficiently thought-through transposition of his own, personally favorable experiences with the scene on the other side of the Channel.

In Britain, then, the kind of progress to be expected from revolutionary nature-knowledge proceeded along lines first drawn by Bacon. Its new visibility as a distinct enterprise of its own was not due, as in France, to an artificial split imposed from the top down. It stemmed rather from the (at least to outward appearances) relatively smooth collaboration between seasoned and less seasoned practitioners as well as sympathetic laymen from a variety of walks of life united in little else but the Baconian message in one or another guise. Due to efforts by Bishop Wilkins and others, what started with rebellious hotheads in the 1640s and 1650s had been steered into the calmer waters of the Restoration, with the new science, the ideology enveloping it, and the reigning powers in broad harmony with each other. True, the harmony thus established need not have lasted; if, following Charles II's death in 1685, the reign of his Catholic brother, James II, had not come unstuck it probably would not have, and the entire religious/political development of England might after all have taken a more Continent-like direction. But the revolutionary replacement of James by the Dutch *stadhouder* Willem III of Orange ensured that the France/Britain contrast did not lose its edge. The question that remains, then, is what originally brought the contrast about. How can we explain it? Ready candidates reside in differences in (1) state policy, (2) social structure, and/or (3) religious outlook.

Regarding *state policy* it is tempting to associate the different ways in which the pursuit of revolutionary nature-knowledge was officially handled with instant recognition of established verities about France's highly formal, centralization-prone, pedantically bureaucratic way of running things versus the stopgap mode of genial improvisation customarily held to characterize the British way of muddling through. The problem with an explanation along such lines is that the last four decades of the 17th century formed the very period when these seemingly timeless features of French government administration first took shape. The split within nature-knowledge effected by the government was itself part and parcel of a package of decisions made by Louis XIV to pursue a sustained policy of consolidation and expansion by means of a centralized, formally operating bureaucracy; decisions in the making, that is, which even in the situation Louis inherited might with a different sort of King have turned out differently.

The French/British contrast finds better causal illumination in a preexisting difference in *social structure*, which in England was more fluid than on the Continent. Men with an aspiration to rise in society the way Jonas Moore did had more chances to do so in Britain than anywhere else save the colonies. The informal way in which nobles and gentlemen associated with commoners in the Royal Society, albeit of course tinged by all kinds of status codes, nonetheless would have been inconceivable anywhere on the Continent but in the Dutch Republic (where revolutionary nature-knowledge failed to institutionalize for quite other reasons; p. 439). Decisive in this regard was the collective posture taken by the aristocracy, which in England was led with relative ease to accept taxation in exchange for a substantial share in the government, thus becoming part of society in ways other and more constructive than as a warrior class alone, whereas in France the aristocracy kept itself aloof from all but the jealous maintenance of its neatly hierarchical prerogatives. The rise of a numerically significant group of social climbers broadly sympathetic to revolutionary nature-knowledge while projecting a variety of aspirations upon it could not under such circumstances fail to be far off yet in France.

Differences in *religious outlook*, finally, contributed their own part to that 'failure'. The connection casually pointed at above between the Baconian Ideology and the shared Protestantism of those who created it (most often Calvinist or radically Baptist) and then went on to consolidate it (most often latitudinarian-Anglican) is not really casual at all. It rather represents an indispensable component of the full story of how revolutionary nature-knowledge regained legitimacy from the 1660s onward. Legitimacy of a wholly novel kind, to be sure. With the Baconian Ideology an unprecedentedly novel phenomenon came into the world. For the first time in history, a purely worldly enterprise (for the close link with philosophy-*cum*-theology, forged by Thomas Aquinas, had just been severed) was not only tolerated but actually sanctioned on grounds that derived from the reigning religion and its constituent values. Whence came this extraordinary development?

Sustenance for nature-knowledge in monotheist surroundings – a comparative summing up

No activity can be maintained in society that goes against its core values. In the Old World certainly, new initiatives tended to be perceived *a priori* as a menace to the social order. To have a chance to endure, they required either positive sanctioning or, at a minimum, a neutral stance with respect to the reigning values. In those civilizations which took their core values from a Holy Book, the sustained pursuit of nature-knowledge that stemmed from pagan resources ran two specific risks. The activity might widely be seen as *superfluous*, in that no valid knowledge was recognized beyond Holy Writ itself. Or it might appear as downright *sacrilegious*, in that it was regarded as incompatible with the core values on which society ran.

During the pioneering period of nature-knowledge in Islamic civilization, from the onset of the Baghdad translation movement until the downturn around 1050, no more than neutrality was attained. Abbasid society was remarkably open and tolerant, and such contemporary complaints on grounds of superfluousness as were voiced on occasion remained neatly confined to the margin. But no *positive* legitimation for such worldly pursuits came forward either, in spite of al-Mamun's short-lived effort to have the *mutazila* intellectualize and thereby sanction the 'foreign' learning sponsored by him. Hence, the translation movement and the enrichment that came from it could be quelled with relative ease once the tables turned and, in the wake of the invasions, self-styled representatives of Holy Writ gained the upper hand (p. 65).

During the first pioneering period of nature-knowledge in Christian civilization, which started with the onset of the Toledo translation movement, a positive sanctioning did come forward. To ward off the perceived risk of worldly, philosophic knowledge detracting from God's omnipotence, Thomas Aquinas forged a close alliance between Aristotle and Jesus Christ (p. 93), which quickly marginalized clerical complaints.

The alliance held over the next episode of transplantation and enrichment of pagan learning, which followed the fall of Byzantium and went on unabated up to and including the Golden Age of Renaissance nature-knowledge. But then the speedy downfall of its Aristotelian component in consequence of the onset of the Scientific Revolution unhinged the alliance by the early 1640s. Worse, sacrilege widely perceived to be implied in the teachings that came to replace the predominant philosophy put the pursuit of nature-knowledge in its realist-mathematical and natural-philosophic modes at the most serious risk. What saved the movement from petering out, as the present chapter has been concerned to show, is a number of successful, Europe-wide neutralization efforts, plus the reasoned expectation that revolutionary nature-knowledge would yield important material benefits (France, England), plus that expectation itself being built into an ideology that sanctioned open-minded research by means of an appeal to religious values deeply anchored in society (England specifically).

Key to the ever widening divergence in this regard between the Islamic world and Christian Europe, and to how the entire development culminated in the rise of the Baconian Ideology, is the extraordinary emergence in Latin Christendom of a markedly this-worldly orientation and of a consequent sanctioning of this-worldly pursuits as acceptable pathways to salvation (p. 263).

Religious sanction given to worldly activity generally and to the pursuit of nature-knowledge in particular worked in more than one direction.

It reinforced such outward-bound inclinations as were also induced by Europe's divisiveness and lack of self-sufficiency – its 'roving' qualities (p. 134).

It sustained Europe's enhanced naturalism – a tendency, noticeable already in the medieval period, to give a more prominent place than usual to the investigation of natural phenomena (p. 94).

592   It furthered the mid-15th-century emergence of a third, empiricist mode of nature-knowledge around the edges of the two Greek ones, along with the specifically coercive nature that its empiricism acquired when a vision of human mastery over nature emerged. It also made for an overrepresentation of Protestants in the pursuit of this particular, non-Greek mode of nature-knowledge (p. 265).

It further made for the presence in Europe of men who, while not occupied with the pursuit of nature-knowledge in any direct sense, nonetheless made a living from it. Unlike in Islamic civilization, in Europe a relatively substantial social layer of skilled craftsmen, mariners, mapmakers, city mathematicians, and the like arose at the time, whose activities were rooted in essentially the same, ultimately faith-derived or at least faith-supported values. As the 17th century began and the pursuit of nature-knowledge was first revolutionized, it was in such expanding social layers that what had meanwhile grown into the everyday business of practical mathematics, instrument building, and the like helped revolutionary nature-knowledge look relatively familiar. Well into the 17th century it was also in such social layers that the utopian schemes of Comenius and Hartlib and their ilk struck a sensitive chord and in which, a little later, the Royal Society found part of its sustenance.

It finally made for the Baconian Ideology, the most extreme form so far in which the this-worldly tendencies of Latin Christendom came to the fore. It appealed in particular to the very circles just mentioned – those men, British Protestants chiefly, who perceived their aspirations in social life to coincide with the values embodied in the experimental pursuit of nature-knowledge.

In Catholic France, amid a rising tide of repression of the Huguenots, the Baconian Ideology struck few roots. Here, in the domain of nature-knowledge a this-worldly religious orientation remained confined to the kind of cool, religiously neutral calculation of present and future material costs and benefits of revolutionary nature-knowledge that was undertaken by Louis XIV's government.

In Islamic civilization, something like the Baconian Ideology would have been altogether inconceivable. Among values underlying the civilization at the time one searches in vain for any that might seem fit to sustain the – world-historically speaking, quite extreme and utterly unprecedented – kind of sentiments expressed by Francis Bacon and elaborated and varied after him by men like Hartlib, Boyle, Hooke, Sprat, and Glanvill and all those other promoters of the conquest of nature through a scientific insight into her operations.

All of which is meant to serve at long last as the full grounds for the bald assertion, made on p. 72 when I first addressed the question of why Europe underwent a Scientific Revolution whereas its elder stepbrother in nature-knowledge, Islamic civilization, did not, that even if a Galileo-like revolutionary transformation had taken place in the latter, as quite conceivably it might have, its chances for long-term survival would have been near zero. Fierce accusations of sacrilege would, in a culture similarly saturated with veneration for a

Holy Book, have been inevitable. Nothing but a sustained development of underlying values 593
in a pronounced, self-consciously this-worldly direction might have made for ideological
arms of sufficient strength to counter such accusations. In Islamic civilization, for reasons
not bound up necessarily with its faith or with anything else, but due above all to the vicissitudes of history, the this-worldly orientation that had manifested itself in Abbasid times leveled off and was then reversed at a critical juncture (p. 65). In Europe, in likewise contingent
historical ways, such an orientation went on and on, reaching its Reformed climax at about
the same time as (in a development starting with the fall of Byzantium) the civilization came
into its own scientifically. No built-in, essential, as it were God- or Allah-given necessity accounts for the vast difference in outcome; the net result of the inquiry is grounds for neither
shame nor pride, but just a case of value orientations grown ever more widely apart.

The road to atheism.   Not only did revolutionary nature-knowledge benefit hugely
from the this-worldly orientation of Europe's religious development, to the point of owing
its very survival to it by midcentury, but once returned to full bloom by the 1660s, revolutionary nature-knowledge quickly began in return to give that orientation a boost of its
own. Within a mere four decades of Voet opening his campaign against Descartes, Galileo,
Kepler, and the menace to proper philosophy and proper religion that these men jointly represented (p. 423), theological stances like those that Voet and his ilk had foreseen and sought
to ward off began indeed to emerge from the resources of revolutionary nature-knowledge.
No man in the 17th century was to delve more deeply into the intellectual legacies of these
three pioneers than Isaac Newton did. Equally, no man was to think more deeply through
the consequences that revolutionary nature-knowledge entailed for a viable conception of
the Deity. In his published work Newton accepted current views of ongoing divine intervention (p. 710). But in the early 1680s already, in the privacy of his theological studies and the
anti-Trinitarian heresy contained therein he saw himself almost willy-nilly driven toward an
as yet nonexistent, near-Deist position, in such a way as to turn upside down the customary
order of things:

> His religious quest presents itself . . . as an effort to purge Christianity of irrationality. . . . At
> his hand, Christian theology felt the touch of cold philosophy. There was much that failed to
> meet the test. . . . Nowhere did he approach the Bible as the revelation of truths above human
> reason unto life eternal.

And here is what, in Newton's own words, his quest culminated in:

> So then the first religion [which Newton ascribed to Noah and found in essentially similar
> fashion in the records of other civilizations] was the most rational of all others till the nations

594   corrupted it. For there is no way (wthout revelation) to come to ye knowledge of a Deity but by
ye frame of nature.

Note that the two words here put in parentheses were added by Newton later, "as an after-thought".[334]

By the 1680s, then, the primacy of science over religion so familiar to us nowadays had already been reached in the private thought of at least one expert, more capable than anyone at the time of probing the depths of both. It is only in retrospect that we can see this radical outcome as historically in line with the this-worldly value orientation Europe had inadvertently begun to acquire some ten centuries earlier.

NOTES ON LITERATURE USED

Conceptualization. Issues pertinent to the legitimation of the new science have occupied historians of science from Robert K. Merton's *Science, Technology and Society in Seventeenth-Century England*. New York: Harper & Row, 1970 (published originally in *Osiris* 4, 2, 1938, pp. 360-632) onward, with sustained attention to such issues over the past decades given a big boost by Steven Shapin & Simon Schaffer, *Leviathan and the Air-Pump. Hobbes, Boyle, and the Experimental Life*. Princeton UP, 1985. Yet I am aware neither of any coherent treatment of the full issue nor of sufficient conceptual clarity being yet attained over the vital difference between issues of *legitimacy* and of *validity*, which tend in the literature to be conflated in the question-begging neologism 'sociocognitive legitimation'. For treatment in the main text I have drawn on a wide variety of scholarly resources, with views from Merton, Ben-David, Rabb, and many others serving as my point of departure (I discussed them at length in *SRHI*, sections 3.6.1., 5.1.2., 5.2.3., and 5.3.).

NB: To the extent that the present chapter examines 'from the outside' issues considered in earlier chapters 'from the inside', I refer the reader to the 'Notes on Literature Used' for those chapters (13 and 14 chiefly).

European framework. A wonderful panorama of cultural/intellectual change in Europe of the late 17th and early 18th centuries is sketched in Paul Hazard, *La crise de la conscience Européenne (1680-1715)*. Paris: Boivin, 1935; 3 vols. (*SRHI* 3.6.1.). One succinct summing up of the shift of geographic center generally is in R.R. Palmer & J. Colton's ever-refreshed textbook of post-1500 history, *A History of the Modern World*; p. 160 in the 7th edition (first page of ch. 4, 'The Establishment of West-European Leadership').

Measured quantities. Alfred W. Crosby, in the development he sketched in his *The Measure of Reality. Quantification and Western Society, 1250-1650*. Cambridge UP, 1997, has given far wider scope to his argument than I do here. In his view, the drift toward quantification that he engagingly analyzes was the prime marker and origin of Europe's unique historical development in acquiring world dominance and ushering in the modern world. A failure to distinguish properly between quantity and mathematical relation is but one of the defects of his vastly overstretched thesis (as I have argued at greater length in a book review in Dutch).

Trade in scientific instruments. Peter de Clercq, *At the Sign of the Oriental Lamp. The Musschenbroek workshop in Leiden, 1660-1750*. Rotterdam: Erasmus, 1997.

Title pages. Volker R. Remmert, *Widmung, Welterklärung und Wissenschaftslegitimierung. Titelbilder und ihre Funktionen in der Wissenschaftlichen Revolution*. Wiesbaden: Harrassowitz Verlag, 2005 (Wolfenbütteler Forschungen, Band 110).

Princely distance. Mario Biagioli, 'Etiquette, Interdependence, and Sociability in 17th-Century Science'. *Critical Inquiry* 22, 1996, pp. 193-238.

Longitude awards. D.S. Landes, *Revolution in Time. Clocks and the Making of the Modern World*. Cambridge (Mass.; Harvard UP), 1983. C.A. Davids, *Zeewezen en wetenschap*. Amsterdam/Dieren, 1986.

596   **British Interregnum**. Charles Webster, *The Great Instauration. Science, Medicine and Reform 1626–1660*. London: Duckworth, 1975 (where, as I saw belatedly, the expression 'Baconian ideology' occurs on p. 99); also his 'Puritanism, Separatism, and Science'. In: D.C. Lindberg & R.L. Numbers, *God and Nature. Historical Essays on the Encounter between Christianity and Science*. Berkeley: University of California Press, 1986; pp. 192-217.

Anglicanism. I owe the points made in the text about the 'doctrinal minimalism' and lack of patience with religious proof built into the Church of England to John Henry, who made a persuasive argument for this in a chapter 'National Styles in Science: A Possible Factor in the Scientific Revolution?' in D.N. Livingstone & C. W.J. Withers (eds.), *Geography and Revolution*. Chicago: University of Chicago Press, pp. 43-74.

Rabb's thesis, as set forth in Theodore K. Rabb, *The Struggle for Stability in Early Modern Europe*. Oxford UP, 1975, is that revolutionary science itself was instrumental in Europe's post-Westphalian return to a measure of stability. As the analyses in the main text show, there is every reason to agree with his fundamental point that the greatly enhanced visibility and staying power displayed by the new science in the second half of the 17th century are historically extraordinary feats well in need of specific explanation. But Rabb overdoes it – what the kings of France and England wanted and got from the new forms of science was substantially less than having these serve as a major tool for stability. The science policies here examined show no more than the presence of a perception, from the top down, that revolutionary nature-knowledge (in France) could enhance royal glory and offered a vast, partly realized potential for improved warfare and intensified production, and (in Britain) offered a welcome opportunity to keep a bunch of potentially disruptive thinkers innocently occupied at no cost at all.

Merton thesis. In picking up Weber's hint (p. 269) and extending it in his *Science, Technology and Society in Seventeenth-Century England* of 1938, Robert K. Merton sought to make the following points: (1) science in 17th-century England was very much practice-oriented; (2) a noticeable correspondence of values with Puritan, this-worldly asceticism goes some way toward explaining such a stress on practical utility; (3) the correspondence helps explain why disproportionately numerous Englishmen with Puritan leanings, while of course facing other career options as well, in the course of the century flocked to science. Surveying the voluminous literature devoted since to critical review of what rightly or (more often) wrongly has passed for Merton's thesis, a later historical sociologist, Joseph Ben-David, proposed to cut out the third claim, on career choice, as weak and unproven, while seeking to strengthen the remainder in other ways. Taking up hints dropped by Merton along the way, Ben-David focused on the world-historically unprecedented rise of an ideology able to legitimate experimental, practice-oriented, forward-looking, progress-bearing science in times prior to those when, at long last, the scientific-technological fruit could be reaped that was then to make further religious sanction for the scientific enterprise more and more superfluous.

My own views on these matters, as rendered in the main text, are inspired by a selection of the various historical-sociological conceptions just outlined. From Weber I have adopted his point about the extraordinarily this-worldly orientation developed in Latin Christendom and culminating in a

distinctly Protestant ethic, but also his hint about not the new science itself but its exploitability for   597
economic gain being particularly associated with Protestantism (a hint most clearly expressed on the
penultimate page of Weber's 1924 book *Wirtschaftsgeschichte. Abriß der universalen Sozial- und Wirtschaftsgeschichte*, and in the concluding passage of Weber's second rejoinder to the second major critic
of his 'Protestant Ethic', Felix Rachfahl [reprinted in M. Weber (ed. J. Winckelmann), *Die Protestan-
tische Ethik, II. Kritiken und Antikritiken*. Gütersloh: GTB Siebenstern, 1987, p. 324-325]). If Merton
had straightforwardly adopted this extraordinarily perceptive insight of Weber's, rather than weaving
it into a tissue of side issues, he might well have prevented the ensuing debates about his thesis from
being sidetracked time and again from the simple core message. Among these side issues are Merton's
needless, unconvincingly argued emphasis on Puritanism, i.e., near exclusion of Anglicanism, which
may or may not make sense for Weber's own thesis on Calvinism/baptism and the spirit of capitalism, but hardly if at all for issues involving *science* ('nature-knowledge') and religion. Later misguided
efforts to make Merton's thesis cover, not just experimental science in 17th-century England but the
Scientific Revolution as one undifferentiated whole, then contributed even further to obscuring the
basic truth hidden in Weber's hint, by making what passed for the Merton thesis lethally vulnerable to
justified pointers at all those Catholic contributions to innovative 17th-century nature-knowledge or
(even more pointedly) at the very substantial, Jesuit contribution to experimental science (*SRHI* 4.2.3.,
5.1.2., 5.2.3., 5.3.).

Not that my own findings in the present chapter square fully with Weber's hint, which certainly
requires expansion in terms of the Baconian Ideology – a concept that I owe to Ben-David and that I
have defined in the main text as a leap of faith in the power of science. Leaving aside all niceties about
what exactly it was in Protestantism that made its adherents so much more prone to such a leap of
faith than is to be observed in Catholic circles at that or later times, I maintain that, phrased thus
broadly, the connection is empirically unmistakable. This assertion does not, of course, come down
to a simple-minded equation 'where Protestantism, there Baconianism' (few traces of which can be
found, for instance, in the Dutch Republic). But it does mean that at those centers in Europe where the
new science (marked by a sacrilege-countering stress on fact-finding experiment) did find itself institutionalized (Rome, Florence, Paris, London), the one center where an ideology grew around it which
focused on its capacities for the practical improvement of humankind's fate on Earth was the one
where Protestantism, not Catholicism, was predominant. Differences in the relative fluidity of social
structure certainly contributed further to this differential outcome. Still, the bald observation strongly
suggests a ringing affirmation of Weber's basic insight that the Protestant variety of Latin Christendom
is where its this-worldly orientation had found its most thoroughgoing expression yet.

In *SRHI*, 5.1., I have outlined a variety of claims about the place of religion in the Scientific Revolution, with special attention given to pertinent theses by Hooykaas and by Merton.

Newton on the move toward deism. R.S. Westfall, 'The Rise of Science and the Decline of Orthodox
Christianity: A Study of Kepler, Descartes, and Newton'. In: D.C. Lindberg & R.L. Numbers, *God and
Nature. Historical Essays on the Encounter between Christianity and Science*. Berkeley: University of
California Press, 1986; pp. 218-237.

XVIII

NATURE-KNOWLEDGE BY 1684: THE ACHIEVEMENT SO FAR

In August 1684 Halley's visit to Newton provoked the creative outburst that led in just two and a half years to the highest achievement of the Scientific Revolution, the *Principia*. The resources on which Newton could and did draw for the revolutionary transformation that he accomplished in that short period derived from all previous revolutionary transformations so far examined. I shall now recapitulate their attainments and the ways in which these had come about; in the next chapter, we find out how Newton drew many of the lines together. What, by 1684, did the big picture of modern-science-in-the-making look like?

In the survey that follows, there is no need to stop punctually in 1684 in every case. With topics like the funding of the Académie or van Leeuwenhoek's microscopic work, a given line of development has been pursued until near the end of the century or somewhat beyond, as the case might require. Here the main purpose is to review the major developments and also to draw certain conclusions left pending so far:

- What had, by the end of the 17th century, become of the three prerevolutionary modes of nature-knowledge: Alexandrian mathematical science, Athenian philosophy of nature, and Renaissance Europe's control-oriented empiricism?
- To what, if any, extent did the speculative approach to natural phenomena *as such* survive the onslaught of its new rivals, mathematical-experimental and fact-finding experimental science?
- To what, if any, extent was the Greek corpus of nature-knowledge still felt by century's end to be a living presence?
- What about the arts and crafts, in view of claims made throughout the Scientific Revolution for their scientific improvement?
- What forces propelled the Scientific Revolution forward beyond its pioneering stage?
- What did the full achievement add up to if considered from the viewpoint of an outside observer, if we fancy such a person to be charged in 1684 with reporting on its major strengths and weaknesses?

The reporter's imaginary report will reveal, as one major weakness, ongoing conceptual confusion over an issue that kept turning up from Galileo and Descartes onward – is a body's motion brought about or maintained by some force and, if so, how? Since final resolution of that burning question was the very hallmark of what Newton attained in the *Principia*, I end the present chapter by examining what half a century of assiduous thinking on the subject had led to by 1684.

Predecessors on their way out

Revolutions neither result from complete breaks with the past, nor do they bring a rupture about all at once. The three distinct modes of nature-knowledge that by 1600 filled the European scene did not take their exit all at once or of their own accord. In each case remnants lingered on for a while. Here is how they did that.

The fate of pre-1600 modes of nature-knowledge.   In two cases out of three, *absorption* is the main theme. Setting up progressive series of fact-finding experiments was still quite rare by the end of the 16th century – a hundred years later even the most purely descriptive activities, as in collecting and cataloguing and writing natural histories, had become infused with making and reporting on experiments wherever suitable (ch. 13). But this also means, conversely, that prerevolutionary work by, for example, Vesalius and Tycho was far from being fully repudiated. The difference is that work in their lines was now pursued at greater depth and with more precision by means of microscope-supported experimentation and telescope-guided observation. In short, the new approach supplanted the elder while incorporating some of its principal features.

The situation was similar for the revolutionary transformation of Alexandria into Alexandria-plus. By 1684 planetary theory was still affected by the long-standing addiction of practitioners to the sheer saving of phenomena, but for the rest mathematical science had ceased to be handled in the thoroughly abstract manner of Archimedes or of Ptolemy's *Almagest* (ch. 10). Even so, the Alexandrian achievement lived on in the new, realist mode of mathematical science. The properties of light rays were now studied in early attempts at physical optics, consonant intervals were now handled in close conjunction with the vibrations that produced them, equilibrium states were now taken up in the budding disciplines of statics and hydrostatics, and even planetary theory was handled in realist fashion wherever Kepler's celestial physics and its outcomes kept being pursued.

Inevitably, the Athenian picture looks quite different. The very approach of comprehensive speculation did not allow for smooth absorption. To be sure, almost right from the start there were many reasons for kinetic corpuscularianism to look much more appealing than any of its elder philosophic rivals (p. 383). But this did not mean that those rivals just disappeared without leaving a trace. Still, each underwent drastic changes. So, too, did natural philosophy as a mode of nature-knowledge in its own right.

Aristotelianism was by far the most solidly entrenched institutionally among existent natural philosophies. Its reception in medieval Europe had coincided with the rise of the universities. It had captured these as a matter of course, going on for centuries to shape the minds of Europe's intellectual elites. When the scholastic forms in which it had been clad came under grave assault from the humanist movement in the 15th and 16th centuries,

Aristotelians had risen to the challenge. In returning to its Greek sources, in readdressing unresolved questions, and above all in adopting novel modes of presentation, they managed to revitalize the doctrine. Early in the 17th century it came under relentless attack of a principally novel kind, at its most spectacular in clashes centered on immersion (p. 404) and on the void (pp. 410; 484). These and similar attacks upon Aristotle's doctrine were marked by a blend of polemic vigor, highly specific targeting, often fair-minded criticism, sometimes very unfair propaganda, and a general tendency to hold the doctrine up for ridicule. As such, the attacks expressed a genuine inability of advanced thinkers to make proper sense anymore of such basic, Aristotelian concepts as 'substantial form' or the potentiality/actuality definition of local and other change, all of which had for centuries made perfect sense even to the minds of those who had little or no explanatory use for it.

The weakening consequent upon such attacks was reinforced by the tight coherence between diverse components of the doctrine. That coherence had long served to keep the whole together when local leaks appeared (p. 84) but now began to work against it as delle Colombe and his Pisan colleagues in the 1610s already came to realize with such acute alarm (p. 405). In time even the staunchest defenders of the fortress as one impregnable whole, the Jesuits, felt compelled to retreat from so large a number of really indispensable Aristotelian positions as to fall back upon the stratagem of working a selection of remaining positions plus the general tenor of the doctrine into their particle- and experiment-infused, inevitably deeply incoherent synthesis.

The original institutional stronghold of the Aristotelians, academia, was not committed to the forefront of learning the way the Jesuit Order was. Even so the universities felt that they had been put on the defensive as well. Here, too, refuge was sought in all kinds of compromise formulae, as in 'new-old philosophy'. Still, the full extinction in academia, too, of Aristotelian natural-philosophical doctrine was not to come about until far into the 18th century. Neither Galileo nor Descartes nor Bacon delivered the deathblow – Newton did, or rather the textbook tradition that was extracted early in the century from his *Principia* and *Opticks*.

For Aristotelianism as a feasible conception of the natural world this was indeed the end. Not so for Aristotle's expertise with regard to the parts of animals and the functions thereof but also with regard to logic, to poetry, and to politics. To zoologists his pertinent works remained a force to be reckoned with until the mid-19th century; to logicians, until that century's end; to certain literary critics, at least up to and including the Chicago school of the 1950s; to political scientists, they still are.

With *Platonism* it was much easier for adherents to let go of what pieces of natural philosophy the full doctrine had to offer, in view both of its weaker institutional position and of the more eccentric, less critical place of natural philosophy inside the doctrine as a whole. The picture of the world conjured up in the *Timaios*, still such a source of inspiration to

602   Kepler, did little as the years went by but bequeath its world-soul to the aethereal spirit that went into the Baconian Brew (p. 554). Meanwhile, even Henry More and other Cambridge Platonists, so fervently engaged in upholding the autonomy of spirit in the face of the apparent materialism of Descartes' teachings, found themselves unable to do so on other terms than those of kinetic corpuscularianism itself (p. 399). Henceforth, for Platonists the obvious move was to focus on Plato's other teachings, the timeless problems and literary brilliance of which have kept enticing readers up to this day. The elevated knowledge status assigned in Platonism to mathematics has, ever since Kepler and Galileo, been indulged in the sterner frame of the realist-mathematical science created by them; it is, however, true that certain mathematical scientists (notably, Einstein and Heisenberg) were to find in Platonism a source of inspiration for their own budding convictions regarding the world-building (or world-constraining) power of mathematics.

For stoicism even more than Platonism did the center of gravity rest in ethical and political conceptions hardly touched upon by the Scientific Revolution. As a distinct philosophy of nature, stoicism likewise disappeared quietly from the scene, yet not without leaving something of great use to the very doctrines that had dispelled it. Among theories of a universal aether developed in the second half of the 17th century, several, while purely corpuscularian on the surface, came to display many more properties of the continuous pneuma of stoic provenance than of the atomic fragmentation from which this class of conceptions originally sprang. By far the most significant aether conceptions constructed along such hidden, stoic lines were developed in the 1670s by Isaac Newton (pp. 561, 642).

The revival of *atomism* in the late 16th and early 17th century took several guises. Its transformation into kinetic corpuscularianism, with its radically novel focus on the varied motions that particles were taken to be subject to, came to dominate particle thinking throughout the 17th century (ch. 11). Kinetic corpuscularianism as a philosophy of nature was not to survive, however – in the course of the first half of the 18th century Newtonian forces (or, in German lands, Leibniz's 'monads') gained the upper hand over Cartesian whirlpools. Original atomist doctrine had lived on in the meantime, not as a philosophy of nature any longer but rather as a background worldview animating chemical theorizing in particular (p. 469). By the mid-18th century, with kinetic corpuscularianism out of the way, atomism revived for a while as a speculative philosophy of nature, particularly so among certain French Enlightenment philosophes. Atomism definitely became a part of chemical science when, early in the 19th century, John Dalton managed to link up Lavoisier's new approach to chemical elements with atomic doctrine. During the 19th century the scientific status of atomism remained heavily disputed; not until a few decades before Rutherford shot the atom to pieces was informed scientific opposition to the physical existence of atoms to yield to the inevitable. The 'modern' atom, constitutive of the elements in ways inconceivable to Demokritos or Epikouros, has little if anything in common with how Greek

speculative thought originally conceived it. Still, of the five Athenian schools of philosophy 603
it is atomism alone that has managed to keep advanced thinking about the constitution of
matter busy over the full period of its existence.

The *skeptical* doctrine, finally, was not of course a philosophy so much as, rather, a coherent demonstration of its congenital lack of epistemic certainty. Descartes had set out to slay
the skeptical dragon at one stroke, by means of his *cogito* (p. 236). His overstretched argument proved less effective as the Scientific Revolution advanced than the very tenor of the
science to come out of the revolution. That tenor was, as it were, one of skepticism turned
constructive. Practitioners of nature-knowledge in Galileo's wake came more and more to
share with ancient and modern skeptics abandonment of first-principle thinking, lack of
satisfaction with the quick fit, and disdain for seeking refuge for real problems in verbal solutions. But unlike Pyrrhôn and Sextus Empiricus and their latter-day followers, scientists of
various new stripes did not leave it at that critical part. They busily set out to discover ways
and means to build into their various methods such checks upon unbridled fancy as could
help ensure a measure of reality confirmation for their positive findings (pp. 363, 501). No
one has phrased in more pointed fashion the resulting stance of modern science half-way
wholesale certainty and wholesale *un*certainty than Blaise Pascal:

> We have an incapacity for proof which no amount of dogmatism can overcome. We have an
> idea of truth which no amount of skepticism can overcome.[335]

As one major result, skeptical modes of thought were redirected to the domain where, upon
the printing of Sextus' works, they had come to the fore in the first place – theology. Employed previously to undermine the alleged certainty of either Catholic or Reformed tenets
of belief, skeptical arguments now began, at the hands of Spinoza and several other thinkers
inspired by Cartesian doubt, to be directed against the foundations of religious belief itself.
Still, the skeptical mode of thought was not then or later to withdraw from the domain
of science altogether; rather, it was to turn into its steady companion over the centuries.
Skepticism, no longer, as a rule, in the format of a radical denial of the possible validity of
knowledge about nature but rather in forever shifting shades and guises – probabilism, Hu-
mean skepticism, positivism, relativism one way or another – was directed always toward the
exploration of how far the knowledge claims of science may be taken to extend. The skeptical temper has not, therefore, vanished from the world as a result of the advent of modern
science; rather, modern science has provided it with a suitable terrain to exercise its often
considerable, yet always circumscribed, skills.

Natural philosophy on its way out.   The long and short of the foregoing survey is
that the Scientific Revolution brought about a restructuring from top to bottom of the very

604   domain of natural philosophy, which came to involve the latest variety, kinetic corpusculari-anism, as well. In this regard the following broad picture presents itself.

Prior to the Scientific Revolution, natural philosophy appeared in three distinct guises, dogmatic, syncretist, and opportunist. *Dogmatic* (or, in early Royal Society parlance, 'no-tional') is the label that saw increasing use over the 17th century for the original, Athenian knowledge structure of first principles stretched to explain the world in its totality and sus-tained by a modicum of supporting evidence. This mode of natural philosophy predomi-nated over the twenty-odd centuries between the 4th century BCE and the 17th century CE. *Syncretist* has from the 2nd century BCE been one standard expression for those philoso-phers concocting, out of tenets selected at will from any combination of dogmatic philoso-phies, a necessarily incoherent philosophic blend aimed likewise at explaining the world in its totality. *Opportunist*, in significant contrast, is what (using Einstein's phrase) we have called a stance first taken by Ptolemy in his pursuit of mathematical science – without com-mitting to any wholesale explanation of the world he invoked separate tenets of any available philosophy wherever they could be of service as a stopgap argument.

During the Scientific Revolution these three varieties fared as follows. *Dogmatic* natural philosophy retreated, then began to dissolve in ways just specified for each (not counting Cartesianism and the system made out of Leibniz's philosophy, both to harden and survive until far into the 18th century). *Syncretism* (basically unchanged) ran rampant in the Jesuit synthesis as also in all sorts of even less creative, academic compromise philosophy. *Op-portunism* (likewise unchanged) became the hallmark of mathematical scientists who, in Galileo's vein, subordinated their philosophical education to the primacy they gave to the exigencies of realist-mathematical science. But during the Scientific Revolution new ways to handle natural philosophy also emerged: as a hybrid, as an eclectically employed resource of a background worldview, and as a source for specific hypotheses. No more than two *hybrid* philosophies of nature were ever to see the light of day: Kepler's effort to build an *a priori* world upon mathematical science in ways subject to ongoing, empirical *a posteriori* check-ing, and a quite differently tinged variety that Newton created (p. 711). Natural philosophy employed as one possible *resource for background worldviews* was prefigured in, for example, Paracelsus' work and its magical background. It came into its own in fact-finding experi-mental science, where, for instance, Harvey was helped in his experimentation and theoriz-ing on the flow of blood by the Aristotelian notion of *circulatio* (pp. 253, 259). Finally, the kinetic-corpuscularian philosophy of nature was in revolutionary ways turned into a source of quite specific, *hypothetical* mechanisms. Out of this came the Baconian Brew as a histori-cally significant, short-term product (ch. 16) and physical optics and mathematical accounts of generic motion as scientifically very significant, long-term products (ch. 15).

Compared, then, to the temporary supremacy that one or another philosophy of nature had enjoyed in previous eras, an even more radical upheaval had now taken place. With not only its various categories but its very knowledge structure radically overhauled, natural philosophy *as a distinct mode of acquiring knowledge of the external world*, albeit hardly vanquished once and for all, had by the end of the 17th century lost its near monopoly for good.

Cessation of dialogue with the ancients. By the 1640s ongoing dialogue with the ancients had certainly not ceased; by the 1660s, in the Quarrel of the Ancients and Moderns, a sense of distance from the ancients had begun to be expressed. As the century advanced, so did the sense of distance. Take the case of late-16th- and early-17th-century debates on comets. Even open-minded participants in these debates, ready in principle to abandon one view on the supralunar or sublunar nature of comets in favor of another, listed pertinent ideas of ancient and modern authors side by side, as views of equal standing. That habit did not fully come to an end until Newton's *Principia* decisively altered the terms of the debate. Not until then was a full-scale 'cessation of dialogue' attained; not until then did ancient authors begin to be routinely cited only for the occasional celestial observation or in accounts meant to be historical, and not any longer for views worth examining as such.

Another case in point concerns the altered status of music. By 1600 audiences were at least dimly aware that music conveyed the cosmic harmony in which it was enveloped (theoretically *and* practically). By 1700 music was no longer felt to express any such thing but rather to have for its sole purpose to please the senses. In the performance of music, by 1600 Pythagoras was still a living presence; by 1700, at court and in the church but also in the emerging public concert halls, he had vanished from the scene for all but antiquarians or the odd scientist. It was not that scientific discovery was totally severed from ancient resources by the end of the century. Newton came to regard his *dis*covery of universal gravitation diminishing with the square of the distance between bodies as a feat of *re*covery. Pythagoras, so Newton thought, had deliberately hidden this law in the story (rendered falsely so as, according to Newton, the better to hide the true law) of how he had found a similarly inverse-square relation. To alter the pitch of a note emitted by a vibrating string one must either multiply by a given quantity the weight used to stretch it or divide its length by the square root of that quantity (this was actually Vincenzo Galilei's find; p. 145). The economist John Maynard Keynes was the first to rummage through the vast depositories of Newton's papers with other purposes in mind than the reconstruction of his work in 'positive' science. In encountering passages like the one on how Pythagoras hid his early knowledge of universal gravitation, Keynes had good reason in the early 1940s to draw the oft-quoted, attractively overstated yet by no means unfounded conclusion that, rather than "the first of the age of reason", Newton had been

the last of the magicians, the last of the Babylonians and Sumerians, the last great mind which looked out on the visible and intellectual world with the same eyes as those who began to build

our intellectual inheritance rather less than 10,000 years ago . . . the last wonder-child to whom
the Magi could do sincere and appropriate homage.[336]

It is still refreshing to see Newton, the 'first' in so many respects, as the 'last' in the sense of
closing off an entire age. Indeed, by the time Newton had passed away, 'dialogue' in the sense
of routinely, rather than for some specific purpose, consulting ancient authors not only as
conveyors of one or another dogmatic philosophy of nature but as a possible source of timeless wisdom had (at least in the sciences) become a thing of the past.

17th-century props

In our time the permanence of the scientific enterprise is warranted by an ever-widening,
ever-advancing research front and by a science-based technology that keeps supplying us
forever with new, most often science-imbued products. A good part of the investigation
so far has been devoted to finding out whether, and if so to what extent, these two major
props served already in the 17th century to keep the new science going. Did something like a
science-based technology emerge during the Scientific Revolution? And was there an identifiable research front that kept moving forward? In foreseeing the possibility of a negative
answer to one or both of these questions, I have also examined whether perhaps other props
kept the Scientific Revolution going.

A variety of conclusions on these questions has been reached along the way. It is now
time to assemble them and to sketch the broad picture that emerges.

'Thinking with the hands' before and during the Scientific Revolution.   Prior
tothe mid-15th century, a certain amount of intertwinement between craftsmanship and
nature-knowledge marked China far more than it did Europe (p. 137). Not until the corpus of highly intellectualist, Greek nature-knowledge was supplemented by a large range
of empiricist investigations bent on practical improvement did opportunities arise for the
emergence of interfaces between the two, hitherto wholly separate domains.

Indeed, between that time and the wholesale conversion of craft practice into science-
based technology, the arts and crafts served as useful resources for drawing verbal analogies
and as straightforward sources of inspiration for more theory-oriented pursuits. Acquaintance with centuries of craft literature in the vernacular helped Galileo in his Padua days to
abandon the abstract Archimedean approach current in the circle of his patron, Guidobaldo
del Monte (p. 182). On the lookout for a macro-world analogon to some intricate mechanism of moving particles, more often than not kinetic-corpuscularians found it in craft
practice, as for instance in how Beeckman compared the way animal spirits move through

or alongside the nerves with how water may stream both through a conduit and alongside it. Bacon, in seeking to make his case for collaborative research, invoked the example of the artisans (p. 246). When Mersenne wanted to gain an understanding of musical sound, he sought out the makers of violins or trumpets or church organs in their workshops (p. 449). The Royal Society even went so far as to commission a wholesale 'History of Trades' (p. 481). Although it would go much too far to ascribe the revolutionary work of the pioneers to their concern with craft practice alone (p. 634), it is certain that throughout the Scientific Revolution a variety of interaction patterns manifested themselves. This was of course most conspicuously the case with the new scientific instruments. 'Thinking with the hands', so marginal by the century's beginning, had by its end become well-nigh indispensable to the pursuit of revolutionary nature-knowledge.

How did these interaction patterns work out in the reverse direction? Certain craft practices began from the mid-15th century onward to be informed by mathematics or by outcomes of sustained observation. With linear perspective, fortification, and the determination of place on Earth for navigational purposes improvement was indeed attained, with considerable economic profit in the latter case. There were also numerous failed attempts, as with the calculations by means of which Stevin hoped to enhance the draining power of Dutch windmills (p. 131). And there were attainments for which it is impossible to decide whether they actually were of any practical use, as notably with those mineral cures that derived from Paracelsus' extended theorizing about the constitution of the world.

The advent of drastically novel modes of nature-knowledge by the early decades of the 17th century was accompanied by high expectations right from the start. It seemed to highly visible pioneers like, notably, Galileo, Bacon, and Descartes as if the novel insights and approaches now probed in mathematical science, in fact-finding experimental science, and also in natural philosophy could be used to jolt craftsmen out of their reliance on rules of thumb and to subject their trial-and-error methods to orderly, well-thought-through procedures. Jupiter's moons could be pressed into service for resolving the problem of longitude; experiments with sounds and their echoes could show the way toward amplifying sound over many miles. Indeed, as Descartes pithily phrased it, the human race could turn itself into "masters and possessors of Nature".[337]

For all the promise involved in the new approaches, very little of these high expectations was actually fulfilled in the course of the Scientific Revolution. By the end of the 17th century, craftsmanship looked much like it did a hundred years previously. Efforts at mathematization in about a dozen craft domains did not lead to improvements substantially beyond what in practical mathematics had already been attained before, with the sole exception of the telescope and the pendulum clock. Four distinct impediments stood in the way. The world is far messier than mathematical scientists first realized. Even when they began to recognize this, their mathematical techniques were as yet too constricted to be of much help.

608   Also, even in those cases where these techniques were already up to scratch, craftsmen were not yet prepared to master them. All this was exacerbated by the vast social distance between the world of learning and the domain of craft practice – a gap as yet overcome only in the rarest of cases (pp. 325; 482).

For mostly the same reasons, efforts to draw practical improvement of the arts and crafts from outcomes and products of fact-finding experimentation equally yielded little, except, once again, in the domain of the scientific instrument (pp. 333; 473).

All this began to change by the 18th century. By means of the calculus, messiness could be subjected at times to second-order rule. Even more importantly, craftsmen of a new kind emerged, who were prepared to master the theoretical knowledge required for the case at hand, and managed on occasion even to engage with theorists on near-equal footing. More than anywhere else this happened in Britain, where science of a less sophisticated but more practically viable kind than in France was cultivated at exactly that intermediate level of abstraction where bridges with innovative craftsmanship proved capable of being built. That is where, and how, theoretical solutions to some practical problem (as in Papin's cylinder) could be turned into truly practicable solutions (as in Newcomen's fire engine; p. 478). By the middle of the 18th century, the path toward a truly science-based technology had been opened, if not yet on the Continent, then at least in Britain (pp. 584; 729).

But this also means that, as far as the Scientific Revolution is concerned, no prosperity-inducing or warfare-improving products were as yet created that might propel revolutionary science forward.

Forces propelling the Scientific Revolution forward.   What *did* propel the Scientific Revolution forward once the pioneering stage was left behind, were three other forces.

Not indeed the reality of a science-based technology but the sustained expectation thereof and the various aspirations projected upon those expectations served to overcome the midcentury crisis of legitimacy. Major realignments in European politics made it possible. What came out of them took shape as a Baconian Ideology bent upon celebrating the still largely imaginary utility of the new science while giving indispensable religious sanction to it.

But there was a reality, of another kind to be sure, that could and did serve to propel the revolution forward: an ever-expanding research front, not over the full width of the Scientific Revolution, but concentrated at two strategic places. Advance in fact-finding experimental science proceeded in just too haphazard a way to allow relentlessly ongoing progress, nor by the end of the century had sufficiently solid criteria been established to distinguish with confidence between true and spurious facts, between valid and mistaken conclusions (p. 501). Nor did an infusion with mechanisms of particles in motion make enough of a difference in this regard. But in realist-mathematical science relentless progress was attained through the ongoing mathematics/experiment dialectic and the feedback mechanisms enabled thereby (p. 363). This proved to be the case to an even larger extent when barriers between distinct modes of nature-knowledge began to break down. That is when highly productive ways were found to resolve the apparent tension between the Galilean and the Cartesian ways of dealing with problems of motion – work that found its high point in Huygens' *Horologium oscillatorium* (p. 522).

In short, three major forces served to keep revolutionary science going and ensure a vital measure of permanence for it before the rise of a science-based technology. These forces, then, were the power inherent in realist-mathematical science, the breakdown of barriers, and the perceived utility of revolutionary science and the sanctioning thereof in the Baconian Ideology.

What enabled these three 17th-century props to function the way they did is a question that has occupied us at length in previous chapters. Schema 5 provides an overview.

No doubt the manner in which legitimacy was restored provided the primary force. If Europe had not by the late 1640s begun in some last-minute spirit of compromise to retreat from the brink of all-out war fostered by all-round doctrinal infighting, and if no shift of geographic center to the northwest had occurred in addition, the leveling off so noticeable in revolutionary nature-knowledge of the 1650s would in all likelihood have continued unabated, as the forces of devastation and delegitimation would have grown stronger and stronger. Given, however, that these two decisive, Europe-wide developments did occur, and that, as one consequence, the pace of revolutionary nature-knowledge could be regained, its ongoing expansion in width and in depth was mightily enhanced in three significant respects.

Qualitatively, expansion was enhanced by the ensuing, partial breakdown of barriers between standing modes of nature-knowledge, which was fostered likewise by the new spirit of reconciliation. This, in its turn, made possible some degree of fusion between abstract mathematical relation and more concretely tangible measurement. Even more importantly, it also made possible the near-simultaneous occurrence of revolutionary transformations 4 and 5, with moving corpuscles being plugged into mathematical science and into fact-finding experimental science. Out of these two revolutionary transformations emerged in its turn a final one – we shall find soon enough that the revolution wrought in the mid-1680s by Isaac Newton, FRS, Lucasian Professor of Mathematics, benefited to the point of indispensability from such pathbreaking contributions as Isaac Newton, BA, had made in the mid-1660s to *both* preceding ones.

In terms of scope, pace, and direction alike, the driving force built into realist-mathematical science readily proved able to propel it onward over an ever-widening front almost by itself, and in so doing render a badly needed sense of direction to the pursuit of nature-knowledge. This, too, was soon to be reinforced by Newton's second round of wonder years.

Finally, the pace of all this was enhanced by the printing press. It is questionable whether

Schema 5: Three forces propelling the Scientific Revolution forward

1. *Power of realist-mathematical science*; due to (see p. 361):
   (a) Galileo's versatility
   (b) close intertwinement of subjects
   (c) problem/solution/problem chases
   (d) feedback mechanisms arising from search for balance between mathematical theory &
   experimental check

2. *Breakdown of barriers* starting c. 1655-1660; caused by (see p. 511):
   ~~(a) geography (including speed of communication)~~
   (b) techniques of communication in nonteaching situations:
       ~~(i) travel~~
       ~~(ii) rate of book printing~~
       (iii)correspondence networks (involving heightened open-mindedness)
       ~~(iv) scientific journals~~
   (c) scientific societies:
       (i) unprecedented measure of autonomy (London)
       (ii) everyday personal interaction (London, Paris)
   (d) spirit of Westphalia (some measure of Europe-wide reconciliation)

3. *Utility as sanctioned in the Baconian Ideology* (see p. 578):
   (a) broad framework:
       (i) post-Westphalian return to a measure of European stability
       (ii) shift of Europe's geographic center ⇨ new centers Paris + London ⇨ Académie + Royal
       Society
   (b) two principal agents:
       (i) fact-finding experimental science
       (ii) breakdown of barriers (hence, new science perceivable from outside as one whole)
   (c) removal of impediments:
       (i) strangeness mitigated
       (ii) sacrilege insulated
   (d) range of boons, chiefly the utilitarian ones of
       (i) expectation of straightforward material benefit, even if delayed ('engine': Paris)
       (ii) Baconian Ideology = leap of faith in power of science ('symbol': London); causally to be
       associated with
       ~~A) state policy~~
       B) relative fluidity of social structure
       C) religious outlook (Protestant preponderance, out of specifically European this-
       worldliness)

NB: Crossed-out items indicate explanations which have in the main text been considered but rejected.

the printing press made a decisive difference with respect to the making of the Scientific Revolution; no clear-cut reason is apparent for why the revolutionary transformations of the first half of the 17th century could not have taken place, albeit plainly much more slowly, in a script culture (p. 205). Nor is the mid-century breakdown of barriers between modes of nature-knowledge attributable to the printing revolution; the only communication tool to make a distinctive contribution was the regular exchange of handwritten letters (p. 512). But print *was* indispensable for the blow-by-blow exchanges between practitioners at some geographic distance from one another which fostered so much advance in revolutionary nature-knowledge from the 1660s onward – not only through the letters they wrote but also thanks to the articles and books which came to their knowledge with so little delay and which they were so quick to comment on. With the three *original* revolutionary transformations, the principal point of which had been to make a relatively clean break with the past, it was not as a rule the studious examination of the latest literature but rather a decision to lay it aside and make a more or less new start that had counted above all. But with these transformations by and large completed by the 1640s, it was rarely possible any more to make substantial advances in substantial ignorance of what one's fellow practitioners had meanwhile found. Also, it was only a print culture that enabled an undergraduate like Isaac Newton to read his way so quickly and creatively into the Scientific Revolution. In short, the pace of net-productive interaction between practitioners of revolutionary nature-knowledge from roughly the 1660s onward was such as no script culture could possibly have accommodated. Alexandria's library could help sustain a relatively prolonged tradition in mathematical science and a few other pursuits; it could not possibly have sustained a full-fledged Scientific Revolution.

Advances on many fronts: the big picture

Three times so far have we parachuted a fictional observer into the past and charged her with reporting on the present and inferring the future state of things in nature-knowledge. The main aim of the artifice has been to provide historians with a measure to help decide ongoing debates about continuity and break, as the more continuous an event is with what preceded, the better it might have been predicted (p. 141).

At the first occasion, by the turn of the 17th century, for all the forward-looking zest that animated practitioners in the European-colored mode of coercive empiricism the reporter would have been led by her procedure of comparative extrapolation to expect the Golden Age of European nature-knowledge to come to a fairly speedy end in more or less the same manner as its Hellenist, its Islamic, and also its medieval predecessors had (p. 149). Instead, events took a wholly unpredictable turn, and each distinct mode of nature-knowledge

612   underwent revolutionary transformation, with only no. 3 displaying a more gradual overhaul of the predecessor from which it emerged.

By the mid-1640s, with the opening stages of the Scientific Revolution by and large completed, we left our reporter clueless over what would happen next in view of the utter lack of
precedent for what had just occurred (p. 283).

By the 1650s we ascribed to the reporter full awareness of a noticeable slowing down, due
to a looming crisis of legitimacy which arose out of a sense of acute threat arising from perceived strangeness and perceived sacrilege. At that turn of events, broad analogy with how in
another sacred-book civilization the pursuit of nature-knowledge had come to be perceived
as a threat caused the reporter to return to the predictive outcome of her first report and
make it both more alarming and more specific: an already noticeable sapping of the will
toward continued, drastic innovation might well increase proportionally and thereby speed
up the natural ending that in c. 1600 she had already foreseen (p. 438; 566).

Actual resolution of the crisis of legitimacy by the early 1660s has proven the reporter
wrong again. The various turns of history that made the survival and renewed flourishing of
revolutionary nature-knowledge possible after all were, once again, truly unpredictable.

In view of so many predictive failures we shall assign a more modest task to our reporter,
now reinstated one final time. We no longer invite her to foresee future events, which have
turned out to be so disconcertingly unpredictable. We charge her instead with filing a report
on the state of European nature-knowledge by 1684 by listing its *strengths* and its *weaknesses*
from the viewpoint of a sympathetic outsider. That is, if one places oneself in or around the
year 1684 and makes an inventory of what appear to be the principal outcomes of the revolutionary transformations that have taken place so far, what does the broad picture look like
of advances attained and weak points still remaining?

Strengths first. The reporter starts off with a list of the most decisive substantive changes
brought about by the pioneers.

Pioneers' basic achievement.   Kepler: a specific arrangement of world harmony encompassing three empirically checked, mathematical regularities governing planetary orbits, embedded in their turn in a tentative notion of forces behind them; further, a new
conception of vision grounded in a partly novel analysis of light rays.

Galileo: mathematical rules for falling and projected bodies; a triplet of novel conceptions of motion, centered on a principle of motion retained; a host of further probings into
the mathematization-cum-experimental-validation of phenomena encountered either in
nature directly or in craft practice; further, a range of telescopic discoveries made to serve
his campaign for heliocentrism.

Beeckman + Descartes + Gassendi: speculative construction of a world made up of particles in motion in such a way that motion is somehow retained.

Gilbert: experiment-sustained investigation of magnetism and electricity, in a vitalist con-text.

Van Helmont: experiment-sustained restructuring of inherently vitalist, Paracelsian chemistry.

Harvey: experiment-sustained establishment of blood circulation and of general features of reproduction, in a vitalist context.

Research cores.   In elaborating and expanding the work of the pioneers, their successors effectively organized their own researches around certain cores. Here is a list of them.

In both mathematical and fact-finding experimental science sustained efforts were made at *absorption* of the respective predecessor modes. In mathematical science these took shape as enhancement of the reality content of the five Alexandrian subjects. Treatment of consonant intervals, light rays, and planetary trajectories was greatly altered thereby, whereas the Archimedean treatment of equilibrium states was placed in the wider frames of the increasingly distinct fields of statics and hydrostatics. In fact-finding experimental science, even in the case of the production of natural histories sheer unaided observation gave way where possible to observation experimentally sustained.

In both mathematical and fact-finding experimental science attempts were made to improve *craft practices* of the most varied kinds, from enhancing the efficacy of machine tools to use of wood juice for producing artificial manure. Even though the effort yielded hardly any tangible results, it was vastly instructive in teaching investigators important lessons about how the natural world's messiness and, every so often, sheer caprice may stand in the way of subjecting natural phenomena and man-made operations alike to mathematical and/or experimental rule and order.

In both mathematical and fact-finding experimental science, *instruments* of a new kind came to the aid of investigators. Major attainments included the discovery of the previously invisible, like planetary satellites and spermatozoa; exploration of the behavior of objects and beings in the void; vastly enhanced accuracy in the measurement of time; establishment of the variables that determine the properties of musical sound; empirical reinforcement of the conception of a Sun-centered universe.

In both mathematical and fact-finding experimental science, it was sought to reduce arbitrariness and attack otherwise insoluble problems by shoring up theorizing with *moving particles*. In mathematical science this led to clarification of four genera of motion: motion in impact, motion retained, circular motion, and pendular motion. In fact-finding experimental science this led to conceiving of diverse aethers, all employed to explain a wide variety of experimentally found properties.

Further, in mathematical science alone certain scholastic ideas were *reconceptualized* such that, duly transformed, they became vastly more productive. This was true notably of

614   the medieval concept of impetus and the technique of graphic representation. More generally speaking, a habit arose of taking certain results out of their original context and reinserting them in other ones with a view to enhancing their productivity. An early, very fine example is how Horrocks managed to pry Kepler's three laws of planetary motion loose from their context of harmonic speculation.

In fact-finding experimental science alone, the best-focused efforts were directed at a quite circumscribed range of subjects, to wit, electricity, magnetism, light, the structure and function of bodily parts, and the fine structure of matter.

In kinetic corpuscularianism, finally, one can hardly speak of research in the sense meant here, of individuals out to discover phenomena and properties of nature as yet unknown. What counted was to explain phenomena, not to discover them. Since all but low-level details were held already to be known and understood the speculative way, empirical research, if undertaken at all, remained confined to such low-level details, as, for instance, with Descartes' animal dissections.

Phenomena investigated.   Upon their revolutionary transformation, in all three modes of nature-knowledge there was a vast expansion of topics subjected to scrutiny. In realist-mathematical science alone, the range of phenomena subjected to its specific procedures was still easily countable by the end of the 17th century. The list compiled in Schema 6 may well come near exhausting their number.

By contrast, in view of the world-encompassing claims of the natural philosophy of particles in motion, the number of phenomena its practitioners subjected to explanation ran into the tens of thousands. So did, albeit for other reasons, the sheer number of phenomena investigated by fact-finding experimental scientists. Here solid achievement ran the perennial risk of drowning in phenomenal overflow, as no limit was built into their observational and experimental procedures. Among phenomena given in nature, those newly investigated ranged from everyday ones like wind or the speed of sound to arcane niceties like the blue opalescence of nephritic wood. Among phenomena produced artificially, the range of investigation ran from the length/width ratios of diverse organ pipes to craft practice in the dyeing of textiles.

Methods.   The term 'method' carries meaning on three distinct levels of generality.

At the *highest* level, it is 'the scientific method' as such – that delusion of so many armchair methodologists. Two big, well-thought-through methodologies accompanied the Scientific Revolution from the start. Of these, Descartes' top-down method was not really about research at all, so concerned was he instead to nail down the certainty of his own variety of speculative philosophy of nature. Bacon's refined procedures for a bottom-up research method of comparative listing so as to arrive at higher-order generalities quickly de-

Schema 6: Phenomena treated over the 17th century the realist-mathematical way   615

	'Given' phenomena	Artificially produced phenomena
Planets	trajectories; order and measure of solar system	
Solids	equilibrium; light and heavy	five simple machines; STRENGTH OF MATERIALS
Liquids	• *at rest*: floating; sinking; suspension; buoyancy (idem with AIR) • *in flow*: STREAM; OUTFLOW	• *at rest*: COMMUNICATING VESSELS; HYDRAULIC PRESS • *in flow*: FOUNTAINS
Light	• *image formation*: eye • *refraction*: rainbow; ICELAND SPAR	• *image formation*: LENS (both single and in telescopic configuration) • *refraction*: PRISMATIC COLORS
Sound	consonant intervals	division of octave
Place	place on Earth	
Time	local time	
Nothing		void
Motion	• vertical descent • projectile motion • IMPACT • ROTATION	• INCLINED DESCENT • MOTION RETAINED • PENDULAR MOTION • FORCE OF IMPACT
Tools		• linear perspective • *warfare*: gunnery; fortress building • PERFORMANCE OF MACHINES • OPERATION OF SCIENTIFIC INSTRUMENTS

The horizontal separation is between the five Alexandrian subjects and the chief categories added to them during the 17th century.

The headings 'given' and 'artificially produced phenomena' correspond to a distinction made on p. 247.

Phenomena printed in SMALL CAPITALS are those hardly or not at all examined before the Scientific Revolution in any way.

generated, even among those who professed to follow his detailed recipes, into those rarely completed lists compiled over much of the 17th century in catalogs and natural histories. Nor can any methodology (i.e., generalized method) of later invention and/or prescription be unproblematically projected back upon the bulk of research activities performed during the Scientific Revolution.

At the *lowest* level of generality, 'method' may stand for certain special research techniques. In mathematical science these were focused on the taming of the infinite and the infinitesimal and led to algorithms of the calculus. In fact-finding experimental science they were focused on case-by-case refinements in the practice of experimentation and in the everyday use of instruments, as with the illumination of objects of microscopic research or the fastidious cleaning of one's equipment. Methods of such a kind count among the most significant and enduring products of the Scientific Revolution.

Methodologically more intricate is, finally, the question of whether the practice of revolutionary nature-knowledge was governed, at an *intermediate* level of generality, by certain broad guidelines to research. Here it is wise to disregard practitioners' own statements. These tend to confuse us rather than help us find out what actually happened. Those to think about method at all were rarely good at making their prescriptions match their actions, as with Newton whenever he found it convenient to pose as a Baconian. Or they might phrase novel prescriptions in the language of some no longer fitting methodology, as with Galileo whenever he fell back on the vocabulary of Aristotle's *Analytica posteriora*.

Here another useful distinction to make is the well-known one between 'context of discovery' and 'context of justification', that is, between the making and the checking of purported discoveries.

Methods for *making* discoveries were quite different in the two major research modes to come out of the Scientific Revolution, mathematical-experimental and fact-finding experimental.

In the former mode, the making of discoveries involved first of all methods of *abstraction*, which stripped a given phenomenon down to what were thought to be its essentials. *Concepts* were employed or freshly minted to clarify and delimit those essentials. *Analogies* might be invoked to provide the investigator with a clue for what precise mathematical relation might obtain in a given case (p. 336). Theoretical outcomes might be *played around with*, preferably but by no means necessarily until some desired measure of correspondence with experimental results had been attained. All this might be done in any number of reiterations, in any order, and directed most of all by sudden flashes of more or less brilliant insight.

By almost total contrast, in the mode of fact-finding experimental science the making of discoveries involved as a rule either (1) the more or less haphazard encounter, in the course of some experiment, with phenomena unknown and unsuspected but subsequently recognized as such (if shown to contemporary satisfaction not to be spurious) or (2) the explanation, whether more or less ad hoc, of experimentally established phenomena in light of some background worldview.

But this is about as far as the definition of 17th-century research methods can go. True, for the historical episode of the Scientific Revolution one cannot quite speak of a Feyerabend-like 'anything goes' posture taken by its protagonists. Yet within the broad confines just sketched practitioners enjoyed a very large measure of individual freedom of invention, as in fact they have ever since.

Finally, methods were sought for the *checking* of purported discoveries. In the 17th century these took shape in a protracted search for ways and means to rein in our human propensity for fancying plausible things rather than seeking to nail our suspicions down in solid proof. What, if any, criteria for proof did they find?

Checks upon possible fancy. Francis Bacon opened the preface to his *Novum organum* with a plea to get rid of the two broad conceptions of nature-knowledge adopted by the Greeks, the dogmatic and the skeptical. Dogmatists, in "daring to speak of nature as a thing already sufficiently explored . . . have been effective in quenching and stopping inquiry." Skeptics may have had some good reasons for "asserting that absolutely nothing can be known", yet in making that claim "zeal and affectation have carried them much too far." What is needed instead is "a position between these two extremes, – between the presumption of pronouncing on everything, and the despair of comprehending anything", since "whether or no anything can be known [is a question] to be settled not by arguing but by trying."[338]

Not only Bacon, but everyone in the 17th century who sought nature-knowledge in a new vein faced that question, and 'try' is precisely what they did in seeking to resolve it. Their exploration of viable ways and means to make assertions about facts and properties of natural phenomena stick, and to rein in fancy and arbitrariness, naturally took different forms in different modes of revolutionary nature-knowledge (p. 284).

In *realist-mathematical science* a delicate and variously weighted interplay emerged between, on the one hand, mathematical relation (functional dependence expressed as equality or proportion) and, on the other, experimental outcomes meant to confirm it. The interplay yielded valuable pointers toward possible correction and/or falsification, in that for the first time in history it provided systematic feedback against the very realities of nature (p. 364).

Insofar as *kinetic corpuscularianism* kept being handled in the dogmatic way of the Greeks, four 'internal' checks served to some small extent as bridles upon unlimited fancy: foundational certainty, consistency, analogy with the macro-world, and 'physical intuition' (p. 378). Further, after the breakdown of barriers between modes of nature-knowledge, both mathematical analysis and fact-finding experimentation yielded, for a few highly important topics of inquiry, ways to keep kinetic-corpuscularian fancies in check from the outside as well (p. 554).

In *fact-finding experimental science*, inevitably, nature's whimsy held the upper hand, and practitioners found themselves burdened with the task of dealing with it as best they could. Here feasible checks upon outcomes proved to reside in little beyond (1) systematic efforts at purification of substances and at compensation for apparently consistent measurement errors, (2) ways to ensure actual and/or 'virtual' witnessing, and (3) such coherence as was

618   yielded by a rather tenuous blend of background worldview, explanatory theory, and experimental outcome (p. 501).

Behind this variety of checks lurks the core issue of how best to balance the neatness of the (most often mathematical, but in any case simplifying) model with the (where possible, experimentally reduced) messiness of the full world meant to be pictured in the model and subsequently needed to check it. There are no hard and fast rules for such balancing acts, as if the feedback gained from the messy world could be taken into account just automatically, as a routine matter. Yet taken into account it can be. *How to do that in the best possible way is one core issue first explored during the Scientific Revolution.* When that revolution was over, the realm of the mathematical model, on the one hand, and the realm of the phenomenal world and its messiness and experimentally apparent whimsy, on the other, were still largely separate. Barring Newton's unique case (next chapter), it was not until the early 19th century that the two realms were made to fuse to such an extent as to become regularly susceptible to somewhat smoother and more routinely applicable ways to accomplish the balancing.

Two visible overhauls.   What is the world-out-there like? And what are we ourselves like? During the Scientific Revolution, the answers given to these two basic questions changed radically. The new answers were provided by practitioners of mathematical and fact-finding experimental science, whereas the spread of these new answers among the academically educated from roughly midcentury onward was aided greatly by the rapidly increasing popularity of kinetic-corpuscularian philosophies.

Starting with Ptolemy, the received answer to the first question had been: The universe with its basic appurtenances (Moon, Sun, five planets, numerous stars) reveals itself by naked-eye observation alone. If measured from the Earth at the center to the outer bounds marked by the sphere of the fixed stars, the universe has a radius about 20,000 times as large as the radius of the Earth. By the end of the Scientific Revolution, this answer had been given up by every expert and by significant numbers of the educated as well. It was replaced by either a wholly different answer or a less or more watered-down variety thereof. In its most radical version it ran: The universe is infinite in size and requires sustained telescopic observation and measurement for reliable knowledge of its star-, planet-, moon-, and comet-studded makeup. It contains our own solar system among many others. The one where we live measures c. 200,000 Earth radii from the Sun at its center to its sixth and outermost planet, Saturn. Its distance to even the nearest star covers more space than is traversed in six weeks by light, which takes a few seconds to reach us from our Sun.

At least from Galen onward, the received idea of human beings had been of an indissoluble unity of body and mind (even to express it thus yields too much to the Scientific-Revolution-born separation between them). Working their effects upon body and mind alike, a variety of substances (blood and the other humors, spirits, waste products, etc.) were

regarded as flowing freely through the bodily vessels. Harvey unhinged this hoary conception at one critical point. Thus, he unwittingly brought about a radical revision in how the
human body and its fluids were viewed, along with a momentous relocation of mental functions to the nervous system and the brain. Prior to the Scientific Revolution it was customary in the Greek tradition to regard the human being as a scaled-down mirror image of the
cosmos, that is, as a microcosm furnished with numerous profound analogies to the macrocosm enveloping it. By the end of the 17th century, among the educated little remained of
such a conception of ourselves in the apparent absence of any other analogy in the cosmos
than with the animals, our nearest relations on Earth after all.

So much for the two most spectacular outcomes of one century of revolutionary nature-
knowledge. Naturally, these outcomes served to enhance its visibility. Even apart from the
startling theological implications, the agent of such drastic reversals in our idea of ourselves
and of the heavens around us acquired a more distinct profile thereby than nature-knowledge had ever attained in any of its prerevolutionary modes. No longer was its pursuit just
a marginal activity, shorn of wider repercussions. But its enhanced visibility did not stem
solely from the sources just mentioned. A range of independent developments in the institutional sphere led to a significant increase in both autonomy and visibility by the second half
of the 17th century.

Autonomy and visibility.   Here the organization of revolutionary nature-knowledge
in the two major scientific societies was largely responsible. In the French as in the British
variety, the deliberate abstention (imposed in the former; taken upon itself in the latter)
from discussion of religious, philosophic, and political issues led to unintended consequences fraught with significance for the future. The pursuit of revolutionary nature-knowledge
came to be seen as an activity in its own right and also (in spite of bouts of British ridicule)
as an occupation that serious people, members of a prestigious, royally stamped, and (in
the French case) generously funded institution, could seriously devote their careers to. In
the British case this went together with self-regulation and a large amount of freedom to
pursue research irrespective of where it might appear to lead – freedom bought, to be sure,
at the expense of absence of state-funding. In the French case, autonomy in the sense of no
binding restrictions placed upon research outcomes was granted likewise. Here, however,
the flip side of royal largesse and of concomitant absorption into the royal bureaucracy and
its policy guidelines was a perennial diversion of research aims in the direction of fairly narrowly defined, practically useful results.

Visibility and autonomy were still circumscribed in many ways. Still, compared to the
institutional state of pre-revolutionary nature-knowledge this was a big step forward into
the as yet uncharted waters of self-regulated pursuit of nature-knowledge.

620   *Two weaknesses discerned.*   We have now reviewed the long list of strengths filed in the report of our imaginary observer. But from her viewpoint of 1684 she is bound to perceive certain *weaknesses* as well. Two observable states of affairs capable of weakening the healthy pursuit of revolutionary nature-knowledge beyond 1684 suggest themselves, one specific and from the outside and the other diffuse and from the inside.

The weakness from the outside derived from the inherently time-bound nature of the Baconian Ideology. Not that this ideology, the way it had taken shape in Interregnum Britain and had grown institutional roots with the Restoration, was just a fleeting fashion, bound to go up in smoke with the next hype. For that to happen, both its social backing and the two interrelated sensibilities to which it appealed – the expectation of economic rewards to come from the pursuit of revolutionary nature-knowledge, and the values underlying this-worldly Christianity – were much too solid. Still, limits set to that solidity could easily be foreseen. The expectation could sour; Christianity could turn otherworldly after all; Christianity could lose its appeal altogether. The second contingency was a fairly remote one, after centuries of movement in the opposite direction (even though a re-Catholization of Europe, such as for half a decade seemed possible with James II's 1685 succession, might have gone some way to reverse this). The third contingency, even if considered in light of the numerous theological problems already provoked by revolutionary nature-knowledge, might also seem fairly remote – very few people were yet around in Europe seriously to question their faith. But the first contingency was not nearly so remote; there are after all limits to patience. By 1684 Louvois (Colbert's successor) had begun to slash the budget for the Académie by a vast amount, and even though it is true that, by their very nature, ideologies are less sensitive to prompt delivery of promises, the Baconian Ideology could hardly be expected to remain a living source of legitimacy and moral support forever. As our observer may well have ended this section in her report, at some point or other over the next decades the new science had better deliver, or else . . .

The other weakness, the diffuse one from the inside, emerges from an awkward question we allow our reporter to pose but not any longer to respond to in person (since the full answer requires a measure of prescience no contemporary observer could possibly possess): Where, for all its strengths, is all this tremendous effort leading us? Where, if anywhere, is it heading?

In part, that question might be answered by pointing (the way Kant was to do about a century later; cf. p. 363) at the built-in open-endedness of the scientific enterprise.

A more specific answer would be that the kinetic-corpuscularian philosophy of nature was not heading anywhere in that, as a speculative philosophy, it was meant to be all-embracing and definitive. Insofar as it was nonetheless heading somewhere, it could only do so on the coattails of something else, be it of mathematical-experimental or of fact-finding experimental science.

But what, more alarmingly so, about a certain lack of direction observable with these two? With fact-finding experimental science in its 17th-century guise a lack of direction was well-nigh inevitable – its uphill struggle with the whimsy of nature left it no other option in this regard. But with realist-mathematical science it was not necessarily so. On the one hand, quite a constricted number of practitioners had managed to advance over a quickly widening front with discoveries of ever increasing depth and sophistication. In so doing, these few men gave some badly needed coherence to the pursuit of nature-knowledge. Even so, their pursuit lacked a central problematic – there was no unifying viewpoint to bind a large variety of results together. The place to find that viewpoint would have been the subject of motion, which was after all the most distinctive issue of the entire Scientific Revolution (p. 521). By 1684 no such central problematic had yet been found.

True, from the early 1650s to the late 1660s both Huygens and the young Newton made major advances on all the relevant genera of motion: impact, motion retained, centrifugal motion, to-and-fro motion. But Huygens had not meant his efforts to contribute to some comprehensive science of motion capable of covering all genera alike – his deliberate abstention from any concept of force in view of its apparent intractability doomed him to treating these *genera* one by one, as mutually connected yet not jointly derivable from some shared core or unifying principle. By contrast, Newton did mean his efforts to contribute to a generalized science of motion. Only, in seeking it he got stuck. He fell into the very trap Huygens had foreseen and chosen to avoid almost right from the start. Newton's diffuse handling of a variety of really incompatible concepts of force was what made him reach the dead end that in 1668 caused him to abandon his quest (p. 539). What, then, made force so tough a concept, as Huygens had intuited, and yet so indispensable to the mathematical analysis of motion, as the young Newton had sensed?

The force knot

So far I have spoken of force only in passing, so as to be able to treat in close succession the force knot and its first major, large-scale unraveling between 1684 and 1687. The expression 'the force knot' occurs in Richard S. Westfall's 1971 *Force in Newton's Physics. The Science of Dynamics in the Seventeenth Century*. It captures the central theme of that book, conveniently summed up in two passages. In the first, the subject is Torricelli's attempt to find in static force a measure for the force of percussion. There Westfall observes that "the confusion on this point . . . was the confusion of the entire century in its jumbling together of various irreconcilable concepts under the heading of 'force.'" In another passage he speaks of "the incredible capacity to mislead that intuitive ideas of force possessed".[339] What, then, did that incredible capacity stem from?

622 Force is inherently abstract and intangible. In wider senses than the experience of our body being hit by fist or foot, force is hard to picture in a concrete sort of way and hard to
conceptualize in a precise sense. Not by chance, the term is used in everyday language in
many different ways, with no clear-cut distinction made between them. For instance, in a
deliberate effort to come to grips with the concept Leonardo da Vinci probed the following
definition:

> What force is.
> I say that force is a spiritual virtue, an invisible power which, by means of accidental, ex-
> ternal compulsion, is caused by motion and located and infused into bodies which are drawn
> and forced out of their natural state of being; in giving these an active life of wonderful power,
> it constrains all created things to change shape and place; it runs with fury toward its death,
> and takes different guises as the case requires.[340]

Such outpourings are as inspired as they are ultimately opaque. A variety of modern conceptions of force may be associated with these and similar utterances in Leonardo's notebooks. Is potential energy what he is addressing (the concept most readily associated with
the passage quoted)? Or rather kinetic energy? Or work? Or action? Or momentum? Or
static force? Or motive force? Or some or even all of these?

No doubt, the latter. A definition like his neither was nor could be satisfactory in any
way. It rather expresses the inevitable confusion that arises from a conceptual knot holding together a range of as yet undifferentiated strands. Retrospectively aware as we are that
to dispel confusion minimally requires definition in the language of the calculus, we can
further see that for as long as intuition-guided conceptualization was the best that could be
undertaken, investigation of force was fraught with traps. In lesser measure, this was true
of the concept of speed, too. In its customary, pre-revolutionary usage it might indiscriminately cover average speed, terminal speed, and/or speed at one given instant. The Scientific
Revolution was the event in the course of which these various senses were gradually unrav-
eled and eventually given not just intuited but unambiguous definition by means of the calculus (p. 357). Just so was it to be with the concept of force, albeit in an even more laborious,
complicated, and drawn-out process, not to be completed until far into the 19th century.

Beside everyday indistinctness and the indispensability of mathematical definition, dynamic intuitions could also lead one astray due to the retrospectively identifiable circumstance that different concepts were often marked by different dimensions. For us it is easy
to see that one cannot, for instance, weigh an instantaneous force. In the 17th century, with
mathematical relations still expressed in proportions rather than in equations, it was not
nearly so evident.

The customary, wholly tacit conception of force as a *substance* served to obscure further the presence of really incompatible concepts. Rather than being embedded in a structure of space and time, force was held capable of being transferred from one body to another the way one pours water from one vessel into another (p. 308). It was likewise held capable of appearing (as Leonardo noted) in many different guises without losing thereby its substantial unity.

Finally, insofar as force was employed in everyday life to push or draw a cart, no observer could miss that, as the force ceases, so does the movement. Hence, it seemed self-evident to regard everyday force as first producing and then sustaining motion. Gradual acceptance, in Galileo's wake, of some principle of motion retained (p. 528) upset this commonsense conception – considered in the absence of all apparent impediments, a body requires no force to pursue its uniform motion if it has any. Galileo himself drew numerous important consequences from the principle. Still, in retrospect, it cries out for the insight that force comes in, not indeed to sustain, but rather to *alter* that state of uniform motion. Newton, in the course of a prolonged struggle with orbital motion, was finally to draw that ultimate conclusion (p. 653). But in the meantime, in the half-century between the *Discorsi* of 1638 and the *Principia* of 1687, for any given kind of *non*uniform motion the true role of force did not become one whit clearer by the new principle of motion retained. It could not even help clarify these things for those few who, upon adopting the principle for uniform motion in a straight line, then managed to stick to it consistently.

For the Scientific Revolution in its ongoing advance, three points sum up the net result of all these distinct barriers on the road to conceptual clarity regarding force.

In the *first* place, between Galileo and Newton/Leibniz only two comparatively sharply delimited modes of dynamic action were regarded as unproblematic: those involving either good old static force or the newfangled 'force of a body's motion' arising from collisions of particles.

Static force, operative in equilibrium situations that from Archimedes onward could be managed well, was the most frequently employed source of analogy to fall back on. Time and again attempts were made to measure entities in nonequilibrium situations by reducing those situations to those that are governed by the law of the lever. Galileo's effort to measure the force of impact is a case in point (p. 337). More often than not, both the dimension problem just mentioned and too careless discrimination made between the proper entities (distance, time, speed) to put into one's resulting proportionalities blocked advances on which others could build. At times 'physical intuition' could overcome these impediments, as with a distinction Descartes made between two varieties of static force. Calling the one 'action' and the other 'puissance', he used them to keep separate from each other a continuously exerted force that presses or pulls and something akin to the modern concept of work. Even so, the same law of the lever did lead Descartes astray when providing him with the sole available model for deriving rules of impact (p. 523).

624   The conception of the world as made up of moving particles that had made derivation of those rules imperative for Descartes involved *another* concept of force as well. He dubbed this 'the force of a body's motion'. It was the force exerted by moving particles when they bumped into other moving particles. It was measured by a body's size and its velocity, the product of which was held by Descartes to be conserved in impact. It served those who, in preference to the law of the lever, were to take impact for the prime model of their dynamic investigations. Acting through contact alone, this force represented a major constraint upon dynamic thinking at the time. The constraint was derived from the quite justified suspicion that forces of attraction and of repulsion were tainted with magic, with vitalist conceptions, in short, with everything that smacked of the occult. Forces like this were habitually invoked by noncorpuscularians to account for such phenomena as sympathetic resonance or magnetic action. Obviously, explanations along such lines were incompatible with the various motions of the particles our non-occult but rather (in the final analysis) wholly transparent world was exclusively made up of in the view of any serious kinetic corpuscularian. That is, forces operating not through immediate contact but at a distance, as active principles, were officially shunned in that philosophy, even though on occasion they were smuggled in through the backdoor (p. 552). *In the conceptual frame of kinetic corpuscularianism* (whether taken dogmatically or hypothetically makes no difference here) *no formal introduction of forces as active principles was at all possible.* To the extent that it was attempted nonetheless, the effort ended in irresolvable ambiguity; witness the various aether conceptions tried out by Hooke and by the young Newton (p. 560; 564).

   Second, whether the law of the lever or the collision of bodies was taken as the point of departure, between the time of Galileo's *Discorsi* and Newton's *Principia* numerous men bent upon investigating motion the dynamic way (be it in the mode of realist-mathematical science or of kinetic corpuscularianism) indulged in arrays of concepts of force stretching from the moderately indefinite to the fully intuitive. The way they sought to put these to work has been exhaustively analyzed by Westfall, who derived from such analyses his conclusion about their "incredible capacity to mislead". Whether labeled 'force' or 'strength' or 'power' or 'moment' or 'impetus' or 'energy' or 'virtue' or 'propensity to motion' or 'endeavor' (yes, terminological chaos contributed its own part to the confusion) and whether investigated by Mariotte or Borelli or Wallis (to name a few of the, comparatively, most insightful), no concept of force came even near fulfilling what was needed for a general treatment of motion. One powerful example of how even the best of the lot could be led astray this way is Robert Hooke's effort to employ the most sophisticated among his various concepts of force (with the uniform acceleration of falling bodies serving as his starting point) for teaching craftsmen a lesson or two. One such lesson involved a device to make lamps burn steadily; another, an effort to demonstrate, by means of well-grounded scientific insight deemed by Hooke vastly superior to any groping, trial-and-error method, how the way sailors trimmed their sails from time immemorial was dead-wrong. In the latter exemplary case, the dynamic confusion is as rampant as it was inevitable (Hooke not being Newton, as we shall have ample occasion to observe in the next chapter). Incidentally, the very failure in this regard of this particular man, so knowledgeable about the latest science and also an inventor of almost matchless ingenuity, serves to reinforce all conclusions drawn earlier about the lack of capacity of revolutionary nature-knowledge as yet to improve contemporary craftsmanship. Among retrospectively ascertainable grounds for that lack of capacity, the inherent deficiencies of pre-Newtonian concepts of force definitely had a part of their own to play.

Third, the sharpest thinker on motion between Galileo and Newton/Leibniz, Christiaan Huygens, was more acutely aware than his contemporaries of the presence of some profound force knot. He responded to it by letting go, in all but static cases, of any concept of force whatsoever. He thus withdrew into the defensive but, as such, secure fortress of a merely kinematic treatment of various genera of motion at the willingly accepted expense of giving up beforehand any unified treatment of these diverse cases. In this he was preceded by Galileo's handling of his own major case of motion, falling bodies (p. 192). Huygens withdrew into pure kinematics the very first time he came to handle a Cartesian problem the Galilean way. Seeking at first to analyze the problem of speeds in impact dynamically, he quickly got rid of any notion of a force of collision, so as instead to make a boat in a canal do all the virtuoso analytical work needed (p. 525). Huygens was not henceforth to swerve from this kinematic path in his handling of any specific case of motion in nonequilibrium situations. When in a 1669 Académie session Roberval attacked him, Huygens insisted that such attractive forces as his opponent preferred to invoke were not to be allowed in any sound approach to the phenomena of nature:

> To discover a cause of weight that is intelligible, it is necessary to investigate how weight can come about, while assuming the existence only of bodies made of one common matter in which one admits no quality or inclination to approach each other but solely different sizes, figures, and motions.[341]

Everything said so far about force and the knot it was entangled in came together in particularly focused manner in two standing enigmas. One was uniform acceleration in vertical or inclined descent, coupled to Galileo's other big find on the subject, that for all falling bodies, irrespective of what they are made of, the rate of acceleration is the same. The other was the role of force in orbital motion.

The enigma of falling bodies.

In retrospect, it is all so very simple. It already appeared that way to Newton in the first edition of his *Principia*. Famously and quite understandably yet wrongly he attributed his own immortal find, the second law of motion, which

626   introduces force as that which generally engenders uniform acceleration (or at least change of motion), to Galileo, as if the latter had derived his law of falling bodies that way rather than in purely kinematic fashion. In historical reality, the road toward the second law was extremely arduous, with Galileo's law of falling bodies alternating between acknowledged enigma, a starting point in the absence of much of an idea on how to advance from there, and a major stumbling block. Time and again did the varied sources of conceptual obscurity, confusion, and deliberate confinement intervene to bog one seeker after another down in a dynamic morass.

If, on the one hand, one started from some idea of force, however far one might perhaps get with other cases of motion, no way to include free fall presented itself – time and again its inexplicable acceleration stood in the way.

If, on the other hand, one set out by taking uniform acceleration in free fall rather than impact or the law of the lever as one's model for dynamic action, no idea of force presented itself that was suitable to generalize the special case of free fall. Clearly, specific forces, like magnetic attraction, would not do (even apart from what made them so reprehensible from a kinetic-corpuscularian point of view), unless and until they were shorn of all specificity, which, indeed, was between 1684 and 1687 to become Newton's decisive step. But the only universal forces yet available, static force and the force of a body's motion, would not do, either. The former, in that static force is about equilibrium cases alone; not for nothing had Galileo found the bridge from statics to kinematics through his idea of indifference to motion, not through any explicitly dynamic consideration (p. 193). The latter, in that no force brought about *by* motion could possibly serve to account for the action of a force applied to a moving body from the outside. Thus, along either pathway the case of uniform acceleration in vertical or inclined descent was bound to remain an enigma. This was the more true as this case alone was governed by the body's weight, which is naturally unable to play a motive role in any horizontal motion uniformly accelerated. That is to say, the proper role of weight had to be figured out as well. As only the wisdom of hindsight informs us, a clear-cut distinction between weight and mass, such as was still wholly outside Galileo's purview, was required to make it possible to get from uniform acceleration as a special case in free fall to any generalized, dynamic account of motion.

If, finally, for reasons of general intractability one excluded dynamic considerations altogether, uniform acceleration had to arise somehow from the sheer movement of particles.

The various possibilities here outlined in principle took illuminating guises in historical reality. Prior to the *Principia* (or at least unaffected by this specific aspect of that book) Torricelli, Huygens, and Leibniz were the acutest thinkers on the subject. Here is how they dealt with falling bodies.

In a way, *Torricelli* came closest. Along a path already considered in another context (p. 308) he managed, by means of a frankly dynamic analysis of Galileo's account of falling bodies, to pry loose from the force knot one of its numerous constituent concepts, to wit, something closely akin to our modern concept of momentum, that is, of a constant force considered in its operation over time. A workable dynamics might have emerged from further pursuit of this route, which was different from the path Newton was to take and was foreclosed by Torricelli's early death and by long-delayed publication of his most promising results.

In another way, *Huygens* came closest. Not, to be sure, in the 'Discours de la cause de la pesanteur' (Discourse on the Cause of Weight), to which he treated his fellow Académiciens in 1669. Here, Huygens sought to repair mathematically another of Descartes' perceived failures – in *Principia philosophiae* he had explained vertical descent without taking its acceleration into account (p. 408). With Galileo, but against Descartes, Huygens accepted the uniform acceleration of falling bodies. With Descartes, but against Roberval (who had been Descartes' opponent on the subject, now to become Huygens' as well), he accepted no other force than the force of a body's motion. Thus, Huygens' task became the impossible one of making some mechanism of particles in motion produce the uniformly accelerated motion of heavy bodies down to Earth. To that end, he amended Descartes' ingenious mechanism of a flat vortex composed of particles tending through the force of their whirling motion to press gross bodies down to Earth. Huygens replaced this with a *spherical* vortex, so as to bring about downward pressure in all planes rather than in Descartes' one plane, which, alas, was bound to be perpendicular to the Earth's axis. On this somewhat improved setup Huygens based a calculation of the speed of the subtle matter of his vortex sufficient to result in at least rough approximation of uniform acceleration at the values actually observed. This, then, was the best he could do to reconcile the Galileo in his mind with the sole feature of Descartes' kinetic corpuscularianism he had committed himself to forever. But it was not good enough to satisfy Roberval, who, unimpressed, kept appealing to some specific attraction between the Earth and smaller bodies of the same makeup ('like attracts like'). Two decades later Huygens was to turn his manuscript lecture into his major vehicle for determining in public his own position vis-à-vis Newton's meanwhile published solution to the same problem, and to a few more (p. 677).

Not in his *Discours* of 1669/1690 but in a brief manuscript from the mid-1670s that he abandoned incompleted, Huygens came close to devising in outline a concept of force that, if pursued to its final conclusion and then published, would have made us speak ever since, not of Newtonian 'force', but of Hugenian 'incitation'. The choice of term looks deliberate – a concept handled so opaquely and, on the whole, unproductively by his contemporaries deserved a fresh start marked by an unspoilt label and a clear-cut definition. Inspired directly by the case of free fall, and with the principle of uniform, persisting rectilinear motion settled in his mind more fully than with any of his contemporaries, he defined incitation as that which acts to put a body at rest in motion or *alters* the speed of a moving body. Its mea-

628   sure is given by "the force that must be employed, in the place where the body is and in the direction it can take, to prevent it from beginning to move." After an example, Huygens went on to make the universalizing intent behind his concept unmistakably clear: "the incitations of a body can be equal to each other although they are caused by different causes, as weight, elasticity, wind, attraction of a magnet, or something else."[342] He further sketched how this incitation might be applied to one specific problem. And that is how far he got before he laid the paper, which comprises two manuscript pages in a bound workbook, to rest forever.

Not only with force did Huygens allow his considerable yet, on the whole, intuitive dynamic insight to get the better of his kinematic restraint for a little while. He did so also with the concept of mass, toward the distinction of which from weight he was also groping in other unpublished work of his, notably when polishing his treatise on impact originally composed in the mid-1650s.

In yet another way, *Leibniz* came closest. The substantial conceptual sharpening that he attained, rather than preceding Newton's second round of wonder years, 1684-1687, as with Torricelli in the 1640s and Huygens in the 1660s/1670s, ran largely in parallel not only in time but *qua* content, too. In differentiating quite another concept (the one eventually to take shape as modern kinetic energy) out of the force knot than Newton was simultaneously seeking to unravel, Leibniz can be seen in retrospect to have hit upon a concept fraught with developmental possibilities scarcely to begin to unfold until decades later, far past the Scientific Revolution properly delimited. Beyond its barest outlines it must remain untreated here. In the very act of prying it loose Leibniz resolved in a way alternative to Newton's a range of quandaries that had beset the concept of force for at least half a century. In so doing Leibniz caused a veritable upheaval in current thinking about force.

The kernel out of which the upheaval grew, was contained in a concise contribution he made in 1686 to the recently founded, Leipzig-seated journal *Acta eruditorum* (Records of the Learned) under the nicely provocative title 'Brief Demonstration of a Memorable Error Committed by Descartes and Others concerning a Law of Nature'. Here Galileo's law of falling bodies was directed against Descartes' law of conservation of quantity of motion defined as the product of a body's size and its velocity (in modern rendition, *mv*). Leibniz started from two assumptions hard to dispute: (1) in falling from a given height, a body gains force sufficient to carry it back to the same height (if otherwise, perpetual motion would be possible); (2) if a given force can lift a body of 1 pound to a height of 4 feet, it can also lift a body of 4 pounds to a height of 1 foot. So much granted, Leibniz went on to argue that,

> if the four pounds were divided into four units of one pound, lifting them one after another to a height of one foot would clearly be equivalent to lifting one pound one time after another through four stages of one foot each. Now set at one unit the velocity that a body acquires in falling one foot. Four pounds falling one foot must acquire a force of four units; and by

Descartes' formula, that force transferred to the body of one pound will give it a velocity of   629
four. Once again Galileo stood in the way of Descartes, for Galileo had proved that if a body
projected upwards with one unit of velocity rises one foot, a body projected upwards with four
units of velocity will rise sixteen feet.[343]

Leibniz concluded that the full effect of a force at work on a body of given size is not to be
measured by its velocity – Descartes had conflated motive force and quantity of motion. The
example itself suggested what quantity is really conserved in dynamic action. It is the product of a body's size and its velocity *squared* (in modern rendition, mv^2). Young Leibniz had
made its acquaintance under the tutelage of Huygens, to whom it was and forever remained
a measure devoid of physical content (p. 526). In coming into his own as a philosopher bent
on clarifying principles and examining preconceptions down to their very core, Leibniz was
to turn a mere quantity into a cosmic force, which he called *vis viva* ('living force'). Due to
the conservation of this force the universe could run on and on as a self-sustaining machine,
which it never could if Descartes' conservation law held. Indeed, avoidance of any conception of God as occupationally concerned with stopping gaps in the engine He had created
Himself was among the prime motives of Leibniz's entire philosophy.

Leibniz's thinking about force from the 1680s onward was drenched in the calculus he
had (essentially independent from Newton) been developing since 1672. Dynamic 'living
force', associated with motion itself, had its counterpart in static 'dead force' as mere 'solicitation' to motion. In 1699 he defined their relation thus:

the impression of living force is related to bare solicitation as the infinite to the finite, or as
lines to their elements in my differential calculus. . . . Meanwhile as the law of equilibrium is
found always to apply to differentials or increments, so also in the integrals by the admirable
art of nature the same *vis viva* is found to be conserved according to the law of equivalence
[of cause and effect]. . . . Consequently, in the case of a heavy body which receives an equal
and infinitely small increment of velocity in each instant of its fall, dead force and living force
can be calculated at the same time: to wit, velocity increases uniformly as time, but absolute
force as distance or the square of time, that is, as the effect. Hence according to the analogy of
geometry or of our analysis, solicitations are as dx, velocities as x and forces as xx or $\int x dx$.[344]

The distinction, however vital for his thinking on the level of phenomena, was not the primary one in his full philosophy. It was one of Leibniz's leading concerns harmoniously to
subordinate a physics of contact action to a metaphysics of substance – not Aristotelian substance, to be sure, but substance thought of as kernel of activity transmitted by force ("the
mark of substances is to act").[345] Many of the riddles of his final conception, his peculiar
doctrine of 'monads', were to stem from so admirable yet reckless an ambition.

630   Several major issues were clarified in Leibniz's 'dynamics' (his word). One he attained through his recognition that Descartes' memorable error was due to his using for a model the law of the lever. Out of the recognition (which enabled him to distinguish between 'living' and 'dead' force) came a clear-cut, once-and-for-all distinction between statics and dynamics, between forces, that is, in equilibrium and in nonequilibrium situations. Other pieces of clarification resulted from Leibniz's *forte*, his refined thinking about continuity and discreteness and how to resolve the paradoxes that ensue when we keep cutting up pieces of matter or space or time or motion. Leibniz was the first stubbornly and consistently to think through Zeno's paradoxes in light of the ongoing mathematical taming of the infinite to which he himself contributed so much. He subjected Descartes' still influential rules for impact to a searching critique from the point of view of continuity in a way Huygens (whose rules Leibniz adopted as a matter of course) may at best have been groping at (p. 524). Also, Leibniz's dynamic approach to impact allowed him to reconceive it in terms of force, that is, of the perfect elasticity demanded by both his conservation rule and his principle of continuity. Finally, and most important of all, Leibniz faced openly rather than with ambiguous circumspection (as with Hooke and young Newton in the mode of the Baconian Brew) or not at all (as with Huygens all his life) the trouble kinetic corpuscularians were in the habit of evading when employing none but the motion of particles to account for activity in nature. What Leibniz's philosophy implied was not so much a wholesale overturning of kinetic corpuscularianism, which he always accepted without reservation, in part because it scorned the concept of 'occult' forces acting at a distance. Rather, he restructured it so that moving particles became the phenomenal manifestation of ultimate substances embodying the activity in nature that mere moving particles could not produce out of themselves. What Leibniz (up to 1687, unwittingly so) shared with post-1684 Newton was an awareness that some concept of force was required to repair this (in both his and Newton's perception) basic deficiency of kinetic corpuscularianism. What kept the two men separated forever was not only what concept to apply but also how (but for both men's usage of the calculus) to do the applying.

The above and more is what Leibniz derived, over the late 1680s and 1690s, from the head-on confrontation set up by him in 1686 between Galilean uniform acceleration in free fall and the Cartesian conception of motion and force. The vast conceptual clarification of whole ranges of forces that emerged from that confrontation was confined, however, to linear action, notably descent and impact. Nearly negligible, amid all this, was Leibniz's contribution to the *other* current enigma.

The role of force in orbital motion.   During his wonder months in late 1659, Huygens let himself be guided by a profound analogy between two tendencies toward motion – as provoked by heaviness and by being swirled around – that he then found to counterbalance

in the conical pendulum. Along this route he established the basic proportionalities governing centrifugal force (pp. 343; 537). He felt free to call it 'force' in spite of his commitment
to the kinematic approach. This was because, in conceptualizing rotational motion thus,
he remained true to any kinetic corpuscularian's broad conception of ongoing restoration
of equilibrium between the constraint to which a body is subjected by the ongoing particle
pressure that keeps it whirling around and its perennial endeavor to fly off at a tangent. Huygens failed to publish his full account, in the end adding none but the bare proportionalities
to his *Horologium oscillatorium* of 1673. In the meantime, unknown to anyone else the young
Newton derived the same result from dynamic reasoning. He then went on to extend his
result to a mathematical comparison between bodies falling to Earth, on the one hand, and
lunar and planetary orbital motion, on the other, without, however, ceasing to conceive of
the fundamental setup of orbital motion the way Huygens did – as an endeavor of orbiting
bodies to recede from the center (p. 533).

Over these very decades of the 1660s and 1670s orbital motion became rather a topic among certain fellows of the Royal Society. Newton did not become one until, in 1672,
he went public with his sensational reflector; nor did he more than incidentally attend its
weekly sessions, let alone the London coffee houses. But three other fellows, when bumping
into each other, brought up from time to time the subject of orbital motion. They looked at
it from another viewpoint than the one shared as a matter of course between Huygens and
(prior to 1679) Newton.

These fellows were Robert Hooke and Christopher Wren from the 1660s onward, joined
by the much younger Edmond Halley in the early 1680s. Their viewpoint was shaped by two
specific notions. One was an awareness (shared on the Continent by Borelli) that Kepler's
dynamics of planetary motion (p. 173), in being by near-unanimous consent untenable, had
to be replaced with something better. In line with Gilbert's book on the Earth as a magnet,
the three Britons joined to this sentiment a sense of the Earth exerting a magnetic pull that,
with gradually diminishing intensity, operates as far out as its 'sphere of activity' extends
(p. 250).

Most of the creative thinking on the subject was done by Hooke. He brought to it all his
assets – thorough command of the literature, good hunches about where to go for the next
step, unceasing, sometimes brilliant flashes of insight, ability to write up an at least superficially coherent argument. But the investigation was marred at the same time by the sum
total of his liabilities – lack of geometry beyond Galileo's, lack of capacity to think an insight
through to its ultimate conclusion, and, consequently, lack of conceptual consistency, of
systematic follow-up, and of a sense of the power of rigorously mathematical science. An
inclination to boasting that his difficult social status furthered helped seal one of history's
most inspired failures.

In 1674 Hooke added three 'suppositions' to an argument meant to prove the motion of

632   the Earth. In supposition 1 Hooke proposed that every celestial body, the Earth of course included, has an "attraction or gravitating power" that not only keeps its own parts together but also stretches out to all other celestial bodies so as to exert "a considerable influence" upon them as far as its "sphere of activity" extends.[346] It is impossible to find in anybody's writing from before the spring of 1685, which is when Newton worked his way to the conception of universal gravitation in full, a statement that comes closer to it than the one lengthy sentence in which Hooke phrased his first supposition. Still, what did he mean exactly? This appears from similar passages in somewhat later work of his, which reveal more of the ambiguities that beset Hooke's core conception of congruity & incongruity (p. 560). Considered in their light, the two apparent qualifiers ("considerable influence"; limitation of gravitational effect to width of sphere of activity) must be taken seriously in revealing at least remnants of specificity in his conception, which still betray its origin in magnetic attraction in the first place. But the most significant difference with Newton's later, definitive conception is surely that Hooke characteristically left his definition of what he had in mind as enticing as he left it vague. Here as elsewhere, Hooke's tragedy was his inclination to miss his own point.

About the mode of operation of the attraction he had more to say in the two following suppositions.

In supposition 2 Hooke proposed that if any body in uniform, rectilinear motion be subjected to an "effectual power", it will be deflected thereby into some curved orbit, be it a circle, an ellipse, or some more complex curve. At an earlier occasion, Hooke had chosen the conical pendulum for an illustration. In his hands its treatment suffered lethally from the conflation of two quite different concepts of force (modern 'Newtonian' force and modern 'work'), as also from his inability to carry the mathematics involved to successful conclusion. Even so supposition 2 made a seminal contribution to the dynamics that, albeit out of reach of the man who stood at the beginning of this particular road, was soon to prove to lie at its end. Hooke was the first to draw from the Galileo-inspired principle of uniform, rectilinear motion retained the retrospectively obvious conclusion that the proper way to conceptualize orbital motion is as a force-induced *deflection away* from uniformly rectilinear motion rather than (the other, Cartesian way round) as an ongoing balancing act between the constraint exerted upon a moving body to follow a curved orbit and its endeavor to recede from the center. In short, "if the principle of inertia [the definitive name by which Newton was soon to call it] is given, the question is what constrains a body to follow a curved path and not the tendency to recede that it exhibits when so constrained."[347] In 1679, half a decade after phrasing this very insight as his second supposition, Hooke was to jump at an occasion to employ it for teaching Newton a lesson (p. 647).

In supposition 3 Hooke proposed "that these attractive powers [i.e., what he had just called 'effectual powers'] are so much the more powerful in operating, by how much the nearer the body wrought upon is to their own Centers."[348] In exactly what proportion, then, 633 does the attraction decrease with distance? Not here but in other papers at the time Hooke suggested that the decrease would be in squared proportion. While a variety of possible analogies presented themselves (e.g., with the intensity of light), Hooke is likely to have arrived at the conclusion by applying a current yet abortive version of Kepler's area law to his own conception of a discretely acting force proportional to the square of velocity.

Could Hooke's suspicion of an inverse-square relation be proved? Retrospectively, it could not by any means available to Hooke – apart from using the wrong area law, he lacked all sense of such mathematical limiting procedures as were required to get from force acting discretely to force acting continuously. In January 1684 the question of proof was put to Hooke by Wren and Halley. Both men had independently arrived at the same suspicion along a somewhat different pathway, more similar to the one eighteen years earlier taken in solitude by Newton. Hooke's response was that, yes, he could prove it, but that he had decided to withhold his proof until the inability of everybody else to prove it had become sufficiently evident. This failed to satisfy Wren, who went on to offer a book worth 40 shillings as a prize for the missing proof if supplied within two months (in retrospect, as a prize for fame everlasting). Hooke's response failed likewise to satisfy Halley, who, while not in a hurry to see the question resolved, found it important enough to keep it in mind. The chief purpose of Halley's trip to Cambridge half a year later, in August 1684, is unknown; but whoever it was that he wished to see in the first place, with the inverse-square riddle firmly in mind he used the occasion to pay a side visit as well.

NOTES ON LITERATURE USED

Aristotelianism: *SRHI* 3.5.2., 4.4.3.

Platonism: The Cambridge Platonists have lately come under intensified scrutiny; my own view still owes most to E.A. Burtt's perceptive treatment in his *The Metaphysical of Modern Physical Science. A Historical and Critical Essay*. London: Routledge & Kegan Paul, 1924 (*SRHI*, 2.3.4.).

Stoa: S. Sambursky, *Physics of the Stoics*. Princeton UP, 1987 (1st ed.: London: Routledge & Kegan Paul, 1959), especially pp. 37-38. Gad Freudenthal, 'Clandestine Stoic Concepts in Mechanical Philosophy: The Problem of Electrical Attraction', in: Judith V. Field & Frank A.J.L. James (eds.), *Renaissance and Revolution. Humanists, Scholars, Craftsmen and Natural Philosophers in Early Modern Europe*. Cambridge UP, 1993, pp. 161-172. Papers contributed by Peter Barker, Betty Jo Teeter Dobbs, and Jamie C. Kassler to Stephen Gaukroger (ed.), *The Uses of Antiquity. The Scientific Revolution and the Classical Tradition*. Dordrecht: Kluwer, 1991, and to Margaret J. Osler (ed.), *Atoms, Pneuma and Tranquillity. Epicurean and Stoic Themes in European Thought*. Cambridge UP, 1991.

Atomism: K. Lasswitz, *Geschichte der Atomistik vom Mittelalter bis Newton*. Leipzig: Voss, 1890 (2nd ed. 1926).

Skepticism: *SRHI* 4.4.5.

Dialogue with the ancients. Tabitta van Nouhuys, *The age of two-faced Janus. The comets of 1577 and 1618 and the decline of the Aristotelian world view in the Netherlands*. Leiden: Brill, 1998; p. 15-25. I owe the idea of the contrast drawn in the main text between music at the start and at the end of the 17th century to personal communication from Penelope Gouk. She has treated Newton's ideas on Pythagoras in her *Music, Science and Natural Magic in Seventeenth-Century England*. New Haven / London: Yale UP, 1999; pp. 251-254, with due reference to a classic article by J.E. McGuire and P.M. Rattansi, 'Newton and the "Pipes of Pan"'. *Notes & Records of the Royal Society* 21, 1966, pp. 108-143.

Thinking with the hands. The expression 'penser avec les mains', often used by R. Hooykaas, is due to Denis de Rougemont (1935).

In the 1940s Edgar Zilsel developed the following thesis: due to the breakdown around 1600 of social barriers between skilled artisans, on the one hand, schoolmen and humanists, on the other, such rules of thumb as used to be practiced by the former could now be formalized into scientific law by men skilled in logical argument who were drawn from the latter circles. I have discussed the merits, limitations, and historiographical ramifications of this thesis in *SRHI*, 5.2.4., 5.2.6. In view of conclusions reached in the present book, I evaluate Zilsel's thesis thus. To speak of rapprochement between practitioners of nature-knowledge and the crafts does make sense, not to be sure for some imaginary, wholesale breakdown of social barriers around 1600 but for an earlier period starting by the mid-15th century. Also, that *rapprochement* remained confined to those active in the coercively empiricist mode of nature-knowledge. It is further true that Galileo owed inspiration to the literature that came out of that *rapprochement*. Still, only in the rarest of cases (notably with some work by Zilsel's chief witness, William Gilbert) can findings attained in any of the three revolutionized modes of nature-knowledge

be regarded as straightforwardly due to sheer formalization and logical rearrangement of what arti-   635
sans already 'knew' by their hands and by their trial-and-error methods alone. In a wider sense, too,
the hoped-for improvement of craft practice through revolutionary science failed to become manifest
until the early 18th century. As an overall fact of life, then, by the end of the 17th century 'rule of thumb'
and 'scientific law' were still wide apart.

Domenico Bertoloni Meli, in his *Thinking with Objects. The Transformation of Mechanics in the
Seventeenth Century*. Baltimore: Johns Hopkins UP, 2006, has dealt in a nicely concrete manner with
how 17th-century mathematical scientists began to think with their hands.

Print. In *SRHI* 5.2.9. I have critically examined Elizabeth L. Eisenstein's numerous, strongly worded
claims in her *The Printing Press as an Agent of Change. Communications and Cultural Transformations
in Early-Modern Europe*. 2 vols. Cambridge UP, 1979. Thanks to Adrian Johns' *The Nature of the Book.
Print and Knowledge in the Making*. Chicago: University of Chicago Press, 1998, I have since become
impressed with the questionable state of something that Eisenstein took for granted – the alleged
'fixity' of texts printed prior to the mid-19th century. It remains true that the instability of textual
traditions in a script culture (due to fire, theft, mice, copying errors, etc.) differs greatly from defective
fixity in the first centuries of print culture (due, among much more that Johns also details, to near-
ubiquitous pirate editions).

Method. *SRHI* 3.1.1.-2.

Pre-Newtonian force. Central to my treatment is Richard S. Westfall, *Force in Newton's Physics. The
Science of Dynamics in the Seventeenth Century*. New York/London: Elsevier/MacDonald, 1971. I have
also used with profit his 'Robert Hooke, Mechanical Technology, and Scientific Investigation'. In: J.G.
Burke (ed.), *The Uses of Science in the Age of Newton*. Berkeley: University of California Press, 1983;
pp. 85-110 (*SRHI* 3.4.3.); further E.J. Dijksterhuis, *Val en worp. Een bijdrage tot de geschiedenis der me-
chanica van Aristoteles tot Newton*. Groningen: Noordhoff, 1924 (*SRHI* 2.3.2.; a shortened version of his
specific argument (ch. 2, section 2.1.) on Leonardo's conceptualization of force is in his *The Mechanization of the World Picture*. Oxford UP, 1961, III 42–44); J. Christiaan Boudri, *What Was Mechanical
About Mechanics. The Concept of Force between Metaphysics and Mechanics from Newton to Lagrange*.
Dordrecht: Kluwer, 2002; Catherine Wilson, *Leibniz's Metaphysics. A Historical and Comparative Study*.
Manchester UP, 1989.

XIX

THE SIXTH TRANSFORMATION:
THE NEWTONIAN SYNTHESIS

Halley had made his first acquaintance with Isaac Newton two years earlier, in 1682. The Lucasian Professor of Mathematics, known outside Cambridge as the author of a few remarkable papers on light and color and as a secretive mathematical prodigy, might perhaps be of help in disentangling the riddle, so Halley now looked him up to ask him "what he thought the Curve would be that would be described by the Planets supposing the force of attraction towards the Sun to be reciprocal to the square of their distance from it."[349] To which query Newton answered right away that it would be an ellipse. Halley, "struck with joy & amazement", asked in return how his host knew that. To which query Newton answered that he had calculated it; which answer both men perfectly understood to mean that, rather than just the lucky guess it had so far seemed at best, this was a property amenable indeed, as Hooke had vainly boasted and as Wren and Halley still hoped, to rigorous proof.

The calculation had been made by Newton five years earlier, after an exchange with Hooke over the dynamics of orbital motion, in the course of which Hooke taught him a painful yet seminal lesson about its proper conceptualization. The exchange had been the first significant, albeit still fleeting, occasion for Newton to return to problems of motion, which he had abandoned inconclusively a decade earlier (p. 536).

Who, then, was this man Isaac Newton, and what have we over past chapters learned about him already?

The Newton knot

Newton frankly defies any facile attempt to catch him whole. Why is that? What makes Newton and his achievements, published and unpublished, so hard to grasp? Was he not almost unique among protagonists of the Scientific Revolution for leaving to posterity an almost complete record of everything he ever noted down?

In the first, as in the final, analysis, Newton was a *sans pareil* genius. Here is how a man more intimately familiar than most of us with what it means to be a genius, John Maynard Keynes, defined the peculiar nature of Newton's creativity:

> I believe that the clue to his mind is to be found in his unusual powers of continuous concentrated introspection. . . . I fancy his pre-eminence is due to his muscles of intuition being

638

the strongest and most enduring with which a man has ever been gifted. Anyone who has ever attempted pure scientific or philosophical thought knows how one can hold a problem momentarily in one's mind and apply all one's powers of concentration to piercing through it, and how it will dissolve and escape and you find that what you are surveying is a blank. I believe that Newton could hold a problem in his mind for hours and days and weeks until it surrendered to him its secret.[350]

Consequently, even with a written record as rich as the one near-integrally preserved until 1936 and then (in spite of Keynes' efforts) by auction spread all over the globe, those to examine its contents have tended to feel rather humble in comparison. To the present author, who seeks to come to terms here with what he takes to be the best work of those who have examined it firsthand, the same rule applies in squared proportion. This is by no means to deny that those dedicated to the task may approach the record and come to a measure of historically responsible understanding of a Newton whose picture has deservedly altered over half a century of 'Newton industry' in ways almost unrecognizable two generations ago. The point, however, is that it is bound to be defective in a deeper sense than ordinary when historians seek to come to grips with the past.

A further cause of opacity manifold rests in the nature of the record. It has over the past half-century appeared to be much like the proverbial iceberg, with the two tips that rise above the surface representing almost the full Newton to posterity until (starting with Keynes) the record began to be examined in earnest and over its full breadth. Historians face the inherent difficulty of interpreting alchemical texts and the niceties of 17th-century theology and biblical chronology. They face the rarity of dates appended, thus sending them back to watermarks, the evolution of writing hands, and often dubious third-party recollections. But above all they face the circumstance that the habits of a person writing notes and tracts meant for private use alone are different from those of an author intent on making his ideas clear to prospective readers. The loner often lets his unintended reader guess at the very why of a text: In what context to place the piece? What did its author have in mind when taking up a given subject? What made him go in this direction rather than that? In throwing himself *in medias res* as was his wont, Newton did little to explain himself, intent as he was on prey yet unknown to him while clear in his own mind about why he wanted to catch it.

In regard to the published work, sources of opacity abound, too. Not that Newton's writing style is obscure in any customary sense, or that the organization he gave to *Principia* and *Opticks* is anything other than carefully thought out. Still, most among those who approach these works find themselves in trouble. There are specific sources of trouble, such as, for example, in *Principia,* identification of the core argument itself and its peculiar, soon-outmoded mathematical language and, in the more accessible *Opticks,* the full meaning and purpose of the 'Queries' at the end. Neither book is a marvel of didactic exposition, lacking markers to guide the uninitiated, nor is either book leavened by any charm or human touch, be it Galileo's wit or Kepler's catchy metaphors. But what most of all stands in the way of quick understanding is the density of Newton's arguments. One may easily read a paragraph and feel that one has caught the point, then return to it and find that there is still much more to capture there. Or one may have worked one's way through a whole range of successive arguments and still wonder what, in the end, they add up to. All this is in addition to such more customary problems as are raised by the variety of later editions and/or translations or by the need to rid oneself of the load of later interpretations provided at a time when little but the two books remained in the public eye, with little hint that they were just the tip of an iceberg underneath which hid other, quite surprising, not to say shocking, work.

Finally, there is Newton's versatility. Not only is it humanly impossible to become a Newton specialist in the full sense of equal command of his mathematics, his optics, his dynamics, his chemistry and alchemy, his theology, his chronology, and his ideas on the constitution of the world (to name only the most important and also to leave aside his position in Cambridge, his public service in Parliament and later at the Mint, and his dealings with his fellow men in general and the Royal Society in particular). More problematic still is the question of whether and, if so, how these varied pursuits hang together. Here modern historians' positions range from continued insistence on the primacy of the first three in the list (the traditional items, and without any doubt the principal grounds for calling him a *sans pareil* genius) to the claim that a unified conception behind it all can indeed be identified.

Whatever one may think of this highly controversial issue of 'the whole Newton' (I set forth my own view on p. 711), the sheer variety of Newtons encountered in previous chapters already testifies to his versatility. What have those chapters meanwhile taught us about him?

In chapter 15 (p. 531), we saw how the 'Waste Book' marked this Cambridge undergraduate's quickly intensifying acquaintance with the ongoing Scientific Revolution. Newton's reading notes gave way from 1664 onward to increasingly pointed questions, so as in 1665/6 to explode in one original investigation after another – the essence of the calculus; colors as primary and sunlight as compounded of them; work on motion culminating in a quantified, mathematical comparison between celestial orbits and free fall on Earth – all of which he kept to himself.

Near the end of chapter 10 (p. 355) we met him in the first domain where his notes moved from absorption to creation, the mathematical analysis of infinites, which he single-handedly turned into the calculus. It remains to show how, a quarter century later, Newton employed his calculus in the *Principia*.

Earlier in chapter 10 (p. 300) and then again in chapter 15 (p. 543) we found him grappling, in first one mode, then in another, with strange refraction in prisms – the principal subject of his first and (up to 1687) only publications. In the present chapter we shall deal with the whole of his work on light and color (p. 678).

640   In chapter 13 we encountered him as a highly gifted experimenter, at work on electric phe-nomena (p. 491) and on the fine structure of matter as investigated in chemical/alchemical reactions (p. 467). What remains to be said about these subjects concerns the conclusions Newton drew from all this in his full conception of nature, and what those conclusions were to entail for his mature dynamics.

In chapter 15 I hinted in passing at the 'fiery religiosity' that in the late 1660s provoked Newton's rupture with Cartesianism (p. 536), whereas at the end of chapter 17 (p. 594) I men-tioned how his anti-Trinitarianism and his science had jointly driven him to an essentially deist stance by the early 1680s. In what follows we come back to both Newton's private and his public conceptions of the Deity.

In chapter 16 (p. 561) we encountered in his 1675 'Hypothesis of Light' a conception of the world in which experimental outcomes, corpuscular mechanisms, and ambiguous hints at active principles were combined in such a manner as to form the high point of the Baconian Brew. What became of those active principles forms the very crux of the present chapter.

In chapter 15 (p. 532) we watched Newton at work between 1665 and 1668 on several *gen-era* of motion, in ongoing comparison with how Huygens had come to grips with them. Due to the lack of clarity of the various concepts of force that he employed, his quest terminated in a dead end. I attempted to clarify in chapter 18 what made those concepts so enigmatic (p. 621) and moved ahead to Halley's visit of 1684. I shall now jump back to where we left Newton in 1668, so as to be able to explain what had in the meantime enabled him so self-assuredly to tell his guest 'that he had calculated it', and then to examine the creative outburst unpredictably unleashed by that answer – or rather, by Halley's response to it.

Toward the Principia

Between 1668, when growing awareness of the force knot compelled him to give up his search for a general science of motion, and 1684, when he set out to unravel the knot, New-ton faced it in two different ways. The confrontation with the force knot occasioned by an exchange of letters with Hooke in 1679 was brief and momentous. Another confrontation, which involved an ever more acute conviction of the irreparable shortcomings of kinetic corpuscularianism even if pushed to its limits and beyond, was drawn out over the full period but not, for that, one whit less momentous. Experts dispute the dating of specific insights and even the precise order in which Newton attained them. They also dispute the extent to which the 1679 correspondence with Hooke accelerated them or influenced them in other ways. So much is certain that along the way Newton passed various milestones, not necessarily with full consciousness of their retrospectively obvious consequences until after Halley's visit. These milestones minimally include the following:

- ongoing concern with those phenomena that were the hardest to account for in a conception of the world as made up solely of moving corpuscles;
- consequently, prolonged investigation of the fine structure of matter by means, primarily, of chemical/alchemical researches;
- elaboration of a succession of conceptions of the natural world in which moving corpuscles *and* active principles were made to account for its principal phenomena as made manifest in experiment;
- increasing tendency to conceive of those 'active principles' as 'forces', due to
  (1) waning confidence in their reducibility to contact action in some mechanism of particles in motion;
  (2) increasing awareness that physically operative forces (whether thus reducible or not) are in any case suitable for mathematical treatment in ways aethereal mechanisms are not;
- ongoing wavering over whether or not those forces operate solely on the microscopically short range pertinent to chemical/alchemical reactions or also over the much longer ranges that mark, notably, falling bodies and the Moon's orbit;
- experimental proof that at least in the terrestrial region no aether exists;
- adoption, without apparent questioning, of Hooke's conception of some attractive force left unspecified but for the rate of its variation with distance.

Although the order of milestones here outlined, however plausible logically, may not correspond to that in which Newton experienced them, for better or worse I shall now consider them in this order.

Exploring the limits of kinetic corpuscularianism.   From his 'Waste Book' days onward, Newton displayed an almost perverse inclination to focus on such natural phenomena as the doctrine of kinetic corpuscularianism, so enthusiastically yet critically embraced by him at the time, was least equipped to handle. The pioneers themselves had already sensed that certain phenomena did not fit too well. Indeed, the surreptitious invocation of pre-Revolutionary, really magic-infused notions irreconcilable with the basic principle of explanation by means of moving particles alone began already with Descartes and Gassendi (p. 552). By the 1660s a felt need to widen such conceptual expansion and make it more explicit led to a hypothetical, rather than dogmatic, approach to the doctrine as effected in Revolutionary fashion by Boyle and by Hooke. Spurred on by his assiduous reading of their works, in 1664/5 Newton set out to widen the breach. Time and again, throughout his entire career really, he was to return to four issues in particular: "the cohesion of bodies, capillary action, surface tension of fluids, the expansion and pressure of air."[351] For instance, what he found astonishing in the conduct of air as demonstrated in recent but well-known experiments with the air pump was the apparent lack of limit set to its capacity to expand, as also

642   the variety of causes thereof. A decrease in outside pressure makes it expand; so does heat; so does the presence of other bodies, as apparent from capillary action arising from differences in pressure. Indeed, surface tension in fluids suggests in particularly outspoken fashion that not just particles of air but all bodies tend to avoid each other. How to explain such phenomena of mutual avoidance and the apparent generality thereof by means of nothing but intricate corpuscular mechanisms of the kind strenuously yet implausibly invoked by their principal student, Boyle?

Basically the same question presented itself to Newton's mind when, starting around 1666 and perhaps motivated by it, he turned to chemical reactions. There, too, numerous phenomena were hard to reconcile with the reigning conception; there, too, Boyle heroically sought to reconcile them. Why, so Newton asked himself, is it that the reaction of certain substances at room temperature produces heat powerful enough to burst the tube? Why is it that *aqua fortis* (by its modern name, nitric acid) dissolves silver but not gold, whereas *aqua regis* (nitric acid mixed with hydrochloric acid) dissolves gold but not silver? Why is it that certain substances fail to mix (e.g., water and oil, in contrast to water and salts)? Why is it that certain substances that likewise fail to mix (e.g., molten lead and copper) do mix as soon as a third substance (e.g., tin) is added?

The 'Hypothesis of Light' as a stage in Newton's thinking. By the early 1670s the insights gained by asking questions like these precipitated in a detailed conception of the natural world that we are already familiar with. The 'Hypothesis of Light', read on Newton's behalf to the Royal Society on 9 December 1675, was the high point of the 'Baconian Brew' – the culmination point, that is, of a mode of nature-knowledge pioneered in less elaborate fashion by numerous speculative philosophers of British stock and in less radical fashion by Boyle and Hooke, both of whom had preceded Newton in the experimental bent given to it (p. 554). But the 'Hypothesis' must also be seen as one passing stage in the ongoing development of Newton's forever fluctuating conception of nature. Examination of the milestones that came next makes it possible to gauge how in the sequel he managed to *transcend* it – that is, to transform the Baconian Brew into something radically different. The main points Newton was concerned to make in the 'Hypothesis' were as follows.

- All known phenomena of light can be explained by means of an all-pervasive, extremely subtle aether. Its particles vibrate swiftly and aggregate to locally greater density than in free aethereal space as the pores of the gross bodies in which the aether stands are bigger.
- In this as in other respects, the aether, albeit of course much subtler than air, is closely similar to it.
- The actual existence of the aether is empirically demonstrated by the fact that the rate of dampening of a pendulum is nearly independent of whether it swings in air or in the void (i.e., hardly the air but overwhelmingly the aether is responsible for the resistance manifest in the dampening).

- In ongoing circulation all over the universe, the aether particles press down in showers   643
of varying density upon gross bodies on or near the Earth, from which they then rise up
in exhalation, thus perhaps producing such phenomena as gravity, electricity, capillary
action, sensory perception, or animal motion.
- A 'secret principle of unsociableness' is needed to explain further how it is that certain
stuffs fail to mix, even though their particulate, hence, porous structure would seem sufficiently subtle to allow mutual penetration.

This secret principle was much like Hooke's principle of congruity and incongruity. As with
Hooke, Newton left it unclear whether it was still reducible in the end to some mechanism
of particles in motion. Had he been exploring the outer limits of kinetic corpuscularianism,
or had he already transcended them? He left the audience guessing – not for nothing had he
asked Oldenburg, who read the piece to the assembled fellows, to drop the phrase "the whole
frame of Nature may be nothing but aether condensed by a fermental principle". To the
historian, that phrase, and in addition the entire content of Newton's private paper 'On the
Vegetation of Metals', leave no doubt that already by 1669 Newton felt sure that "mechanicall coalitions or seperations of particles" did not suffice to explain activity in nature – "wee
must have recourse to som further cause".

Active principles. Speaking in the most general of terms, that further cause was 'active
principles'. But what active principles precisely, and what to think of their ultimate reducibility to some underlying mechanism of particles in motion? On the former question Newton was not to settle his mind for good until 1684/5; on the latter he never would, or at least
not quite. What underlay both questions was his conviction, arrived at by the early 1670s at
the latest, that whole ranges of phenomena in nature display degrees of activity such as no
ordinary mechanism of fundamentally passive matter in motion can possibly account for.

This, indeed, was the big issue held of old against atomism by both Aristotelians and
stoics, with the latter providing the most pointed answer to the various conundrums of atomism (besides activity in nature, the question of what makes particles cohere for a while in
gross bodies, and the relation between matter and spirit). Newton was an avid reader of the
stoic fragments, and his aether, as well as that of other British thinkers, deliberately blended
the subtle matter of orthodox kinetic corpuscularianism with stoic pneuma, as also with the
Platonic world-soul. Two features made Newton's (and Hooke's) aether different from whatever was conceived by armchair philosophers like Petty or Hale: its much greater specificity
and differentiation (especially the density gradient) and the demonstration of its apparent
capacity to explain experimentally produced phenomena of nature of the most varied kind.

Speaking now more specifically, in terms of Newton's alchemy in particular, that 'further
cause' was 'vegetative spirit' (or at times some kindred expression for the same thing). By the
'vegetation' of metals Newton meant all kinds of lifelike phenomena to be associated with

644 them. In his 1669 tract by that name he listed phenomena like the (on occasion) treelike growth of a variety of metals in ores, his observation "that in the yᵉ first days of yᵉ stone green is yᵉ only permanent colour & so in yᵉ least mature vegetables", and the circumstance "that nothing has so great power on animalls as mineralls witnes not only the Alkahest to destroy & yᵉ Elixir to conserve but their operations in common chymicall physick & in springs".

In the conception here developed Newton made a clear-cut distinction between phenomena like these, which require for full understanding the vegetative spirit, and other phenomena, such as heaviness, which are liable to explanation by means of some mechanism of particles in motion. Or rather, mechanical and vegetative processes find themselves in interaction all the time. Here is what the natural process subjected to examination in 'On the Vegetation of Metals' came down to:

> the metalline spirit ascends from the bowels of the earth, is fixed into salts and minerals when it meets water, and is thereby alienated from its metallic nature. The alienation arises because concretion is not a process of vegetation but only "a gross mechanicall transposition of parts". If the spirits can be freed from their fixed compositions, they can again "receive metallick life & by degrees recover their pristine metallick forme".

With respect to this vegetative spirit:

> so far as by vegetation such changes are wrought as cannot bee done wᵗʰout it wee must have recourse to som further cause. And this difference is vast & fundamentall because nothing could ever yet bee made wᵗʰout vegetation wᶜʰ nature useth to produce by it. . . . There is therefore besides yᵉ sensible changes wrought in yᵉ textures of yᵉ grosser matter a more subtile secret & noble way of working in all vegetation which makes its products distinct from all others & yᵉ immediate seate of thes operations is not yᵉ whole bulk of matter, but rather an exceeding subtile & inimaginably small portion of matter diffused through the masse wᶜʰ if it were seperated there would remain but a dead & inactive earth.
>
> [and elsewhere in the tract, when speaking of the aether:] This is the subtil spirit, this is Natures universall agent, her secret fire, yᵉ onely ferment & principle of all vegetation. The material soule of all matter wᶜʰ being constantly inspired from above pervades & concretes wᵗʰ it into one form & then if incited by a gentle heat actuates & enlivens it.

The vegetative spirit was marked in Newton's view by two further features: its possible identity with 'the body of light' and its divine nature. In regard to the former, Newton wrote:

> This spt perhaps is yᵉ body of light 1 becaus both have a prodigious active principle (both are perpetual workers) 2 because all things may bee made to emit light by heat.[352]

He went on from there to outline a range of further parallels and relations between light, 645
the vegetable spirit, and the "gentle heat" required to make the spirit inform matter with life.
Much of this was to come back in the 'Queries' of *Opticks* – indeed, light was never far from
Newton's mind when brooding over the constitution of the world (p. 704).

'On the Vegetation of Metals' also gave a place to the deity. A few years earlier, at the
end of his first run of private papers on generic motion, Newton had come to discern the
atheist implications of Descartes' radical separation between body and spirit (p. 535). The
dependence of the material world on God could be maintained, so he had argued there in an
outburst of fury, only by getting rid of Descartes' identification of essentially passive matter
with extension. Within a few years Newton hit upon the positive solution. The vegetative
spirit not only mingles with passive matter all the time, and vivifies it, but is itself of divine
origin.

> Ultimately the cause [of vital phenomena] is God, and within the realm of vegetable chemis-
> try one may find an area of continuing divine guidance of the world and of matter, an area of
> providential care. It is God's will that directs the motion of the particles of matter and guides
> them into their designed arrangements. The vital Stoic and alchemical agent, the subtle spirit
> of life, the secret fire in the earth's aethereal breath is thus simply the natural agent God uses in
> directing the motion of the passive particles of matter.[353]

All the way through 'On the Vegetation of Metals', then, the tension between passive matter
and active spirit is almost palpable. How far could it be maintained in Newton's thinking
until it snapped?

From active principles to forces. In 1679 Newton set out to write a treatise on air and
aether, in which he amended aspects of the conception of nature set forth four years earlier
in the 'Hypothesis of Light'. The treatise displays a Newton clearly on the move from active
principles to forces. He now declared himself unable to account for the immense expansion
of air, and for the mutual distance kept between things generally, by any other means than
an appeal to some large, repulsive force operating (by implication) at a distance. He also set
out to describe the aether as generated by the fragmentation of air but quickly broke off the
treatise; quite possibly he realized that he was heading for an infinite regress, with properties
of the air reduced to a subtle aether whose similar properties were to require a further reduction along the same lines to an even subtler aether. Whether or not this is what made him
stop, there is good reason to assume that it was about this time, with his thinking directed
toward an aether ever more ambiguous and problematic in the properties assigned to it, that
he decided to examine all over again what he had always regarded as proof of its existence,
to wit, how the dampening of a pendulum in the void was so similar to its dampening in the

646   air. With his knack for experimental design and for precision measurement, he thought of a pendulum experiment that could exploit the principal difference between the resistance offered by air and by aether – air operates only on a body's outside, whereas aether, in penetrating a body's pores, ought to offer resistance at its inside as well. To that end, he tied a box to a pendulum and let it swing first when empty (i.e., filled with air), then when filled with heavy metal. Eliminating as many sources of friction and of stretching as he could, he first marked the respective heights reached by the pendulum with the empty box on completion of its first, second, and third full swing. Next, he weighed very carefully how much heavier the box became by filling it up and found a factor of 78. On making the heavy box swing to and fro, he found that for it to reach any height first attained by the empty box required seventy-seven swings from one mark to the next. He then reasoned as follows:

> Let A therefore designate the resistance of the box on its external surface, and B the resistance of the empty box on its internal parts; then, if the resistances of equally swift bodies on their internal parts are as the matter, or the number of particles that are resisted, 78B will be the resistance of the full box on its internal parts; and thus the whole resistance A + B of the empty box will be to the whole resistance A + 78B of the full box as 77 to 78, and by separation A + B will be to 77B as 77 to 1, and hence A + B will be to B as 77 x 77 to 1, and by separation A will be to B as 5,928 to 1. The resistance encountered by the empty box on its internal parts is therefore more than 5,000 times smaller than the resistance on the external surface. This argument depends on the hypothesis that the greater resistance encountered by the full box does not arise from some other hidden cause but only from the action of some subtle fluid upon the enclosed metal. But I believe the cause is quite another. For the periods of the oscillations of the full box are less than the periods of the oscillations of the empty box, and therefore the resistance on the external surface of the full box is greater than that of the empty box in proportion to its velocity and the length of the space described in oscillating. Hence, since it is so, the resistance on the internal parts of the box will be either nil or wholly insensible.[354]

Unfortunately, the paper in which he described this important experiment is among the few he lost, so we know of it only from the *Principia,* where he carefully recounted it from memory. By that time (in 1685/6, that is) Newton had completed his move from aethereal mechanisms to the operation of forces, and it is not at all certain whether the experiment caused him to give up belief in an aether unambiguously and at one stroke. Remnants of aethereal mechanisms do make their passing appearance in utterances of his up to 1685, and this may well have been so because, even with terrestrial aether showers discarded, at least two big questions remained.

   One was whether the operation of forces could not after all be reduced, in the final analysis, to some aethereal mechanism (even though, given the outcome of the revised pendulum experiment, that aether had to lack resistance). Whereas the terms on which he conceived of the question were radically to alter indeed, Newton was never to resolve it to his own full satisfaction.

The other question was whether what he might hold true of the short-range forces he found himself accepting in vegetative processes could be extended without more ado to the longer range required, notably, if some force were to account for heaviness and its effects. On this Newton may have changed his mind – whether indeed and, if so, how often remain unclear. In the days when he was still thinking in terms of active principles ultimately operating through contact action in however unspecified and ambiguous a manner, he had accounted for heaviness in more than one way. In the 'Hypothesis of Light', as earlier in 'On the Vegetation of Metals', he had made his aether shower press gross bodies down to Earth as, in its ongoing circulation, the aether went on to vivify the Earth itself. Four years later, in a letter to Boyle and perhaps in view of its recipient only (though likewise an alchemical adept), Newton devised a distinct aethereal mechanism for heaviness and thus detached it from the action of the vegetative spirit. Quite possibly, his ongoing wavering (if that is what it was) over whether or not extrapolation from the short-range forces of alchemical action to any long-range force was warranted indeed came to an end due to uninvited outside intervention starting on 24 November 1679.

Hooke and Halley intervene.   To whatever extent Newton had by 1679 come to conceive of active, aether-guided principles as irreducible forces, he still conceived of orbital motion in the same manner as almost everybody else tainted by kinetic corpuscularianism did – as circular motion arising from particle pressure and counterbalanced by a tendency to recede from the center.

'Almost everybody' indeed – Hooke alone had come to conceive of orbital motion as arising from the deflection, due to some attractive force, of a body away from the uniform movement in a straight line it tends to persist in (p. 632). Hooke served the Royal Society right from the start as its curator of experiments, and when in 1677 Oldenburg died he acquired his job in addition. In that capacity, he sat down on 24 November 1679 to write Newton a letter of the kind secretaries the world over write to activate dormant members. He used as bait his own, five-year-old conception of orbital motion. The exchange comprises four plus two letters altogether, of a nature galling at least to Newton (Hooke's anger was aroused later, as we shall see). These led Newton, in the privacy of his chambers, to adopt Hooke's setup (whether for good or for the pleasure of the challenge only is impossible to tell for sure), invert the original problem, and work it out. In so doing Newton demonstrated mathematically by infinitesimal means forever foreclosed to Hooke himself that, yes, Hooke had it right – if a body is deflected from its uniform, rectilinear path into an ellipse, the attractive force that from one focus does the deflecting is such as to diminish in proportion to the square of the distance. Here is how he did it:

648 Newton began . . . by demonstrating Kepler's law of areas. Using the law of areas and accepting Hooke's definition of the dynamic elements of orbital motion, he showed first that the force varies inversely as the square of the distance at the two apsides [endpoints of the major axis] of an ellipse, and then that the same relation holds for every point on an ellipse. If the inverse-square relation initially [in 1666, that is; see p. 533] flowed from the substitution of Kepler's third law into the formula for centrifugal force under the simplifying assumption of circular orbits, the demonstration of its necessity in elliptical orbits far excelled in difficulty what had been a simple substitution.[355]

This was the private proof Newton had in mind when, five years later, he told Halley that 'he had calculated it'. He also told Halley that he could not find the calculation. More likely, he was not prepared to show it without having a chance to look it over first. And that is how, in response, Halley could not satisfy his burning curiosity except by asking for (and indeed by receiving) Newton's promise to recover it and send Halley a copy soon after his guest's return to London.

What, not days but months later, Halley did receive in London was something rather different. It was a nine-page-long tract entitled 'De motu corporum in gyrum' (On the Motion of Orbiting Bodies). Halley took care at once to have it registered with the Royal Society. It contained the expected proof for sure, yet the proof came enveloped in an argument of far wider scope – a budding, general dynamics conceived with mathematical rigor. After Halley's visit, all that from the 'Waste Book' days onward Newton had picked up *and* taught himself about forces, spurred on both by his growing mistrust of any aether conception whatsoever and by Hooke's conceptualization of orbital motion, began to gestate and spawned, first, the short tract and then, within two more years spent, to the exclusion of almost everything else, in utter transport and possession, the *Principia*. Gone were the days when he believed to have shaken hands with problems of motion for good (as in 1679 he had grimly informed Hooke). Gone, for a long time to come, were the days when he spent his working hours on private, esoteric-alchemical and heretical-theological endeavors enveloped in almost complete secrecy. Gone were the days when he littered his chambers with private papers overflowing with grandiose conceptions and pathbreaking calculations of the most varied kind but never yet considered conjointly or thought through to their final conclusion.

Not, to be sure, that Halley's intervention ceased to be of service. Immediately after receiving 'De motu corporum in gyrum' Halley returned to Cambridge, now with the sole purpose of visiting Newton. He proposed to his host that he should rework it into publishable form under the aegis of the Royal Society. Deeply impressed as he was with 'De motu', the huge promise of which he grasped at once, Halley was nonetheless quite unprepared for what nine months later, in August 1685, Newton sent him by way of a duly reworked version. It had in the meantime assumed book-length proportions.

Perhaps we can trace to that reading Halley's extraordinary behavior in 1686 and 1687, when he sacrificed all his own activities to insure the publication of the *Principia*. And perhaps we can trace to that reading the tone of awe with which he henceforth referred to the still growing masterpiece.[356]

Thus, Halley took upon himself an ever more arduous job, alleviated only by his sense of awe, his generosity, and his diplomatic skill – the job of midwife to Newton's labors. These labors were marked in the first instance by a succession of huge conceptual steps that Newton made between August 1684, when he began to turn his 1679 proof into the short tract 'De motu', and the spring of 1685, when he decided that nothing less than a book would be able to contain his full investigation – what would in the end become the *Principia*, to be completed in April 1687.

'On the Motion of Orbiting Bodies' – the decisive difference.

Already in the first draft sent to Halley, 'De motu corporum in gyrum' introduced four basic concepts:

- A *centripetal* (i.e., center-seeking) *force* of attraction that operated from outside a body to alter its uniform motion in a right line and that was always proportional to the rate of change. Coined in allusion to Huygens' centrifugal force, this new concept captured Hooke's conceptualization of orbital motion. It expressed in more general and also mathematically more rigorous terms the idea of force as altering a body's motion that Newton had in his 'Waste Book' days employed solely when trying to grasp impact (p. 532).
- An *inherent force*, operating from inside a body to maintain its uniform motion in a right line. In Newton's mind this force had taken the place of the principle of uniform, rectilinear motion retained once, by the late 1660s, he had come to reject Cartesian relativity of motion (p. 535). As such, it formed a retreat from Hooke's arrangement, where uniform motion persisting in a right line was taken to be self-explanatory rather than in need of a force to ensure persistence.
- A force of *resistance*, which was irrelevant to the principal thrust of the tract but which Newton needed for deriving the last few theorems, which dealt with motion in resisting media.
- A *parallelogram of forces*, to make it possible to evaluate the outcome of the simultaneous operation of a multiplicity of forces. Set up in conscious analogy to the standard, Galilean tool for compounding a variety of motions, this new parallelogram affirmed Newton's newly gained conviction of the primacy of force over motion.

The first two forces and the parallelogram then served Newton to derive those *theorems* that jointly were the principal thrust of the tract. Not in so many words but by way of tacit supposition the tract implied that planets move without the intermediary benefit of an aether. The forces that regulate their movements act at a distance rather than through the contact action that alone was admitted in the doctrine of kinetic corpuscularianism.

Among these theorems Kepler's three laws of planetary motion stand out. To derive the second law (the area law) Newton employed a straightforward limiting procedure. The derivation (Figure 19.1) showed the law to be valid for any centripetal force, irrespective of whether it alters with distance and, if so, at what rate.[357]

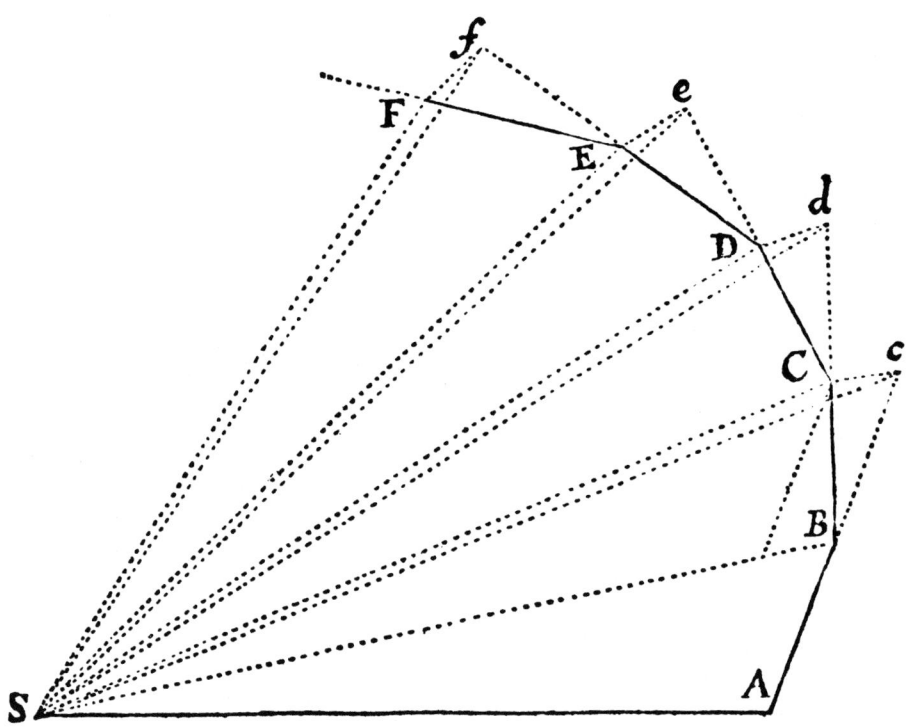

"Force [is treated] as a series of discrete impulses at equal intervals of time. Using the parallelogram of forces and the elementary geometry of triangles, Newton showed that the areas swept out by the radius vector in successive intervals are equal, and so are they also equal when the triangles are infinitely small and the polygon approaches a curve. Hence Kepler's second law holds for the path of a body under the action of any centripetal force."

Figure 19.1: Newton's derivation of Kepler's area law

For the first law (elliptical orbit) and the third (proportionality between period squared and mean distance cubed) Newton rather worked with the version of his calculus that he currently preferred. By its means he showed both laws to follow indeed if the centripetal force decreases in squared proportion to the distance (the core of his derivation is shown in Figure 19.2).

He made it clear that he regarded all of Kepler's laws as concretely applicable to any pair of planets or satellites in the solar system. And in a side passage which, for the wide vista opened by it, must have whetted Halley's appetite even further, he declared them applicable to comets, too.

This, then, is the state that Newton's thinking had attained by November 1684, when he sent the tract to Halley. From the point of view of the principal task still to be performed, it had not advanced nearly far enough. Above all, in order definitely to throw open the gates to a workable dynamics the tension had to be resolved between the two really incompatible concepts defined at the head of the tract: force inducing change of motion (and/or acceleration) and force required to preserve motion. But if considered from the point of view

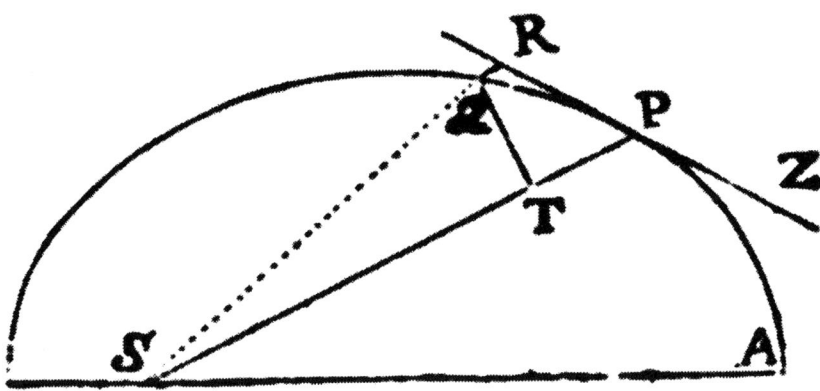

Newton let a centripetal force (with center S for Sol = sun) deflect a body continuously from its uniform motion in a right line. Due to this force the body follows in P (Planeta), not the tangent ZPR, but the arc PQ. So QR marks the resulting deviation. In accordance with Kepler's second law just proven, the area SPQ is proportional to time passed. Now draw QT perpendicular to SP. Between all these line segments there are certain proportionalities which tend to the required ratios when (in accordance with Newton's calculus) QP is taken ever smaller and when P's orbit around S is given the shape of an ellipse. Newton demonstrated the converse along similar lines: given the ratios, to find the orbit.

Figure 19.2: The core of Newton's derivation of Kepler's first and third laws

652 of how Newton had managed to come this far, the number and nature of the milestones he passed on the road toward it is nothing if not impressive. Among those milestones, two stand out: his *experimental* demonstration of the nonexistence of the aether and the possibility of *mathematization* of forces, which had engaged him between 1665 and 1668 and to which his 1679 exchange with Hooke had momentarily recalled him. For all the captivating richness of the picture of the natural world sketched in 'On the Vegetation of Metals' and the 'Hypothesis of Light', and for all the empirical sustenance Newton gained from both his fact-finding experimentation and his wide reading, in the final analysis it is these two basic attainments of the new science, mathematization and validating experiment (particularly if capped by careful measurement), that were decisive in enabling Newton to transcend at almost one and the same time revolutionary transformations 4 and 5, so as single-handedly, between 1684 and 1687, to produce no. 6. Take these two basic attainments away from the account, and what remains is grandiose but, in the end, barren speculation. Nor, on looking back, was Newton to see it differently. Given as he was throughout his career to reasoned and, to some extent, controlled speculation about the whole frame of nature, in the end he knew that if his wonderfully fertile fancy were to produce enduring results, he had to keep it rigorously in check. Mathematics had been the first subject to attract him when in 1664 he set out to discover what the ongoing Scientific Revolution had produced so far. Take away Newton's mathematical proclivities and his quantitative-experimental bent, and we would be left (as we would with Kepler likewise) with just one idiosyncratic philosophy of nature among so many others at the time.

Contradictions removed.   In the seminal half year after he had sent 'De Motu' to Halley, Newton set to work to revise its crude points of departure. In so doing he eliminated, or at least smoothed out, their increasingly apparent, inner contradictions. Salient points covered along the way were (a) his envelopment of 'centripetal force' in 'impressed force', (b) his elimination of the incompatibility between impressed force and inherent force, and (c) his clarification of the relation between resistance and quantity of matter.

In coining the expression 'impressed force' for any force operating from the outside, centripetal force of course included, Newton indicated that he was now ready to subject the concepts somewhat casually thrown down in 'De motu' to a measure of rigor so far absent from either his own or anybody else's handling of concepts of force. As he moved from one revision of 'De Motu' to the next, 'impressed force' became his general expression for an outside force acting upon a body so as to effect a proportional alteration of its motion in speed *or in direction*. Indeed, the analytical power of the full-blown concept of impressed force rested in the supreme act of imagination in the construction of modern dynamics, the recognition that uniform circular motion is dynamically equivalent to its apparent opposite, uniformly accelerated motion in a right line. *All curvilinear motion could now be subjected to the principles of a generalized dynamics adequate to every change of an inertial state.*[358]

Still, in the juxtaposition of this immensely fertile concept with the 'inherent force' that Newton had defined as the force required to keep a body in its state of uniform motion in a right line, that analytical power was thwarted almost from the start. Not entirely so – usage of the parallelogram of forces enabled Newton, albeit in retrospectively unwarranted ways, to compound the effects of two forces which really had little but the word 'force' in common, so as to derive from their joint operation Kepler's laws. As such, this of course represented the huge advance that Halley recognized at once. But much farther than this Newton would not be able to go without entangling himself in contradictions no longer capable of even surface resolution. Absorbed wholly in the investigation to the point of frequently forgetting to eat or sleep, between November 1684 and the spring of 1685 Newton transformed his inherent force from one that *usurped* the place by rights belonging to the principle of inertia into one that, in expressing the resistance of a resting body to its being put in motion, *complemented* the principle.

The transformation did not go smoothly, given the need for Newton to solve in another manner than so far attempted the problem that had in 1668 made him renege upon the principle of uniform rectilinear motion retained: its apparent corollary, the principle of relativity of motion. As Newton became aware that reinstatement of the former principle was indispensable for unleashing the full analytical powers of his centripetal force, he also began to see that this required as a preliminary step that the empirically demonstrable presence in our universe of absolute motion be established in some other way. That is, he felt bound to determine at least one kind of motion independent of a given frame of reference that could serve as a benchmark for all other frames. In the end Newton found that 'other way' in certain varieties of centrifugal motion. Given the high religious stakes of the issue as conceived by Newton, no other way out was left to him.

For all the dispute his peculiar solution was to provoke, this is what enabled him, grudgingly no doubt, to reinstate the principle that, in the absence of any force impressed upon it, a body persists in its state of rest or of uniform motion in a right line. Adopted earlier by Torricelli and Huygens in unambiguous fashion (p. 528), Newton, in the very act of reinstatement, was the man to provide it with its near-definitive formulation, as also with its definitive name, inertia. The principle of inertia, then, made it possible for Newton to realize the wide vistas first opened by means of his concept of impressed force – to treat change of direction in full equivalence with the more familiar treatment of change of speed.

What this transformation of the role of inherent force in Newton's budding dynam-

654 ics presupposed was clarification of the 'indifference to motion' originally proclaimed by Galileo. Independently of Leibniz but roughly simultaneous with him, Newton now made unmistakably clear what Huygens and Mariotte had intuited before. One may rightly speak of indifference in the sense of relative to what particular frame of reference motion is being considered, but this does not imply that the resistance which any quantity of matter exerts against being put in motion may therefore be overlooked. To the contrary, resistance grows with the quantity of matter that exerts it. This insight brought Newton to formulate a principle of equality between 'action' and 'reaction' – absent such equality, the proportionality between impressed force and the change of motion generated by it would make no sense.

The same insight brought Newton formally to define quantity of matter rather than weight as the proper entity to which impressed force is ever proportional (again, Huygens had already distinguished between mass and weight, albeit in less rigorous fashion). Mass, Newton now asserted in an inevitably circular way, is given by a body's volume multiplied by its density. The circularity rests in the ultimate impossibility of defining density other than through mass. It did not plague Newton, because he conceived of matter as packed, more or less densely, with fundamentally identical particles.

By way of successive revisions of 'De motu', the above points taken together mark Newton's advance by the spring of 1685. But the advance sums up by no means all the work by then accomplished. In tandem with the momentous opening of the path toward a dynamics capable of treating any curvilinear orbit generated by any centripetal force acting to alter the orbiting body's motion at any rate (i.e., not just uniformly), Newton began to think more and more of this centripetal force as embodied in physical reality in an attraction not specific but truly universal in the sense of applying to all of matter.

The road toward universal gravitation.   Hesitatingly at first for a variety of very good reasons, Newton set out on the revolutionary transformation of what in 1666 had been to him no more than a significant analogy, not even sufficiently underwritten by calculation, between the fall of an apple on Earth and the Moon's orbit around Earth. In 'De Motu' of 1684 still, the analogy amounted to no more than assertion of the general presence of inverse-square attractions in the solar system. Transformation of the analogy into universal law involved the following major steps: (1) to move from specific to universal attraction and (2) to eliminate three big obstacles. Two of these obstacles concerned the Earth-Moon correlation, to wit (2a) its exact quantitative determination and (2b) how to extend attraction to the inside of the Earth. A final obstacle rested in (2c) confirmation of the proportionality between mass and weight.

Specific attractions were nothing new or particularly remarkable. The idea that like attracts like goes back to antiquity; for instance, that was the way Plato explained why heavy bodies fall down on Earth. In Europe in the late 15th and 16th centuries, with the rise of an elaborate worldview of natural magic, forces of attraction and repulsion had multiplied, with magnetic attraction as a prime example. Most often such forces took shape as magically sympathetic or antipathetic, in short, as 'occult', or hidden, in the dual sense of tinged with magic and recognizable not as such but in their effects only. Forces of that kind had always been incompatible with the particle-infested world of the atomists. With the advent of kinetic corpuscularianism, when Beeckman and then Descartes began vehemently to combat the supposed operation of any occult forces, the incompatibility turned into a program. Originally this proved an act of intellectual liberation in many respects – recall one final time Beeckman's account of sympathetic resonance (p. 375). But by the 1660s, notably in Britain, so monolithic an explanation of nature's phenomena through the exclusive operation of particles in a variety of motions began to be perceived by numerous adherents as more of a straitjacket, to be loosened up by means of a liberal dose of active principles. Hooke and young Newton, who were the two men to do so in most outspoken fashion, then found themselves willy-nilly on the move from active principles (notably Hooke's principle of congruity and its look-alike, Newton's secret principle of sociability) toward playing with attractive forces. With both Hooke and Newton the question of their ultimate reducibility to some intricate mechanism of particles in motion was drenched in ambiguity as we have seen. Their unusually thorough anchoring in experimental findings notwithstanding, these were still specific forces harking back to those 'occult forces' of old. Without going all the way as we have also seen, up to 1685 Hooke was the one farthest advanced on the pathway – less smooth in any case than he could suspect – from specific attractions to a universal one. But early that year Newton overtook him. Propelled forward no doubt by growing awareness of the mathematical powers built into the centripetal force he had recently conceived and was now ready to unleash, Newton began to consider in earnest the final step – to attribute attractive force to all of matter.

How odd such a step was bound to appear in the context of contemporary thinking may be gauged from the response by John Flamsteed (the first Astronomer Royal) when late in 1684 Newton approached him for quantitative data about the latest conjunction between Jupiter and Saturn. Flamsteed rightly surmised that Newton had in mind a mutual influence between the two planets leading to a deviation from their elliptical orbits best measurable (if present indeed) at their point of closest proximity. He did not need to think twice to reject the idea out of hand. His reason was not the notion of attraction per se, which he was prepared to accept as others in Hooke's surroundings were. The point was rather that he found it inconceivable for the sphere of activity of these planets to extend over so huge a distance as one amounting to no less than twice the diameter of the Earth's annual orbit around the Sun. Why was that distance much too large for Flamsteed? Because in proportion to the planets' sizes it exceeds by far the proportion of the size of an ordinary magnet to its sphere of activity, which had been found to be limited to about one hundred yards.

656   Certain stimuli furthered Newton's revolutionary move from specificity to universality. In the 'Hypothesis of Light' he had already gone to great lengths in conceiving a universal (as yet aethereal) mode of operation for the whole frame of nature, even though his secret principle of sociability, devised especially for chemical phenomena, did not seem to take part in those aethereal operations. More generally speaking, both then (in 1675) and now (early 1685) Newton may have felt the pull of the unity of matter. That unity had marked the atomist doctrine of old. It had turned both programmatic and far more outspoken in the laws of motion Descartes proclaimed for all of matter. Unified matter versus specific forces, then, had for decades been the standard opposition, as between Huygens and Roberval (p. 627). Now, early in 1685, the opposition was being resolved in a true synthesis, of matter unified in the attractive force exerted by every particle composing it.

   Further incentives may well have entered the mix of stimuli and motives that drove Newton toward universal attraction to the extent of taking the idea seriously enough to pursue it down to the level of compelling demonstration if indeed he could. And there was of course the promise that universal attraction offered on account of its apparent readiness to be subjected to mathematical treatment. Still, Newton felt bound to face squarely three major obstacles to thus unprecedentedly linking up celestial and terrestrial motions. Two of these involved the Earth-Moon correlation.

   A quantitative problem had presented itself from 1666 onward, when Newton's first attempt to correlate the fall of a body on Earth with the orbit of the Moon in hoped-for accordance with the inverse-square rule that he had found in a comparison between planetary orbits had no more than 'answered pretty nearly' (p. 534). The comparison of uniform acceleration in free fall with the Moon's 'endeavor to recede' had yielded a ratio of 4,000:1, whereas feeding the commonly accepted measure of 60 Earth radii for the distance between the Earth and the Moon into the inverse-square rule yields a ratio of 3,600:1. Into the original comparison had gone a measure for the circumference of the Earth that Newton had picked up from Galileo's *Dialogo*. Armed in the meantime with the more precise value that Picard had come up with while remeasuring France by order of the king, Newton at some time in 1685 found the correlation to fit his comparison with the highest exactitude then possible. Consequently, he could lay his tentative interim solution – that the difference could be accounted for in terms of some small whirlpool of matter – to rest for good.

   An even bigger problem was conceptual and, as such, entirely overlooked by Hooke when, informed in 1686 of the budding *Principia*, he began to complain that Newton had abused their 1679 exchange to steal universal gravitation from him. The whole setup of the Earth/Moon correlation makes sense only if the distance between the apple and the Earth is taken to be, not the, say, five meters over which it falls to the ground, but the entire radius of the Earth. It was one thing to think of the Earth's attraction extending beyond the apple to the Moon, or perhaps (as Newton now began to consider in earnest) of an attraction exerted

by every body in the solar system upon every other body. It was still something else to think
of every single particle of matter *inside* the Earth as both attracting and attracted by every
other particle, such that every gross body, composed of particles thus attracting, may itself
be taken to attract other gross bodies *as if its particles were lumped together at one point situated at its center of gravity*. To convince himself of the feasibility of the former assertion (i.e.,
of attraction being truly universal), Newton had first to suspect, then convince himself of,
the truth of the latter. This he did to his own satisfaction in a range of mathematical demonstrations of (even for his powers) formidable difficulty, with a solid sphere that stands
for the Earth taken as composed of homogeneous, spherical shells made up in their turn of
attracting particles.

Universal gravitation naturally implies a conception of heaviness as due to the attractive
pull exerted by all of matter upon all other matter, in proportion to the mass by which the
particles are lumped together. Final acceptance therefore required prior demonstration of
the proportionality between mass and weight. If demonstrable, this would make definitive
sense of the equal speed with which bodies of whatever size or composition, if released from
the same height, fall to the ground. Galileo had not been able to do more than suspect the
truth of this wholly counterintuitive proposition. Boyle with his air pump had shown that
yes, a lead ball and a bird's feather do fall through the void with equal speed. But Newton,
always on the lookout for perfection, sought and found more exact confirmation in experiments with pendula, ideally suited to the test because tiny deviations add up over successive
swings:

> Using equal pendula, let the oscillations of two bodies of the same weight be counted, and the
> bulk of matter in each will be reciprocally as the number of oscillations made in the same time.
> When experiments were carefully made with gold, silver, lead, glass, sand, common salt, water,
> wood, and wheat, however, they resulted always in the same number of oscillations.[359]

Empirical confirmation galore.   With the conceptual equipment for a mathematically rigorous dynamics firmly in hand, and with the three big obstacles to the physical
embodiment of centripetal force, universal gravitation, overcome in the ways just surveyed,
Newton was ready to widen his view to include phenomena beyond those obstacles. Among
his early explorations of (not yet universal) gravitation as mentioned briefly in 'De Motu'
and as implied in his subsequent exchange with Flamsteed were comets and the Jupiter/
Saturn pair. Not only were the required deviations from perfectly elliptical shape and from
exact obedience to Kepler's third law readily yielded (or so Newton thought) by Flamsteed's
data on their latest conjunction, but the orbits of Jupiter's satellites fell in line, too. Take further Newton's achievement with comets. Several years earlier, in 1681, Flamsteed had come to
the inspired hunch that the comet then flaming the sky and the comet of the previous year

658   were really one and the same comet just before and just after it reversed its trajectory near the Sun. Due at least in part to Flamsteed's shaky underpinning for his idea, Newton had rejected it as everybody else had; but in the fall of 1684, in the context of his budding theory of universal gravitation, it began to make sense to him. From time immemorial comets had been regarded as quite irregular phenomena, be they in the Earth's atmosphere according to Aristotle or in the heavens as upheld by the Stoa and, since Tycho, by most astronomers, too. It was a striking confirmation of inverse-square attraction and its proportionality to mass to show, as Newton managed in the end to do, that even comets obey the same laws as planets obey in their orbits around the Sun, albeit in much more elongated conics.

These demonstrations being confined to the heavens, they did not of course imply the universality of gravitational attraction. This found its most glaring empirical confirmation in a phenomenon produced by heavenly and terrestrial matter conjointly – the tides. Galileo had employed for his explanation the dual motion of the Earth; Kepler, an influence exerted by the Moon; and Descartes, a local condensation in the aethereal whirlpool between the Earth and the Moon. None of these ideas proved to be really satisfactory, and it was now up to Newton to demonstrate, with quantitative precision grounded in mathematical law, the alternation between ebb and flood to be due, in its fundamental regularity, to the attractive forces the Sun and the Moon jointly exert upon the seas and oceans of the Earth. With the twin concepts of centripetal force and its real-world embodiment, universal gravitation, by now well prepared, Newton saw that no possible contribution to the *Philosophical Transactions*, even if of extraordinary length, would be able to contain all the many things that unleashing these concepts was bound to produce. He was ready to write the *Principia*.

Principia

The title *Philosophiae naturalis principia mathematica* (Mathematical Principles of Natural Philosophy) served Newton to announce the new ideal of science the work was to embody more and more as he went along. A more content-oriented title would have run something like the Latin equivalent of 'Mathematical investigations, sustained by experimental and other measurements, into a variety of orbital motions due (under idealized conditions) to forces acting either (1) in the void or (2) in variously resisting media, and (in physical reality) to (3) a force of universal gravitation acting over distances as far as our solar system, and even beyond'. Such a title would have the added benefit of neatly capturing Newton's division into the three successive 'books' that follow upon the enunciation of the very foundation stone of his dynamics, his laws of motion. This way of organizing the material – idealized at first; next, semi-idealized by factoring in resistance; fully physical at long last, by applying results thus obtained to what Newton called 'the system of the world' – is not only satisfactory from a logical point of view, but also serviceable to a reader who wishes to follow the author into all the nooks and crannies of his exploration of a great variety of orbital motions. But Newton was well aware that his own delight in exploring a vast terrain hitherto hidden from anybody's view had carried him far beyond what was needed for his central argument, which was to derive universal gravitation and demonstrate that it entails the elliptical orbits of the planets and their satellites. This is why in the preface to Book III he suggested a shortcut to readers who wanted to grasp the central argument alone. Properly executed, that shortcut first takes us from the laws of motion as the obvious starting point to a range of abstract propositions derived therefrom in Book I. The reader is then, in Book III, sent to an array of 'hypotheses', consisting first of some general rules for proper scientific procedure and then Kepler's second and third laws presented as outcomes of up-to-date measurements. Finally, in the same Book III, the reader moves to a range of propositions about the real world. All this is interspersed with a few dips back into propositions that appear in Book I or II. To complicate the picture further, several *other* propositions, corollaries, lemmata, and scholia in Books I and II appear to be relevant to the central argument as well, though not quite so directly.

I shall now discuss the principal contents of the *Principia*, with my principal objectives being to mark what was truly revolutionary about the book and to explain what enabled Newton to take those revolutionary steps. To that end, I shall successively

- outline the central argument;
- list the main subjects in each book, with comments on some;
- consider how certain propositions in Books I and II impinge on Book III issues in ways not immediately pertinent to the central argument, culminating in a discussion of possible and impossible universes and consequent demonstration of God's mathematical powers;
- examine the ideal of science that Newton took his book to embody;
- seek possible causes of the revolutionary transformation he thus brought about in the two and a half years between August 1684 (Halley's first visit) and April 1687, when he sent Halley the definitive manuscript of Book III; and
- survey how the book was received by those among Newton's contemporaries knowledgeable enough to pass expert judgment upon it.

The central argument.   Creative Newtonians in 18th-century France were particularly keen to identify the central argument, which has recently been set forth in all required detail by Dana Densmore. Naturally, her book starts with the three laws of motion the reader encounters right after Newton's highly programmatic preface. Laws 1 (principle of inertia) and 3 (rule of action and reaction) require no further comment here; but the retrospective ambiguity built into law 2 as defined and handled by Newton throughout the *Principia* still

660 does. The ambiguity pertains to proper axiomatization alone; in terms of net results attained by means of the second law it proves generally innocuous. The point, however, is that impressed force was at times taken by Newton to produce accelerated motion (anachronistically, $F = ma$), whereas at other times he took it to produce change of motion ($F = \Delta mv$). Nor did Newton acknowledge a difference between the two concepts. The ambiguity betrays a remnant of the impact way of thinking from which Newton arrived in the end at his mature dynamics. It was also powerfully suggested by the still current conception of force as a substance, the transfer of which to a moving body inevitably looks like a unitary action at bottom, thus further obscuring that passing from discrete, instantaneous impulses to a continuous force that takes time involves more than sheer summation. Finally, the language of geometric proportion in which Newton clad the mathematics he employed to accomplish such integrations worked in the same direction of effectively masking the dimensional incompatibility.

To introduce that mathematical language ('of first and last ratios') is the burden of the first section of Book I, which follows at once upon enunciation of the three laws of motion. Figure 19.2 displays Newton's mode of proof in action, which is the form then given by him to his calculus (p. 651). His decision to express in Euclid's language of geometric proportions advanced techniques utterly foreign to Euclid stemmed from his conviction that this was the best way to avoid the paradoxes of the infinite (p. 357). To that end he accepted the consequence that he had to derive proof for every theorem anew, rather than just letting a few algorithms do the work for him. Consequently the proofs given throughout the *Principia* are not really full-blown proofs in the literal sense. Rather, they are more or less detailed pointers at what, assisted by the lemmata that make up section 1, other competent mathematicians (and it is to these alone that the *Principia* directs itself) would readily recognize as the way toward complete proof with no step, however elementary, left out.

Section 2, then, is the first to display centripetal force in action. We are already familiar with proposition 1 – derivation in the abstract of the area law from the reiterated (and, upon summation, continuous) action of a centripetal force upon a body moving uniformly in a right line. Schema 7 gives an idea of how the central argument then unfolds.[360]

Quite unmistakably Newton's original response to Hooke's challenge has meanwhile undergone a vast expansion in depth and in breadth, so as to make derivation of the elliptical orbit of the planets from some universally operating, inverse-square attractive force truly and fully conclusive. Even so the organization of the core argument still harks back to that challenge as Newton conceived it. It is designed from start to finish to show that to have a bright idea is one thing, to prove it quite another – not just to 'reason about these motions from the phenomena' but, 'now that the principles of the motions are known, [to] infer from these the celestial motions *a priori*'. No longer are Kepler's laws arbitrary-looking pieces of information about planetary orbits, devoid of either a rationale (barring Kepler's own) or

Schema 7: The central argument of Newton's Principia   661

I	P 1	derivation of area law for any centripetal force (Figure 19; p. 650)
	P 2	converse of P 1
	P 3	extension of P 2 involving a body at the center of force
	P 4	relations between centripetal forces on two bodies moving uniformly in circles at various distances
	P 6	core property of centripetal force on a body moving in any orbit around center at rest
	c 1	general expression for any centripetal force (Figure 19.2; p. 651)
	P 7	law for centripetal forces on bodies in eccentric circular orbits
	P 9	centripetal force tending toward center of spiral with constant angle between radius and tangent is in inverse-cube proportion to distance
	P 10	centripetal force tending toward center of ellipse is directly proportional to distance
	c 1	converse of P 10
	c 2	equality of periodic times for pairs of ellipses with the same center
	P 11	centripetal force tending toward focus of ellipse is in inverse-square proportion to distance
	P 12	same for hyperbola
	P 13	same for parabola
	c 1	converse of P 11/12/13: centripetal force in inverse-square proportion to distance tends toward focus of a conic section
	P 14	property of orbits of several bodies around same center (focus of some conic section), with centripetal force in inverse-square proportion to distance
	P 15	under givens of P 14, periodic times in ellipses are proportional in 3/2 power of their major axes
	P 16	property of the velocities of bodies revolving under givens of P 14
	P 17	how, given centripetal force in inverse-square proportion to distance plus its absolute quantity, plus certain givens about body on which force operates, to find orbit of that body
III	Ph 1	reliable measurements indicate that Jupiter's satellites obey Kepler's second and third laws
	[Ph 2]	[Ph 1 likewise for Saturn's satellites]
	Ph 3	reliable measurements indicate that Mercury, Venus, Mars, Jupiter, and Saturn enclose Sun in their orbits
	Ph 4	reliable measurements indicate that planets of Ph 3 plus either Sun or Earth obey Kepler's third law
	Ph 5	reliable measurements indicate that planets of Ph 3 obey Kepler's second law if orbiting Sun, but by no means if orbiting Earth
	Ph 6	reliable measurements indicate that Moon obeys Kepler's second law
	P 1	forces which keep Jupiter's satellites in orbits are inversely proportional as the square of their respective distances to Jupiter
	P 2	P 1 likewise for planets and Sun (NB: the tiniest deviation would be noticeable as motion of line of apsides)

	P 3	force which keeps Moon in orbit is inversely proportional as the square of their distance to Earth
	P 4	Moon gravitates toward Earth and is kept in orbit by force of gravity
	P 5	all heavenly bodies that orbit Jupiter, Saturn, or the Sun gravitate toward their central body and are kept in orbit by the force of gravity
II	P 24	masses of bobs of pendula with equal string lengths that swing in void vary as weights multiplied by periodic times squared
	c 1	"Therefore if the times are equal, the quantities of matter in individual bodies will be as the weights".
	c 7	Thus, possibility opened up of comparing the masses of different bodies or of comparing the weight of one body at different places, "to discover the variation of gravity"; "moreover, by the most accurate experiments I have found that the quantity of matter in individual bodies is always proportional to their weight".
III	P 6	"That all bodies gravitate to each of the planets, and that their weights toward whichever particular planet you please, at equal distances from the center of the planet, are proportional to the quantity of matter in each".
I	P 69	in a system of more than two bodies, absolute weights of bodies attracting each other in inverse-square proportion to the distance from the attracting body are as their respective masses
III	P 7	"That gravity is given in bodies universally, and that it is proportional to the quantity of matter in each".
I	P 71–75	attraction by gross bodies takes place as if concentrated at center point (p. 656)
III	P 8	weights of two homogeneously composed globes in mutual gravitation are inversely proportional to square of distance between respective centers
	c 1	computation, by means of periodic times, of weights on various planets
	c 2	computation of masses of planets
	P 13	"The planets move in ellipses having their focus at the center of the sun, and by radii drawn to that center describe areas proportional to the times". NB: The whole point of this proposition is rendered by Newton in the line that follows: "Hitherto we have reasoned about these motions from the phenomena. Now that the principles of the motions are known, we infer from these the celestial motions *a priori*". Moreover, deviation from pure ellipticity due to mutual attractions is noticeable in Saturn's orbit at conjunction with Jupiter.

The first column indicates the book number.

In the second column, P = proposition (numbered anew for every book); c = corollary (numbered anew for every proposition); Ph = phenomenon (in book III only).

The third column sums up the principal point made in those propositions, corollaries, and phenomena that together make up the central argument according to Newton/Densmore. Their numbering follows the third edition, which has since become standard.

NB: Since Flamsteed, consulted by Newton in 1685, had yet to be persuaded by Cassini's reports of his discovery of more Saturn satellites than Huygens' Titan alone, Phenomenon 2 is not present in the first edition. There, too, the 'Phenomena' of the second and third editions go under the original name of 'Hypotheses'.

mutual coherence. In Newton's treatment they follow from the interplay of three laws of motion plus the specific force relation which God (with little choice really left to Him, as we shall see) has selected for our universe. The utmost care has been given throughout to let no one escape the compelling force of the argument, in that nothing is taken for granted and (barring mathematical details) not one step is omitted that is required to arrive at the end point of the investigation – derivation of the planets' elliptical orbit. True, only a super-human could have made the argument proceed without a flaw or two. For instance, in an effort to make III-4, which is about the Moon gravitating toward the Earth, fully clinching, Newton invoked a proposition in Book I which deals (in the abstract, of course) with *mutual* attractions – a point not to be made about the Earth and the Moon until III-7.

In terms of the core argument, then, Books I and II are no more than preparatory to the real-world applications performed in Book III. For most propositions and corollaries this is transparent enough. Not so, to be sure, for propositions I-P7, I-P9, and I-P10, which pertain to the central argument in a more roundabout manner (p. 670). Further, the joy of find-ing the proportionality between weight and mass confirmed by his pendulum experiments caused Newton in II-P24-c7 to make a statement about the real world that ought strictly to have been reserved for Book III. Considering the structure as a whole, however, it is quite remarkable how adroitly the preparatory ground has been laid and how quickly, on arriving at propositions 1–13 of Book III, Newton could now dispose of these. With their demonstra-tions requiring little beyond a monotonous 'this is so by propositions I-x, -y, and -z, and by phenomena p, q, and r', the set of propositions 1–13 of Book III occupies a mere fifteen pages, compared to the approximately seventy pages needed for their more abstract counterparts in Books I and II.

Books I and II as early exercises in dynamics.　　If, in contrast, the *Principia* is read from cover to cover, Books I and II appear to perform in addition a function of scarcely lesser import. In Book I Newton did two principal things. He brought to order just about everything that had from Galileo onward been discovered about generic motion of bodies taken to move in the void. To that he added what his second law of motion and his calculus now allowed him to add, an examination of bodies moving in more complicated ways than just uniform or uniformly accelerated. In partial contrast, Book II, on motion under condi-tions of resistance, a theme that Newton had begun to broach in a few theorems at the end of 'De motu', was a pioneering effort almost entirely. Schema 8 lists the successive section titles of Books I and II and also the subjects successively handled in Book III once carried beyond the conclusion listed at the end of Schema 7.[361]

I shall now examine the revolutionary steps that were involved in Books I and II and then highlight certain topics not yet treated that come up in these books or in Book III.

I shall first consider *Book I*. Prior to August 1684, the highest achievement in the study of

664   generic motion rested in the revolutionary transformation brought about almost exclusively by Huygens and by the young Newton. Independently, in different ways, and leaving their pertinent work in manuscript mostly (Huygens) or entirely (Newton), they harmonized the incompatible legacies of Galileo and Descartes at least insofar as impact, motion retained, rotation, and oscillation were concerned. Before giving up, Newton struggled in vain for a unification Huygens felt to be out of reach anyway. Two decades later, Book I of the *Principia* did provide such unified treatment, thus giving birth to what Newton in his general preface called 'rational mechanics'. The expression has stuck, as with so many coined by Newton. The mathematical rules for the four *genera* thus far established, and of course Galileo's laws of falling bodies as well, now found themselves taken up into a system that held far-reaching consequences for the status of motion in impact. That is, impact lost the centrality that it necessarily has in a Cartesian world of contact action. In the paper of c. 1666 meant to sum up his results thus far, Newton duly treated impact as the very key to motion; in Book I of the *Principia*, with motions derived from forces, the laws of impact take a subordinate place, on a level with numerous other *genera*. *In the domain of motion, then, revolutionary transformation 4 had within thirty years been consumed entirely by transformation 6.*

In addition to such reorganization (and, here and there, extension) of essentially known properties of moving bodies, Newton broke new ground in many respects and in many directions. Among the most important expansions was a systematic inquiry into orbits produced by the action of a large variety of centripetal forces – constant, directly proportional to distance, or inversely proportional to distance in accordance with a variety of powers, from the first through the sixth. Newton undertook these investigations with three distinct aims in mind: (1) to probe what his new conceptual apparatus was up to; (2) to investigate whether there were viable alternatives for the universe as we know it, and (3) to lay the mathematical groundwork for forces of other kinds.

Considered in the abstract sphere of Book I alone, here was terrain ready-made to test the full worth of the mathematical and dynamic equipment his exertions had now put him in possession of. One example is his analysis of vertical descent under any force law, leading to establishment of the mathematical equivalent of what has since become known as 'the work-energy equation'. It testifies to the real-world significance that Newton's conception of force had meanwhile come to have for him that the expression mv^2 that appears here, which Leibniz (calling it 'living force'; p. 629) was busily expanding at the time into a cosmic principle, remained for Newton what it had been for Huygens, a mathematical expression of no particular meaning in the real world. Another outstanding example is Newton's treatment of what has since become known as simple harmonic motion. He demonstrated what Hooke had no more than suspected: isochronous vibration as effected, for example, in cycloidal pendula is governed by an accelerative force that increases in proportion to distance. He went on from there to derive the relation between force, displacement, acceleration, velocity, and time during a single vibration.

Schema 8: Contents of Newton's Principia

I	1	The method of first and last ratios
	2	To find centripetal forces
	3	The motion of bodies in eccentric conic sections
	4	To find elliptical, parabolic, and hyperbolic orbits, given a focus
	5	To find orbits when neither focus is given
	6	To find motions in given orbits
	7	The rectilinear ascent and descent of bodies
	8	To find the orbits in which bodies revolve when acted upon by any centripetal force
	9	The motion of bodies in mobile orbits, and the motion of the apsides
	10	The motion of bodies on given surfaces, and the oscillating motion of simple pendula
	11	The motion of bodies drawn to one another by centripetal forces
	12	The attractive forces of spherical bodies
	13	The attractive forces of nonspherical bodies
	14	The motion of minimally small bodies, that are acted on by centripetal forces tending toward each of the individual parts of some great body
II	1	The motion of bodies that are resisted in the ratio of their velocity
	2	The motion of bodies that are resisted in the squared ratio of their velocity
	3	The motion of bodies that are resisted partly in the ratio of their velocity and partly in the squared ratio of their velocity
	4	The revolving motion of bodies in resisting media
	5	The density and compression of fluids, and hydrostatics
	6	The motion of simple pendula and the resistance to them
	7	The motion of fluids and the resistance encountered by projectiles
	8	Motion propagated through fluids
	9	The circular motion of fluids
III	10	Beyond the core argument, from proposition 14 onward, the following subjects come up for treatment: • more-detailed properties of planetary motion • irregularity in the shape of the Earth • irregularities in the motion of the Moon • tides • orbits of comets

The first column indicates the book number, the second lists section numbers, and the third lists section titles. Book III has no sections. As in Schema 7, the list follows the third edition.

666   But such abstract investigations into a variety of motions under a variety of proportions of force to distance carried real-world implications as well. They were indispensable for finding out for each case what the universe would have looked like if it operated by another rule for universal gravitation than the inverse-square law (p. 670). Further, now that Newton had decided that his beloved active principles actually were (or at least ought to be treated as) forces rather than aethereal mechanisms, his entire worldview became populated with physical forces other than universal gravitation alone. And those forces, about which Newton was remarkably reticent in the first edition of the *Principia*, might well prove to operate under rules other than inverse-square. That is, the setup of Book I provided Newton with a splendid opportunity to run in advance through the properties each of these rules would then prove to entail. In other words again, the groundwork was tacitly being laid here for an as yet missing Book IV about such other forces as Newton now felt certain governed vital aspects of our world, for example, in chemical/alchemical reactions or in magnetic or electric action. This is a very involved issue. I shall come back to it when discussing the programmatic significance that Newton attached to his *Principia* (p. 672), and again when I consider the status of the *Principia* in terms of Newton's ongoing thinking about the entire frame of nature (p. 710).

Next, *Book II*. This was itself a prime example of exploration of territory hitherto untrodden. In treating of motion under conditions of resistance it brings closer to the real world the state of motion in Book I, where motion is treated as if taking place in the void. To Galileo the resistance that moving bodies undergo in the air had presented itself as nothing but an impediment to the ideal phenomenon of motion in the void, that is, as a deviation not capable any further of being handled mathematically (p. 195). Newton, now in possession of a generalizable dynamics and of calculus-inspired techniques adequate to it, saw how to subject motion under conditions of resistance to mathematical treatment after all:

> Recognizing the exponential nature of such motion, he seized on a means of expressing it mathematically by means of the rectangular hyperbola. With this device, he was able to derive the consequences for various assumed functional relations of velocity and resistance.
>
> Newton was not satisfied, however, with a purely hypothetical treatment of motion under resistance. He was convinced that the principal component in the resistance of a medium derives from the inertia of the medium itself which must be set in motion in order for a body to pass. This component must be proportional to the density of the medium, the square of the velocity of the moving body, and the area of the body's cross-section (or the square of its diameter). By experiments with bodies falling through air and through water and with pendulums swinging in various fluids, he convinced himself that the reality of resistance is as his theory predicts. The climax of his consideration of motion through resisting media is found in Corollary 3 to Proposition XXXVIII, Book II.

If there be given both the density of the globe and its velocity at the beginning of the motion, and the density of the compressed quiescent fluid in which the globe moves, there is
given at any time both the velocity of the globe and its resistance, and the space described
by it.

This was to do for motion through a resisting medium exactly what Galileo had done for motion under idealized conditions from which resistance is eliminated.[362]

As one consequence of these (in retrospect) mostly mistaken points of departure, Newton
followed up his prior analysis of pendular motion in the void with an investigation of how
oscillations decay in air. Another significant example of Newton's treatment of motion under conditions of resistance is the propagation of waves. He first investigated surface waves,
then how waves propagate through an elastic medium, going on to check the resulting model for the speed of sound against his own measurement thereof. The satisfactory outcome of
the check was not to survive later measurements, leading in later editions to some extensive
fudging (p. 494).

Finally, *Book III*. Here the Earth's deviation from perfect sphericity requires comment.
The axial rotation of the Earth appeared to Newton to carry gravitational consequences, to
wit, unequal distribution of gravity in such a way that the Earth bulges at the equator and is
flattened at the poles. He derived for the equator a 17 miles' excess of length over the polar
diameter. He pointed at what this entails for the length to give to a pendulum meant to make
exactly one oscillation every second. This, after all, depends on gravity and is therefore not
constant but varies with geographic latitude if Newton's argument is sound. Nor did he fail
to cite here Richer's reports from Cayenne about precisely such an effect (p. 454). Within
a year of publication of the *Principia* Huygens was to seize on this very issue (vital in any
case for his preferred solution to the problem of longitude) in a sustained effort to shore up
empirically his rejection, not of gravitational attraction *per se,* but of its alleged universality
(p. 677).

Requirements for a stable universe.   The three subjects to be addressed now – whirlpools, perturbations, and God's mathematical wisdom – are all connected to the viability of
the universe as Newton conceived it. Each was prepared in the abstract in Book I and/or II.
In each case a real-world point, first made in Book III, appeared to be involved that was of
consequence for (albeit not quite at the heart of) the central argument of the *Principia*.

Book II culminates in a section on *whirlpools*. Whirlpools of subtle matter had been
around ever since Descartes deduced their omnipresence from his first principles so as
to make them central to the explanation of nature's phenomena. These explanations had
never gone beyond the qualitative (with the partial exception of Huygens' as yet unpub-

668   lished Académie lecture on heaviness; p. 627). Certainly no one had sought to investigate
the mathematical properties of motion of a gross body in a whirlpool of subtle matter. The
starting point for Newton's investigation was his general point about motion in resisting
media that the density of the medium, not the subtlety of its parts, is the key factor. Adding some further assumptions that allowed him to build what seemed to him a well-fitting
model, he found that model to entail a whole range of consequences incompatible with the
predominant current in the natural philosophy of kinetic corpuscularianism, the Cartesian
one. Most devastating of all were those conclusions about motion in whirlpools of which the
validity did not depend on any special assumptions – that is, conclusions no Cartesian could
escape from by proposing other assumptions instead.

> On the one hand, the variation of velocity with distance in the orbit of an individual planet, as
> demanded by Kepler's second law, could never be reconciled to the variation of velocity with
> distance among the planets demanded by his third. On the other hand, and in the end most
> compelling of all, a vortex cannot be a self-sustaining system. In order for it to continue, "some
> active principle is required from which the globe [that moves in the whirlpool] may receive
> continually the same quantity of motion which it is always communicating to the matter of the
> vortex". The hypothesis of vortices, he concluded, "is utterly irreconcilable with astronomical
> phenomena, and rather serves to perplex than explain the heavenly motions".[363]

The *first* conclusion reflects once more the primacy realist-mathematical science had acquired in Newton's mind over whatever exigencies might be posed by natural philosophy.
For Newton (unlike Huygens) there had eventually proven to be no way harmoniously to
resolve the (to both men) apparent tension between 'Galileo' and 'Descartes' – for falling
bodies you could have uniform acceleration or you could have whirlpools but not both. But
now 'Kepler' and 'Descartes' proved similarly incompatible – you could have planets obeying
Kepler's laws or you could have them move in whirlpools but not both. Newton's usage of
forces as his prime tool of analysis throughout the *Principia*, then, constituted more than an
implicit challenge to the reigning conception of the world; in his section on whirlpools he
openly defied its adherents to get rid of what to most of them was its central feature.
    Meanwhile Newton's *second* conclusion about whirlpools, that their failure to sustain
themselves appears to preclude a stable universe, implied a challenge to himself as well –
what about the stability of a universe governed instead by a force attracting in inverse-square
proportion to distance?
    A handle on that question could be acquired by examining the conditions for, and
mathematical properties of, *perturbations*. As Newton demonstrated in Book I, even in the
relatively simple case of two bodies attracting each other in accordance with the inverse-
square rule, deviations from pure ellipticity are bound to arise. He further demonstrated

that, whereas the inverse-square rule itself produces stable orbits, the tiniest deviation from inverse-square puts them on the move (by way of a forever-shifting orientation of the line of apsides). What, then, did the mutuality of gravitational attraction as Newton conceived it, and the actual presence throughout the solar system of bodies thus attracting each other, jointly entail? The possible consequences were of two opposite kinds. If shown indeed to exist, deviations from pure ellipticity, which repeat themselves in cycles but do not add up, would naturally serve as empirical affirmation of the ubiquity, throughout the solar system, of inverse-square attraction. So would the observationally known motion of some celestial body's orbit, if it could be proven to fit the case for mutual attraction on the inverse-square rule. However, orbital shifts add up with every next revolution and, if present throughout the solar system, would damage its stability sooner or later. Hence, the observational absence so far of such secular shifts in the revolutions of the planets was in need of being mathematically demonstrated to be compatible with mutual inverse-square attraction – if not, the universality of inverse-square attractions had to be questioned after all.

Jupiter/Saturn conjunctions and also the orbit of the Moon seemed to Newton ready candidates for 'affirmative' perturbations. With less justice than has retrospectively proven warranted, he sought to strengthen the case for universality by invoking for Saturn's orbit a noticeable deviation from Keplerian ellipticity due to the attracting mass of Jupiter – a phenomenon that Flamsteed had forthwith held to be inconceivable (p. 655). For Newton the case of the Moon, attracted as it is by the Earth and the Sun alike in his theory, appeared much more complicated. The Moon provided the principal reason for Newton's tackling of what has since become known as the 'three-body problem' – how to determine the orbit of a body attracted by two others. The problem admits of no general solution – a mathematical fact of life Newton was the first to run up against. Approximation was what he had to settle for, both in the abstract treatment of Book I and in the Book III application to the orbit of the Moon. Its gross irregularities (notably, ongoing shift in orbital orientation) had been notorious to mathematical astronomers from Ptolemy or even Hipparchos onward. Newton was the first to grope for more than ad hoc solutions. His struggle with the problem, the only one ever to give him a headache, was rewarded at least insofar as to elevate it to a wholly new level of understanding. Still, in the end he had to acknowledge partial defeat, in that his theoretical model yielded no more than half the principal irregularity observed. Thus was the ground laid, not only for his future ignoble treatment of Flamsteed as sole possessor yet reluctant communicator of up-to-date lunar observations, but also for substantial changes to be made in the second and third editions of the *Principia*.

Both planetary conjunctions and the orbits of some planetary satellites under the additional attraction exerted by the Sun posed self-contained problems, inconsequential as such for the stability of the solar system. But there might also be perturbations all over the universe that, on investigation, were to prove to reinforce themselves over time, the way that

670   in the ongoing swings of a pendulum tiny deviations from some regularity quickly add up to ever more deeply perturbing ones. That is, if inverse-square attractions were to involve perturbations of a damagingly cumulative kind, our own, apparently stable universe was bound to run on some other force rule after all. Considerations like these led Newton to investigate what, if any, alternative universes were at all possible. Three distinct building principles for the universe were around: *eccentric* (Ptolemy, Copernicus, Tycho), *whirlpool* (Descartes), and *ellipse* (Kepler, Newton himself). And of course there might be still others.

For the *whirlpool universe*, which by its very nature does not run on any force rule, there was Newton's mathematical demonstration at the end of Book II that (beside being incapable of producing orbits that obey Kepler's laws) it is inherently unstable (p. 668).

For the benefit of those astronomers still thinking on pre-Kepler lines, in proposition 7 of Book I Newton derived by what force rule *eccentric* orbits are produced (the stratagem central to Ptolemy's *Almagest*; Figure 1.2, p. 14). In confirmation of Kepler's hunch, he found the force to require some sort of foreknowledge of the orbiting body's future course. "Using an entirely different physical theory [from Kepler's, Newton] nonetheless comes to the same conclusion: there seems to be something unnatural about eccentric circular motion."[364]

Further, Newton investigated in Book I what orbits are generated by attractions in accordance with a host of *other* proportions to distance than inverse-square (p. 664). Among the ends for which he undertook the investigation was the question of whether a viable universe could run on any of these alternative rules. With one partial exception, the answer turned out to be 'no'. Thus, proposition 9 showed that a force diminishing with distance in inverse-*cube* proportion produces a certain kind of spiral orbit incapable of sustaining a conceivable universe. For instance, at the surface of a sphere composed of particles thus attracting, the attraction becomes infinite, so that only aerial beings would stand a chance to begin with.

Finally, in the most challenging and enlightening case by far, Newton probed the depths of the partial exception just hinted at. Compare, in Schema 7 (p. 661), propositions 10 and 11 – inverse-square centripetal forces directed to a focus are not the only ones capable of producing elliptic orbits; so are centripetal forces directed to the center and increasing in direct proportion with distance (i.e., the very force that Newton showed elsewhere in the *Principia* to govern simple harmonic motion). Moreover, as Newton went on to establish, whereas under the former rule attractions between many bodies lead to perturbations, under the latter rule pure ellipticity is always preserved.

Was this advantage of direct over inverse-square proportionality decisive? To show that it is not, Newton had to prove two things. In the first place, the gradual destabilization of planetary orbits due to a cumulative deviation from inverse-square that is itself due to attractions among the planets proved to be as negligible as their observational absence already suggested; for example, for Mars Newton calculated a shift in orbital orientation of less than a degree in a century. Second, the rule of direct proportionality, even though not giving rise

to any perturbation at all, does in other respects entail properties no Creator in His right 671
mind might wish to inflict upon the inhabitants of a universe thus construed. For this was
where, to Newton, the crux of the matter truly resided. Not whether our universe is run on
inverse-square or on direct proportionality could be in serious doubt to Newton, but whether the Creator had any real choice on this score was what he desired to know. In view of the
further circumstance that "in an infinite universe attraction in direct proportion to distance
entails the impossible consequences of infinite forces, infinite accelerations, and infinite velocities",[365] it ought to be clear for all to see (or so Newton felt) that the Being to preside over
the creation of the universe had to be thought of as a skilled Mathematician, fully aware of
the tight constraints the mathematical principles of natural philosophy imposed upon Him.

A new program for science. 'The mathematical principles of natural philosophy', indeed! Not only is the title Newton gave to his masterpiece a sly dig at Descartes, who had
thought he could establish the 'Principles of Philosophy' without knowing how to do it
mathematically. But there was also a positive meaning, succinctly announced in the preface.
Here Newton told his prospective readers both what he felt he had accomplished in the work
that followed and what still remained to be done:

> We put forward this work of ours as mathematical principles of philosophy [i.e., in the present
> context, 'physical science']. For the whole difficulty of philosophy seems to reside in this, that
> from the phenomena of motions we investigate the forces of nature and then from these forces
> demonstrate the other phenomena. It is to these ends that the general propositions in books 1
> and 2 are directed. Then, in book 3, we have put forward as an example of this an explanation
> of the system of the world. For there, by means of propositions demonstrated mathematically
> in the earlier books, we derive from celestial phenomena the gravitational forces by which bod-
> ies tend toward the sun and the individual planets. Then the motions of the planets, the com-
> ets, the moon, and the sea are deduced from these forces by propositions that are also math-
> ematical. If only we could derive the other phenomena of nature from mechanical principles
> by the same kind of reasoning! For many things lead me to suspect that all phenomena may
> depend on certain forces by which the particles of bodies, by causes not yet known, either are
> impelled toward one another and cohere in regular figures, or are repelled from one another
> and recede. Since these forces are unknown, philosophers have hitherto made trial of nature in
> vain. But I hope that the principles set down here will shed some light on either this mode of
> philosophizing or some truer one.[366]

672   As is customary with Newton's highly condensed sentences, a variety of claims (and one significant disclaimer) were packed together in this oft-quoted passage.

In the widest of its numerous senses, it involves the claim, directed not only against Descartes but against all of natural philosophy past and present, that science ought to be mathematical, not verbal. In this wide sense it is a ringing endorsement of the kind of realist-mathematical science first explored by Kepler and Galileo, whose principal findings Newton set out in the *Principia* to derive from a mathematical regularity capable of encompassing them all.

In a narrower sense the passage addresses one issue crucial to the knowledge structure of realist-mathematical science: the problem of how exactly to establish the validity of whatever connection is being made between some mathematical theory and its purported, real-world embodiment. Unceasing interplay between mathematical regularity and effort at experimental validation, guided by analogy and other means, is what from Galileo to Huygens had become the customary way of proceeding (p. 616). How much tighter did Newton in the *Principia* manage to make the connection! It rested in the ongoing back-and-forth between the propositions of Book I (on occasion II, too) and the 'phenomena' and propositions of Book III. There is a distinct pattern here of decreasing idealization leading to increasingly close approximation, which in our time has been recognized as a specifically 'Newtonian style'. Book I advances from single bodies attracting, to two bodies attracting one another, to three bodies attracting, to a multiplicity of mutual attractions. In the advance, ever more elements of idealization are dropped, with an eye to such real-world complications (shifting orientation of the Moon's orbit, shape of the Earth, orbits of comets, etc.) as are bound to call for treatment in Book III. In so doing, the Newtonian style opened a path toward a more closely knit ensemble of mathematical theory and real-world application than had seemed within reach before.

In the passage quoted, Newton further suggests that he has first inferred gravitational force from various motions of bodies, so as to go on to demonstrate its presence and properties for other phenomena. This is indeed how he organized the *Principia*. As a historical statement about his personal process of discovery, which took him to forces along quite another pathway, it cannot of course stand. In this and similar passages (notably, the 'rules of philosophizing' at the head of Book III), as also decades later in the entire setup of the *Opticks*, Newton helped reinforce the Baconian misconception that science, even mathematical science, operates by way of generalization from the less to the more inclusive.

What, finally, about the one real-world force apparently laid bare in the *Principia*, and what about those other forces from which, as he confidently foresees here, all remaining phenomena are in due time to be derived? What *is* their status in Newton's own view?

Even if no more evidence were available than the quite outspoken 'Conclusion' to the *Principia* that he drafted *but in the end filed away*, there is no mistaking Newton's position at this stage in his career. He had just thrown out aethereal mechanisms and replaced them with forces, whose mathematical fertility had just proved to exceed all known bounds, to the point of enabling him to gauge the Divine Mind at creation. Consequently, he had a higher regard for the unadulterated physical reality of forces than ever before or since (p. 652). What worked in the opposite direction, however, was that, in bitter remembrance of the unceasing polemics over his 1672 article on colors, Newton, not yet the august authority the very *Principia* was to turn him into, was by 1687 even less ready to face criticism than at any other time. And nothing was surer to come up for biting criticism than a full-blooded affirmation of the physical reality of attracting forces – the nemesis of the crowd of card-carrying kinetic corpuscularians.

So, instead of sounding a ringing proclamation of Newton's full view of the world, that the various motions undertaken by their beloved particles are guided throughout by a variety of attracting and repelling forces, the *Principia* is about as tacit on the subject of the reality of forces as can be. In the passage just quoted Newton acknowledges at once that these forces – but for universal gravitation itself, not even hinted at elsewhere in the *Principia* – are still in need of explanation. Over a few other passages inserted at a late date hangs the artful suggestion that 'force' in general might be no more to Newton than a handy analytical tool devoid of real-world significance. Few readers were taken in by such incongruous disclaimers. Still, in a formal sense Newton's genuine view on the physical reality of the force of gravity could be derived only from the very serious constraints he imposed upon the sole alternative available, aethereal mechanisms. Not only would these require hosts of ad hoc assumptions, they also had to be thought of as consisting of principally other stuff than gross matter is made of; and, on top of that, they had to lack resistance. In short, when the first edition of the *Principia* came out in 1687, Newton was as unambiguously convinced of the primacy of forces in nature as he was concerned to hide that conviction.

One final reason for wishing to hide it rested in the partial nature of his accomplishment – on his own admission in the preface, hosts of forces were still awaiting discovery. Only an intimate among his readers (and Newton had no intimates) might suspect where they could actually be found, which was in the dozens and dozens of notes and unpublished treatises, on chemical/alchemical phenomena chiefly, that Newton had for decades been accumulating in his desk drawer. Torn between his forever rival urges to give his speculative fancy free rein and keep it in check by rigorous mathematical and/or experimental proof, Newton in 1687 kept silent about his private speculations, filled as these were with aethereal mechanisms reinterpreted as forces but not (as with gravity) subjected to mathematical treatment yet. Where, then, does the *Principia* stand in respect to Newton's full conception of the frame of nature? Pending fuller treatment of this vexed issue (p. 710), I note here only that Newton's achievement in the *Principia*, even if a small step compared with what he felt still had to be accomplished, nonetheless represented a stunning leap forward compared with where

674   the next-best practitioners, Huygens and Hooke, stood. Now that we have discerned the components, and gauged the dimensions, of Newton's revolutionary accomplishment, such a comparison opens the door toward resolving the question of what had made it possible.

Causes of the revolution.   In seeking causes for really the only one-man revolutionary transformation among the six that together constitute the Scientific Revolution, three dimensions must be kept analytically distinct: the structural, the coincidental, and the strictly personal. Leaving aside Newton's genius, the peculiar nature of which was perceptively probed by Keynes (p. 637), I shall first address Hooke's and Halley's timely interventions. After all, it is quite conceivable that in the absence of either of these highly personal events the *Principia* would have remained unwritten. Insofar as these were just coincidences, serving to redirect Newton's mind in particularly fruitful ways and in the end leading Halley to underwrite publication of the *Principia*, they might at first blush seem to defy causal analysis. At most, they alert us to what in history remains irreducibly personal (chiefly, in the present case, Halley's extraordinary generosity). However, the likelihood of such happy coincidences is greatly enhanced by the presence of institutional arrangements that favor them. Precisely such had come into being in the 1660s for the cultivation of science in England, as also in France (p. 514). These men *knew* each other; they talked in the coffeehouse, corresponded, paid each other visits, much of all this with what was new in science for their prime subject. This was the case to such an extent that not even Newton's reclusiveness, directed 'chiefly to decline my acquaintance' as in 1670 he brutally expressed it in response to an eager correspondent,[367] sufficed fully to seal him off from such interaction. In one particular sense did Newton unreservedly partake of ongoing exchanges, in that even at the height of his withdrawal, between 1678 and 1684, he rarely missed a publication, news of which the *Philosophical Transactions* and its extensive accounts and digests of what was going on at home and abroad kept laying on his doorstep. If in the *Principia* he showed himself in full mastery of Flamsteed's data for the Jupiter/Saturn conjunction of 1683, of Richer's report on pendular behavior in the tropics, and of Picard's measurements on the Earth's circumference, it was due to such communication networks as two decades earlier had come into full bloom.

What such communication networks do not begin to explain, however, is the magnitude of Newton's achievement, which, albeit sustained by up-to-date information and also by quantitative-experimental confirmation, was guided chiefly by the inner logic of conceptual advance. It is true that Newton was not the only one at the time to probe the capacity of some concept of force to show the way out of the quandary presented by particles in motion and their ever more apparent inability to account for activity in nature. At the very time when Newton was toiling day and night at his investigation of orbital motion under the guidance of forces, unbeknownst to him Leibniz was busily seeking to enrich (rather than replace) the reigning natural philosophy of kinetic corpuscularianism with a conception of force, even if an almost wholly different one (p. 628). Still, as Halley's astonishment at re- ceipt of even so retrospectively crude a piece as 'De motu corporum in gyrum' testifies, it is impossible to regard the revolutionary transformation single-handedly effected by Newton as the obvious next step which, if he had not taken it, someone else would have. True, in the fullness of time someone (or rather a succession of someones) is likely to have, but not then and there – for that, Newton's scientific stance was too singular by far.

In the singularity of his stance rests the prime structural cause of the revolution New-ton brought about between August 1684 and April 1687. That cause was his unique prior participation in *both* revolutionary transformations 4 and 5, *both* of which he now went on to transcend. What enabled Newton to go beyond Huygens in adding forces to the world's assemblage of moving particles so as to make them their prime agents, was that he was also a kind of Hooke. Just like Hooke he allowed the tension inherent in the Baconian Brew (i.e., its built-in ambiguity, in Newton's personal case urged upon him by his chemical/alchemical concerns above all) to drive him toward free exploration of the outer bounds of what aeth-ereal mechanisms still could accomplish. Huygens once spent a day or two contemplating an accelerative force of 'incitation' (p. 627), but there was nothing in his intellectual makeup to incite him to pursue the idea any further; for that Huygens (scarcely connected to this Brit-ish movement in any case) was just too Hooke-less. And what, conversely, enabled Newton to go beyond Hooke in moving their shared search for active principles from some specula-tive and, at best, guess-quantitative hand waving at real-world forces toward a sustained ex-ploration of where these were to lead once anchored in mathematical rigor was that he was also a kind of Huygens, that is, a man whose prime allegiance lay with realist-mathematical science and its stern exigencies. By blending all that was fertile in the scientific inclinations and best achievements of the two men to dominate the most advanced research of the 1660s through mid-1680s, Newton made a breakthrough of a kind out of reach to either.

Reception.   How a book is first received more often than not determines its future desti-ny. In the case of the *Principia*, Newton was once again lucky in having Halley as a supporter. Not only did Halley see the book through the press and pay for the printing costs out of his own pocket, but he also made sure to trumpet its virtues as early and as widely as he could.

He did so in an 'Ode to Newton' that disfigures the first pages of the *Principia,* in a lengthy advance review for the *Philosophical Transactions,* and in an array of letters care-fully placed both in Britain and on the Continent, but first of all through word of mouth. News of the principal ideas to receive treatment in the *Principia* began to make the rounds of London's coffeehouses by 1686, thus causing Hooke indignantly to raise with Halley and other fellows Newton's alleged 1679 theft of universal gravitation. In the squabble that en-sued, with Halley caught in the middle, Newton went out of his way to belittle Hooke's real contribution to his own accomplishment, which contribution did not so much reside in the

676 discovery of universal gravitation as in Hooke's making Newton face the proper parameters of orbital motion. Still, in view of the whole of Newton's accomplishment, which Hooke was congenitally unable to fathom, it is hard to discount the truth of Newton's oft-quoted complaint:

> Now is not this very fine? Mathematicians that find out, settle & do all the business must content themselves with being nothing but dry calculators & drudges & another that does nothing but pretend & grasp at all things must carry away all the invention as well of those that were to follow him as of those that went before.[368]

It may well be that Hooke's ire was enhanced by a dark foreboding that the *Principia* marked the end of the peculiar approach to nature here distinguished as the Baconian Brew. Its pioneers, Boyle and Hooke, lived on to 1691 and 1703, respectively, yet their claims to fame stem almost exclusively from pre-*Principia* days – the book that at one stroke sent these men from the forefront of the Scientific Revolution to the second rank. So it did (albeit in smaller measure) with Huygens, whose strictly geometric treatment of individual *genera* of motion was likewise turned by the *Principia* into a time-bound curiosity, an aberration from the historical pathway that seemed in retrospect to lead straight from Galileo to Newton.

This is not to say that Huygens' response to the *Principia* was as narrow-minded as that of Hooke (whose dependent status was directly threatened by it). To the contrary, Huygens showed himself deeply impressed. What enchanted him in particular was Newton's determination of the weight of Jovians and of Martians – a kind of measurement by proxy no one would have thought possible, so he noted with genuine awe. The story further goes that when John Locke, the philosopher, inquired with Huygens whether the mathematics of the *Principia* could be trusted, Huygens put his worries to rest without reserve. Huygens even let himself be convinced by Newton's conception of gravity insofar as, by way of a powerful mathematical tool, it could account for the celestial bodies' orbits in accordance with Kepler's laws.

Two major aspects failed to gain his allegiance, however. Even though he felt compelled by the force of Newton's argument to acquiesce in the destruction of Cartesian whirlpools, whirlpools of another kind had to replace them. These were the spherical whirlpools he had already broached in his 1669 Académie lecture on the cause of heaviness (p. 627). Kinetic corpuscularian that he was and remained, he kept requiring of Newton's force of gravitation a reduction to some mechanism of particles in motion. Pending such a reduction, he regarded any real-world claim for it (which he felt sure Newton was hiding from his readers' view) as a fanciful absurdity. He further refused to go along with Newton's conception of gravitation as *universal*, that is, as operating on Earth the way Huygens granted it to do in the heavens. On this score Huygens, with his penchant for the telling detail rather than the

sweeping vision, perceptively saw that the *Principia* yielded no more than one phenomenon
fit as yet to serve as empirical testing ground: not the tides, which suffer from too many local
particularities, but the shape of the Earth.

To set forth in public his *first* major point of disagreement, Huygens need do no more
than retrieve from among his manuscripts that Académie lecture of eighteen years back and
prepare it for the press under the title *Discours sur la cause de la pesanteur* ('Discourse on
the Cause of Weight'), which in 1690 he appended to another lecture now finally published,
Traité de la lumière. To make his *second* point clinching, he had to enlarge his original text
with conclusions to be drawn from an account of an expedition that, as luck would have it,
had just returned from a test of Huygens' preferred solution to the problem of longitude, his
marine chronometer with cycloidal cheeks (p. 332). Huygens succeeded on retrospectively
mistaken yet at the time plausible grounds in showing that the data produced en route from
the Cape to Holland came closer to the sole deviation from constant gravity on Earth acknowledged by him than to the dual deviation Newton had just derived in the *Principia* and
regarded as confirmed by Richer's measurements (p. 667). Despite these two major disagree-
ments with the *Principia*, which cannot but look in retrospect as rearguard actions, Huygens
paid Newton the implicit compliment of filling the major part of the correspondence of
his last eight years with subjects Newton had in 1687 placed on the agenda. This was true in
particular of Huygens' exchanges with his most gifted disciple, Leibniz.

Also in 1687, Leibniz paid Newton a big compliment of another kind. As soon as he heard
of the contents of the *Principia* he hastily (and "ever mindful of his intellectual capital",[369]
which he needed for a living no less than Hooke did) produced three papers on subjects
treated therein and had them published forthwith in *Acta eruditorum*. Leibniz's enduring objections to what Newton was up to in the *Principia* stem in good part from his own
concerns with force (p. 629). At times he phrased those objections as radical opposition,
notably in his notorious 1715/16 exchanges with Newton's stooge, Samuel Clarke, in which
philosophical-theological issues were paramount. At other times he rather suggested that a
certain aether mechanism of his own devising accomplished all that Newtonian attractions
between heavenly bodies did, and more. Notably, he felt that by means of that particulate
mechanism he could explain what in Newton's treatment remained an accidental circumstance, to wit, that all planets orbit the Sun in roughly the same plane. Also, in 1715 Leibniz
still regarded the alleged universality of gravitation as empirically refuted by Huygens' appeal to pendular motion on board ship at different latitudes.

So much for the three men (Hooke, Huygens, Leibniz) whose own scholarly concerns
were most narrowly intertwined with the large range of issues solved or at least raised in
the *Principia*. Albeit with individual variations in the amount of openness with which they
approached the book, each man was too settled in his own trusted approach to jump on
board – Baconian Brew in the case of Hooke; sheer geometrization of corpuscular motion
with Huygens; philosophy-drenched, living-force-centered dynamics in Leibniz's case.

678   Jump on board was what younger men did. One major example is Pierre Varignon in Paris. He quickly dedicated himself to 'translation' of Newton's calculus-based yet geometry-drenched proofs into algorithms expressed in the calculus variety to which, as Varignon rightly foresaw, the future belonged, Leibniz's (p. 360). In Britain, too, it was the mathematically skilled, in particular, who (on Halley's example) began to flock to Newton forthwith. In later years, following Newton's 1703 appointment as president of the Royal Society, this was to lead to the rise of Newtonianism as concocted and heatedly defended by a cabal of intellectual servants swearing by every word of their master (not counting a few who were more independently minded). By then, that master himself was still redoubtable and energetic but no longer creative, the well-established and adulation-enjoying shadow of the ardent young man who, within two years of making his acquaintance with the ongoing Scientific Revolution, had joined in one revolutionary transformation (the one accomplished by Huygens) in the mid-1660s, capped another (Boyle's and Hooke's) in 1675, so as finally to pull off entirely on his own the most revolutionary transformation of them all in the mid-1680s.

Toward the Opticks

And then there was light.

   With light, as with the calculus and with motion, Newton's major discoveries stem from his wonder years of 1665–7. As with the calculus, but unlike with motion (sole subject of his second creative outburst, the one just examined), he never changed his mind about his fundamental conceptions regarding light and color. Newton's optical work forms a blend of mathematical science, fact-finding experimental science, and underlying mechanisms derived from his successive philosophies of nature. In its final presentation, the mathematical component fell to the wayside. Newton's ongoing failure to find a satisfactory dispersion law obliged him in the end, when finally ready in 1704 to publish *Opticks*, to emphasize its fact-finding experimental component (p. 544). This gave a distinctly Baconian flavor to his final presentation. Since that was precisely how decades earlier he had chosen to present one portion to the world and how in most of his private papers he had treated the other portions as well, *Opticks* eventually differed little from a comprehensive albeit more compact treatise he had written back in 1672 but did not publish.

   In discussing *Opticks* and the path toward it, I shall first survey how problems of light and color were treated in all five revolutionary modes of nature-knowledge prior to Newton. I shall then examine to what extent and in what respects he went beyond them. Finally, I shall investigate in narrow connection with the *Principia* what enabled him to take these optical steps forward and to what extent and in what respects it makes sense to regard these, too, as revolutionary.

Light and color prior to Newton.   In the history of the investigation of light up to
and including the Scientific Revolution three successive periods can be distinguished: (1)
from Ptolemy and his predecessors onward until the first, 'perspectivist' synthesis of optical knowledge; (2) the perspectivist tradition itself, as established by Ibn al-Haytham and
left unscathed both in Islamic civilization and in Europe until Kepler; (3) the disintegra-
tion of the perspectivist tradition during the first stage of the Scientific Revolution, followed by some groping after a fresh optical synthesis in accord with ongoing discovery. Ibn
al-Haytham's synthesis served to unite outcomes of geometric ray tracing with speculative
accounts of the nature of light and available knowledge of the structure of the human eye
(p. 59). Before Kepler and Descartes the perspectivist tradition was centered on the problem
of vision; starting with them it was centered more and more on the problem of the nature of
light. As such it provided the framework in which, up to *and including* Huygens' and Newton's respective first struggles with cases of strange refraction, issues of light and vision kept
being tackled. Among the major discoveries and reorientations that served to undermine the
viability of the perspectivist tradition were the following.

(a) In the abstract *Alexandria* mode: Discovery of the sine rule of refraction by Harriot,
Snel, and Descartes. Attempted proof given by the latter, along perspectivist lines. Later
(thanks to the sine rule), sophisticated ray tracing by Barrow.

(b) In the *Alexandria-plus* mode of enhanced realism: 'Visual cone' account of vision replaced by Kepler with a 'retinal image' account. Determination of foci and image formation for a variety of lenses, with special attention given to what actually happens when
light hits the surface of a refracting medium (Kepler) and/or to optimization of lens
configurations in telescopes (Kepler, Huygens).

(c) In the *Athens-plus* mode of kinetic corpuscularianism: Conception of the nature of light
corpuscularized by Descartes (light as particle pressure passed on instantaneously) and
by Gassendi (light as streams of emitted particles). Color no longer regarded as a quality of objects but as our perception of originally white light weakened in the shadow or
modified by some interposed medium.

(d) In the combined mode of *corpuscular motion geometrized*: Strange refraction in Iceland
spar accounted for by Huygens according to the mathematical properties of rectilinear
propagation, reflection, and refraction both ordinary and strange (but not for color phenomena) by means of a specific wave mechanism, thus getting rid of the last remnants
of the perspectivist tradition. Anomalous refraction in prisms identified by Newton as
color sifting and accounted for eventually by means of specific mechanisms of particles
and forces far removed likewise from the perspectivist tradition yet not in his own view
sufficiently adequate.

All this just recapitulates points made before. In addition, certain discoveries and/or reorientations arose in those two revolutionary modes of nature-knowledge where neither

680   mathematical derivation nor first-principle speculation of limited empirical import held sway but rather a deliberate search for phenomena as yet unknown.

(e) In the *fact-finding experimental* mode, both Boyle and a range of Jesuits examined phenomena of light and color in a broadly Baconian setting. As usual, Jesuit authors enveloped their observations in some personal variety of their syncretic worldview. In his all-encompassing *Ars magna lucis et umbrae* of 1646 (Figure 17.1; p. 573) Kircher, with his predilection for phenomena associated with natural magic, dealt with fluorescence, opalescence, and similar curiosities apparent, for example, in nephritic wood. Francesco Grimaldi SJ sought to reinforce his own conception of light by demonstrating that a tiny, sharp object like a nail, if placed in the path of an extremely narrow beam of light, causes it to cast a shadow much wider and more diffuse than it should. He named the phenomenon 'diffraction'. Particularly puzzling was that it called the rectilinear propagation of light into question and was also inexplicably accompanied by small color bands. His favored conception of light was as either itself a fluid substance or a modality of some primary fluid. The outcome of his experiment made Grimaldi draw an analogy between diffraction and how the regularity of water waves gets disturbed by a small midstream obstacle. He published his finds in 1665 in his book *Physico-mathesis de lumine, coloribus, et iride* (Physico-mathematics of Light, Colors, and the Rainbow). A review appeared in the *Philosophical Transactions* of 1672, which quickly drew the attention of Hooke and Newton. Both were intrigued by the color bands and by the apparent incompatibility of diffraction with what (unlike Grimaldi) both men regarded as the rock-bottom property of light, its rectilinear propagation.

Unlike with the Jesuits, Boyle's experimental investigation of color phenomena was not bound up with a worldview but with a specific issue. In his ongoing campaign against Paracelsian conceptions of matter (p. 555) he was concerned to refute the idea, current in those quarters, that what color a body has is determined by how it partakes of the three constitutive principles of iatrochemistry, 'sophic' mercury, sulphur, and salt. His refutation took shape as the natural-historical listing of numerous observations (some straightforwardly observational, others mediated by experiment) that appeared to go against such an account. One example, gratefully borrowed from Kircher and then extended, is the opalescence of nephritic wood, which he also turned to chemical purposes. Another example is a range of color phenomena in thin layers, such as the rainbow colors that appear on the surface of thin glass or of soap bubbles. These colors appeared to him to vary with the angle of incidence and with the plane of vision.

(f) To the combined mode of the *Baconian Brew*, finally, belongs a lengthy passage on light and color in a book published one year after Boyle's, Hooke's *Micrographia* of 1665. Citing partly the same phenomena as Boyle did, Hooke placed them in a wider and more intricate frame – not just experimental but experimental in such a way as to sustain, and

in turn to be reinforced by, hypothesized mechanisms of corpuscles in motion. Hooke's
starting point was likewise color phenomena in thin layers, which he, too, used for purposes of refutation. His target was Descartes' explanation of color as arising from the differential axial rotation (quickest for red; slowest for blue) that aether corpuscles allegedly
undergo upon refraction. Descartes had acknowledged this to imply that, unlike with
rainbow-producing water droplets or rainbow-like color fringes near prisms, no colors
will appear if light passes through two parallel surfaces. The reason is that in that case
refractions are reversed and each primary rotation is nullified by its secondary counterpart. Here, then, is the opening Hooke exploited to refute Descartes by means of a variety
of mostly experiment-induced observations on light effects in "plates of muscovy glass
(mica), liquids of various kinds pressed between two plates of ordinary glass, liquids and
glass blown into bubbles, and the surfaces of metals".[370] With flakes of mica first, Hooke
found such thin layers to display colored rings under his microscope. In deference to
the man to examine them more thoroughly, these have come to be known as 'Newton's
rings'. Among their properties Hooke listed the order of appearance of their various
colors. Having thus shown colors to appear where on Descartes' account they could not,
Hooke went on to explain how they truly originate. Inspired by Hobbes' earlier account
of light as pulses, Hooke's conception of light took shape by way of an extension of
Descartes'. He employed an analogy with the current account of sound, altogether in
line, that is, with his views on the generally vibrational constitution of matter (p. 558).
Thus was Cartesian particle pressure turned into a succession of pulses which, through
the aether, pass the vibrational movements of the luminous body on to the retina. When
these pulses fall obliquely on a refracting surface, the light gets confused and colors are
produced, such that we perceive as blue an oblique pulse whose weaker part precedes and
whose stronger part follows (Figure 19.3). With red it is the other way round, and with all
other colors the degrees of weaker and stronger are intermediate.
Within a year of Hooke's publication, Newton's voyage of optical discovery was to com-
mence with a flat-out contradiction of this particular account of light and color. This came
to pass as follows.

Newton sets off on his own. Barring only Grimaldi's discovery of diffraction and Huygens' and Barrow's still-ongoing work on image formation with lenses, all the literature produced by the developments listed above under (a), (b), (c), and (e) found itself duly digested
in 1664/5 in the 'Waste Book', in which the twenty-one-year-old Isaac Newton set out, in the
solitude of a self-imposed reading program, to master the Scientific Revolution in its current state. As yet, rejection of Descartes' conception of light on a range of semi-experimental
grounds was the chief outcome of his highly critical questioning (e.g., "Light cannot be by
pression &[c] for y^n wee should see in the night as wel or better y^n in y^e day").[371] One or two

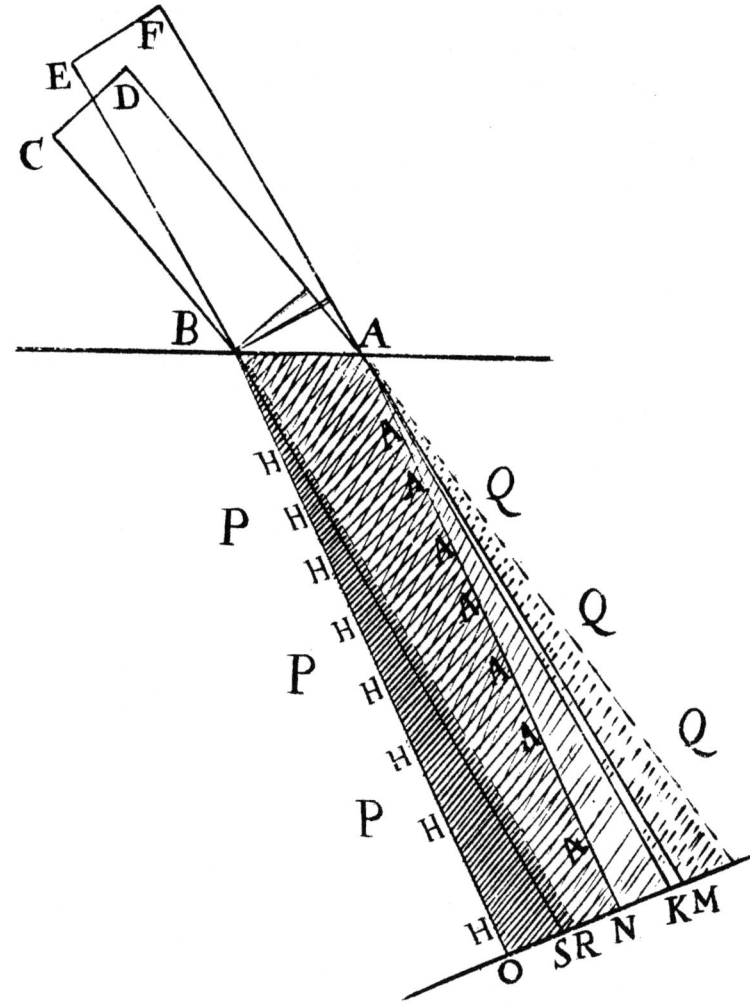

Light ray EFB[L] is white. When the left side of the pulse is refracted at B and passes into another medium, the right side has not yet arrived at A. The difference makes for what Hooke called the 'confusion' that manifests itself as the various colors.

Figure 19.3: Hooke: The making of color

years later, his apprenticeship clearly over, he returned to the 'Waste Book' to enter further 683
notes. Hooke had meanwhile published his account of light and color in *Micrographia*, and
that is where Newton set out on a voyage of discovery all his own:

> The more uniformly the globuli move ye optick nerves ye more bodys seme to be coloured red
> yellow blue greene &c but ye more variously they move them the more bodys appeare white
> black or Greys.[372]

That is, light is not pulses propagated but corpuscles emitted, and colors are not confused
white light. Rather, white light (the diminishing intensity of which yields gray or black, not
colors) is itself confused, with its individual components giving rise to the perception of
individual colors if made to fall separately on the retina. The *Opticks* Newton was to publish
almost forty years later contains little but this one lapidary statement, elaborately extended,
however, in a threefold direction:

(1) He drew ever more radical consequences, in ongoing intertwinement with an ever more
thorough, at times mathematical but chiefly quantitative-experimental investigation of
the two specific modalities (refraction, partial reflection) in which the separation of colored rays could be effectively investigated (1666, 1669/70).

(2) He learned the hard way how in public to present the primary nature of colors separately
from the emission account of light that had given rise to it (1672–1678).

(3) He wove his account of light and color and their measured properties into his successive
conceptions of the world (from c. 1669 onward until close to his death in 1727).

Here, in highly simplified rendition, is how he did all that.

Color sifting and its consequences.   Moving from the rhetorical experimentalism of
his 1664 'Waste Book' questions to the real thing, Newton subjected his newfangled, utterly paradoxical idea of the primacy of color to the test. Unequal refraction of distinct
colors should cause a thread that is blue at one end and red at the other, if held against a
shady background and viewed through a prism, to appear broken halfway. Newton already
knew from Boyle's description of the experiment that indeed it does appear so, although he
made sure to repeat it for himself. For a while this is where he left it. So paradoxical was his
tentative contradiction of what everybody had always taken to be self-evidently the case –
the natural primacy of pure, white light – that a good deal more than an idea and its first
empirical confirmation was needed even for its originator to let the consequences sink in.
A first consequence occurred to Newton when seeking to grind hyperbolic lenses. These,
as shown in Descartes' *Dioptrique*, would, unlike spherical ones, converge rays to a perfect
focus (p. 328). But suddenly Newton realized that his grinding, even if successful, would still
be a failure – with spherical aberration gotten rid of, telescopic images would still remain

684 blurred due to differential color refraction (which as he was soon to demonstrate yields an aberration much larger than its spherical counterpart). Newton never returned to lens grinding, deciding instead to pursue his original idea further. This he began to do, teeth and claws sunk into the problem as always with him, in a paper properly entitled 'Of Colours' (one of the 'wonder years' papers).

He was aware from the start that the colors we see all around us come about in overwhelming majority through preferential reflection: a surface exposed to white light (i.e., to a range of colors blended) displays the color it reflects best. He was also aware that two artificial means presented themselves actually to investigate the separation of colors out of white light. One, the prism, sifts colors through differential refraction, as his first experiment was soon to confirm. The other, thin layers of whatever composition, produce one color on one side, due to reflection of part of the light falling upon it, and another color on the other side, due to the passage of the remainder. With partial reflection he confined himself at first to assembling what he had found in the literature. One example concerned nephritic wood. Jesuit missionaries encountered it in the jungles of Mexico and found that it was used by the natives for medicinal purposes. Samples were therefore sent through their network to Kircher's museum in Rome. There the wood was investigated by Kircher for its magical opalescence. Encountered by Boyle in Kircher's *Ars magna lucis & umbrae*, Boyle acquired some for himself and put it to work as a standard discriminator between acids and alkalis and also exploited it for its opalescence to confute the Paracelsians. This versatile tropical wood was now used by Newton to explain why an infusion looks blue on the reflection side and yellow on the other (the opposite of gold leaf).

Once he had exhausted for the time being the range of examples of thin layers culled from the literature, Newton took up the other means to investigate the separation of colors, the prism. Why had Descartes and Boyle, the two men who preceded him in its optical use, failed to observe the deviance from the sine law of refraction that Newton ex-

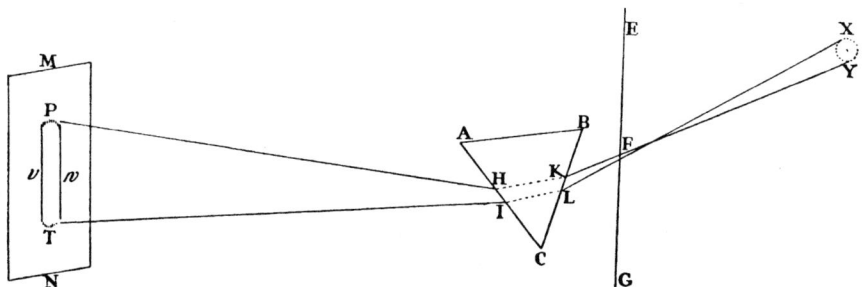

Figure 19.4: Newton's first demonstration of anomalous refraction in prisms

pected it to display? Newton realized that, whereas none but colored fringes appear near the prism, the colored rays that emerge from it upon refraction (each, then, by a somewhat different amount) require space duly to separate. If allowed to form at a proper distance, the spectrum does not display the same circular shape as the original pencil of light that falls on the prism but rather appears *elongated*. Newton equipped his room back home (for in Cambridge the plague reigned) in the manner sketched in Figure 19.4.

Indeed, the color spectrum that emerged turned out to be semicircular at both ends and oblong, five times wider than high. This was his first experimental measurement, with many more to follow. Newton was also aware from the start that the experiment requires the light rays to be refracted equally at both faces of the prism. More than mathematician enough to see at once that the condition for that is also the condition for minimal deviation, he only had to turn his prism until it projected the spectrum at its lowest possible point.

Newton's essential insight, that the oblong shape of the spectrum indicated a significant violation of the sine rule due to the compound nature of white light, could of course still be undermined in many ways, which he went on systematically to eliminate. For example, since the hole in the window through which the sunlight fell comprised an angle larger than negligible, did the elongation not follow naturally from the rays in the pencil of sunlight running less than exactly parallel? No, it did not – a further narrowing of the beam (in the end Newton managed to use Venus rather than the Sun for a light source!) caused the elongation to increase rather than vanish. He also tackled head on the most important of all anticipated objections: was not the elongation, along with the colors that made their appearance, itself an alteration brought about likewise by the interposed prism? For an answer, Newton devised in stages his celebrated 'crucial experiment' (Figure 19.5).[373]

This is how, at least to his own satisfaction, Newton threw out of court the current conception of color being white light modified. He went on to gain further clarity over what exactly color sifting implied. At first he remained indifferent to the range of colors in between the blue and red ends of the spectrum. In line still with Hooke's two-color idea, for a while he kept regarding these two as basic. It took time for it to dawn upon him that white light is not just composed of two colors, red and blue, which may then combine in various ways to form all other colors, but that there are really innumerably many colors, each a shade different, each immutably marked by a somewhat different degree of refrangibility, and together composing white light. Newton found it henceforth convenient, in analogy with how many diatonic notes there are in the octave, to treat the spectrum as if made up of seven colors (red, orange, yellow, green, blue, indigo, violet). Even so, he was convinced that this expresses no more than how we subjectively perceive and intersubjectively discuss bundles of rays that are each endowed with a somewhat different index of refraction. Vast improvement on Descartes' account of the rainbow (p. 299) now also proved possible. More exact, empirically confirmed determination of the angles for the primary and secondary rainbows, as well as

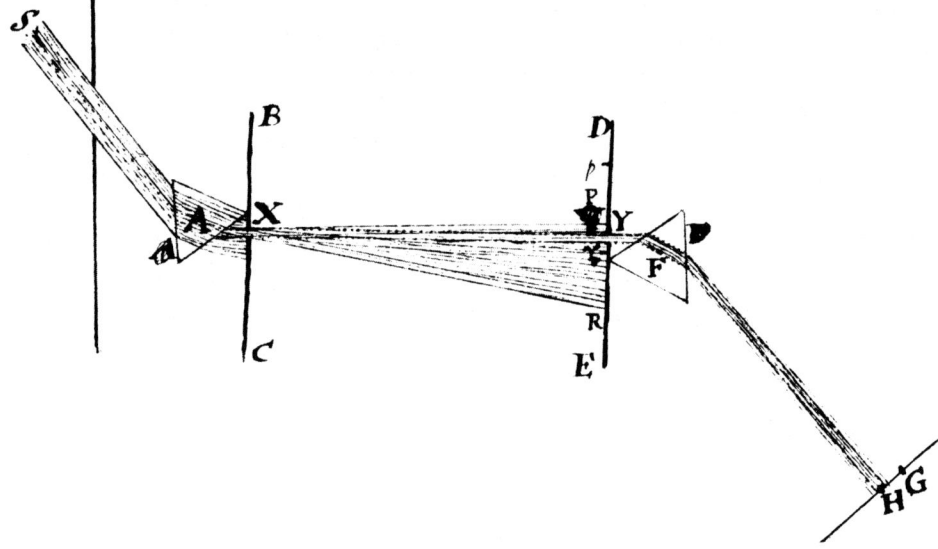

"Behind the prism, he set up a board with a small hole so placed that by rotating the prism slightly on its axis he could project different parts of the spectrum through the hole. Halfway across the room another board with a hole allowed the beam to pass. Since the two boards were fixed in place, they defined a constant path for the beam and a constant angle of incidence on a second prism set in a fixed position behind the second board. When the red end of the spectrum was projected through the holes and onto the second prism, it was refracted there at an angle corresponding to its refraction by the first prism; blue was refracted at a greater angle, again corresponding to that of the first prism. In no case did the second prism cause further dispersion."

Figure 19.5: Newton's experimentum crucis

the respective widths of their arcs and the distance between them, readily emerged from equally exact determination of the ratio between the indices of refraction at both ends of the spectrum – 108:81 for red, 109:81 for violet. Newton even went so far as to calculate the dimensions of as yet unseen, higher-order rainbows.

The growth of his insight into the fundamental heterogeneity of white light went in tandem with ranges of experiments of increasing sophistication meant to reconstitute white light after its separation. Half a century earlier, the pharmacist Angelo Sala decomposed copper vitriol into copper ash, acid spirit, and water but refused to find the feat of actual decomposition sufficiently persuasive until he managed to recombine these materials into the original vitriol. Just so did Newton now operate to make the primacy of color clinching. And indeed, with white light reconstituted and then separated once more, colors proved to reemerge as before – colored rays are truly immutable (Figure 19.6).

As noted, Newton was originally moved to get to the bottom of his original idea of the compound nature of white light when a sudden awareness dawned upon him of the apparent futility of using lenses of whatever shape for telescopes in view of the apparent inevitability of chromatic aberration. Consequently, he took another approach and built a reflector instead (p. 329). Consequently, too, he began to search for the regularity that the variously colored rays obey in each refracting to a slightly different focus. At first he did so in a purely mathematical vein still tacitly governed by perspectivist presuppositions. Later he aimed at the same objective in hoped-for accordance with some mechanism consistent with his own emission conception of light. The approach was quite novel, but he failed to find such a dispersion law (p. 544).

So much for Newton's findings on the separation of light by means of refraction in prisms and lenses. Armed thus with full awareness of the heterogeneity of white light, Newton now set out to examine in equal depth and in likewise quantitative fashion its separation due to partial reflection in thin layers. Some observations by Boyle and by Hooke gave him a welcome head start.

Hooke's investigation of what in 1665 were still definitely Hooke's rings was directed at the patient finding of facts about them by means of microscope-aided experimentation. To explain them, he came up in his customary way with an inventive hypothesis, grounded in his conception of light as vibrational pulses and of colors as their varied confusions. What, then, were these 'rings'?

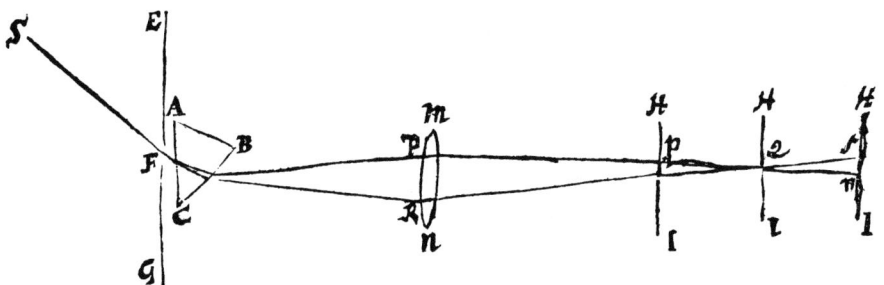

HI is a sheet of white paper held at first in the focal plane, then beyond it.

Figure 19.6: Newton's experimental reconstitution of white light

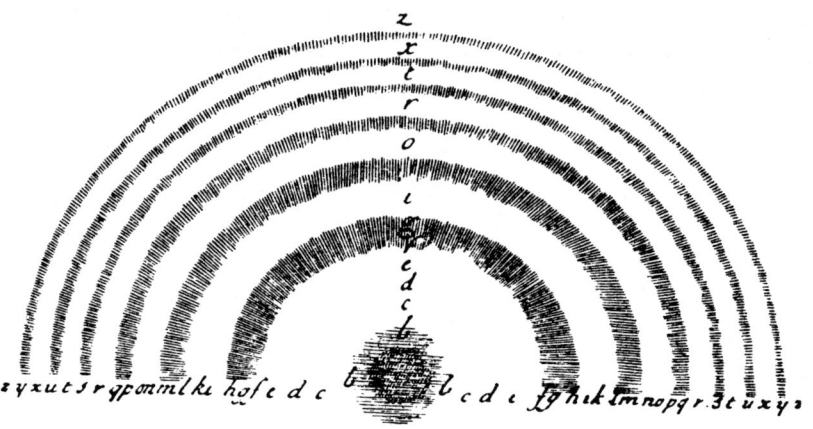

Figure 19.7: Newton's (at first Hooke's) rings

By splitting the muscovy glass with a needle into thin flakes [Hooke] observed that there were several white specks or flaws in some parts of the flake, while other parts appeared tinged with all the rainbow colours. These colours appeared under the microscope to be arranged in rings, circular or irregular according to the shape of the spot which they terminated. The order of the colours was, from the middle, as follows: blue, purple, scarlet, yellow, green; and this series repeated itself around every spot from six to nine times. He called these systems of colours *rings*, *lines* and *irises* – counting all the gradations between the ends of each series for one.[374]

Hooke further observed the following phenomena. The rings seem to repeat themselves at regular intervals; that is, they are periodic. The central spot is mostly of one color only. Pressure exerted upon the thin plate causes the colors to change places. Definite upper and lower limits are set to the thickness of the layer for the rings to appear at all. The last-mentioned phenomenon was of particular significance for Hooke's explanatory mechanism (Figure 19.8).

He imagines a pulse impinging in an oblique direction on the first surface of a lamina having the minimum requisite thickness. Some of the light is reflected by that surface while the rest is transmitted into the lamina. When the refracted light reaches the second surface, some of it will be reflected back to the first where it will be again transmitted out of the lamina. Two pulses will thus proceed from the first surface: a pulse that has been reflected at that surface, followed by another which has undergone one reflection at the second surface and two refractions at the first. Owing to these two refractions and the time spent in traversing the thickness of the lamina twice, the succeeding pulse will – according to Hooke – be weaker than the one preceding it. But as the distance between the surfaces of the lamina is very small, the impression of these two pulses on the eye will be rather that of *one* pulse whose stronger part precedes and whose weaker part follows.[375]

And so we are back at Hooke's general explanation of colors. He has now suitably refined it by noting that the order of strength and weakness, which in his general account determines what specific color appears, depends in its turn on the exact thickness – within the upper and lower bounds outside which no rings appear at all – of the layers in question. But here was also the rub: in view of their very thinness Hooke acknowledged that he was incapable of measuring it.

Upon making their acquaintance with this particular passage in *Micrographia*, two attentive readers, one in Paris and one in Cambridge, saw at once how to measure it nonetheless. Almost simultaneously it occurred to both Huygens and the young Newton that, while direct measurement was indeed out of the question, it could be done indirectly by taking a lens of known curvature and pressing it upon a flat glass plate, thus producing a thin layer of air in which the rings duly appeared. By means of the same geometric property of the circle that both men had used earlier to calculate centrifugal force, they could now calculate how thin the layer was at any one ring by measuring the diameter of the colored circles, which could be done by means of just a pair of compasses.

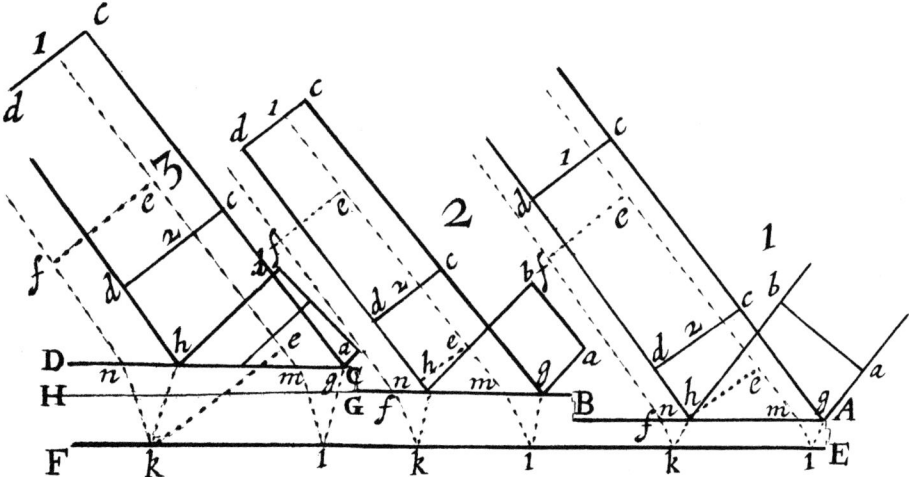

Figure 19.8: Hooke on the thickness of layers and the confusion of light

690   The degree of precision eventually attained by each man was rather different, though. Already with his first attempt, in 1666, the obscure Cambridge student did better than Louis XIV's scientific crown jewel, only to surpass him by huge lengths some four years later, when he sat down to pursue the issue with that peculiar, dogged tenacity in which he excelled. Recording outcomes in another private paper, 'Of ye coloured circles twixt two contiguous glasses', he devised suitable techniques for tightly pressing a lens with a radius of curvature of 50 feet upon a flat piece of glass. Here is an instructive comparison:

> As with [Newton's] prismatic observations, the powerful experimental imagination which conceived the means to reduce thin films to measurement cannot fail to impress. ... Newton's skill in performance ... outran that of Huygens from the beginning, and in the more sophisticated experiments of 1670 he simply eclipsed his unsuspected rival. What he demanded of his measurements tells us much about the man. Measuring with a compass and the unaided naked eye, he expected accuracy of less than one-hundredth of an inch. With no apparent hesitation, he recorded one circle at 23 ½ hundredths in diameter and the next at 34 1/3. When a small divergence appeared in his results, he refused to ignore it but stalked it relentlessly until he found that the two faces of his lens differed in curvature. The difference corresponded to a measurement of less than one-hundredth of an inch in the diameter of the inner circle and about two-hundredths in the diameter of the sixth. "Yet many times they imposed upon mee," he added grimly to his successful elimination of the error. No one else in the seventeenth century would have paused for an error twice that size. Newton was confident enough in his technique that he used his results to correct the radius of curvature of the lens.[376]

To both men the comparison of the diameter of the first ring with those of the others kept coming out incompatible with periodicity (reappearance of all distinctly visible circles at equal intervals) by a seemingly inexplicable factor of 2½. As a consequence, Huygens abandoned his entire effort. Newton, being Newton, did not, and in the end he identified deformation of the lens due to an overdose of pressure as the culprit. Insofar as a then/now comparison is feasible, it discloses that Newton's final results do "not ... diverge from modern measurements, an extraordinary achievement for a pioneering investigation."[377]

    Confirmation of his capacities as an experimenter naturally figured in the investigation, even if it did so only to himself, publication being as yet far from his mind. But Newton made it serve numerous other ends as well. Boyle had discovered the variation that colored circles display when viewed at diverse obliquities. This was important for Newton in view of his own corpuscular conception of light (yet to be separated from his account of color phenomena in and of themselves). The matter led to another array of virtuoso micromeasurements, guided by mathematical law in this particular case. Periodicity was another concern. Huygens, as Newton in 1666, had assumed it as Hooke had; now, in 1670, Newton first established the periodicity of some optical phenomena by careful measurement. From the geometry of the circle, the thickness of the film between the lens and the flat sheet is proportional to the square of the diameter of the colored circles. Newton measured dark circles between colored circles, squared their diameters, and found a simple arithmetic progression. If the thickness of the first dark circle were set at 2 units, successive circles appeared at thicknesses of 4, 6, 8, 10 and 12 units; colored circles appeared between them at thicknesses of 1, 3, 5, 7, 9, and 11 units, becoming less distinct until they merged completely into whiteness [as according to Newton's account of color they should]. No matter what the obliquity, no matter what the medium, air or water, the same progression held. There could be no question about the periodicity of the circles. Whereas Huygens simply forgot his measurements when periodicity became an embarrassment to his treatment of light, Newton's measurements etched its reality so deeply on his consciousness that he could not forget it even though it eventually became an even greater embarrassment to his treatment.[378]

Finally, Newton addressed the rings in terms of his own account of color, as changed from a two-color account to the immutability of all colors. That is, he went on to subject to measurement, not the thickness of his thin plates from one ring to another but, inside a given ring, from one color to another.

Measuring them proved to be more difficult. Instead of attempting to measure the colors on the circles themselves, where only the innermost and least reliable circle was free from overlap, he tried to cast colors separated by the prism on his apparatus. Limitations of space dictated his method. There was no place where he could project a spectrum down vertically far enough to let the colors separate. Hence he projected the colors onto a white sheet of paper which reflected the light down on his lens. Apparently the intensity was not great enough to allow measurements of circles, but Newton's ingenuity rose to the challenge. Simply laying the lens on the sheet of glass, he cast purple on the paper. As he rotated the prism through the spectrum, circles expanded and new ones appeared in the center. The number of circles from purple to red corresponded to the number of extra half-pulses [see p. 704 for the account in terms of vibrations to which this refers] for purple at that thickness. When he then pressed the lens down on the glass, more circles appeared in the center and expanded until the glasses came into contact. The number of red circles corresponded to the number of half-pulses for red at that thickness. The ratio of the total number of circles to the number of red circles gave the ratio between the half-pulses for purple and red. Newton set it at 14:9 or 20:13. This ratio remained the empirical foundation of Newton's quantitative treatment of colors in solid bodies. Bodies are composed of transparent particles the thickness of which determines the colors they reflect. He had demonstrated with the prism that ordinary sunlight is a heterogeneous mixture of rays, each with its own immutable degree of refrangibility. "And what is said of their refran-

692 gibility may be understood of their reflexibility; that is, of their dispositions to be reflected, some at a greater, and others at a less thickness of thin plates or bubbles, namely, that those dispositions are also connate with the rays, and immutable ...” Hence all the phenomena of colors derive from processes of analysis, whether refraction or reflection, which separate individual rays from the mixture. In 1666, Newton laid out the program and carried it through for refractions. Only about in 1670 did he fully work out the details for the colors of solid bodies.

With 1670, Newton's creative work in optics virtually came to an end. He had worked out the implications of his initial insight, answering to his own satisfaction the questions he had set himself. Though he would devote considerable time to the exposition of his theory, first in 1672, later in the 1690s, and carry out some minor experimentation, he had effectively exhausted his interest in the subject. Never again was it able to command his undivided attention.[379]

Color sifting and the nature of light.   As much as could be done (with periodicity it cannot), the account of Newton's discoveries on light and color has so far been rendered only in terms of phenomena and their properties as Newton established them. That, however, is not how he conceived them. After all, the first independent sentence he wrote down on the subject was “the more uniformly the globuli move ye optick nerves ye more bodys seme to be coloured red yellow blue greene”. That is, how he conceived of color and how he explained light took shape right from the start as one, as yet inseparable blend.

The preference for an emission account of light Newton that settled upon in his first ‘Waste Book’ notes of 1664 went along with his rejection of the only kinetic-corpuscularian alternative then known to him, Descartes’ pressure account. When one year later Hooke in *Micrographia* turned Cartesian pressure into a succession of pulses, Newton stuck to emission of corpuscles in view of what, much to the distress of posterity, was forever to remain his principal objection to any such account, Huygens’ later wave front mechanism included. For Newton kept arguing that if, like sound, light were pulse- or wavelike, it should likewise bend in the shadow. The counterobjection that emission of light corpuscles entails the occasional collision of light rays from opposite directions, hence, their mutual blocking or at least diffusion, never exerted any force upon him.

Newton's conception of light as corpuscles emitted, which he appears to have visualized more vividly than its latter-day reviver, Gassendi, helped him conceive of his new account of color:

> The notion of strong and weak rays in which he couched his original insight readily clothed itself in the images of large and small corpuscles. His geometrization of colors found its counterpart in the paths of single corpuscles, whereas the wave theory of light had always to think of a physical beam of finite dimensions. As the immutability of rays impressed itself on his understanding, it became a further argument for the corpuscular conception, finding its physical

basis in the definition of an atom. Thus there were many filiations binding the corpuscular conception of light to the new theory of color. His initial experiments with thin films in 1666 served to reinforce them. He observed that the circles of colors between a lens and a flat piece of glass appeared to grow in diameter the more obliquely he observed them. He immediately interpreted their size in terms of the ray's blow on the film. A stronger blow allowed the rays to pass through more readily in a circle of smaller diameter, while a weaker blow caused it to pass through less readily and hence to appear as a larger circle.[380]

At a later stage, Newton found that the consequences of this particular speculation about blows of subvisible particles conflicted with what his experiments with refraction led him to conclude. It is typical of Newton that, rather than glossing over the conflict as of course he could have, he recognized it as offering a choice. It is no less typical of him that what he then went on to drop was the speculation rather than a view he felt to be dictated by experiment.

In any case, Newton's account of colors as settled upon for good by 1670 was thoroughly intertwined with the emission account of light he had settled upon for good in 1664. And it was in such a vein that he presented, not the whole, but only the prismatic portion of his investigations to the world. Upon Oldenburg's eager acceptance, 'New Theory about Light and Colors' appeared in the *Philosophical Transactions* of 19 February 1672.

With that publication Newton burst upon the budding scene of Europe's ongoing revolution in science. The paper has with justice (albeit in retrospect only) come to be regarded as one of its landmarks. Here, as with the fixed centrality of the Earth, was another assumption challenged that had tacitly been held through the ages about one of nature's outstanding phenomena. Moreover, unlike the step-by-step unfolding that occurred between Copernicus and Kepler/Galileo, the tacit assumption found itself here being challenged whole, in one fell swoop. Here it was stated flatly and argued forcefully that colors are primary and that pure, radiant sunlight is composed of them – what better empirical confirmation could be had of that seemingly speculative tenet underlying the approach of realist-mathematical science and the kinetic-corpuscularian view of things alike that the world as it really is differs in kind from how we perceive it? It was in full awareness of its significance that, in submitting the paper to Oldenburg, Newton spoke of "the oddest if not the most considerable detection wch hath hitherto beene made in the operations of Nature", later to remark as well that "I perswade my selfe that this assertion [of the heterogeneity of white light] above the rest appears *Paradoxicall*, & is with most difficulty admitted".[381]

It certainly was admitted with difficulty. The act of challenging so intimately held a conviction, if understood at all, was challenged in its turn. Newton responded to those challenges with vehemence. After all, although he himself had needed four years for the consequences of his original insight to sink in fully, by the time of its publication he had held it for six years, so as to tinge it with a sense of self-evidence no reader could possibly share with

694   him yet. Another part of his vehemence rested in natural impatience with the slower-witted, particularly if coming up with objections he had already countered in advance. And there was this near incapacity of his to suppress the rage that tended to spring up in him at just about every criticism that was then or later to come his way. The debate to ensue after his first publication went on for six years and was terminated by Newton in the most ungracious manner possible. The silence that thus began in 1678, and which he never ceased to look back upon in later years as an act of liberation, persisted for six more years, not to be terminated until Halley's memorable visit.

For our purposes there is no profit in examining the exchanges with the Jesuit fathers Pardies (who merely sought clarification of points as yet unclear to him) and Hall (Latinized Linus; the only opponent to question Newton's experiments rather than how to interpret them) plus two posthumous defenders of Linus' honor. Instead, I shall focus on exchanges with Huygens and with Hooke, and what these did for Newton's later presentation of his unaltered views.

Huygens' response was dictated in good part by his own concerns at the time. In possession of a freshly discovered configuration of lenses able to get rid of spherical aberration, Huygens approached both Newton's 'New Theory' and the near-simultaneous account of his reflecting telescope (which he acknowledged at once to be a splendid invention) from a telescope-governed point of view. Originally full of praise, Huygens became ever more critical as the upshot of Newton's paper – the heterogeneity of white light – sunk in. In his final substantive response he asked why, if the composite nature of white light be granted, it was not enough to regard it as composed of just blue and yellow, out of which all other colors could after all be composed just as well? In response, Newton pointed to his assiduous experimentation as showing conclusively that, whatever anybody's idea of how things could or should be, nature had ineluctably arranged them otherwise. Moreover, so Huygens argued, to declare white light composite left wide open the question of what specific mechanism of particles in motion could be adduced to explain it. Note here the similarity with Huygens' later response to universal gravitation. In the course of the exchange Huygens crossed out his latest 'eureka' in view of what he now took to be the proven inevitability of much larger chromatic aberration. Also, he was sufficiently in awe of this newcomer's apparent gifts not to feel lethally offended by the impatient, know-all tone of voice of Newton's responses. Still, he was far from convinced. This had much to do with the impression he gained from a phrase in Newton's paper to the effect that, in view of his discoveries on colors, it could "be no longer disputed . . . whether Light be a Body".[382] That is, Huygens saw Newton's conception of color as bound up with an emission conception of light he was prepared to share even less when, a few years hence, he was to conceive of his own wave front mechanism.

Even more than Huygens was Hooke inclined to read Newton's 'New Theory' in terms of his own concerns and presuppositions. Within three to four hours of receiving a copy of

Newton's manuscript he sat down to pen a report for the benefit of the Royal Society. He declared Newton's experiments to cover no new ground since in previous years Hooke had himself tried them all – the statement testifies at once to Hooke's superficial reading and to the possessive attitude he took toward the subject. True, so Hooke granted, Newton's 'hypothesis' about the nature of color was ingenious, yet the experiments advanced to sustain it were in no way incompatible with other plausible hypotheses, be it Hooke's own 'confused pulse' account, which he went on to expound all over again, or an alternative account Hooke devised on the spot. In all these alternative hypotheses the standard idea of colors arising from white light modified was taken for granted. As with Huygens for more than a little while, Hooke missed entirely Newton's key point, differential refrangibility.

The response to Hooke that Newton wrote at first covered his views on light and color wholly, his measurements on colored circles in thin layers now also included. Since a summary of his causal speculations was included as well, the response came close to a synopsis-in-advance of his eventual *Opticks*, and if published in 1672 it would "have advanced the science of optics by thirty years". There is no need to dwell on the "viciously insulting" tone of voice, "filled with hatred and rage",[383] that Newton used in the response he finally did publish, which was a reiteration of his original paper done at greater length and (where needed, as in his all-too-brief account of the *experimentum crucis*) greater clarity. More to the point, there were two intertwined lessons that Newton derived from the whole of these exasperating (or so he felt about them) exchanges.

What the exchanges brought home to him in the first place was that his opponents held a different view on the proper role of causal hypotheses in science. True, all three were agreed that phenomena had to be secured as best one could, that is, the experimental way, and also (at this stage of Newton's thinking) that it was desirable to explain them by means of nothing but motions of corpuscles. Even so, Newton could agree neither with Huygens' strict *requirement* (which Hooke shared) that an account of phenomena experimentally established ought to be accompanied by some causal mechanism nor with Hooke's loose dealing with such causal hypotheses as more or less interchangeable. A response to Pardies carried the core of what was to remain Newton's prime methodological message:

For the best and safest method of philosophizing seems to be, first to enquire diligently into the properties of things and to establish these properties by experiments, and then to proceed more slowly to hypotheses for the explanation of them. For hypotheses should only be employed for explaining the properties of things, not abused for determining them; unless so far as they may furnish experiments. For if one were from the sheer possibility of hypotheses to make a conjecture of the true state of things, I see not in what manner one can determine anything in any science with certainty, since one may always think up other and again other hypotheses, which shall then appear richly to supply new difficulties. Which is why I have

696   judged that consideration of hypotheses ought here to be abstained from, as inappropriate to the argument.[384]

Given Newton's speculative proclivities, the message was directed as much to himself as to his opponents. True, he had deliberately neglected to weave his ideas on particle blows of different strengths or on any other corpuscular mechanisms into his public account of what his experiments with prisms had taught him. Still, he had felt at liberty to present the corpuscular nature of light as confirmed by them. However convinced he remained of its underlying reality, he now saw that he had needlessly contributed to his opponents' misreading of his core point by drawing his emission account in, and he decided henceforth to present the two – the experimentally established properties and the speculations into which he was to keep weaving them over a lifetime – separately from one another.

Following Newton in that design, I shall now first complete the account of his findings on the phenomenal level as eventually published in Books I–III of *Opticks* and only then examine the 'Queries' he attached at the end of Book III, plus a few instructive things still to be picked up from the late 1660s onward.

Opticks

The book completed by 1694 at the latest yet not to be published until 1704, when Hooke had finally passed away and Newton had reestablished contact with the Royal Society due to his appointment to its presidency, was entitled *Opticks: or, a Treatise of the Reflections, Refractions, Inflections, and Colours of Light*. Its ostensible structuring into Definitions, Axioms, Propositions, Theorems, and Problems makes the reader expect at first glance a mathematical treatise not unlike the *Principia*. But soon enough *Opticks* appears, much less forbiddingly, to be organized rather along Baconian lines. A sequential listing of Observations and Experiments is consistently made to precede the propositions, most glaringly so in the setup of Book II. The accessibly conversational, 'I first did this, then that . . .' tone of voice reinforces the suggestion that Newton's discovery of the properties of light and color had come about through prior observation of facts found through a spontaneous and, as it were, neutral search for them. As in the *Principia*, Newton was aiming in *Opticks* for the maximally attainable amount of demonstrative rigor. However, he had learned the hard way that a verbal account (larded here and there with a mathematical demonstration) in which to string his experiments, precision measurements, and conclusions together was the farthest he could get in this regard. Even so, the book is vintage Newton. It did not quite, however, become the twin pillar to the *Principia* that he intended to make it when in the early 1690s he sat down to put together in three books what he had already drafted decades before, while adding a few optical subjects examined anew. He further intended to write a Book IV that would contain his full argument for a world populated not only with the one macroforce investigated in the *Principia* but also with the various microforces he believed he had encountered over many decades of research in optics and in chemistry/alchemy alike. In or around 1693, his creative powers exhausted, as marked by his months-long mental breakdown of that year, he gave up on the plan. Glimpses of what remained of Book IV and the ideas underlying it were in the end to surface in the 'Queries' (p. 704).

In *Opticks* as eventually published in 1704, then, the reader is made to work his way through three books, each divided into parts. Its principal subjects are listed in Schema 9.[385]

Schema 9: Contents of Newton's Opticks

I	1	· axioms and definitions meant to sum up what was already known before · heterogeneity of white light; colors differently refrangible and reflexible · sine rule applies to every ray individually · consequences for telescopes
	2	· colors do not come about by way of modification of white light · colors are immutable · composition of the spectrum · TO COMPOSE COLORS OUT OF COLORS · TO COMPOSE WHITE, GRAY, AND BLACK OUT OF COLORS · explanation of the rainbow · explanation of the permanent colors of natural bodies
II	1	"Observations concerning the Reflexions, Refractions, and Colours of thin transparent Bodies"
	2	"Remarks upon the foregoing Observations"
	3	"Of the permanent Colours of natural Bodies, and the Analogy between them and the Colours of thin transparent Plates" · FITS OF EASY REFLECTION AND EASY TRANSMISSION
	4	"OBSERVATIONS CONCERNING THE REFLEXIONS AND COLOURS OF THICK TRANSPARENT POLISH'D PLATES"
III	1	"Observations concerning the Inflexions of the Rays of Light, and the Colours made thereby" · i.e., DIFFRACTION "Since I have not finish'd this part of my Design, I shall conclude with proposing only some Queries, in order to a farther search to be made by others." · QUERIES

Subjects in SMALL CAPITALS are those added by Newton after his major round of optical researches in 1666-1670.

698   As the schema shows, after 1670 there were only four subjects which Newton examined for the first time (in the early 1690s mostly) or on which he arrived at novel conclusions.

The first is *color mixing*. In 'New Theory' of 1672 he emphasized the infinite number of colors in the spectrum. Without giving up that view, in *Opticks* he did his best to meet halfway such objections as stemmed from the habit painters had developed over past centuries to regard only a few colors (including at least red, blue, and yellow) as primary and the remainder as composed of them (this, of course, concerned pigments, not rays, but not until Helmholtz was their mixing found to obey distinct rules). After all, as Huygens had pointed out, white light may be composed of fewer colors than all of them together. Although obliged to concede the point while insisting that the white that results from the mixing of only a few colors is still different from the white light that emerges from the Sun, in *Opticks* Newton went ahead to elaborate his earlier, sevenfold division of colors (in analogy with the octave) in the direction of rules for how and to what effect to combine them.

Next, *'fits'*. In *Opticks* Newton recounted the tenacious measurements by means of which in 1670 he established the periodic nature of 'Newton's rings'. He now went on as follows:

> Every Ray of Light in its passage through any refracting Surface is put into a certain transient Constitution or State, which in the progress of the Ray returns at equal Intervals, and disposes the Ray at every return to be easily transmitted through the next refracting Surface, and between the returns to be easily reflected by it ... the returns of the disposition of any Ray to be reflected I will call its *Fits of easy Reflexion*, and those of its disposition to be transmitted its *Fits of easy Transmission*, and the space it passes between every return and the next return, the *Interval of its Fits*.[386]

Here as in later passages Newton treated these fits as a matter-of-fact property inferred directly from phenomena, not by way of an explanation on the microlevel. In historical reality they owed their introduction to his abandonment of his original explanation of periodicity. The methodological strictness he had in 1672 begun to impose upon himself as upon others prevented him from discussing causal hypotheses on the same level as the phenomena they were meant to explain. So the drastic alteration that occasioned the appearance of these fits remained hidden below the surface. Actually, the momentous switch he made between 1679 and 1685 from an aether to forces was responsible for it (p. 645).

Diffraction, which Newton called 'inflection', was subjected in *Opticks* to an investigation yielding detailed measurements far beyond Grimaldi's original discovery of the phenomenon, with special attention given to the color bands that arise in the shadow projected by small, nail-like objects. Unlike with the other subjects treated in *Opticks*, Newton drew no inferences on the investigative plane intermediate between the bare facts as observed and/

or measured and such explanatory hypotheses as he felt free to divulge, if at all, only with a question mark.

Finally, *thick transparent polished plates*. The appearance of Newton's rings in spherical mirrors has failed to evoke much interest in the years since the *Opticks* but for a late acknowledgment of Newton's pertinent researches as the very culmination point of his investigative powers. Our concern here is rather the application of those powers to the entire *Opticks*. We shall now examine Newton's optical researches in sustained comparison with the *Principia* and also, as before, with the approach to the subject taken by the best among Newton's immediate predecessors, Huygens and Hooke.

Opticks *revolutionary as well?*

How does *Opticks* stand in relation to revolution 6, the one that Newton brought about by moving from aether to force and by applying his new concept of force to his new concept of universal gravitation? The *Principia* emerged as the product of revolutions 4 and 5 jointly transcended – in producing it Newton out-Hooked Huygens and out-Huygensed Hooke. Does an analysis along similar lines justice to *Opticks*? If so, did *Opticks reinforce* the revolutionary transformation thus attained or even *widen* it or just *adorn* it?

In a fleeting mood of generosity Newton granted to Hooke in 1676 that, if in matters optical he himself had seen farther, it was due to his standing on the shoulders of giants. And indeed, Hooke's account of light and color was what Newton had taken his original cues from – the general account, so as doubly to contradict it, and the thinness of colored layers, so as to measure it.

What Newton did in measuring it evidently went beyond the outer bounds of the approach that marked the Baconian Brew, which in its qualitative slant, too, Hooke's entire account exemplified as well as anything in his varied writings did. Huygens' identical response to the measurement issue – a mathematical device is at hand to circumvent the difficulty – serves to confirm that Newton's readiness to transcend this particular boundary of revolution 5 stems from his simultaneous participation in 4. The further historical fact that, in carrying out the measurement, Newton outdid Huygens by large stretches is due to more contingent circumstances. On a wholly personal level there is the incredible persistence in going to the bottom of things that singles Newton out, not just among his contemporaries, but among the vast majority of achievers of all times and places. But also, for Newton much more was at stake in the issue. To Huygens, motivated as always by a sense of 'I can do that better', all there was to the issue was to succeed where Hooke had declared himself beaten. To Newton, in contrast, the quantitative handle on optical issues that the different refrangibility of light had provided him with was quickly turning into the very core of his enterprise.

More circumspection is needed in establishing how Newton's rejection of Hooke's general account of light and color is related to the approach that marked the Baconian Brew.

700 Newton's adoption of forces among the ultimate constituents of the world, such as worked out in the *Principia* for universal gravitation, went ipso facto beyond the bounds of revolution 5. But with Newton's emission conception of light it was not so. Be it Cartesian pressure or Hooke-like pulses or (later) Huygens-like wave fronts or Gassendist or (now also) Newtonian emission of particles, the first principles of kinetic corpuscularianism were spacious enough to encompass all such conceptions. Characteristically, Newton's ever more ambiguous aether was in no respect more orthodox than how in 1675 he used it to explain the properties of light. His later switch to forces, while an act of liberation from a *Principia* point of view, rather acted as a setback to what was to become *Opticks*. Further, while it remains true that his emission account helped him hit upon his great discovery of the primacy of colors, neither did it *predispose* him to do so (after all, no earlier adherent had discovered it) nor was the find, once made, *necessarily* incompatible with a wavelike account. There is of course no way to know for sure what, on this specific issue, a joining of forces between Newton and Hooke would have accomplished – these two men who were on the same track so often, yet were so antagonistic in research style and in personality. Their imagined collaboration could not possibly have led to the modern theory, which depends crucially on Young's and Fresnel's discovery of interference. But with some luck it might have led to elimination of the retrospectively weakest aspects of the theories of each. In that manner a fruitful blend might have been attained between different refrangibility fearlessly accepted, periodicity squarely faced, and some wave mechanism underlying it all. What stood in the way of such an unrealized yet historically possible development was not so much the experiments adduced by Newton to confirm the immutability of colors – Hooke very characteristically believed that he had already performed them all. What stood in the way more than anything else was the big leap of the imagination required to attain or even to adopt Newton's paradoxical insight to that effect.

So much for the limited extent to which Newton went beyond the Baconian Brew considered from the point of view of its *content* – *qua* measurement, certainly; *qua* kinetic corpuscularianism, not at all. With the *knowledge structure* of revolution 5, as also with revolution 4, things are different. The whole point of revolution 5 pioneered by Boyle and Hooke and capped by the young Newton was to provide often haphazard, fact-finding experimentation with more of a sense of direction and also, conversely, to enhance the plausibility of explanatory mechanisms by linking them up with experiments. Similarly, the whole point of revolution 4 undertaken by Huygens and the young Newton alike was to harmonize the realist-mathematical and the kinetic-corpuscular approaches to those issues where the two proved incompatible – that is, where possible, to enrich a mathematical regularity with some definite mechanism of corpuscles in motion, or the other way round. *In both cases, the combination still permitted a freedom of speculation that came to strike Newton more and more as too fanciful, at least if exercised in public.* Newton offered something else instead.

That 'something else' was *quantity-driven, maximally coherent experimentation consistently*
reinforced, where at all possible, by precision measurement. Not that precision measurement
was quite unheard of at the time. Cassini's and Flamsteed's fact-finding investigation of
heavenly bodies by means of the telescope (p. 451) and also Starkey's earlier and Newton's
own, slightly later chemical/alchemical researches (p. 466) were cases in point. Newton's
optical work added two things to precision measurement of such a kind. One was the imaginative ingenuity he displayed in squeezing the maximum attainable precision out of experimental situations well suited to sending their creator beyond the merely qualitative. The
other was his readiness to regard their outcomes as indissolubly bound up with such more
general inferences about experimentally investigated states of the macroworld as he was
moved to draw from them.

Negatively, that readiness was born of the very impatience with which he regarded such
out-of-control speculation as he found rampant among his contemporaries, with Hooke
serving as its perennial exemplar and Huygens as the best among a flock of unrepentant adherents. It was this sentiment that made Newton dismiss as just too fanciful any hypothesis
accepted solely (as he wrote to Oldenburg in a draft version of his 'New Theory' that did
not make it into the *Philosophical Transactions*) "because it satisfies all phaenomena (the
Philosophers universall Topick)."[387]

Positively, Newton's readiness to regard inferences from his precision measurements as
conclusive rather than just plausible found expression in phrases like the following, written
to Oldenburg half a year later:

> I told you that the Theory wch I propounded was evinced to me, not by inferring tis thus
> because not otherwise, that is not by deducing it only from confutation of contrary supposi-
> tions, but by deriving it from Experiments concluding positively & directly.[388]

What Newton desired of a scientific assertion was *not* to have it bound up with some inherently unascertainable and, at its frequent worst, wholly fanciful microworld mechanism.
Instead, he wanted to have it ascertained *either*, where feasible, by means of mathematical
demonstration experimentally tested *or* by means of precision measurement experimentally
undertaken as the only acceptable alternative. From the start of his optical investigations, the
unheard-of property of different refrangibility which he first conceived theoretically, then
established empirically, put Newton in possession of a ready means to go beyond the purely
experimental and turn it into precision measurement. The revolutionary transformation he
brought about was wider than the one embodied in the *Principia* alone; what he added to it
in *Opticks* was another, subsidiary yet independent component of valid science.

The differences between the two books are correspondingly vast. Finding a satisfactory
dispersion law would have enabled Newton to turn the latter into more of a mathemati-

702   cal treatise than in the end he could make it. His failure to find that law compelled him to fall back on the Baconian style of presentation that he had adopted in 1672 to enhance the eligibility of his 'New Theory' for publication in *Philosophical Transactions*. But even if he had found it, *Opticks* would still have remained a less rigorous treatise than its predecessor. That is why Newton preferred the earlier book over the later, as apparent from the care he bestowed on their respective reeditions. What the two books (and these alone among the multitude of treatises resting unpublished in his desk drawer) shared nonetheless was a capacity to make their respective cases clinching. As far as Newton was concerned, the age of fanciful science was over.

Two fragments and one whole

The problem was, many of those treatises resting unpublished in his desk drawer were filled to the brim with fanciful science. And so pressing was Newton's urge to speculate on what constitutes our world and holds it together that he felt unable fully to stick to his stern methodological requirement of keeping a science of properties separate in public from a natural philosophy of unproven connections and underlying mechanisms. This inability manifested itself in a variety of ways, now to be listed on a scale of increasingly public communication, as follows:

- *incidental, private communication*, as in the sketch of his current worldview that he sent to Boyle in 1679;
- *ongoing drafting, followed by cancellation*, of prefaces and conclusions to his two books (or, in one case, the drafting of Book IV of *Opticks*, followed by its abandonment while incomplete), with phrases from and allusions to these drafts sometimes making it into the published books after all;
- *having his current worldview (larded with 'perhapses' and 'it-may-be-thats') read in formal session* of the Royal Society, yet expressly *not* published in the *Philosophical Transactions* (the 1675 'Hypothesis of Light');
- *ostensibly noncommittal publication* of vital aspects of his still evolving view of the world in the format of 'Queries' (final portion of the *Opticks*).

Since the mid-1970s the pioneering labors of a generation of historians of Newtonian science have alerted the scholarly world to the full scope of Newton's work of a lifetime. The *Principia* and *Opticks* appear as two fragments of a much larger whole, now visible in a detailed manner foreclosed to either his contemporaries or posterity. But what contemporaries and posterity, faced with hardly more than the two books alone, did have a chance to catch was the intriguing fragment *of* the whole they encountered on the final sixty-or-so pages of the *Opticks* and in one celebrated passage inserted in the second edition of the *Principia* (the

'Scholium generale'). What they could pick up most clearly from among the 'Queries' was
the mechanisms Newton thought to underlie the properties of light and color he had been
treating in the preceding pages of *Opticks* – naturally, the 'Queries' were meant primarily to
elucidate the book's principal topic. Still, in one particular Query he added in 1706 to the
first Latin edition, finally to be numbered Query 31, Newton went far beyond optical phenomena and revealed more elaborately than anywhere else in his published work his current
worldview as reinforced by a generous listing of phenomena he felt to sustain it. What still
remained hidden from sight until the second half of the 20th century, however, was the extent to which both his alchemical researches and his theological views that derived from his
rejection of the Trinity impinged on his evolving conception of the world.

The threefold task that remains, then, is to examine successively
- *light and color in the world* (i.e., Newton's views on how light acts in the world and his
  successive micro-world explanations of the properties of light and colors);
- *the whole AND its fragments* (i.e., the totality of Newton's life's work, considered both as
  such and in relation to the two books to rise above the surface);
- *the fragments AS a whole* (i.e., what the Newtonian synthesis, as made up of *Principia* and
  Opticks alone, looked like in its own time).

Newton's unceasing vacillation over how best to deal with his speculations, plus the fundamental revision these underwent between 1679 and 1685, plus the very long gestation period
of *Opticks*, in which to some limited extent his speculations finally saw the light of day,
greatly confuse the chronology of his work. Schema 10 serves to clarify things.

Schema 10: Newton's speculations on the whole frame of nature

	up to 1679 (aether)	*after 1685 (forces)*
Private communication	• Letter to Boyle (1679; p. 647)*	• Letters to Bentley (1692/3)*
Canceled drafts	• Full response to Hooke (1672; p. 695)	• Conclusion to *Principia* (1686; p. 672) • Book IV, *Opticks* (1691-3; p. 697)
Public reading alone	• 'Hypothesis of Light' (1675; pp. 561; 642)*	
Noncommittal publication		• 'Queries' in *Opticks* (1704–1717) • 'General Scholium' in *Principia* (1713)

insofar as reaching out (prior to the 1930s) to other people than Newton himself.
* Texts listed with an asterisk were published posthumously by the mid-18th century.

704   *'Light and Bodies'.*   Query 5 offers Newton's general expression for light as a phenomenon in the world:

> Do not Bodies and Light act mutually upon one another; that is to say, Bodies upon Light in emitting, reflecting, refracting and inflecting it, and Light upon Bodies for heating them, and putting their parts into a vibrating motion wherein heat consists?[389]

In a range of further Queries Newton elaborated this lapidary statement. Take first the connection between light and heat, and recall that in his 1669 account of 'the vegetation of metals' he had hesitatingly identified the vegetative spirit (the alchemical principle he introduced as a necessary complement to particle action) with "ye body of light". In support he had argued there that light and the vegetative spirit share "a prodigious active principle" and are alike "perpetual workers"; also, "all things may bee made to emit light by heat" (p. 644). That is, light is not just a product of certain particle motions but is among the active principles in the world that help sustain life ('vegetation').

Carefully shorn now of its alchemical connotations, this conception of light as "the chief principle of activity in [bodies]"[390] found more elaborate, public expression in a range of Queries focused on a variety of connections between heat and light as marked by fire and flame, in particular. What he went on to outline in some empirical detail was all kinds of experiment-induced or everyday events, between which he jumped indiscriminately to and fro. Examples are the exceptional proneness to burning of bodies composed in part of sulfur, the light that comes from the back of a cat rubbed in the dark, or the extraordinarily vehement explosion of salt of tartar when mixed with gunpowder and then gently heated. Treatment of the issue culminated in Query 30, which begins by 'asking' (here as with all Queries, the evidently intended answer is 'yes'),

> Are not gross Bodies and Light convertible into one another, and may not Bodies receive much of their Activity from the Particles of Light which enter their Composition?

Amid the empirical examples of such conversions that he went on to give, he allowed himself one passing allusion to their hidden alchemical background when arguing that

> the changing of Bodies into Light, and Light into Bodies, is very conformable to the Course of Nature, which seems delighted with Transmutations.[391]

Next, we examine those 'vibrations' that were also mentioned by Newton in Query 5, and the explanatory uses to which he put them. At first, in his 1675 'Hypothesis of Light', he explained by their means those properties of refraction and reflection that he had investigated so thoroughly between 1666 and 1670 (p. 561).

Central to that hypothesis was a very subtle aether that he held to pervade everything and to be made up of particles that vibrate swiftly and that aggregate in such a way that the larger the pores of gross bodies are in which they find themselves, so much the denser is the aggregate. Newton urged his audience to take neither this aether nor its vibrations for light itself. Still, whatever light might be (corpuscles emitted or something else), the aether serves in any case to explain its experimentally manifest properties by acting as an intermediary between light and the bodies encountered along its way. The density gradient serves to explain refraction, in that a larger amount of aether pressure arising from denser aggregation makes the ray bend more; also, in similar fashion, the density gradient explains diffraction (which Newton always regarded as a special kind of refraction). Since reflecting surfaces are never quite smooth, reflection does not proceed through mere impact of the substance of light upon the surface particles but is mediated by the aether that stands in the pores between particles. Finally, the vibrations set up in the aether by (as Newton saw it in private) the blows of light corpuscles upon it serve to explain all those phenomena of light and color Newton had sensed and/or confirmed to be periodic, notably 'Newton's rings'. These explanations were not just ad hoc inventions or *a posteriori* justifications. They had actually been of great help to Newton in coming to grips with the phenomena themselves, most notably so in his ingenious measurement of the thickness of a thin film at the red and violet endpoints of just one colored ring (p. 691).

Inherently satisfying as all these explanations were to Newton, there was one problem with them. When between four and ten years later his sophisticated pendulum experiment gave him the final push and he dropped the aether, he *ipso facto* lost the carrier of those vibrations of his. For his explanation of refraction, diffraction, and reflection his otherwise momentous shift from aether to forces was really innocuous – what pressures and smoothing out accomplished in 1675, a variety of attractions and repulsions accomplished in the Queries. For instance, he now explained the difference between colors as follows:

> Nothing more is requisite for producing all the variety of Colours, and degrees of Refrangibility, than that the Rays of Light be Bodies of different Sizes, the least of which may take violet the weakest and darkest of the Colours, and be more easily diverted by refracting Surfaces from the right Course; and the rest as they are bigger and bigger, may make the stronger and more lucid Colours, blue, green, yellow, and red, and be more and more difficultly diverted.[392]

For colored rings, however, and for the vibrations Newton had from early on held to underlie them, no such solution was anywhere in sight. From 1685 onward, four ways out of this quandary stood open to him.

He could conveniently forget the property of periodicity for which he had introduced those vibrations in the first place. This was the course subconsciously taken by Huygens

706   when, in the advance toward his eventual *Traité de la lumière*, he felt compelled for quite different reasons to deny periodicity to the wave pulses central to his explanatory wave front mechanism. To Newton, who had plunged himself so much more deeply into the measurement of colored rings, that way out was barred (p. 691).

Or he could drop his emission account of light and exchange it for a pulse account, with the pulses (unlike with either Huygens or Hooke) made periodic. That way was barred, too, due to his standing objection that a pulse account is incompatible with rectilinear propagation (a point he took care to make once again in the course of a range of Queries directed specifically at contradicting Huygens' *Traité de la lumière*).

Or he could assign his vibrations, no longer to the vanished aether, but, however implausibly, to the transparent medium. This he did in one proposition in part 3 of Book II.

Or – and this is the principal line he took when in the mid-1690s he was completing the text of *Opticks* – he could clothe desperation in a label by adorning the periodic phenomena he was no longer able to refer to a proper vehicle with the dignified name of 'fits of easy reflection and easy transmission'. In parts 3 and 4 of Book II he handled their appearance on the phenomenal level, going on in Query 29 to account for them as follows:

> Nothing more is requisite for putting the Rays of Light into Fits of easy Reflexion and easy Transmission, than that they be small Bodies which by their attractive Powers, or some other Force, stir up Vibrations in what they act upon, which Vibrations being swifter than the Rays, overtake them successively, and agitate them so as by turns to increase and decrease their Velocities, and thereby put them into those Fits.[393]

If still in his years of creativity, he might well have taken the very oddity of those fits, and the impossibility of making the vibrations giving rise to them vibrate in anything at all, for an occasion to rethink his underlying conception of light from scratch. Instead, it stimulated him only to come up in the end with an aether of another kind – one at least nominally compatible with three cornerstones of the *Principia*, to wit, (a) his force of universal gravitation and the measure thereof, (b) his sophisticated pendulum experiment (the final conclusion of which he characteristically dropped in later editions), and (c) his sustained argument against Cartesian whirlpools.

The aether, then, with which the seventy-five-year-old Newton filled his last eight Queries in 1717, was born, in part from ongoing worries over those fits of easy reflection and easy transmission, in part from experiments with electricity in the 1710s that Francis Hauksbee undertook for the Royal Society under Newton's presidential guidance and that impressed him greatly. The biggest incentive, though, was a compromising spirit lately come to the erstwhile solitary rebel now comfortably established in London's high circles. The principal task of this new aether was belatedly to accomplish what Huygens, Leibniz, and other critics

of the *Principia* had required him to do from the outset: to account for his forces in terms
of an underlying mechanism of particles in motion. As such, however, the new aether could
not but be a dismal failure. Due to requirements (a)–(c) just listed, the ratio of its formidable
elasticity to its minimal density that he duly calculated came out such as in the end to make
the new aether account for the motions of particles by means of their mutual repulsion –
hardly the way to reduce the action of forces to sheer matter in motion!

The above account covers all but one of the Queries in the various editions that *Opticks*
went through during Newton's lifetime. That one remaining Query, the very last in number
though not in age, was the most intriguing of them all. It attractively opened thus:

> Have not the small Particles of Bodies certain Powers, Virtues, or Forces, by which they act at
> a distance, not only upon the Rays of Light for reflecting, refracting, and inflecting them, but
> also upon one another for producing a great Part of the Phaenomena of Nature?[394]

And that is how in Query 31 Newton set out over some thirty pages to outline his conception of the world, to the extent that he felt that conception to be fit for public consumption
at all, even with a question mark added. That is, we are now ready to examine the larger
Newtonian synthesis.

The whole AND *the fragments: The larger Newtonian synthesis.*   Query 31 was
never more than just an abstract of the never completed Book IV of *Opticks*, with esoteric
alchemy and heretical theology carefully left out to boot. Even so it provides a useful guide
through some intricacies of Newton's conception of the world and the alterations it underwent along the way.

Take, first, Newton's lifelong focus on phenomena at bottom incompatible with kinetic
corpuscularianism in any of its standard versions: surface tension in fluids, capillary action,
the cohesion of bodies, the expansion and pressure of air, elective affinities, explosive heat
produced in certain chemical reactions. In the 1675 'Hypothesis of Light' these phenomena
served as his principal evidence for his aether (p. 561). Now, in Query 31, they came to serve
as his principal evidence for such microforces as he now held the world to be filled with –
some specific, others universal; some attracting, others repelling; but all operating over short
distances only. The cohesion of bodies is now explained through mutual attractions of the
particles composing them, whereas the almost unlimited expansion of air is now due to mutual repulsion. And here is how Newton now accounts for the capacity of those two slightly
different acids to dissolve gold and silver, respectively, in such curiously inverse ways:

> When *Aqua fortis* dissolves Silver and not Gold, and *Aqua regia* dissolves Gold and not Silver,
> may it not be said that *Aqua fortis* is subtil enough to penetrate Gold as well as Silver, but wants

the attractive Force to give it Entrance; and that *Aqua regia* is subtil enough to penetrate Silver as well as Gold, but wants the attractive Force to give it Entrance?[395]

If almost literally the same evidence (rendered only in more intricate detail in the Query than in the Hypothesis) could within thirty years serve to draw radically different conclusions, how different, then, were these conclusions really? What, in the shift from aether to forces, had really changed? Just labels? Or anything substantial after all?

What had substantially changed in the first place was Newton's conception of matter. True, in 1675 as in 1706 and beyond, Newton found it "probable" that (the way he phrased it in Query 31)

> God in the Beginning form'd Matter in solid, massy, hard, impenetrable, moveable Particles, of such Sizes and Figures, and with such other Properties, and in such Proportion to Space, as most conduced to the End for which he form'd them; and that these primitive Particles being Solids, are incomparably harder than any porous Bodies compounded of them; even so very hard, as never to wear or break in pieces; no ordinary Power being able to divide what God himself made one in the first Creation. While the Particles continue entire, they may compose Bodies of one and the same Nature and Texture in all Ages: But should they wear away, or break in pieces, the Nature of Things depending on them, would be changed.[396]

But the big difference that the abolition of the aether carried with it in this respect was the immense rarity, not only of space between heavenly bodies, but of all bodies. In paraphrase, "the matter of bodies is highly textured – not a pile of stones, but a crystal lattice," or, in Newton's own, highly significant analogy, matter is formed "more retium" ('on the pattern of nets'). That is, the 'net' that had become central to Newton's alchemical experimentation (p. 468) came in due time to govern his conception of matter as such. As evidence for it, Newton invoked the transparency from every angle of an incompressible substance, water, itself far more attenuated already than the densest substance, gold (which in view of its capacity to absorb mercury cannot be entirely massive). Light going through water would on striking a solid particle be deviated from its path. "How bodies may be sufficiently porous to give free passage every way in right lines to the rays of light is hard to conceive but not impossible," Newton argued in one manuscript passage out of which Query 31 was eventually composed. Some calculation of the ratio of pores to particles led him to regard matter as (in another striking paraphrase) "diaphanous threads tenuously strung in a three dimensional net".[397]

If so-called massive bodies themselves are rare already, so much the rarer is outer space between heavenly bodies. With the aether exchanged for forces, those forces are bound to act at a distance. In the very first sentence of Query 31 Newton phrased their action in precisely

this way. But what did it mean? How is action at all possible in the absence of an apparent   709
carrier for it?

In 1692, one year after Robert Boyle died, a formidable Cambridge theologian/philologist, Richard Bentley, was charged with delivering the first of the Boyle lectures. These lectures were meant by the man who bequeathed the funds for them to have atheism confuted annually by means of the argument from design (p. 437). Bentley decided to marshal Newton's *Principia* for the task. Daunted a little by the circumstance that he could not understand the book, he directed himself to Newton with some pointed questions, interesting chiefly for the well-known answer Newton gave to one of them:

> Tis unconceivable that inanimate brute matter should (wthout ye mediation of something else wch is not material) operate upon & affect other matter wthout mutual contact. ... That gravity should be innate inherent & essential to matter so y^t one body may act upon another at a distance through a vacuum wthout the mediation of anything else by & through wch their action or force may be conveyed from one to another is to me so great an absurdity that I beleive no man who has in philosophical matters any competent faculty of thinking can ever fall into it. Gravity must be caused by an agent acting constantly according to certain laws.[398]

The precise nature of the immaterial agent here invoked was in the end revealed by Newton in fleeting passages in Queries 28 and 31 and also, in a hardly less obscure manner, in the General Scholium he added to the second edition of the *Principia*. Serving truly as "the ultimate foundation of Newton's conception of nature,"[399] the immaterial agent proved to be none other than God, enabled to the task by infinite space being His sensorium. Now what does that mean?

As Newton, always so shy to share his most intimate thoughts with others, made much clearer in some of his private papers, the analogy is with human perception and locomotion:

> Life & will are active Principles by wch we move our bodies, & thence arise other laws of motion unknown to us. And since all matter duly formed is attended with signes of life & all things are framed wth perfect art & wisdom & Nature does nothing in vain: if there be an universal life & all space be the sensorium of a thinking being who by immediate presence perceives all things in it ... the laws of motion arising from life or will may be of universal extent.[400]

That is, all those attractions and repulsions between bodies that serve to make our world more than a mere heap of inert matter take place, neither immediately (for no greater absurdity is possible) nor mediated by a material aether (for the sophisticated pendulum experiment and the argument against Cartesian whirlpools have conclusively shown that no such aether exists), but mediated by infinite, omnipresent God. The God Who thus at long last made His fleeting yet so much the more significant appearance in Newton's published work is not the God Newton had portrayed in his private, anti-Trinitarian treatises. Still, in His reason-drenched, prime attributes of omnipresence and omnipotence this God is a strikingly Christ-less God, Who if asked could well do without the Trinity Newton begrudged Him in private. Theology apart, what the above implies is that "the universal presence of God performed all the functions that Newton ascribed to the aether in the 'Hypothesis of Light.'"[401] Which sober observation provokes all over again and in somewhat different form the question of what, in the switch from a material to an immaterial aether underlying all physical action, has *really* changed.

The question sends us back to the nature and attributes of those forces that from 1685 onward had become the primary agents in the world as Newton conceived it.

In the first place (to phrase a query about a Query), is it not evident that, in their very specificity, all those forces conjured up in Query 31 suffer from an innate tendency to become as numerous as the phenomena they are invoked to explain? Take, as one example, Newton's explanation of the taste of acids – "do not the sharp and pungent Tastes of Acids arise from the strong Attraction whereby the acid Particles rush upon and agitate the Particles of the Tongue?"[402] Compare this with Lucretius' explanation by means of sheer particle shape, and with Descartes' by means of the 'agitation' that likewise spike-shaped particles bring about in stinging the tongue (p. 228). What has been gained other than that one plausible-looking explanation has been exchanged for another?

The letter to Bentley further confirms that Newton never regarded those forces of his as self-explanatory. Near the end of Query 31 Newton faced squarely the objection that Huygens and Leibniz placed before him long ago: he had perversely reintroduced occult forces into a mode of nature-knowledge happily liberated from them. To this Newton countered that he left the underlying causes of those forces to be discovered by posterity, but that he need not in their absence abstain from here and now inferring their reality from phenomena. From what phenomena? For the macro-force of universal gravitation that attracts over large distances, from the phenomena of motion handled in *Principia*. For the micro-forces that attract or repel at shorter distances, from those adduced in Query 31. Still, even apart from their sizes and their ways of acting, there was of course one big difference between these forces, marked by the fact that *Principia* is a mathematical treatise and Query 31 is a verbal disquisition. What singled out universal gravitation from all other forces that Newton believed to exist was that *he had found the mathematical law obeyed by it.*

Not that Newton had failed to make due mathematical preparations for those micro-forces. He had done so in Book I of the *Principia*, by means of the variety of possible force rules he established far beyond the inverse-square rule that he needed for universal gravitation (p. 666). Nor had he failed to seek to subject those micro-forces to quantitative regular-

ity. With two micro-forces he even managed to derive a crude measure from such motions   711
as he held to come from them. One was the attraction exerted by light corpuscles, which he
estimated "by comparing measured velocities and radii of curvatures of inflected paths".
The other was capillary action in the case of a drop of orange juice in its accelerated motion
toward the joined endpoint of two glass sheets slightly tilted toward each other.[403] However,
the bulk of Newton's micro-forces, be they magnetic or electric or chemical or whatnot,
withstood with retrospective inevitability any such attempt at quantification however crude,
let alone at catching any one of them in the net of any one among his ready-made force rules.
Still, none of this is the main point. What in Newton's view distinguished forces (whether
successfully subjected to mathematical and/or quantitative treatment or not) from aether
of any kind was their capacity to be subjected to such treatment *in principle*. "What the new
philosophy of nature made possible was less the concept of force than its precise mathematical formulation."[404]

Does it make sense to adorn these conceptions of Newton's with the name of 'philosophy
of nature'? Earlier I called Johannes Kepler the creator of a 'hybrid philosophy of nature'. By
this I meant lots of *a priori* speculation unusual in his own time solely for its idiosyncratic
slant being followed up in wholly unprecedented manner by assiduous *a posteriori* checking
(pp. 211; 702). In that particular sense, Newton's philosophy of nature in its successive guises
was likewise a hybrid. He and Kepler shared a vision of the Creator as supreme Mathematician and a consequent desire to investigate the mathematical constraints set to Creation.
One difference was that this was Kepler's prime objective, whereas for Newton it was an
added boon at the endpoint of where his mathematical researches had carried him. Also different was that Newton had very little time for the archetypal harmonies so dear to Kepler. A
further difference was that with Kepler the assiduous *a posteriori* checking was observational
only, whereas with Newton it was also experimental. Still, both men held in common that
speculation was closely intertwined with the mathematical outcomes. But what was different above all was that, with Kepler, their decoupling was performed by posterity (more
specifically, by Horrocks), whereas with Newton, in reluctant obedience to his own methodological strictures, it was performed by and large by himself.

So much for the whole of Newton's conception of the world and how *Principia* and
Opticks express two portions of that whole. Little doubt is possible that, in the assumed absence of the underlying whole, the two portions would not have come into being. Wherever
the paths of discovery led Newton, experimental investigation and mathematical derivation
were spurred on by his aether speculations first (light and colors) and then by the shift
from aether to forces (*Principia*). Here, too, lies the deepest reason for all the many things
that set Newton apart from even the very best of his near contemporaries, Huygens, Boyle,
and Hooke. Distinctive of Newton is the scope and tenacity of his search for an integrated
conception of the world. To him, light as such and motion as such carried so much deeper

712 and richer meanings than (as with Huygens, in particular) just serving as food for exercising one's mathematical and/or experimental ingenuity. The 'positive' outcome of Newton's assiduous speculation about the whole frame of nature, then, is well-nigh inconceivable in the imagined absence of the fertile yet retrospectively muddy soil on which it grew. To that extent Keynes (p. 605) was right – the full handwritten legacy does serve to identify Newton as 'the last of the Babylonians'.

But only to that extent.

For what (other than for the historian) counts in the end is the neat separation Newton's methodological strictness compelled him, most of the time, to maintain between the speculative soil and the scientific fruit. That methodological strictness left him with two criteria only, those of mathematical law experimentally confirmed and of precision measurement experimentally established. Together, these two criteria made for what rightly goes by the name of 'the Newtonian synthesis'.

The fragments AS a whole: The smaller (but historically most significant) Newtonian synthesis.   Newton's published accomplishment may sensibly be called a synthesis in at least four respects: (a) heaven and Earth, (b) matter and force, (c) revolutionary transformations 4 and 5, (d) checks upon fancy. Items (a) and (b) involve *Principia* alone; items (c) and (d) *Principia* and *Opticks* alike. Each item has been treated here at length, so all we need still is a summing up placing it all in the wider frame of the Scientific Revolution.

Heaven and Earth were generally regarded in pre–Scientific Revolution times as an Earth-centered unity, even though in the most detailed and most widely accepted doctrine of heaven and Earth, Aristotle's, considerable divergences were maintained between the supralunar and the sublunar. Copernicus' proposal to regard Earth as one of the planets, if taken in the fully realist sense almost no one took it in prior to Kepler and Galileo, called for a unity both more closely knit than Aristotle's and reconstituted on a non-Earth-centered basis. Such a reconstitution was achieved in a qualitative vein by Descartes; yet to those ready to accept Kepler's laws for planetary trajectories and Galileo's rules for bodies falling to the Earth, reconstitution actually receded farther from sight because any apparent coherence between those mathematical rules was lacking. In deriving both the former and the latter from the inverse-square law for universal gravitation, Newton took up the terrestrial and the heavenly domains in one, no longer Earth-centered unity. Among the various respects in which Newton wrought a synthesis, *qua* content this one was the most important.

Further, universal gravitation blended, on the one hand, the *universality* of just one homogeneous matter known of old in atomism and resuscitated in kinetic corpuscularianism and, on the other, those *specific* forces rampant in the magic-drenched mode of coercive

empiricism yet excluded for that very reason from kinetic corpuscularianism (p. 552). Again,
Principia carried the synthesis, a force universal in its action upon all of matter.

As a youngster, Newton joined the pioneers of the two revolutionary modes of pursuit of nature-knowledge embodied in the geometrization of corpuscular motion and in the Baconian Brew, respectively. He quickly went on to excel in everything he touched there. Later, in *Principia* as in *Opticks*, Newton even managed creatively to transcend both modes. This had vast consequences for the later careers and reputations of all four men who from the 1650s onward had done importantly innovative work therein. Newton's own, almost entirely unpublished work in these two modes was now made obsolete. So was Huygens' achievement in what was henceforth called mechanics – it was doomed to live on in a frame quite foreign to how he had originally attained his results. So was the bulk of Boyle's chemical treatises and Hooke's entire life's work, as revolution 5 was wiped out by the Newtonian synthesis entirely. And the only contribution to revolution 4 to find itself neither absorbed nor invalidated, Huygens' wave front explanation of reflection and of refraction both ordinary and strange, was bound to appear henceforth as an anomaly, since the proper frame in which to comprehend it had meanwhile been lost from sight. Only in quite another context, such as arose more than a century later at the hands of Young and Fresnel, could it be revived in a mostly altered frame brought about by their novel concept of interference.

Finally, the methodological strictures to come out of Newton's lifelong campaign against 'fansying' form a synthesis between, on the one hand, the indubitable certainty claimed of old by natural philosophy and, on the other, such checks upon its actually unbounded fancy as had meanwhile been worked out in the revolutionary modes of mathematical-experimental and of fact-finding experimental knowledge of nature. Common in those checks was their groping and (particularly with the latter mode) rather tenuous character (p. 617). What Newton achieved in *Principia* was a more rigorous connection than ever before attained between abstract mathematical rules and their real-world embodiment, by means of the 'Newtonian style' (p. 672). What Newton achieved in *Opticks* was to take much of the customarily haphazard mode of advance arising from the perceived whimsy of nature out of fact-finding experimental science by anchoring it wherever he could in quantity-driven experimentation and precision measurement.

What Newton felt free to claim for both his own mathematical science and his own experimental science was *certainty*. He reluctantly bought that certainty at the cost of a rigorous separation, in that mathematical law bound to phenomena through empirical or even experimental validation, as well as experimental inferences from phenomena, ought to be kept rigorously apart from anything beyond these two, uniquely valid modes of scientific inference. His elder contemporaries, Huygens, Boyle, and Hooke, had sought to rein in the near-unbridled fancying they perceived the natural philosophy of kinetic corpuscularianism

714   to be innately subject to in its customary, dogmatic mode. They had done so by linking hypothesized mechanisms of particles in motion with either mathematical rule (Huygens) or with facts found the experimental way (Boyle, in his chemical work especially; also Hooke). What came out of such exercises was the kind of heightened plausibility each of these men claimed for his own attainments. *To Newton almost from the outset, such plausibility was not enough.* Rather than regarding it as an imperfect yet effective check upon unbridled fancy, as (with Descartes in mind) its pioneers saw it, Newton regarded it as definitely less than the scientific enterprise deserved. What he requested was a reconstitution of certainty – certainty at quite another level, to be sure, than the hoary fake provided by dogmatic natural philosophy. Realist-mathematical science, such as pioneered by Galileo and Kepler, and fact-finding experimental science, such as pioneered by Bacon and others, did offer the means to bring about such certainty, if cultivated in the more pointed manner of *Principia* and *Opticks*, respectively. Everything else ought, however reluctantly, to be scrapped. This was not to be the last methodological word in science, particularly in its insistence on nothing less than certainty. Still, in the manner that it was not just proclaimed but actually embodied in Newton's two books, it did serve posterity as an exemplar and a measure for strictness that posterity, even if unwilling or unable to emulate it to the full, could not in good scientific conscience ignore.

Even disregarding the vicissitudes of customarily short (not to mention nasty and brutish) 17th-century life, there were four moments in Isaac Newton's life when he was almost lost to his native calling, the pursuit of nature-knowledge. Only a perceptive uncle and a likewise-perceptive schoolmaster saved the adolescent from a life spent in the countryside, far away from the learning he craved. In defiance of the rules of seniority that obtained at Trinity College, patronage twice carried the student through the successive elections that decided on distribution of its fellowships. Finally, patronage made the king absolve the Lucasian Professor from taking the obligatory holy vows Newton conscientiously felt to be incompatible with his secret anti-Trinitarianism. What if any of these four events had turned out differently? That is, what would have befallen the Scientific Revolution if Newton had not come to cap it?

There is of course no way to tell for sure. But call to mind for a moment the final conclusion drawn by our outside reporter about two well-perceptible weaknesses in the state of nature-knowledge by 1684 (p. 620), and compare it with the fourfold dimensions of the Newtonian synthesis. We cannot know whether, in the imagined absence of that synthesis, the Scientific Revolution would have petered out after all, as we made our imaginary observer suggest as a serious possibility in view of the lack of a central focus to mathematical-experimental and fact-finding experimental science and also in view of the apparent persistence of the wide gap between the promise of improving human life through science and its actual delivery. But we do know for sure that the added boost the Newtonian synthesis

was bound to give to a more focused and, as such, ever more recognizably modern science 715
and to the forces that propelled it forward did much to keep the enterprise alive over those
upcoming decades when scientific technology was not yet there to cement it.

NOTES ON LITERATURE USED

Sources. Unlike with Kepler, Galileo, Descartes, Bacon, Huygens, Boyle, and numerous lesser contributors to the Scientific Revolution, for Newton (as for Hooke) there are no 'Collected Works'. Still, his *Mathematical Papers* have been edited in 8 volumes by Derek T. Whiteside (Cambridge UP, 1967-1981); his *Optical Papers* in 1 volume to date (two more projected) by Alan E. Shapiro (Cambridge UP, 1984), and his *Correspondence* in 7 volumes by H.W. Turnbull, J.F. Scott, A. R. Hall, and L. Tilling (Cambridge UP, 1959–1977). I. Bernard Cohen edited a collection of facsimile reprints, *Isaac Newton's Papers & Letters on Natural Philosophy*. Cambridge UP, 1958. The *Principia* has found its standard edition in *Philosophiae Naturalis Principia Mathematica. The Third Edition (1726) With Variant Readings Assembled and Edited by Alexandre Koyré and I. Bernard Cohen, With the Assistance of Anne Whitman*: Cambridge UP, 1972. The widespread, unreliable Motte/Cajori translation has been superseded by *The Principia. Mathematical Principles of Natural Philosophy. A New Translation by I. Bernard Cohen and Anne Whitman*. University of California Press, 1999. The manuscript papers on motion that preceded the *Principia*, as well as translations thereof into English, are best accessible through John Herivel, *The Background to Newton's Principia. A Study of Newton's Dynamical Researches in the Years 1664–1684*. Oxford: Clarendon, 1965. The *Opticks* is still being read as a rule in an edition (based on the fourth and last one still revised by Newton himself: London, 1730) prepared and prefaced by I. Bernard Cohen (New York: Dover, 1952). Of Newton's chemical/alchemical and theological manuscripts relatively little has been printed yet (but see http://www.newtonproject.sussex.ac.uk).

'Newton knot'. To the extent that I am familiar with the vast literature to emerge from half a century of 'Newton industry', the (in most respects like-minded) interpretations by Richard S. Westfall and by Betty Jo Teeter Dobbs have come to appear to me as providing the most convincing picture of the man and his work. Westfall's picture is in his *Never at Rest. A Biography of Isaac Newton*. Cambridge UP, 1980 (summed up in his lemma 'Newton' in W. Applebaum (ed.), *Encyclopedia of the Scientific Revolution from Copernicus to Newton*. New York/London: Garland, 2000); for the *Principia*, the run-up to it, and Newton's successive conceptions of the world the two final chapters of his *Force in Newton's Physics. The Science of Dynamics in the Seventeenth Century*. London: Macdonald, 1971, remain decisively important, too. Dobbs' picture is in her *The Janus Faces of Genius. The Role of Alchemy in Newton's Thought*. Cambridge UP, 1991 (summed up in part 1 of Betty Jo Teeter Dobbs and Margaret C. Jacob, *Newton and the Culture of Newtonianism*. Amherst NY: Humanity Books, 1998). More specific literature that I have used includes:

- **Principia**. Dana Densmore, *Newton's* Principia*: The Central Argument. Translation, Notes, and Expanded Proofs*. Santa Fe: Green Lion Press, 1995. I. Bernard Cohen, *The Newtonian Revolution. With Illustrations of the Transformation of Scientific Ideas*. Cambridge UP, 1980. J. Christiaan Boudri, *What Was Mechanical About Mechanics. The Concept of Force between Metaphysics and Mechanics from Newton to Lagrange*. Dordrecht: Kluwer, 2002; sections 2.5.4.–2.7. George E. Smith & Eric Schliesser, 'Huygens's 1688 Report to the Directors of the Dutch East India Company on the

Measurement of Longitude at Sea and its Implications for the Non-Uniformity of Gravity'. *De*   717
zeventiende eeuw 12, 1, 1996; p. 198-214. [NB: Not to saddle the reader with avoidable complications, I have in the main text ignored the intermediate stage of Newton's 'Lectiones de Motu' as if, after improving 'De Motu', he set out at once to write the *Principia* itself.]

•   ***Opticks***. A.I. Sabra, *Theories of Light from Descartes to Newton*. Cambridge UP, 1981 (1st ed.: 1967). Fokko Jan Dijksterhuis, *Lenses and Waves. Christiaan Huygens and the Mathematical Science of Optics in the Seventeenth Century*. Dordrecht: Kluwer, 2004. Casper Hakfoort, *Optics in the Age of Euler. Conceptions of the Nature of Light, 1700-1795*. Cambridge UP, 1995. Carl B. Boyer, *The Rainbow. From Myth to Mathematics*. New York: Sagamore Press, 1959 (reprinted in 1987 by Princeton University Press). Richard S. Westfall, 'Uneasily Fitful Reflections on Fits of Easy Transmission'. *Texas Quarterly* 10, 3, 1967; pp. 86-102. Alan E. Shapiro, *Fits, Passions, and Paroxysms. Physics, Method, and Chemistry and Newton's Theories of Colored Bodies and Fits of Easy Reflection*. Cambridge UP, 1993; idem, 'Artists' Colors and Newton's Colors'. *Isis* 85, 4, December 1984, pp. 600-630; idem, 'Newton's "Experimental Philosophy"'. *Early Science and Medicine*, 9, 3, pp. 185-217.

•   **Fragments and whole**. The question of whether Newton's life's work, the full handwritten legacy included, does constitute a unified whole is still deeply controversial among Newton scholars. Treatment here is mostly along lines drawn by Westfall (in particular pp. 377-400 of *Force in Newton's Physics*) and by Dobbs. On God as mediator between gravity and matter see also John Henry, "'Pray Do Not Ascribe That Notion To Me': God and Newton's Gravity", in J.E. Force & R.H. Popkin (eds.), *The Books of Nature and Scripture: Recent Essays on Natural Philosophy, Theology and Biblical Criticism in the Netherlands of Spinoza's Time and the British Isles of Newton's Time*. Dordrecht: Kluwer, 1994; pp. 123-47.

EPILOGUE

A DUAL LEGACY

Soon after his visit to the Academy of Lagado (p. 481), Captain Lemuel Gulliver availed himself of the opportunity granted him on the isle of Glubbdubdrib to converse with the dead. In 1726, one year before Newton's demise, he reported as follows:

> I then desired the Governor to call up Descartes and Gassendi, with whom I prevailed to explain their systems to Aristotle. This great philosopher freely acknowledged his own mistakes in natural philosophy, because he proceeded in many things upon conjecture, as all men must do; and he found, that Gassendi, who had made the doctrine of Epicurus as palatable as he could, and the *vortices* of Descartes, were equally exploded. He predicted the same fate to *attractions*, whereof the present learned are such zealous asserters. He said, that new systems of nature were but new fashions, which would vary in every age; and even those who pretended to demonstrate them from mathematical principles would flourish but a short period of time, and be out of vogue when that was determined.[405]

As before with the extraction of sunbeams from cucumbers, Jonathan Swift had it both right and wrong. With unfailing precision he now pointed at the very thing that, *pace* the captain, *had* altered for good between Aristotle and Newton, or even between Descartes and Newton. For the first time in history, the pursuit of nature-knowledge was no longer helplessly subject to change of fashion. 'Attractions', as embodied in Newton's discovery of the first of just four forces that ultimately underlie all nature's phenomena, have not 'flourished but a short period of time'. Other than with Aristotle's 'mistakes' or Descartes' 'vortices', universal gravitation has come to stay. So have the 'mathematical principles' that, in Newton's book of that name, establish 'attraction' and define its measure. It has now been more than three centuries since mathematical principles, in ongoing interplay with rigorous experimentation, have proven themselves capable of deciding in the end what in scientific research is just a matter of 'vogue' and what transcends fashion such that it can provide us with enduringly valid knowledge of nature.

'Enduringly valid knowledge' – is that article actually available on the market of ideas?

The 1980s have seen a return to ghostly 'Aristotle's' conception of scientific change as running on fashion. As subjective preferences shift, and the power relation between negotiating practitioners alters, so, allegedly, do our views of how the natural world is constituted

(p. xxi). This particular variety of the 'fashion' conception was first developed by 'sociologists of scientific knowledge', with an extreme wing claiming that the rules that scientists follow are just as arbitrary as those of baseball. Out of this came the so-called Science Wars, fought out between those who adopted this conception, albeit most often in some watered-down variety, and practicing scientists who felt their commitment to a lifetime of sometimes successful struggle with the very realities of nature to be unjustifiably wronged thereby.

Numerous historians of science for their part have since preferred in sophisticated accounts of some artfully contextualized episode to take a seemingly neutral position and just bracket the possible validity of the claim to truth so often made by our own protagonists, the scientists whose achievement we investigate. Unfortunately, the practice entails our losing sight of a distinctive, not to say decisively important aspect of our chosen subject: the apparent human ability to gain a reliable grasp of natural reality. After all, as Steven Shapin observed on the final page of his 1996 *The Scientific Revolution*, albeit without drawing any consequences from the observation, science is "certainly the most reliable body of natural knowledge we have got".[406] Is it not our historians' duty, then, to find out when and how science became so reliable in the first place?

Indeed, it is about time once again to look our protagonists' claim to truth squarely in the face. *A priori* reaffirmation of the validity of that claim, however, will not do. In the present book I have deliberately refrained from accepting without more ado the position of either party in the Science Wars. Instead, I have chosen to *historicize* the issue. For each of those six revolutionary transformations that I have found to make up together the sum total of the Scientific Revolution — Alexandria-plus, Athens-plus, fact-finding experimentalism, and three post-1660 blending efforts — I have investigated the empirical way whereby protagonists actually sought in their own practice to make their claims about natural reality stick. The outcome has turned out to be highly differentiated among these revolutionary episodes. Notably, far more robust procedures emerged in realist-mathematical than in fact-finding experimental science (p. 619). The culmination point of both was attained by Newton, in his ongoing fight against others' and his own 'fansying', in *Principia* and in *Opticks* (p. 713).

The differences between the various ways in which practitioners of each revolutionary mode of nature-knowledge sought to curb fancy are vastly intriguing. The inevitably groping manner in which the pioneers and their immediate successors probed such ways reveals the Scientific Revolution to have served, as it were, as a laboratory in which this wholly unknown terrain was first explored. How, between the Scylla of dogmatic assertion and the Charybdis of all-pervasive, skeptical doubt, to ferret out ways and means to anchor their novel findings in reality? And yet, for all the vast differences between these ways and means that remained over the 17th century, some ground was shared between them. Out of the Scientific Revolution came something of still wider import: *procedurally and institutionally established self-correction*.

Significant cases of self-correction manifested themselves right from the start of the Scientific Revolution. Very well known are Kepler's one-man decision to drop his own unprecedentedly accurate, painstakingly arrived-at hypothesis on the orbit of Mars in view of the tiniest of empirical discrepancies (the famous eight minutes of arc; p. 171) and Galileo's one-man realization that his original supposition of speed in vertical descent increasing with distance ran into a dead end and had to be replaced with an increase in proportion with time (p. 194). These events, to be sure, did not reflect changes of mood, let alone changes of fashion; rather, both men drew radical consequences from a dilemma that their own investigations presented them with. They were of course at liberty to face the dilemma or evade it or even ignore it, whatever way out might please them best, yet if faced it compelled the very correction both men went on to apply.

What was still a matter of private decision with the pioneers (albeit prepared to some extent in the handling of models in Ptolemaean astronomy) quickly became accepted procedure as revolutionary transformation caught on. Not that self-correction need remain confined to the investigator himself, as with Torricelli's vacillations over water outflow, taken up by Huygens and then again by Varignon (p. 340). Gilbert's large-scale winnowing of facts and lore regarding amber and the lodestone is an early example of systematic correction of possibly erroneous facts maintained by predecessors, in ways that did not (as with the scholastics) oppose opinion to counteropinion but genuine fact to apparently spurious fact.

Here, too, there were incidental precedents (as with Castro), but this time the practice, now followed more systematically, caught on at once. Within decades of the onset of the Scientific Revolution correction of oneself and/or others was built into regular procedure and subsequently enshrined in the first institutions of science, where 'provando e riprovando' (the Accademia del Cimento's motto; p. 487) became standard practice.

What made these 17th-century developments so decisively important is the dual makeup of human cognition. We are blessed with a trenchant capacity to grasp realities with which the world confronts us. But that capacity is accompanied by an at least equally powerful capacity to enshrine the ideas in which we seek to capture those realities in allegedly timeless, dogmatically held truths or (with Swift's ghostly 'Aristotle') to move from one 'vogue' to the next or – riskiest of all – to get lost in flights of fancy. The agents that enable us time and again to lose sight of reality have been identified by Francis Bacon as the 'idols' that so easily come to bewitch our minds (p. 486). We are subject to bias of the most varied kinds, from self-serving distortions to an even more powerful inclination to take our wishes for reality. In no human endeavor has our capacity to grasp reality in any of its varied aspects exceeded the level of a rare conjunction of individual gifts – except for recognizably modern science. Nowhere else have viable ways been found yet to turn the possible correction of our fondest assertions into standard practice that may serve, not just the exceptionally gifted and reality oriented, but everyone with a solid training in those procedures.

722   To be sure, a solid training in procedures does not guarantee the making of discoveries – for that, solid training must be accompanied by exceptional gifts. Nor does institutionally and procedurally established self-correction lead to valid results without fail. Recall how Newton, feeling in good empirical conscience compelled to adjust his mathematical model to new measurements of the speed of sound, spoiled the original, straightforward simplicity of the model by means of a 'fudge factor', and how it took almost a century until Laplace found the true grounds of the discrepancy to reside in the heat that sound waves engender (p. 494). The point is rather that, as one major outcome of the Scientific Revolution, institutionally and procedurally established self-correction has come to be built into how practitioners routinely operate.

Of course, with practitioners' growing familiarity with the ins and outs of procedural self-correction, the practice has over time become more and more refined. The Scientific Revolution did not in any way *complete* the search for how to achieve self-correction in the most rigorous and productive manner possible – it only marks the very onset of that never-ending search.

And so it is in just about every other respect as well. To be sure, the Scientific Revolution was, first and foremost, a *historical* phenomenon. It was a clearly circumscribed episode, coherently made up of six distinct yet variously intertwined revolutionary transformations. With the completion of the sixth one, Newton's, the episode came to a well-marked end. But from the point of view of well-nigh everything else, the Scientific Revolution set in motion a train of events that has yet to cease. With respect to the methods of science; its contents; its professional status; its institutional bearings; its societal preconditions and underpinnings; its capacity to revolutionize the arts and crafts; its expansion worldwide; claims made on its behalf for a monopoly on rational thought and for leading inexorably to a truly scientific worldview; and also the countercurrents provoked by science – in all these and many more respects vast extensions and huge changes have kept occurring. Together these have resulted in the Scientific Revolution bequeathing to the generations that followed a dual legacy: a legacy of gains attained and conquests made but also of losses suffered. To treat the legacy at the length it deserves would at least triple the length of this book and turn it into a survey of the entire history of science. But a few remarks on the major extensions and changes just listed may serve as a fitting epilogue to the conception of the Scientific Revolution here put forward.

Expanding modern science

Take first of all the way in which the prime culmination point of the Scientific Revolution, Newton's dynamics and the law of universal gravitation as expounded in the *Principia*, has

over time undergone several major alterations. Starting in the 1720s, Newton's works gave 723
rise to an extended textbook tradition. Even earlier Varignon in France and a few British
scholars with a penchant for mathematics began to reorganize the main message of the
Principia and to modernize its mathematical language, with the latter sticking to Newton's
calculus of fluxions and the former preferring Leibniz's differential calculus (p. 360). Later
in the 18th century mathematical scientists like d'Alembert, Euler, and Lagrange made big
strides in the same direction, while in addition pursuing a further differentiation between
a variety of concepts of force (p. 622). This led among many other things to shearing 'force'
of its remaining substantive connotations and enveloping it for good in a more abstract
structure of space and time. The effort culminated in the mid-19th century in the coining of
a clear-cut concept of energy destined to replace the primacy that Newton had envisaged for
all those forces he mathematically prepared for in the *Principia* (p. 666).

But Newtonian force was still much more than a key topic for mathematical scientists.
By the late 1720s it acquired wider meanings as the contested cornerstone of a closed system of natural philosophy, loaded with ideological connotations, that went by the name of
Newtonianism. As such it played its part in a renewed process of compartmentalization.
From the early 18th century onward, and due in part to the increasing consolidation and
prominence of the states of England, France, and Prussia, mathematical science, fact-finding
experimentalism, and speculative natural philosophy were once again pursued by and large
in mutual isolation, with the latter being split up along national lines in three systematized
philosophies, Newtonianism in England, Cartesianism in France, and, in German lands, a
concoction of ideas first thought up by Leibniz. Not until the turn of the 19th century, when
the French Revolution put so much else in flux as well, were the fences broken down all over
again but now for good. This was just one major portion of a process of drastic change on
many fronts that well deserves the label the 'Second Scientific Revolution' recently given to it.
As that revolution unfolded, the term 'science' came to stand for the consolidated product of
decompartmentalization, accompanied in its turn by its practitioners becoming 'scientists'
and their vocation turning into the full-fledged profession it has remained ever since.

Regarding universal gravitation, for a long time all predictions entailed by the theory
that came up for possible confirmation were confirmed indeed, as notably with the discovery of Neptune consequent upon the detection of small apparent deviations of Uranus'
actual from its calculated trajectory. Not until the early 20th century, when the advent of
relativity theory at Einstein's hands revealed universal gravitation to be a property of space-
time curvature, did it turn out, at velocities near the speed of light, to be no more than
approximately valid. Even so, the enduring validity of the mathematical law of universal
gravitation but for extreme cases was resoundingly confirmed all over again when in the
1950s rockets became powerful enough to overcome the limitations imposed by the escape
velocity that the law entails and to start the exploration of outer space. Take, for instance,

724   the calculated launch date and the fourfold 'gravity assist' that enabled a spacecraft named after Cassini and Huygens, which was launched on 15 October 1997 from Cape Canaveral, to swing by three planets in succession so as to accelerate it sufficiently to reach Saturn within seven years:

> The primary launch period for Cassini, based on the alignment of the planets and the capabilities of the Titan IV/Centaur launch vehicle, was in October 1997. The launch boosted the spacecraft into a Venus-Venus-Earth-Jupiter Gravity-Assist (VVEJGA) trajectory toward its final destination of Saturn.[407]

This feat was a fitting demonstration of the power and precision of what Isaac Newton began by 1685 to consider in earnest and then managed to maintain against three possibly lethal objections of his own devising (p. 656). When on the predicted date of July 1, 2004, *Cassini* smoothly slipped into its well-planned orbit around Saturn, even Jonathan Swift might have chosen to reconsider his 'fashion' judgment . . .

The scientific enterprise has kept expanding in other ways as well. Recall how speedily topics that prior to 1600 used to be treated in the wholly abstract manner of prerevolutionary Alexandria or of prerevolutionary, as yet scarcely experimental empiricism were absorbed into Alexandria-plus and into the practice of fact-finding experimentation (p. 600; 614). Since then, in stark contrast with the 'world of the more-or-less' from before the Scientific Revolution, ever more domains of natural phenomena and of human experience have come to be absorbed by what Alexandre Koyré dubbed 'the universe of precision' (p. 160). Its prime manifestation has been by means of numbers. For instance, the phenomenon of color, which natural philosophers like Aristotle treated as irredeemably qualitative, has proved analyzable in terms of wavelengths, with each color corresponding to exactly so many angstroms.

In due course, ever more natural phenomena, like those treated in chemistry and then biology, have similarly proven amenable to being handled in rigorously quantitative terms with great success. In at least partial contrast, in the human sciences it has become a perennial bone of contention whether across-the-board quantification can usefully supplement, or perhaps even replace, more qualitative analyses, or whether instead quantification tends to mask or even call into being the very events and states of affairs it is meant only to pinpoint. On the one hand, there is the perennial temptation eloquently expressed in 1793 by the marquis de Condorcet, and still adopted without much reflection by many a present-day scientist dabbling in the social sciences and the humanities:

> If man is capable of predicting, with almost full assurance, the phenomena of which he knows the laws; if even when these are not known to him he is capable from past experience to foresee

with great probability the events of the future; why then would we regard it as a chimerical 725
enterprise to draw with a certain amount of likelihood the table of future destinies of humankind from the results of its history? The sole foundation of belief in the natural sciences is this
idea that the general laws (be they known or not) that rule the phenomena of the universe are
necessary and constant; and for what reason would this principle be less true for the development of the intellectual and moral faculties of man than for the other operations of nature?[408]

On the other hand, there is the 'revolt against positivism' around the turn of the 20th century, engaged in by social thinkers like Weber and Durkheim and Croce and Freud, out of
which came an armory of arguments against such a lawlike approach to social reality and
also a range of tools for how instead to come to grips with the inner world and the culture
and society that intentional human beings create in their own right. Overall, however, the
prestige of the natural sciences and their successes on their own terrain are such as in our
day to have their standards and methods bureaucratically enforced upon humanistic disciplines that by the very nature of their subject matter can profit from those standards and
methods but little.

Outside academia, in everyday life, the scope of the quantitative likewise keeps expanding. Nowadays we no longer just spend (or idle away) time as students; rather, we fulfill a
prescribed number of internationally standardized, administratively serviceable units called
'study points' or 'credits'. Similarly, we no longer just contemplate nature but take part in
the organized enjoyment of measurable 'nature values'. It is a matter of personal preference
whether in such quantification measures we value the orderliness and the semblance of unadulterated objectivity or rather resent the effacing of individual difference and the increase
of control. Yet so much is certain that the invention and global spread of a wonderfully
handy machine which in its obstinately binary way treats the entire universe as if it were
composed solely of os and 1s has given a big boost to the predominance of the 'universe of
precision' that was instigated in Europe some four centuries ago.

In tandem with such expansion into our daily lives of the apparent ways of modern science has gone an equally incisive, geographic expansion. In our time science has definitely
become an operation spanning the globe, but the first vestiges of the process appear already
in the very period of the Scientific Revolution or its immediate aftermath. What is meant
here is not so much the 'cultural transplantation' of nature-knowledge of Greek provenance
treated in chapters 2–5 but rather efforts undertaken in other civilizations to come to terms
with certain wished-for attainments of Europe's Scientific Revolution. Three instructive
early examples concern the Ottoman Empire, China, and Russia.

The Ottoman Empire shared borders with Europe, and diplomats, merchants, and fugi-
tives kept moving both ways, so contacts were close enough to enable the court at Istanbul to
acquaint itself already in the 16th century with a carefully made selection of possibly useful

726 advances in, notably, gunnery and mapmaking as cultivated in Europe. The first product of the Scientific Revolution to be translated in this way into Arabic and then Turkish was a set of detailed tables compiled c. 1640 by a French astronomer on a Copernican basis. The translator, a native from present-day Hungary who in the 1660s presented his manuscript to the chief astronomer at the Ottoman court, rendered the book in accordance with a centuries-old tabulation genre, known as *zij,* to which al-Battani in the 10th (3rd) century and many others contributed and which was concerned almost exclusively with the ongoing correction and improvement of stellar and planetary data and parameters (p. 58). In his introduction, the translator did not address the conceptual issues involved but rather, in line with that tabulating tradition, treated Copernicus, Tycho, and Kepler as latter-day *zij* authors.

When in 1683 the army besieging Vienna was decisively defeated, the Ottoman perception of Europe changed. Apparently, Europe had more to offer than some useful bits and pieces, as it had meanwhile acquired a general superiority in the area of science and even more so in the military and naval arts. Consequently, translation and study began to cover a wider terrain, albeit still with a marked tendency to confine the science to its utilitarian and instrumental over its theoretical and conceptual aspects (p. 69). Early in the 19th century educational institutions were founded to supply their students with secular learning not on offer in the *madrasa*s. Even so, the unbroken reliance on the centuries-old Golden Age of Islamic nature-knowledge made it possible for the principal of the Imperial Engineering School in 1834 to "explain the systems of Ptolemy, Tycho Brahe and Copernicus at length and in full technical detail", all the while regarding heliocentrism as "quite possibly mistaken". [409] An even more telling example of how much the new kept as yet being taken up in the framework of the old is the manner in which the same authority handled basic findings of post-Lavoisier chemistry; in his own words:

> Some earlier scholars held that there were only four elements, but modern scholars of chemical science claim that there are more than four, and that water, air, and earth, previously considered to be elements by old scholars, actually consist of more than one element. [410]

Unlike with the Turks, acquaintance with products of European nature-knowledge came to China by way of the Jesuit mission, which the emperor first allowed to enter Beijing in 1601. In line with the current conversion policy (p. 495), its head, Matteo Ricci SJ, aimed to persuade the local intellectual elite of the unmistakable superiority of mathematics and astronomy as cultivated in Europe and to use this superiority, once acknowledged, as a ready means to convert the court mandarins, and then the population at large, to the Catholic faith. Together with an early convert he translated the first six books of Euclid's *Elements* into Chinese, preceded by an extensive preface in which, following Clavius but also with a shrewd eye to the predominantly practical orientation of Chinese nature-knowledge (p. 45),

he extolled the many advantages for everyday life that a knowledge of demonstrative geometry had to offer. But Ricci's prize asset was astronomy. The Ming 'season-granting system'
(p. 41) had remained unaltered for the entire duration of the dynasty, and the prediction of
eclipses was off by more than half an hour. Eager to point this out and to offer improvement
from European resources, Ricci's successor, Adam Schall von Bell, attained the very directorate of the emperor's Astronomical Bureau (p. 42).

The job, however, also entailed responsibility for the establishment of lucky days. Accusations of having selected an inauspicious day and site for the burial of an infant prince,
coupled to the political troubles that went with the forceful replacement of the Ming by the
Ch'ing (Manchu) dynasty, caused the fall of 'the second Adam' in 1661. However, when seven
years later the second Manchu emperor, young Kang Shi, in a move to get rid of his regents,
wished to reinstall the Jesuits as reformers and guardians of the season-granting system, he
set up a competition for the directorate between Schall's successor as head of the mission,
Ferdinand Verbiest, and native astronomers, who had meanwhile regained control over the
bureau. Of course, ever since 1616 the Jesuits were compelled to work under the handicap of
the official prohibition of Copernicus, which led to some planetary double-dealing that did
not escape their Chinese opponents. Also, in his later account of the event Verbiest acknowledged that in executing the range of tests set for him he had been the recipient of several
strokes of good luck (e.g., bad weather masked errors consequent upon his less-than-opti-
mal tables). However, and this is a key point, for all the vast complexities of the local context
in which the 'duel' took place, under the fair conditions the emperor set for it the superiority
of the new astronomy, even if in a mutilated guise, was plain to see.

Not that the Jesuits' mission succeeded in its principal aim of large-scale conversion.
Among other setbacks, repeated accusations from the Jesuits' own ranks that their errant
brethren in faraway China were dangerously 'going native' proved decisive in the end for
the Roman headquarters calling the mission back home. Also, the local repercussions of the
far-reaching reforms instituted by the Jesuits in the Ming season-granting system proved too
disturbing for ongoing adoption, so much did the system govern everyday life for the ordinary Chinese. All in all, the modes of nature-knowledge embodied in organic materialism,
on the one hand, and in Alexandria and Athens in any of their varieties, on the other, proved
too incompatible for large-scale adoption under the conditions of relative equality that as
yet obtained. Not until the 19th century, when the pressure of Europe's military superiority
was brought home to China the hard way, did large-scale adoption begin in earnest.

Russia's introduction to products of the Scientific Revolution began with the translation, under Czar Peter the Great's personal auspices, of a posthumous work by Christiaan
Huygens, *Kosmotheoros*. Before it came out in 1717, the latest exposition known in Russia of
how the natural world is constituted was a 6th-century tract by a monk intent on explaining
Genesis. *Kosmotheoros* (1696) is a running speculation about life on other planets, driven by

728   persistent analogical reasoning that takes its point of departure in the view of the universe Huygens settled on during the final years of his life – filled with spherical whirlpools, yet governed mathematically by what Huygens took to be the acceptable portion of Newton's inverse-square law (p. 676). Czar Peter seized on the book as an engaging way to make "the discoveries of Western science clear and intelligible to the still minuscule reading public."[411] The czar is likely to have met Newton in person during his 1698 stay in England, probably at the Mint. At that time Newton served as its warden, and Peter consulted him primarily on monetary matters so as to prepare for the large-scale reforms of the Russian polity he began executing once he was back home in his newly founded capital of Saint Petersburg. The abiding interest that sent the czar on his exploratory voyage to the Netherlands and England was in advanced techniques (such as notably for shipbuilding) as the best means to drive his planned reforms. Yet his lively interest in the latest science was not just derivative. In Amsterdam he arranged for the purchase of a well-stocked *Kunstkammer* (cabinet of curiosities). In England he had the man who was to translate *Kosmotheoros*, a courtier by the name of Iakov Vilimovich Bruce (of Scottish descent), purchase the *Principia* and other books with mostly mathematical content and also construct reflecting telescopes and other scientific instruments.

Bruce's collection of instruments and books, soon enriched with many more works on science and also with translations specifically ordered done by Czar Peter, was in due time to serve as one kernel of the Russian Academy of Sciences that Peter founded but whose first session he did not live to see. Benefiting from Leibniz's counsel among others, he and his successors on the throne managed to attract to Saint Petersburg leading luminaries from all over Europe, such as Euler and several Bernoullis. By the late 1720s the respective merits of Newton, Descartes, and Leibniz were discussed in Saint Petersburg as elsewhere on the European continent – within less than ten years, that is, of the attempted sabotaging of the Russian translation of *Kosmotheoros* by the man whom Peter dragooned into printing it, a task that the printer detested in view of Huygens' clearly apparent "Satanic perfidy".[412]

The three cases just discussed have enough in common to invite an instructive comparison. In each case, policy considerations largely determine the course of events from the outset – in the Chinese case, policies on the receiving, as well as the supplying, end. In each case, an autocratic court initiates events and keeps running the show, sometimes at one remove, sometimes directly, yet invariably present and always decisive in the end. In two cases out of three, both what portions of recent European science are selected for introduction and how these are actually received are matters very much determined by local circumstances and by constraints set by the locally predominant conceptual framework. Adaptation of European work to the legacy of the Islamic Golden Age comes with considerable substantive impoverishment; the contrast with organic materialism even proves so vast as in the end to lead to almost wholesale ejection. In the Russian case, by contrast, there is a blank slate. There,

Newtonian science is parachuted into a scholarly wasteland, and in the absence of any rival conceptual framework the science imported is very much received on its original terms.

For all the differences between the local contexts in each specific case, *the superior substantive quality of the science adopted is recognized in all three*. In the Ottoman Empire and in Russia, European scientific superiority goes without saying, or is at least accepted without more ado by the respective autocrats involved (in the former case only after a crushing military defeat). In China the ruler deliberately sets up a competition and forthwith follows up its confidently foreseen outcome with the appropriate measure: reinstallment of an expert European as head of the emperor's Astronomical Bureau. To be sure, in each case local acceptance of European superiority concerns *only* the portions of science actually adopted – not science as such, let alone the culture as such (here, too, Czar Peter's Russia is a bit of an exception). Even so, the acceptance is all the more telling because all three episodes took place a long time before the as yet latent capacity of the science that came out of the Scientific Revolution to alter the arts and crafts from the ground up and thus help instigate the rise of the modern world became at all apparent. Neither to the Ottoman Empire nor to China or Russia did these early products of Europe's Scientific Revolution come on the coattails of modern imperialism or on anything at all but their own substantive merits in the eyes of the mutually so different beholders.

Indeed, the Baconian vision, enthusiastically proclaimed in the 17th century, did not begin to work out in a manner visible to the public at large until the early decades of the 19th. As I have insisted at many places in the present book, the promised rise of a science-based technology, the obstacles to which were first encountered and then gropingly explored over the period of the Scientific Revolution, went through a creative incubation period during much of the 18th century. That is when 'craftsmen of a new type', like Newcomen, Harrison, Smeaton, and Watt, succeeded in overcoming all those obstacles arising from complex mathematics, from underestimated messiness, from basic miscommunication, etc. (p. 325). To be sure, the achievement of these men who managed to make novel science incarnate in novel machines forms only a necessary, not a sufficient condition for the onset of the Industrial Revolution. There was definitely more to the onset of industrialization than modern science alone. This is so in two major respects, quite distinct from each other.

In the first place, not all techniques that jointly made for the world's first industrialization process owed much to science. The steam engine that came from Watt's drastic transformation of Newcomen's fire engine surely did. It was imbued to the very core with revolutionary scientific insights into, notably, atmospheric pressure (Galileo, Torricelli), the void (Torricelli, Pascal), latent heat (Black, Watt), and the expansion of steam (Watt). It was quite otherwise with the range of inventions that jointly made possible the mass fabrication of cotton (and soon enough other fabrics as well): the spinning machines invented by Hargreaves, Arkwright, and Crompton, plus Cartwright's power loom. None of these machines

730   owed anything to modern science but the Baconian Ideology itself, even though by the turn
of the 19th century the mass bleaching and dyeing that quickly came to present a production
bottleneck was to benefit soon enough from contemporary work in advanced chemistry. In
short, science did come in *after* the fact without fail (as also with the onset of thermodynam-
ics in Sadi Carnot's reflections on the motive power of steam), but only selectively as part
and parcel of the original invention.

However, and this is my second major point, all this technical ingenuity would still have
been wasted if technical *invention* had not been turned into veritable economic *innovation*
due to a range of bold decisions actually to *invest* in these inventions and *market* the products to come out of those investments. What entered into the original investment decisions
made by men like Boulton for the steam engine or Arkwright for the water frame that goes
by his name was not just their capital or their visionary entrepreneurial foresight (although
these particular components of the full equation are not to be neglected) but really an entire
economic climate. In Britain the relevant economic parameters displayed certain special
features, such as (to name but a few) a none-too-constricting patent law; an above-average
public awareness of what was going on in science and in the crafts; reasonably dependable
safeguards against arbitrary expropriation; a regularized money-lending system; and a unified national market (and the potential for an international market in the United States and
in India) not encumbered by prohibitive interior or exterior tariffs. It is, then, in the seemingly miraculous *confluence* of, on the one hand, British craftsmen of a new type managing
to overcome all the obstacles to a radical, science-based overhaul of the crafts and, on the
other, British entrepreneurs seizing on the chance thus arising to invest, with hopes for vast
long-term profit, in large-scale machines of a kind quite unheard of that the true riddle
resides of the onset of our modern world in early-19th-century Britain.

In whatever manner that vast historical enigma is in the end to be resolved, early industrialization did give England, and soon enough all of western Europe and the United States,
a decisive edge and then a hold on the world at large that in many ways endure to the present
day. The making and the running of colonies in Asia and Africa came and went, yet in many
respects Western predominance is still a fact of global life, and science has kept contributing to that predominance in large measure and without cease. This is not, however, a state
of affairs that is bound to remain with us in any case. One major point of the present book
has been that the events out of which modern science arose – a feat absolutely indispensable for the modern world to come into being at all – took place in Europe due to events
and circumstances that, while well-explicable up to a point, nonetheless might have taken a
somewhat different course in which the Scientific Revolution might not have happened at
all or have occurred somewhere else or at another time (pp. 72; 204; 271). Just as there was
no inherent necessity for modern science to emerge in the West, just so there is no inherent
reason why its prime manifestations should forever remain an attribute of the West. There

is nothing *a priori* unlikely about a vision, recently promulgated by Kishore Mahbubani (a diplomat from Singapore), of Asian civilizations regaining their longtime predominance after an altogether fairly brief, two- to three-centuries-long interruption by the West.

For such a grand reversal actually to come about, it would nonetheless seem that – at least where science is concerned – certain conditions of an as yet unsettled status ought to be met. For how does modern science fare in civilizations other than the one that has given rise to it? The specifics of local context are highly consequential, decisive even, as the radically different outcomes of the early Chinese and Russian efforts at importing portions of revolutionary European science show – nearly wholesale ejection in the one case, almost wholesale adoption on its own terms in the other. Over the larger part of the post-1700 expansion of science, however, the Ottoman effort to adopt portions of science in the curtailed guise of ready-made usefulness at the cost of conceptual depth seems more of a harbinger of the future. To have just the science while leaving everything else intact has been a well-understandable reflex on the part of rulers who desire the useful goods but resist the wholesale modernization and its numerous discontents that all too easily come with it. But is such a strictly utilitarian manner of handling the cultivation of science really feasible? Does not modern science come with certain value-laden attributes which, although not directly involved in how research is actually done on a day-to-day basis, nonetheless co-determine the ongoing flourishing of the enterprise as such?

Science and values

Take first those values that have over time become anchored in stable institutions. Although scientific societies and academies remained significant sites of original research until some time into the 19th century, and although the Royal Society is still alive and kicking, laboratories maintained by universities and by private companies with R&D departments of their own have replaced them as the principal institutions housing scientific research. In addition, just as military and commercial interests were co-constitutive of the very foundation of the two earliest societies (p. 580), so have the 20th-century industrialization of warfare and the rise of the pharmaceutical industry, in particular, caused commissioned research to become a growing part of the landscape of science.

Consequent upon all this, one indispensable ingredient of scientific research, its autonomy in the sense of the tenability and scientific standing of an investigator's conclusions being subject only to the judgment of his or her peers, has come to face novel challenges. Indeed, nothing remains of the very specific constellation of large-scale European power realignment, rapidly emerging royal confidence, Baconian optimism, and religious sanctioning out of which an unprecedented amount of autonomy for the investigation of nature emerged

732 in the 1660s. Suddenly, within ten to twenty years of revolutionary nature-knowledge running a grave risk of losing momentum for good, a second generation of practitioners could be seen strolling in the Jardin des Plantes or seated in London's coffeehouses to discuss the latest findings made under the auspices of the two royal societies newly founded in the two principal cities of Europe (p. 513; 619). The scope and the underpinnings of this newly attained and still rather circumscribed autonomy have since changed almost beyond recognition. Even so, a serviceable amount of freedom of research irrespective of where it may lead has remained, by and large undisputed (except for a range of totalitarian onslaughts), in quite different constellations.

In the late 1940s the manner in which Nazi and communist regimes appeared to handle certain disciplines (efforts at constructing an 'Aryan' physics and the slightly later imposition of Lysenko's ideas on heredity are well-known examples) inspired the sociologist of science Robert Merton to articulate the ethos of the scientific enterprise, which he took to be defined by four values. When the pursuit of science is made dependent on anything other than individual merit (such as race, class, or gender); when scientific knowledge remains secret; when results are affected or even determined by the investigator's personal stake (glory, money) in the outcome; or when limits are set to research in view of what any given society may hold sacred – in all such cases (which appear to be less rare than one might hope for and also at present on the increase) practitioners collectively feel in their bones that a binding norm is being trespassed. What is minimally required for the ongoing cultivation of science even on a day-to-day basis is the presence of safeguards against wholesale infringement of the self-critical procedures on which the scientific enterprise thrives, and which it cannot do without in the long run. Moreover, the general public's trust in science, while bound up to a large extent with an as yet unceasing readiness to acquire the goods that scientific research is widely taken to hold in store for us, is ultimately dependent on a certain confidence in precisely these self-critical procedures. For as long as we sense that, *in the end*, nothing needs to be taken on trust and that anyone who decides to take the trouble and acquire the needed expertise may confirm or repudiate for himself the validity of a scientist's assertions, our trust seems sufficiently well founded. If, on the contrary, science were widely seen as overall up for grabs by the highest bidder or as giving in to the loudest accusations of trespassing some sacred principle, it would spell its effective demise. The machinery might still rumble on for quite a while, but the lifeblood would meanwhile seep away.

From the 1940s to the 1970s it used to be maintained without much risk of contradiction that only a democratic environment is capable of providing the safeguards embodied in the scientific ethos. In his thought-provoking book, Mahbubani suggests that the home-grown cultivation of top-notch science may also thrive in the absence of Western-style democracy:

the 'magical' development that propelled Western science and technological research ahead in the past few centuries has now penetrated the Asian cultural fabric. Asians no longer believe that they are inferior to the West in science and technology research. They believe that they can do equally well on their own.[413]

As yet it is an open question whether or not 'the Asian cultural fabric', permeated as it is for the largest part by comparatively mild forms of authoritarian rule and behavior, would be sufficiently capable of ensuring the institutionalization of regular, objectivized feedback procedures, although the example of Japan (for all its democratic elections a deeply hierarchical, authority-run society) seems to confirm it.

A closely related, perhaps more worrisome question is whether science can enduringly flourish in an environment that has not or not yet interiorized the principal values of the Enlightenment. Surely there are intimate *historical* connections between the rise of modern science and the emergence of the 18th-century movement known by that name. In his 1784 essay 'What Is Enlightenment?', Kant famously defined it thus:

> Enlightenment is man's emergence from his self-imposed immaturity [*Unmündigkeit*]. Immaturity is the inability to use one's understanding without guidance from another. This immaturity is self-imposed when its cause lies not in lack of understanding but in lack of resolve and courage to use it without guidance from another. *Sapere aude* [dare to know]! "Have the courage to use your own understanding!" – that is the motto of enlightenment.[414]

A century and a half earlier, Galileo's *Dialogo* and those deeply ambiguous 'everything up for doubt' passages in Descartes' *Discours de la méthode* and *Meditationes* had already begun, if not literally to express, then at least exude this very admonition to think for oneself unencumbered by anybody else's prescriptions however much imbued with authority (pp. 282; 385; 435). Also, one fundamental presupposition shared by just about all protagonists of the Scientific Revolution save only the dogmatic natural philosophers was that, in spite of what the vast majority of scholars knew for sure, the constitution of the natural world was not at all known yet but was open to fresh discovery by mostly novel, as yet hardly tried-out means. The 17th century, with its religious wars and its persecution of witches, was as little an enlightened age as any preceding century was; but to the coming about of such an age the rise of modern science contributed greatly. In his *Lettres philosophiques* of 1734, Voltaire, in creating the *persona* of the 'philosophe' and wittily taking sides in thirty years' worth of debate in the Académie on the problems of Newtonian attraction, initiated the French Enlightenment movement all the while seizing on Newton's published work. He adopted it in its current, mostly British/Dutch interpretation as primarily experimental rather than mathematical in its approach and its implied methodology, as nearly materialist and deist in its ideological

734   underpinnings, and as a close ally somehow in the upcoming fight against all that the 'ancien régime' stood for. Newton, already dead for seven years, would have been horrified. To be sure, other 16th- and 17th-century events, notably the emergence of a public sphere and the rise of biblical criticism out of new methods of philology, also contributed their fair share to the coming about of the Enlightenment – as with the later Industrial Revolution, the rise of modern science is only part of the story, albeit certainly an indispensable part.

This is equally true of one more portion of enlightened thought: the aimed-at eradication of superstition in all its many guises. Here, too, what protagonists of the Scientific Revolution contributed forms rather a mixed bag. Debate about the very possibility of miracles was affected to say the least by the idea and the confirmation of the constant action of inexorable laws of nature like, notably, Newton's; yet denial of miracles could and did emerge as well out of Protestant polemics against what was felt to be a popish overdose of miraculous events and cures. Belief in the reality of witchcraft and of demonic possession was by no means confined to *opponents* of the new currents of nature-knowledge such as, for instance, Voet; yet the remarkable shift in the debate that took place by the second half of the century owed much to a consistent failure to render urgently searched-for, experimental proof of demonic action in any of its numerous varieties (p. 480). Similarly, as debates for and against the portentous meaning of comets continued through the seventeenth century and beyond, opponents appealed less and less to just common sense endorsed by ancient authorities and more and more to a century's worth of astronomical investigation that had revealed the regular calculability of their appearance in the sky and, indeed, of their entire trajectory, due, once again, to the inexorable operation of dependable laws of nature.

Thus was the beneficial rule of enlightened 'reason' opposed to the tyrannical rule of irrational authority on many fronts at a time, and scientists were quick to avail themselves of so advantageous an opposition. To what extent were they, and are they, entitled to do so?

They are very much so entitled insofar as Enlightenment feats and values like those of unencumbered inquiry and of a securely institutionalized public sphere which leaves ample room for critical, open-ended debate form the very lifeblood of the scientific enterprise. Even in totalitarian states attempts have been made to create carefully encapsulated domains where, in maximum insulation from society at large, scientists were given privileges along such 'enlightened' lines. Alexandr Solzhenitsyn's novel *V kruge pervom* (The First Circle) provides a haunting, firsthand picture of such a privileged site during Stalin's reign, while also intimating that artificial hothouses like these are doomed in the long run to collapse under their own contradictions. Whether, in the far milder circumstances of run-of-the-mill authoritarian regimes in Asia, the surely smaller amount of insulation needed there would be sufficiently limited to allow truly groundbreaking scientific research to flourish and survive in the long term is an as yet open question, albeit one to which within a few decades we are likely to find out the answer.

So much for the 'freedom and critical openness' side of the Enlightenment in connection with the scientific enterprise. But what about its 'rationality' and its 'liberation from superstition' aspects?

Surely scientists deservedly lay claim to rationality in the sense of a proven capacity to train the human powers of reason-guided action and understanding in a manner optimal for coming to grips with the realities of our natural environment – precisely there lies the greatest triumph of the Scientific Revolution.

But does it follow that scientists have a monopoly on rationality? On hearing some of their self-appointed representatives, one would almost believe it. For instance, in his book *The God Delusion* Richard Dawkins makes no secret of his conviction that any riddle in the world, if it can be solved at all, can be solved by the strictly rational, strictly empirical means which science alone is in a position to bring to bear on it. A delightful sentence in Dawkins' book provides a neat measure for the degree of rationality and empiricism that mark his own argument, viz. "Newton did indeed claim to be religious".[415] That is, Newton said so, but he cannot have meant it, for how possibly could that prime icon of science have been a true Christian believer? And so Dawkins feels at liberty to ignore the tons of evidence in Newton's own manuscripts and actions so carefully assembled and analyzed by historians of science over the past four decades. Did Newton risk his entire career for his religious convictions just on a whim? Was it just to mask his utter unbelief that he appended to the second edition of his *Principia* a 'Scholium generale' in which he set forth the publishable portion of the highly personal conception of the Deity that he had arrived at after decades of assiduous wrestling with theological issues (p. 710)? In short, if the facts of history do not fall in line with what Dawkins already knows for sure on sheer *a priori* grounds, then so much the worse for the facts …

Or take a cosmologist by the name of Vincent Icke, who recently, in his ongoing fight against superstition, wrote an op-ed piece attacking astrology.[416] In effect the piece came down to propaganda *for* astrology – if there are really no better arguments against astrology than vulgar name-calling and misinformed accusations dished up without adducing one shred of evidence, then perhaps there may be something to it after all? For instance, Icke jeered at unnamed astrologers who thought they could 'prove' that there were only seven planets. As everyone with an inkling of the history of astrology knows, there has been no more than one astrologer intent on proving that there could only be a determinate number (six, not of course seven) of planets, and his name was Johannes Kepler (p. 164).

What all this amounts to is not of course that scientists are incapable of rational thought and action or of taking empirical realities into account. Quite to the contrary – even if not in person, then collectively they may rightly lay claim to one highly specific and as such most successful and admirable form of rationality. It means only that scientists, as soon as they feel at liberty to expound *ex cathedra* beyond the proper boundaries of science, are prone

736   to all the distortions that Bacon, in his analysis of the four idols, eloquently warned against (p. 486). Outside their own field of competence, they may all too easily fall back into the very habits of our naturally biased thinking that so often mar progress in other domains than the natural sciences – taking one's wishes for reality and ignoring or even forgoing readily available evidence.

Behind such arrogance on the part of numerous self-appointed spokesmen of science often resides a sentiment far more interesting than mere arrogance, and a culturally very significant one to boot – *a sense of unity lost*. This, then, is my final topic.

Among all the many changes due to the Scientific Revolution was the loss of a by-and-large unified conception of the world. Human concerns, spiritual agents, and natural givens all formed coherent parts of one overarching whole, be that whole thought of as a seamless web held together by correlations, as in China, or rather as a multilevel construction held together by logic of the 'if this, then that' type, as predominantly in the Greek tradition. Albeit not without its scary aspects, this was a cosmos one could feel at home in. But the rise of modern science changed all that.

On the smaller scale of the 'nature' portion of that overarching unity, some practitioners adapted without any apparent difficulty to the apparent lack of a unified structure for all of nature. For all Galileo's 'opportunism' in matters philosophical (p. 211), his work still displays a lively interest in how the whole of nature hangs together. One generation later, Huygens is the prime exemplar of a virtuoso practitioner who scarcely cared for anything beyond just the piecemeal and the readily graspable. His near-contemporary Isaac Newton was of a partly different temperament, or rather of two temperaments at once. In the preceding chapter I argued at some length that the very tension between his urge to go for the whole and his self-enforced confinement to the mathematically and experimentally securely proven was in good part responsible for making him an even greater discoverer than his so often unwitting rival Huygens (p. 675; 711). A little more than two centuries later, Lorentz and Einstein were to replicate much the same opposition, which indeed seems to run (albeit not as a rule in quite so plain a fashion) through the entire history of science.

As specialization keeps remorselessly increasing, the longing for the whole has more and more difficulty finding a responsible outlet. It comes to the fore in admirable ways in the Santa Fe Institute, dedicated to integrating a variety of scientific disciplines by throwing "a crude look at the whole" (in Murray Gell-Mann's wonderful catchphrase).[417] But it also comes to the fore in adorning the aimed-for unification of the four basic forces of nature, which would of course be a formidable achievement in its own right, with the silly label 'Theory of Everything'. It is as if using that theory we could henceforth not only explain relativity theory but also determine the causes of cancer and warfare and even what exactly made Mozart a composer superior even to Haydn – 'everything' after all meaning nothing less than, indeed, everything.

Nor have all scientists withstood the urge for reunification on even grander scale than just all of nature. Particularly prominent in our time are efforts to reduce without more ado all human phenomena, from ethical rules and conduct to literary criticism, to Darwinian evolution in one or another interpretation thereof. But proponents of a superficially quite other-directed mode of unification, which usually goes by the name of 'New Age', really drink from the same source. Scientists like Capra, Bohm, Prigogine, and Lovelock have aimed to show how this or that piece of scientific endeavor, from quantum physics to dissipative structures in molecular chemistry, constitutes nothing less than proof for some all-encompassing conception of the world and humankind – each to be sure a different conception.

In short, many scientists are united in the belief that there exists something like a scientific worldview, which most place in square opposition to religious belief, whereas others find the two readily compatible. But is there really a *scientific* view of the world? What can be said, at most, is that the advance of science over, by now, some four centuries has, if not straightforwardly ruled out, then at least made quite unlikely the tenets of any religious creed in its literal sense. As noted before (p. 430), I go along with Steven Weinberg's idea of an ongoing retreat of the conception of a caring Deity in the face of what science has successively revealed to be the case. Whether that retreat is full scale in the sense of in the end leaving no legitimate room for any religious belief at all is another, ultimately personal matter, which goes far beyond the scope of the present book. Within that scope, however, fits the observation that to many upholders of a 'scientific worldview' – Dawkins is a prime exemplar – science has taken all the trappings of a worldly religion, just as has happened to many pronounced atheists of quite other stripes. Here, too, nostalgia for a lost unity seems to be at work.

What it all comes down to in the end is that certain limits are set to scientific knowledge. This is hardly a novel revelation – Kant made a formidable first step in laying down those limits. Since then, it has been sought time and again to overstep them, most often under the self-invoked aegis of science. One obvious example is Comtean positivism, with its ultimate aim of providing "one body of homogenous doctrine."[418] Another, much less known yet even more instructive example is Schopenhauer's Kantian or, rather, brilliantly pseudo-Kantian effort to bring the latest science in harmony with an all-encompassing conception of the world at large.

The appeal that efforts to overstep Kant-like limits have nonetheless kept exerting on so many people, be they themselves scientists or not, testifies to the presence and the acuteness of the sense of loss that the Scientific Revolution has brought about. From a human point of view, the world that science reveals to us looks bleak – its richness rests solely in the incredible vastness and precision of understanding that modern science has enabled us to attain. The world that we ourselves experience on a day-to-day basis is much richer still, and science seems to do nothing but impoverish it.

738   All this is reflected in the modern ways of life that, due in good part to the coming into the world of modern science, many of us have become accustomed to. No one who has tasted the ways of modern life wishes to give them up, for very good reasons (higher life expectancy, better health, greater wealth, a greatly enlarged choice of consumer goods). Even so, modern life comes with a vengeance – it tends to lock us up in an alienation-inducing technopolis that endangers our very survival in more ways than one. The challenges are new – the Scientific Revolution and a range of events that more or less inadvertently followed in its wake have called them up. Yet the predicament is as old as humanity itself. Ever since the rise of consciousness we have been the objects *and* subjects of nature's greatest experiment to date. What exactly that experiment is meant to prove, and whether it is meant to prove anything at all, or even whether it has been meant by Someone or something in the first place is hard or even impossible to tell – quite possibly forever.

Here is a final image of science, of its triumphs as well as its limitations. I take the image from music – that perennial source of illustration throughout my account of how modern science has come into the world. For the longest part of history, music was just music. Or rather, music was far more than music – it was communal sharing and bonding; it was sacred ritual; it was medical cure; it was accompaniment to dance, warfare, worship. Music was all these things, and of course pure relaxing or energizing beauty to boot, yet it was not science. By now it is. Or rather, in addition to all these things that it still is, it has become science as well. We know (yes, dear reader, we *know*) how the consonant and dissonant intervals that music is made of are mathematically and physically linked up with the vibrations of the air that make up their constituent sounds. We know the anatomy and physiology of our sense of hearing, from the eardrum, across the hammer, anvil, and stirrup to the oval window, the cochlea, and the auditory nerve. We are learning more and more about those specific places inside the brain that, on investigating musicians' heads, light up in the test subject's MRI scan. We further know how musical notes if sounded together produce beats and combination tones, such that the man to make the single biggest stride in this domain of inquiry, Hermann Helmholtz, could show that the spacing of chords in Palestrina's masses was precisely the optimal one from the viewpoint of what his own theory predicted.

But none of all this gets us (nor to be sure does it claim to get us) any closer to what transcends even Palestrina's expert craftsmanship – the capacity of his music to bring us to tears fully four centuries after he first composed it.

In Palestrina's day, music was still a vital part of the cosmos. It even served as the prime expression of the harmony that was taken to bind humans and the cosmos together. That unifying harmony has gone forever.

In Palestrina's day, musical composition was given to exploiting the finest varieties between intervals tuned as purely as the elementary arithmetic of tuning and temperament

allows. Those subtle differences have gone forever. We at present experience every day how   739
scientific knowledge of musical and other sound has homogenized our experience of music:
that most deadening of all possible tuning systems, equal temperament, has now spread
throughout the globe. Meanwhile technologies originating in scientific knowledge of the
radio wave have blessed us with the opportunity to listen to the loveliest music we know of,
whenever and wherever we wish for it. They have also given us the ubiquity of those really
submusical sounds that go by the name of muzak, and by means of boom boxes they have
made true Narcissus Marsh's confident 1684 prediction of the advent of "microphones . . .
contriv'd after that manner, that they shall render the most minute Sound in nature distinctly audible, by Magnifying it to an unconceivable loudness".[419] The Scientific Revolution, in
short, has truly come with a dual legacy. It is one we had better face upright, with a mixture
of delight, admiration, and revolt, and, above all, with all the courage we can muster.

NOTES ON LITERATURE USED

Makeup of the human mind. Rob Wentholt, manuscript provisionally entitled 'Selfish, Unselfish, And Much More Besides; A Treatise On The Nature Of Human Nature'.

Newton and Newtonianism. J. Christiaan Boudri, *What Was Mechanical About Mechanics. The Concept of Force between Metaphysics and Mechanics from Newton to Lagrange*. Dordrecht: Kluwer, 2002, treats the differentiation of Newtonian force in the 18th century. J.B. Shank, *The Newton Wars and the Beginning of the French Enlightenment*. Chicago: University of Chicago Press, 2008, gives a fresh account of the origins of Newtonianism, which is treated by Margaret C. Jacob from a different viewpoint in a book co-written with B.J. Teeter Dobbs, *Newton and the Culture of Newtonianism*. Amherst NY: Humanity Books, 1998.

'Universe of precision'. Alexandre Koyré, 'Du monde de l' "à-peu-près" à l'univers de la précision'; in: idem, *Études d'histoire de la pensée philosophique*. Paris: Armand Colin, 1961 (pp. 341-361 of the 1971 reprint by Gallimard). Theodore M. Porter, *Trust in Numbers. The Pursuit of Objectivity in Science and Public Life*. Princeton UP, 1995.

'The Second Scientific Revolution'. An attempt at across-the-board conceptualization was undertaken by five speakers during an instructive round-table discussion 'The Birth of the Sciences in the Age of Revolutions? Was Knowledge Reconfigured c. 1800, And If So How And Why?' at The Sixth Three Societies Conference: Connecting Disciplines' held at the University of Oxford 4-6 July 2008.

'Revolt against Positivism' is the core issue of H. Stuart Hughes, *Consciousness and Society. The Reorientation of European Social Thought 1890-1930*. New York: Knopf, 1958.

Expansion beyond Europe. OTTOMAN EMPIRE: Ekmeleddin Ihsanoglu, *Science, Technology and Learning in the Ottoman Empire*. Aldershot: Ashgate Variorum, 2004. CHINA: Peter M. Engelfriet, *Euclid in China. The Genesis of the First Translation of Euclid's* Elements *in 1607 and its Reception up to 1723*. Leiden: Brill, 1998; idem, essay review of Noël Golvers (transl.), T*he Astronomia Europea of Ferdinand Verbiest, S.J.*; in: T'oung Pao 63, 1996, pp. 206-220. Nathan Sivin, 'Copernicus in China or, Good Intentions Gone Astray': ch. 4 of his *Science in Ancient China. Researches and Reflections*. Aldershot: Variorum, 1995 (first published in 1973). RUSSIA: Valentin Boss, *Newton and Russia. The Early Influence, 1698-1796*. Cambridge (Mass.): Harvard UP, 1972.

Scientific Revolution and Industrial Revolution. A more-detailed exposition of my views on the subject is my 'Inside Newcomen's Fire-Engine, or: The Scientific Revolution and the Rise of the Modern World'. *History of Technology* 25, 2004; pp. 111-132 (a later, more succinctly phrased version is 'The Rise of Modern Science as a Fundamental Pre-Condition for the Industrial Revolution'. In: P. Vries (ed.), *Global History. Österreichische Zeitschrift für Geschichtswissenschaften* 20, 2, 2009; pp. 107-132; both pieces derive from my article in Dutch 'Het ontstaan van onze moderne wereld: wat natuurwetenschap en techniek ermee van doen hadden'. *Theoretische Geschiedenis* 25, 4, 1998; pp. 322-349. On Watt's investigation of the expansion of steam, and on Carnot's theoretical comparison between water power and the steam engine: Donald S.L. Cardwell, *Turning Points in Western Technology. A Study of Technology, Science and History*. New York: Science History Publications, 1972; pp. 85-93; 127-138.

Science and values. Merton's four values: *SRHI* 3.5.1. The evidence that Kishore Mahbubani, in his *The New Asian Hemisphere. The Irresistible Shift of Global Power to the East.* New York: Public Affairs, 2008, provides for the thesis cited in the main text (pp. 57-66, section 'Science and Technology') is too anecdotal, and too much based on statistical evidence which does not sufficiently differentiate between foundational and more run-of-the-mill research, to be truly convincing for the present state of things. Even so his vision of the near future is both intriguing and refreshing.

Enlightenment. A wide-ranging albeit one-sidedly Spinozist history of the onset of the Enlightenment is Jonathan Israel, *Radical Enlightenment: Philosophy and the Making of Modernity 1650-1750*. Oxford UP, 2001. Wijnand W. Mijnhardt, in 'Urbanization, Culture and the Dutch Origins of the European Enlightenment', *Bijdragen en Mededelingen betreffende de Geschiedenis der Nederlanden* 125, 2-3, 2010, pp. 137-173, rather insists on the prime significance for the Enlightenment of the 18th-century emergence of chances for open, public debate in an urban setting.

Efforts at unification. A nicely balanced view of what is and what is not at stake in the future discovery of a unified theory of the four forces is Steven Weinberg, *Dreams of a Final Theory. The Scientist's Search for the Ultimate Laws of Nature.* New York: Vintage, 1992. On Lorentz and Einstein compared: Nancy J. Nersessian, 'Why wasn't Lorentz Einstein? An Examination of the Scientific Method of H. A. Lorentz'. *Centaurus* 29, 1986, pp. 205-242. In *Copernicus is ziek. Een geschiedenis van het New-Agedenken over natuurwetenschap.* Delft: Eburon, 1996. Tomas Vanheste has analyzed in detail the arguments used by a variety of New Age oriented scholars and scientists (Mumford, Roszak, Capra, Bohm, Prigogine, Lovelock, Sheldrake, Wilber, Varela) to connect science with their various, holistic conceptions of the world. On Schopenhauer: Paul F.H. Lauxtermann, *Schopenhauer's Broken World-View. Colours and Ethics Between Kant and Goethe.* Dordrecht: Kluwer Academic Publishers, 2000.

Music and science. I have treated the subtleties of tuning and temperament, and Helmholtz's analysis of consonance, in my *Quantifying Music*. Dordrecht: Reidel, 1984.

ENDNOTES

1   As quoted in A.C. Kors, *Atheism in France 1650–1729. Vol. 1: The Orthodox Sources of Disbelief*. Princeton UP, 1990; pp. 334 (passage referred to a letter of 18 September 1676 by Dom Robert Desgabets OSB to an unknown correspondent: *Oeuvres complètes de Malebranche* 18, p. 122).

2   S. Drake used the expession (most often in the variety 'recognizably modern *physics*') in many of his works, e.g., on p. 98 of his *Galileo: Pioneer Scientist*. Toronto: University of Toronto Press, 1990.

3   H.F. Cohen, *The Scientific Revolution. A Historiographical Inquiry*. Chicago & London: University of Chicago Press, 1994; p. 9.

4   J. Needham, *The Grand Titration. Science and Society in East and West*. London: Allen & Unwin, 1969; p. 216.

5   I owe the phrase to Dick van Lente, whom I once heard using it in conversation.

6   T.E. Huff, *The Rise of Early Modern Science. Islam, China, and the West*. Cambridge UP, 1993; p. 321.

7   M. Kundera, 'Les chemins dans le brouillard'. *L'Infini* 40 (November 1992); pp. 42-64; p. 64.

8   F. Furet, *Le passé d'une illusion. Essai sur l'idée communiste au XXᵉ siècle*. Paris: Laffont, 1995; p. 16.

9   H. Trevor-Roper, 'History and Imagination'. In: H. Lloyd-Jones, V. Pearl & B. Worden (eds.), *History and Imagination: Essays in Honour of H. Trevor-Roper*. London: Duckworth, 1998; pp. 356-369; p. 364.

10   Francis Bacon, *Works* 1, p. 261 (*Novum organum*, second book, aphorism 20).

11   A. Koyré, 'Les philosophes et la machine'. In: idem, *Etudes d'histoire de la pensée philosophique*, pp. 305-339; p. 316; p. 339.

12   See endnote 273.

13   C.B. Boyer, *The Rainbow. From Myth to Mathematics*. New York: Sagamore Press, 1959; p. 252 in the 1987 reprint by Princeton UP.

14   E.J. Dijksterhuis and R.J. Forbes, *Overwinning door gehoorzaamheid*. Zeist: De Haan, 1961; pp. 51–52 (English edition: *A history of science and technology*: Harmondsworth, Middlesex : Penguin, 1963).

15   Ibidem, p. 42.

16   As quoted in S. Sambursky, *Physics of the Stoics*. London: Routledge & Kegan Paul, 1959, p. 93 (referred to fragments quoted by Plutarch in his *Moralia*).

17   Ibidem.

744   18   A. Koyré, 'Du monde de l'"à-peu-près" à l'univers de la précision', in: idem, *Études d'histoire de la pensée philosophique*. Paris: Armand Colin, 1961, on pp. 341–361 of the reprint published in 1971 by Éditions Gallimard.

19   D. Pingree, 'Hellenophilia versus the History of Science', *Isis* 83, 1992, pp. 554–563; p. 560.

20   Section I:1 in the *Almagest* (p. 36 in G.J. Toomer's translation: *Ptolemy's Almagest* (London: Duckworth, 1984)).

21   Usage of the expression 'opportunist' in this context goes back to Einstein: see note 120 and appended text.

22   A. Van Helden, *Measuring the Universe: Cosmic Dimensions from Aristarchus to Halley*. Chicago: University of Chicago Press, 1985 ; p. 26.

23   J.-F. Revel, *Histoire de la philosophie occidentale*, vol. 1, *De l'antiquité à la Renaissance*. Paris: Stock, 1969; pp. 233–234 (here translated with some slight transposition in the order of sentences).

24   In the abridged version of Needham's multivolume *Science and Civilisation in China* prepared under his supervision by C.A. Ronan: *The Shorter Science and Civilisation in China*, vol. 1. Cambridge UP, 1978; pp. 165, 167–168.

25   G.E.R. Lloyd and N. Sivin, *The Way and the Word: Science and Medicine in Early China and Greece*. New Haven: Yale UP, 2002; p. 192.

26   N. Sivin, in: J. Needham et al., *Science and Civilisation in China*, vol. 5, part 4. Cambridge UP, 1980; pp. 222–223.

27   G.E.R. Lloyd & N. Sivin, *The Way and the Word: Science and Medicine in Early China and Greece*. New Haven: Yale UP, 2002; pp. 198–199.

28   N. Sivin, in: J. Needham et al., *Science and Civilisation in China*, vol. 5, part 4. Cambridge UP, 1980; pp. 298.

29   Ibidem, p. 236.

30   Ibidem, pp. 228; 230.

31   N. Sivin, lemma 'Shen Kua' in *DSB* 12, pp. 369–393 (passage quoted from section 'Quantity and Measure' in a slightly revised version on Sivin's personal website).

32   Ibidem (passage quoted from section 'Astronomy' in a slightly revised version on Sivin's personal website).

33   N. Sivin, *Granting the Seasons: The Chinese Astronomical Reform of 1280, with a Study of Its Many Dimensions and an Annotated Translation of Its Records*. Berlin: Springer, 2008; p. 31.

34   Ibidem.

35   As quoted in N. Sivin, lemma 'Shen Kua' in *DSB* 12, pp. 369–393 (section 'Formation of the Earth' in a slightly revised version on Sivin's personal website).

36   As quoted in N. Sivin, in: J. Needham et al., *Science and Civilisation in China*, vol. 5, part 4, p. 299.

37   A. C. Graham, *Later Mohist Logic, Ethics and Science*. Hongkong: Chinese University Press, 1978; p. 54, 55.

38   Ibidem, pp. 388–389.

39 G.E.R. Lloyd and N. Sivin, *The Way and the Word: Science and Medicine in Early China and*   745
Greece. New Haven: Yale UP, 2002; p. 241.

40 D.S. Landes, *Revolution in Time: Clocks and the Making of the Modern World*. Cambridge, Mass.,
Harvard UP, 1983. The first chapter is entitled 'A Magnificent Dead-End'.

41 D. Gutas, 'The Social and Political Context of the Graeco-Arabic Translations and the Origins
of Arabic Science and Philosophy' (abstract for symposium 'Crossing Boundaries', 20th International Congress of History of Science, Liège, 20–26 July 1997).

42 As quoted in S.L. Montgomery, *Science in Translation. Movements of Knowledge Through Cultures
and Time*. Chicago: University of Chicago Press, 2000; p. 125 (passage referred to D.M. Dunlop,
Arab Civilization to A.D. 1500. New York: Praeger, 1971; p. 228).

43 Ibidem, p. 120 (passage referred to P. Kunitzsch, *Der Almagest: Die* Syntaxis Mathematica *des
Claudius Ptolemaeus in Arabisch-Lateinischer Überlieferung*. Wiesbaden: Harrasowitz, 1974, p. 68).

44 K.P. Moesgaard, entry 'Astronomy' (section 2.7.) in *Companion Encyclopedia of the History and
Philosophy of the Mathematical Sciences* (ed. I. Grattan-Guinness (2 vols.); London: Routledge,
1994); vol. 1, pp. 244–245.

45 G.A. Russell, 'The Emergence of Physiological Optics'. In: R. Rashed (ed.), *Encyclopedia of the
History of Arabic Science* (3 vols.). London: Routledge, 1996; vol. 2, pp. 672–715; pp. 686–687.

46 A.I. Sabra, entry 'Ibn al-Haytham' in *DSB* 6, pp. 189–210; p. 196.

47 E.J. Kennedy, entry 'al-Biruni' in *DSB* 2, pp. 147–158; p. 155.

48 R. Arnaldez & A.Z. Iskandar, entry 'Ibn Rushd' in *DSB* 12, pp. 1–9; p. 3.

49 J.J. Saunders, 'The Problem of Islamic Decadence'. *Journal of World History* 7, 1963, pp. 701–720; p.
716.

50 A. Sayili, *The Observatory in Islam and Its Place in the General History of the Observatory*. Ankara,
1960; p. 331.

51 As quoted in idem, p. 408.

52 S.L. Montgomery, *Science in Translation. Movements of Knowledge Through Cultures and Time*.
Chicago: University of Chicago Press, 2000; p. 147.

53 Dante Alighieri, *Divina Commedia*, Canto 4, line 131.

54 *Acutissimi philosophi reverendi magistri Johannis Buridani subtilissime questiones super octo physi-
corum libros Aristotelis* as printed in 1509 by Johannes Dullaert from Ghent (facsimile ed.: Frank-
furt am Main: Minerva, 1964); ninth question on the fourth book: folio lxxiiii[r].

55 Ibidem, folio lxxiiii[v] / lxxv[r].

56 Nicole Oresme (A.D. Menut & A.J. Denomy, eds., intr., trans., comm.), *Le Livre du ciel et du
monde*. Madison: University of Wisconsin Press, 1968; p. 522/3 (NB: In this and following passages
I have somewhat adapted the translations there given).

57 Ibidem, pp. 524-5.

58 Ibidem, pp. 536-7.

59 Ibidem.

746   60   Ibidem.

61   M. Clagett, *Nicole Oresme and the Medieval Geometry of Qualities and Motions*. Madison: University of Wisconsin Press, 1968, p. 112 (this book is an edition and translation of Oresme's treatise on configurations, with extensive comments).

62   Michael McVaugh, entry 'Frederick II of Hohenstaufen' in *DSB* 5, p. 146-148; p. 147.

63   al-Biruni, 'The History of India' (translated by E.C. Sachau as *Alberuni's India*), opening passage of ch. 14 (p. 152 in the abridged version prepared by A.T. Embree (New York: Norton, 1970)). NB: On my suspicion that Sachau's expressions 'science' and 'research' are anachronistic, Jan Hogendijk was kind enough to look the passage up in the original; I have accordingly replaced 'science' by 'learning' (*ilm*) and 'research' by 'new beginning' (*naash*).

64   J.L. Heilbron, *The Sun in the Church. Cathedrals as Solar Observatories*. Cambridge (MA): Harvard UP, 1999, p. 27.

65   N.M. Swerdlow, 'Science and Humanism in the Renaissance: Regiomontanus' Oration on the Dignity and Utility of the Mathematical Sciences'. In: P. Horwich (ed.), *World Changes. Thomas Kuhn and the Nature of Science*. Cambridge, Mass.: MIT Press, 1993, pp. 131-168; p. 149.

66   Galileo Galilei, *Opere* 1, p. 300; p. 368.

67   On http://faculty.fullerton.edu/cmcconnell/Planets.html Craig Sean McConnell has created a range of moving models that make visualization of these complex matters much easier.

68   N.M. Swerdlow & O. Neugebauer, *Mathematical Astronomy in Copernicus's De Revolutionibus*. New York: Springer, 1984, p. 59.

69   M. Kemp, J. Roberts, & P. Steadman, *Leonardo da Vinci*. Yale University Press, 1989, p. 14.

70   Ibidem, p. 123; p. 128.

71   D. Laurenza, *Leonardo on Flight*. Baltimore: Johns Hopkins UP, 2004; p. 84.

72   L. Reti's contribution to the entry 'Leonardo da Vinci' in *DSB* 8, p. 210.

73   Latin text (in capital letters on the balustrade on which Derich is shown leaning) transcribed from the painting on display at a 2003 exhibition of Holbein paintings at the museum 'Mauritshuis' in The Hague; I have adopted the translation (even if not quite literal) from the accompanying leaflet.

74   E.J. Dijksterhuis, 'Ad Quanta Intelligenda Condita (Designed for Grasping Quantities)'. *Tractrix* 2, 1990, pp. 111-125 (originally Dijksterhuis' inaugural address at Leyden University on 21 January 1955); p. 120 (my italics).

75   As quoted in A.G. Debus, *The Chemical Philosophy. Paracelsian Science and Medicine in the Sixteenth and Seventeenth Centuries*. 2 vols. New York: Science History Publications (Watson), 1977; vol. 1, p. 77 (passage referred to Thomas Tymme, from 'the dedication to Sir Charles Blunt in Joseph Duchesne (Quercetanus), *The Practise of Chymicall, and Hermeticall Physicke, for the preservation of health*, trans. Thomas Tymme (London: Thomas Creedle, 1605)'.

76   F.A. Yates, *The Rosicrucian Enlightenment*. London: Routledge & Kegan Paul, 1972; introduction to Paladin edition of 1975, p. 17 (my italics).

77   J. Needham, *The Grand Titration. Science and Society in East and West*. London: Allen & Unwin, 1969; pp. 119-122.

78   Luis de Camões, *Lusiadas,* canto V.

79   Francis Bacon, *Works* 1, p. 191 (*Novum organum* I, aphorism 84; passage found through R. Hooykaas, 'The Rise of Modern Science: When and Why?' *British Journal for History of Science* 20, 4, 1987, p. 453-473; p. 470).

80   N. Sivin, lemma 'Shen Kua' in *DSB* 12, p. 369-393 (passage quoted from section 'Attitudes toward Nature' in a slightly revised version on Sivin's personal website).

81   This expression was Needham's preferred way to render that celebrated phrase (*SRHI*, pp. 384, 449, 474).

82   R. Hooykaas, *Science in Manueline Style. The Historical Context of D. João de Castro's Works.* Coimbra: Academia Internacional da Cultura Portuguesa, 1980; p. 127 (section 4.2.a.) (This is a separate edition of pp. 231-426 in vol. 4 of *Obras Completas de D. João de Castro* (eds. A. Cortesão & L. de Albuquerque), Coimbra, 1968-1980; 4 vols.).

83   Ibidem, p. 137.

84   C.V. Palisca, 'Scientific Empiricism in Musical Thought', in: H.H. Rhys (ed.), *Seventeenth Century Science and the Arts.* Princeton UP, 1961, pp. 91-137; p. 129-130.

85   P. Dear, *Mersenne and the Learning of the Schools.* Ithaca/London: Cornell UP, 1988; p. 221.

86   J.M. Forrester & J. Henry, 'Jean Fernel and the Importance of His *De Abditis Rerum Causis.*' Introduction to idem (ed., transl.), *Jean Fernel's On the Hidden Casuses of Things.* Leiden: Brill, 2005; p. 43; 49.

87   Johannes Kepler, *Gesammelte Werke* 3, p. 141.

88   Idem, *Gesammelte Werke* 6, pp. 104-105; p. 223.

89   Johannes Kepler to Herwart von Hohenburg, 14 September 1599: *Gesammelte Werke* 14, p. 73, l. 441-443.

90   Nicolaus Copernicus, *De Revolutionibus Orbium Coelestium*, book I, chapter 10 (p. 10 in the first edition).

91   V.E. Thoren, *The Lord of Uraniborg. A Biography of Tycho Brahe.* Cambridge UP, 1990, p. 126.

92   Johannes Kepler, *Gesammelte Werke* 3, p. 178. (I prepared the translation given in the main text with glances at those by M. Caspar on p. 166 of his translation *Neue Astronomie.* München: Oldenbourg, 1929; by E.J. Dijksterhuis on pp. 339-340 of his *De mechanisering van het wereldbeeld.* Amsterdam: Meulenhoff, 1950, and by W.H. Donahue on p. 286 of his translation *New Astronomy.* Cambridge UP, 1992; I further acknowledge gratefully C.A. Heesakkers' expert advice).

93   Johannes Kepler to Christoph Heydon, October 1605: *Gesammelte Werke* 15, p. 233, l. 91-92.

94   See note 2.

95   Galileo Galilei, *Opere* 1, p. 298-299. A rather less literal translation by Drabkin is on p. 65 of: I.E. Drabkin & S. Drake: *Galileo Galilei on Motion and on Mechanics.* Madison: University of Wisconsin Press, 1960.

96 Idem, *Opere* 7, pp. 141-142 (translation adopted from Drake's (on which more in the Notes on Literature Used appended to the present chapter): *Dialogue*, p. 116, with one small addition).

97 Ibidem, p. 171 (translation adopted from Drake's: *Dialogue*, p. 145).

98 E.J. Dijksterhuis, *The Mechanization of the World Picture*. Oxford UP, 1961; section IV 111 (passage retranslated in accordance with the original, Dutch edition of 1950).

99 Federico Cesi to Galileo Galilei in Firenze; Roma, 21 July 1612; *Opere* 11, pp. 365-367; p. 366: "se la via de' pianeti è elliptica, come vol Keplero."

100 Galileo Galilei, *Opere* 7, p. 369 (translation adopted from Drake's: *Dialogue*, p. 341).

101 "assai concludenti" (*Opere* 7, p. 487) was rendered by Drake (*Dialogue*, p. 462) as 'very convincing'. I suspect that Drake, whose command of Italian was superb, did so (see my pertinent remarks in the Notes on Literature Used for this chapter) to make Galileo look less 'guilty' of trespassing the 1616 ban on arguing about Copernicanism in terms of conclusiveness. Remarkably, one page further down, where the same term, '*concludente*', is used by Simplicio, Drake does render it as 'conclusive'.

102 Galileo Galilei, *Opere* 7, p. 29. The authorship of such phrases is uncertain, however, as the preface, the title page, and possibly other passages are known to be a coproduction between the author and his none-too-knowledgeable ecclesiastical censors. Whichever party thought of this and similar phrases, each must have read them in a very different way, with the intended or unintended irony escaping the censors.

103 Ibidem, p. 489 (Drake, *Dialogue*, p. 464, rather ignores the dual meaning of '*mirabile*' and leaves out '*veramente*', thus producing 'admirable and angelic doctrine').

104 Idem, *Opere* 8, p. 202-203 (translation adopted with two minute changes from Drake's: *Two New Sciences*, p. 158-159).

105 Idem, *Opere* 1, p. 299 (translation adopted from I.E. Drabkin & S. Drake: *Galileo Galilei on Motion and on Mechanics*. Madison: University of Wisconsin Press, 1960; p. 66).

106 Idem, *Opere* 2 , p. 179-180.

107 P. Dear, *Mersenne and the Learning of the Schools*. Ithaca/London: Cornell UP, 1988; p. 138.

108 Galileo Galilei, *Opere* 8, p. 209 (translation adopted with one minute alteration from Drake's: *Two New Sciences*, p. 166).

109 Ibidem, p. 212-213 (translation adopted from Drake's: *Two New Sciences*, p. 169-170).

110 This oft-quoted phrase is not to be found in the *Opere*. It blends Galileo's famous statement (*Opere* 6, p. 232) "La philosophia è scritta in questo grandissimo libro che continuamente ci sta aperto innanzi a gli occhi (io dico l'universo), ... Egli è scritto in lingua matematica ..." in his *Saggiatore* with what Viviani, in his biography of Galileo, says of him (*Opere* 19, p. 625), "che la libertà della campagna fossa il libro della natura ...; dicendo che i caratteri con che era scritto erano le proposizioni, figure e conclusioni geometriche, per il cui solo mezzo potevasi penetrare alcuno delli infiniti misterii dell'istessa natura".

111 Galileo Galilei, *Opere* 7, pp. 232-234 (translation adopted from Drake: *Dialogue*, pp. 206-208, with some minor alterations).

112 Idem, *Opere* 8, p. 296 (translation adopted from Drake: *Two New Sciences*, p. 245).

113 Ibidem, p. 274 (translation adopted from Drake: *Two New Sciences*, p. 223)

114 Ibidem, p. 276 (my italics; translation adopted from Drake: *Two New Sciences*, p. 225).

115 S. Shapin, *The Scientific Revolution*. Chicago: University of Chicago Press, 1996; p. 1 (The notion of the Scientific Revolution as "a coherent, cataclysmic, and climactic event", to be sure, is precisely what Shapin in the second line of his book *rejects*; my point here is that this is a caricature of even the most radically 'discontinuist' position ever put forward in serious historiography).

116 D. Gutas, *Greek Thought, Arabic Culture. The Graeco-Arabic Translation Movement in Baghdad and Early Abbasid Society (2nd–4th / 8th–10th centuries)*. London/New York: Routledge, 1998; p. 192 (final paragraph of the Epilogue; in the appended footnote Gutas refers to the late expert on Aristotle Werner Jaeger for making this particular point).

117 E.L. Eisenstein, *The Printing Press as an Agent of Change. Communications and Cultural Transformations in Early-Modern Europe*. 2 vols. Cambridge UP, 1979; passim, e.g. pp. 517-519.

118 Johannes Kepler, *Gesammelte Werke* 6, p. 279.

119 Johannes Kepler to Herwart von Hohenburg, 26 March, 1598: *Gesammelte Werke* 13, p. 193.

120 Albert Einstein, 'Remarks Concerning the Essays Brought Together in This Co-operative Volume'. In: P.A. Schilpp (ed.), *Albert Einstein: Philosopher-Scientist*. LaSalle (Illinois): Open Court, 1949; p. 684 (passage found through a course syllabus prepared by my late colleague Casper Hakfoort).

121 B. Stephenson, *Kepler's Physical Astronomy*. Princeton UP, 1994, p. 14.

122 Isaac Beeckman, *Journael* 1, p. 244.

123 Idem, *Journael* 4, p. 41.

124 Lucretius, *De rerum natura,* verses 2.398-407; 2.426-430 (English taken from W.H.D. Rouse's translation in the Loeb Classical Library, p. 127; p. 129; the passage on magnetism summarized slightly farther down in the main text comprises verses 6.906-1089; "primordia ferri" is in line 1006.)

125 René Descartes, *Oeuvres* 11, p. 146 (*Traité de l'homme*; the account in the main text blends this passage with one from *Les Météores*: *Oeuvres* 6, p. 237-238). The passage on magnetism summarized next is in *Oeuvres* 8, pp. 291-292, and 9, pp. 286-287 (*Principia Philosophiae*, part 4, article 152).

126 Idem, *Oeuvres* 1, p. 264.

127 J.A. Schuster, entry 'Descartes, René', in W. Applebaum (ed.), *Encyclopedia of the Scientific Revolution from Copernicus to Newton*. New York/London: Garland, 2000; pp. 184-188; p. 187.

128 Christiaan Huygens, *Oeuvres Complètes* 10, p. 403.

129 René Descartes, *Oeuvres* 10, p. 213 ('Cogitationes Privatae').

130 Idem, *Oeuvres* 8, p. 329 (*Principia philosophiae*: part 4, article 207; French translation of 1647: *idem* 9, p. 325).

131 Ibidem, p. 203 (*Principia philosophiae*: part 4, caption for article 1; French translation of 1647: *Oeuvres* 9, p. 201).

750   132   Idem, *Oeuvres* 2, p. 501 (Letter to Mersenne, 9 February 1639).

133   Christiaan Huygens, *Oeuvres Complètes* 10, p. 403.

134   René Descartes, *Oeuvres* 7, p. 22 ("genium aliquem malignum"); p. 34 ("ego sum res cogitans"); *Oeuvres* 9, p. 17 ("mauvais genie"); p. 27 ("ie suis une chose qui pense") (end of first and beginning of second meditation of the *Meditationes de prima philosophia*).

135   Isaac Beeckman, *Journal* 1, p. 25.

136   H. Butterfield, *The Origins of Modern Science 1300 - 1800*. London: Bell, 1957; p. 118 (first edition: 1949).

137   A concept coined, for almost the same purpose, by G. Sorel, *Les préoccupations métaphysiques des physiciens modernes*. Paris, 1905 ('Cahiers de la Quinzaine'; Série VIII, Numéro 16), esp. p. 58-9: "Le but de la science expérimentale est donc de construire une *nature artificielle* (si on peut employer ce terme) à la place de la *nature naturelle* ..."

138   Francis Bacon, *Works* 3, p. 156 (*The New Atlantis*).

139   Translation, published in 1577 by John Frampton, of a book by Nicolás Bautista Monardes entitled *Dos libros ... que trata de todos las cosas que traen de nuestras Indias Occidentales* (1565, with additional parts in 1571 and 1574; all this as set forth in Allen Debus, *Man and Nature in the Renaissance*. Cambridge UP: 1978, pp. 47-48).

140   J. Bylebyl, entry 'Harvey, William', in W. Applebaum (ed.), *Encyclopedia of the Scientific Revolution from Copernicus to Newton*. New York/London: Garland, 2000; pp. 285-288; p. 287.

141   As quoted in R. Porter, *The Greatest Benefit to Mankind. A Medical History of Humanity from Antiquity to the Present*. London: Fontana, 1997, pp. 215-216 (passage referred to the 1653 translation of *De Motu Cordis*).

142   As quoted in R.S. Westfall, *The Construction of Modern Science. Mechanisms and Mechanics*. Cambridge UP, 1971, p. 92.

143   Idem, p. 98.

144   As quoted in A.G. Debus, *The Chemical Philosophy. Paracelsian Science and Medicine in the Sixteenth and Seventeenth Centuries*, 2 vols. New York: Science History Publications, 1977; vol. 2, p. 319 (passage referred to Jean Baptiste van Helmont, *Ortus medicinae*. Amsterdam: Elsevier, 1648; p. 108-109, and to its English translation *Oriatrike*. London: Lloyd, 1662; p. 109).

145   W.R. Newman & L.M. Principe, *Alchemy Tried in the Fire. Starkey, Boyle, and the Fate of Helmontian Chymistry*. Chicago: University of Chicago Press, 2002, p. 90.

146   A.G. Debus, *The Chemical Philosophy. Paracelsian Science and Medicine in the Sixteenth and Seventeenth Centuries*, 2 vols. New York: Science History Publications, 1977; vol. 2, p. 341.

147   E. Reeves, *Galileo's Glassworks. The Telescope and the Mirror*. Cambridge, MA: Harvard UP, 2008; p. 5.

148   J.-F. Revel, *Pourquoi des philosophes, suivi de La cabale des dévots*. Paris: Laffont, 1976, p. 364: "Montaigne n'aurait jamais dû écrire une ligne, faute de pouvoir être à la fois et sans délai Galilée, Newton, Lavoisier ... Quant à nous, aujourd'hui, ce que nous avons à faire est suffisamment pénible: il est des époques où *ne pas* penser certaines choses exige déjà un énorme effort."

149 Galileo Galilei to Belisario Vinta, 19 March 1610: *Opere* 10, p. 298 (letter 277; translation adopted
     with some changes from M. Biagioli, *Galileo Courtier. The Practice of Science in the Culture of
     Absolutism*. Chicago: University of Chicago Press, 1993; p. 135).

150 René Descartes, *Oeuvres* 6, p. 72 (*Discours de la méthode*: part 6).

151 Francis Bacon, *Works* 3, p. 518 ('De Interpretatione Naturae Prooemium'; passage found through
     W. Durant, *The Story of Philosophy*. New York: Simon & Schuster, 1926; section 'The political
     career of Francis Bacon'; p. 108 in the Cardinal pocketbook edition of 1952).

152 J.A. Schumpeter, *Capitalism, Socialism and Democracy*. New York: Harper, 1942; ch. 11 (p. 124 in
     the third edition of 1950).

153 Christiaan Huygens, *Oeuvres Complètes* 16, p. 302, where the editors refer to Horace (*Epistolarum
     liber* I, no. 19, vs. 21), the verse quoted by Huygens on a manuscript page called by them "Appen-
     dix I to 'De vi centrifuga'".

154 Johannes Kepler, *Gesammelte Werke* 6, p. 290 (final lines of his 'Prooemium' for book V of *Har-
     monice Mundi*).

155 Galileo Galilei, *Opere* 8, p. 296 (translation adopted from Drake: *Two New Sciences*, p. 245).

156 Idem, *Opere* 7, p. 229 (translation adopted from Drake: *Dialogue*, p. 203, but with 'the questions
     of nature' instead of Drake's 'physical problems', and some other, smaller alterations as well).

157 Idem, *Opere* 8, p. 190.

158 I phrased this sentence with a close look at E.J. Dijksterhuis, *The Mechanization of the World
     Picture*. Oxford: Oxford University Press, 1961; section IV, 167.

159 F.J. Dijksterhuis, *Lenses and Waves. Christiaan Huygens and the Mathematical Science of Optics in
     the Seventeenth Century*. Dordrecht: Kluwer, 2004, pp. 27-28.

160 David Fabricius to Johannes Kepler, 20 January 1607: Johannes Kepler, *Gesammelte Werke* 15, p.
     377 (passage found through Arthur Koestler, *The Sleepwalkers*. London: Hutchinson, 1959; p. 352
     in the Penguin edition of 1964).

161 Peter Crüger to Phillip Müller, 1 July 1622 (old style): Johannes Kepler, *Gesammelte Werke* 18, p. 92
     (passage found through idem, p. 353 in the Penguin edition of 1964).

162 Isaac Newton, *Correspondence* 1, p. 297-303 (there dated "c. 1669"; passage found (with two mini-
     mal changes in the translation) through R.S. Westfall, *Never at Rest*. Cambridge UP, 1980; p. 152).

163 R.S. Westfall, *The Construction of Modern Science. Mechanisms and Mechanics*. New York: Wiley,
     1971; p. 124 (in all later editions with Cambridge UP).

164 J.C. Boudri, *What Was Mechanical About Mechanics? The Concept of Force between Metaphysics
     and Mechanics from Newton to Lagrange*. Dordrecht: Kluwer, 2002; p. 68.

165 Christiaan Huygens, *Oeuvres Complètes* 20, p. 133.

166 Benedetto Castelli, *Della misura dell'acque correnti*, Rome, 1628; p. 48 (reproduced on p. 49 of C.S.
     Maffioli, *Out of Galileo. The Science of Waters 1628-1718*. Rotterdam: Erasmus Publishing, 1994,
     with a translation on p. 48 which I adopt here with one minor alteration).

167 As quoted in M. Blay, 'Le développement de la balistique et la pratique du jet des bombes en

752   France à la mort de Colbert'. In: L. Godard de Donville (ed.), *De la mort de Colbert à la révocation de l'édit de Nantes: un monde nouveau?* Marseille: Centre Méridional de Rencontres sur le XVIIᵉ siècle, 1984, p. 33-51: p. 45 and p. 47, respectively; passages referred to Surirey de Saint-Rémy, *Mémoires d'Artillerie* of 1697, and to De Ressons, 'Méthode pour tirer les bombes avec succès'. *Mémoires de l'Académie des sciences.* 1716, p. 79.

168   D.S.L. Cardwell, *Turning Points in Western Technology. A Study of Technology, Science and History.* New York: Neale Watson, 1972; pp. 42-43; 38-39; idem, *The Fontana History of Technology.* London: Fontana Press, 1994; p. 87.

169   Christiaan Huygens to Constantijn Huygens, Jr., 11 May 1668: *Oeuvres Complètes* 6, p. 216 (passage found through K. Andersen, 'Stevin's Theory of Perspective: The Origin of a Dutch Academic Approach to Perspective'. *Tractrix* 2, 1990, pp. 25-62; p. 41).

170   A. Keller, lemma 'Fortification' in: W. Applebaum (ed.), *Encyclopedia of the Scientific Revolution from Copernicus to Newton.* New York/London: Garland, 2000; p. 239.

171   J.A. Bennett, *The Divided Circle. A History of Instruments for Astronomy, Navigation and Surveying.* Oxford: Phaidon/Christie's, 1987, p. 40.

172   As quoted from Leonard Digges (edited by his son Thomas), *A geometrical practise, Named Pantometria.* London, 1591; p. 55 (passage referred to in J.A. Bennett, *The Divided Circle. A History of Instruments for Astronomy, Navigation and Surveying.* Oxford: Phaidon / Christie's, 1987, p. 46).

173   J.A. Bennett, *The Divided Circle. A History of Instruments for Astronomy, Navigation and Surveying.* Oxford: Phaidon / Christie's, 1987, p. 62.

174   Christiaan Huygens, *Oeuvres Complètes* 9, p. 272-291; p. 289 (report to the Dutch East India Company of 1688).

175   As quoted in M. Blay, 'Le développement de la balistique et la pratique du jet des bombes en France à la mort de Colbert'. In: L. Godard de Donville (ed.), *De la mort de Colbert à la révocation de l'édit de Nantes: un monde nouveau?* Centre Méridional de Rencontres sur le XVIIᵉ siècle, 1984; p. 33-51: p. 49 ; passage referred to Surirey de Saint-Rémy, *Mémoires d'Artillerie* of 1697: preface, p. 2.

176   Constantijn Huygens, draft letter to H. de Beringhen, 14 October 1683. In: J.A. Worp (ed.), *De briefwisseling van Constantijn Huygens (1608-1687)*; 6 vols. Den Haag: RGP, 1911-1917: vol. 6, p. 450-1 (passage found through R. Hooykaas, *Experientia ac Ratione: Huygens tussen Descartes en Newton.* Leiden: Museum Boerhaave, 1979; p. 34).

177   J.C. Boudri, *What Was Mechanical About Mechanics. The Concept of Force between Metaphysics and Mechanics from Newton to Lagrange.* Dordrecht: Kluwer, 2002; p. 44-45.

178   Galileo Galilei, *Opere* 8, p. 149, as quoted in Weston's 1730 translation of the *Discorsi.*

179   As quoted in M. Blay, *Les raisons de l'infini. Du monde clos à l'univers mathématique.* Paris: Gallimard, 1993, p. 114 (passage referred to Evangelista Torricelli, *De motu gravium naturaliter descendentium et projectorum libri duo.* Firenze, 1644; book 2 'De motu aquarum', p. 191).

180   As quoted ibidem ('De motu aquarum', p. 192).

181   Christiaan Huygens, *Oeuvres Complètes* 19, p. 171.

182   As quoted in M. Blay, *Les raisons de l'infini. Du monde clos à l'univers mathématique*. Paris: Galli-
        mard, 1993, p. 117 (passage referred to the archival collections of the *Académie Royale des Sciences*,
        vol. 14, folio 94 vᵒ).

183   Christiaan Huygens, *Oeuvres Complètes* 16, p. 392 (passage found through J.G. Yoder, *Unrolling
        Time. Christiaan Huygens and the Mathematization of Nature*. Cambridge UP, 1988, p. 195, note
        16).

184   C.B. Boyer, *The History of the Calculus and Its Conceptual Development*. New York: Dover, 1959, p.
        51-52.

185   Ibidem, pp. 57-58.

186   H.J.M. Bos, 'Tradition and Modernity in Early Modern Mathematics: Viète, Descartes and Fer-
        mat', p. 192 (with Viète's own claim, and its Cartesian counterpart, being quoted on p. 187).

187   As quoted in C.B. Boyer, *The History of the Calculus and Its Conceptual Development*. New York:
        Dover, 1959, p. 154 (no place mentioned).

188   As quoted ibidem, p. 110 (likewise no place mentioned).

189   Isaac Newton, *Correspondence*, 2, pp. 179-180 (passage found through R.S. Westfall, *Never at Rest.
        A Biography of Isaac Newton*. Cambridge UP, 1980, p. 138).

190   As quoted in R.S. Westfall, *Never at Rest. A Biography of Isaac Newton*. Cambridge UP, 1980, p. 131
        (passage referred to *Add* MS 3968.41, f. 86ᵛ, dated about 1714).

191   Galileo Galilei, *Opere* 8, p. 208 (in Drake's *Two New Sciences*: p. 165).

192   Ibidem, p. 208-9 (translation adopted with one tiny alteration from Drake: *Two New Sciences*, p.
        165-6).

193   S. Weinberg, *Facing Up. Science and Its Cultural Adversaries*. Cambridge: Harvard UP, 2001; p. 6.

194   René Descartes, *Oeuvres* 6, p. 64 (*Discours de la méthode*: part 6).

195   Ibidem, p. 68; p. 69.

196   Isaac Beeckman, *Journael* 4, pp. 214-215.

197   Idem, *Journael* 1, p. 177.

198   René Descartes to Marin Mersenne, mid-January 1630 (passage reproduced by C. de Waard in
        Isaac Beeckman, *Journael* 4, p. 177).

199   Isaac Beeckman, *Journael* 3, p. 67.

200   Ibidem, p. 69.

201   'Réponse de Blaise Pascal au très bon Réverend Père Noël' (29 October 1647; p. 373 in Chevalier's
        'Pléiade' edition of Pascal's *Oeuvres Complètes*).

202   Christiaan Huygens, *Oeuvres Complètes* 10, p. 403.

203   A. Van Helden, 'The Birth of the Modern Scientific Instrument, 1550-1700'. In: J.G. Burke (ed.),
        The Uses of Science in the Age of Newton, 1983, pp. 49-84.

204   Walter Charleton, *Physiologia Epicuro-Gassendo-Charltoniana*. London: Thomas Heath, 1654
        (facs. reprint, with introduction and indices by R.H. Kargon; New York: Johnson Reprint, 1966

754   ('The Sources of Science, vol. 31')); p. 116 (passage found through C. Wilson, *The Invisible World. Early Modern Philosophy and the Invention of the Microscope*. Princeton UP, 1995; p. 57-58).

205   Henry Power, *Experimental Philosophy*. London, 1664, p. 193 (final sentence of the book; I found this passage as well as the preceding references to Malebranche and Patrick through O. Mayr, *Authority, Liberty and Automatic Machinery in Early Modern Europe*. Baltimore: Johns Hopkins UP, 1986, p. 68 and 83).

206   Christiaan Huygens, *Oeuvres Complètes* 10, p. 406.

207   K. Lasswitz, *Geschichte der Atomistik vom Mittelalter bis Newton*. Leipzig: Voss, 1890; vol. 2, p. 498 in the second ed. of 1926.

208   As quoted ibidem, p. 488 in the second ed. of 1926; referred to Claude François Milliet Dechales, *Cursus seu mundus mathematicus*. Lyon, 1674; no page mentioned.

209   K. Lasswitz, *Geschichte der Atomistik vom Mittelalter bis Newton*. Leipzig: Voss, 1890; vol. 2, p. 518 in the second ed. of 1926 (the one obscure name in May's list of modern authors is David de Rodon, mentioned in passing by Lasswitz on p. 500).

210   Ibidem, p. 301.

211   I felt honored when, soon after writing this sentence, a quite similar observation made by Leibniz in a letter to Malebranche in 1679 caught my eye on p. 238 of D.M. Clarke, *Occult Powers and Hypotheses. Cartesian Natural Philosophy Under Louis XIV*. Oxford UP, 1989.

212   K. Lasswitz, *Geschichte der Atomistik vom Mittelalter bis Newton*. Leipzig: Voss, 1890; vol. 2, p. 520 in the second ed. of 1926.

213   From Aristotle's *Peri Ouranou*, IV:6, in the translation by W.K.C. Guthrie, *On the Heavens*. London: Heinemann, 1939; Loeb Classical Library (pp. 367-369).

214   Galileo Galilei, *Opere* 4, p. 385; 391 (passage found through, and translation adopted from, M. Biagioli, *Galileo Courtier. The Practice of Science in the Culture of Absolutism*. University of Chicago Press, 1993; pp. 207-208).

215   Ibidem, p. 436 (passage found through, and translation (but for one phrase) adopted from, M. Biagioli, ibidem, pp. 192-193).

216   René Descartes to Marin Mersenne, 12 September and 11 October 1638: *Oeuvres* 2, p. 355; p. 380.

217   The expression (in frequent 17th-century usage) 'to live at the bottom of a sea of air' was first used, more or less, by Giovan Battista Baliani, in a letter to Galileo of 24 October 1630: *Opere* 14, letter 2075, p. 159: "Lo stesso [namely, phenomena at the bottom of the sea] mi è avviso che ci avvenga a noi nell'aria, che siamo nel fondo della sua immensità, nè sentiamo nè il suo peso che la compressione che ci fa da ogni parte ..." (passage found through C.S. Maffioli, *Out of Galileo. The Science of Waters 1628-1718*. Rotterdam: Erasmus Publishing, 1994, pp. 92).

218   Blaise Pascal, Proposition no. 4 in the second part of his *Expériences nouvelles touchant le vide*. Paris, 1647.

219   Gabriel Daniel SJ, *Voyage du monde de Descartes*. Paris, 1691, p. 188 (passage found through P. Mouy, *Le développement de la physique Cartésienne 1646-1712*. Paris: Vrin, 1934, p. 169).

220   Blaise Pascal, *Pensées*, n° 72 in Brunschvicg's arrangement. 755

221   As quoted in A.C. Duker, *Gisbertus Voetius* (4 vols.). Leiden: Brill, 1897-1915 (reprint Leiden: Groen, 1989); vol. 2, p. xlv-xlvi.

222   The full text of the condemnation is quoted on pp. 466-467 of vol. 1 of F. Bouillier, *Histoire de la philosophie Cartésienne*. Paris, 1868 (third ed. in 2 vols.; facs. reprint: Genève: Slatkine Reprints, 1970).

223   As quoted ibidem, p. 469, with reference to *Quaedam recentiorum philosophorum ac praesertim Cartesii propositiones damnatae et prohibitae*. Paris: 1705.

224   D.M. Clarke, *Occult Powers and Hypotheses. Cartesian Natural Philosophy Under Louis XIV*. Oxford UP, 1989, p. 16.

225   Ibidem, p. 27.

226   As quoted in F. Bouillier, *Histoire de la philosophie Cartésienne*. Paris, 1868 (third ed. in 2 vols.; facs. reprint: Genève: Slatkine Reprints, 1970; vol. 1, p. 470 (note 1).

227   Blaise Pascal, 'Mémorial' (of his night of religious ecstasy, Monday 23 November 1654; on pp. 552-553 in Chevalier's 'Pléiade' edition of his *Oeuvres complètes*).

228   E.A. Burtt, *The Metaphysical Foundations of Modern Physical Science. A Historical and Critical Essay*. London: Routledge & Kegan Paul, 1972 (reprint of second edition of 1932; first edition: 1924); pp. 298-299.

229   R.S. Westfall, *Science and Religion in Seventeenth-Century England*. New Haven: Yale UP, 1958 (on pp. 219-220 in the second edition; Ann Arbor: University of Michigan Press, 1971).

230   L. Jardine, *Ingenious Pursuits. Building the Scientific Revolution*. London: Little & Brown, 1999; p. 312.

231   René Descartes, *Oeuvres* 5, p. 548 (letter, possibly to Constantijn Huygens, written around 1642).

232   P. Findlen, *Possessing Nature. Museums, Collecting, and Scientific Culture in Early Modern Italy*. Berkeley / Los Angeles: University of California Press, 1994; p. 236.

233   Marin Mersenne, *Harmonie Universelle*. Paris, 1636/7 (facs. ed.: Paris: Centre National de la Recherche Scientifique, 1975); p. 347 of the 'Livres des instrumens'.

234   P. M. Gouk, *Music, Science and Natural Magic in Seventeenth-Century England*. New Haven/London: Yale UP, 1999; p. 191.

235   J. Hollander, *The Untuning of the Sky. Ideas of Music in English Poetry 1500-1700*. Princeton UP, 1981.

236   Francis Roberts, 'Concerning the Distance of the Fixed Stars'. *Philosophical Transactions* 18, 209; 1694; pp. 101-103; p. 103 (passage found through A. Van Helden, *Measuring the Universe*. Chicago: University of Chicago Press, 1985, p. 159).

237   M. Fournier, *The Fabric of Life. Microscopy in the Seventeenth Century*. Baltimore/London: Johns Hopkins UP, 1996; p. 58.

238   Jan Swammerdam (Herman Boerhaave ed.), *Biblia naturae; sive historia insectorum ... .* Leiden, 1737-1738, p. 664 (passage found through M. Fournier, *The Fabric of Life. Microscopy in the*

756   *Seventeenth Century*. Baltimore/London: Johns Hopkins UP, 1996; p. 67; with a few changes in the translation).

239 My onetime colleague M.J. Sparnaay considers Huygens an early specimen of this 'aha; but I can improve on that' type of scientific investigator Sparnaay himself came to recognize early on in his teacher, the Dutch chemical physicist and Nobel prize winner Peter Debye.

240 Samuel Pepys' diary for that day.

241 J.L. Heilbron, *Electricity in the 17th and 18th Centuries. A Study of Early Modern Physics*. Berkeley / Los Angeles: University of California Press, 1979, p. 179.

242 Ibidem, p. 3.

243 W.R. Newman & L.M. Principe, *Alchemy Tried in the Fire. Starkey, Boyle, and the Fate of Helmontian Chemistry*. Chicago: University of Chicago Press, 2002; p. 154.

244 R.S. Westfall, *The Construction of Modern Science. Mechanisms and Mechanics*. Cambridge UP, 1971; p. 79.

245 J.M. Engelbrecht, *De onttovering van de waanzin. Wetenschap, het bovennatuurlijke en de opkomst van een psychologisch mensbeeld* (to appear in 2011; end of ch. 5).

246 T.M. Brown, entry 'Lower, Richard' in *DSB* 8, pp. 523-527; p. 526.

247 J.M. Engelbrecht, *De onttovering van de waanzin. Wetenschap, het bovennatuurlijke en de opkomst van een psychologisch mensbeeld* (to appear in 2011; ch. 7).

248 P.M. Gouk, 'Music in the Natural Philosophy of the Early Royal Society' (unpublished doctoral dissertation, Warburg Institute, University of London, 1982), p. 102.

249 Narcissus Marsh, 'An Introductory Essay to the Doctrine of Sounds'; *Philosophical Transactions* 14, 1684; pp. 472-488; p. 482 (passage found through P.M. Gouk, *Music, Science and Natural Magic in Seventeenth-Century England*. New Haven/London: Yale UP, 1999, p. 185).

250 Frederick Slare, 'A Further Confirmation of the Above Said Contagion'. *Philosophical Transactions* 13, 1683, pp. 94-95 (passage found through C. Wilson, *The Invisible World. Early Modern Philosophy and the Invention of the Microscope*. Princeton UP, 1995; p. 158).

251 C. Wilson, *The Invisible World. Early Modern Philosophy and the Invention of the Microscope*. Princeton UP, 1995; pp. 171-173.

252 Ibidem, p. 235.

253 M.J. Sparnaay, *Adventures in Vacuums*. Amsterdam: North Holland, 1992; pp. 40-41.

254 Christiaan Huygens collection of papers in Leyden University Library: p.241 of v.22, midpage (coming from HUG 2, f.163r).

255 K. Neal, 'On the 'Attractions' and 'Uses' of Gilbert's Magnetic Philosophy: Gilbert and the Practical Mathematicians' (unpublished lecture delivered at Oberwolfach, 6 January 2003).

256 As quoted in B. Shapiro, *Probability and Certainty in Seventeenth-Century England*. Princeton UP, 1982; p. 221, with reference to Joseph Addison in the *Spectator* of July 11, 1711, no. 117 (passage found through J.M. Engelbrecht, 'De onttovering van de waanzin', ch. 7).

257 R. Porter, *The Greatest Benefit to Mankind. A Medical History of Humanity from Antiquity to the Present*. London: HarperCollins, 1997; p. 231.

258 K.H. Ochs, lemma 'Histories of Trades' in W. Applebaum (ed.), *Encyclopedia of the Scientific Rev-*
     olution from Copernicus to Newton. New York/London: Garland, 2000; p. 296.

259 Jonathan Swift, *Gulliver's Travels* (1726): third voyage, ch. 5 'The author permitted to see the grand
     Academy of Lagado. The Academy largely described. The arts wherein the professors employ
     themselves'.

260 Thomas Hobbes (ed. W. Molesworth), *Opera* 2, pp. 243-4; p. 236 (passage found through, and
     translation adopted from, Schaffer's translation in S. Shapin & S. Schaffer, *Leviathan and the Air-*
     Pump. Hobbes, Boyle, and the Experimental Life. Princeton UP, 1985, p. 353 and p. 347; cp. p. 115).

261 Francis Bacon, *Works* 3, p. 529 ('Temporis Partus Masculus', first section; passage found through
     J.M. Engelbrecht, *De onttovering van de waanzin. Wetenschap, het bovennatuurlijke en de opkomst*
     van een psychologisch mensbeeld).

262 Idem, *Works* 1, p. 643 ('De Augmentis Scientiarum', liber 5, caput 4; passage found through J.M.
     Engelbrecht, *De onttovering van de waanzin. Wetenschap, het bovennatuurlijke en de opkomst van*
     een psychologisch mensbeeld; I have adopted the English translation, with several changes, from
     Works 4, p. 431).

263 Robert Hooke, *Micrographia*. London, 1665; 6th and 7th pages of the Preface (or, if one follows
     the lettering underneath, the second and third pages marked 'b').

264 In a different meaning the phrase comes from Dante's *Paradiso*, canto 3, verses 1–3.

265 P. Findlen, *Possessing Nature. Museums, Collecting, and Scientific Culture in Early Modern Italy*.
     Berkeley / Los Angeles: University of California Press, 1994; p. 236-237.

266 J.L. Heilbron, *Electricity in the 17th and 18th Centuries. A Study of Early Modern Physics*. Berkeley
     / Los Angeles: University of California Press, 1979, p. 219.

267 Robert Hooke, *Micrographia*. London, 1665; 24th page of the Preface (or, if one follows the letter-
     ing underneath, the page just before the one marked 'g'; passage found through C. Wilson, *The*
     Invisible World. Early Modern Philosophy and the Invention of the Microscope. Princeton UP, 1995;
     p. 221).

268 J.L. Heilbron, *Electricity in the 17th and 18th Centuries. A Study of Early Modern Physics*. Berkeley
     / Los Angeles: University of California Press, 1979; pp. 4-5 (the passage quoted from Newton is in
     Correspondence 1, p. 364-365).

269 A. Stroup, lemma 'Botany' in W. Applebaum (ed.), *Encyclopedia of the Scientific Revolution from*
     Copernicus to Newton. New York/London: Garland, 2000; p. 98.

270 M. Hunter, *The Royal Society and its Fellows 1660–1700. The Morphology of an Early Scientific*
     Institution. Oxford: BSHS Monographs (vol. 4), 1994 (second ed.; originally 1982); p. 27.

271 R. Hooykaas: 'The Rise of Modern Science: When and Why?'. *British Journal for History of Science*
     20, 4, 1987, pp. 453-473; p. 470.

272 J.-F. Revel, *Histoire de la philosophie occidentale*, vol. 2, *La philosophie pendant la science*. Paris:
     Stock, 1970, p. 229.

273 Isaac Newton to Robert Boyle, 28 February 1678/9: *Correspondence* 2, p. 288 (passage found

758   through R.S. Westfall, *Force in Newton's Physics. The Science of Dynamics in the Seventeenth Century*. London: MacDonald, 1971; p. 370).

274   Blaise Pascal, *Pensées*, n° 79 in Brunschvicg's arrangement.

275   'Réponse de Blaise Pascal au très bon Réverend Père Noël' (29 October 1647; p. 374 in Chevalier's 'Pléiade' edition of Pascal's *Oeuvres Complètes*).

276   P. McLaughlin, 'Force, Determination and Impact'. In: S. Gaukroger, J. Schuster, J. Sutton (eds.), *Descartes' Natural Philosophy*. London: Routledge, 2000; pp. 81-112; p. 101.

277   René Descartes, *Oeuvres* 9, p. 93 (authorized, French translation of *Principia philosophiae*: part 2, section 53; the corresponding, more succinct passage in the Latin original is in *Oeuvres* 8, p. 70).

278   R.S. Westfall, *Force in Newton's Physics. The Science of Dynamics in the Seventeenth Century*. London: MacDonald, 1971, p. 150.

279   Christiaan Huygens, *Oeuvres complètes* 16, p. 132 (passage found through R.S. Westfall, *Force in Newton's Physics. The Science of Dynamics in the Seventeenth Century*. London: MacDonald, 1971, p. 153).

280   R.S. Westfall, *Force in Newton's Physics. The Science of Dynamics in the Seventeenth Century*. London: MacDonald, 1971; pp. 154-155.

281   René Descartes, *Oeuvres* 8, p. 62; p. 63 (*Principia philosophiae*: part 2, sections 37 and 39; translation adopted from D.E. Garber, *Descartes' Metaphysical Physics*. Chicago: University of Chicago Press, 1992; p. 201).

282   J.G. Yoder, *Unrolling Time. Christiaan Huygens and the Mathematization of Nature*. Cambridge UP, 1988, pp. 42-43.

283   R.S. Westfall, *Force in Newton's Physics. The Science of Dynamics in the Seventeenth Century*. London: MacDonald, 1971, p. 344 (my italics).

284   As quoted ibidem, p. 345 (referred to Newton's 'Waste Book', folio 12^v, and to the text as edited by J. Herivel, *The Background to Newton's Principia. A Study of Newton's Dynamical Researches in the Years 1664–1684*. Oxford: Clarendon, 1965; p. 157).

285   As quoted ibidem, p. 346 (referred to Newton's 'Waste Book', folio 12, and to J. Herivel, *The Background to Newton's Principia. A Study of Newton's Dynamical Researches in the Years 1664–1684*. Oxford: Clarendon, 1965; p. 150).

286   R.S. Westfall, *Force in Newton's Physics. The Science of Dynamics in the Seventeenth Century*. London: MacDonald, 1971; p. 348.

287   Idem, *Never at Rest. A Biography of Isaac Newton*. Cambridge UP, 1980, p. 155 (sentence slightly altered to fit better the construction of my own sentence).

288   As quoted ibidem, p. 143 (referred to *Add MS* 3968.41, folio 85).

289   Ibidem, p. 302 (the two quoted passages are referred to Isaac Newton (A.R. & M.B. Hall, eds.), *Unpublished Scientific Papers*. Cambridge UP, 1961; p. 144; p. 124, with the original Latin on p. 110; p. 92).

290   Idem, *The Construction of Modern Science. Mechanisms and Mechanics*. Cambridge UP, 1971; p. 134.

291 Christiaan Huygens, *Oeuvres complètes* 13/II, p. 741 (passage found through F.J. Dijksterhuis, *Lenses and Waves. Christiaan Huygens and the Mathematical Science of Optics in the Seventeenth Century.* Dordrecht: Kluwer, 2004, p. 112). 759

292 Isaac Newton, *Optical Papers* 1, pp. 170-171 and 312-313 (translation adopted from the editor's (A.E. Shapiro) with a few minor alterations; passage found through F.J. Dijksterhuis, 'Once Snel[l] Breaks Down: From Geometrical to Physical Optics in the Seventeenth Century'. *Annals of Science* 61, 2004, pp. 165-185; p. 174).

293 F.J. Dijksterhuis, 'Once Snel[l] Breaks Down: From Geometrical to Physical Optics in the Seventeenth Century'. *Annals of Science* 61, 2004, pp. 165-185; pp. 183-184.

294 As quoted from the manuscript source in R.H. Kargon, *Atomism In England From Hariot to Newton.* Oxford: Clarendon Press, 1966, p. 69.

295 R.S. Westfall. 'Newton and the Hermetic Tradition', in: A.G. Debus (ed.), *Science, Medicine and Society in the Renaissance. Essays to Honor Walter Pagel.* 2 vols. New York: Science History Publications, 1972: vol. 2, pp. 183-198; p. 187.

296 As quoted from the manuscript source in J. Henry, 'Occult Qualities and the Experimental Philosophy: Active Principles in Pre-Newtonian Matter Theory'. *History of Science* 24, 1986, pp. 335-381; p. 351.

297 Ibidem, p. 342.

298 As quoted from the manuscript source in R-M. Sargent, *The Diffident Naturalist. Robert Boyle and the Philosophy of Experiment.* Chicago: University of Chicago Press, 1995, p. 164.

299 Robert Boyle, as quoted in Birch's introduction to Boyle's *Works* 1, p. xlix (passage found through A.G. Debus, *The Chemical Philosophy. Paracelsian Science and Medicine in the Sixteenth and Seventeenth Centuries,* 2 vols. New York: Science History Publications, 1977, p. 474).

300 Idem, *Works* 3, p. 426 (passage found ibidem, p. 480).

301 As quoted in T.S. Kuhn, 'Robert Boyle and Structural Chemistry in the Seventeenth Century'. *Isis* 43, 1952, p. 12-36; p. 22 (passage referred to Boyle, *Works* 2, p. 474).

302 Ibidem, p. 28.

303 Ibidem, p. 31.

304 Robert Hooke, *Posthumous Works*, p. 3 (passage found through J. Henry, 'Robert Hooke, the Incongruous Mechanist'. In: M. Hunter & S. Schaffer (eds.), *Robert Hooke. New Studies.* Woodbridge: Boydell Press, 1989; pp. 149-180; p. 163).

305 Ibidem, p. 165 (passage found ibidem).

306 Isaac Newton, *Correspondence* 1, p. 376 (i.e., in the 'Hypothesis of Light' of 1675; used later as Rule III of the 'Regulae Philosophandi' in the final version of the *Principia*).

307 Robert Hooke, *Micrographia.* London, 1665; p. 28 (passage found through J. Henry, 'Robert Hooke, the Incongruous Mechanist'. In: M. Hunter & S. Schaffer (eds.), *Robert Hooke. New Studies.* Woodbridge: Boydell Press, 1989; pp. 149-180; p. 162).

308 Idem, 'Cometa'. In: idem, *Lectures and Collections.* London: John Martyn, 1678 (passage found ibidem).

760   309   Idem, *Micrographia*. London, 1665; p. 15 (all italics removed; passage found through P.M. Gouk, 'The Role of Acoustics and Music Theory in the Scientific Work of Robert Hooke'. *Annals of Science* 37, 1980, pp. 573-605; p. 586).

310   Idem, *Posthumous Works*, p. 184 (passage found through J. Henry, 'Robert Hooke, the Incongruous Mechanist'. In: M. Hunter & S. Schaffer (eds.), *Robert Hooke. New Studies*. Woodbridge: Boydell Press, 1989; pp. 149-180; p. 157).

311   Idem, *Micrographia*. London, 1665; p. 16 (all italics removed; passage found ibidem).

312   Ibidem, p. 31-32 (all italics removed; passage found ibidem).

313   Idem, *Posthumous Works*, p. 172 (passage found ibidem, p. 151).

314   Idem, *Micrographia*. London, 1665, p. 127 (all italics removed; passage found ibidem).

315   Isaac Newton, *Correspondence* 1, p. 367.

316   Ibidem, p. 371.

317   Ibidem, p. 364.

318   Ibidem.

319   Ibidem, p. 366.

320   Ibidem, p. 368.

321   As quoted from the manuscript source in R.S. Westfall, *Never at Rest. A Biography of Isaac Newton*. Cambridge: Cambridge UP, 1980, p. 307.

322   T.K. Rabb, *The Struggle for Stability in Early Modern Europe*. Oxford UP, 1975, p. 112 (word order adapted to fit my text).

323   Thomas Sprat, *The History of the Royal-Society of London*. London, 1667 (facs. reprint: London: Routledge & Kegan Paul, 1966); p. 56 (passage found through S. Shapin & S. Schaffer, *Leviathan and the Air-Pump. Hobbes, Boyle, and the Experimental Life*. Princeton UP, 1985; p. 306).

324   Francis Bacon, *Works* 3, p. 218 (second page of first chapter of his fragmentary treatise, written under the pseudonym 'Valerius Terminus', *Of the Interpretation of Nature*; passage found through C. Webster, *The Great Instauration. Science, Medicine and Reform 1626–1660*. London: Duckworth, 1975, p. 22).

325   Christiaan Huygens, *Oeuvres complètes* 6, p. 95.

326   L. Jardine, *Ingenious Pursuits. Building the Scientific Revolution*. London: Little & Brown, 1999; p. 157.

327   In the last years of his life R.S. Westfall conducted a large-scale survey of patronage in the 16th and 17th centuries. Some preliminary results appeared in his 'The Background to the Mathematization of Nature'. In: J.Z. Buchwald & I.B. Cohen (eds.), *Isaac Newton's Natural Philosophy*. Cambridge (MA): MIT Press, 2001; pp. 321-339. On p. 331 he mentioned in this connection that "a young man might commit himself to mathematics with confidence." In a letter to me of 15 June 1995 he wrote in response to my pertinent query: "About patronage: I am uncertain, but on the whole I incline to a seller's market – never forgetting that ultimate power lay with the patron."

328   D.S. Landes, *Revolution in Time. Clocks and the Making of the Modern World*. Cambridge (Mass.; Harvard UP), 1983; p. 112.

329   Thomas Sprat, *The History of the Royal-Society of London*. London, 1667 (facs. reprint: London: 761
Routledge & Kegan Paul, 1966); all passages here quoted are on pp. 365-369.

330   G. Rees, lemma 'Baconianism' in W. Applebaum (ed.), *Encyclopedia of the Scientific Revolution from Copernicus to Newton*. New York/London: Garland, 2000; pp. 70-71.

331   J. Ben-David, *The Scientist's Role in Society. A Comparative Study*. Englewood Cliffs, NJ: Prentice Hall, 1971; p. 66.

332   As quoted in W.H. Austin, lemma Glanvill, *DSB* 5, pp. 414-417; p. 416 (passage referred to Glanvill's *Plus Ultra*).

333   John Dryden, *Of Dramatick Poesie, An Essay*; vol. 1 (p. 12 in M. Summers' 1931 edition of Dryden, *The Dramatic Works*).

334   R.S. Westfall, 'The Rise of Science and the Decline of Orthodox Christianity: A Study of Kepler, Descartes, and Newton'. In: D.C. Lindberg & R.L. Numbers, *God and Nature. Historical Essays on the Encounter between Christianity and Science*. Berkeley: University of California Press, 1986; pp. 218-237; p. 231-233 (passage referred to a manuscript known as 'Yahuda MS 41', folio 7).

335   Blaise Pascal, *Pensées*; n° 395 in Brunschvicg's arrangement.

336   J.M. Keynes, 'Newton, the Man'. In: *The Royal Society: Newton Tercentenary Celebrations 15–19 July 1946*. Cambridge UP, 1947; pp. 27-34; p. 27.

337   René Descartes, *Oeuvres* 6, p. 62 (*Discours de la méthode*: part 6).

338   Francis Bacon, *Works* 1, p. 151 (translation adopted from *Works* 4, p. 39, with the first phrase altered).

339   R. S. Westfall, *Force in Newton's Physics. The Science of Dynamics in the Seventeenth Century*. London: Macdonald, 1971, p. 136, p. 256.

340   As quoted in E.J. Dijksterhuis, *Val en worp*. Groningen: Noordhoff, 1924, p. 147 (referred to Leonardo da Vinci, manuscript A, p. 34^v in vol. 1 of *Les manuscrits de Léonard de Vinci* (ed. C. Ravaisson-Mollien). 6 vols. Paris: Quantin, 1881–1891; I thank Paolo Gozza for helping me translate this passage properly).

341   Christiaan Huygens, *Oeuvres complètes* 19, p. 631 (passage found through, and translation adopted from, R. S. Westfall, *Force in Newton's Physics. The Science of Dynamics in the Seventeenth Century*. London: Macdonald, 1971, p. 186).

342   Idem, *Oeuvres complètes* 18, p. 496; 497 (passage found through, and translation adopted from, R. S. Westfall, *Force in Newton's Physics. The Science of Dynamics in the Seventeenth Century*. London: Macdonald, 1971, pp. 177-179).

343   R. S. Westfall, *The Construction of Modern Science. Mechanisms and Mechanics*. Cambridge UP, 1971, p. 135.

344   As quoted in idem, *Force in Newton's Physics. The Science of Dynamics in the Seventeenth Century*. London: Macdonald, 1971, p. 298 (passage referred to Gottfried Wilhelm Leibniz, *Philosophische Schriften* 2, p. 154-156: letter to Burchard de Volder, January 1699).

345   As quoted in J.C. Boudri, *What Was Mechanical About Mechanics. The Concept of Force between*

762   *Metaphysics and Mechanics from Newton to Lagrange.* Dordrecht: Kluwer, 2002; p. 89 (passage referred to Gottfried Wilhelm Leibniz, 'Specimen dynamicum'. *Acta eruditorum*, April 1695, p. 235).

346   Robert Hooke, *Attempt to prove the motion of the Earth.* London: John Martyn, 1674, pp. 27-28. In: idem, *Lectiones Cutlerianae, or a Collection of Lectures: Physical, Mechanical, Geographical, & Astronomical.* London: John Martyn, 1679.

347   R. S. Westfall, *Force in Newton's Physics. The Science of Dynamics in the Seventeenth Century.* London: Macdonald, 1971, p. 210.

348   As note 346.

349   As quoted in R. S. Westfall, *Never at Rest. A Biography of Isaac Newton.* Cambridge UP, 1980, p. 403, from a manuscript source ('Joseph Halle Schaffner Collection, University of Chicago Library, MS 1075-7') containing Abraham de Moivre's account of how Newton later recalled the story.

350   J.M. Keynes, 'Newton, the Man'. In: *The Royal Society: Newton Tercentenary Celebrations 15–19 July 1946.* Cambridge UP, 1947; pp. 27-34; p. 28.

351   R. S. Westfall, *Never at Rest. A Biography of Isaac Newton.* Cambridge UP, 1980, p. 96.

352   Ibidem, pp. 305-307 (the quoted passages are referred to 'Burndy MS 16, folio 3'; edited and printed since by B.J. Teeter Dobbs as 'Appendix A' in her *The Janus Faces of Genius. The Role of Alchemy in Newton's Thought.* Cambridge UP, 1991; pp. 256-270). The phrase "both are perpetual workers" is a later insertion by Newton.

353   B.J. Teeter Dobbs & M.C. Jacob, *Newton and the Culture of Newtonianism.* Amherst NY: Humanity Books, 1998; p. 30.

354   Isaac Newton, *Philosophiae Naturalis Principia Mathematica*, Book II, end of General Scholium at the end of Section 6 (taken from idem, *The Third Edition (1726) With Variant Readings Assembled and Edited by Alexandre Koyré & I. Bernard Cohen, With the Assistance of Anne Whitman*: Cambridge UP, 1972; vol. 1, pp. 462-463; translation adopted from idem, *The Principia. Mathematical Principles of Natural Philosophy. A New Translation by I. Bernard Cohen & Anne Whitman.* University of California Press, 1999; p. 723, but for the final three sentences, which Newton left out of the second and third editions, and for which I adopt the translation in R.S. Westfall, *Never at Rest. A Biography of Isaac Newton* Cambridge UP, 1980; pp. 376-7).

355   R. S. Westfall, *Never at Rest. A Biography of Isaac Newton.* Cambridge UP, 1980; p. 387-388.

356   Ibidem, p. 436-437.

357   Ibidem, p. 412.

358   Idem, *Force in Newton's Physics. The Science of Dynamics in the Seventeenth Century.* London: Macdonald, 1971; p. 433 (italics mine).

359   As quoted on pp. 318-319 of the edition of one of Newton's revisions of 'De Motu ...' provided by J. Herivel in his *The Background to Newton's Principia. A Study of Newton's Dynamical Researches in the Years 1664–1684.* Oxford: Clarendon, 1965 (passage found through, and translation adopted

from, R. S. Westfall, *Never at Rest. A Biography of Isaac Newton*. Cambridge UP, 1980; pp. 421-2).   763

360   Compiled (with a few corollaries that pertain rather more tangentially to the central argument left out) from D. Densmore, *Newton's* Principia: *The Central Argument. Translation, Notes, and Expanded Proofs*. Santa Fe: Green Lion Press, 1995 (for the quoted passages I have adopted W.H. Donahue's translations in the same book).

361   Isaac Newton, *Philosophiae Naturalis Principia Mathematica* (taken from idem, *The Third Edition (1726) With Variant Readings Assembled and Edited by Alexandre Koyré & I. Bernard Cohen, With the Assistance of Anne Whitman*: Cambridge UP, 1972; translations adopted with a few minor changes from idem, *The Principia. Mathematical Principles of Natural Philosophy. A New Translation by I. Bernard Cohen & Anne Whitman*. University of California Press, 1999).

362   R. S. Westfall, *Force in Newton's Physics. The Science of Dynamics in the Seventeenth Century*. London: Macdonald, 1971; p. 494.

363   Ibidem, p. 511 (the first passage quoted from the *Principia* is from corollary 4 to proposition 52 of Book II; the second, from the scholium to proposition 53, is the penultimate sentence of Book II).

364   D. Densmore, *Newton's* Principia: *The Central Argument. Translation, Notes, and Expanded Proofs*. Santa Fe: Green Lion Press, 1995; p. 155 (italics for 'unnatural' removed).

365   R. S. Westfall, *Never at Rest. A Biography of Isaac Newton*. Cambridge UP, 1980; p. 441 (a few words from the previous sentence inserted).

366   The translation of this passage, which ends the first paragraph of the 'Auctoris praefatio' to the first edition, has been adopted with some alterations from Isaac Newton, *The Principia. Mathematical Principles of Natural Philosophy. A New Translation by I. Bernard Cohen & Anne Whitman*. University of California Press, 1999; pp. 382-393.

367   Isaac Newton to John Collins, 18 February 1670: *Correspondence* 1, p. 27 (passage found through R.S. Westfall, *Never at Rest*, p. 224; untransposed, the sentence runs thus: "It [i.e., public esteem] would perhaps increase my acquaintance, ye thing wch I cheifly study to decline".)

368   Isaac Newton to Edmond Halley, 20 June 1686: *Correspondence* 2, pp. 435-440; p. 438.

369   R. S. Westfall, *Never at Rest. A Biography of Isaac Newton*. Cambridge UP, 1980; p. 472.

370   A.I. Sabra, *Theories of Light from Descartes to Newton*. Cambridge UP, 1981; p. 322.

371   R. S. Westfall, *Never at Rest. A Biography of Isaac Newton*. Cambridge UP, 1980; p. 91 (passage referred to Add MS 3996, f. 103^v).

372   Ibidem (passage referred to Add MS 3996, f. 122).

373   R. S. Westfall, *The Construction of Modern Science. Mechanisms and Mechanics*. Cambridge UP, 1971; p. 57-58.

374   A.I. Sabra, *Theories of Light from Descartes to Newton*. Cambridge UP, 1981; p. 323.

375   Ibidem, p. 325-326.

376   R. S. Westfall, *Never at Rest. A Biography of Isaac Newton*. Cambridge UP, 1980; p. 217 (the quotation referred to is Add MS 3970.3, f. 352^v).

377   Ibidem, p. 218.

764 378 Ibidem, p. 220.

379 Ibidem, pp. 221-222 (the passage referred to is in the 'Discourse of Observations', as reproduced in I. Bernard Cohen (ed.), *Isaac Newton's Papers & Letters on Natural Philosophy*. Cambridge UP, 1958; p. 224).

380 Ibidem, pp. 171-172.

381 Isaac Newton to Henry Oldenburg, 18 January 1671/2 and 11 June 1672: *Correspondence* 1, pp. 82-83; p. 183 (passages found through R. S. Westfall, *Never at Rest*. Cambridge UP, 1980; p. 237, p. 170).

382 Isaac Newton, 'New Theory about Light and Colors'. *Philosophical Transactions*. February 19, 1672/1, p. 3085 (as reproduced in I. Bernard Cohen (ed.), *Isaac Newton's Papers & Letters on Natural Philosophy*. Cambridge UP, 1958; p. 57).

383 R.S. Westfall, *Never at Rest. A Biography of Isaac Newton*. Cambridge UP, 1980; p. 244, 247.

384 'Mr. Newtons Answer to the foregoing Letter'. In: *Philosophical Transactions*. 1672, p. 4014 (as reproduced in I. Bernard Cohen (ed.), *Isaac Newton;'s Papers & Letters on Natural Philosophy*. Cambridge UP, 1958; p. 99; the translation is as reproduced there from an 18th century translation (p. 106), but with some alterations I have adopted from R.S. Westfall, *Never at Rest. A Biography of Isaac Newton*. Cambridge UP, 1980; p. 242-243, and with some further alterations made by myself).

385 Listing prepared with some help from the 'Analytical Table of Contents' compiled by D.H.D. Roller on pages lxxix-cxvi of the standard modern edition of *Opticks* (New York: Dover, 1952) by way of a reprint of the fourth: London, 1730.

386 Isaac Newton, *Opticks*, Book II, Part 3, Proposition 12 and Definition (pp. 278, 281 of the New York: Dover, 1952 edition).

387 Isaac Newton to Henry Oldenburg, 6 February 1671/2: *Correspondence* 1, pp. 96-97.

388 Isaac Newton to Henry Oldenburg, 6 July 1672: *Correspondence* 1, p. 209.

389 Isaac Newton, *Opticks*, Book III, Part 1 (p. 339 of the New York: Dover, 1952 edition).

390 As quoted in R.S. Westfall, *Force in Newton's Physics. The Science of Dynamics in the Seventeenth Century*. London: Macdonald, 1971; p. 389 (no manuscript source mentioned).

391 Isaac Newton, *Opticks*, Book III, Part 1 (p. 374 of the New York: Dover, 1952 edition).

392 Ibidem, p. 372.

393 Ibidem.

394 Ibidem, p. 377.

395 Ibidem, p. 382-383.

396 Ibidem, p. 400.

397 R.S. Westfall, *Force in Newton's Physics. The Science of Dynamics in the Seventeenth Century*. London: Macdonald, 1971; pp. 385-386 (the passages quoted from Newton are referred to as Add MS 3965.6, f. 266^v, and Add MS 3970.3, f. 296, respectively).

398 As quoted ibidem, p. 396 (the manuscript text differs somewhat from the one printed in 1756: I. Bernard Cohen (ed.), *Isaac Newton's Papers & Letters on Natural Philosophy*. Cambridge UP, 1958; p. 302-3).

399 Ibidem, p. 396.

400 As quoted ibidem, p. 397 (passage referred to Add MS 3970.9, f. 619).

401 Ibidem, p. 397.

402 Isaac Newton, *Opticks*, Book III, Part 1 (pp. 385-386 of the New York: Dover, 1952 edition).

403 R.S. Westfall, *Force in Newton's Physics. The Science of Dynamics in the Seventeenth Century*. London: Macdonald, 1971; p. 399.

404 Ibidem, p. 398.

405 Jonathan Swift, *Gulliver's Travels* (1726): third voyage, ch. 8 'A further account of Glubbdubdrib. Ancient and modern history corrected'.

406 S. Shapin, *The Scientific Revolution*. Chicago: University of Chicago Press, 1996; p. 165.

407 http://sci.esa.int/science-e/www/object/index.cfm?fobjectid=31240 (illustrated with the trajectory of the spacecraft).

408 Marquis de Condorcet, *Esquisse d'un tableau historique des progrès de l'esprit humain* (1793): first lines of the 10th 'époque', 'Les progrès futurs de l'esprit humain'.

409 E. Ihsanoglu, *Science, Technology and Learning in the Ottoman Empire*. Aldershot: Ashgate Variorum, 2004; p. V-246-7.

410 Ibidem, p. V-247; the passage quoted (without a place being given) is taken from the principal in question, Ishak Efendi.

411 V. Boss, *Newton and Russia. The Early Influence, 1698-1796*. Cambridge (MA; Harvard UP), 1972; p. 51.

412 Ibidem, p. 61.

413 K. Mahbubani, *The New Asian Hemisphere. The Irresistible Shift of Global Power to the East*. New York: Public Affairs, 2008; p. 61.

414 Immanuel Kant, first lines of his 'Beantwortung der Frage: Was ist Aufklärung?' (1784; *Berlinische Monatsschrift*).

415 R. Dawkins, *The God Delusion*. London: Transworld Publishers, 2006; p. 124.

416 V. Icke, 'Astrologie is geen lieve maar gevaarlijke onzin' ('Astrology Is Not Gentle But Dangerous Nonsense'). In: *NRC/Handelsblad*, 3 September 2008.

417 M. Gell-Mann, *The Quark and the Jaguar. Adventures in the Simple and the Complex*. New York: Freeman, 1994, p. xiv.

418 Auguste Comte (eds. M. Serres, F. Dagognet, A. Sinaceur), *Philosophie première. Cours de philosophie positive. Leçons 1 à 45*. Paris: Hermann, 1975; p. 39.

419 See note 249.

NAME INDEX

Abbas (Mohammed's uncle), 54
Addison, Joseph, 480
Agricola (Georg Bauer), 131, 250
Agrippa (Cornelius von Nettenheim), 133, 139, 472
Aiton, E.J., 216, 401
Al Ashari, 72
Al Battani, 64, 68, 726
Al Birjandi, 69, 71, 91, 160, 214
Al Biruni, xxix, 55, 59, 60-65, 68-72, 78, 91-92, 136, 149, 160, 204, 213
Al Farabi, 56-58, 61, 64
Al Ghazali, 56, 66, 72, 209, 240
Al Khwarizmi, 57, 64, 95
Al Kindi, 55-58, 61, 64, 70, 91, 149
Al Mamun, 54, 64, 93, 591
Al Mansur, 54-55
Al Qushji, 69-71, 91
Al Razi, 62-64, 88
Al Urdi, 68, 452
Albert the Great (Albertus Magnus), 79, 91, 149
Alberti, Leon Battista, 131
Aldrovandi, Ulisse, 130, 137, 446
Alexander of Aphrodysias, 8, 10, 56
Alexander the Great, 5, 10, 27, 30, 46
Alhazen, 55, 82, 146, 207, 294, 297, 299, 524. See also Ibn al-Haytham
Almeida, O.T., 154-155
Andersen, K., 369, 371
Anne Boleyn, 587
Apianus, Petrus 22
Apollonios, 10, 29, 57, 102-103, 173, 192, 345, 356, 366
Applebaum, W., xxxix, 244, 268, 369-370, 506-508, 546, 716
Aquinas, Thomas, 79-80, 93-94, 146, 427-428, 590-591

Archimedes
    in Greece, 10-13, 18-19, 29, 47, 346-350, 357, 404
    in Islamic civilization, 57, 59
    in Middle Ages, 82, 84, 88
    in Renaissance, 101-103, 131, 139, 143, 150-151
    in 17th century, 179-181, 199, 204, 207, 282, 293, 336, 353-354, 366, 404-405, 407, 411, 524, 600, 623
Aristarchos, 15, 25, 29, 106, 209, 395
Aristotle. See Subject Index s.v. Aristotelianism
Aristoxenos, 25, 64, 144-145, 247
Arkwright, Richard, 729-730
Artigas, M., 442
Augustine, St., 79, 93, 112, 420-421, 427-428, 438
Auzout, Adrien, 575
Averroes. See Ibn Rushd
Avicenna. See Ibn Sina

Bacon, Francis, **245-248**, 261, 266, 281-284, 378, 392, 440, 490, 578-579, 586, 589, 592, 607, 714. *See also in Subject Index* Baconian Brew *and* Baconian Ideology
    against Gilbert, 249, 251
    against Greek-style nature knowledge, 45, 239, 382, 601, 617
    fact-finding experiment, 245-248, 499, 576. *See also below*, B., sound
    general reform, xvi, 137, 245-247, 472, 566, 585
    idols, 486, 721, 736
    natural history, method and proposed application, 246-248, 259, 446, 503, 614-615
    self-confidence, 276
    sound, 247-248, 448-450, 473-474
    truth and error, xxix
    worldview, 247-248, 550
Bacon, Roger, 82, 102
Bagrow, L., 370

768

Bala(subramaniam), A., 219
Baliani, Giovan Battista, 408
Barker, A., 51
Barker, P., 634
Barnes, J., 50
Barrow, Isaac, 299, 329, 335, 366, 541, 679, 681
Bartholinus, Erasmus, 300, 366, 514, 539, 541-542
Barton, T., 51
Beeckman, Isaac, 221-222, **224-227**, 230-233,
    237-241, 261, 272, 276, 281, 301-302, 367, 374-381,
    391-393, 397, 418, 445, 521, 528-529, 552, 606,
    612, 655
    and Descartes, 221-222, 226-227, 231, 241, 378-
    381, 393
Bellarmino, Cardinal Roberto, 189, 420-422, 428,
    433
Ben-David, J., xxii-xxxix, 50-51, 287, 443, 595-597
Benedetti, Giovanbattista, 143-144, 147-150, 159-
    160, 181-182, 203, 300, 317
Benedict, St., 264
Bennett, J.A., 216, 370
Bentley, Richard, 703, 709-710
Berkel, K. van, 243, 369, 442
Bernoulli (family), 728
Bertoloni Meli, d., 368, 635
Bessarion, Cardinal, 101
Biagioli, M., 219, 244, 441-442, 507, 595
Black, Joseph, 729
Blackwell, R.J., 371, 441
Blaeu, Willem, 129, 322
Blay, M., xxxix, 370-371
Blondel, François, 314, 583
Boas-Hall, M., 506-508
Bock, Hiëronymus, 128, 136, 149, 266
Boethius, Anicius Manlius Severinus, 32, 48, 77,
    81, 145
Bohm, David, 736
Boileau, Nicolas, 436
Bono, M. di, 155
Boorstin, D.J., 154
Borelli, Giovanni Alfonso, 305, 393-394, 398, 496,
    503, 510, 516, 624, 631
Born, Derich, 128
Bos, H.J.M., 370-371, 546
Boss, V., 740

Boudri, J.C., 369-371, 635, 716, 740
Bouillier, F., 442
Boulliau, Ismaël, 304, 409, 439
Boulton, Matthew, 730
Boyer, C.B., xxxiii, 350, 369-371, 717
Boyle, Robert, 239, 386, 392, 446, 464, 476, 480,
    498, 509, 531, 560, 592, 647, 676, 678, 700, 702-
    703, 709, 711
    air, pump, and void, 414, 459-462, 476, 484,
        485, 558, 576, 658
    chemistry and alchemy, 466-469, **554-557**, 713
    clash with Hobbes, xviii, 483-484, 571
    corpuscles, views on, xxxvi, 221, 230, 285, 503,
        512, 549, 554-557, 587, 641-642
    law of, 462, 501, 516
    light, 680, 683-684, 687, 690
    practice of experiment, views on, xxxiii, 414,
        487-491, 554-557, 571
Bradwardine, Thomas, 86
Brock, W.H., 506
Bruce, Iakov Vilimovich, 728
Brunfels, Otto, 128, 136, 139, 149, 266
Bruno, Giordano, 159, 348, 396
Bukofzer, M.F., 156
Buonamici, Francesco, 404, 405
Burch, C.B., 546
Bürgi, Jost, 330
Buridan, Jean, 83-85, 88-89, 130, 180, 184, 239, 258,
    307-308, 404-405, 489, 527
Burnett, C., 51, 97, 155
Burtt, E.A., 401, 437, 564, 634
Butterfield, H., 244
Bylebyl, J., 268

Cabeo, Niccolò, 464, 495
Capra, Fritjof, 736
Cardano, Girolamo, 148, 249, 494
Cardwell, D.S.L., 155, 369, 740
Cartwright, Edmund, 729
Caspar, M., 216, 369
Cassini, Giovan Domenico (Jean-Dominique),
    452-454, 488, 496-497, 516, 568, 575, 579-580,
    583, 662, 701, 724
Castelli, Benedetto, 312, 420, 439
Castro, João de, xxix, xxxiii, 139-141, 149, 245, 261,

265, 446, 487-489, 509, 511, 721

Cavalieri, Bonaventura, 353

Cavendish, Charles, 550, 578

Cavendish, Margaret, 550-551, 553

Cavendish, William, 550, 578

Cesi, Federico, 187

Charles I, 252, 490, 549

Charles II, 461, 497, 500, 571, 577-580, 589

Charleton, Walter, 385, 509, 551-553

Chen Shao-Wei, 39

Cheng Ho, 136, 265

Cheng I, 43

Chilmead, Edmund, 448, 450

Chottin, A., 75

Christina (queen of Sweden), 373

Christina de' Medici, 420

Chrysippos, 8-9, 20, 29

Chuang Tzu, 35, 138

Clagett, M., 50

Clarke, D.M., 244, 443, 677

Clarke, Samuel, 677

Clauberg, Johann, 383

Clavius, Christoph, xxix, 143, 146-150, 180, 212-
   213, 222, 232, 495, 726

Clercq, P. de, 595

Clerselier, Claude, 425, 435, 566

Cohen, I.B., 716

Colbert, Jean-Baptiste, 497, 579-580, 589, 620

Collins, John, 354

Colombe, Lodovico delle, 405-407, 419, 485, 601

Colombo, Realdo, 252-253

Colton, J., 595

Comenius (Jan Amos Koménsky), 586, 592

Comte, Auguste, 737

Condorcet, Nicolas de, 724

Confucius, 33, 35

Cook, H.J., 506

Copernicus, Nicolaus, xvii, xxiv, **105-113**, 128,
   149-150, 160, 208, 253, 348, 387, 396, 452-453,
   515, 572, 670, 693
   in China, 727
   early adherents, 159
   and Galileo, 159, 181, 186-190
   and Kepler, 159, 161-164, 167-174, 177
   objections, 111-113, 129, 187-189

'opportunist' stance, 24, 109, 259

   in Ottoman Empire, 726

   and Ptolemy, 13, 58, 105-110

   realism, ambiguity of, 107-109,, 112-113, 208-
      210, 213, 303, 395, 417, 712

   De Revolutionibus, Books I and II-VI con-
      trasted, 108-110, 113, 163, 209

   Tusi couple, 68, 106, 207

Cosimo II de' Medici, 417

Croll(ius), Oswald, 256

Crombie, A.C., xxviii

Crompton, Samuel, 729

Cromwell, Oliver, 514

Crosby, A.W., 569, 595

Crüger, Peter, 303

Dalton, John, 602

Dante Alighieri, 81, 386

Darby, Abraham, 481

Daston, L., xxxix, 506

Daumas, M., 370

Davids, C.A., 595

Dawkins, R., 735, 737

Dear, P., xxxix, 153, 244, 441, 507, 546

Debus, A.G., 153-154, 268, 564

Dechales, Claude François Milliet, 389-392

Delalande, Michel-Richard, 579

Demokritos, 9, 19, 221, 223, 381, 384-386, 392, 602

Densmore, D., 659, 662, 716

Derham, William, 494

Desargues, Girard, 319

Descartes, René, xviii, xxviii, xxx, **226-241**, 258,
   276, 284, 399, 431, 445, 447, 471, 510, 512, 550,
   599, 601, 607, 641, 658
   'Alexandria', 59, 202, 233, 306, 366
   and Beeckman, 221-222, 226-227, 231, 241,
      378-381, 393
   atomism, 227-230, 710
   certainty, skepticism, and 'cogito', 235-237, 379,
      435, 603
   clash over faith, xxiv, 390, 418, 422-424, 432,
      439, 464, 535, 568, 593
   color, 299
   critized, 233, 237, 382-383, 388-389, 415, 423,
      497, 517, 535, 543, 602, 628, 630, 645, 668,

770 671-672, 714, 728

disciples, 234-235, 383, 393, 400, 423, 425-426, 435, 551, 566, 574-575, 579, 728

dualism, 227-228, 381, 434, 645

force, 623-624

geometry, 233, 351-354, 356

heliocentrism, 227, 386-387, 396-397, 423, 712

imagination, 234, 714

kinetic corpuscularianism, xvi, 227-228, 284, 367, 373-374, 380, 388-389, 391, 416-417, 534, 551-552, 576, 587, 612, 655, 667, 670, 719

light, nature of, 233, 299, 541, 679, 681, 692

magnetic action, 229, 463, 553

methodology, 614

motion, 228, 231, 238, 241, 345, 386, 411, 521, **522-532**, 535, 623, 627, 629, 656, 664

musical intervals, 226, 301, 377-378

novelty, 238, 282, 384

on Galileo, 241, 408, 484

'physics and mathematics', 221, 226, 232

refraction, 59, 202, 205, 296-299, 306, 328, 398, 539, 543, 683, 685

thinking for oneself, with safeguards, 385-386, 435-436, 588, 733

two modes of nature knowledge simultaneously, 299, 509

youthful choices, 231-238

Dickreiter, M., 216

Digby, Kenelm, 392, 398-399, 504, 551

Digges, Leonard, 321

Digges, Thomas, 322, 348, 396

Dijksterhuis, E.J., xxiii, xxxvi, xl, 50-51, 97, 152, 218, 243-244, 368-369, 441, 546, 635

Dijksterhuis, F.J., 368-370, 546, 717

Diophantos, 29-30, 57, 67, 70, 91, 149

Dioskorides, 139

Dobbs, B.J., 546, 634, 716-717, 740

Donahue, W.H., 216

Dorn, H., xxiii, xxxix

Dorstenius, Johann Daniel, 570

Drabkin, I.E., 217, 368

Drake, S., 153-155, 179, 217-218, 368

Dryden, John, 588

Duccio di Buoninsegna, 128

Duhem, P., xix, xxx, 19, 50, 97, 203

Dürer, Albrecht, 131, 319

Dury, John, 586

Dyck, W. van, 216

Eamon, W., 154, 507

Ehrlich, M.E., 564

Einstein, Albert, 210, 602, 604, 723, 736

Eisenstein, E.L., xix, xxvi, 635

Elisabeth of Bohemia, 373

Elizabeth I, 249

Engelbrecht, J.M., 401, 507

Engelfriet, P.M., 156, 371, 740

Epikouros, 7-9, 29, 31, 221, 223, 226, 235, 385, 551, 602, 719

Erasmus, Desiderius, 100, 300, 514

Eratosthenes, 321, 515

Euclid, 10-13, 17-18, 25, 29-31, 46, 55-59, 81-82, 101-102, 131, 146, 150, 164, 208, 295, 345, 349, 354, 360, 660, 726

Euler, Leonhard, 723, 728

Evelyn, John, 586

Fabri, Honoré, 409, 425, 464, 495

Fabrici d'Aquapendente, Girolamo, 252-253

Fabricius, David, 214, 303-304

Fakhry, M., 75

Farrington, B., 50

Favaro, A., 103, 217

Feldhay, R., 442

Fend, M., 51, 155

Ferdinando II de' Medici, 496, 578

Fermat, Pierre de, 103, 351-356

Fernel, Jean, 143, 146-148, 150, 222

Ferrier, Guillaume, 328

Fibonacci (Leonardo di Pisa), 95

Ficino, Marsilio, 240

Field, J.V., 51, 216, 634

Findlen, P., 154, 506, 508

Finocchiaro, M.A., 441

Flamsteed, John, 451-453, 498, 516, 578-580, 655-657, 662, 669, 674, 701

Fontenelle, Bernard le Bovier de, 425

Forbes, R.J., 368

Forrester, J.M., 156

Fournier, M., 506

Fracastoro, Girolamo, 148, 250
Francesco di Giorgio, 131
Freiberg, Dietrich von, 82, 299
Fresnel, Augustin-Jean, 700, 713
Freudenthal, G. (Gad), 634
Friedrich II, 88-89, 92, 95, 130, 137, 261
Fuchs, Leonhardt, 121, 128, 136, 149, 266
Furet, F., xxvii

Gabbey, A., 546
Gabriel, Daniel, 414
Galen, 16, 59, 61, 64, 128, 139, 147, 252-258, 379,
 470-472, 618
Galilei, Galileo, 143, **178-200**, 205, 207, 215, 217-
 219, 251, 281, 284, 293, 396, 413, 445, 494, 510,
 531, 572, 574, 593, 599, 612, 631, 639, 656, 676,
 693
 Aristotelianism, views on, 109, 180-181, 186-
  187, 210, 239, 382, 601, 616
 common sense, 111, 182-184
 crafts, 182, 208, 309-310, 312-317, 322, 324-326,
  334-335, 472, 606-607
 clash with Rome, xxiv, **189-191**, 227, 258, 278,
  418-422, 426, **428-432**, 434, 496
 consonant intervals, 301-302, 339-340, 377
 disciples of, 202, 292, 294, 309, 326, 353, 403,
  411, 439, 496-497, 528
 falling bodies, 90, 103, 150, 180-181, **191-199**,
  232, 307-308, 336-340, 342, 357-360, 367, 403,
  408-410, 461, 504, 509, 513, 625, 627-629, 657,
  668, 712, 721
 floating bodies, 294, 403-408, 419, 485
 heliocentrism, 159, 178, 181, 187-191, 345, 395,
  417-423, 434, 436, 533, 658
 and Kepler, 167, 175, 178-179, 185-188, 303, 327,
  362
 magnetic action, 266
 mathematical humanism of, 151, 179-182, 203,
  206, 242
 mathematics/experiment relation, **194-200**,
  363-364, 414
 mathematization of nature, xvi, 201
 motion, xxvii-xxviii, 70, 159-160, 175, **182-186**,
  192-194, 210, 221, 225, 241, 272, 337, 367, 521,
  523-526, 528-531, 536, 556, 623, 626-627, 632,
  654, 663-664, 666-667

natural philosophy, 24, 211, 237, 391, 603-604,
 736. *See also above* G., Aristotelianism
 novelty, 282-283, 385, 733
 patronage. *See* G., social position
 pendulum, 330-331, 337-339, 343, 537
 possibly similar figure elsewhere, 72, 204, 213,
  261, 592
 projectiles, 192, 313
 realist-mathematical science, 26, 151, **178-215**,
  222, 231, 233, 238, 259, 271, 291, 323, 378, 400,
  407, 416-417, 484, 602, 672, 714
 self-confidence, 275-277
 social position of, 101, 103, 179-181, 206, 212,
  406-407, 415, 417-418, 429
 telescopic observations, 173, 186, 275, 278, 317,
  327-328, 387, 395, 406, 448, 516, 579
 versatility of, 361-364, 397, 610
 void, 410, 461, 729
Galilei, Vincenzio (Galileo's son), 330
Galilei, Vincenzo (Galileo's father), 143-146, 148-
 149, 182, 245-247, 261, 300, 311, 446, 510, 605
Garber, D.E., 243-244, 546
Gassendi, Pierre, 222, 226, 235, 238, 241, 261, 281,
 367, 391-392, 398, 408, 411, 418, 439, 445, 452,
 504, 512, 527-531, 550-552, 560, 568, 587, 612,
 641, 679, 692, 719
Gaukroger, S.W., xxxix, 244, 401, 442, 546, 634
Gauß, Carl Friedrich, 88
Geber, 88
Gellibrand, Henry, 478, 487
Gemma Frisius, 316, 321-322, 332
Gerard of Cremona, 77-78
Gilbert, William, 173, 229, 245, **249-252**, 259-261,
 266, 281, 316, 440, 446, 463-465, 478, 500-503,
 510, 553, 566, 613, 631, 721
Gille, B., 153
Gingerich, O., 216-217
Giotto di Bondone, 128
Glanvill, John, 571, 576, 585-587, 592
Glauber, Johann Rudolf, 479
Goethe, Johann Wolfgang von, 138
Goldstone, J., xl
Gouk, P.M., 51, 155, 268, 506, 564, 634
Gould, S.J., 429, 443
Gozza, P., 97

772

Graaf, Reinier de, 459
Graham, A.C., 50
Grant, E., 74, 97
Grattan-Guinness, I., 51, 74, 97, 370-371
Grazia, Vincenzo di, 406, 415, 484
Grégoire de Saint Vincent, 356
Gregory, James, 314
Grew, Nehemiah, 454-457, 509
Grimaldi, Francesco, 680-681, 698
Grunebaum, A. von, 74-75
Guericke, Otto von, 459-461, 464, 476, 487-488, 504
Guglielmini, Domenico, 313, 324
Guicciardini, N., 371
Guido d'Arezzo, 95
Guidobaldo dal Monte, 103, 179, 181, 194, 317, 336, 606
Guldin, Paul, 212
Gunter, Edmund, 322, 478
Gutas, D., 74-75

Hakfoort, C., 717
Hale, Matthew, 553-554, 643
Halley, Edmond, 306, 445, 514, 535-536, 599, 631, 633, 637, 640, 647-653, 659, 674-675, 678, 694
Hargreaves, James, 729
Harriot, Thomas, xxix, 59, 202, 205, 275, 297, 306, 314, 366, 511, 679
Harrison, John, 326, 333, 729
Hartlib, Samuel, 481, 586, 592
Harvey, William, xviii, 245, **252-255**, 259-261, 266, 281, 440, 446, 456, 459, 463, 470-471, 480, 490, 501-503, 512, 566, 604, 613, 619
Hauksbee, Francis, 706
Hazard, P., 595
Heilbron, J.L. , 97, 268, 465, 506-508
Heisenberg, Werner, 602
Helmholtz, Hermann von, 302, 451, 473, 698, 738
Helmont, Jean Baptiste van, **245-261**, 266, 281, 418, 439-440, 446, 463-470, 501-503, 512, 553, 555, 566, 568, 613
Henry VIII, 587
Henry, J., xxvi, xxxix, 154-156, 564, 596, 717
Herakleitos, 9
Herivel, J., 716

Hermes Trismegistos, 24, 133, 139, 240, 258
Herôn, 13, 130
Hesse, M., 268
Heuraet, Hendrik van, 354
Hevelius, Johannes, 329, 334-335, 451, 516
Hipparchos, 13-14, 29-31, 70, 91, 106, 149, 188, 669
Hippokrates, 16, 147, 474
Hobbes, Thomas, xviii, 239, 399-400, 461, 484-485, 490, 512, 550-553, 571, 681
Hohenburg, Herwart von, 167
Holbein, Hans the Younger, 128
Hooke, Robert, 285, 471, 481, 494, 503, 504, 549, 554, 592, 624-625, 652, 655, 656, 660, 664, 678, 713
  air pump, 460-461, 476
  corpuscular thinking, 557-562, 630, 642-643
  experimental practice, 485-488, 498, 571
  light and color, 561, 680-695, 706
  microscope, 454-457
  and Newton, xxiv, 536, 637, 640-641, 647-650, 694-696, 699-703, 711
  orbital motion, 305, 631-633, 637, 640, 647-650
  sound, 450
Hooykaas, R., 97, 153-155, 243, 368, 441, 508, 546, 597, 634
Horrocks, Jeremiah, 304-305, 439, 614, 711
Howitt, P., 75
Hudde, Johan, 354-356
Huff, T.S., xxiii, xxv, xxxix
Hughes, H.S., 740
Hulagu Khan, 68
Hunter, M., 508, 564
Huygens, Christiaan, 70, 239, 282, 391, 494, 499, 662
  Académie, relations with, 326, 341-342, 364, 439, 452-453, 497, 514, 568, 575, 579-580, 583, 589, 721
  air pump, 461-462, 476, 483
  'Alexandria', 294, 366
  compared with Newton, xxiv, 531-539, 543, 544, 621, 649, 656, 664, 667-668, 672, 674-678, 689-692, 694-695, 698-701, 705-706, 710-711, 736
  cosmological views, 727-728
  crafts, xxxiii, 312, 317, 324, 334

Descartes and Cartesianism, 233-234, 237, 383-384, 388, 398, 464, 510, 524, 527, 630
electric action, 464, 510
faith, 536
force. *See under* H., motion
gunpowder engine, 476-477
harmonic speculation, 304
improving others' work, 457, 461, 476, 504, 699
infinitesimal ratios, 345, 353-354, 360
isochrony and pendulum clock, 331-333, 343-346, 360, 530, 537-538
light and vision, 299, **539-543**, 689-691, 694-695, 698-699, 705-706. *See also under* H., telescope
longitude. *See under* H., crafts *and* H., isochrony
missile named after, 724
motion, 314, 340, 410, **521-539**, 609, 621, 625-627, 631, 640, 653-654
musical intervals, 302, 312
pioneer of hypothetical natural philosophy, 285, 413, **521-528**, 557, 678, 713-714
telescope, 299, 328-329, 334, 451, 516, 583, 679, 681
wonder months (fall of 1659), 342-346, 362, 502, 530
Huygens, Constantijn (Christiaan's brother), 317
Huygens, Constantijn (Christiaan's father), 282, 335, 410

Ibn al-Haytham, 25-26, 55, 59, 62-65, 70-72, 82, 91, 102, 139, 149, 204, 207, 213, 294-300, 306, 679
Ibn as-Shatir, 67-70, 78, 89, 91, 149
Ibn Qutaiba, 66
Ibn Rushd, 56, 62, 67, 70, 78, 91, 149
Ibn Sahl, 59, 202, 205, 297
Ibn Sina, xxix, 56-58, 61, 64-65, 70, 84, 91, 149, 258, 509-511
Ihsanoglu, E., 75, 740
imaginary observer, 141-143, 149-151, 201, 238, 253, 283, 286, 566, 599, 611-612, 620, 714
Ishaq ibn Hunayn ibn Ishaq, 55
Israel, J., 741

Jabir ibn Hayyan, 62, 88
Jacob, J.R., xxxix
Jacob, M., 716, 740
James I, 246
James II, 589, 620
Jardine, L., xxxix, 506-508
Jardine, N., 152
João III, 140
Johannes Philoponos, 31, 57, 67, 70, 84, 91, 149
Johns, A., 635
Jones, A., 74
Jordanus de Nemore, 336

Kamal al-Din, 60
Kandel, David, 128
Kant, Immanuel, 363, 620, 733, 737
Kargon, R.H., 564
Kassler, J.C., 546, 634
Kemp, M., 115-116, 153
Kepler, Johannes, 195, 203, 206-207, 276, 281, 291-293, 353, 362, 366-367, 395, 423, 429, 434, 439, 445, 593, 614, 639, 693, 712
celestial physics, xxxvi, 168-169, 172-174, 210, 417, 600, 612, 631
counterevidence, how dealt with, 169-172, 363, 451, 721
faith of, 161-162, 211-212, 418, 421, 433
and Galileo, 167, 175, 178-179, 185-188, 303, 327, 362
harmony, 162, 167, 174-177, 179, 301, 416, 612
hybrid philosophy of nature, 211-212, 240-241, 604, 652, 711
laws, 70, 167, 173-174, 177, 304, 521, 534, 612-614, 633, 648-653, 657-661, 668, 670, 676
light and vision, 59, 102, 205, 295-299, 327-329, 335, 362, 451-452, 524, 612, 679
magnetic action, 173, 251, 266, 510
mathematical humanism of, 26, 151, 161, 203
planetary theory, xvii, 108, 150, 151, **159-178**, 208, 271, 302-306, 316, 362, 452, 496, 504, 612, 658, 726, 735
possibly similar figure elsewhere, 72, 213-214
realist-mathematical science, xvi, 26, **160-178**, 200-201, 215, 225, 239, 261, 271, 360, 602, 672, 714

774   and Tycho, 159, 167-171, 174, 177, 188, 200, 208, 212, 327
Keynes, J.M., 605, 637-638, 674, 712
King, D.A., 75
Kircher, Athanasius, 447-449, 473-475, 487-488, 495, 509, 513, 517, 572-573, 680, 684
Kissinger, H.A., 519
Knowles Middleton, W.E., 508
Koestler, A., 217, 369, 442
Kors, A.C., 442
Koyré, A., xix, xxx, 51, 154, 201, 218-219, 244, 370, 716, 724, 740
Ktesibios, 13, 130, 195
Kubilai Khan, 41
Kuhn, J.R., 154
Kuhn, T.S., xx, xxii, 152, 268, 368-370, 441, 564
Kundera, M., xxvii

La Mettrie, Julien Offray de, 435
Lana Terzi, Francesco, 464, 501
Landes, D.S., 47, 52, 370, 595
Lao Tzu, 33, 35, 138
Laplace, Pierre-Simon, 494, 722
Lasswitz, K., 243, 287, 392, 400-401, 634
Lattis, J.M., 156, 442
Lauxtermann, P.F.H., 155, 401, 741
Lavoisier, Antoine-Laurent, 467-469, 602, 726
Le Cazre, 367, 398, 408-409
Leclerc, Sébastien, 581
Leeuwenhoek, Antoni van, 392, 454-459, 498-499, 509, 568, 571, 599
Leezenberg, M., 75
Leibniz, Gottfried Wilhelm, 464, 488, 524, 568, 602, 604, 623, 677-678, 723, 728
    against attractions, 677-678, 706-707, 710
    calculus, 342, 354, 359-360, 391, 494
    force, 308, 625-630, 654, 664, 674-675
Leonardo da Vinci, 114-119, 128, 131, 137-141, 149, 245, 261, 265, 313-315, 446-447, 472, 622-623
Leopold, J.H., 370
Leopoldo de' Medici, 496, 577-578
Leybourne, William, 322
Lieburg, M. van, 243
Lindberg, D.C., xxx, xxxix, 50-51, 74, 97, 153, 368, 596-597

Linus (Hall), Francis 461, 694
Lipperhey, Hans, 274-275, 327
Livingstone, D.N., 154, 596
Lloyd, G.E.R., 50-52
Locke, John, 399, 676
Lorentz, Hendrik Antoon, 736
Louis XIV, 335, 425-426, 431, 440, 497, 567, 577-580, 587-589, 592, 690
Louis XVI, 514
Loukianos, 30, 100
Louvois, François-Michel, 335, 497, 620
Lovelock, James, 736
Lower, Richard, 471, 480
Lucretius, 7, 31, 32, 223-224, 228-230, 469, 537, 558, 710
Lully, Jean-Baptiste, 579
Luther, Martin, 113

Maffioli, C., 369, 441
Mahbubani, K., 730-732, 741
Maier, A., 97
Malebranche, Nicolas, 386
Malpighi, Marcello, 454-459, 475, 480, 499, 501, 504, 568
Maricourt, Pierre de, 88-89, 95, 130, 134, 137, 139, 229, 249-250, 261
Mariotte, Edme, 315-316, 326, 439, 497, 527, 575, 624, 654
Marr, John, 478
Marsh, Narcissus, 474, 739
Mästlin, Michael, 159, 162, 167, 303
Maurits of Orange, 221, 274, 275, 319
Maurolyco, Francesco, 103
May, Heinrich, 391, 393
Mayr, O., 401
McClaughlin, T., 401
McGuire, J.E., 634
McLaughlin, P., 546
McMullin, E., 442
McNeill, W.H., 52
McVaugh, M., 97
Megenberg, Conrad, 124
Mehmet II, 99
Melsen, A. van, 243
Mercator, Gerard, 129, 265, 321, 322

Mercator, Nicolaus, 305
Merrett, Christopher, 481
Mersenne, Marin, 366, 513, 550-551, 569
    correspondence network, 373, 513
    experimentation, 266, 342, 408-410, 449, 462,
        494, 501, 509, 516
    falling bodies, 342-343, 345, 408-410, 494, 509,
        515, 530, 537
    music, harmony, and sound, 301-302, 311, 376-
        378, 448-450, 462, 501, 516, 537, 559, 607
Merton, R.K., xix, xxxiii, 269, 507, 595-597, 732,
    741
Mijnhardt, W.W., 741
Mo Ti, 33, 37, 44
Moerbeke, Willem van, 82, 103
Mohammed, 54, 93
Molière, Jean Poquelin, 425, 579
Molyneux, William, 329
Montaigne, Michel de, 100, 235
Montanari, Geminiano, 313
Montgomery, S.L., 51, 74-75, 97, 152
Montmor, Henri Louis Habert de, 439
Moore, Jonas, 498, 578-582, 589
More, Henry, 398-399, 504, 553, 602
More, Thomas, 100
Mouy, P., 401, 442
Münster, Sebastian, 126-127
Musschenbroek (family), 483, 570
Mydorge, Claude, 550

Neal, K., 507
Needham, J., xx, xxv, xxx, 27, 50-51, 135, 138, 154-
    155, 219
Nersessian, N.J., xxxix, 370, 741
Netz, R., 51
Neugebauer, O., 152
Newcomen, Thomas, 478, 482, 608, 729
Newman, W.R., 97, 268, 506, 564
Newton, Isaac, xvii-xviii, xxxiv, xxxvi, 70, 72, 178,
    211, 284-286, 293, 302, 382, 391, 413, 439, 445,
    479, 483, 491-494, 501, 503-505, 549, 554, 558,
    572, 574, 599, 601, 606, 609, 611, 616, 618, 637-
    640, 722, 728, 733-734
    active principles, 641, 643-647
    calculus, 314, 346, 349, **354-362**, 651, 660

checks upon 'fansying', xxxii, 516-517, 672,
    700-701, 713-714
chemistry and alchemy, **466-470**, 501, 561-562,
    643-644, 704
comparisons, xxiv, 531-539, 674-675, 690-692,
    699-702, 706, **711-712**, 736
corpuscular conceptions, 561-562, 641-643,
    645-647
electricity, 499-500
faith, 437, 593-594, 709-710, 735
force, 169, 308, 462, 621-633, 641, **647-674**, 707-
    711, 719, 723. *See also below under* N., motion,
    nature of
laws of motion, 522, **649-654**, 659-660, 734
light and color, 329, 334-335, 414, 451, 455, 465,
    539-544, 561-562, 678, **681-707**
motion, nature of, 305, 314, 521, **527-536**, 664-
    667
outside comments, 675-678, 694-696
synthesis, 565, 604, 707-714
telescope, 334-335, 451, 483, 631, 683-684, 687
universal gravitation, 306, 316, 454, 605, **654-**
    663, 667, 723-724
Noël, Étienne, 413-415, 461, 517
Norman, Robert, 250
Nouhuys, T. van, 243, 287, 634
Nunes, Pedro, 137-140, 149, 265, 320

O'Brien, P.K., 268
Ochs, K.H., 507
Ogilvie, B., 153
Oldenburg, Henry, 485-487, 491, 494, 497-498,
    513, 571, 586, 643, 647, 693, 701
Olschki, L., 154, 219
Oresme, Nicole, 85-91, 111-112, 147-149, 205, 307,
    358
Ornstein, M. [Bronfenbrenner], 507
Ørsted, Hans Christian, 463, 500
Orta, Garcia de, 136, 149, 265, 446
Osiander, Andreas, 109, 113, 395
Osler, M.J., 243, 634

Pagel, W., 268
Palestrina, Giovanni Pierluigi da, 738
Palisca, C.V., 155

776   Palissy, Bernard, 131
Palmer, R.R., 595
Palmerino, C.R., 441
Paltrow, G., 75
Panofsky, E., 219
Papin, Denis, 477, 482, 608
Pappos, 31, 103, 293, 351-352
Paracelsus (Theophrastus Bombastus von Ho-
    henheim), 88, 132-134, 137, 149, 246, 255-258,
    266, 472, 479, 552, 604, 607
Pardies, Ignace, 391, 514, 541-543, 694-695
Park, K., xxxix
Parmenides, 5-6, 9, 19, 45
Pascal, Blaise, 213, 294, 349, 353, 359, 382, 392, 395,
    397, 403, 412-415, 436, 491, 493, 517, 527, 603,
    729
Patrick, Simon, 386
Patrizi, Francesco, 251
Pesic, P., 268
Peter the Great, 727
Petit, Pierre, 440, 575
Petty, William, 550, 553-554, 643
Philo of Alexandria, 32, 80, 93
Picard, Jean, 452-453, 488, 514, 656, 674
Plato. *See* Subject Index s.v. Platonism
Pliny the Elder, 31-32, 139-140, 258
Plotinos, 32, 56, 240
Pontchartrain, Louis de, 497
Popkin, R.H., 244, 717
Porta, Gianbattista della, 248-249, 275, 487
Porter, R., 216, 268, 506
Porter, T., 740
Poseidonios, 32
Poullain de la Barre, François, 436, 588
Poussin, Nicolas, 579
Power, Henry, 387, 457, 509
Praetorius, Michael, 449
Prigogine, Ilya, 736
Principe, L.M., 97, 268, 506-507, 564
Proklos, 31, 67, 70, 91, 149
Ptolemy (Klaudios Ptolemaios), 24, 27, 29-31, 46,
    64, 67, 70, 89-91, 136, 139, 149-151, 204, 207, 600
    bridge-building by, 13-14, 23-27, 48, 49, 62, 210,
      294, 306
    determination of place on Earth, 26, 126-127,
      129

harmony, 25-26, 58-59, 102, 144, 161, 176
light and vision, 18, 59, 202, 297-299, 679
measure of universe, 25, 452-453, 618
natural philosophy, views on, 24-25, 62, 101,
    109, 209, 259, 391, 604
planetary theory, 24-25, 57-58, 61-62, 68-69,
    81-82, 105-107, 110-112, 163-164, 170, 187-188,
    214, 395, 451, 515, 572, 669-670, 726
Ptolemy (successive kings of Egypt), 10, 16, 30,
    429
Pumfrey, S., 268, 506
Pyrrhôn, 8, 29, 603
Pythagoras, 11, 26, 64, 139, 144-145, 161, 537, 605

Qutb al-Din al-Shirazi, 82

Rabb, T.K., 519, 566, 595, 596
Racine, Jean, 579
Ragep, F.J., 75
Ramus, Petrus, 130, 149, 208, 266
Rashed, R., 74
Rathborne, Aaron, 322
Rattansi, P.M., 634
Ray, John, 447
Redondi, P., 219
Rees, G., 268
Reeves, E., 279
Regiomontanus (Johannes Müller), 101, 105, 146,
    149, 203, 209, 291
Régis, Pierre-Sylvain, 383, 425-426, 588
Remmert, V.R., 595
Rescher, N., 371
Ressons, Jean-Baptiste Deschiens de, 314
Reti, L., 153
Reve, K. van het, 52
Revel, J.-F., 51
Rheede tot Drakestein, Hendrik Adriaan van,
    447
Rheticus, Joachim, 112
Ricci, Matteo, 726
Riccioli, Giambattista, 342-343, 367, 408-409,
    509, 515, 530, 572
Richer, Jean, 454, 516, 667, 674, 677
Robartes, Francis, 450, 501
Roberts, L., 153

Roberval, Gilles Personne de, 382, 440, 550, 625-627, 656
Rochberg, F., 50
Rodon, David de, 392
Rohault, Jacques, 383, 392-393, 425, 570, 574-575, 588
Rømer, Ole, 453, 510, 541-542
Ronan, C.A., 155
Rose, P.L., xxiv, 97, 152, 155, 564
Rossi, P., xxxix, 154
Rougemont, D., de, 634
Roy, Henrick de (Regius), 390, 423-424, 432
Rumphius, Georg Everhard, 447
Rutherford, Ernest, 602

Sabra, A.I., 74-75, 219, 717
Sachau, E.C., 72
Sacrobosco (John of Holywood), 82, 139, 418
'Sagredo', 184, 189, 191, 199, 291
Sala, Angelo, 686
Saliba, G., 74
Salinas, Francisco, 102, 144, 295, 300
'Salviati', 183-184, 188, 190-192, 197-199
Sambursky, S., 50-51, 519, 634
Sargent, R.-M., 564
Sarton, G., 219
Saunders, J.J., 74-75
Sauveur, Joseph, 494
Sayili, A., 74-75, 219
Sbaraglia, Girolamo, 475, 480
Schaffer, S., xxxii, 153, 506-508, 564, 595
Schall von Bell, Adam, 727
Schliesser, E., 716
Schöffer, Peter, 120
Schooten, Frans van, Jr., 354-356, 524
Schopenhauer, Arthur, 737
Schott, Caspar, 449-450, 459-461, 476, 504
Schumpeter, J.A., 277
Schuster, J.A., 244, 401, 442, 519, 546
Sebokht (Bishop), 48
Selin, H., 50, 74, 219
Sen, A., 52
Senguerd, Wolfgang, 389, 392, 397
Severinus, Petrus (Peder Sørensen), 256, 552
Sextus Empiricus, 31, 100, 483, 603

Shadwell, Thomas, 501
Shank, J.B., 740
Shapin, S., xxxii, xxxix, 506-508, 595, 720
Shapiro, A., 547, 716-717
Shapiro, B., 757
Shea, W.R., 218, 243, 442
Shehadi, F., 75
Shen Kua, 34, 41-43, 47, 137-138
Shiloah, A., 75
'Simplicio', 183-184, 188-190, 198-199
Simplikios, 56
Sivin, N., 40, 50-52, 740
Slare, Frederick, 474
Sloane, Hans, 445
Smeaton, John, 326, 729
Smith, G.E., 716
Snel, Rudolf, 241
Snel, Willebrord, 59, 202-205, 297, 306, 321, 328, 366, 452, 522, 539, 679
Sosigenes, 95
Soto, Domingo de, 90, 307
Sparnaay, M.J., 506
Spinoza, Baruch de, 435, 588, 603
Sprat, Thomas, 571, 576, 584-587, 592
Starkey, George, 466-470, 501, 505, 516, 701
Steno, Nicolaus, 487
Stephenson, B., 216
Stevin, Simon, 103-105, 131-132, 140, 149, 159, 203, 232, 241, 293, 301, 309, 317-320, 336, 366, 569, 607
Streete, Thomas, 305
Stroup, A., 508
Stückelberger, A., 51, 243
Su Sung, 47
Surirey de Saint-Rémy, Pierre, 314, 325
Swammerdam, Jan, 454-458, 570
Swerdlow, N.M., xxiv, 152
Swift, Jonathan, 481-483, 500, 514, 719-721, 724
Sydenham, Thomas, 472

Tartaglia, Niccoló, 131, 313, 494
Taton, R., 369, 401
Thabit ibn Qurr, 55, 58-61, 64, 68, 293
Thoren, V.E., 154, 217-219
Toomer, G.J., 51

778 Topper, D., 218
Torricelli, Evangelista, 294, 308, 313, 340-342, 345-347, 362-366, 403, 411-415, 461, 526-529, 621, 626-628, 653, 721, 729
Tournefort, Joseph Pitton de, 447
Trevor-Roper, H., xxvii
Truesdell, C.A., 546
Tusi (Nasir ed-Din al-Tusi), 67-70, 78, 82, 91, 106, 149, 207
Tycho Brahe, **128-129**, 134, 149, 170, 172, 177, 179, 206, 208, 213, 266, 297, 303, 418, 453, 494, 515, 600, 726
   comet (of 1577), 164-166, 222, 658
   compromise system, 129, 166, 395, 397, 670
   Kepler's research director, 159, 167-171, 174, 177, 188, 200, 208, 212, 327
   nova (of 1572), 125, 274
   observational accuracy, 15, 129, 171, 214, 451, 515-516

Umar al-Khayyami, 57, 348
Urban VIII, 190, 422, 429, 431
Ursus (Reymers Bär), 494

Van Helden, A., 51, 75, 217, 370, 442, 506-508
Vanheste, T., 741
Varignon, Pierre de, 341, 357, 360, 363-364, 678, 721-723
Vasco da Gama, 136, 265
Vaughan, D., 370
Verbeek, T., 442
Verbiest, Ferdinand, 727
Vermij, R., xxxix, 401
Vesalius, Andreas (Andries van Wesel), xxiv, 123, 128, 139, 149, 252, 258, 265, 297, 455, 600
Viète, François, 103, 351-352
Vigevano, Guido de, 122
Vives, Luís, 130
Viviani, Vincenzo, 496
Voelkel, J.R., 216, 369
Voet, Gijsbert, 390, 422-426, 431-435, 439, 593, 734
Volder, Burchard de, 570, 574
Voltaire, 733

Waard, C. de, 243

Waerden, B.L. van der 50
Walbridge, J., 75
Walker, D.P., 155
Wallis, John, 354-356, 527, 624
Watt, James, 326, 478, 729
Weber, M., 50, 75, 135, 263-264, 268-269, 584, 596-597, 725
Webster, C., 596
Weiditz, Hans, 128
Weinberg, S., 363, 371, 430, 443, 737, 741
Wendelin, Godefroy, 367, 408
Wentholt, R., xl, 371, 740
Werckmeister, Andreas, 312, 324
Westfall, R.S., xxii, 218, 268, 287, 368-371, 437, 506-507, 519, 522, 546, 564, 597, 621, 624, 635, 716-717
Westman, R.S., 152-153, 216--219
Whiteside, D.T., 371, 716
Wilkins, John, 585-586, 589
Willem III (king William), 589
Willis, Thomas, 471-472
Willoughby, Francis, 445
Wilson, C. (Catherine), 401, 442, 506, 635
Wilson, C. (Curtis), 369, 401
Witt, Johan de, 354-356
Wren, Christopher, 305, 450, 527, 631-633, 637
Wright, Edward, 250

Yates, F.A., xix, 154, 244, 279
Yoder, J.G., 370-371, 546
Young, Thomas, 713

Zarlino, Gioseffo, 26, 102, 144-145, 149, 167-168, 226, 294, 300-301, 306, 311, 375
Zeno (of Elea), 19, 346, 357, 630
Zeno (of Kitheion), 7
Zilsel, E., xix, xxxiii, 219, 268, 634

SUBJECT INDEX

Académie Royale des Sciences, 326, 439-440, 497, 513, 567, 574-580, 583, 586, 625, 627, 676, 731-733

Accademia del Cimento, 487, 496, 575, 577, 579

acoustics. *See* musical intervals *and* sound

action at a distance. *See* force

active principles, 549-563, 624, 643-647, 704

actors' categories, xxi, xxxvi

aether (Cartesian and post-Cartesian subtle matter), 558-562, 601, 642-647, 652, 658, 668-669, 677, 705

air
    pressure, 462, 641
    pump. *See* void

alchemy, 39-40, 63-64, **466-470**, 479-480, 562, 643-645, 704, 707-708. *See also* chemistry

'Alexandria'
    abstract nature of, 18, 21-22, 60
    in Greece, **10-27**
    in Islamic civilization, **57-62**, 90-92
    in Middle Ages, **81-83**
    in Renaissance, **100-113**, 143-144, 151
    in 17th century Europe, 202, 233, 282-283, 366, 524, 600, 615, 679

'Alexandria-plus', 26-27, **160-215**, 231-235, 282-283, **291-367**, 404-418, 504, 600, 603, 606-608, 612-621, 623-629, 672, 678-679
    checking procedures practiced in, 199-200, 363-365, 516-518, 652
    combined with kinetic corpuscularianism, **521-545**, 652
    sacrilege possibly involved in, 426-440, 572-577
    strangeness of, 403-417, 429, 568-571

algebra, 57, 64. *See also* calculus

analogy. *See* motion, analogies of, *and* Scientific Revolution, checking procedures

animals, 86-87, 131, 423, 434, 447, 455-458, 461, 471, 474-475, 485

Aristotelianism, 197, 552, xxvii-xxviii
    in Greece, 6-7, 21-22
    in Islamic civilization, 56-57, 61-62
    in Middle Ages, 77-81, 83-87, 99
    in Renaissance, 100, 139, 143, 146-148, 150
    in 17th century Europe, 173, 186-187, 222, 239-240, 252-253, 382-383, 404-408, 411-415, 424, 483, 531, 600-601, 613-614. *See also* (in Name Index) 'Simplicio'

astrology, 23-25, 54-55, 68, 735

astronomy. *See* planetary trajectories *and* observatory *and* heliocentrism

'Athens'
    in Greece, 4-10, 15-27
    in Islamic civilization, 56-57, 60-62, 83-87
    in Middle Ages, 79-81, 90-92
    in Renaissance, 99-100, 139, 143, 146-148, 150
    in 17th century Europe, 208-212, 282-283, 600-605

'Athens-plus', **221-242**, 282-283, 300, **373-400**, 418, 484-485, 504, 600, 607
    checking procedures practiced in, 234-237, 378-381, 386, 389-390, 392-394, 398, 516-518, 558
    sacrilege possibly involved in, 426-440

atomism, 7, 56, 81, 223-226, 228-229, 240-241, 247, 384-385, 554, 602-603

attraction, 560, 631-633, 654-656, 705, 709-710, 719, 733. *See also* force, magnetic action, *and* universal gravitation

Baconian Brew, **554-563**, 677, 680

Baconian Ideology, 583-593, 609, 620

780 balance problems
    fluid, 11, 13, 82, 84, 103, 105, 232, 293-294, 393-394, 411, 600
    solid, 11-12, 44, 82, 103-104, 146, 179-181, 293, 336-338, 404-408, 419, 523, 600, 623-626, 630
bias, 486-487, 721, 735. *See also* self-correction
biology. *See* animals *and* plants
body and bodily processes, 16, 59, 64, 122-123, 147-148, 252-255, 295-297, 376-378, 381, 398-399, 456-459, 461, 470-472, 475-476, 480, 613, 618-619
botany. *See* plants

calculus, 20, **346-360**, 608, 622, 629
    and motion, 349-350, 357-360, 651, 657, 660, 664, 678
Cartesianism, 424-426, 435-436, 440, 588, 723. *See also* kinetic corpuscularianism
censorship and self-censorship, 418, 432, 438-440, 568
certainty. *See* Scientific Revolution, checking procedures
chemistry, 132-133, 255-258, 461, **465-470**, 479, 554-557, 562, 580, 613, 642-643, 656, 680, 697, 707-708. *See also* alchemy
China. *See* nature-knowledge, in China
clockwork universe, 230, 386-387, 436-437
collections, 129-130, 448-450
comparison, xxiv-xxvi
    cross-cultural, xxiv-xxvi, 44-47, 70, 90-96, 135-138, 142, 160, 202, 205-207, 213-214, 278-279, 427, 432-433, 590-593
consonances. *See* musical intervals
continuity/break, xxx-xxxi, 141-142, 202-201, 238, 260, 282, 600
corpuscles. *See* atomism *and* kinetic corpuscularianism
crafts, 95, 130-132, 137, 182, 285, **309-335**, 450, **472-483**, 490, 580, 606-608, 613, 624-625, 729
cultural transplantation, **46-50**
    absence from Chinese world, 46-48
    to Islamic civilization, 54-57
    to medieval Europe, 77-78
    to Renaissance Europe, 99, 152
    to Great Britain, 549-551

dialogue with ancients, 282, 605-606
dynamics. *See either* Scientific Revolution, dynamics of, *or* force

Earth,
    correlation with Moon, 305, 533-535, 656-657
    daily rotation of, 85-86, 91, 304. *See also* heliocentrism
    determination of place on, 26, 316-317, 319, 321-323, 344, 580, 583
    shape of, 454, 677
Easter, date of, 88, 95, 134, 265
electric action, 249-251, 462-465, 479, 491-492, 499-500, 613
empiricism, 4, 45
    coercive, **113-142**, 144-146, 151, 208, 245, 260-261, 600
Enlightenment, 263, 419, 436, 588, 601, 605, 733-735
error, handling of, 140, 488-489, 494
Eurocentrism, xxii-xxiii, 261-263
Europe
    authority in, 430-431, 438-439
    center of gravity of, 285, 567-568
    peculiarities of, 134-138, 263-266, 273-278. *See also* thisworldliness
experiment, 407-408, 410-415
    artificiality of, 247, 568
    checking procedures practiced in, 501-503, 516-518, 554
    drift toward, 574
    in early ranges, 117, 140, 145, 245, 600
    education in, 571
    experience and, 84, 246-247, 407, 491, 493
    fact-finding, **245-267**, **445-505**, **554-563**, 571, 600, 603, 606-608, 613-614, 617-618, 620, 652, 678, 680
    in natural philosophy, 392-393, 404, 574
    neutrality of, 574-577

facts, 171-172, 259-260, 448, 462. *See also* collections *and* natural history
    problematic nature of, **483-494**
faith, 735, 737
    Christian, 79-81, 85, 93-94, 112, 161-162, 189-191, 258, 263-266, 278, 390, 412, **417-440**,

535, 551-552, 575, **584-594**, 629, 645, 653, 709, 735. *See also* Reformation
    Islamic, 65-67, 93-94, 427, 432
falling bodies, 85, 87, 143, 180-181, 191-199, 231, 307, 358-360, 408-410, 530, 612, 625-630, 647, 657, 721
fancy. *See* Scientific Revolution, checking procedures
fictional observer, 141-143, 149-151, 201, 238, 253, 283, 286, 566, 599, 611-612, 620, 714
fire. *See* chemistry *and* alchemy
force, 173, 306-307, 523, 526, 532-534, 599, **621-633,** 640, **645-678,** 697, **706-711.** *See also* motion

geometry, 10, 57, 81-82, 101-103, 146, 161-167, 207, 233, 236-237, 319, 344-346, 394, 516, 518. *See also* calculus
global history, 262-263
graphic representation, 86-87, 91, 307

heat, 246, 496, 559-560, 642, 645, 704
heliocentrism, 14-15, 106-113, 159, 162-178, 181, 187-191, 209-211, 302-304, 348-349, 386, 394-397, 417, 419-424, 429, 434, 452, 618, 712, 726-729
history, conceptions of, xxvii, 260-261, 277. *See also* Scientific Revolution, non-occurrence
human sciences, 724-725
hydrostatics. *See* balance problems, fluid

illness. *See* body and bodily processes
impact. *See under* motion, nature of
impetus, 84-85, 239-241, 307-309
inertia, principle of ('motion retained'). *See under* motion, nature of
infinite
    divisibility of matter, 230, 391
    mathematical, 19-20, 346-351, 391, 630, 648, 650. *See also* calculus
    space, 111, 119, 230, 348-349
Inquisition, 258, 421-422, 431-432, 438, 565-567, 575
instrument trade, 570-571
inverse-square relation, 533-535, 632-633, 647-652, 663, 665, 670-671

Jesuit order, 146-147, 212, 382, 390, 431, 439, 495-496, 517, 572-574, 579, 601, 680, 726-729

kinetic corpuscularianism, 221-222, **224-242,** **373-400,** 411, 416-417, 516-518, 568, 617, 620, 623-625, 630, 641-647, 668-669, 707, 714
    combined with 'Alexandria-plus', **521-545**
    combined with fact-finding experimentalism, **549-563**
knowledge structure, xxviii, 16-18, 195-201, 227, 237-238, 406-408, 416, 485, 517-518, 603-605, 624, 641, 672, 700

laboratory, 459, 466-470, 490, 495-498, 504
lever, law of. *See* balance problems, solid
lifelike depiction, 114-128
light and vision, 13, 25, 44, 59, 82, 102, 246, 295-300, 327-329, 415, 453, 455, 600, 612, 642, 644, **678-715**
    color, 299-300, 543-545, 561-562, **679-702,** 724
    corpuscular, 539, 541, 561-562, 679, 700
    diffraction, 680-681, 698
    emission account of, 300, 561-562, 683, 690, 692-693, 700, 705-706
    periodicity, 688-692, 698, 705
    perspectiva, 82, 296-300, 539, 542, 679
    rainbow, 82, 299, 681, 685
    reflection, 13, 44, 59, 687-692
    refraction, 25, 59, 296-300, 455, 539, 541-545, 561-562, 683-687. *See also* telescope
    pulse/wave account of, 300, 541, 679-680, 700, 704-707
longitude. *See* Earth, determination of place on

machine efficiency, 117-119, 315-316
magic, xxxvii, 133-134, 148, 240, 247, 450, 495, 549, 552, 560, 605, 641
magnetic action, 43, 89, 249-251, 463, 478-479, 553, 613, 624
maps, 126-127, 129. *See also* Earth, determination of place on
mass, concept of, 654, 657
mathematical science
    abstract. *See* 'Alexandria'
    realist. *See* 'Alexandria-plus'

782   mathematization of nature, xvi, 201

measurement, 39, 42, 69, 165, 202, 214, 222, 251, 257, 311, 317, 321, 327-328, 343, 406, 409-410, 439, 470, 515-516, 569-570, 609. *See also* telescope

experimental, 451-454, 462, 466-468, 646, 689-692, 699-701

mechanics

expression avoided, xxxvi

rational (Newtonian), 522, 664

medicine. *See* body and bodily processes

microscope, 384-385, 392, **454-459**, 474-476, 600, 687-689

mixed mathematics, 146-147, 150, 208, 212, 221-222, 232, 313, 366, 408, 509, 515, 537, 574

modes of nature-knowledge

breakdown of barriers between, 285, **509-518**, 609, 723

bridge-building between, 13-14, 24-27, 48-49, 62, 102, 210, 294-295, 306

clashes between, **403-415**, 484-485

concept of, xxviii-xxix

developmental potential of, 47-48, 203, 239, 260-261, 271, 277

separation between, 18-27, 61-62, 88, 138-140, 143, 242, 282, 511-512, 723

tension between, 521-545, 624

motion

analogies of, 338-346, 523

and calculus, 349-350, 357-360

laws of, 230-231, 528-529, 625-626, 656, 659-660

nature of, 44, **182-186**, 192-194, 225, 228, 405, 408, **521-539**, 550, 612-613, 621, 622, 663-667

musical intervals, 11, 20-21, 25-26, 58-59, 81-83, 102, 144-146, 167-168, 175-177, 182, 226, 300-302, 311-312, 339-340, 375-378, 448-449, 451, 600, 605, 738. *See also* sound

natural history, 246-248, 446-448, 503

natural philosophy, 9-10, 16-18, 237. *See also* 'Athens' *and* 'Athens-plus'

clashes, 403-415

debate in, 8, 100, 222-223, 407-408, 719, 721

dogmatic to hypothetical, 517, **521-545**, **549-563**, 603-604, 624, 641, 714

eclectic, 259, 509, 604

hybrid, 210-212, 604, 703-712

opportunist stance toward, 24-25, 62, 101, 109, 209, 259, 391, 509, 604

predominance in, 32-33, 383-384, 604-605

resource for worldviews, 259, 604

succession through changing fashion, 719-721

syncretist, 32, 388, 391-392, 572, 604

naturalist drift, 94-96, 591. *See also* thisworldliness

nature-knowledge. *See also* Scientific Revolution

beginnings of, 3-4

in China, 4, **33-47**, 44-48, 55, 726-727, 729, 736

in Greece, **4-33**, 44-49

in Byzantium, 53-54, 69, 99

in Islamic civilization, **54-75**, 90-96, 207-208, 566, 725-726, 729

in Middle Ages, **77-96**

in Renaissance, **99-151**

aftermath, 29-30, 60, 67-70, 89, 277

autonomy of, 577, 610, 619, 731-732

communication in, 282, 490-494, 498-499, 512-514, 675, 684, 702. *See also* printing press

counterintuitive, 111, 113, 238, 316, 407, 416-417, 528, 569, 657

and crafts. *See* crafts

decline of, 27-33, 64-70, 89-90, 148-151, 611

and faith. *See* faith

Golden Age of, 28-30, 64-65, 89, 142-149, 276-277, 611, 726

legitimation of, 93-94, 309-310, **403-440**, 565-594, 612

limits set to, 737

marked by the culture, 62-64, 87-89, 95-96, 113-141. *See also* European peculiarities

petrifaction of, 69, 71, 90-92, 277

social anchoring of, 15-16, 42, 60, 92-93, 139, 282, 407, 417, 504, 511, 619-620, 731-733. *See also* patronage *and* university

unified conception of, 233, 384, 387, 540, 702-703, 707-712, 735-738. *See also* natural philosophy

visibility of, 589, 619
Newtonianism, 678, 722-723
novelty, 151, 251, 274-275, 277, 282-283, 383-384, 400, 419-420, 579

observatory and observational astronomy, 68, 124-125, 128-129, 177, 495-496, 498. *See also* telescope
optics. *See* light and vision

patronage, 16, 90-93, 206, 419, 428-429, 567-568, 577-583
Peace of Westphalia, 285, 431, 514-515, 566-568, 587
pendulum, 337-340, 342-346, 537-539, 559, 561, 646, 657
  clock, 329-335, 343-344, 538, 677
perspectiva. *See under* light and vision
perspective (linear), 131, 317, 319
plagiary, 492-494
planetary theory, 40-41
  in Greece, 13-15, 21, 24
  in Islamic civilization, 57-58, 68-69
  in Middle Ages, 81-82
  in Renaissance, 105-113, 129
  in 17th century, 161-178, 302-306, 534-535, 600, 612, 648, 651, 657-658, 660-671, 721, 726-729
plants, 115, 120-121, 128, 447-448, 455-456
Platonism, 5-6, 56, 77, 81, 99-100, 161, 197-199, 240, 383-385, 554, 601-602, 643
presentism, how to avoid, xxi, xxv
printing press, xxvi, 101-102, 205, 222, 611
projectile motion, 84-85, 192-193
publication. *See* printing press *and* Scientific Revolution, communication in

qibla, 63, 66, 95
quantities, 515-516, 569-570. *See also* measurement

rationality, xxxvii, 734-737
Reformation, 264-266, 590-592. *See also under* faith, Christian
religion. *See* faith

replication, 140, 478, 487-488
Restoration, 566, 585-586
Royal Society, 445, 497-499, 513-514, 567, 576-578, 584-589, 631, 639, 647, 731-732
Russia, nature-knowledge in, 727-729

Scientific Revolution. *See also under* nature-knowledge
  revolutionary transformation (1), **159-215**. *See also* 'Alexandria-plus'
  revolutionary transformation (2), **221-242**. *See also* 'Athens, plus'
  revolutionary transformation (3), **245-267**. *See also* experiment, fact-finding
  revolutionary transformation (4), 517, **521-545**, 609, 652, 664, 699-701, 713
  revolutionary transformation (5), 517, **549-563**, 609, 652, 699-701, 713
  revolutionary transformation (6), 599, 609, **637-678, 681-715**, 664
  absent necessity of, 23, 215, 271, 273, 730-731
  in brief, xiii, xv-xvi, 521, 569, 611-619, 621, 734
  causes of, xxvi-xxvii, xxix, xxxiii, 47-48, 160-161, 201-215, 238-242, 260-267, 271-286, 284, 674-675, 730
  checking procedures practiced during, xxxii, 284, 516-518, 603, **617-618, 700-701, 713-714**, 720-722. *See for specifics under* 'Alexandria-plus', 'Athens-plus', *and* experiment
  chronology of, 284-285, 403, 509-510
  conceptualization of, xxi-xxiv
  dual legacy of, **722-739**
  dynamics of, 284-286, 291-293, 306, 360-367, 397-400, 499-505, 608-611
  expansion beyond Europe, 725-729
  historiography of, **xv-xx**, xxii, **xxxiii-xxxviii**, xxxvii, 19, 50-52, 74-75, 153, 218, 323, 441-442, 466, 507, 566-567, 569, 595-597, 621, 634-635, 638-639, 693, 719-720. *See also* the Name Index
  legitimacy of, **403-440, 565-594**, 609, 620
  method in, 614-617, 695, 701, 714
  non-occurrence outside Europe of, 69, 71-73, 261-266, 592-593, 730

784   principles underlying selection of facts
              about, xxiii, xxxiv-xxxv, 510
         provisional end of, 722
         Second, 723
         solutions found to problem of, xxix-xxxiii
         staying power of, xv-xvi, 276-278. *See also*
              Scientific Revolution, dynamics *and*
              nature-knowledge, legitimation of
         superiority of, 263, 729
         utility perceived to rest in, 579-594, 609, 620,
              726, 731
    self-confidence, 66-67, 275-277
    self-correction, 169-172, 494, 721-722. *See also*
         error, handling of, *and* Scientific Revolution,
         checking procedures
    skepticism, 8, 56, 100, 209, 235-236, 240, 435, 483,
         602. *See also under* Scientific Revolution,
         checking procedures
    sound, 221, 247-248, 448-451, 473-474, 559, 667,
         738-739
    spirit, 247-248, 399, 549, 551-554, 560, 647
    statics. *See* balance problems, solid
    Stoa, 7-8, 81, 240, 248, 301, 383-385, 396, 552, 554,
         602, 647
    sympathetic resonance, 39, 133, 221, 225, 375, 552,
         559, 624

    technology. *See* crafts
         science-based, 3, 28, 117, 132, 284-285, 308-
              309, 335, 478, 545, 608, 715, 729-730
    telescope, telescopic observation and measure-
         ment, 186-189, 205, 299, **327-329**, 333-335, 385,
         419, 451-454, 474, 516, 600, 612, 683-684, 687.
         See also light and vision
    terminology, xviii, xxxv-xxxvii, 258
    theory involved in present analysis, xxvii-xxix,
         9-10, 46-49, 70-71, 141-142, 171-172, 258-260,
         261-263
    thinking
         for oneself, 385-386, 435-436, 588, 733
         with the hands, 265, 606-608. *See also* crafts
    thisworldliness/otherworldliness, 49, 263-265,
         591-593
    translation, 31-32, 48, 53-55, 77-78, 102
    trustworthiness, 140, 489-491, 732

    universal/local, xviii, xxxi-xxxii, 204, 719-720,
         723-724
    universal gravitation, 533-535, 605, 654-659, 675-
         677, 694, 712-713, 723-724
    universe
         measure of, 25, 110-111, 452-453, 618. *See also*
              infinity, space
         stability of, 667-671
    'universe of precision', 23, 160, 724-725
    universities, 80-81, 139, 382, 397, 418, 439, 601

    values underlying nature-knowledge/science,
         485, 495, 731-739
    vibration. *See under* motion, nature of, *and*
         pendulum
    vitalism, 251, 254, 257, 259, 555
    void, 83-84, 130, 186, 410-415, 459-462, 476-478,
         484-485, 491, 493, 558, 657
    Voyages of Discovery, 135-141, 265-266

    warfare, 37, 48, 54, 65, 78, 99, 277, 285, 313-315,
         318-320, 430-431, 438-439, 479, 566, 580, 583,
         585-586, 609, 726, 729
    water, 140, 410, 256, 556
         outflow, 340-342
         streaming, 312-313
    waves. *See under* motion, nature of
    whirlpools. *See under* kinetic corpuscularianism
    world harmony. *See under* musical intervals
    'world of the more-or-less', 21, 23, 111, 724
    worldview, 9-10, 36-38, 247-248, 258-260, 462

    zoology *see* animals